CENTURY OF RICKETTSIOLOGY

Emerging, Reemerging Rickettsioses, Molecular Diagnostics, and Emerging Veterinary Rickettsioses

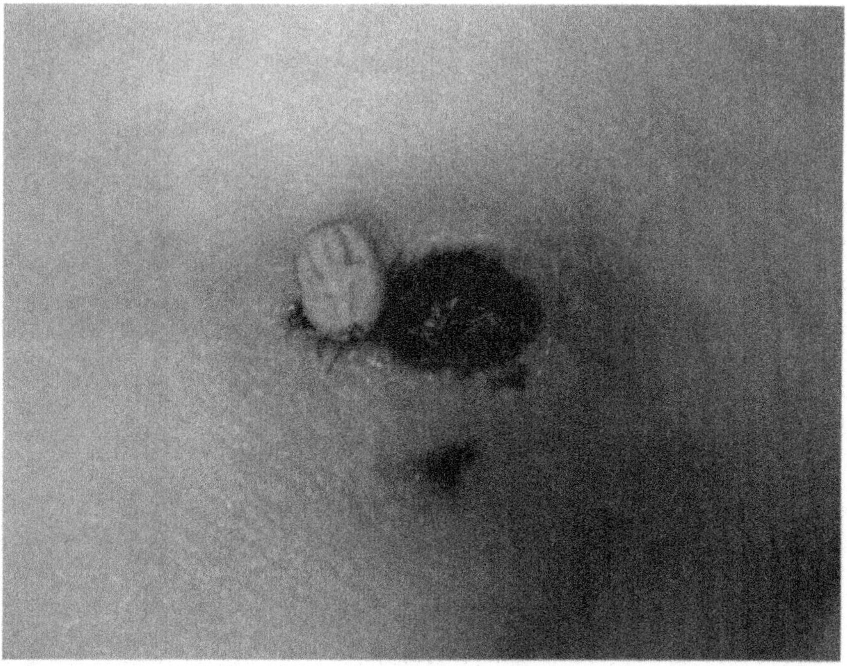

An engorged female *Dermacentor marginatus* tick was present near an eschar found on an 85-year-old man, from Logroño, Spain, complaining of mild pain on his left scapula for eight days. Polymerase chain reaction of the tick was positive for the gltA gene. Photograph courtesy of José A. Oteo and José R. Blanco.

ANNALS OF THE NEW YORK ACADEMY OF SCIENCES
Volume 1078

CENTURY OF RICKETTSIOLOGY
Emerging, Reemerging Rickettsioses, Molecular Diagnostics, and Emerging Veterinary Rickettsioses

Edited by Karim E. Hechemy, José A. Oteo, Didier A. Raoult, David J. Silverman, and José R. Blanco

Published by Blackwell Publishing on behalf of the New York Academy of Sciences
Boston, Massachusetts
2006

Library of Congress Cataloging-in-Publication Data

International Conference on Rickettsiae and Rickettsial Diseases (4th : 2005 : Logroño, Spain) Century of rickettsiology : emerging, reemerging rickettsioses, molecular diagnostics, and emerging veterinary rickettsioses / edited by José A. Oteo ... [et al.].

p. ; cm. – (Annals of the New York Academy of Sciences, ISSN 0077-8923 ; v. 1078) "4th International Conference on Rickettsiae and Rickettsial Diseases in Logroño, (La Rioja), Spain, on June 18-21, 2005"–Pref.

Includes bibliographical references and index.
ISBN-13: 978-1-57331-639-2 (pbk. : alk. paper)
ISBN-10: 1-57331-639-3 (pbk. : alk. paper)
1. Rickettsial diseases–Congresses. 2. Rickettsia–Congresses. 3. Rickettsial diseases in animals–Congresses. I. Oteo, José A. II. New York Academy of Sciences. III. Title. IV. Series.

[DNLM: 1. Rickettsia Infections–Congresses. 2. Bartonella–Congresses. 3. Coxiella burnetii–Congresses. 4. Q Fever–Congresses. 5. Rickettsia–Congresses. W1 AN626YL v.1078 2006 / WC 600 I585c 2006]

QR201.R59158 2005
616.9'22—dc22

2006019882

The *Annals of the New York Academy of Sciences* (ISSN: 0077-8923 [print]; ISSN: 1749-6632 [online]) is published 28 times a year on behalf of the New York Academy of Sciences by Blackwell Publishing, with offices located at 350 Main Street, Malden, Massachusetts 02148 USA, PO Box 1354, Garsington Road, Oxford OX4 2DQ UK, and PO Box 378 Carlton South, 3053 Victoria Australia.

Information for subscribers: Subscription prices for 2006 are: Premium Institutional: \$3850.00 (US) and £2139.00 (Europe and Rest of World).
Customers in the UK should add VAT at 5%. Customers in the EU should also add VAT at 5% or provide a VAT registration number or evidence of entitlement to exemption. Customers in Canada should add 7% GST or provide evidence of entitlement to exemption. The Premium Institutional price also includes online access to full-text articles from 1997 to present, where available. For other pricing options or more information about online access to Blackwell Publishing journals, including access information and terms and conditions, please visit www.blackwellpublishing.com/nyas.

Membership information: Members may order copies of the *Annals* volumes directly from the Academy by visiting www.nyas.org/annals, emailing membership@nyas.org, faxing 212-888-2894, or calling 800-843-6927 (US only), or +1 212 838 0230, ext. 345 (International). For more information on becoming a member of the New York Academy of Sciences, please visit www.nyas.org/membership.

Journal Customer Services: For ordering information, claims, and any inquiry concerning your institutional subscription, please contact your nearest office:
UK: Email: customerservices@blackwellpublishing.com; Tel: +44 (0) 1865 778315; Fax +44 (0) 1865 471775
US: Email: customerservices@blackwellpublishing.com; Tel: +1 781 388 8599 or 1 800 835 6770 (Toll free in the USA); Fax: +1 781 388 8232
Asia: Email: customerservices@blackwellpublishing.com; Tel: +65 6511 8000; Fax: +61 3 8359 1120
Members: Claims and inquiries on member orders should be directed to the Academy at email: membership@nyas.org or Tel: +1 212 838 0230 (International) or 800-843-6927 (US only).

Mailing: The *Annals of the New York Academy of Sciences* are mailed Standard Rate. **Postmaster:** Send all address changes to *Annals of the New York Academy of Sciences*, Blackwell Publishing, Inc., Journals Subscription Department, 350 Main Street, Malden, MA 01248-5020. Mailing to rest of world by DHL Smart and Global Mail.

Annals are available to subscribers online at the New York Academy of Sciences and also at Blackwell Synergy. Visit www.annalsnyas.org or www.blackwell-synergy.com to search the articles and register for table of contents e-mail alerts. Access to full text and PDF downloads of *Annals* articles are available to nonmembers and subscribers on a pay-per-view basis at www.annalsnyas.org.

The paper used in this publication meets the minimum requirements of the National Standard for Information Sciences Permanence of Paper for Printed Library Materials, ANSI Z39.48_1984.

ISSN: 0077-8923 (print); 1749-6632 (online)
ISBN-10: 1-57331-639-3 (paper); ISBN-13: 978-1-57331-639-2 (paper)

A catalogue record for this title is available from the British Library.

Digitization of the *Annals of the New York Academy of Sciences*

An agreement has recently been reached between Blackwell Publishing and the New York Academy of Sciences to digitize the entire run of the *Annals of the New York Academy of Sciences* back to volume one.

The back files, which have been defined as all of those issues published before 1997, will be sold to libraries as part of Blackwell Publishing's Legacy Sales Program and hosted on the Blackwell Synergy website.

Copyright of all material will remain with the rights holder. Contributors: Please contact Blackwell Publishing if you do not wish an article or picture from the *Annals of the New York Academy of Sciences* to be included in this digitization project.

ANNALS OF THE NEW YORK ACADEMY OF SCIENCES
Volume 1078
October 2006

CENTURY OF RICKETTSIOLOGY
Emerging, Reemerging Rickettsioses, Molecular Diagnostics, and Emerging Veterinary Rickettsioses

Editors
KARIM E. HECHEMY, JOSÉ A. OTEO, DIDIER A. RAOULT, DAVID J. SILVERMAN, AND JOSÉ R. BLANCO

This volume comprises part of the proceedings of the **4th International Conference on Rickettsiae and Rickettsial Diseases**, held June 18–21, 2005, in Logroño (La Rioja), Spain. The first volume was published in 2005, *Rickettsioses*: *From Genome to Proteome, Pathobiology, and Rickettsiae as an International Threat* (Volume 1063).

CONTENTS

Part II. Clinical Aspects and Diagnosis

Part III. Vectors

Part IV. Veterinary Rickettsiology

Part V. Isolation, Cell Culture, and Diagnostics

Financial assistance was received from:

General Sponsors

- Gobierno de la Rioja, Fundacíon Rioja Salud
- European Society of Clinical Microbiology and Infectious Diseases (ESCMID)
- American Society for Rickettsiology (ASR)

Supporters

- National Institutes of Health, USA
- Fundacíon Logroño Turismo, Ayuntamiento de Logroño
- Consejo Regulador DOC Rioja
- Asociacíon para Estudio de Patógenos Especiales
- Sociedad Española de Enfermedades Infecciosas y Microbiología Clínica (SEIMC)

Corporate Supporters

- Abbott
- Boehringer Ingelheim
- Bristol-Myers-Squibb
- Focus Diagnostics
- Gilead
- Glaxosmithkline
- Ibercaja
- Merck
- Pfizer
- Sanofi-Aventis
- Schering-Plough
- Vircell
- Vitro
- Wyeth

A Century of Rickettsiology: Emerging, Reemerging Rickettsioses, Clinical, Epidemiologic, and Molecular Diagnostic Aspects and Emerging Veterinary Rickettsioses

An Overview

KARIM E. HECHEMY,[a] JOSÉ A. OTEO,[b] DIDIER RAOULT,[c] DAVID J. SILVERMAN,[d] AND JOSÉ RAMÓN BLANCO[b]

[a]*Division of Infectious Disease, Wadsworth Center, New York State Department of Health, Albany, New York 12208, USA*

[b]*Area de Enfermedades Infecciosas, Complejo Hospitalario San Millan, San Pedro de La Rioja, Hospital de La Rioja, 26001 Logroño (La Rioja), Spain*

[c]*Université de la Mediterranée, Faculté de Médecine, Unité des Rickettsies, 13385 Marseille Cedex 05, France*

[d]*University of Maryland School of Medicine, Department of Microbiology/Immunology, Baltimore, Maryland 21201, USA*

ABSTRACT: This overview summarize the salient features of advances in the epidemiology, vectors, and clinical and laboratory diagnoses of rickettsiology. Presentations on veterinary rickettsiology highlight the importance of the rickettsiae in animal husbandry.

KEYWORDS: epidemiology; vectors; laboratory diagnosis; clinical diagnosis; veterinary rickettsiology.

This is the second of two volumes encompassing the presentations that were given at the Fourth International Conference on Rickettsiae and Rickettsial Diseases in Logroño, (La Rioja), Spain, on June 18–21, 2005. The first volume was published in 2005 (Volume 1063 of the *Annals*) as *Rickettsioses: From Genome to Proteome, Pathobiology, and Rickettsiae as an International Threat*. This volume dealt with the molecular aspects of rickettsiae, from genomics to

Address for correspondence: Karim E. Hechemy, 5 Windham Ct. Clifton Park, New York 12065, USA.
e-mail: khechemy@nycap.rr.com

Ann. N.Y. Acad. Sci. 1078: 1–14 (2006).
doi: 10.1196/annals.1374.001

proteomics; the pathobiology of query (Q) fever, a potential bioweapon agent, and the possible use of other rickettsiae in bioterrorist activities; and the pathobiology of rickettsiae, including the prospects for development of vaccines against these organisms, especially if antibiotic resistance in rickettsiae becomes a major issue. This second volume focuses on the epidemiology, vectors, and clinical and laboratory diagnoses of rickettsioses and "associated" organisms that were formerly included in the Order *Rickettsiales*: Q fever, which is caused by *Coxiella burnetii*, Family *Coxiellaceae*, Order *Legionellales*, and members of the Family *Bartonellaceae*, Order *Rhizobiales*. In this volume, we also introduce a new subdiscipline in the discipline of Rickettsiology, namely Veterinary Rickettsiology, which encompasses the pathobiology of these and associated organisms on vertebrates and invertebrates, as well as addressing the economic effects of these rickettsiae on domestic and peridomestic animals.

In this volume, we also commemorate the centennial of the discovery of the "bacillus of Rocky Mountain spotted fever" by Howard Taylor Ricketts in 1906. His studies led to the discovery of the causative agent of Rocky Mountain spotted fever (RMSF). It was named *Rickettsia rickettsii* after Dr. Ricketts' untimely death in Mexico City in 1910 of a laboratory-acquired typhus fever. As we approach their centenaries, it is also a good occasion to revisit two other milestone discoveries in the rickettsial field in 1909, namely, Stanislaus von Prowazek's discovery of the causative agent of epidemic typhus fever, *Rickettsia prowazekii*, and Charles Nicolle's discovery that epidemic typhus fever is transmitted by the body louse. Subsequently, the latter discovery helped to draw a clear distinction between the classical louse-borne epidemic typhus and murine typhus, which is conveyed to man by the rat flea. Wolbach, who studied further the causative agent of epidemic typhus, named it *R. prowazekii* after its discoverer. Of these three pioneers, only Nicolle survived the rickettsiae unscathed and went on to win the Nobel Prize in 1928. The other two died prematurely, both of typhus infections. Until the discovery and availability of chloramphenicol and the tetracyclines in the early 1950s, Nicolle's discovery of the vector formed the only basis for measures taken against these diseases.

Our understanding of the epidemiology of rickettsiae and associated organisms, namely, the former rickettsiae *C. burnetii*, and some species of the family *Bartonellaceae* formerly in the Order *Rickettsiales*, is evolving rapidly, thanks to the ongoing discoveries and dramatic advances in molecular epidemiology. At this conference, a minisymposium to review the worldwide aspects of the epidemiology of rickettsioses and associated organisms was presented.

In Europe, where only one spotted fever group (SFG) rickettsia, *Rickettsia conorii* (which causes Mediterranean spotted fever, MSF), had been recognized until recently, a large number of rickettsiae have now been discovered and their pathogenicity to the human recognized; one of these is *Rickettsia slovaca*. Some rickettsiae have been imported through the travel of humans, (e.g., *Rickettsia aeschlimannii*); and for others, migrating birds are the likely source, either carrying the imported tick vector or infecting native ticks after having been infected elsewhere (e.g., *Rickettsia sibirica mongolotimonae*). Murine typhus appears to be present mainly in the coastal areas of southern Europe. Murine typhus-like *Rickettsia felis*, which is a SFG rickettsia transmitted by cat fleas, has been described in several European countries. The diagnosis of *R. felis* should be considered in patients presenting with fever and/or rash, if there is a history of cat contact or fleabite. Many other rickettsiae have been isolated but not yet implicated in human diseases.

In North Africa, rickettsioses resemble those that are found in southern Europe, with some quantitative variations. Among healthy populations, seroprevalence of antibodies to *R. conorii*, the presumed main rickettsial disease in this region, range from 5 to 8%. It is also believed that the number of murine typhus infections is underreported. Recent studies showed that murine typhus infections were erroneously considered to be *R. conorii* or viral infections. Surveillance of typhus and SFG rickettsiae needs to be reevaluated in light of the molecular advances in the field of rickettsiology as tools to reassess the extent and types of infections, and the pathology due to these infections exhibited. In another report, the highest prevalence of antibodies is against several SFG rickettsiae, *C. burnetii*, *Bartonella* spp., and *Ehrlichia* spp, among patients with acute febrile illness of unknown origin. Prospective studies performed in Oran, Algeria, to assess the frequency of MSF and to describe the clinical and epidemiological characteristics of the disease, show the existence of severe forms of MSF characterized by the occurrence of complications in about half of the patients and three deaths. Another important finding is the presence of multiple eschars in these patients. There is no clear explanation for the occurrence of multiple eschars; it could be due to other coinfecting rickettsial strains, or due to a change in the epidemiology/behavior of the *Rhipicephalus sanguineus* tick, the vector for *R. conorii*.

Five tick-borne SFG rickettsioses are currently known to occur in sub-Saharan Africa. One rickettsia, *Rickettsia africae*, the agent of African tick-bite fever (ATBF), is differentiated from *R. conorii*. Various species of the *Amblyomma* ticks are recognized as the vector and reservoir of *R. africae*. Because *Amblyomma* readily bites humans, cases of ATBF often occur in clusters, and patients often present with multiple inoculation eschars. Another SFG rickettsia, *R. sibirica mongolotimonae* carried in *Hyalomma* ticks, causes multiple eschars and draining lymph nodes, and a lymphangitis that extends from the inoculation eschar to the draining node. The particular clinical features of this new rickettsiosis have led to the appellation "lymphangitis-associated rickettsiosis" (LAR). *Rickettsia typhi* infections occur mostly in coastal areas. In contrast, *R. prowazekii* outbreaks have occurred in refugee camps necessitated by such upheavals as civil wars or volcanic eruptions in Central Africa, under conditions of high crowding. Little is known regarding human infections in Africa by members of the family *Anaplasmataceae*. One recent infection in a human by *Ehrlichia ruminantium* in South Africa was reported in Volume 1063 of the *Annals*.

At present, two rickettsioses are classified officially as reportable in the Russian Federation. The first is epidemic typhus and Brill–Zinsser disease, both caused by *R. prowazekii* and the second is Siberian tick typhus, caused by *R. sibirica*. Q fever is also under compulsory official registration. During the previous two decades two new rickettsioses have been described. They are Astrakhan spotted fever, caused by a rickettsia recently classified as *R. conorii* subsp. *caspia*, and Far Eastern tick-borne rickettsiosis, caused by *Rickettsia heilongjiangensis*. Another paper shows that the *Haemaphysalis* ticks transmit the new rickettsia *R. heilongjiangensis.* A subsequent presentation describes the extensive heterogeneity and the coexistence of multiple species of nonpathogenic rickettsiae in the same ticks. It was assumed that the specific/particular geographic area influences both the epidemiologic activity of natural foci and also the morbidity due to tick bites of humans or animals. Studies on zoonotic infections, including Q fever in humans and animals, showed that livestock are infected with *C. burnetii* in most of the districts of the northwestern region of Russia. It was further reported that most human infections are caused by the presence of the bacterium in cattle. In present day Russia *C. burnetii* is thought to be underreported. Another study describes the geoinformation mapping of tick rickettsiosis zoonoses in the Altai Krai region of the Russian Federation.

Three rickettsial diseases are known to exist in Japan currently: Japanese spotted fever (JSF), Tsutsugamushi disease (scrub typhus), and Q fever. Tsutsugamushi disease was the only known rickettsiosis in Japan until 1984, when the first clinical cases of JSF were reported. A Q fever patient was reported in 1989. In addition to Tsutsugamushi disease, these two emerging infectious diseases are designated as notifiable diseases. Clinical and epidemiological information on JSF and Q fever are limited, because outbreaks have been sporadic, and the numbers of reported cases are low. The vectors of JSF appear to be *Haemaphysalis flava* and *Haemaphysalis hystericis*.

Foci of Tsutsugamushi disease on Pingtan Island, China, shows that it tends to be seasonal, occurring in spring time. Such information may help experts to develop prevention strategies targeted to this time of the year.

Australia has a range of rickettsial diseases that include typhus group rickettsia, *R. typhi*, SFG rickettsiae, *Rickettsia australis* and *Rickettsia honei*, the scrub typhus group, and Q fever. In addition, a new SFG species (or subspecies) detected in the eastern half of Australia is tentatively named "*Candidatus* Rickettsia marmionii." This rickettsia causes both acute disease and a chronic illness. The significance of the chronic illness is under investigation. In western Australia, ticks were collected from people and then screened using polymerase chain reaction (PCR) assay. The rickettsial species was then cultured, and its novelty and phylogenetic position were examined. The infecting rickettsia is divergent enough to be classified as a new species. Sequence data suggest that the evolution of Australian rickettsiae was not routed through a recent common ancestor. The pathogenic potential of the new species is unknown.

In Europe, seroprevalence studies in northern and northeastern Poland showed evidence of contact with the human granulocytic anaplasmosis (HGA) agent, *Anaplasma phagocytophilum*, although none of the seropositive persons developed clinical manifestations of the disease. Similarly, Slovenian forestry workers exposed to tick bites did not develop clinical manifestations of HGA. Molecular epidemiology of human and bovine anaplasmoses in southern Europe showed that the molecular epidemiology of bovine and human anaplasmosis in regions where *Anaplasma marginale* and *A. phagocytophilum* coexist indicates a common reservoir, hosts, and vectors. The increasing contact between wildlife, domestic animals, and human populations will undoubtedly increase the risk of outbreaks of human and bovine anaplasmoses. In Bosnia and Herzegovina, the numbers and distribution of Q fever–seropositive individuals suggest that Q fever is an endemic disease in this country. In certain areas of the Basque region of Spain, Q fever endemicity is lower than in other regions of Spain. In northeastern Spain, in rural and urban areas, cases of murine typhus and *R. felis* infections in humans are reported. It was postulated that the classical murine typhus cycle seems to be replaced in some regions by the peridomestic animal cycle. In the Spanish Mediterranean area of Valencia, findings suggest a low incidence in human cases of MSF, but a higher incidence of tick-borne lymphadenopathy (TIBOLA) caused by *R. slovaca*.

In Portugal, a study indicates the first evidence of human exposure to *A. phagocytophilum* or an antigenically related bacterium. Climate changes in the last two decades have influenced vector-borne diseases. In Portugal, where the average temperatures in the last 10 years have been the warmest on record, coupled with low rain falls, the number of cases of boutonneuse fever (BF) caused by the *R. conorii*, and Israeli spotted fever doubled in the last two years.

Rickettsioses in Italy are primarily due to infection with *R. conorii*. Only sporadic cases of murine typhus are reported. Other observed rickettsioses appear to have been imported from abroad. In northeastern Italy, studies showed that HGA should be included in the serologic testing of patients with a tick bite history, and it should also be considered in patients symptomatic for or confirmed as having Lyme disease. In Denmark, Danish elite orienteerers, who have the potential for massive exposure to ticks, and are thus theoretically highly susceptible to the acquisition of rickettsial diseases, rarely show positive results in a battery of serologic tests for rickettsiae. An explanation for this is that the orienteerers wear protective clothing. In the European region of Turkey, mainly elderly patients were observed with acute and sometimes lethal MSF infection.

In Central and South America, a first report (from Venezuela) of *Ehrlichia canis* infection in human patients showing clinical signs of human monocytic ehrlichiosis (HME) is presented. Patients with clinical symptoms of an infection with *R. felis*, a rickettsia that has characteristics of both SFG and typhus group rickettsiae, are diagnosed in Brazil and Mexico. Brazilian spotted fever (BSF), the most prevalent tick-borne disease in Brazil, is caused by *R. rickettsii* and transmitted by the ixodid tick *Amblyomma cajennense*. Since

1985, there has been a large increase in the number of diagnosed cases. The highest incidence is from May to October. The capybara, a large rodent, is thought to be a reservoir for the tick vector in São Paulo state; control of the capybara population could help in the control of BSF.

In the United States, an apparent increase in incidence of HME and HGA may be related to enhanced awareness and reporting, and/or to increased human–vector interactions. Both infections may be significant opportunistic infections and should be strongly considered by any physician treating a patient with a clinically compatible illness and possible tick exposure. The increased incidence of RMSF in the period 1997–2002 is likely related to enhanced awareness and reporting of RMSF, as well as to increased human–vector interactions. The persistence of RMSF mortality, despite a low incidence due to timely antibiotic treatment, underscores the need for physician vigilance in considering a diagnosis of RMSF for febrile individuals who have potentially been exposed to ticks. Further it stresses the importance of treating such persons regardless of the presence of a rash, and before receiving the laboratory test results.

Clinical and microbiological characteristics of a newly recognized tick-borne disease in Spain (DEBONEL/TIBOLA) caused by *R. slovaca* are described. All ticks removed from the Spanish patients were PCR-positive for SFG rickettsiae. By the sequencing procedure, eight of them were identified as a *R. slovaca* and two as rickettsial genotypes RpA4 and DnS14. The pathogenicity in humans of the latter genotypes is not known. The clinical presentations include an eschar at the site of the tick bite, surrounded by an erythema, and painful regional lymphadenopathy. The disease appears during the colder months and its vector is *Dermacentor marginatus.* The clinical, epidemiological, and therapeutic aspects of 15 patients with MSF from the island of Crete, Greece, are described. Among them two became ill during the winter months, indicating that the disease there is endemic.

A clinical study on BSF, based on a retrospective review of a series of cases including medical records and case notification files of hospital-admitted patients, describes clinical manifestations in patients with a 30% case fatality rate. The authors theorize that the high fatality rate was largely due to delays in initiating appropriate antibiotic treatment. A study of Brazilian patients from various municipalities identifies the causative organism as *R. rickettsii* on the basis of the application of immunohistochemical staining and sequencing methods to fixed tissues or blood clots. In another study from Brazil, the authors discuss the possible efficacy of vector control measures adopted in a certain area of the country, as well as the role of dogs and horses as sentinels of infection by rickettsiae.

In this volume, data on the epidemiology, diagnosis, prevention, and treatment of tick-borne diseases in the homeless population are presented. Homeless people, whose numbers are increasing in the inner cities, are frequently exposed to ectoparasites. The low standard of living conditions and the crowding in shelters provide ideal conditions for the spread of "infected" vectors such as lice, flea, ticks, and mites. These arthropods carry rickettsiae and associated organisms, and may transmit them to humans. Great increases in outdoor recreational activities conducted without proper protective measures have also increased human exposure to pathogens that previously cycled almost exclusively within natural, nonhuman enzootic hosts.

A review of the clinical diagnosis and treatment of HGA is presented. For the most part, it is a mild or even asymptomatic illness. However, older individuals and patients who are immunocompromised by natural disease processes or immunosuppressant medications may develop an acute, influenza-like illness characterized by high fever, rigor, generalized myalgia, and severe headache.

Detection and treatment of pericarditis remain a challenging problem and the unknown etiology remains high, between 40 and 85%. A 6-year study attempting to reduce this proportion via development of a thorough diagnostic strategy has suggested the systematic use of a battery of noninvasive tests to diagnose benign pericardial effusion. The study shows that *C. burnetii* is found in almost 5% of the cases of idiopathic pericardial effusion.

Studies on vectors, which are obviously a crucial element in the maintenance and transmission of the pathogens, are presented. The results indicate that the vector may contain more than one species of rickettsia and therefore can no longer be considered to be the specific carrier for a specific species of microorganism. The understanding and prediction of habitat suitability for ticks together represent a step toward the development of strategies to minimize tick-borne infections. The techniques described use information that captures the abiotic features of the habitat, for example, bioclimate envelopes that are defined as constituting the climatic component of the ecological niche. The authors assert that they are now able to predict the habitat suitability and the density of the various stages of ticks. This predictive capability provides health authorities with information on the risk in a given local for human-to-tick exposure.

In Southern Spain, *R. slovaca* and SFG rickettsiae of the closely related genotypes RpA4, JL-02, DnS14 , and DnS28, have been detected principally in *De. marginatus*. Double infections have not been detected. Another presentation describes the high incidence and distribution pattern of *R. felis* in peridomestic fleas. This suggests that people with pets are exposed to *R. felis*. In northern Spain, prevalence of SFG rickettsiae in ticks shows that the *De. marginatus* tick is the major species of tick that carries the SFG rickettsiae. Another study reports the molecular identification of SFG rickettsiae in ticks removed from human subjects in a 7-year study. The near-absence of *R. conorii* in the ticks, despite the frequent diagnosis of MSF cases in that area for years, suggests that many MSF or MSF-like cases attributed to *R. conorii* have been actually caused by other rickettsiae, such as *R. aeschlimannii* and *R. slovaca*. Alternatively, other SFG rickettsiae, whose pathogenicities are not yet known, may be the agents. A study shows *R. slovaca* in *De. marginatus* ticks that were removed from European wild boar during the winter of 2004 to be the predominant rickettsia carried by that vector. In another study, *R. felis*-like bacterium, *Bartonella clarridgeiae*, and *Bartonella henselae* are detected in *Ctenocephalides* fleas infesting peridomestic animals in northern Spain. The authors recommend that physicians consider rickettsia and bartonella as potential causes of fever in cat or dog owners, as well as in patients bitten by fleas. In Croatia, the vector that carries *R. felis* is *Haemaphysalis sulcata* tick collected from domestic animals. Another study describes the prevalence of ticks, tick-borne rickettsiae, and *C. burnetii* in a Greek island. Nine species of ticks are identified. Other than, *Rhipicephalus* spp. ticks, the ticks of greatest importance in the epidemiology of SFG rickettsiae in this Greek island are *Hyalomma* spp. *C. burnetii* and four SFG rickettsiae, including *R. conorii, R. massiliae, R. rhipicephali*, and *R. aeschlimannii* are detected by molecular methods. Double infection with *R. massiliae* and *C. burnetii* was found in one of the positive ticks.

A study to determine the production of certain rickettsial proteins in ticks, depending upon the stage of the tick cycle and/or influenced by temperature and feeding, shows that the expression of rickettsial outer-surface membrane proteins rOmpA and rOmpB in *R. massiliae* in the vector *Rhipicephalus turanicus* is not influenced by the temperature. However, rOmpA production is influenced by the life stage of the tick, although rOmpB is not. Another study describes the role of the vector in maintaining and transmitting pathogens to vertebrate hosts, information that is crucial to control the vector, and hence the transmission of the organism. The study characterizes several tick-derived molecules, including a histamine release factor, serine proteases, and lysozymes. The identification and expression analysis of immune-like and stress-response molecules, as well as host immunomodulatory molecules, suggest that ticks are actively responding to rickettsial infection, rather than merely serving as transmission vehicles. This fact obviously requires that the SFG rickettsiae be able to adapt in response to the host cell environment, as has been observed for other rickettsiae.

R. felis is the second species of rickettsia identified in Uruguay in the past two years. *R. parkeri* was detected in *Amblyomma triste*. In studies from Brazil, one shows the prevalence of *R. felis* in the fleas *Ct. felis* and *Ct. canis*. This indicates that rickettsiosis due to *R. felis* should be suspected in the differential diagnosis of a zoonotic infection. Another Brazilian study reports that *R. belli* infects the *Ixodes loricatus* tick, and that an efficient transovarial transmission, and transstadial survival, of this rickettsial species occur in the tick, with no vertebrate animal required as an amplifier host.

In a Russian province, the bacterial agents "*Candidatus* Rickettsia tarasevichiae," *Borrelia*, and Montezuma, a rickettsia-like endosymbiont, were the agents most frequently detected in *Ix. persulcatus* ticks. Ehrlichiae and rickettsiae frequently share a tick host with *Borrelia burgdorferi* (*sensu lato*), so the potential for cotransmission and mixed infections in vertebrate hosts, including humans, exists. In another study, four pathogens recognized in humans were detected in hard ticks, the pathogenic *R. sibirica*, *R. heilongjiangensis*, *Rickettsia helvetica*, and *A. phagocytophilum*; rickettsiae and ehrlichiae of unknown pathogenicities were also detected. In the Ukraine, a study describes the use of pediculicide lotions produced from medicinal plants to decrease lice in the child population. A study from Poland indicates variation in the prevalence of *A. phagocytophilum* in the *Ix. ricinus* tick. In Slovakia, *A. phagocytophilum*, *C. burnetii*, *Rickettsia* spp., and *B. burgdorferi* are found in *Ix. ricinus* ticks; all but the fourth of these three organisms are also found in mice. In a similar study of a region of France, three bacteria *A. phagocytophilum*, *Rickettsia* spp, and *B. burgdorferi* (*sensu lato*), while having similar prevalences in questing *Ix. ricinus* ticks, appear to have different life cycles. Pastures and fields as well as woods are high-risk areas for humans and animals. The authors concluded that there is a need for parallel investigation on human and animal occurrences of the diseases caused by all three bacteria in the region.

Many arthropod species are vectors for rickettsial agents of human disease. Some rickettsial species have evolved a stable means of transovarial maintenance within their arthropod hosts, whereas other rickettsiae are acquired when the arthropod vector feeds on infected vertebrates (horizontal transmission). It is widely accepted that an initial infection of an arthropod by one rickettsial species prevents the acquisition and transmission of a secondary rickettsial form. The cause of this phenomenon, referred to as interference, is unknown. However, a tick had been found to be infected with multiple rickettsial forms, *R. bellii*, *Rickettsia montanensis*, and *R. rickettsii* after a screen of rickettsia-infected *De. variabilis* ticks were collected in Ohio in the United States. The tick was hemolymph-positive for *Rickettsia* spp. and positive for a direct fluorescent antibody test specific to *R. rickettsii*. It is postulated that multiply-infected vectors exist in nature and may serve a vital (if only intermediate) role in the maintenance of these rickettsiae. In another study, RMSF is generally reported to be rare in Arizona because of a climate unsupportive of the typical *R. rickettsii* tick vectors *De. variabilis* or *De. andersoni*. Thirteen cases of RMSF, including one death, have recently identified in Arizona and a heavy environmental infestation of *Rh. sanguineus* at an endemic site has been observed. Although *Rh. sanguineus* is not normally the vector for *R. rickettsii*, the authors propose that that under certain circumstances, *Rh. sanguineus* transmits *R. rickettsii* to humans. It thus may merit reconsideration as a vector in other geographic areas as well.

Ticks collected from animals in Egypt show the presence of *Anaplasma* and *Ehrlichia* spp. *Bartonella* spp. *C. burnetii*, and *R. typhi* are detected in fleas. SFG rickettsiae (*R. aeschlimanii* and an unnamed *Rickettsia* spp.) are detected in both fleas and ticks. A study from Algeria indicates that there is a dual risk of exposure to tick-borne SFG and flea-transmitted SFG rickettsiae. The epidemiological aspects of the two types of exposure as well as prevention aspects are different and need careful consideration. In another study from east Africa, the soft tick *Ornithodoros moubata* is a vector for a rickettsia species yet to be determined. Also to be established is whether this species can infect man, or whether it is merely harbored by man.

An entire section of this volume is devoted to Veterinary Rickettsiology. Rickettsiology is gaining importance in the veterinary sciences on account of several factors, for example, increased awareness of the veterinary community as to the extent of infections by the rickettsiae and associated organisms in animals, the scientific advances, and the economic factors. It may help us understand the pathogenicity of these bacteria in vertebrates that become infected with rickettsiae via the vector or by ingestion by the vertebrate of rickettsiae-infected invertebrates. We hope and expect that this subdiscipline will grow, so as to assist our understanding of the effects of rickettsiae and associated organisms on animal vertebrates.

Persistence of *Anaplasma* spp. in the animal reservoir host is described. Such persistence is required for efficient tick-borne transmission of these pathogens to animals and humans. With *A. marginale* infection used as a model, persistent infection can be shown to reflect cycles in which antigenic variants emerge, replicate, and are brought under control by the immune system. Long-term persistent infection is dependent on the generation of complex mosaics in which segments from different unique hypervariable region sequences recombine into the expression site. The resulting combinatorial diversity generates the numbers of variants that have been both predicted and shown to emerge during persistence. The authors conclude that the recombination of immunodominant outer-membrane protein variants in the membrane creates novel quaternary structural epitopes; this process could add a novel mechanism of posttranslational combinatorial diversity to the suite of mechanisms that induced generation of surface-coat variation. In addition, recent studies raise the possibility that infection-induced regulation of the immune response is synergistic with the defined mechanisms of antigenic variation, thereby extending the period during which recombination can generate successful escape variants. Both the immune regulatory mechanisms and the specific requirement for generation of closely related variants to trigger these mechanisms are as yet obscure and they should provide exciting areas for future investigation.

In a theoretical consideration of the "ultimate" vaccine for the control of anaplasmosis, the conclusion is that it will be one that reduces infection in, and transmission of the pathogen by, the tick vector. Various approaches are presented. One is the characterization of *A. marginale* adhesins, which shows potential as a candidate vaccine antigen for the control of bovine anaplasmosis. Another approach identifies and characterizes tick-protective antigens; this approach has produced vaccine candidates that reduce tick infestation, molting, and oviposition and that affect *Anaplasma* infection levels in ticks. However, transmission-blocking antigens have not been identified from either the tick vector or the pathogen. In another presentation, the authors describe an evaluation of *E. ruminantium* genes in mice, as potential DNA-construct vaccine candidates for control of heartwater. Heartwater is an acute and fatal tick-borne disease of domestic and some wild ruminants. The authors concluded that certain of their DNA-construct vaccines offered partial protection against lethal challenge, as demonstrated by the reduction in mortality compared to the control groups. Furthermore, protection is increased when DNA-primed mice are boosted with the corresponding homologous recombinant protein.

A review of the new findings and advances in research of members of the family *Anaplasmataceae* of veterinary importance is providing us with insights into the evolution, reservoir, and transmission of these organisms in nature and their pathogenesis in natural and accidental hosts. Our ability to now apply molecular approaches to research on *Anaplasmataceae* is not only advancing our diagnostic capabilities, but also significantly advancing our understanding of these organisms and their evolutionary relationship, as well as the pathogenesis of the diseases that they cause. It is through these types of studies that surveillance, diagnosis, preventive measures, and treatment of ehrlichioses, in both animals and humans, will be improved. Another paper reviews *A. phagocytophilum* in ruminants in Europe as the causative agent of what is known in Europe as tick-borne fever (TBF). Serological evidence of infection of humans has been demonstrated in several European countries, creating a renewed interest and increased awareness of the zoonotic potential of TBF variants. It remains to be established whether the variants causing HGA in Europe are genetically and biologically different from those causing TBF in ruminants.

An epidemiological survey of *Ehrlichia*, the causative agent of canine monocytic ehrlichiosis, and related species, in dogs in eastern Sudan, shows that the tick *Rh. sanguineus* transmits the agent. The infection rates of dogs with *E. canis* in Sudan are far higher than those reported in other countries. Also, the authors show first evidence of coinfection with *E. canis* and hemoplasma. A similar survey on seroprevalence of *E. canis* in dogs from the Ivory Coast and Gabon also shows a very high seroprevalence.

A study in France done to determine the virulence and the pathogenicity of *E. canis* strain Borgo 89 in experimentally infected dogs showed that not all dogs were symptomatic as a result of the infection. The authors conclude that the prevalence of the natural disease is likely underestimated.

Results obtained by gene sequencing of isolates of *R. rickettsii* from naturally infected dogs and people in North Carolina show that a high degree of homology exists in the dog and human isolates. Clinical manifestations are strikingly similar in people and in dogs. Illness in dogs can precede illness in people in the same household. The clinical and temporal relationships of the naturally occurring disease in dogs and people suggest that dogs can serve as sentinels for natural infection.

In a review on *Bartonella*, felids, including free-ranging wild and domestic felids, represent a major reservoir for several *Bartonella* species. Prevalence of infection is highest in the warm and humid climates optimal for the survival of cat fleas, which is essential for transmission of the infection. Flea feces are the likely infectious substrate. Cats are usually asymptomatic, but uveitis, endocarditis, neurological signs, fever, necrotic lesions at the inoculation site, lymphadenopathy, and reproductive disorders have been reported in naturally or experimentally infected cats. Molecular studies of a given species (e.g., *B. henselae* isolated from various wild animals) indicate the coexistence of various subspecies of the bacterium.

The laboratory diagnosis of rickettsioses and diseases caused by associated organisms is still retrospective and therefore confirmatory. Isolation of rickettsiae and their maintenance in cell cultures remain daunting tasks, even though a number of new techniques and improvements on old techniques have been described. There is an urgent need to increase the efficiency of isolation and culture of the rickettsiae. Also, improvements of laboratory serodiagnostic and antigen diagnostic methodologies are needed, requiring the development of new platforms or improvement of established platforms, so as to decrease the turnaround time for test results. We need to increase test sensitivity, to lower the minimum number of bacteria that can be practically detected, and to increase our ability to identify the isolated bacterium at the species level. In this volume, various papers attempt to address these issues. Methods of isolation and cultivation of new rickettsiae are described; by such methods, we can carry out a complete study of all variant rickettsiae within a population, including the nonvirulent rickettsiae. A study in Brazil provides a description of the first cell-culture isolation of *R. rickettsii* and *R. bellii*, from *Amblyomma aureolatum* ticks.

The efficacy of a *Rickettsia* genus-specific and a *R. prowazekii* species-specific quantitative real-time PCR (qPCR) cassette was assessed, using experimentally infected human body lice. The rickettsial burden in infected lice demonstrated an initial decrease in total number of rickettsia after treatment followed by a robust increase coincident with louse death.

Another paper describes the isolation of the Ap-Variant 1 strain of *A. phagocytophilum* from laboratory-infected goats and from field-collected *Ix. scapularis* ticks. The isolation was made in the *Ix. scapularis*–derived cell line. A 10-year study to isolate rickettsial species from blood samples using the shell vial cell culture assay showed that, at present, this technique is the most suitable method to obtain rickettsiae from patient blood samples and/or for use in a PCR assay to identify the rickettsia at the species level. A paper describes the PCR procedure for the detection of *A. phagocytophilum*. The authors indicate that early in the infection, negative serologic test for *A. phagocytophilum* is frequent in patients with HGA and positive PCR assay. However, a positive PCR test is uncommon in human patients who have positive serologic test results for the HGA agent. Another study substantiates the use of PCR to confirm the clinical diagnosis of HGA. A multiplex detection of *Ehrlichia* and *Anaplasma* pathogens in vertebrate and tick hosts by real-time RT-PCR is described. The authors claim that this molecular test is useful to rapidly diagnose single infection or coinfection by up to five tick-borne rickettsial pathogens. The test serves as a tool to monitor rickettsial infections in dogs and ticks. It may also be a useful tool for experimental infection studies using single infection or coinfection to assess the disease outcome, and to evaluate vaccines and therapeutics. An evaluation of a real-time PCR assay to detect *C. burnetii* shows that the test is highly sensitive, with reproducible detection limits of approximately 10 copies per reaction; it is thus at least 100 times more sensitive than capture enzyme-linked immunosorbent assay (ELISA) when performed on infected placental material. Another *C. burnetii* PCR study showed that use of the insertion sequence IS1111 provides greater sensitivity than does targeting of the 27-kDa outer-membrane protein COM1. Molecular typing of novel *R. rickettsii* isolates from *Rh. sanguineus* in Arizona is presented. The authors indicate that variable nucleotide tandem repeat (VNTR) sequences, and probably other sequence variants, provide novel molecular epidemiologic tools for further investigating the origin of RMSF in Arizona. The use of VNTR sequences in an automated platform is also described in another study, which showed promising results in identifying the rickettsia to the species level and for use in epidemiologic surveillance. The use of molecular techniques is described to identify the SFG rickettsiae *Rickettsia sibirica* to the subspecies level.

A study using an immunoblot platform with monoclonal antibodies as the first antibody to identify and characterize *C. burnetii* strains and isolates is described. Such a testing procedure revealed unique characteristics of the strain of *C. burnetii* studied, as well as the variability in its structural components. In another study, test results for four commercially available assays for the detection of IgM phase II antibodies to *C. burnetii*, in the diagnosis of acute Q fever, were found to be equivalent. A report on the evaluation of IgG antibody response against *R. conorii* and *R. slovaca* in patients with DEBONEL/TIBOLA indicated that the species-level specific identification of the SFG rickettsiae studied cannot be accomplished by serologic means. However, a study indicated that the identification of the rickettsia at the species level could be done when the sera that initially yielded positive results were cross-absorbed and the absorbed sera retested. Early diagnosis of rickettsioses by electrochemiluminescence is promising, when whole rickettsiae are used to detect antibodies in the acute phase of the disease. A study presents a description of a multiplexed serological test in atypical pneumonia, which may be caused by *C. burnetii* and/or other microorganisms. The authors conclude that the development of a protein microarray could provide many advantages in the determination of specific etiological pathogens involved in atypical pneumonia. Use of such an array would provide the ability to select the optimal drug for treatment, thereby potentially reducing the course of the disease and the occurrence of antibiotic resistance and adverse drug reactions, as well as the cost of treatment. In the same vein, another study describes the successful use of a corpuscular antigenic microarray for the serodiagnosis of blood culture–negative endocarditis. A comparison of the immune response against *O. tsutsugamushi*, in a 4-week-old and a 10-week-old scrub typhus–infected laboratory mouce, showed by ELISA that older mice mount a response against *O. tsutsugamushi* faster than do younger mice.

Clearly, the full scope of rickettsial pathogens and rickettsia of unknown pathogenicity is still being defined. This upsurge in isolation/identification of new rickettsiae, the discovery that vectors can carry pathogenic rickettsiae not known to be harbored by them, and the resurgence of RMSF in various parts of the world (e.g., the United States and Brazil) may be indicative of the increase of numbers and maybe the severity of outbreaks of rickettsioses. In the United States, cases of RMSF cycle are increasing and then gradually decreasing. The question that recurred constantly at the meeting was, "Are we truly in an upswing in the number of cases of rickettsioses?" Only time will tell. Veterinary rickettsiology is growing as a subdiscipline, as more veterinarians become aware of the rickettsial threat to domestic and peridomestic animals, and the economic impact on animal husbandry.

We, the editors, thank the authors of the papers in these volumes for the time and effort that they have put into writing their manuscripts. We also greatly thank the reviewers. The New York Academy of Sciences is to be thanked for publishing these proceedings of the fourth International Conference. We are also grateful to Linda Mehta and Steve Bohall of the Academy, as well as to Julia Keller of Blackwell Publishing, for their efforts in seeing this volume through the press.

Insights into Mechanisms of Bacterial Antigenic Variation Derived from the Complete Genome Sequence of *Anaplasma marginale*

GUY H. PALMER,[a] JAMES E. FUTSE,[a] DONALD P. KNOWLES, JR.,[b] AND KELLY A. BRAYTON[a]

[a]*Program in Vector-Borne Diseases, Washington State University, Pullman, Washington 99164-7040, USA*

[b]*Animal Diseases Research Unit, USDA-ARS, Pullman, Washington 99164-7030, USA*

ABSTRACT: Persistence of *Anaplasma* spp. in the animal reservoir host is required for efficient tick-borne transmission of these pathogens to animals and humans. Using *A. marginale* infection of its natural reservoir host as a model, persistent infection has been shown to reflect sequential cycles in which antigenic variants emerge, replicate, and are controlled by the immune system. Variation in the immunodominant outer-membrane protein MSP2 is generated by a process of gene conversion, in which unique hypervariable region sequences (HVRs) located in pseudogenes are recombined into a single operon-linked *msp2* expression site. Although organisms expressing whole HVRs derived from pseudogenes emerge early in infection, long-term persistent infection is dependent on the generation of complex mosaics in which segments from different HVRs recombine into the expression site. The resulting combinatorial diversity generates the number of variants both predicted and shown to emerge during persistence.

KEYWORDS: *Anaplasma*; *A. marginale*; antigenic variation; immune evasion; gene conversion; functional supergenes

INTRODUCTION: THE IMPORTANCE OF PERSISTENT INFECTION IN *ANAPLASMA* TRANSMISSION

Tick-borne transmission of *Anaplasma* spp. requires continual presence of an infected mammalian reservoir host to serve as a source for ticks to acquire

Address for correspondence: Guy H. Palmer, Department of Veterinary Microbiology and Pathology, Washington State University, Pullman, WA 99164-7040. Voice: 509-335-6033; fax: 509-335-8529. e-mail: gpalmer@vetmed.wsu.edu

**Ann. N.Y. Acad. Sci. 1078: 15–25 (2006). © 2006 New York Academy of Sciences.
doi: 10.1196/annals.1374.002**

and subsequently transmit the pathogen to susceptible animals, including humans. The ticks themselves cannot maintain *Anaplasma* between subsequent generations as transovarial transmission does not occur.[1–3] As a consequence, each generation of ticks must acquire the organism—which, in turn, is dependent on the presence of infected reservoir hosts. Natural reservoir hosts for *Anaplasma* spp. maintain a persistent infection, thus maximizing the opportunity for the appropriate tick stage, which is present only during a limited period of the year, to acquire the pathogen.[4–15] Importantly, ticks are able to efficiently acquire *Anaplasma* during feeding on the low bacteremia levels associated with long-term persistence.[4,5,14]

ANTIGENIC VARIATION AS A MECHANISM OF PERSISTENT *ANAPLASMA* INFECTION

Studies over the past decade using infection of *Anaplasma marginale* in its natural cattle reservoir host have revealed that persistence reflects a classic model of antigenic variation (FIG. 1).[2,16–18] Although *A. marginale* is an obligate intracellular parasite, as are all bacteria in the order *Rickettsiales*,[19] the presence of cyclic bacteremia is clearly reminiscent of the cycles that occur in relapsing fever and African trypanosomiasis, caused by, respectively, an extracellular bacterium (*Borrelia hermsii*) and an extracellular protozoon (*Trypanosoma brucei*).[16,20] This similarity among genetically widely disparate organisms reflects the strong evolutionary pressure for development of a mechanism to generate antigenic variants and highlights the importance of antigenic variation in efficient vector-borne transmission.[20,21] In *A. marginale* infection, each cycle reflects the emergence of one or, more commonly, multiple clones that express a unique immunodominant outer-membrane protein, designated major surface protein (MSP)-2.[6,17,18] These variants express a unique surface-

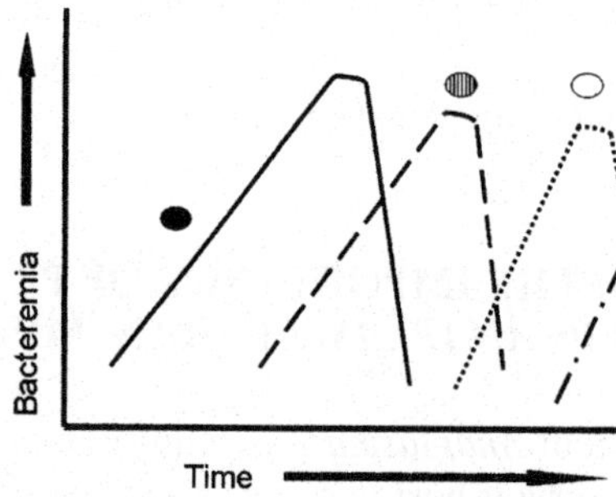

FIGURE 1. *Anaplasma marginale* persistent infection is characterized by sequential emergence, replication, and immune control of organisms expressing antigenically variant MSP2. Although only four variant cycles are shown, cyclic bacteremia persists for at least 6 years and appears to be lifelong in infected ruminants. (Modified from Palmer[21] with permission.)

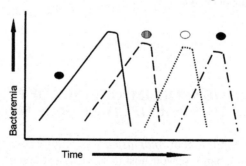

exposed domain, designated the hypervariable region (HVR) flanked by highly conserved N- and C-terminal domains.[6,17,18] These represent true "escape variants" as they are not recognized by antibody present at the time of emergence and are controlled concomitant with the development of IgG2 directed against the specific HVR.[2,18] This cycle of emergence and control continues unabated, allowing life-long persistent *A. marginale* infection.[2,16,18,22]

GENERATION OF MSP2 ANTIGENIC VARIANTS VIA HOMOLOGOUS RECOMBINATION

The question as to how *A. marginale*, a small (approximately 1.2 Mb) genome pathogen, generates this remarkable MSP2 variation has been the subject of ongoing investigation. Initial studies by Barbet *et al.* identified a single operon-linked *msp2* expression site,[23] which was subsequently confirmed by whole genome sequencing[24,25] and the operon defined by transcriptional analysis[26] (FIG. 2). The *A. marginale* expression locus, which is syntenic with the *msp2* (also designated p44) expression site in *A. phagocytophilum*,[27–29] is composed of a transcriptional regulator, followed by, in the direction of the polycistronic transcript, five genes encoding outer-membrane proteins: outer-membrane protein (*omp*)-1, operon-associated genes (*opag)* 1-3, and *msp2*.[23,25,26,28,30] Unlike the *omp-1* and the *opags*, which are highly genetically stable during acute and persistent infection, the expressed *msp2* is highly variable, reflecting the diversity of sequential variants that arise, are recognized

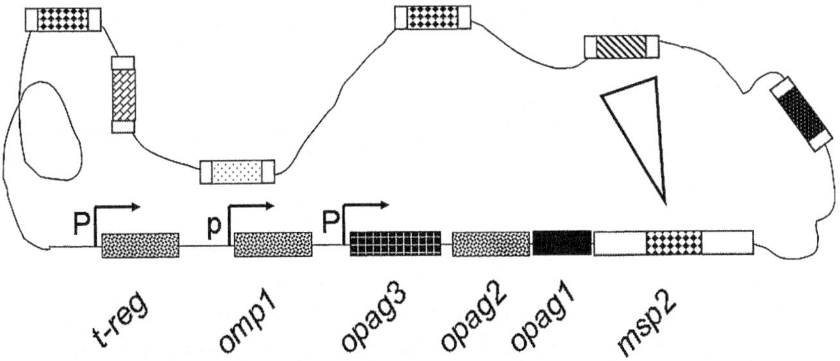

FIGURE 2. Structure of the single *msp2* expression site and the multiple *msp2* functional pseudogenes in the *A. marginale* chromosome. The *msp2* expression locus is within an operon that includes transcripts initiated from multiple promoters. The direction of transcription is indicated by the *arrows* of the promoters. "P" represents the two stronger promoters, 5′ to the putative transcriptional regulator (*t-reg*) and 5′ to *opag3*; "p" represents the weak promoter 5′ to *omp1*.[26] The HVR in both the expression site *msp2* and the pseudogene copies are flanked by identical 5′ and 3′ domains, represented in *light gray*, which direct the recombination (indicated by the *triangle*).

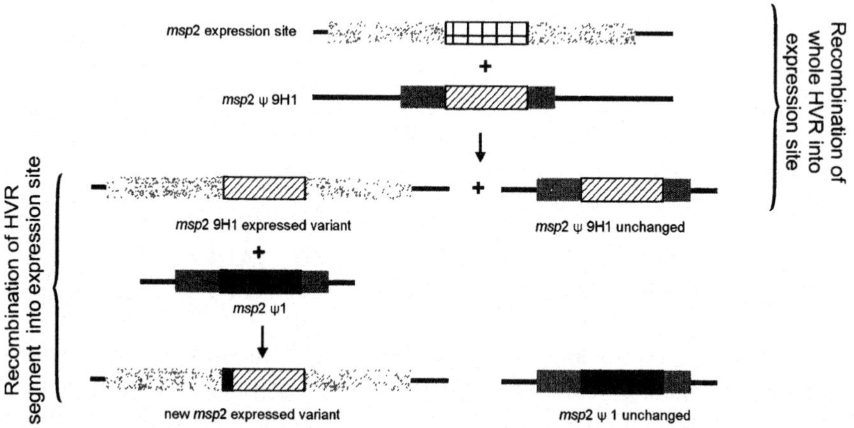

FIGURE 3. MSP2 variants are generated by gene conversion utilizing either the intact HVR or an oligonucleotide segment of the *msp2* pseudogenes. In either recombination event, the original pseudogene is unchanged. Ψ1 and Ψ9H1 are representative pseudogenes identified in the St. Maries strain of *A. marginale*.[25] The identical 5′ and 3′ regions flanking the HVR are indicated in gray. The figure has been modified from Reference 31, with permission.

by the immune system, and are subsequently cleared.[6,17,23,24,28,30–32] The source of this diversity is the presence of multiple partial *msp2* chromosomal sequences (FIG. 2). These sequences resemble the complete *msp2* but are truncated on both the 5′ and 3′ ends, encoding both the N- and C-termini.[23–25,31] Although these are not complete genes and cannot be expressed from their chromosomal loci, they encode unique HVRs that serve as a template for homologous recombination into the single *msp2* expression site. Thus, while they are pseudogenes in their current loci, these truncated sequences are also functional in the context of the organism's biology, and have been designated "functional pseudogenes."[24,31] They are not true "silent gene copies," a terminology used in other microbes that generate antigenic variation through recombination, as only a small part of the *msp2* pseudogene, the HVR, and the identical immediate flanking regions, are used in recombination. The recombination mechanism itself is a unidirectional gene conversion in which the pseudogene HVR is recombined into the expression site, but is also retained unchanged within its nonexpression site.[24,31] The pre-existing expression site copy is lost as the new *msp2* variant is expressed (FIG. 3).

USE OF SEGMENTAL GENE CONVERSION TO GENERATE MSP2 ANTIGENIC VARIANTS

The recombination of pseudogene HVRs, tethered by their immediate flanking regions, which are identical to those in the expression site copy, as a mech-

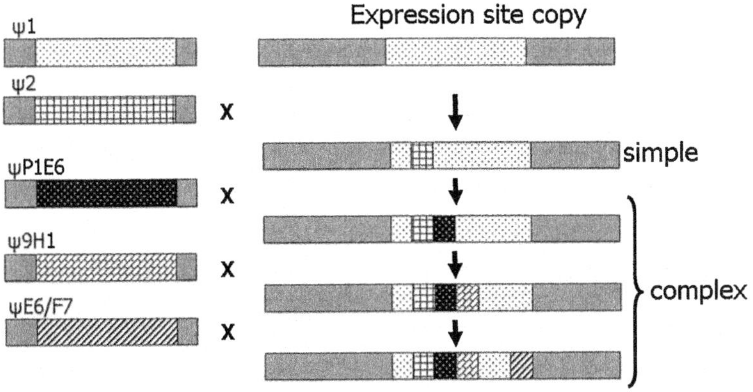

FIGURE 4. Generation of complex MSP2 variants. The repertoire of *msp2* functional pseudogenes with unique HVRs in the St. Maries strain of *A. marginale* is shown on the left and the expression site *msp2* on the right. Simple variants can be generated by a single recombination event utilizing either a whole HVR (see FIG. 3) or an oligonucleotide segment of the pseudogene HVR, as shown in the first recombination (indicated by "**X**"). Sequential recombination events utilizing segments derived from different pseudogene HVRs generates progressively complex expression site HVR mosaics.

anism to generate new antigenic variants is limited by the number of pseudogenes with unique HVRs. Complete genome sequencing of the St. Maries strain of *A. marginale* has identified seven copies, of which only five encode unique HVRs (FIG. 4).[25] Targeted sequencing of *msp2* pseudogene loci in other strains supports that the number of pseudogenes is limited, less than 10/genome in all strains examined to date.[25,33] While this number is clearly inadequate to generate the number of variants either predicted or actually identified during persistent infection, examination of sequential variants demonstrated the presence of expressed HVRs that were composed of oligonucleotide segments derived from different pseudogene HVRs (FIGS. 3, and 4).[31] This process, termed "segmental gene conversion," results in a dramatic increase in the number of possible variants, from <10 using only the whole HVR of the pseudogenes to thousands generated by the combination of one of several oligonucleotide segments from each of the unique pseudogene HVRs.[31]

As recently reported by Futse *et al.*,[34] these two related gene conversion mechanisms, recombination of whole HVRs and HVR segments, account for almost all of the variation (98.8%) observed during more than a year of tracking over 1,300 variants during infection. The remaining 1.2% represent relatively short oligonucleotide deletions, insertions of oligonucleotides not represented in the genome, or stretches of substitutions.[34] Notably, all of these changes occurred at recombination sites, consistent with their introduction during mismatch repair during the process of gene conversion. These repair-associated changes may well occur with greater frequency than what is actually detected in emergent organisms; however, the majority of changes, specifically those that would result in a change in reading frame or premature introduction of a stop codon, would not be expected to generate viable organisms. Regardless, the number of variants derived using these alternative mechanisms is relatively small and does not appear to be a major contributor to MSP2 antigenic variation and *A. marginale* persistence.

PROGRESSION OF VARIANT COMPLEXITY DURING PERSISTENT INFECTION

Initial studies by Brayton *et al.* identified the preferential emergence of "simple variants," those defined as being generated by a single recombination event, either using a whole HVR or a single HVR segment (FIG. 3), early in infection using the South Idaho strain of *A. marginale*.[24] The availability of the genome sequence of the St. Maries strain,[25] which provided the complete pseudogene complement, allowed subsequent studies of acute through persistent infection in which each expressed variant could be mapped precisely to its pseudogene template. The data, generated from the study of infected animals during one year of infection, supported the preferential emergence of simple variants early in infection with "complex variants," those representing mosaics with sequences derived from at least three different pseudogenes (FIG. 4), becoming predominant during persistent infection and increasing with duration of infection (FIG. 5A).[34] The use of whole HVRs decreased dramatically during infection, consistent with the development and maintenance of an immune response capable of recognizing and clearing organisms expressing whole HVRs (FIG. 5B). These findings underscore the importance of segmental gene conversion and consequent generation of HVR mosaics in long-term persistent infection. The expression site HVR mosaics are not represented by any single pseudogene HVR (FIG. 4) and with each subsequent change allow generation of additional unique variants that emerge unrecognized by the existing immune response, resulting in the subsequent bacteremic cycle. There are, at least, two explanations for the preferred sequential use of whole HVRs early in infection versus developing complex mosaics on a backbone of a single HVR and then later switching to a second HVR with subsequent mosaic generation, etc. The first assumes that the recombination of pseudogene whole HVRs, sharing identical 5′ and 3′ flanking regions with the expression site copy, occurs at a substantially greater frequency than that of segments.[34] Thus, variant emergence early in infection simply reflects the frequency of recombination as all generated variants, simple or complex, would be unrecognized by the existing immune response. The second explanation supposes that early in infection, in the absence of the strong selective pressure of the immune system, an *A. marginale* clone expressing a MSP2 HVR derived from a pseudogene whole HVR would have a growth advantage over clones expressing complex mosaics,

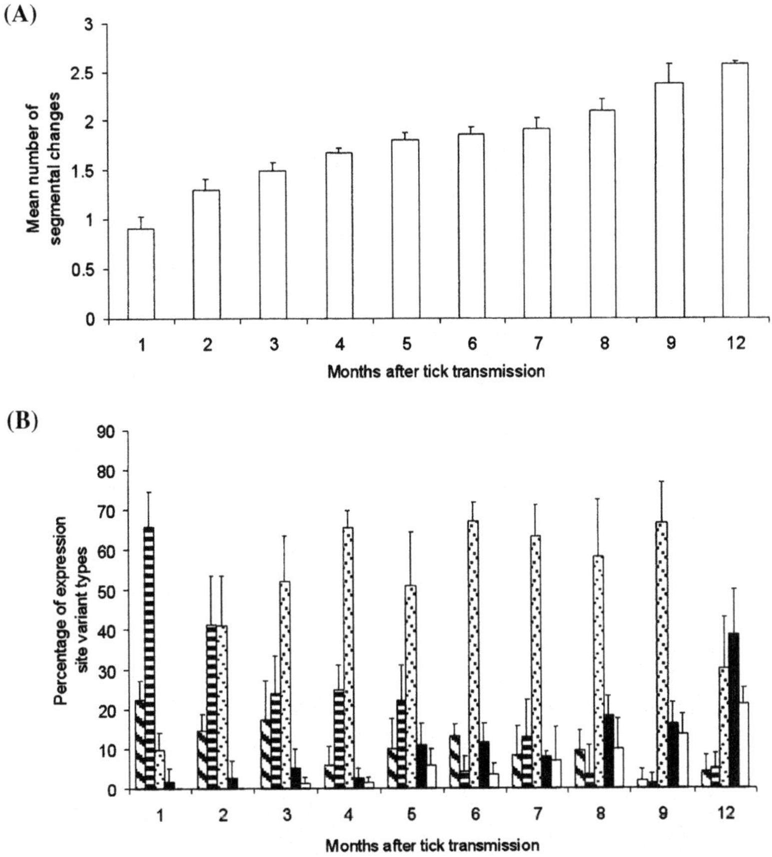

FIGURE 5. Increasing complexity of MSP2 variants during persistent infection. Greater than 1,300 variants were sequenced from animals persistently infected for >12 months and each sequence mapped to its parent *msp2* pseudogene. (**A**) Change in mean complexity of the expressed variants during infection. Complexity was scored as follows: 0, represents variants derived from recombination of a whole HVR; 1, variants with a single segmental change; 2, variants with two segmental changes derived from different pseudogenes; 3, variants with three segmental changes each derived from different pseudogenes, *etc*. The bars represent the mean complexity ± 1 SD. (**B**) Change in predominance of simple versus complex variants during infection. *Crosshatched bars* represent variants derived from recombination of a whole HVR; *horizontal bars*, single segmental changes; *stippled bars*, two segmental changes; *black bars*, three segmental changes; white bars, four segmental changes. Percentage of total variants examined ±1 SD. (From Futse *et al.*[34] Reproduced by permission.)

based on MSP2 function. This would signify that the *msp2* pseudogene HVR sequences have evolved for optimal growth and that the mosaics are at a growth disadvantage until the immune response develops to provide the strong selective pressure. Both explanations remain open at this point.

CONCLUSIONS AND FUTURE DIRECTIONS

Combinatorial diversity using assembly of oligonucleotide segments into complex HVR mosaics is responsible for the generation of MSP2 variants during long-term *A. marginale* persistent infection. This mechanism also appears to be manifest in generating variation in MSP3, a related immunodominant protein.[35] Similar to MSP2, MSP3 variants are generated using recombination of functional pseudogenes into a single expression site, and MSP2 and MSP3 variants emerge and are cleared simultaneously.[35,36] Whether the combination of MSP2 and MSP3 variants in the membrane, where they associate "nearest-neighbors,"[37–39] creates novel quaternary structure epitopes is intriguing and, if identified, would add a novel mechanism of post-translational combinatorial diversity to the generation of surface coat variation. In addition, recent studies using an immunization model of immunity to MSP2 have uncovered loss of CD4+ T lymphocyte responses upon infection,[40] raising the possibility that infection-induced regulation of the immune response is synergistic with the defined mechanisms of antigenic variation, extending the period for recombination to generate successful escape variants. Both the immune regulatory mechanisms as well as the requirement for generation of closely related variants in triggering this mechanism are unknown and will provide exciting areas for investigation.

ACKNOWLEDGMENTS

This primary research reviewed in this manuscript was supported by NIH R01 AI44005 and AI 45580; and USDA-CSREES NRI 2001-52100-11342; USDA-ARS-CRIS 5348-32000-016-00D and 58-5348-3-212. FIGURE 1 is modified from a figure originally published in *Veterinary Immunology and Immunopathology*.[21] FIGURE 3 is modified from a figure originally published in *Molecular Microbiology*[31] and FIGURE 5 is reproduced from its original publication in *Molecular Microbiology*.[34] Figures are published with permission of the authors and publishers.

REFERENCES

1. STICH, R.W. *et al.* 1989. Transstadial and attempted transovarial transmission of *Anaplasma marginale* by *Dermacentor variabilis*. Am. J. Vet. Res. **50:** 1377–1380.
2. PALMER, G.H., W.C. BROWN & F.R. RURANGIRWA. 2000. Antigenic variation in the persistence and transmission of the ehrlichia *Anaplasma marginale*. Microbes Infect. **2:** 167–176.
3. KOCAN, K.M. 1995. Targeting ticks for control of selected hemoparasitic diseases of cattle. Vet. Parasitol. **57:** 121–151.

4. ERIKS, I.S., D. STILLER & G.H. PALMER. 1993. Impact of persistent *Anaplasma marginale* rickettsemia on tick infection and transmission. J. Clin. Microbiol. **31:** 2091–2096.
5. FUTSE, J.E. *et al.* 2003. Transmission of *Anaplasma marginale* by *Boophilus microplus*: retention of vector competence in the absence of vector–pathogen interaction. J. Clin. Microbiol. **41:** 3829–3834.
6. RURANGIRWA, F.R. *et al.* 1999. Restriction of major surface protein 2 (MSP2) variants during tick transmission of the ehrlichia *Anaplasma marginale*. Proc. Natl. Acad. Sci. USA **96:** 3171–3176.
7. ZAUGG, J.L. *et al.* 1986. Transmission of *Anaplasma marginale* Theiler by males of *Dermacentor andersoni* Stiles fed on an Idaho field-infected, chronic carrier cow. Am. J. Vet. Res. **47:** 2269–2271.
8. CHRISTENSEN, J.F. *et al.* 1960. Persistence of latent *Anaplasma marginale* infection in deer. J. Am. Vet. Med. Assoc. **136:** 426–427.
9. ZAUGG, J.L. 1986. Experimental anaplasmosis in American bison: persistence of infections of *Anaplasma marginale* and non-susceptibility to *A. ovis*. J. Wildl. Dis. **22:** 169–172.
10. ZAUGG, J.L. 1988. Experimental anaplasmosis in mule deer: persistence of infection of *Anaplasma marginale* and susceptibility to *A. ovis*. J. Wildl. Dis. **24:** 120–126.
11. STUEN, S., E.O. ENGVALL & K. ARTURSSON. 1998. Persistence of *Ehrlichia phagocytophila* infection in lambs in relation to clinical parameters and antibody responses. Vet. Rec. **143:** 553–555.
12. STUEN, S. & K. BERGSTROM. 2001. Persistence of *Ehrlichia phagocytophila* infection in two age groups of lambs. Acta Vet. Scand. **42:** 453–458.
13. STUEN, S., R. DJUVE & K. BERGSTROM. 2001. Persistence of granulocytic *Ehrlichia* infection during wintertime in two sheep flocks in Norway. Acta Vet. Scand. **42:** 347–353.
14. TELFORD, S.R., 3RD, *et al.* 1996. Perpetuation of the agent of human granulocytic ehrlichiosis in a deer tick-rodent cycle. Proc. Natl. Acad. Sci. USA **93:** 6209–6214.
15. PALMER, G.H. *et al.* 1998. Persistence of *Anaplasma ovis* infection and conservation of the *msp-2* and *msp-3* multigene families within the genus *Anaplasma*. Infect. Immun. **66:** 6035–6039.
16. KIESER, S.T., I.S. ERIKS & G.H. PALMER. 1990. Cyclic rickettsemia during persistent *Anaplasma marginale* infection of cattle. Infect. Immun. **58:** 1117–1119.
17. FRENCH, D.M. *et al.* 1998. Expression of *Anaplasma marginale* major surface protein 2 variants during persistent cyclic rickettsemia. Infect. Immun. **66:** 1200–1207.
18. FRENCH, D.M., W.C. BROWN & G.H. PALMER. 1999. Emergence of *Anaplasma marginale* antigenic variants during persistent rickettsemia. Infect. Immun. **67:** 5834–5840.
19. DUMLER, J.S. *et al.* 2001. Reorganization of genera in the families *Rickettsiaceae* and *Anaplasmataceae* in the order *Rickettsiales*: unification of some species of *Ehrlichia* with *Anaplasma*, *Cowdria* with *Ehrlichia* and *Ehrlichia* with *Neorickettsia*, descriptions of six new species combinations and designation of *Ehrlichia equi* and "HGE agent" as subjective synonyms of *Ehrlichia phagocytophila*. Int. J. Syst. Evol. Microbiol. **51:** 2145–2165.
20. BARBOUR, A.G. & B.I. RESTREPO. 2000. Antigenic variation in vector-borne pathogens. Emerg. Infect. Dis. **6:** 449–457.

21. Palmer, G.H. 2002. The highest priority: what microbial genomes are telling us about immunity. Vet. Immunol. Immunopathol. **85:** 1–8.
22. Eriks, I.S. *et al.* 1989. Detection and quantitation of *Anaplasma marginale* in carrier cattle by using a nucleic acid probe. J. Clin. Microbiol. **27:** 279–284.
23. Barbet, A.F. *et al.* 2000. Antigenic variation of *Anaplasma marginale* by expression of MSP2 mosaics. Infect. Immun. **68:** 6133–6138.
24. Brayton, K.A. *et al.* 2001. Efficient use of a small genome to generate antigenic diversity in tick-borne ehrlichial pathogens. Proc. Natl. Acad. Sci. USA **98:** 4130–4135.
25. Brayton, K.A. *et al.* 2005. Complete genome sequencing of *Anaplasma marginale* reveals that the surface is skewed to two superfamilies of outer membrane proteins. Proc. Natl. Acad. Sci. USA **102:** 844–849.
26. Barbet, A.F. *et al.* 2005. Identification of functional promoters in the msp2 expression loci of *Anaplasma marginale* and *Anaplasma phagocytophilum.* Gene **353:** 89–97.
27. Lin, Q. *et al.* 2003. Mechanisms of variable p44 expression by *Anaplasma phagocytophilum*. Infect. Immun. **71:** 5650–5661.
28. Löhr, C.V. *et al.* 2004. Characterization of the *Anaplasma marginale msp2* locus and its synteny with the *omp1/p30* loci of *Ehrlichia chaffeensis* and *E. canis.* Gene **325:** 115–121.
29. Barbet, A.F. *et al.* 2003. Expression of multiple outer membrane protein sequence variants from a single genomic locus of *Anaplasma phagocytophilum.* Infect. Immun. **71:** 1706–1718.
30. Löhr, C.V. *et al.* 2002. Expression of *Anaplasma marginale* major surface protein 2 operon-associated proteins during mammalian and arthropod infection. Infect. Immun. **70:** 6005–6012.
31. Brayton, K.A. *et al.* 2002. Antigenic variation of *Anaplasma marginale* msp2 occurs by combinatorial gene conversion. Mol. Microbiol. **43:** 1151–1159.
32. Barbet, A.F. *et al.* 2001. Antigenic variation of *Anaplasma marginale*: major surface protein 2 diversity during cyclic transmission between ticks and cattle. Infect. Immun. **69:** 3057–3066.
33. Rodriguez, J.L., G.H. Palmer, D.P. Knowles, & K.A. Brayton, 2005. Distinctly different *msp2* pseudogene repertoires in *Anaplasma marginale* strains that are capable of superinfection. Gene **361:** 127–132.
34. Futse, J.E. *et al.* 2005. Structural basis for segmental gene conversion in generation of *Anaplasma marginale* outer membrane protein variants. Mol. Microbiol. **57:** 212–221.
35. Meeus, P.F. *et al.* 2003. Conservation of a gene conversion mechanism in two distantly related paralogues of *Anaplasma marginale.* Mol. Microbiol. **47:** 633–643.
36. Brayton, K.A. *et al.* 2003. Simultaneous variation of the immunodominant outer membrane proteins, MSP2 and MSP3, during *Anaplasma marginale* persistence *in vivo.* Infect. Immun. **71:** 6627–6632.
37. Tebele, N., T.C. McGuire & G.H. Palmer. 1991. Induction of protective immunity by using *Anaplasma marginale* initial body membranes. Infect. Immun. **59:** 3199–3204.
38. Vidotto, M.C. *et al.* 1994. Intermolecular relationships of major surface proteins of *Anaplasma marginale.* Infect. Immun. **62:** 2940–2946.

39. Brown, W.C. *et al.* 1998. CD4(+) T-lymphocyte and immunoglobulin G2 responses in calves immunized with *Anaplasma marginale* outer membranes and protected against homologous challenge. Infect. Immun. **66:** 5406–5413.
40. Abbott, J.R. *et al.* 2005. Rapid and long-term disappearance of CD4+ T lymphocyte responses specific for *Anaplasma marginale* major surface protein-2 (MSP2) in MSP2 vaccinates following challenge with live A. *marginale*. J. Immunol. **174:** 6702–6715.

Rickettsiosis in Europe

J.R. BLANCO AND J.A. OTEO

Área de Enfermedades Infecciosas, Complejo San Millán –San Pedro–de La Rioja, Hospital de La Rioja, Logroño (La Rioja), Spain

Abstract: In Europe, rickettsioses are long-known infectious diseases. Until recently, it was thought that Mediterranean spotted fever due to *Rickettsia conorii* was the only tick-borne rickettsiosis in Europe. In the last decade new *Rickettsia* spp. have been implicated in human pathology (*R. slovaca, R. sibirica mongolotimonae, R. helvetica*). Furthermore, cases of infection due to flea-borne rickettsioses (*R. typhi, R. felis*) have been described. Finally, although no outbreak of epidemic typhus has been reported yet in central and southern Europe, we should be aware of the possibility of reemergence of this disease in Europe. Other rickettsioses exist that have not yet been implicated in human pathology. We should consider that climate changes and other factors could contribute to the emergence and reemergence of other new diseases.

Keywords: rickettsia; rickettsioses; tick; flea; louse; Europe

In Europe, rickettsioses are long-known infectious diseases. Although their epidemiology and etiology have changed with history, to date they still an important threat to the community. The oldest recorded epidemic that was thought to be typhus (*Rickettsia prowazekii*) was the Plague of Athens, which occurred around 429 B.C.[1,2] The earliest military report concerning typhus described more than 17,000 deaths in the Catholic kings, Army during the reconquest of Granada in 1489.[2] It is very feasible that this disease was introduced in Spain from sailors who returned from Crete, where the disease was endemic. Later, this disease reappeared in Europe during the Napoleonic Wars and the World Wars. After that the improvement of hygienic conditions and practices, the use of repellents, rodent control, and field sanitation have controlled outbreaks of these and other vector-borne disease. However, we should be aware of its possible reemergence.

Historically, the genus *Rickettsia* was divided into three groups on the basis of phenotypic criteria: the spotted fever group, the typhus group, and the scrub typhus group (this one is absent in Europe). In this review, we will describe them according to their vector (tick, flea, and louse).

Address for correspondence: José R. Blanco, Área de Enfermedades Infecciosas, Complejo San Millán–San Pedro–de La Rioja, Hospital de La Rioja, Avd. Viana 1, 26001, Logroño (La Rioja), Spain. Voice: +34-941-297-273; fax: +34-941-297-267.
e-mail: jrblanco@riojasalud.es

Ann. N.Y. Acad. Sci. 1078: 26–33 (2006).
doi: 10.1196/annals.1374.003

TICK-BORNE RICKETTSIOSES

Until recently it was thought that Mediterranean spotted fever (MSF), also known as "*botonneuse fever,*" "*Marseilles fever*" or "*Escaro-nodular fever,*" was the only tick-borne rickettsiosis in Europe, but new *Rickettsia* spp. are continuously being isolated and implicated in human pathology.

MSF is caused by *R. conorii* and some genetic variants (*R. conorii* Israeli strain and *R. conorii* Astrakhan strain). All of them are transmitted by *Rhipicephalus sanguineus* (brown dog tick).[3–5] The disease is endemic in southern Europe (Mediterranean area),[6–10] but has been described in other European countries as well.[11] Most of the cases happen during the hot months (end of spring and summer). After an incubation period of around 7 days, the onset of MSF is abrupt and patients have high fever, malaise, headache, rash (typically involving palms and soles) (100%) and frequently, a local black eschar (or *tache noire*) at the site of tick bite (70%). In reference to the genetic variants of *R. conorii*, in 1999, *R. conorii* Israeli strain was associated with human disease in Portugal and in Sicily.[12,13] It seems that the infection for this genetic variant is more severe than classical MSF. The presence of eschar is less common (35%).[14] In 2001, *R. conorii* Astrakhan strain was detected in one *R. sanguineus* taken from a French United Nation soldier in Kosovo.[15] This was the first detection of this rickettsia out of Russia.

During the last decade, other newly recognized tick-borne diseases have been described throughout Europe. One of these is *R. slovaca* infection. The implicated vector is *Dermacentor marginatus*, and wild boar is the main host. In Hungary this illness is named TIBOLA (tick-borne lymphadenopathy)[16] and DEBONEL in Spain (*Dermacentor*-borne-necrosis-erythema-lymphadenopathy).[17,18] *R. slovaca* has been identified in *D. marginatus* and *D. reticulatus* in a great majority of European countries.[19–22] In 1997, the first documented case of infection due to *R. slovaca* was reported.[23] Since then, several human cases have been described in Europe, mainly in Hungary, France, and Spain.[16,20,21,24–28] Most of the cases happen during the cold months (winter). The disease is characterized by a necrotic black lesion surrounded by erythema at the site of tick bite and regional painful lymph nodes. Fever and rash are uncommon.[18,20,21,28] DEBONEL, in patients with, tick-bite is more common in scalp (<0.001) than in MSF patients.[24,28] At this point the patient commonly develops alopecia.

In 1996, the first human case of infection due to *Rickettsia sibirica mongolotimonae* was reported.[29] Since then, new cases have been described, mainly in France.[30–32] This rickettsia was first isolated from *Hyalomma asiaticum* ticks collected in Mongolia in 1991.[33] In Europe, most of the cases occur in spring. In contrast with other tick-borne rickettsioses, in this disease the presence of enlarged lymph nodes, lymphangytis, and multiple eschars, is common,[32] these authors prompting to denominate this disease as LAR (lymphangitis-associated rickettsiosis). Because of the absence of this tick in Europe, it is possible that another tick could be the vector. The implication of birds in the dissemination of this infection (bird-carried ticks) seems to be very possible.[30,32]

In 1999, *R. helvetica* was implicated in a case of fatal perimyocarditis in a young patient in Sweden.[34] It has also been implicated as a cause of unspecific febrile illness[35,36] and in sarcoidosis,[37] although its role is unclear in the latter. This rickettsia has been detected in *Ixodes ricinus* in France, Sweden, Slovenia, Portugal, Italy, Switzerland, and more recently in Spain.[36,38–41] The infection is present during the hot months as a mild disease and is associated with fever, headache, arthralgia, and myalgia, but not with a cutaneous rash.[36]

More recently, *R. aeschlimannii* has been implicated in human pathology in a patient from France who returned from Morocco.[42] This rickettsia is mainly isolated from *Hyalomma marginatum*, although in other ticks such as *Rhipicephalus* sp. and *Ixodes* sp. it has been identified. *R. aeschlimannii* has been identified in *H. marginatum* in Portugal, Spain, Croatia, and the Corsica region in France[13,22,40,43,44] and symptoms are similar to those of MSF. In our area this tick commonly bites humans. We have demonstrated the presence of this rickettsia in patients bitten by *H. marginatum*, but none of them developed illness. It is possible that *R. aeschlimannii* is less pathogenic than the others.[43]

Finally, European clinicians should keep in mind other tick-borne rickettsioses described throughout the world, as in the case of African tick bite fever caused by *R. africae* in patients returning mainly from sub-Saharan Africa.[45–49]

FLEA-BORNE RICKETTSIOSES

In reference to flea-borne rickettsioses, the most important, and one of the most prevalent human rickettsioses is murine typhus (or endemic typhus or urban typhus). Rats and other rodents act as reservoir, and the rat flea (*Xenopsylla cheopis*) is the main vector of *R. typhi*. Humans are infected by inoculation through damaged skin by infected flea feces. Murine typhus occurs worldwide, mainly in warm areas, particularly in port cities and coastal regions, where rodents are present. In Europe, this rickettsiosis is prevalent in Portugal, Spain (included the Canary Islands), and Greece.[50–54] In southern Spain, murine typhus is considered an important cause of fever of intermediate duration.[51] Most of the cases occur in summer and fall. Patients have fever, constitutional symptoms, and often a poorly visible maculopapular exanthema on the trunk. Sometimes it is difficult to consider the implication of *R. typhi* on account of the absence of rats and its fleas. Further, the diagnoses of human cases were made by serology (absence of culture) and it is known that cross-reaction among rickettsias is the rule. So, it is very feasible that the disease could be due to other "murine typhus-like rickettsiosis" such as *R. felis* infection. *R. felis* is a spotted fever group rickettsia transmitted by cat fleas (mainly *Ctenocephalides felis*). *R. felis* has been described in fleas in Spain,[55,56] France,[57] and the United Kingdom.[58] In Europe, its pathogenic role has been demonstrated (by serology and/or polymerase chain reaction [PCR]) in patients from France[59] and Germany.[60] We should consider this diagnosis in patients with fever and/or rash and a history of cat contact or flea bite.

LOUSE-BORNE RICKETTSIOSES

Epidemic typhus is a disease caused by *R. prowazekii* and transmitted by the body louse (*Pediculus humanus corporis*). The main reservoir is humans. In Europe epidemic typhus is known by several names such as *Fleckfieber* (German), *typhus exanthematique* (French), or *tabardillo* (Spain). This disease is always closely allied with war, poverty, natural disasters, and hunger. Although the infection can apparently self-resolve, the bacteria can persist for life in humans, and under stress conditions (war, poor living conditions, natural disaster, lapses in public health) reappear as a Brill–Zinsser disease. Because this clinical form is bacteriemic, it can initiate an outbreak of epidemic typhus when body louse infestation is prevalent in the population. Although no outbreak of epidemic typhus has been reported to date in central and southern Europe, we should be aware of the possibility of reemergence of this disease in Europe (i.e., imported typhus).[61] Recently, an autochthonous case of *R. prowazekii* infection has been reported in a homeless patient from Marseille (France).[62]

Finally, there are other rickettsioses not implicated in human pathology yet, but it is very feasible that in the next few years these will be finally implicated as human pathogens.

REFERENCES

1. Roux, V. & D. Raoult. 1999. Body lice as tools for diagnosis and surveillance of reemerging diseases. J. Clin. Microbiol. **37:** 596–599.
2. Kelly, D.J., A.L. Richards, J. Temenak, *et al.* 2002. The past and present threat of rickettsial diseases to military medicine and international public health. Clin. Infect. Dis. **34:** S145–S169.
3. Gilot, B., M.L. Laforge, J. Pichot, *et al.* 1990. Relationships between the *Rhipicephalus sanguineus* complex ecology and Mediterranean spotted fever epidemiology in France. Eur. J. Epidemiol. **6:** 357–362.
4. Oteo, J.A., A. Estrada-Pena, C. Ortega-Pérez, *et al.* 1996. Mediterranean spotted fever: a preliminary tick field study. Eur. J. Epidemiol. **12:** 475–478.
5. Giammanco, G., S. Mansueto, P. Ammatuna, *et al.* 2003. Israeli spotted fever Rickettsia in Sicilian *Rhipicephalus sanguineus* ticks. Emerg. Infect. Dis. **9:** 892–893.
6. Raoult, D., P.J. Weiller, A. Chagnon, *et al.* 1986. Mediterranean spotted fever: clinical, laboratory and epidemiological features of 199 cases. Am. J. Trop. Med. Hyg. 35: 845–850.

7. RAOULT, D., P. ZUCHELLI, P.J. WEILLER, *et al.* 1986. Incidence, clinical observations and risk factors in the severe form of Mediterranean spotted fever among patients admitted to hospital in Marseilles 1983–1984. J. Infect. **12:** 111–116.
8. ANTON, E., B. FONT, T. MUNOZ, *et al.* 2003. Clinical and laboratory characteristics of 144 patients with Mediterranean Spotted Fever. Eur. J. Clin. Microbiol. Infect. Dis. **22:** 126–128.
9. DE SOUSA, R., S.D. NOBREGA, F. BACELLAR, *et al.* 2003. Mediterranean spotted fever in Portugal: risk factors for fatal outcome in 105 hospitalized patients. Ann. N. Y. Acad. Sci. **990:** 285–294.
10. SARDELIC, S., P.E. FOURNIER, P. PUNDA, *et al.* 2003. First isolation of *Rickettsia conorii* from human blood in Croatia. Croat. Med. J. **44:** 630–634.
11. LAMBERT, M., T. DUGERNIER, G. BIGAIGNON, *et al.* 1984. Mediterranean spotted fever in Belgium. Lancet **2:** 1038.
12. BACELLAR, F., L. BEATI, A. FRANCA, *et al.* 1999. Israeli spotted fever rickettsia (*Rickettsia conorii* complex) associated with human disease in Portugal. Emerg. Infect. Dis. **5:** 835–836.
13. PAROLA, P. & D. RAOULT 2001. Tick-borne bacterial diseases emerging in Europe. Clin. Microbiol. Infect. **7:** 80–83.
14. SOUSA, R., N. ISMAIL, S. DÓRIA-NÓBREGA, *et al.* 2005. The presence of eschars, but not greater severely, in Portuguese patients infected with Israeli spotted fever. Ann. N. Y. Acad. Sci. **1063:** 197–202.
15. FOURNIER, P.E., J.P. DURAND, J.M. ROLAIN, *et al.* 2003. Detection of Astrakhan fever rickettsia from ticks in Kosovo. Ann. N. Y. Acad. Sci. **990:** 158–161.
16. LAKOS, A. 1997. Tick-borne lymphadenopathy—a new rickettsial disease? Lancet **350:** 1006.
17. OTEO, J.A. & V. IBARRA. 2002. DEBONEL (*Dermacentor*-borne-necrosis-erythema-lymphadenopathy): a new tick-borne disease? Enferm. Infecc. Microbiol. Clin. **20:** 51–52.
18. OTEO, J.A., V.V. IBARRA , J.R. BLANCO, *et al.* 2003. Epidemiological and clinical differences among DEBONEL-TIBOLA and other tick-borne diseases in Spain. Ann. N. Y. Acad. Sci. **990:** 391–392.
19. BACELLAR, F., M.S. NUNCIO, M.J. ALVES, *et al.* 1995. *Rickettsia slovaca*: an agent of the group of exanthematous fevers, in Portugal. Enferm. Infecc. Microbiol. Clin. **13:** 218–223.
20. RAOULT, D., A. LAKOS, F. FENOLLAR, *et al.* 2002. Spotless rickettsiosis caused by *Rickettsia slovaca* and associated with *Dermacentor* ticks. Clin. Infect. Dis. **34:** 1331–1336.
21. OTEO, J.A., V. IBARRA, J.R. BLANCO, *et al.* 2004. Dermacentor-borne necrosis erythema and lymphadenopathy: clinical and epidemiological features of a new tick-borne disease. Clin. Microbiol. Infect. **10:** 327–331.
22. PUNDA-POLIC, V., M. PETROVEC, T. TRILAR, *et al.* 2002. Detection and identification of spotted fever group rickettsiae in ticks collected in southern Croatia. Exp. Appl. Acarol. **28:** 169–176.
23. RAOULT, D., P. BERBIS, V. ROUX, *et al.* 1997. A new tick-borne disease due to *Rickettsia slovaca*. Lancet **350:** 112–113.
24. IBARRA, V., J.A. OTEO, A. PORTILLO, *et al.* 2005. DEBONEL/TIBOLA: is *Rickettsia slovaca* infection. Ann. N. Y. Acad. Sci. This Volume.
25. LAKOS, A. 2002. Tick-borne lymphadenopathy (TIBOLA). Wien. Klin. Wochenschr. **114:** 648–654.

26. KOMITOVA, R., A. LAKOS, A. ALEKSANDROV, *et al.* 2003. A case of tick-transmitted lymphadenopathy in Bulgaria associated with *Rickettsia slovaca*. Scand. J. Infect. Dis. **35:** 213.
27. CAZORLA, C., M. ENEA, F. LUCHT, *et al.* 2003. First isolation of *Rickettsia slovaca* from a patient, France. Emerg. Infect. Dis. **9:** 135.
28. OTEO, J.A., V. IBARRA, J.R. BLANCO, *et al.* 2003. Epidemiological and clinical differences among *Rickettsia slovaca* rickettsiosis and other tick-borne diseases in Spain. Ann. N. Y. Acad. Sci. **990:** 355–356.
29. RAOULT, D., P. BROUQUI & V. ROUX. 1996. A new spotted fever group rickettsiosis. Lancet **348:** 412.
30. FOURNIER, P. E., H. TISSOT-DUPONT, H. GALLAIS, *et al.* 2000. *Rickettsia mongolotimonae*: a rare pathogen in France. Emerg. Infect. Dis. **6:** 290–292.
31. PSAROULAKI, A., A. GERMANAKIS, E. SCOULICA, *et al.* 2005. Detection, isolation and molecular identification of spotted fever group rickettsiae derived from patients' specimens in Crete, Greece. Presented at the 4th International Conference on Rickettsiae and Rickettsial Diseases. Logroño, La Rioja, Spain, June 18-21.
32. FOURNIER, P. E., F. GOURIET, P. BROUQUI, *et al.* Lymphangitis-associated rickettsiosis, a new rickettsiosis caused by *Rickettsia sibirica mongolotimonae*: seven new cases and review of the literature. Clin. Infect. Dis. **40:** 1435–1444.
33. YU, X., Y. JIN, M. FAN, *et al.* 1993. Genotypic and antigenic identification of two new strains of spotted fever group rickettsiae isolated from China. J. Clin. Microbiol. **31:** 83–88.
34. NILSSON, K., O. LINDQUIST & C. PAHLSON. 1999. Association of *Rickettsia helvetica* with chronic perimyocarditis in sudden cardiac death. Lancet **354:** 1169–1173.
35. FOURNIER, P.E., F. GRUNNENBERGER, B. JAULHAC, *et al.* 2000. Evidence of *Rickettsia helvetica* infection in humans, eastern France. Emerg. Infect. Dis. **6:** 389–392.
36. FOURNIER, P.E., C. ALLOMBERT, Y. SUPPUTAMONGKOL, *et al.* 2004. An eruptive fever associated with antibodies to *Rickettsia helvetica* in Europe and Thailand. J. Clin. Microbiol. **42:** 816–818.
37. NILSSON, K., C. PAHLSON, A. LUKINIUS, *et al.* 2002. Presence of *Rickettsia helvetica* in granulomatous tissue from patients with sarcoidosis. J. Infect. Dis. **185:** 1128–1138.
38. PAROLA, P., L. BEATI, M. CAMBON, *et al.* 1998. First isolation of *Rickettsia helvetica* from *Ixodes ricinus* ticks in France. Eur. J. Clin. Microbiol. Infect. Dis. **17:** 95–100.
39. BAUMANN, D., N. PUSTERLA, O. PETER, *et al.* 2003. Fever after a tick bite: clinical manifestations and diagnosis of acute tick bite-associated infections in northeastern Switzerland. Dtsch. Med. Wochenschr. **128:** 1042–1047.
40. FERNANDEZ-SOTO, P., A. ENCINAS-GRANDES & R. PEREZ-SANCHEZ. 2003. *Rickettsia aeschlimannii* in Spain: molecular evidence in *Hyalomma marginatum* and five other tick species that feed on humans. Emerg. Infect. Dis. **9:** 889–890.
41. BENINATI, T., N. LO, H. NODA, *et al.* 2002. First detection of spotted fever group rickettsiae in *Ixodes ricinus* from Italy. Emerg. Infect. Dis. **8:** 983–986.
42. RAOULT, D., P.E. FOURNIER, P. ABBOUD, *et al.* 2002. First documented human *Rickettsia aeschlimannii* infection. Emerg. Infect. Dis. **8:** 748–749.
43. OTEO, J.A., A. PORTILLO, J.R. BLANCO, *et al.* 2005. Low risk of developing human *Rickettsiae aeschlimannii* infection in the north of Spain. Ann. N. Y. Acad. Sci. **1063:** 349–351.

44. MATSUMOTO, K., P. PAROLA, P. BROUQUI, *et al.* 2004. *Rickettsia aeschlimannii* in *Hyalomma* ticks from Corsica. Eur. J. Clin. Microbiol. Infect. Dis. **23:** 732–734.
45. RAOULT, D., P.E. FOURNIER, F. FENOLLAR, *et al.* 2001. *Rickettsia africae*, a tick-borne pathogen in travelers to sub-Saharan Africa. N. Engl. J. Med. **344:** 1504–1510.
46. FOURNIER, P.E., V. ROUX, E. CAUMES, *et al.* 1998. Outbreak of *Rickettsia africae* infections in participants of an adventure race in South Africa. Clin. Infect. Dis. **27:** 316–323.
47. JENSENIUS, M., G. HASLE, A.Z. HENRIKSEN, *et al.* 1999. African tick-bite fever imported into Norway: presentation of 8 cases. Scand. J. Infect. Dis. **31:** 131–133.
48. JENSENIUS, M., T. HOEL, D. RAOULT, *et al.* 2002. Seroepidemiology of *Rickettsia africae* infection in Norwegian travellers to rural Africa. Scand. J. Infect. Dis. **34:** 93–96.
49. OTEO, J.A., A. PORTILLO, J.R. BLANCO, *et al.* 2004. Infección por *Rickettsia africae*: tres casos confirmados por reacción en cadena de la polimerasa. Med. Clin. (Barc.) **122:** 786–788.
50. ANON. Murine typhus, Portugal. 1998. Wkly. Epidemiol. Rec. **73:** 262–263.
51. BERNABEU-WITTEL, M., J. PACHON, A. ALARCON, *et al.* 1999. ANON Murine typhus as a common cause of fever of intermediate duration: a 17-year study in the south of Spain. Arch. Intern. Med. **159:** 872–876.
52. GIKAS, A., S. DOUKAKIS, J. PEDIADITIS, *et al.* 2002. Murine typhus in Greece: epidemiological, clinical, and therapeutic data from 83 cases. Trans. R. Soc. Trop. Med. Hyg. **96:** 250–253.
53. TSELENTIS, Y., A. PSAROULAKI, J. MANIATIS, *et al.* 1996. Genotypic identification of murine typhus *Rickettsia* in rats and their fleas in an endemic area of Greece by the polymerase chain reaction and restriction fragment length polymorphism. Am. J. Trop. Med. Hyg. **54:** 413–417.
54. HERNÁNDEZ-CABRERA, M., A. ANGEL-MORENO, E. SANATANA, *et al.* 2004. Murine typhus with renal involvement in Canary Islands, Spain. Emerg. Infect. Dis. **10:** 740–743.
55. MARQUEZ, F.J., M.A. MUNIAIN, J.M. PEREZ, *et al.* 2002. Presence of *Rickettsia felis* in the cat flea from southwestern Europe. Emerg. Infect. Dis. **8:** 89–91.
56. BLANCO, J.R., L. PÉREZ-MARTÍNEZ, M. VALLEJO, *et al.* 2005. Prevalence of *Rickettsia felis*-like and *Bartonella* spp. in *Ctenocephalides felis* and *Ctenocephalides canis* from La Rioja (Northern Spain). Presented at the 4th International Conference on Rickettsiae and Rickettsial Diseases. Logroño, La Rioja, Spain, June 18–21.
57. ROLAIN, J.M., M. FRANC, B. DAVOUST, *et al.* 2003. Molecular detection of *Bartonella quintana*, *B. koehlerae*, *B. henselae*, *B. clarridgeiae*, *Rickettsi felis*, and *Wolbachia pipientis* in cat fleas, France. Emerg. Infect. Dis. **9:** 338–342.
58. KENNY, M.J., R.J. BIRTLES, M.J. DAY, *et al.* 2003. *Rickettsia felis* in the United Kingdom. Emerg. Infect. Dis. **9:** 1023–1024.
59. RAOULT, D., B. LA SCOLA, M. ENEA, *et al.* 2001. A flea-associated *Rickettsia* pathogenic for humans. Emerg. Infect. Dis. **7:** 73–81.
60. RICHTER, J., P.E. FOURNIER, J. PETRIDOU, *et al.* 2002. *Rickettsia felis* infection acquired in Europe and documented by polymerase chain reaction. Emerg. Infect. Dis. **8:** 207–208.

61. NIANG, M., P. BROUQUI & D. RAOULT. 1999. Epidemic typhus imported from Algeria. Emerg. Infect. Dis. **5:** 716–718.
62. BADIAGA, S., P. BROUQUI & D. RAOULT. 2005. Autochthonous epidemic typhus associated with *Bartonella quintana* bacteremia in a homeless person. Am. J. Trop. Med. Hyg. **72:** 638–639.

Epidemiology of Rickettsioses in North Africa

AMEL LETAÏEF

Infectious Diseases Unit, Farhat Hached University Hospital, 4000 Sousse, Tunisia

ABSTRACT: The first description of Mediterranean spotted fever (MSF) was made by Conor and Brush in 1910 in Tunisia, where, at the same time, Nicolle described the role of lice in transmission of epidemic typhus. However, along this century, there have been few and fragmentary reports about ecology and epidemiology of rickettsioses in North Africa. This region was always considered, for these diseases, like other Mediterranean regions. The most human tick-borne rickettsiosis known to occur in North Africa is MSF caused by *R. conorii* and transmitted by the brown dog tick, Rhipicephalus sanguineus. Recent studies showed that other arthropode-transmitted rickettsiae are prevalent in North Africa: *R. aeschlimannii*, *R. massiliae,* and *R. felis*. Moreover, *R. felis* and *R. aeschlimannii* human infection were respectively confirmed, by serology in Tunisia, and by PCR in Morocco. The seroprevalence of *R. conorii* among healthy population was ranging from 5% to 8% in most of the countries. Epidemiological and clinical features are frequently resumed in an eruptive fever with eschar occurring in hot season in rural areas. Typhus group rickettsioses, caused by *R. typhi* and *R. prowazekii* are less frequently reported than in the 1970s. Seroprevalence of *R. typhi* among blood donors was from 0.5% to 4%. In Algeria about 2% of febrile patients had *R. prowazekii* antibodies. Moreover, reemerging threat of epidemic typhus should be considered, after the two cases recently diagnosed in the highlands of Algeria. Murine typhus, considered as "benign" typhus, is underestimated. When *R. typhi* was inserted in serologic tests, murine typhus became more frequently confirmed. In a recent study in Central Tunisia, we confirmed an emergence of murine typhus mistaken for *R. conorii* or viral infection. In addition to typhus surveillance, future studies have to determine which spotted fever group rickettsiae are prevalent in vectors and in human pathology.

KEYWORDS: rickettsia; *Rickettsia conorii*; *Rickettsia felis*; *Rickettsia typhi*; *Rickettsia prowazekii*; epidemiology; Africa, Northern

Address for correspondence: Amel Letaief, M.D., Service de Médecine Interne et Maladies Infectieuses, CHU F. Hached, 4000 Sousse, Tunisia. Voice: (+216) 73 21 11 83; fax: (+216) 73 21 11 83.
e-mail: amel.letaief@famso.rnu.tn

Ann. N.Y. Acad. Sci. 1078: 34–41 (2006).
doi: 10.1196/annals.1374.004

INTRODUCTION

The advent of 16s RNA molecular analysis resulted in reclassification of several rickettsial species. Hence, Rickettsiae tribe includes only the genus *Rickettsia*.[1] This genus is subdivided into the typhus group (TG), the spotted fever group (SFG), and scrub typhus (*O. tsutsugamushi*), which was not described in North Africa.

The classic TG pathogens are *R. typhi* and *R. prowazekii*. The three major and classic SFG representatives are *R. conorii*, *R. rickettsii,* and *R. sibirica* with wide territorial distribution. Other SFG pathogens are recently identified in humans and/or vectors, with unknown epidemiological distribution and pathogenesis in several cases. But even isolated solely in ticks or other reservoirs these rickettsiae should be considered a potential human pathogen.

In North Africa, the two rickettsial groups (TG and SFG) were described since the beginning of the 20th century.[2,3] However, information on the epidemiology of rickettsioses is fragmentary. Studies concerned especially clinical topics and few reports were interested in identifying, by new diagnostic tools, of variant rickettsial pathogens in humans and in arthropods.

ARTHROPODS AND RICKETTSIAE IN NORTH AFRICA

The geographical distribution of Rickettsia spp. is determined by the incidence of its arthropod host. The seasonal incidence of diseases parallels tick activity (adult and immature population). The most frequent ticks circulating in North Africa are Rhipicephalus sanguineus and Hyalomma genus.[4]

The prevalence and distribution of rickettsiae among vectors and reservoirs are not known in North Africa. Only some data are available from fragmentary studies. In early studies the recognized vector of African SFG rickettsioses, attributed to infection with solely *R. conorii*, was the dog tick Rhipicephalus sanguineus. However, in 1950s the presence rickettsia-like organisms were reported in Hyalomma ticks of Morocco.[5] In 1984, antibodies anti-*R. prowazekii* and *R. typhi* were reported in lice and mice in Algeria.[6] In Egypt, spotted fever group rickettsioses (SFGR) were detected, by polymerase chain reaction (PCR), in both R. sanguineus (5.4%) and Hyalomma species (2.7%).[7] In 1997, Beati reported the first isolation of *R. aeschlimannii* from H. marginatum (15% infected) collected in Morocco.[8] Later, Raoult documented the first human infection by PCR in a patient returning from Morocco.[9] More recently, *R. conorii*, *R. aeschlimannii,* and *R. massilliae* were first detected by PCR in ticks and *R. felis* in fleas from Algeria.[10,11]

In a study conducted on the ticks of African countries, PCR failed to demonstrate *Rickettsia spp.* and *Ehrlichia spp.* from 42 ticks (H. impeltatum) in Mauritania.[12] However, *Ehrlichia spp.* was detected in 12–25% of Ixodes adult ticks collected, respectively, in Morocco and Tunisia (TABLE 1).[13]

TABLE 1. Current epidemiological data on rickettsioses in North Africa

	Mauritania	Morocco	Algeria	Tunisia	Libya	Egypt
SFGR	No data	*R. conorii*	*R. conorii*	*R. conorii*	no data	*R. conorii* undetermined SFGR
Among ticks		*R. aeshlimannii*	*R. massilliae*			
			R. aeschlimannii			
Fleas			*R. felis*			
Sero prevalence						
R. conorii	13.5%	6%	6% (50%)*	8% (40%*)	no data	1% (10%*+)
R. africae	19.6%					
R. typhi	1.7%	3%	0.5%	3.6% (20–60%+)	-	19% (40%*+)
R. prowazekii	-	-	2%	-	-	-
SFG rickettsioses	MSF	MSF	MSF	MSF	MSF	MSF
		R. aeshlimannii	-	*R. felis***	-	-
TG rickettsioses	no data	no data	*R.typhi*	*R. typhi*	no data	*R. typhi*
			R. prowazekii			

SFGR : Spotted fever group rickettsioses; MSF : Mediterranean spotted fever; TG: Typhus group.
*Among patients with fever and rash; + acute fever with undetermined etiology; *+ among garbage workers; ** serological evidence.

SEROPREVALENCE OF RICKETTSIAL INFECTION IN NORTH AFRICA

1/ SFG Rickettsiae

Sero survey of rickettsial infection showed that seroprevalence of *R. conorii* antibodies among blood donors was 1% in Egypt [14] and 5–8 % in Tunisia, Algeria, Morocco, and Mauritania.[15–18] In a study of 300 hospitalized febrile patients conducted in Central Tunisia, seroprevalence of *R. conorii* antibodies was 23%.[19]

In Egypt seroprevalence of *R. conorii* ranged from 1% to 15% in, respectively, garbage workers and patients with fever of undetermined etiology.[14] In another study among school children in the Nile river delta, using enzyme immunoassay tests, antibodies anti-*R. conorii* were present in 37% cases.[20]

2/ TG Rickettsiae

Although thought to have disappeared in North Africa, TG rickettsial infection is still present, documented by seroprevalence to *R. typhi* in almost all studies conducted in this region. In fact *R. typhi* antibodies were present from 0.5% to 4% in general population.[15–18] Among Egyptian garbage and rodent control workers the seroprevalence was 19%,[14] and among febrile patients reached 15.6% and 40%, respectively, in Egypt and Tunisia.[14,18] Antibodies anti-*R. prowazekii*, rarely added to rickettsial panel serology, were found in 2% of sera from blood donors and hospitalized patients in a recent study in Algeria.[16]

RICKETTSIAL DISEASES DESCRIBED IN NORTH AFRICA

1/ SFG Rickettsioses

A spotted fever was first reported in Tunisia in 1910 by Conor, who described the clinical pattern, proposed the name of Mediterranean spotted fever (MSF), and evoked a vector inoculation role.[3] The etiologic agent, *R. conorii*, and the tick vector R. sanguineus were subsequently identified.[1] Clinical descriptions of MSF were frequently reported from Tunisia and Morocco. This infection is reported more during these last few years probably because it is better known by physicians and on account of better serological confirmation tools, rather than reemergence.[19,21–23] The most common epidemiological and clinical characteristics are occurrence in the rural areas, in hot seasons, and the clinical triad: acute fever, generalized maculo-papular eruption (in >95%), and a single inoculation eschar (in 75%).[21,24] Because of the frequent lack of several classical clinical features, the diagnosis score, established by

Raoult, was validated for clinical and epidemiological features to facilitate the diagnosis of this disease.[24] Although MSF is considered a nonsevere disease, complications and malignant forms of the infection were reported in 5.6% among hospitalized patients, and was 0.55% in the largest study conducted in Tunisia [21, 22, unpublished data]. *R. felis* infection was recently described in Tunisia in patients with acute fever and rash without inoculation eschar.[25] *R. aeschlimannii* infection could mimic MSF.[9]

2/ TG Rickettsioses

TG rickettsial infection is thought to have disappeared since the 1970s in North Africa.[23,26]

Murine Typhus

Recent studies suggested persistence of the infection in Tunisia and Egypt.[14,19,25,27] We demonstrated that this mild infection, often misdiagnosed for a viral infection or MSF, occurred preferentially in hot season and lacked eruption, a major symptom of rickettsioses in 60% of the cases.[25,27] These findings were noted in studies from Greece, where murine typhus is endemic and frequently reported.[28]

Epidemic Typhus

Typhus had been epidemic in North Africa, where the role of lice as vectors was first described by Nicolle in Tunis.[2] Subsequently, the disease was endemic until the seventies.[23,26] Later, using nonspecific serologic tests, no confirmed cases of epidemic typhus were reported.[23] However, we reported, in the 1980s, seven cases of serologically confirmed epidemic typhus that occurred in elderly patients in cold season, with more severe illness than MSF; Brill–Zinsser could not be eliminated.[29] More recently, in 1999, *R. prowazekii* was isolated in France from blood sample in a patient with severe eruptive fever on returning from travel to Algeria.[30,31] In 2004 another case of epidemic typhus was reported in the same region of Algeria.[32] In the recent serological study in Tunisia among febrile patients, all positive sera against *R. prowazekii* were corresponding to cross-reactions with other SFG rickettsiae or *R. typhi*.[25]

CONCLUSIONS

Rickettsioses in North Africa are still endemic. Epidemiology of SFG rickettsial infections could rather be compared to European Mediterranean countries than African ones.[33] MSF, on account of *R. conorii*, is the most frequently recognized and reported SFG rickettsiosis. However, more than one SFG rickettsia is prevalent. In addition, other vectors and reservoirs than brown dog tick could have a role in SFG rickettsioses.

Murine typhus remains a poorly known disease that is underestimated because of nonspecific symptoms leading to misdiagnosing for MSF or viral infection. On the other hand, unlike sub-Saharan African countries,[34,35] epidemic typhus was rather disappearing in North Africa. However, the reemergence of this severe typhus should be considered after the two recent cases reported from Algeria.

Finally, in order to control rickettsioses in North Africa, epidemiological survey of typhus and determination of the most prevalent SFG rickettsiae in vectors and in human pathology are necessary.

REFERENCES

1. Raoult, D. & V. Roux. 1997. Rickettsioses as paradigms of new or emerging infectious diseases. Clin. Microbiol. Rev. **10:** 694–719.
2. Nicolle, C., C. Comte & E. Conseil. 1909. Transmission expérimentale du typhus exanthématique par le pou de corps. C. R. Acad. Sci. **149:** 486–489.
3. Conor, A. & A. Brush. 1910. Une fièvre boutonneuse observée en Tunisie. Bulletin de la société de Pathologie Exotiques et Filiales **8:** 492–496.
4. Bouattour, A. 2002. Dichotomous identification keys of ticks (Acari: Ixodidae), livestock parasites in North Africa. Arch. Inst. Pasteur Tunis **79:** 43–50.
5. Giroud, P., J. Colas-Belcour, R. Pfister & P. Morel. 1957. Amblyomma, Hyalomma, Boophilus, Rhipicephalus of Africa are carriers of rickettsial and neorickettsial elements and occasionally of the two types of agents. Bull. Soc. Pathol. Exot. Filiales **50:** 529–532.
6. Dumas N. 1984. Rickettsiosis and chlamydiosis in Hoggar (Republic of Algeria): epidemiological sampling. Bull. Soc. Pathol. Exot. Filiales **77:** 278–283.
7. Lange, J.V., A.G. el Dessouky, E. Manor, *et al.* 1992. Spotted fever rickettsiae in ticks from the northern Sinai Governate. Egypt Am. J. Trop. Med. Hyg. **46:** 546–551.
8. Beati, L., M. Meskini, B. Thiers & D. Raoult. 1997. Rickettsia aeschlimannii sp. nov., a new spotted fever group rickettsia associated with Hyalomma marginatum ticks. Int. J. Syst. Bacteriol. **47:** 548–554.
9. Raoult, D., P.E. Fournier, P. Abboud & F. Caron. 2002. First documented human Rickettsia aeschlimannii infection. Emerg. Infect. Dis. **8:** 748–749.
10. Bitam, I., P. Parola, K. Dittmar de la Cruz, *et al.* 2006. First molecular detection of *Rickettsia felis* in fleas from Algeria. Am. J. Trop. Med. Hyg. **74:** 532–535.
11. Bitam, I., P. Parola, K. Matsumoto, *et al.* 2005. First molecular detection of R. conorii, R. aeshlimannii, and R. massilliae in ticks from Algeria. (Fourth ICR RD, Logrono June)
12. Parola, P., H. Inokuma, J.L. Camicas, *et al.* 2001. Detection and identification of spotted fever group Rickettsiae and Ehrlichiae in African ticks. Emerg. Infect. Dis. **7:** 1014–1017.

13. Sarih, M., Y. M'Ghirbi, A. Bouattour, *et al.* 2005. Detection and identification of Ehrlichia spp. in ticks collected in Tunisia and Morocco. J. Clin. Microbiol. **43:** 1127–1132.
14. Botros, B.A., A.K. Soliman, M. Darwish, *et al.* 1989. Seroprevalence of murine typhus and fievre boutonneuse in certain human populations in Egypt. J. Trop. Med. Hyg. **92:** 373–378.
15. Letaief, A.O., S. Yacoub, H.T. Dupont, *et al.* 1995. Seroepidemiological survey of rickettsial infections among blood donors in central Tunisia. Trans. R. Soc. Trop. Med. Hyg. **89:** 266–268.
16. Tebbal, S., A. Benyahia, R. Ait Hamouda, *et al.* 2004. Sero-epidemiological study of rickettsioses in the Aures. RICAI 631/92P (unpublished).
17. Meskini, M., L. Beati, A. Benslimane & D. Raoult. 1995. Seroepidemiology of rickettsial infections in Morocco. Eur. J. Epidemiol. **11:** 655–660.
18. Niang, M., P. Parola, H. Tissot-Dupont, *et al.* 1998. Prevalence of antibodies to Rickettsia conorii Ricketsia africae, Rickettsia typhi and Coxiella burnetii in Mauritania. Eur. J. Epidemiol. **14:** 817–818.
19. Omezzine-Letaief A., H. Tissot Dupont, F. Bahri, *et al.* 1997. Etude séroépidémiologique chez 300 malades fébriles hospitalisés dans un service de médecine et maladies infectieuses. Méd Mal. Infect. **27**:RICAI, 663–666.
20. Corwin, A., M. Habib, J. Olson, *et al.* 1992. The prevalence of arboviral, rickettsial, and Hantaan-like viral antibody among school children in the Nile river delta of Egypt. Trans. R. Soc. Trop. Med. Hyg. **86:** 677–679.
21. Jemni, L., H. Hmouda, M. Chakroun, *et al.* 1994. Mediterranean spotted fever in Central Tunisia. J. Travel Med. **1:** 106–108.
22. Alioua, Z., A. Bourazza, H. Lamsyah, *et al.* 2003. Neurological feature of Mediterranean spotted fever: a study of four cases. Rev. Med. Interne. **24:** 824–829.
23. Kennou, M.F. & E. Edlinger. 1984. Current data on rickettsial diseases in Tunisia. Arch. Inst. Pasteur Tunis **61:** 427–433.
24. Letaief, A., J. Souissi, H. Trabelsi, *et al.* 2003. Evaluation of clinical diagnosis scores for Boutonneuse fever. Ann. N. Y. Acad. Sci. **990**: 327–330.
25. Kaabia, N., J.M. Rolain, M. Khalifa, *et al.* 2005. Serologic study of rickettsioses among acute febrile patients in Central Tunisia. (Fourth ICR RD, Logrono June)
26. Le Corroller, Y., R. Neel & R. Lecubarri. 1970. Exanthemic typhus in the Sahara. Arch. Inst. Pasteur Alger. **48:** 125–130.
27. Letaief, A., N. Kaabia, M. Chakroun, *et al.* 2005. Clinical and laboratory features of Murine Typhus in Central Tunisia : a report of 7 cases. Int. J. Infect. Dis. **9**(6): 331–334.
28. Gikas, A., S. Doukakis, J. Pediaditis, *et al.* 2002. Murine typhus in Greece: epidemiological, clinical, and therapeutic data from 83 cases. Trans. R. Soc. Trop. Med. Hyg. **96:** 250–253.
29. Ernez, M., M. Chakroun, A. Letaief & L. Jemni. 1995. Clinical and biological features of exanthematous typhus. Presse. Med. **24:** 1358–1359.
30. Niang, M., P. Brouqui & D. Raoult. 1999. Epidemic typhus imported from Algeria. Emerg. Infect. Dis. **5:** 716–718.
31. Birg, M.L., B. La Scola, V. Roux, *et al.* 1999. Isolation of Rickettsia prowazekii from blood by shell vial cell culture. J. Clin. Microbiol. **37:** 3722–3724.
32. Mokrani, K., P.E. Fournier, M. Dalichaouche, *et al.* 2004. Reemerging threat of epidemic typhus in Algeria. J. Clin. Microbiol. **42:** 3898–3900.

33. DUPONT, H.T., P. BROUQUI, B. FAUGERE & D. RAOULT. 1995. Prevalence of antibodies to Coxiella burnetti, Rickettsia conorii, and Rickettsia typhi in seven African countries. Clin. Infect. Dis. **21:** 1126–1133.
34. PERINE, P.L., B.P. CHANDLER, D.K. KRAUSE, *et al.* 1992. A clinico-epidemiological study of epidemic typhus in Africa. Clin. Infect. Dis. **14:** 1149–1158.
35. RAOULT, D., J.B. NDIHOKUBWAYO, H. TISSOT-DUPONT, *et al.* 1998. Outbreak of epidemic typhus associated with trench fever in Burundi. Lancet **352:** 353–358.

Rickettsioses in Sub-Saharan Africa

PHILIPPE PAROLA

Unité des Rickettsies, CNRS UMR 6020 IFR 48, WHO Collaborative Center for Rickettsial Reference and Research, Faculté de Médecine, 13385 Marseille Cedex 5, France

ABSTRACT: Although rickettsioses are among the oldest known vector-borne zoonoses, several species or subspecies of rickettsias have been identified in recent years as emerging pathogens throughout the world including in sub-Saharan Africa. To date, six tick-borne spotted fever group pathogenic rickettsias are known to occur in sub-Saharan Africa, including *Rickettsia conorii conorii*, the agent of Mediterranean spotted fever; *R. conorii caspia*, the agent of Astrakhan fever; *R. africae*, the agent of African tick-bite fever; *R. aeschlimannii*; *R. sibirica mongolitimonae*; and *R. massiliae*. On the other hand, fleas have long been known as vectors of the ubiquitous murine typhus, a typhus group rickettsiosis induced by *R. typhi*. However, a new spotted fever rickettsia, *R. felis*, has also been found to be associated with fleas, to be a human pathogen, and to be present in sub-Saharan Africa. Finally, *R. prowazekii* the agent of louse-borne epidemic typhus continues to strikes tens to hundreds of thousands of persons who live in Sub-Saharan with civil war, famine and poor conditions. We present an overview of these rickettsioses occurring in sub-Saharan Africa, focusing on the epidemiological aspects of emerging diseases.

KEYWORDS: rickettsioses; ticks; fleas; lice; *Rickettsia africae*; *Rickettsia conorii*; *Rickettsia aeschlimannii*; *Rickettsia sibirica mongolitimonae*; *Rickettsia prowazekii*; *Rickettsia felis*; *Rickettsia massiliae*

Rickettsial diseases are zoonoses caused by obligate intracellular bacteria grouped in the order *Rickettsiales*. Although bacteria of this order were first described as short, gram-negative rods that retained basic fuchsin when stained by the method of Gimenez, the taxonomy of rickettsias has undergone significant reorganization in the last decade. For example, *Coxiella burnetii*, the agent of Q fever has been removed recently from the *Rickettsiales*.[1] The classification within the *Rickettsiales*, including at the species level, continues to be modified as new data become available.[2] To date, three groups of diseases are still commonly classified as rickettsial diseases. These include (*a*) rickettsioses on

This paper was presented as an oral communication at ICRRD in Logrono.

Address for correspondence: P. Parola, Unité des Rickettsies, CNRS UMR 6020 IFR 48, WHO Collaborative Center for Rickettsial Reference and Research, Faculté de Médecine, 27 Bd Jean Moulin, 13385 Marseille Cedex 5, France. Voice: 33-4-91-32-43-75; fax: 33-4-91-83-03-90.
e-mail: philippe.parola@medecine.univ-mrs.fr

**Ann. N.Y. Acad. Sci. 1078: 42–47 (2006). © 2006 New York Academy of Sciences.
doi: 10.1196/annals.1374.005**

account of bacteria of the genus *Rickettsia*, including the spotted fever group and the typhus group rickettsiae, (*b*) ehrlichioses and anaplasmoses on account of bacteria within the family *Anaplasmataceae* that has been reorganized, and (*c*) scrub typhus on account of *Orientia tsutsugamushi*. As scrub typhus is prevalent in the Asia–Pacific region, it will not been discussed in this article focusing on sub-Saharan Africa.

Six tick-borne spotted fever group rickettsioses are currently known to occur in sub-Saharan Africa. Indeed, it has now been almost 15 years since *Rickettsia africae*, the agent of African tick-bite fever (ATBF) was rediscovered in sub-Saharan Africa, and definitely distinguished from Mediterranean spotted fever (MSF) on account of *R. conorii conorii*.[3] In southern Africa *Amblyomma hebraeum*, a tick of large ruminants and wildlife species is a recognized vector and reservoir of *R. africae*. *R. africae* has also been detected in *A. variegatum* throughout west, central, and eastern sub-Saharan Africa, and in *A. lepidum* from the Sudan.[4] Infection rates of the ticks are remarkably high. Because *Amblyomma* readily bite humans, cases of ATBF often occur in clusters and patients often present with multiple inoculation eschars. However, despite high seroprevalence to *R. africae* among native Africans, nearly all acute cases of ATBF described in the literature (more than 250) have occurred in European or American travelers.[4] Recently, however, cases of ATBF were documented by serology and molecular techniques among indigenous patients in Cameroon.[5] Although MSF is endemic in the Mediterranean area where it is transmitted by the brown dog tick *Rhipicephalus sanguineus*, *R. conorii conorii* has been poorly detected in sub-Saharan Africa (South Africa, Zimbabwe, and Kenya). However, a strain of the closely related *R. conorii caspia*, the agent of Astrakhan fever, was obtained recently from a patient from Chad.[6]

Three more tick-borne pathogenic rickettsiae are now known to occur in sub-Saharan Africa. *R. aeschlimannii* had been first characterized as a new spotted fever group rickettsia following its isolation from *Hyalomma marginatum marginatum* ticks in Morocco in 1997.[7] Since that time, it has also been detected and/or isolated in *H. m. rufipes* in Zimbabwe, Niger, and Mali, as well as in *Hyalomma m. marginatum* in southern Europe. Preliminary data have suggested that transstadial and transovarial transmission of the rickettsia occurs in these ticks, suggesting that the geographic distribution of *R. aeschlimannii* would be, at least that of *H. m. marginatum* (southern Europe and northern Africa) and *H. m. rufipes* (Sahelian and southern Africa).[8] Other tick species including *Rh. appendiculatus* have also been implicated in the transmission of *R. aeschlimannii*.[9,10] The pathogenic role of this rickettsia was achieved in 2002, when the first human infection was reported in a patient returning from Morocco to France.[11] A second case was reported in a patient returning from a hunting and fishing trip in South Africa.[10]

R. sibirica mongolitimonae (formally named *R. mongolotimonae*) was reported in sub-Saharan Africa in 2001, when it was detected in *Hyalomma truncatum* in Niger.[12] This rickettsia had been first isolated in *Hyalomma asiaticum* ticks collected in Inner Mongolia in China in 1991. In 1996 the first human cases were reported in Marseille, southern France, where the vector is still unknown.[13] In 2004 the first proven human infection with *R. sibirica mongolitimonae* in sub-Saharan Africa was reported in a construction worker, working in South Africa's Northern Province.[14] Although migratory birds have been suggested to play a role in the epidemiology of this emerging disease, detection of *R. sibirica mongolitimonae* in *Hyalomma spp.* in Mongolia and Africa also suggests a possible association of this rickettsia with ticks of this genus. On the basis of evaluation of a total of nine cases (France, Algeria, and South Africa), specific characteristics include the occasional findings, alone or in combination, of multiple eschars and draining lymph nodes, and a lymphangitis that extends from the inoculation eschar to the draining node. These particular clinical features of this new rickettsiosis have led to the moniker, "lymphangitis-associated rickettsiosis" (LAR).[13] Finally, *R. massiliae* has recently been recognized as a human pathogen.[15] This rickettsia, which was first isolated in Europe, has been also detected in Africa in *Rh. munsamae*, *Rh. lunulatus*, and *Rh. sulcatus* in the central African Republic and in *Rh. muhsamae* collected on Cattle in Mali.

On the other hand, *Rickettsia prowazekii* the agent of louse-borne epidemic typhus continues to strikes tens to hundreds of thousands of persons who live in sub-Saharan with civil war, famine, and poor conditions. The most recent outbreak (and the largest since World War II) has been observed in Burundi in the 1990s during the civil war.[16] The *Unité des Rickettsies* was involved in the surveillance of the outbreak. Lice were collected on three occasions (1998, 2000, and 2001) after the outbreak had been controlled.[17] They were shipped to the laboratory and tested by polymerase chain reaction (PCR). Although they were negative for *R. prowazekii* DNA in 1998 and 2000 as a result of the administration of doxycycline to patients, the persistence of the vector enabled the spread of *R. prowazekii* from human carriers back into the louse population. Indeed, in 2001, 21% of lice from refugee camps in the same areas of Burundi as sampled earlier were shown to be positive by PCR for *R. prowazekii*. Further samples thereafter submitted to the *Unité des Rickettsies* indicated that a typhus outbreak was developing in refugee camps in Burundi. Also, *R. prowazekii* was detected in 7% of body lice collected in 2001 from a jail in Rwanda.[17] At that time, the country was host to 300,000 refugees from the January 2002 eruption of the Nyiragongo volcano.

Fleas are also known vectors of rickettsioses in sub-Saharan Africa. They have been historically associated with the transmission of the ubiquitous murine typhus, a typhus group rickettsiosis induced by *Rickettsia typhi* and transmitted by the rat fleas *Xenopsylla cheopis*.[18] In sub-Saharan Africa, the prevalence of antibodies against *R. typhi* in humans is higher in costal areas where rats are prevalent. More recently, fleas have been involved in the cycle of *Rickettsia felis*, an emerging pathogen belonging to the spotted fever group of *Rickettsia*.[19] Arguments on the pathogenicity of *R. felis* for humans were provided starting 2000.[20] At that time, three patients with fever rash were diagnosed with *R. felis* infection by specific PCR of blood or skin and a seroconversion to rickettsial antigens.[21] Serological evidence of *R. felis* infection has also been shown in patients from France and Brazil. Moreover, molecular documentation was obtained in the serum of one Brazilian patient.[22] In 2002 two typical cases of rickettsial spotted fever including generalized maculopapular rash and a black eschar were reported in an adult couple in Germany. *R. felis* infections were documented by serology for both patients and by detection of *R. felis* DNA in the woman's sera.[23] Finally, the first case in Asia of *R. felis* infection was recently documented by serology in Thailand.[24] However, to date, few confirmed cases of the so-called flea-borne spotted fever have been described throughout the world. *R. felis* has been detected in fleas throughout the world including sub-Saharan Africa (Ethiopia Gabon). Species of fleas that have been associated with *R. felis* include *Ctenocephalides felis*, *C. canis*, *Pulex irritans*, and more recently *Archeopsylla erinacei*.[25] The role of mammals, including cats, dogs, rodents, and hedgehogs, in the life cycle and circulation of *R. felis* is still unclear, in sub-Saharan Africa and elsewhere.

Finally, human ehrlichioses and anaplasmosis have been suspected to occur in sub-Saharan Africa. However, serological cross-reactivity has been found between the agents of human ehrlichioses and members of *Anaplasmataceae* of veterinary importance, which are widely distributed in Africa.[26] To date, then, there is no definitive evidence for the presence of human ehrlichioses or anaplasmosis in Africa.

Still only little is known about rickettsial diseases in sub-Saharan Africa. Most of the sub-Saharan African countries lack specific laboratory facilities for the diagnosis of rickettsial diseases. However, many emerging diseases have been reported in the recent years. Much of the information on the epidemiology of the diseases has been obtained in first-world laboratories that promote international cooperation and have well-developed facilities. Similarly, data on the clinical aspects of the diseases have often been derived from infected visitors to the continent returning to their homes in developed countries. The development of the molecular methods has greatly facilitated collaborative research between rickettsial reference laboratories and laboratories in countries with less-developed facilities for research. It is hoped, however, that as health workers in Africa become increasingly aware of rickettsioses, better descriptions of rickettsial diseases on the continent will become available and in particular the disease situation in the local peoples.

REFERENCES

1. Parola, P., C.D. Paddock & D. Raoult. 2005. Tick-borne rickettsioses around the world: emerging diseases challenging old concepts. Clin. Microbiol. Rev. **18:** 719–756.

2. Zhu, Y., P.E. Fournier, M. Eremeeva & D. Raoult. 2005. Proposal to create subspecies of *Rickettsia conorii* based on multi-locus sequence typing and an emended description of *Rickettsia conorii*. BMC Microbiol. **5**:1–11.
3. Raoult, D., P.E. Fournier, F. Fenollar, *et al.* 2001. *Rickettsia africae*, a tick-borne pathogen in travelers to sub-Saharan Africa. N. Engl. J. Med. **344:** 1504–1510.
4. Jensenius, M., P.E. Fournier, P. Kelly, *et al.* 2003. African tick bite fever. Lancet Infect. Dis. **3:** 557–564.
5. Ndip, L.M., D.H. Bouyer, A.P.A.T. Da Rosa, *et al.* 2004. Acute spotted fever rickettsiosis among febrile patients, Cameroon. Emerg. Infect. Dis. **10:** 432–437.
6. Fournier, P.E., B. Xeridat & D. Raoult. 2003. Isolation of a rickettsia related to Astrakhan fever rickettsia from a patient in Chad. Ann. N.Y. Acad. Sci. **990:** 152–157.
7. Beati, L., M. Meskini, B. Thiers & D. Raoult. 1997. *Rickettsia aeschlimannii* sp. nov., a new spotted fever group rickettsia associated with *Hyalomma marginatum* ticks. Int. J. Syst. Bacteriol. **47:** 548–554.
8. Matsumoto, K., P. Parola, P. Brouqui & D. Raoult. 2005. *Rickettsia aeschlimannii* in *Hyalomma* ticks from Corsica. Eur. J. Clin. Microbiol. Infect. Dis. **23:** 732–734.
9. Fernández-Soto, P., A. Encinas-Grandes & R. Pérez-Sánchez. 2003. *Rickettsia aeschlimannii* in Spain: molecular evidence in *Hyalomma marginatum* and five other tick species that feed on humans. Emerg. Infect. Dis. **9:** 889–890.
10. Pretorius, A.M. & R.J. Birtles. 2002. *Rickettsia aeschlimannii*: a new pathogenic spotted fever group rickettsia, South Africa. Emerg. Infect. Dis. **8:** 874.
11. Raoult, D., P.E. Fournier, P. Abboud & F. Caron. 2002. First documented human *Rickettsia aeschlimannii* infection. Emerg. Infect. Dis. **8:** 748–749.
12. Parola, P., H. Inokuma, J.L. Camicas, *et al.* 2001. Detection and identification of spotted fever group *Rickettsiae* and *Ehrlichiae* in African ticks. Emerg. Infect. Dis. **7:** 1014–1017.
13. Fournier, P.E., F. Gouriet, P. Brouqui, *et al.*2005. Lymphangitis-associated rickettsiosis, a new rickettsiosis caused by *Rickettsia sibirica mongolotimonae*: seven new cases and review of the literature. Clin. Infect. Dis. **40:** 1435–1444.
14. Pretorius, A.M. & R. Birtles. 2004. *Rickettsia mongolotimonae* infection in South Africa. Emerg. Infect. Dis. **10**:125–126.
15. Vitale, G., S. Mansuelo, J.M. Rolain, & D. Raoult. 2006. *Rickettsia massiliae* human isolation. Emerg. Infect. Dis. **12:** 174–175.
16. Raoult, D., J.B. Ndihokubwayo, H. Tissot-Dupont, *et al.* 1998. Outbreak of epidemic typhus associated with trench fever in Burundi. Lancet **352:** 353–358.
17. Fournier, P.E., J.B. Ndihokubwayo, J. Guidran, *et al.* 2002. Human pathogens in body and head lice. Emerg. Infect. Dis. **8:** 1515–1518.
18. Azad, A.F., S. Radulovic, J.A. Higgins, *et al.* 1997. Flea-borne rickettsioses: ecologic considerations. Emerg. Infect. Dis. **3**:319–327.
19. La Scola, B. S. Meconi, F. Fenollar, *et al.* 2002. Emended description of *Rickettsia felis* (Bouyer et al. 2001), a temperature-dependent cultured bacterium. Int. J. Syst. Evol. Microbiol. **52:** 2035–2041.
20. Parola, P., B. Davoust & D. Raoult. 2005. Tick- and flea-borne rickettsial emerging zoonoses. Vet. Res. **36:** 469–492.
21. Zavala-Velazquez, J.E., J.A. Ruiz-Sosa, R.A. Sanchez-Elias, *et al.* 2000. *Rickettsia felis* rickettsiosis in Yucatan. Lancet **356:** 1079–1080.

22. RAOULT, D.B., M. LA SCOLA, P.E. ENEA FOURNIER, *et al.*2001. A flea-associated *Rickettsia* pathogenic for humans. Emerg. Infect. Dis. **7:** 73–81.
23. RICHTER, J., P.E. FOURNIER, J. PETRIDOU, *et al.* 2002. *Rickettsia felis* infection acquired in Europe and documented by polymerase chain reaction. Emerg. Infect. Dis. **8:** 207–208.
24. PAROLA, P., R.S. MILLER, P. MCDANIEL, *et al.* 2003. Emerging rickettsioses of the Thai-Myanmar border. Emerg. Infect. Dis. **9:** 592–595.
25. ROLAIN, J.M., O. BOURRY, B. DAVOUST & D. RAOULT. 2005. *Bartonella quintana* and *Rickettsia felis* in Gabon. Emerg. Infect. Dis. **11:** 1742–1744.
26. PAROLA, P. & D. RAOULT. 2001. Ticks and tickborne bacterial diseases in humans: an emerging infectious threat. Clin. Infect. Dis. **32:** 897–928. Erratum: Clin. Inf. Dis. **33**: 749.

Rickettsial Diseases in Russia

IRINA V. TARASEVICH[a] AND OLEG Y. MEDIANNIKOV[a,b,c]

[a]*Laboratory of Rickettsial Ecology, Gamaleya Research Institute of Epidemiology and Microbiology, Moscow 123098, Russia*

[b]*Department of the Infectious Diseases, Far Eastern State Medical University, Khabarovsk 680000, Russia*

[c]*Unité des Rickettsies, Faculté de Médecine, CNRS UMR 6020 27, Université de la Méditerranée, 13385 Marseille, Cedex 05, France*

ABSTRACT: Currently, several rickettsioses are officially being reported in the Russian Federation. These are epidemic typhus and Brill–Zinsser disease, both caused by *Rickettsia prowazekii* which has a historic prevalence in Russia. Nowadays only single sporadic cases of *R. prowazekii* infection are reported. The last significant outbreak occurred in 1997 in a mental nursing home, where 29 cases were identified. Registered morbidity of typhus in Russia varies from 0 to 0.01 per thousand for the last decade. Siberian tick typhus, caused by *R. sibirica*, is registered on a large territory from Pacific coasts to Western Siberia, and its incidence continuously increases, varying between 2.5 and 4.0 thousand officially registered cases per year. Astrakhan spotted fever, caused by *R. conorii* subsp. *caspia* has been recognized since 1983. Recently, Far Eastern tick-borne rickettsiosis, caused by *R. heilongjiangensis*, has been described. Several other pathogenic spotted fever group rickettsiae have been detected and isolated from ticks in Russia; however, they have not yet been linked with clinical cases in these regions.

KEYWORDS: *Rickettsia*; rickettsiosis; tick-borne rickettsiosis; spotted fever group; epidemic typhus; Russia

INTRODUCTION

The Russian Federation is the largest country in the world, covering more than 17 thousands square km of surface with a population above 143 million. Wide-ranging climatic conditions and a different ecology throughout Russia allow a lot of possible reservoir-vector-pathogenic microorganism combinations in natural foci. Rickettsiae represent an important issue in ecology and

Address for correspondence: Oleg Y. Mediannikov, Laboratory of Rickettsial Ecology, Gamaleya Institute of Epidemiology and Microbiology, ul. Gamalei, 18, Moscow, Russia, 123098. Voice: +7 095 193 43 10; fax: +7 095 193 61 85.
e-mail: olegusss1@mail.ru

Ann. N.Y. Acad. Sci. 1078: 48–59 (2006).
doi: 10.1196/annals.1374.006

in human pathology, causing several diseases, and sometimes presenting with similar manifestations.

Russia has a long-standing tradition of studying rickettsioses that goes back to typhus outbreaks during World War I and the Civil War (1917–1922) of the 20th century.[1–3] These days the number of research groups and laboratories working on epidemiology of obligate vector-borne rickettsioses in Russia is continuously increasing. While it seems to be well-studied these days, our knowledge is constantly being updated by changes in environmental and social factors. Human technogenic influence on the natural foci (and, in fact, on our abilities to study rickettsiae and their environment) and migration of population, so clearly marked in Russia, play their roles in changes of rickettsioses epidemiology. Constantly improving clinical diagnostic tools as well as their availability for routine laboratory testing may increase the number of registered diseases.

Molecular biological methods are now providing a real burst in the investigation of rickettsial ecology, identification of rickettsiae as a cause of novel diseases, or their associations with new vectors. Polymerase chain reaction (PCR) is an effective method for the identification of rickettsial DNA in a variety of specimens, such as different human samples (preferably skin biopsies) and arthropod vectors. It also made it possible to identify multiple microorganisms in one sample, often of a limited DNA quantity something that was hardly possible previously, when only classical, time-consuming methods of identification of rickettsiae were in use. Additionally, most of these method did not allow precise and exact differentiation of close species or strains of the same species. So, identification and differentiation of rickettsiae in the past were based on laboratory (mostly, serological) data, ecology, and epidemiology of these microorganisms.

Two rickettsioses are currently being reported officially in the Russian Federation: epidemic typhus and Brill–Zinsser disease, both caused by *Rickettsia prowazekii* and Siberian tick typhus, caused by *R. sibirica*. Q fever (coxiellosis) is also under compulsory official registration. During the last decades two new rickettsioses have been described: Astrakhan spotted fever,[4] caused by a rickettsia recently classified as *R. conorii* subsp. *caspia*,[5,6] and Far Eastern tick-borne rickettsiosis, caused by *R. heilongjiangensis*.[7]

LOUSE-BORNE (EPIDEMIC) TYPHUS AND BRILL–ZINSSER DISEASE

Epidemic typhus remains a very serious threat for humans. It still exists in South and Central America, low-income African and Asian countries, and highlands with cold climate and low hygienic habits, associated with body louse infestation. Epidemic typhus is transmitted by the body louse, whose role in this process has been studied in many countries including Russia since the middle of 19th century.[2,3] Works on epidemic typhus elucidating pathogenesis, ecology, and epidemiology have been presented. P.F. Zdrodovskii has shown that lice also get a typhus infection and die in 10–18 days because of massive lipis of intestinal cells by rickettsiae. Both naturally (collected on patients with endemic typhus) and experimentally (either by injection of rickettsial into the louse intestine or after feeding on rickettsiemic rabbit) infected lice had the same result.[7]

High-risk group include refugees during local conflicts, homeless people, and the poorest social layers of a population. Patients who have once suffered from acute epidemic typhus and have recovered may develop asymptomatic persistent infection with *R. prowazekii*. Ten to 20% of them may suffer a relapse known as Brill–Zinsser disease, which may be as severe as acute infection but is generally milder. It is thought that humans who recovered from epidemic typhus retain some persisting rickettsiae for the rest of their life. The bacteriemia may then allow feeding body lice to become infected and to start a new epidemic typhus outbreak. The more the lice feed on such person, the more intensive transmission of rickettsia and the larger the new outbreak. The recent outbreak in Burundi is an excellent example: Since January 1997 till September 1997, 45,345 people acquired typhus and mortality was up to 15%.[8,9] Burundi has a repetitive history of outbreaks; here they occur periodically as soon as a young nonimmune population grows up and from time to time becomes infected through frequent relapses and high body louse proliferation.

Small outbreaks also happen in close isolated human communities when both factors (recrudescent typhus and high louse prevalence) are present. An example occurred in 1997 in a mental home in nursing Lipetsk region of Russia, located ~360 km from Moscow.[10,11] The infection was introduced by an elderly nurse badly infested with body lice. Due to inappropriate hygiene, lice quickly spread among the patients, mostly among those who were not able to wash/clean themselves and change their clothes, resulting in 29 cases. Fortunately, there were no fatal cases; most of the cases were not severe; and some were diagnosed only serologically.

Another question concerns the role of head lice in epidemic typhus transmission. Susceptibility of head lice to *R. prowazekii* infection has been determined experimentally.[12] Infected head lice excrete viable rickettsiae and die.[2] Nevertheless, available epidemiological data do not support the idea of head louse as a vector of *R. prowazekii.*

Statistics show that registered morbidity in the Russian Federation is 0.00–0.01 per 100 thousand population during recent years. Except for the above-mentioned outbreak, there are no spontaneous cases of epidemic typhus and statistics reflect 0–10 cases of Brill–Zinsser disease per year, with all cases in patients older than 50 years.

Morphologic characteristics of body and head lice have been intensively studied. Representative collection of lice from different regions of the former USSR and Professor Eichler's collection (Humboldt University in Berlin) with specimens collected in Europe, Asia, and New Guinea were investigated in the laboratory of rickettsial ecology of the Gamaleya Institute of Epidemiology and Microbiology. Morphologic features for both species were stable and definite; there were no specimens that shared features of body and head lice. A comparative table for morphologic identification of these two species has been developed.[13]

Epidemic typhus nowadays represents a big threat for society worldwide, including Russia, because of the high incidence of body lice infestation (about 200–250 cases per 100,000 of the population), and social instability including homelessness and a large number of refugees. Several registered outbreaks and existence of complement-fixing antibodies to *R. prowazekii* detected in healthy blood donors between 25 and 30 years old support the re-emergence of this infection.[11,14,15]

SPOTTED FEVER GROUP RICKETTSIOSES (SFG RICKETTSIOSES)

SFG rickettsiae are evolutionarily the most ancient and ecologically very plastic bacteria. Their natural environment is the organisms of blood-sucking arthropods, mostly ixodid ticks, and their vertebrate hosts. The distribution of rickettsial vectors strongly depends on the territories populated by these animals and arthropods.

Sixteen species of SFG rickettsiae were demonstrated to cause infection in humans; of them seven were described within the last decade and hence are referred to as emerging rickettsioses. Furthermore, at least another 20 SFG rickettsial species and isolates are described, for which pathogenic potential for humans has not been demonstrated yet. It has to be mentioned that Rocky Mountain spotted fever, Mediterranean spotted fever, African tick-bite typhus, Siberian tick typhus, Japanese spotted fever, and other rickettsioses were diagnosed clinically and years passed before identification of their etiology.[2,6,16] In contrast, *R. slovaca*, *R. aeschlimannii*, and *R. parkeri* were isolated from ixodid ticks long before confirmation of their role in human pathology.

Similar clinical presentations and antigenic cross-reactivity of SFG rickettsiae complicate the accurate diagnosis, especially in areas where several rickettsial species may circulate.[17] Differentiation and precise diagnosis are possible by molecular biology methods. However, it is more difficult to identify retrospectively the true prevalence of each rickettsiosis in certain shared areas from epidemiological points of view.

Geographical distribution of rickettsiae, their vectors, and vertebrate hosts is still among the most intensively studied topics for investigation in rickettsiology. Because the distribution of *Rickettsia* spp. is determined by the incidence of its tick host, both the seasonal variations in human morbidity and regional prevalence of the disease also run parallel with it.[16,18–21] Nevertheless, some ecologically unexpected data continuously appear: a case of Mediterranean spotted fever reported in Uruguay,[22] a genetic variant of Astrakhan spotted fever described from a patient from Chad,[23] and a case of *R. aeschlimannii* being detected in *Haemaphysalis punctata* ticks in Kazakhstan in close proximity to the Russian border[21] (although previously these rickettsia were found in *Hyalomma marginatum* in Africa).

PCR-based screening performed on ticks in Russia revealed the presence of several pathogenic for human rickettsiae. These are *R. slovaca* in *Dermacentor marginatus* ticks[24] and *R. aeschlimanni* in *H. punctata*.[21] Although diseases caused by these rickettsiae have not been reported in Russia yet, it is not excluded that cases are misdiagnosed and not identified because physicians are not familiar with clinical manifestations of these infections. Accordingly, *R. aeschlimanii* can be misdiagnosed as a rickettsiosis similar to Mediterranean spotted fever.[25,26] *R. slovaca* causes an illness presenting as lymphadenopathy, erythema, and necrosis on a site of the tick bite and is referred in European countries as TIBOLA (tick-borne lymphadenopathy) or DEBONEL (*Dermacentor*-borne necrosis erythema lymphadenopathy).[27,28] In these countries, the disease often occurs during cold months, is reported in younger patients, and has a shorter incubation period, and the head is the most frequent location of tick attachment. The disease may be misdiagnosed as Lyme disease on account of erythema and migraines. Because *D. marginatus* is a widespread species of tick (existing all across the southern part of European Russia), that very likely bites humans, we may suppose that *R. slovaca* may be also distributed there.

Two close rickettsiae, RpA4 and DnS14, were identified all across Russia (Far East, Eastern; eastern, central, and western parts of Southern Siberia; Ural; and central European Russia).[21,24] This rickettsia (presumably RpA4 and DnS14 represents two subspecies or strains of the same rickettsial species) were suggested to cause Mediterranean-spotted-fever-like illnesses in Castiglia y Leon and La Rioja, Spain. A large screening study carried out in Spain showed that *Rickettsia* sp. DnS14/RpA4 makes up the portion of 14% among all SFG rickettsiae identified in ticks collected from patients.[29] In another study conducted in Spain (La Rioja) six patients were bitten by *D. marginatus* ticks, where the *Rickettsia* sp. DnS14/RpA4 was identified. Two of them developed a disease, clinically identical to DEBONEL/TIBOLA, caused by *R. slovaca*.[30] An isolate of SFG rickettsia similar to sp. DnS14 from *D. silvarum* tick collected in the Russian Far East in 2005 (O. Mediannikov, unpublished data) may help to determine the role of this species in human pathology.

ASTRAKHAN SPOTTED FEVER

An eruptive febrile disease had been being observed in the Astrakhan region on the Caspian Sea (FIG. 1), during the summer time since 1970s. It was apparently unknown before this time; first cases were noticed after construction

FIGURE 1. Distribution of tick-borne rickettsioses in the Russian Federation.

of a petrochemical complex producing a lot of CO_2 in atmosphere. Clinical data suggested the primary name: "viral exanthema." Since 1983 in a small area of Astrakhan region (44.000 sq. km) more than 2,000 cases of the disease have been registered. No gender and age preferences were noticed. The development of the disease is clearly associated with the bite of *Rhipicephalus pumilio* tick. The numbers of this tick species significantly increased since the construction of the petrochemical complex. In the 1930s, specimens of this tick were rarely found in these areas. This tick feeds on dogs, cats, and hedgehogs living in close proximity to humans. Soon, the agent of the disease was isolated from the blood of two patients with the typical clinical picture of rickettsiosis (fever, rash, eschar) and from *Rh. pumilio* ticks collected from dogs, cats, and hedgehogs.[31,32] Biological, antigenic, and immunogenic properties of this rickettsia were similar to those of *R. conorii.*[33] Astrakhan spotted fever was later serologically diagnosed in Kalmykia (Russia); probably also exists in Western Kazakhstan, where *Rh. pumilio* tick is prevalent; however, no clinical cases have been reported yet. Similar rickettsiae were identified as a cause of febrile illness in a patient from Chad[23] and found in *Rh. sanguineus* ticks from Kosovo.[34]

Taxonomic position of this rickettsia has been a subject of discussion since its discovery in 1991.[4] On the basis of genetic parameters, this rickettsia is currently classified as *R. conorii*, subspecies *caspia*.[5,6] Nonetheless, we suggest that epidemiological data, such as isolated area of distribution and specific vector (*Rh. pumilio*), which is genetically far from *Rh. sanguineus* vector, or *R. conorii*,[35] may require re-evaluation of its taxonomic status.

SIBERIAN TICK TYPHUS AND FAR EASTERN TICK-BORNE RICKETTSIOSIS

Siberian tick typhus (STT) is caused by *R. sibirica*. It is reported mostly in the regions of southern Siberia (Altay, Krasnoyarsk, Novosibirsk regions) and Primorie (Maritime territory, with Vladivostok being the capital city) (FIG. 1). Tick-borne spotted fever in Russia was first described in the Far East during 1932–1935.[36] Later, a similar disease was reported in central Siberia.[37] In 1943 the first strains were isolated in Siberia (Krasnoyarsk region) and were named as *R. sibirica*.

Clinical manifestation and epidemiology of STT are well-studied. The recent threefold increase of STT incidence compared with that of the previous decade (FIG. 2) indicates re-emergence of this infection in its classic endemic area.[19] Epidemiology of STT is similar to other tick-borne zoonoses, wherein the risk group are tourists, explorers of natural foci, summer residents in the countryside, and all other persons having close contact with the natural environment of tick vectors. So, the distribution area of STT should, ideally, coincide with a wide area of distribution of the vector, predominantly the *Dermacentor* tick.

R. sibirica has been detected in more than 20 species of Ixodidae ticks. Most of them are believed to be the vectors of STT. Ticks of the genus *Dermacentor* probably play the most important role in transmission to humans, because they readily bite humans and most of these tick species were found to be infected with *R. sibirica*.

R. sibirica subspecies *mongolotimonae* was first obtained as isolate HA-91 from *Hyalomma asiaticum*, a tick collected in Inner Mongolia in 1991.[38] It

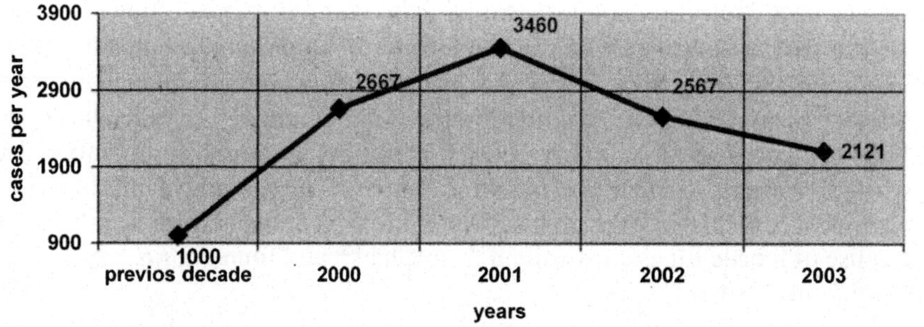

FIGURE 2. Official morbidity data on Siberian tick typhus.

has been implicated as a cause of atypical tick-borne infection associated with lymphadenitis and lymphangitis in southern Europe and Africa.[39] In southern Asiatic Russia, where this tick species is distributed, this infection may also be prevalent.

There is no strict border between Siberia and the Russian Far East (conventionally all regions situated close to the Pacific Ocean coast are called the Far East). So, historically, it has been considered that because of similar clinical and epidemiological features and serological data, rickettsiosis in Siberia and the Far East (from where first cases have been reported) are caused by the same microorganism.[40] Additionally, differentiation of species in rickettsiology was a very difficult task at the time, without modern molecular techniques. Nevertheless, some doubts have been left about the role of *R. sibirica* as a cause of spotted fever in the Far East. Particularly discordant were significant seasonal differences and the mean age of affected population between both regions; also, differences in serological properties were evident. Nevertheless, it has been considered that all strains isolated from patients and ticks in the Russian Far East belong to the same species, *R. sibirica*[2] and PCR-RFLP (restriction fragment length polymorphism)-based characterization of 24 SFG rickettsial isolates including strains from the Prymorye region identified them as *R. sibirica*.[41]

The study of 65 and 42 patients from Khabarovsk region in the Russian Far East has been performed, respectively, during 2002–2003.[7] The patients were selected on the basis of their clinical presentations compatible with Siberian tick typhus, the disease that has been diagnosed in this region since 1934. The major aim of this study was to perform the screening of patients with suggested STT by molecular biology techniques. The disease in Khabarovsk was similar to classical STT, which was distributed in central Siberia. However, specific features, in spite of more southern geography, included a later morbidity peak rising in June–July compared to April–May for typical STT, and relatively often the presence of subcutaneous lymphangitis and an older affected population (mean age: 56 years). All these features were already mentioned since the first description of the disease, but they have been referred to "another form" of "classical" STT and explained by another tick vector and to specific properties of a local strains of *R. sibirica*. It is interesting that morbidity peak in July coincides with population dynamics of *H. concinnae* and *H. japonica douglasi* ticks, so they were thought to be main vectors of rickettsiosis in the Far East, whereas *Dermacentor* sp. ticks, abundant in April–May, are the classical vectors of STT.

PCR-based screening technique has been applied for investigation of samples from these patients. Nested reaction for *gltA* gene has been used for testing DNA extracted from buffy coats and eschar biopsies of acutely ill patients. Positive PCR results were obtained for 11.5% of studied buffy coat samples and for 40% of eschar samples studied. Nucleotide sequence of the DNA fragments amplified from patient specimens had 100% sequence similarity to the sequences of homologous genes from *R. heilongjiangensis*. *R. heilongjiangensis* was previously isolated only once from ticks *D.r silvarum* in Heilongjiang province in northern China.[42] Totally, for portions of three genes (*gltA, ompA*, and *ompB*), *R. heilongjiangensis* has been amplified and sequenced from patients.

This rickettsia has been earlier suggested as a possible human pathogen. Sera of patients with clinical features of rickettsiosis in northern China were tested and found positive in complement-binding studies,[43] but neither the agent had been isolated nor its gene amplified from humans in China.

Serological data from patients from the Russian Far East supported the idea that differences between STT in Siberia and the Far East could be explained by different etiology. In addition, in ticks collected in nature (around Khabarovsk city) and removed from the patients prior to the development of the disease, the same *R. heilongjiangensis* has been found. High infection rate (up to 28%) in *H. concinnae* ticks indicates the main role of this species in transmission. *H. japonica douglasii* ticks also harbor this rickettsia, but it is rarer (only 4.5%).[44]

R. heilongjiangensis was also identified in *H. concinna* ticks in Krasnoyarsk region, where classic STT is reported.[24] Recently, three strains of this rickettsia have been identified among the old-strains collection at the Omsk Institute of Natural Foci Infections (S. Shpynov, unpublished data).

The distinct etiological cause, peculiarities in clinical picture, and epidemiology allowed us to call the rickettsiosis caused by *R. heilongjiangensis*, an emerging disease, Far Eastern tick-borne rickettsiosis. Therefore, the long-known disease in the Russian Far East disease turned out to be a new disease.

The distribution of STT and Far Eastern tick-borne rickettsiosis are close and partly overlay each other, so differential diagnosis should be precisely performed.

CONCLUSION

Nowadays several factors play important roles in rickettsial ecology and epidemiology. For the typhus group rickettsioses, these are economical and hygienic factors, but for the spotted fever group rickettsioses, these are mostly the technogenic influence of human activities on the environment, where arthropod vectors and natural hosts live, and may affect its natural history. Modern methods of isolation and identification of rickettsiae led to the discovery of several new emerging rickettsioses. Nevertheless, precise intimate molecular aspects of parasite–host interactions still remain uncovered. Interesting perspectives are in the field of research of causes and mechanisms of the process of emerging new diseases for the humanity, but, definitely, old for the nature, in order to made better prognosis.

ACKNOWLEDGMENTS

We deeply thank Dr. Marina Eremeeva for generous donations and worthy advice, additions, and corrections to this manuscript and Dr. Karim Hechemy for being patient and benevolent.

REFERENCES

1. Lyskovtsev, M.M. 1968. Tickborne Rickettsiosis: [in Russian]. O.P. Peterson, Ed. Medgiz. Moscow.
2. Zdrodovskii, P.F. & E.M. Golinevich. 1960. The Rickettsial Diseases. Pergamon Press. London.
3. Kramchaninov, N.F. & F.V. Popova. 1966. On the history of the study of typhus in Russia [in Russian]. Zh. Mikrobiol. Epidemiol. Immunobiol. **43:** 152–153.
4. Tarasevich, I.V., V. Makarova, N.F. Fetisova, *et al.* 1991. Astrakhan fever: new spotted fever group rickettsiosis. Lancet **337:** 172–173.
5. Fournier, P.E., Y. Zhu, H. Ogata & D. Raoult. 2004. Use of highly variable intergenic spacer sequences for multispacer typing of *Rickettsia conorii* strains. J. Clin. Microbiol. **42:** 5757–5766.
6. Zhu, Y., P.E. Fournier, M. Eremeeva & D. Raoult. 2005. Proposal to create subspecies of *Rickettsia conorii* based on multi-locus sequence typing and an emended description of *Rickettsia conorii*. BMC Microbiol. **5:** 11.
7. Mediannikov, O.Y., Y. Sidelnikov, L. Ivanov, *et al.* 2004. Acute tick-borne rickettsiosis caused by *Rickettsia heilongjiangensis* in Russian Far East. Emerg. Infect. Dis. **10:** 810–817.
8. Raoult, D., V. Roux, J.B. Ndihokubwaho, *et al.* 1997. Jail fever (epidemic typhus) outbreak in Burundi. Emerg. Infect. Dis. **3:** 357–360.
9. Raoult, D., J.B. Ndihokubwayo, H. Tissot-Dupont, *et al.* 1998. Outbreak of epidemic typhus associated with trench fever in Burundi. Lancet **352:** 353–358.
10. Savelyev, S.I., I.A. Schukina, V.I. Mischuk, *et al.* 1998. Epidemic typhus outbreak in Lipetsk region [in Russian]. ZNISO **1:** 7–11.
11. Tarasevich, I.V., E. Rydkina & D. Raoult. 1998. Outbreak of epidemic typhus in Russia. Lancet **352:** 1151.
12. Murray, E.S. & S.B. Torrey. 1975. Virulence of *Rickettsia prowazeki* for head lice. Ann. N.Y. Acad. Sci. **266:** 25–34.
13. Tarasevich, I.V., A.A. Zemskaya & V.V. Khudobin. 1988. Diagnosis of lice species belonging to the genus *Pediculus*. Med. Parazitol. Mosk. Dis. **3:** 48–52.
14. Tarasevich, I.V. & N.F. Fetisova. 1995. Classical typhus [in Russian]. ZNISO **2:** 9–15.
15. Tarasevich, I.V. 1996. Louse-borne typhus fever in Russia: yesterday, today, and tomorrow. *In* Twelfth Sesqui-Annual Meeting of the ASRRD. Pacific Grove, California.
16. Raoult, D. & V. Roux. 1997. Rickettsioses as paradigms of new or emerging infectious diseases. Clin. Microbiol. Rev. **10:** 694–719.
17. Parola, P. & D. Raoult. 2001. Tick-borne bacterial diseases emerging in Europe. Clin. Microbiol. Infect. **7:** 80–83.

18. TARASEVICH, I.V. 1978. Ecology of rickettsiae and epidemiology of rickettsial diseases. *In* Rickettsiae and Rickettsial Diseases. J. Kazar, R.A. Ormsbee & I.V. Tarasevich, Eds.: 330–349. VEDA. Publishing House of the Slovak Academy of Sciences. Bratislava.
19. RUDAKOV, N.V., S.N. SHPYNOV, I.E. SAMOILENKO & M.A. TANKIBAEV. 2003. Ecology and epidemiology of spotted fever group rickettsiae and new data from their study in Russia and Kazakhstan. Ann. N. Y. Acad. Sci. **990:** 12–24.
20. AZAD, A.F. & C.B. BEARD. 1998. Rickettsial pathogens and their arthropod vectors. Emerg. Infect. Dis. **4:** 179–186.
21. SHPYNOV, S.N., P.E. FOURNIER, N.V. RUDAKOV, *et al.* 2004. Detection of a rickettsia closely related to *Rickettsia aeschlimannii*, "*Rickettsia heilongjiangensis*," *Rickettsia* sp. strain RpA4, and *Ehrlichia muris* in ticks collected in Russia and Kazakhstan. J. Clin. Microbiol. **42:** 2221–2223.
22. CONTI-DIAZ, I.A., I. RUBIO, R.E. SOMMA MOREIRA & G. PEREZ BORMIDA. 1990. Cutaneous-ganglionar rickettsiosis by *Rickettsia conorii* in Uruguay. Rev. Inst. Med. Trop. Sao Paulo **32:** 313–318.
23. FOURNIER, P.E., B. XERIDAT & D. RAOULT. 2003. Isolation of a rickettsia related to Astrakhan fever rickettsia from a patient in Chad. Ann. N.Y. Acad. Sci. **990:** 152–157.
24. SHPYNOV, S.N., P. PAROLA, N.V. RUDAKOV, *et al.* 2001. Detection and identification of spotted fever group rickettsiae in *Dermatocentor* ticks from Russia and central Kazakhstan. Eur. J. Clin. Microbiol. Infect. Dis. **20:** 903–905.
25. RAOULT, D., P.E. FOURNIER, P. ABBOUD & F. CARON. 2002. First documented human *Rickettsia aeschlimannii* infection. Emerg. Infect. Dis. **8:** 748–749.
26. PRETORIUS, A.M. & R.J. BIRTLES. 2002. *Rickettsia aeschlimannii*: a new pathogenetic spotted fever group Rickettsia, South Africa. Emerg. Infect. Dis. **8:** 874.
27. LAKOS, A. 2001. Tick-borne lymphadenopathy (TIBOLA). Clin. Microbiol. Inf. **7:** 31.
28. OTEO, J.A., V. IBARRA, J.R. BLANCO, *et al.* 2004. *Dermacentor*-borne necrosis erythema and lymphadenopathy: clinical and epidemiological features of a new tick-borne disease. Clin. Microbiol. Infect **10:** 327–331.
29. FERNANDEZ-SOTO, P., R. PEREZ-SANCHEZ, R. ALAMO-SANZ & A. ENCINAS-GRANDES. 2005. Spotted fever group rickettsiae in ticks feeding on humans in Northwestern Spain. Is *Rickettsia conorii* vanishing? *In* 4th International Conference on Rickettsiae and Rickettsial Diseases, June 18–21, 2005; Book of abstracts. J.A. Oteo Revuelta, J.R. Blanco Ramos & A. Portillo, Eds.: P–99.
30. IBARRA, V., A. PORTILLO, S. SANTIBANEZ, *et al.* 2005. DEBONEL/TIBOLA: is *Rickettsia slovaca* the only etiological agent? Ann. N.Y. Acad. Sci. **1063:** 346–348.
31. EREMEEVA, M.E., L. BEATI, V.A. MAKAROVA, *et al.* 1994. Astrakhan fever rickettsiae: antigenic and genotypic analysis of isolates obtained from human and *Rhipicephalus pumilio* ticks. Am. J. Trop. Med. Hyg. **51:** 697–706.
32. TARASEVICH, I.V., V.A. MAKAROVA, N.F. FETISOVA, *et al.* 1991. Studies of a "new" rickettsiosis "Astrakhan" spotted fever. Eur. J. Epidemiol. **7:** 294–298.
33. TARASEVICH, I.V. 2002. Astrakhan Spotted Fever [in Russian]. Meditsina. Moscow.
34. FOURNIER, P.E., J.P. DURAND, J.M. ROLAIN, *et al.* 2003. Detection of Astrakhan fever rickettsia from ticks in Kosovo. Ann. N.Y. Acad. Sci. **990:** 158–161.
35. BEATI, L. & J.E. KEIRANS. 2001. Analysis of the systematic relationships among ticks of the genera *Rhipicephalus* and *Boophilus* (Acari: Ixodidae) based on

mitochondrial 12S ribosomal DNA gene sequences and morphological characters. J. Parasitol. **87:** 32–48.
36. MILL, E.N. 1936. Tick-borne fever in Maritime Region. Far East. Med. J. **3:** 34–36.
37. SHMATIKOV, M.D. & M.A. VELIK. 1939. Tick-borne spotted fever. Klin. Med. (Mosk.) **7:** 124–128.
38. YU, X., Y. JIN, M. FAN, *et al.* 1993. Genotypic and antigenic identification of two new strains of spotted fever group rickettsiae isolated from China. J. Clin. Microbiol. **31:** 83–88.
39. FOURNIER, P.E., F. GOURIET, P. BROUQUI, *et al.* 2005. Lymphangitis-associated rickettsiosis, a new rickettsiosis caused by *Rickettsia sibirica mongolotimonae*: seven new cases and review of the literature. Clin. Infect. Dis. **40:** 1435–1444.
40. SAVITSKAYA, E.P. 1943. To the etiology of tick typus in Khabarovsk region. Zh. Mikrobiol. Epidemiol. Immunobiol. **10–11:** 87.
41. BALAYEVA, N.M., M.E. EREMEEVA, V.F. IGNATOVICH, *et al.* 1996. Biological and genetic characterization of *Rickettsia sibirica* strains isolated in the endemic area of the North Asian tick typhus. Am. J. Trop. Med. Hyg. **55:** 685–692.
42. FAN, M.Y., J.Z. ZHANG, M. CHEN & X.J. YU. 1999. Spotted fever group rickettsioses in China. *In* Rickettsiae and Rickettsial Diseases at the Turn of the Third Millennium. D. Raoult & P. Brouqui, Eds.: 247–257. Elsevier. Paris.
43. LOU, D., Y.M. WU, B. WANG, *et al.* 1989. Confirmation of patients with tick-borne spotted fever caused by *Rickettsia heilongjiangi*. Chin. J. Epidemiol. **10:** 128–132.
44. MEDIANNIKOV, O.Y., Y.N. SIDELNIKOV, L.I. IVANOV, *et al.* 2005. Far-Eastern tick-borne rickettsiosis, an emerging spotted fever. *In* 4th International Conference on Rickettsiae and Rickettsial Diseases, June 18–21, 2005; Book of abstracts. Oteo, J.A., J.R. Blanco & A. Portillo, Eds.: Trama Impresores S.A.L.

Rickettsioses in Japan and the Far East

FUMIHIKO MAHARA

Mahara Hospital, 6-1, Aratano, Anan-city, Tokushima, 779-1510, Japan

ABSTRACT: Three rickettsial diseases are known to exist in Japan currently: Japanese spotted fever (JSF), Tsutsugamushi disease (TD; scrub typhus), and Q fever. Since April 1999, the system for infection control and prevention in Japan has changed drastically. JSF, Q fever, and TD, as emerging infectious diseases, are designated as national notifiable diseases.The geographic distribution of JSF patients is along the coast of central and southwestern Japan, whereas TD and Q fever occur almost all over the country. The number of JSF patients reported was 216 cases during 1984998 and 268 cases, under the revised law, in 1999004. About 300000 cases of TD occur every year, and 76 cases of Q fever in 1999004. The number of cases of JSF and its endemic area are gradually increasing. There was only one fatality due to JSF until 2003, whereas two patients died of JSF in 2004, so JSF is still a life-threatening disease in Japan. Treatment of fulminant JSF consists of prompt administration of a combination of tetracycline and quinolone. Recent tick surveys revealed that the most probable vectors of JSF are Haemophysalis flava and Haemophysalis hystericis. In addition to R. japonica, two serotypes or species of spotted fever group rickettsiae have been isolated from ticks in Japan; one is closely related to R. helvetica and the other is a new genotype of unknown genotype AT, which is closely related to a Slovakian genotype. These serotypes are of uncertain clinical significance. Epidemiology of rickettsioses in the Far East is mentioned briefly.

KEYWORDS: Japanese spotted fever; Tsutsugamushi disease; Q fever; rickettsiosis; spotted fever; tick; *Rickettsia japonica*

INTRODUCTION

Three rickettsial diseases are known to exist in Japan currently: Japanese spotted fever (JSF), Tsutsugamushi disease (scrub typhus), and Q fever. Tsutsugamushi disease was the only known rickettsiosis in Japan until 1984, when the author reported the first clinical cases of JSF. Subsequently, a Q fever patient was reported in 1989.

Since April 1999, the system for infection control and prevention in Japan has changed drastically after 100 years of the old infection control law. JSF and

Address for correspondence: Fumihiko Mahara, M.D., Ph.D., Mahara Hospital, 6-1, Aratano, Anan-city, Tokushima, 779-1510, Japan. Voice: +81-884-363339; fax: +81-884-363641.
e-mail:mahara@tokushima.med.or.jp

Ann. N.Y. Acad. Sci. 1078: 60–73 (2006). © 2006 New York Academy of Sciences.
doi: 10.1196/annals.1374.007

Q fever, as emerging infectious diseases, are designated as national notifiable diseases in addition to conventional Tsutsugamushi disease. In the new system, it is the duty of the doctor who diagnoses a notifiable disease to report it immediately to the health authorities.

Clinical and epidemiological information on JSF and Q fever is limited because outbreaks have been sporadic and the numbers of cases small in each hospital. However, information is increasing as a result of the revised law. As for the studies of the vectors, steady progress is being made.

This article mainly aims to describe the recent clinical and epidemiological studies on JSF, as different from Tsutsugamushi disease, and to provide some information about Q fever in Japan and rickettsioses in the Far East.

JSF AND TSUTSUGAMUSHI DISEASE

History

In the 1980s clinicians believed that Tsutsugamushi disease was the only rickettsial disease in Japan, except for sporadic outbreaks of epidemic typhus in the 1950s. In Tokushima Prefecture, located in Shikoku Island in southwestern Japan, neither disease has been reported in the last two decades.

In May 1984, a 63-year-old woman farm worker was hospitalized at Mahara Hospital with high fever and erythematous nonpurpuric skin eruptions all over the body. Antibiotics (β-lactam and aminoglycoside) used for common febrile illnesses were not effective. Two additional patients with similar symptoms were hospitalized in May and July 1984. In both patients, an eschar was observed. Tsutsugamushi disease was suspected; however, the Weil–Felix tests were positive for OX2 serum agglutinins but negative for OXK in all three cases, which did not support Tsutsugamushi disease, but rather OX2-positive infections, that is, spotted fever group (SFG) rickettsiosis.[1] The cases were subsequently confirmed by complement fixation test with antigens of SFG rickettsiae.[2] The name Japanese spotted fever was proposed for these infections[3] and has been commonly used ever since.[4–6] Oriental spotted fever is a synonym for JSF. The causative agent was isolated from the patient and named *Rickettsia japonica (R. japonica).*[7] Tsutsugamushi disease caused by *Orientia tsutsugamushi*[8] was discovered in this area soon afterward.

It was surprising that the first tool used was the Weil–Felix test despite a high level of medical expertise in the country. Investigations of SFG rickettsiosis in Japan date from the recognition of these initial cases.

Clinical Features

Through December 2004, 53 clinical cases of JSF and 23 cases of Tsutsugamushi disease were diagnosed at Mahara Hospital, Tokushima Prefecture.

TABLE 1. Clinical features of Japanese spotted fever

	No. of cases	%
Fever	53	100
Skin rash	53	100
Eschar	50	94
Malaise	47	89
Chills	46	87
Headache	43	81
Confusion	6	11
Hospitalization	43	81
Death	1	1.9
		Total 53 cases

During the same period, 94 cases of human tick bites were recorded in this JSF-endemic area in the same hospital. Clinical features are described on the basis of this experience.

Of the 53 JSF patients, 22 were male and 31 were female. Their ages ranged from 4 to 89 years, but most patients (75.4%) were between the ages of 50 and 80 years. (TABLE 1).

The onset of the disease was 2 to 10 days after work in the fields and was abrupt, with common symptoms of headache (43 patients; 81%), high fever (53; 100%), and shaking chills (46; 87%). Other major objective features of JSF included skin eruptions (53; 100%) and tick bite eschar (50; 94%). In the acute stage, remittent high fever was frequently observed. The maximum body temperature was 38.5–40.9°C (mean 39.7°C), which was higher than that seen 38.5–39.3°C in patients with Tsutsugamushi disease. With abrupt high fever or a few days after mild fever of unknown origin, characteristic erythemas developed on the extremities and spread rapidly (in a few hours) to all parts of the body including the palms and soles, without accompanying pain or itching. These eruptions measured 5 to 10 mm in size and had indistinct margins (FIG. 1).

The erythemas became striking during the febrile period and tended to spread more over the extremities than the trunk. Palmar erythema, a characteristic finding, which does not occur in Tsutsugamushi disease, disappeared in the early stages of the disease.

The erythemas became petechial in 3 to 4 days, peaked in a week to 10 days, and disappeared in 2 weeks. However, in severe cases of petechial, brown pigmentation persisted for 2 months or more. Eschars were observed on the hands, feet, neck, trunk, and shoulders of patients. These eschars generally remained for 1 to 2 weeks but in some cases disappeared within a few days. Eschars of JSF patients are smaller than those seen in patients with Tsutsugamushi disease and may be missed without careful observation (FIG. 2 A and B).

Regional or generalized lymphadenopathy, which is observed in almost all cases of Tsutsugamushi disease, was not remarkable in JSF patients. Enlarge-

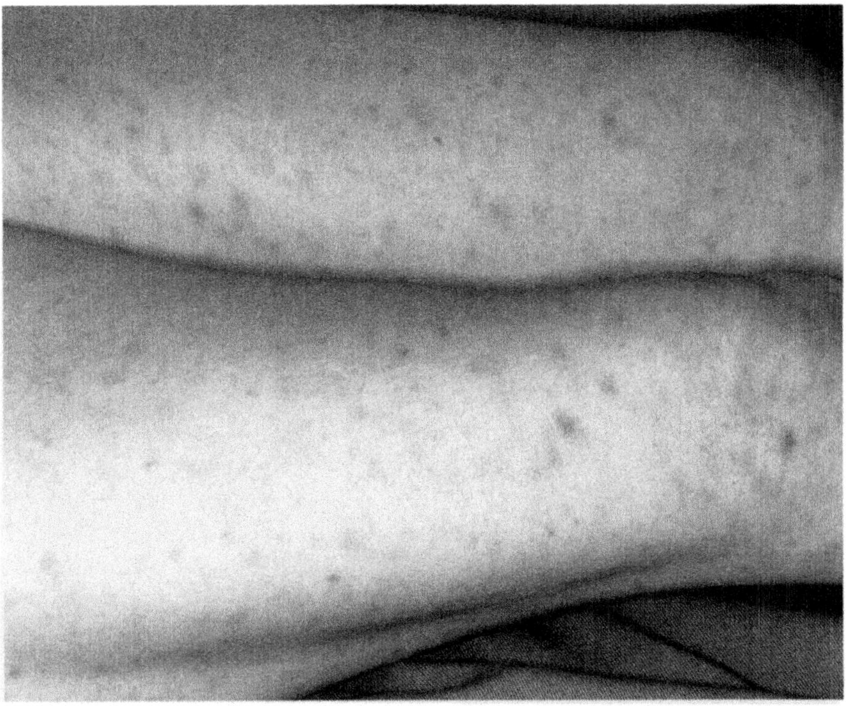

FIGURE 1. Typical erythema.

ment of the liver and spleen occurred in a few patients with cardiomegaly, and six patients had central nervous system involvement. One patient, a 14-year-old male, developed a subshock state on account of acute myocarditis.[48] Four JSF patients with meningitis were reported from other areas.[9]

Laboratory Examinations

The results of laboratory examinations in JSF patients resemble those of patients with common SFG rickettsioses.

During the initial stages of the disease, urinalysis is often slightly positive for protein and occult blood, which may lead to a misdiagnosis of urinary infection.

In the acute stage, leukocytosis or leukopenia may develop (WBC: 18,620–2,810/μL), with a left shift in leukocyte count. Thrombocytopenia (platelets: 2.2–35.3 $\times 10^4$/μL) may also be found. At weeks 1 to 2, leukocyte counts increased slightly, and lymphocyte counts tended to normalize. C-reactive proteins (CRP: 0.3–28.1 mg/dL, mean 10.2 mg/dL) were markedly positive, and liver functions (GOT: 12–305 IU/L; GPT: 7–153 IU/L) were slightly impaired but returned to normal in 2 to 3 weeks.

(A)

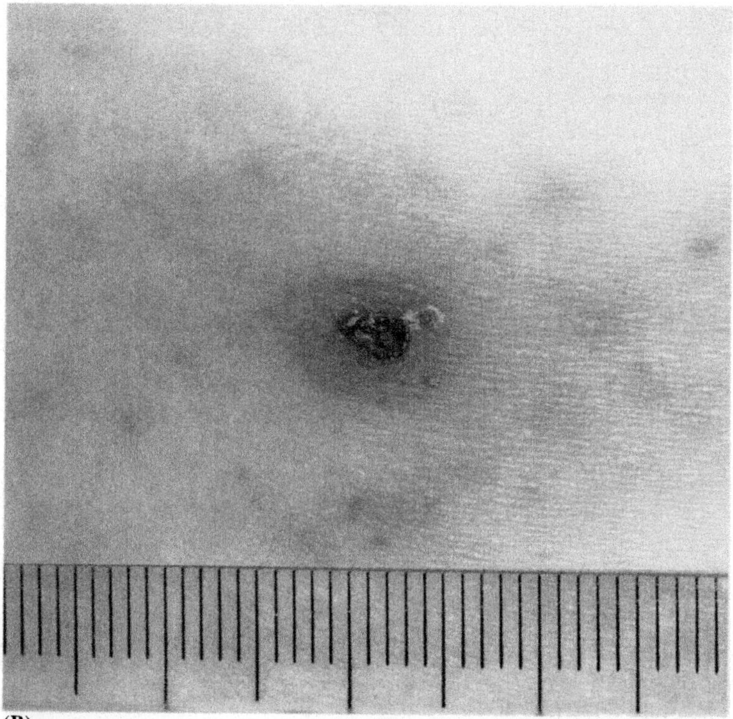

(B)

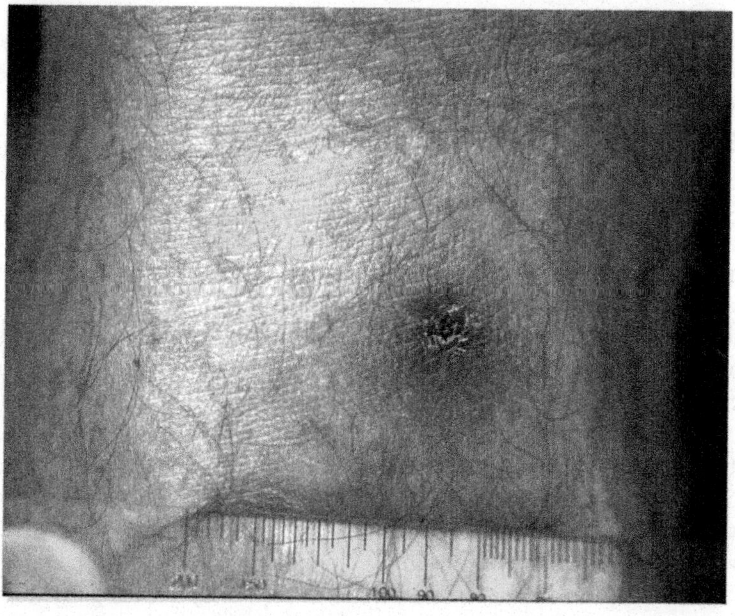

FIGURE 2. **(A)** Eschar of JSF and **(B)** eschar of Tsutsugamushi disease.

Seven cases developed disseminated intravascular coagulation (DIC) and two cases progressed to multiple organ failure (MOF).

Serological Diagnosis

Serodiagnosis for JSF is usually performed by the indirect immunoperoxidase (IP)[10] or immunofluorescence (IF) technique, with antigens prepared from *R. japonica* or other SFG rickettsiae (e.g., *Thai-tick typhus, R. montana*). With the IP test, IgG and IgM antibodies were detected in the sera beginning on day 3 to 5 after the onset of fever.

In the 53 clinically diagnosed cases of JSF at Mahara Hospital, all 52 patients had significant changes in serum IP antibody titers to *R. japonica*, and 27 of 34 (79%) had significant changes in OX2 agglutinin titers by the Weil–Felix technique, but this test has not been used recently.

The polymerase chain reaction (PCR) method, specific to strains isolated from the patients, has become available for the diagnosis of JSF.[11]

One case could not be confirmed serologically because of death within 24 h of hospitalization, although it was diagnosed clinically through major symptoms of the disease. This case prompted us to develop a biopsy-based immunohistochemical method for early diagnosis of JSF.[12]

Treatment

Antibiotics, such as penicillins, β-lactams, or aminoglycosides, commonly used empirically in febrile diseases were completely ineffective, but doxycycline hydrochloride (DOXY) and minocycline hydrate (MINO) were clinically effective in treating the JSF patients.

In an *in vitro* study,[13,14] MINO and DOXY were the most effective antibiotics against *R. japonica*, followed by other tetracycline antibiotics (MICs: 0.04–0.78 μg/dL). In contrast, the sensitivity to β-lactam and penicillin was lower or negligible (MICs: 50–100 μg/dL), but new quinolones were effective (MICs: 0.78–1.56 μg/dL).

Three patients were treated only with a new quinolone (ciprofloxacin) without other antibiotics, which proved effective although the effect was weaker than with MINO. In 1998, we had a case of fulminant JSF and the patient was treated with a combination using MINO and ciprofloxacin hydrate (CPFX) (FIG. 3). This patient developed a shock associated with DIC. MINO was the first choice for this patient; however, the patient's fever recurred and resolved only after the addition of CPFX.[15] An assay of serum cytokines revealed hypercytokinemia. MINO moderated cytokine production, while the administration of CPFX eradicated the *R. japonica.*[16] Since then, seven with fulminant patients JSF have been treated successfully with combined MINO and CPFX.[17]

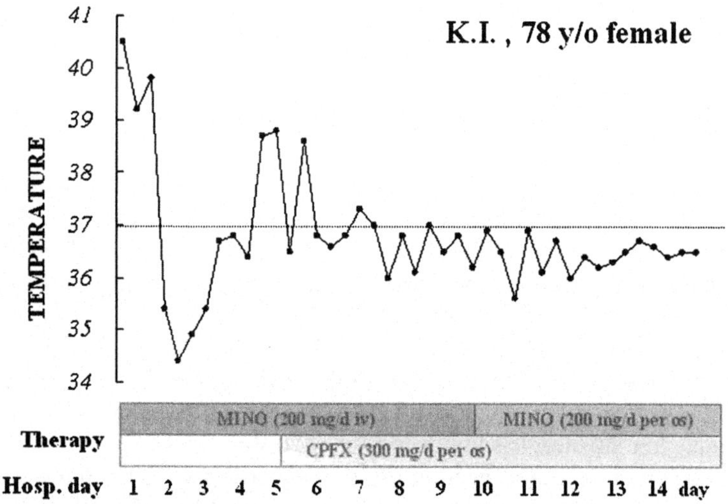

FIGURE 3. Combination treatment (MINO+CPFX).

Pathogen

The etiologic agent was first isolated from a patient in Kochi Prefecture in 1996.[18] The causative rickettsia was also isolated from a JSF patient in Tokushima Prefecture in 1997.[19] The former isolate was the type strain YH (ATCCVR-1363), later named *R. japonica*; the latter strain (Katayama) was the first isolate from outside Kochi. Serotyping by use of the reciprocal cross-reaction of mouse antisera to six human isolates from Tokushima and the type strain YH or *R. japonica* also indicated that these are the same species.[20]

In an electron microscopy study, *R. japonica* was generally recognized as short rods or pleomorphic coccobacillary forms less than 2 μm in length and 0.5 μm in diameter and could be found not only in the cytoplasm, but also in the nuclei of the host cells.[21] A multilayered mesosome-like structure was observed in rickettsiae multiplying in the host cell. This unique structure has not been reported in other species except for *Rickettsia prowazekii*.[22]

After the initial isolation, at least 20 rickettsial strains have been isolated from JSF patients by cell culture techniques or nude mouse passage in Tokushima, Kochi, Hyogo, Chiba, and Wakayama Prefectures. However, it has not been determined whether these strains differ in virulence.

Epidemiology

Geographic distributions (FIG. 3): JSF-endemic prefectures are located along the coast of southwestern and central Japan in a warm climate. The landscape is diverse, including bamboo plantations, crop fields, coastal hills, and forests.

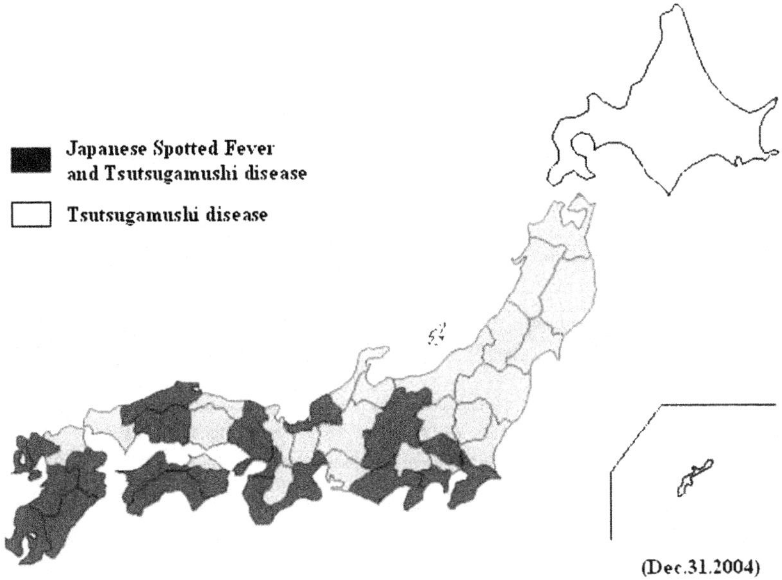

FIGURE 4. Geographical distributions of the cases of JSF and Tsutsugamushi disease in Japan.

On the other hand, Tsutsugamushi disease is observed all over Japan, except for Hokkaido Island in the North. In southern Japan, we can distinguish seasonal differences in the occurrence of JSF and Tsutsugamushi disease. Whereas JSF occurs from spring to autumn (April to October), Tsutsugamushi disease occurs in winter (November to February). However, in northern Japan the prevalent season for Tsutsugamushi disease varies between either spring or autumn. (FIG. 4)

The number of the patients (FIG. 5): From 1984 to 1998, 216 cases of JSF were accumulated by the National Institute of Health in Japan with the voluntary collaboration of investigators and from 1999 to 2004, 268 cases of JSF patients were reported under the revised law. Tsutsugamushi disease presents in about 300–1,000 patients every year. The number of reported cases of JSF is gradually increasing and the endemic areas are spreading.

Vector Study

Like other SFG rickettsioses, JSF is presumed to be transmitted by a tick bite. The high proportion of patients with tick bite eschars supports this hypothesis. Nineteen patients with JSF of 53 (35.8%) recalled tick bites before the onset of their illness; however, the tick had been lost, and no specimens from the patients were available for further study. During the same period, 94 cases of human

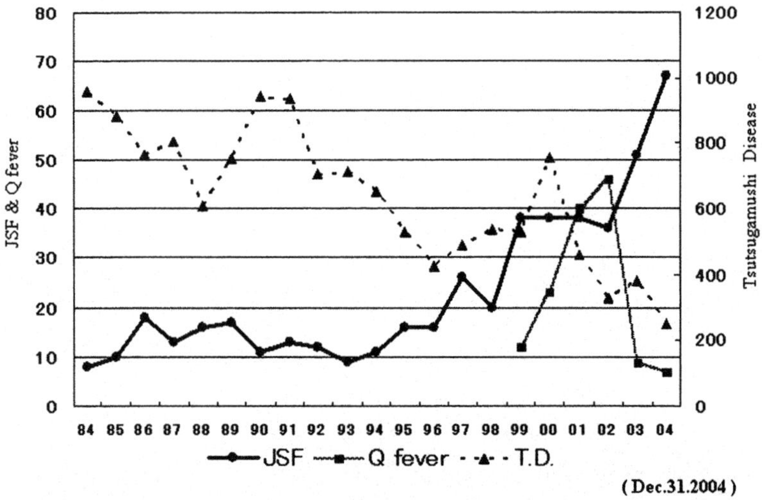

FIGURE 5. The number of patients with rickettsioses in Japan 1984–2004.

tick bites were observed at Mahara Hospital through December 2004. These 94 ticks were broken down into three genera and nine species: *Amblyomma* (*A.*) *testdinarium* 48 (51.1%), *Haemophysalis* (*H.*) *longicornis* 23 (24.5%), *H. flava* 12 (12.8%), *H. kitaokai* 1, *H. hystericis* 1, *Ixodes* (*I.*) *nipponensis* 5 (5.3%), *I. persulcatus* 2 (2.1%), *I. ovatus* 1, and *I. tanuki* 1.

The search for rickettsiae in ticks collected by flagging vegetation revealed four genera and nine species of ticks as positive for SFG rickettsiae in Japan (TABLE 2). *R. japonica* was first isolated from the larva of *H. flava* harvested from the bamboo forest where a JSF patient worked in May 1998,[23] and many strains have been isolated from larvae of the ticks since then. This suggests that the larvae may play an important role in disseminating the pathogen.

Hemolymph samples from *Dermacentor* (*D.*) *taiwanensis*, *H. hystericis*, *H. longicornis*, and *I. ovatus* were positive when tested by the IP technique using a species-specific monoclonal antibody against *R. japonica.*[24]

A polymerase chain reaction (PCR) technique using species-specific primers detected *R. japonica* in *H. hystericis,*[25] *H. flava*, and *I. ovatus.*[26] With amplification and sequencing of the genes encoding 16SrRNA and citrate synthase, the isolates from *D. taiwanensis* and *H. flava* were confirmed to be the *R. japonica* strain.[27]

In summary, these species of ticks are potential vectors of JSF. Of these, *H. flava* and *H. hystericis* are considered to be the most probable vectors of JSF, as they commonly feed on humans in this endemic area. *D. taiwanensis* may have a role as a natural reservoir of the pathogen.

Interestingly, recent tick surveys in Japan have identified two serotypes or species of SFG rickettsial isolates in addition to *R. japonica*; one is closely

TABLE 2. Tick survey for vectors of JSF

SFG Isolated tick species	Human susceptibility	IP(Mab)		PCR amplication		
		SFG	*R. japonica*	gltA	omp A	omp B
H. flava	+++	+	+	+	+	ND
H. hystricis	+	+	+	+	+	ND
A. testudinarium	+++	+	+	+	+	−
H. longicornis	+++	+	±	ND	ND	ND
I. ovatus	++	+	−	+	−	−
H. cornigera	±	+	+	ND	ND	ND
D. taiwanensis	±	+	+	+	+	+
I. persulcatus	++	+	−	+	−	−
I. monospinosus	++	+	−	+	−	−

related to *R. helvetica* and another one is a new genotype of unknown genotype, AT, which was isolated from an *A. testudinarium* tick, which is closely related to a Slovakian genotype. These are of uncertain clinical significance.[27–29]

Q Fever

Serological evidence of Q fever in humans and coxiellosis in animals was reported in Japan in the 1950s; however, systematic studies of the disease did not begin until the isolation in 1989 of *Coxiella (C.) burnetii* from a patient with acute Q fever.[30] Recent investigations on the seroprevalence of *C. burnetii* in animals and humans show that infections are widespread in Japan with 30–40% positive titers in dairy cattle, 20–22% in veterinarians, and 0.8–3.3% in healthy humans.[31]

The Japanese isolates of *C. burnetii* are associated with the acute form of Q fever. These isolates had various degrees of virulence, but are strikingly similar to each other and to reference strains of acute infection with respect to their structure and immunogenicity of proteins. Under the revised surveillance system, clinical cases of Q fever have been reported 7–46 cases since 1999.

Rickettsiosis in Far East

Tsutsugamushi disease has been known to exist widely in Asia and is an increasing problem. Murine typhus and epidemic typhus still exist, but are controlled in each country.

Many SFG rickettsiae have been isolated and identified from ticks and the blood sample of patients.

Korea: Scrub typhus is the major rickettsial disease. An earlier study reported that 34.3% of febrile hospital patients in autumn were seropositive for the disease.[32] The nucleotide sequence from ticks showed high homologies with that of *R. japonica* and *R. rickettsii*, suggesting the possibility of clinical cases in Korea in 2003.[33] Recent investigation of the blood sample of febrile patients suggested that human SFG rickettsia including boutonneuse fever, rickettsial pox, rickettsial felis, and JSF are prevalent in Korea.[34]

Over 200 cases of murine typhus are presumed to occur annually in South Korea, but epidemic typhus cases have not been reported in recent years. *C. burnetii* and Ehrlichia show high prevalence according to animal surveillance.[35,36]

China: In the 1980s five rickettsial diseases existed in the People's Republic of China: scrub typhus, murine typhus, epidemic typhus, Q fever, and one or more spotted fever rickettsioses.[37] In recent years, great progress has been made, although the investigations were carried out only in a limited area. Therefore clinical reports concerning rickettsioses are scarce. At present, eight strains of SFG rickettsiae have been isolated from patients, ticks, tick ova, and rodents. These rickettsial isolates belong to at least four types of SFG rickettsiae including *R. siberica,* Hulin isolate (*R. hulinii*), Inner Mongolian isolate (*R. mongolotimonae*), and Heilongjiang isolate (*R. heilongjiangii*).[38,39] The various tick surveys revealed that a variety of tick-borne ehrlichial agents exist in China.[40]

Russian Far East: An acute tick-borne rickettsiosis caused by *R. heilongjiangensis* was first recognized in 13 patients in May 2002. Clinical symptoms and laboratory data are similar to those of other tick-borne diseases; however, prevalent seasons are different from Siberian tick typhus.[41] Acute granulocytic ehrichiosis was found in the region.[42]

Taiwan: Scrub typhus has been noticed as the major rickettsiosis in Taiwan; however, recently two other diseases, Q fever and murine typhus, have also been noted as an emerging infectious disease.[43,44]

CONCLUSION

JSF is still an emerging infectious disease in Japan. The clinical symptoms of JSF can be summarized as a triad of high fever, skin eruptions, and tick bite eschar. High fever and skin eruptions were observed in 100% of the patients, and tick bite eschars were seen in 94% of the patients after careful observation.

Delayed diagnosis and inadequate treatment led to serious complications, such as cardiac failure, respiratory failure, myocarditis, meningitis, or shock due to disseminated intravascular coagulation.

Under the revised infection control system in Japan since April 1999, 268 cases of JSF patients have been reported and 3 have deaths occurred (1.1%).[45–47] These three patients died shortly after admission (5 h, 23 h, 72 h). Of the three, two cases were diagnosed by PCR or isolation of *R. japonica* from the patient's blood and one was diagnosed clinically. Because all the cases were diagnosed after the fatal outcome we see that prompt diagnosis is crucial.

Treatment of JSF is almost established. The first choice is tetracycline (MINO or DOXY). For fulminant cases, a combination of MINO with a new quinolone is advisable. It is necessary to recognize that JSF is life-threatening if the diagnosis and appropriate treatment are delayed.

The vectors of JSF are indicated to be *H. flava* and *H. hystericis* by serological and epidemiological findings. It is interesting that the tick survey identified an *R. helvetica*-like organism and an unknown SFG rickettsia in Japan.

REFERENCES

1. MAHARA, F. 1984. Three Weil-Felix reaction (OX2) positive cases with skin eruptions and high fever. J. Anan Med. Assoc. **68:** 4–7.
2. MAHARA, F. *et al.* 1985. The first report of the rickettsial infections of spotted fever group in Japan: three clinical cases. J. Jap. Assoc. Infect. Dis. **59:** 1165–1172.
3. MAHARA, F. 1987. Japanese spotted fever: a new disease named for spotted fever group rickettsiosis in Japan. Annu. Rep. Ohara Hosp. **30:** 83–91.
4. EMILIO, W. 1992. Rickettsias. *In* Encyclopedia of Microbiology, Vol. 3. L. Joshua, Ed.: 585–610. Academic Press. Haarcourt Brace Jovanovich. New York.
5. BROUQUI, P. & D. RAOULT. 1996. Clinical aspect of human SFG rickettsiae infection in the era of molecular biology. *In* Proceedings of The Vth International Symposium on Rickettsiae and Rickettsial Diseases. J. Kazar & R. Toman, Eds.: 195–210. Publishing House of the Slovak Academy of Science. Bratislava, Slovak.
6. MAHARA, F. 1997. Synopses: Japanese spotted fever—report of 31 cases and review of the literature. Emerg. Infect. Dis. **3:** 105–111.
7. UCHIDA, T. *et al.* 1992. *Rickettsia japonica* sp. nov., the etiological agent of spotted fever group rickettsiosis in Japan. Int. J. Syst. Bacteriol. **42:** 303–305.
8. TAMURA, A. *et al.* 1995. Classification of *Rickettsia tsutsugamushi* a new genus, *Orientia* gen. nov., as *Orientia tsutsugamushi* comb. nov. Int. J. Syst. Bactriol. **45:** 589–591.
9. ARAKI, M. *et al.* 2002. Japanese spotted fever involving the central nervous system: two case reports and a literature review. J. Clin. Microbiol. **40:** 3874–3876.
10. SUTO, T. 1991. A ten years experience on diagnosis of rickettsial diseases using the indirect immunoperoxidase method. Acta Virol. **35:** 580–586.
11. TANGE, Y. *et al.* 1994. Detection of DNA of causative agent of spotted fever group rickettsiosis in Japan from the patient's blood sample by polymerase chain reaction. Microbiol. Immunol. **38:** 665–668.
12. TSUTSUMI, Y. & F. MAHARA. 2006. Early diagnosis of Japanese spotted fever by immunostaining of skin biopsy. Infect. Agents surveill. Rep. **27:** 38–40.
13. SUTO, T. *et al.* 1989. *In vitro* susceptibility of a strain of rickettsia recently isolated from a case of Japanese spotted fever to chemotherapeutic agents. J. Jpn. Assoc. Infect Dis. **63:** 35–38.
14. MIYAMURA, S. & T. OOTA. 1991. *In vitro* susceptibility of rickettsial strains from patients with Japanese spotted fever to quinolones, penicillins and other selected chemotherapeutic agents. Chemotherapy **39:** 258–260.
15. MAHARA, F. 1999. Rickettsioses in Japan: Rickettsiae and Rickettsial Diseases at the Turn of the Third Millennium. D. Roult & P. Brouqui Eds.: 233–239. Elsevier. Paris.
16. IWASAKI, H. *et al.* 2001. Fulminant Japanese spotted fever associated with hypercytokinemia. J. Clin. Microbiol. **39:** 2341–2343.

17. MAHARA, F. 2004. Japanese spotted fever. Jpn. J. Med. Assoc. **132:** 146–147.
18. UCHIDA, T. *et al.* 1986. Isolation of a spotted fever group rickettsia from a patient with febrile exanthematous illness in Shikoku, Japan. Microbiol. Immunol. **30:** 1323–1326.
19. FUJITA, H. *et al.* 1993. Isolation and serological identification of causative rickettsiae from Japanese spotted fever patients. Asian Med. J. **36:** 660–665.
20. OIKAWA, Y. *et al.* 1993. Identity of pathogenic strains of spotted fever rickettsiae in Shikoku district based on reactivities to monoclonal antibodies. Jpn. J. Med. Sci. Biol. **46:** 45–49.
21. IWAMASA, K. *et al.* 1992. Ultrastructural study of the response of cells infected in vitro with causative agent of spotted fever group rickettsiosis in Japan. APMIS **100:** 535–542.
22. AMANO, K. *et al.* 1991. Electron microscopic studies on the in vitro proliferation of spotted fever group rickettsia isolated in Japan. Microbiol. Immunol. **35:** 623–629.
23. FUJITA, H. 1998. Further observation of tick fauna endemic areas of Japanese spotted fever in Shikoku district. Ann. Rep. Ohara Hosp. **41:** 49. Laboratory, personal communication.
24. TAKADA, N. *et al.* 1992. Vectors of Japanese spotted fever. J. Jpn. Assoc. Infect. Dis. **66:** 1218–1225.
25. FURUYA, Y. *et al.* 1994. Analysis of *Rickettsia japonica* DNA and detection of the DNA by PCR. Organizing Committee of SADI, Acari-Disease Interface, Eds.: 41. Yuki Press. Fukui, Japan.
26. KATAYAMA, T. *et al.* 1996. Spotted fever group rickettsiosis and vectors in Kanagawa Prefecture. J. Jpn. Assoc. Infect. Dis. **70:** 561–568.
27. FOURNIER, P.E. *et al.* 2002. Genetic identification of rickettsiae isolated from ticks in Japan. J. Clin. Microbiol. **40:** 2176–2181.
28. FUJITA, H. *et al.* 1999. List of isolates of spotted fever group rickettsiae from ticks in Japan 1993–1998. Annu. Rep. Ohara Hosp. **42:** 45–50.
29. TAKADA, N. *et al.* 1994. Isolation of a rickettsia closely related to Japanese spotted fever pathogen from a tick in Japan. J. Med. Entomol. **31:** 183–185.
30. ODA, H. & K. YOSHIIE. 1989. Isolation of a *Coxiella burnetii* strain that has low virulence for mice from a patient with acute Q fever. Microbiol. Immunol. **33:** 969–973.
31. HIRAI, K. & T. HO. 1998. Advances in the understanding of *Coxiella burnetii* infection in Japan. J. Vet. Med. Sci. **60:** 781–790.
32. CHOI, M.S. *et al.* 1994. A seroepidemiological survey on the scrub typhus in Korea. J. Korean Soc. Microbiol. **30:** 593–602.
33. LEE, J.H. *et al.* 2003. Identification of spotted fever group rickettsiae detected from *Haemaphysalis longicornis* in Korea. Microbiol. Immunol. **47:** 301–304.
34. CHOI, Y.J. *et al.* 2005. Spotted fever group and typhus group rickettsioses in humans, South Korea. Emerg. Infect. Dis. **11:** 237–244.
35. KOMIYA, T. *et al.* 2003. Seroprevalence of *Coxiella burnetii* infections among cats in different living environments. J. Vet. Med. Sci. **65:** 1047–1048.
36. CHAE, J.S. *et al.* 2003. Molecular epidemiological study for tick-borne disease (*Ehrlichia* and *Anaplasma* spp.) surveillance at selected U.S. military training sites/installations in Korea. Ann. N.Y. Acad. Sci. **990:** 118–125.
37. FAN, M.Y. *et al.* 1987. Epidemiology and ecology of rickettsial diseases in the People's Republic of China. Rev. Infect. Dis. **9**: 823–840.

38. FAN, M.Y. *et al.* 1999. Spotted Fever Group Rickettsioses in China at the Turn of the Third Millennium. D. Roult & P. Brouqui Eds. : 247–257. Elsevier. Paris.
39. ZHANG, J.Z. *et al.* 2000. Genetic classification of " *Rickettsia heilongjiangii* " and " Rickettsia hulinii ."Two Chinese spotted fever group rickettsiae. J. Clin. Microbiol. **38:** 3498–3501.
40. WEN, B. *et al.* 2003. Ehrlichiae and ehrlichial diseases in China. Ann. N. Y. Acad. Sci. **990:** 45–53.
41. MEDIANNIKOV, O.Y. *et al.* 2004. Acute tick-borne rickettsiosis caused by Rickettsia heilongjiangensis in Russian Far East. Emerg. Infect. Dis. **10:** 810–817.
42. SIDELNIKOV, Y.N. *et al.* 2002. Clinical and laboratory features of human granulocytic ehrlichiosis in the south of Russian Far East. Epidemiologia i Infectsionnye Bolezni **3:** 28–31.
43. LEE, H.C. *et al.* 2002. Clinical manifestations and complications of rickettsiosis in southern Taiwan. J. Formos. Med. Assoc. **101:** 385–392.
44. WANG, J.H. *et al.* 1993. Acute Q fever :first case report in Taiwan. J. Formos. Med. Assoc. **92:** 917–919.
45. KODAMA, K. *et al.* 2002. Fulminant Japanese spotted fever definitively diagnosed by the polymerase chain reaction method. J. Infect. Chemother **8:** 266–268.
46. WADA, K. *et al.* 2005. A case report of the fluminant Japanese spotted fever. J. Jpn. Assoc. Infect. Dis. **79:** 253.
47. MAHARA, F. *et al.* 2005. Early diagnostic method for Japanese spotted fever. J. Jpn. Assoc. Infect. Dis. **79:** 254.
48. FUKUTA, Y. *et al.* 2005. Case of Japanese spotted fever complicated with acute myocarditis. J. JSEM **8:** 170.

Rickettsioses in Australia

STEPHEN GRAVES, NATHAN UNSWORTH, AND JOHN STENOS

Australian Rickettsial Reference Laboratory, The Geelong Hospital, Geelong, Victoria 3220, Australia

ABSTRACT: Australia, an island continent in the southern hemisphere, has a range of rickettsial diseases that include typhus group rickettsiae (*Rickettsia typhi*), spotted fever group rickettsiae (*R. australis*, *R. honei*), scrub typhus group rickettsiae (*R. tsutsugamushi*), and Q fever (*C. burnetii*). Our knowledge of Australian rickettsiae is expanding with the recognition of an expanded range of *R. honei* (Flinders Island spotted fever) to Tasmania and southeastern mainland Australia (not just on Flinders Island), and the detection of a new SFG species (or subspecies), tentatively named "*R. marmionii*" in the eastern half of Australia. This rickettsia causes both acute disease (7 cases, recognized so far) and is also associated (as a "*R. marmionii*" bacteriaemia) with patients having a chronic illness. The significance of the latter is under investigation. It may be a marker of autoimmune disease or chronic fatigue in some patients.

KEYWORDS: *rickettsiae*; *Australia*; *R. marmionii*

INTRODUCTION

The first rickettsial disease recognized in Australia was louse-borne epidemic typhus (*Rickettsia prowazekii*). Louse-infested convicts began to arrive by ship from England in 1788 (the first permanent settlement of Europeans in Australia). This condition prevailed for about 100 years and many outbreaks of epidemic typhus were recorded. However, the disease never became established in Australia, probably because of the sunny climate and the opportunity to wash clothes regularly.

The Portuguese who occupied the nearby colony of East Timor had been granted permission by Pope Alexander VI, as part of an agreement with Spain (the treaty of Tordesillas in 1494), to occupy any land to the east of longitude 129°E (Greenwich). The land to the west of this line ("the Pope's line") was "granted" to Spain.[1] Hence, when the English settled in eastern Australia in 1788 they only claimed the "Spanish" part of the Australian land mass (as Spain

Address for correspondence: Dr Stephen Graves, Australian Rickettsial Reference Laboratory, The Geelong Hospital, PO BOX 281, Geelong, Victoria 3220. Voice: 61-2-4921-442; fax: 61-3-5226-3183. e-mail: Stephen.graves@hnehealth.nsw.gov.au

Ann. N.Y. Acad. Sci. 1078: 74–79 (2006). © 2006 New York Academy of Sciences.
doi: 10.1196/annals.1374.008

was then a British enemy) and did not claim the western part of the Australian land mass (as Portugual was a British ally). This line of demarcation still exists to this day as the current border between Western Australia and the rest of Australia. The Portuguese did not settle in Western Australia and it was eventually colonized (in 1829) by the British.

TYPHUS GROUP (TG) RICKETTISAE

Murine typhus (*Rickettsia typhi*) was recognized as a separate disease from the louse-borne epidemic typhus by two groups working independently—Hone,[2] a public health doctor, working in Adelaide, South Australia, and Maxcy and Havens, working in Montgomery, Alabama.[3] However, Hone's article appeared a year earlier (1922) than the other (1923). Thus murine typhus was first described in Australia.

The original Australian reports were in men who were loading bags of wheat onto ships, in rodent-infested circumstances. Later cases were associated with persons living in rodent-infested houses, especially after a disturbance of the rodent nests. This action is now known to generate infectious aerosols of *R. typhi*-infected rodent flea feces. These are inhaled by the patient. Murine typhus outbreaks have also been associated with mouse plagues in rural Queensland. These days, intermittent cases occur regularly in all parts of Australia (especially Queensland and Western Australia), so that the disease is considered to be Australia-wide. Unfortunately, no Australian isolates of *R. typhi* have been made. Diagnosis of cases has been by serology (initially by Weil-Felix test and later by micro-immunofluorescence).

SPOTTED FEVER GROUP (SFG) RICKETTSIAE

There are two definite SFG rickettsiae in Australia: *R. australis* (Queensland tick typhus), and *R. honei* (Flinders Island spotted fever).

A recently discovered third rickettsia ("*R. marmionii*") is currently being investigated as a new species or new subspecies of *R. honei.*

A. *R. australis* is transmitted by two species of ticks, *Ixodes holocyclus* and *I. tasmani*. The disease occurs down the eastern seaboard of Australia from the Torres Strait islands in the north to the southeastern corner of the Australian land mass (Wilson's Promontory).

The rickettsia is a very atypical SFG rickettsia and is a phylogenetic outlier of this group. It has probably undergone a long period of independent evolution in Australia.

The disease (Queensland tick typhus) was first recognized among soldiers training in the Australian jungle during World War II.[4] Several were rickettsiaemic and *R. australis* was isolated by animal inoculation.

This is now recognized as a fairly common disease in urban areas of eastern Australia (especially Sydney and Brisbane), where native bush is present and ticks commonly bite people.

The disease is a typical SFG illness, except that the rash may be vesicular and the disease misdiagnosed as chicken pox. There have been a few reported deaths, usually due to misdiagnosis or late diagnosis.

Native mammals such as the bandicoot and various rodent species are the normal vertebrate hosts of *R. australis*.

B. *R. honei*. On Flinders Island, in Bass Strait, the body of water between mainland Australia and Tasmania, the sole medical practitioner (Stewart) recognized a summer syndrome of fever, headache, and rash. Many patients recalled a tick bite.[5] It was subsequently shown to be a spotted fever group rickettsial infection[6] and named *R. honei* in honor of Frank Hone, the Australian discoverer of murine typhus.[7] Further studies revealed that *R. honei* was associated with the reptile tick, *Aponomma hydrosauri*.[8] This was a surprising observation, as reptile ticks had not previously been implicated in the transmission of human rickettsial diseases. *A. hydrosauri* ticks on blue-tongue lizards, tiger snakes, and copperhead snakes all contained *R. honei*. Humans are accidental hosts.

R. honei has now been shown to cause human disease in Tasmania and southeast mainland Australia,[9] as well as other parts of the world (Thailand, Texas [USA], and Sicily [Italy]). It may well have a worldwide distribution. Other tick species are likely to be involved.

Phylogenetically, *R. honei* is a mainstream SFG rickettsia. The disease it causes is relatively mild and no deaths have been reported.

C. *R. honei* subsp. *marmionii* or "*R. marmionii*." This is a new SFG rickettsia detected in Australia. Seven cases of acute illness have been detected so far. The main features are fever and headache. Only three patients had a rash and only two of these had an eschar. These cases occurred in widely diverse regions of Australia—Torres Strait islands (×3), north Queensland (×2), Tasmania (×1), and near Adelaide (×1). All cases were confirmed by PCR (7/7), culture (5/7), seropositivity (5/7), or serconversion (2/7). The polymerase chain reaction test was a real-time assay. Targeting the citrate synthase gene of *Rickettsia sp.* it was extremely sensitive (1–10 copies detectable) and extremely specific (no reactivity with *Orientia sp.*, *Ehrlichia sp.*, *Bartonella sp.*). All tested members of the SFG and TG were detected, with the exception of *R. bellii*.

Amplified gene sequence comparisons (16S, citrate synthase, 17kDa, ompA) showed that *R. honei* and *R. japonica* were its closest relatives. It remains to be seen whether "*R. marmionii*" is a new species of SFG rickettsia or a subspecies of *R. honei*. It was named in honor of Barrie Marmion, an Australian rickettsiologist, whose lifetime of work on Q fever has led to many advances, including the development of a Q fever vaccine.

It appears to grow best in the laboratory at 28°C in XTC-2 (amphibian) cells. The tick vector(s) is (are) not yet known, although one patient was bitten by *Haemophysalis novaeguineae* in north Queensland.

"*R. marmionii*" has also been associated with chronic illness in 14 patients. Many hundreds of patients with a variety of chronic illnesses (some undiagnosed) were tested for rickettsial disease by their doctor. Rickettsial serology, PCR, and culture were performed. Three percent of the group were rickettsaemic by real-time PCR targeting the citrate synthase gene (the assay mentioned previously). Of these 14 patients, 5 were also positive by conventional PCR targeting the 17 kDA antigen gene of *Rickettsia sp*. Sequencing of this amplified gene, in all 5 patients showed all to have 100% homology with "*R. marmionii*." A group of 400 control patients, from a different medical practice, were negative for rickettsiaemia, and this difference was statistically significant ($P <$ 0.001).

The 14 rickettsiaemic patients had a variety of chronic diseases, including autoimmune disease (rheumatoid arthritis ×5; Hashimoto's thyroiditis ×2; polymyalgia rheumatica ×1), chronic pain syndrome ×7, and chronic fatigue ×7; the total was >14 due to same symptoms in multiple patients.

The interpretation of these results is difficult. There appear to be three possibilities:

1. All data are spurious due to DNA contamination in our lab. This is unlikely as meticulous extraction and amplification protocols were in place, the master-mix contained UDG to prevent carryover contamination, and all control patient blood specimens were negative.
2. "*R. marmionii*" causes chronic illness. This is a possibility but the data are insufficient to make such a claim at this time.
3. "*R. marmionii*" becomes latent in a patient after primary infection. At some stage later in their life patients develop an immunosuppressive illness or are therapeutically immunocompromised. This allows the latent "*R. marmionii*" to reactivate and establish a low-level rickettsiaemia, which was detectable by our assays.

It is not known whether the circulating "*R. marmionii*" is responsible for any of the patients' symptoms or whether rickettsaemia is just a marker of patient immunosuppression.

This work needs to be repeated with another similar patient cohort in another rickettsial laboratory.

SCRUB TYPHUS (*ORIENTIA TSUTSUGAMUSHI*)

The northern third of Australia is tropical. In the early days of European colonization of Australia (19th century), febrile illness was a major impediment to people living and working in tropical Australia. Among the various causes of fever, not differentiated at that time, however, was scrub typhus.

The mite *Leptotrombidium deliense* is endemic in tropical Australia and cases have been detected in the Kimberley region of Western Australia, the Top End of the Northern Territory (especially Litchfield Park), coastal tropical Queensland, and the islands of the Torres Strait (especially Darnley Island). In many of these parts of Australia human population density is very low, so cases are not common or there are few doctors to diagnose the cases.

The disease also occurs in Papua New Guinea from where the KARP serotype originated from a soldier during World War II.

Darnley Island, a small island with only 150 people, has an extremely high incidence of rickettsioses, including *O. tsutsugamushi*, *R. australis*, and "*R. marmionii*." Each year there are several cases on the island.

Deaths have occurred from scrub typhus due to the patients' remote location, delayed medical attention, slow evacuation to hospital, or inappropriate antibiotics.

Q FEVER (*COXIELLA BURNETII*)

II Q fever disease is common in central eastern Australia: There are about 500 cases reported in Australia per year. With Australia's population of 20 million, this gives a Q fever incident rate of approximately 25/100,000 population per year.

Most cases occur in adults in the workforce, with certain occupations (associated with sheep, cattle, or goats) being at most risk.

Q fever was first described in Australia in 1937[10] among abattoir workers in Brisbane. It is still a significant health hazard in those abbatoirs that do not immunize their staff against Q fever.

Q fever also occurs among farmers and the farming community. It occurs more commonly in drought conditions since the dust, containing *C. burnetii* in a dessicated but viable form, is blown long distances from the infected source animals.

Many sero-positive persons do not recall an illness compatible with Q fever, so it is possible that asymptomatic sero-conversion can occur. Several species of Australian ticks carry *C. burnetii*, although virtually all human cases are thought to occur via infected aerosols. Pneumonia is a rare manifestation of Q fever in Australia and patients with acute disease usually present with fever, myalgia, headache, and sometimes hepatitis. Chronic Q fever in Australia occurs mainly as endocarditis or post-Q fever chronic fatigue.

CONCLUSION

Australia has a full range of rickettsial diseases, with the exception of louse-borne epidemic typhus (*R. prowazekii*).

Q fever (*C. burnetii*) is the most common, but Queensland tick typhus (*R. australis*), Flinders Island spotted fever (*R. honei*), murine typhus (*R. typhi*), and scrub typhus (*O. tsutsugamushi*) also occur regularly.

A newly recognized SFG rickettsia ("*R. marmionii*") appears to cause both an acute rickettsial illness and to be associated (as a rickettsaemia) in some patients with various chronic (and presumably immunosuppressive) diseases. Its role in chronic illness is not yet clear.

REFERENCES

1. McIntyre, K.G. 1977. The Secret Discovery of Australia: Portuguese Ventures 200 Years Before Captain Cook. Souvenir Press. London, U.K.
2. Hone, F.S. 1922. A series of cases closely resembling typhus fever. Med. J. Aust. **7:** 1–13.
3. Maxcy, K.F & L.C. Havens. 1923. A series of cases giving a positive Weil-Felix reaction. Am. J. Trop. Med. **3:** 495–507.
4. Andrew, R., J. Bonnin & S. Williams. 1946. Tick typhus in north Queensland. Med. J. Aust. **24:** 256–258.
5. Stewart, R. 1991. Flinders Island spotted fever: a newly recognised endemic focus of tick typhus in Bass Strait. Part 1. Clinical and epidemiological features. Med. J. Aust. **154** : 94–99.
6. Graves, S.R. *et al.* 1993. Spotted Fever Group rickettsial infection in south-eastern Australia: isolation of rickettsiae. Comp. Immun. Microbiol. Infect. Dis. **16:** 223–233.
7. Stenos, J. *et al.* 1998 *Rickettsia honei* sp. nov, the aetiological agent of Flinders Island spotted fever in Australia. Int. J. Systematic Bacteriol **48:** 1399–1404.
8. Stenos, J. *et al.* 2003 *Aponomma hydrosauri*, the reptile associated tick reservoir of *Rickettsia honei* on Flinders Island, Australia. Am. J. Trop. Med. Hyg. **69:** 314–317.
9. Unsworth, N.B. *et al.* 2005 Not only "Flinders Island" spotted fever. Pathology. **37:** 242–245.
10. Derrick, E.H. 1937 "Q" fever, a new fever entity: clinical features, diagnosis and laboratory investigation. Med. J. Austr. **21:** 281–299.

Far Eastern Tick-Borne Rickettsiosis

Identification of Two New Cases and Tick Vector

OLEG MEDIANNIKOV,[a,b,c] YURI SIDELNIKOV,[b] LEONID IVANOV,[d] PIERRE-EDOUARD FOURNIER,[c] IRINA TARASEVICH,[a] AND DIDIER RAOULT[c]

[a]*Laboratory of Rickettsial Ecology, Gamaleya Research Institute of Epidemiology and Microbiology, Moscow, 123098, Russia*

[b]*Department of Infectious Diseases, Far Eastern State Medical University, Khabarovsk, 680000, Russia*

[c]*Unité des Rickettsies, Faculté de Médecine, CNRS UMR 6020 27, Université de la Méditerranée, 13385 Marseille, Cedex 05, France*

[d]*Khabarovsk Plague Control Station of Ministry of Health of Russian Federation, Khabarovsk, 680000, Russia*

ABSTRACT: We recently reported the first documented cases of a new rickettsial disease caused by *Rickettsia heilongjiangensis* in the Russian Far East (Far Eastern tick-borne rickettsiosis). Here we report the amplification of DNA of *R. heilongjiangensis* from both the skin biopsy of an acutely ill patient and the tick removed from him prior to the disease development. The tick has been identified as *Haemaphysalis* spp. The clinical picture was that of a spotted fever group rickettsiosis and a seroconversion was noted with *R. heilongjiangensis* antigen. Screening testing of both species of *Haemaphysalis* ticks inhabiting Russian Far Eastern regions showed that up to 28.13% of *H. concinnae* and 4.48% of *H. japonica douglasii* ticks harbor *R. heilongjiangensis.* It has been concluded that *H. concinnae* may serve as the main vector for the transmission of *R. heilongjiangensis. H. japonica douglasii* ticks harbor several varieties of rickettsiae. DNA of "*Candidatus* Rickettsia tarasevichiae," previously found in *Ixodes persulcatus* ticks, was amplified from one male tick. Two sequenced complete gltA genes belong to the novel spotted fever group rickettsial species provisionally called here "*Candidatus* Rickettsia principis" variants Hjd54 and Hjd61. The rate of infection has been found to be not higher than 1.5%.

KEYWORDS: *Rickettsia heilongjiangensis;* tick-borne rickettsiosis; Russia; Far East

Address for correspondence: Oleg Y. Mediannikov, Laboratory of Rickettsial Ecology, Gamaleya Institute of Epidemiology and Microbiology, ul. Gamalei, 18, Moscow, Russia, 123098. Voice: +7-095-193-43-10; fax: +7-095-193-61-85.
e-mail: olegusss1@mail.ru

**Ann. N.Y. Acad. Sci. 1078: 80–88 (2006). © 2006 New York Academy of Sciences.
doi: 10.1196/annals.1374.010**

INTRODUCTION

The Russian Far East is a geographical, economical, and political unit within the Russian Federation. It consists of the smaller administrative areas (regions) located on or close to the Asian Pacific coast. Since 1932,[1] tick-borne fever has been registered in the Khabarovsk region. After the first isolation of *Rickettsia sibirica* in the nearest regions of Siberia,[2] tick-borne spotted fever in the Russian Far East is thought to be caused by *R. sibirica*. *Rickettsia heilongjiangensis* has been isolated in China in 1982.[3] It is closely related to *Rickettsia japonica* on the basis of serotyping, and on the genetic-based definition it is considered to be a new species.[4]

Recently, acute rickettsiosis caused by *R. heilongjiangensis* (Far Eastern tick-borne rickettsiosis) has been described in the Khabarovsk region of the Russian Far East. DNA of this rickettsia was found in all polymerase chain reaction (PCR)-positive samples obtained from acutely ill tick-bitten patients in this region.[5] Type strain (HLJ-054) of *R. heilongjiangensis* was isolated from *Dermacentor sylvarum* tick. Indeed, this species of tick is quite rare in the southern territories of the Russian Far East. Tick-bitten patients rarely report the bites of these ticks and, additionally, seasonal activity peak of *Dermacentor* spp. ticks (April–May) does not correspond to the seasonal morbidity of rickettsiosis caused by *R. heilongjiangensis*.[5]

Two species of the *Haemaphysalis* genus are prevalent in the same territory where Far Eastern tick-borne rickettsiosis has been described.[6,7,8] They are *Haemaphysalis concinnae* and *Haemaphysalis japonica* ssp. *douglasii*. Epidemiological data suggest that these species may be vectors of rickettsiosis. Recently, DNA of *R. heilongjiangensis* has been amplified from one *H. concinnae* tick collected in the Krasnoyarsk region in Central Siberia.[9]

The aim of our study was to identify the vectors of the *R. heilongjiangensis* infection and to evaluate the role of *Haemaphysalis* ticks in transmitting and harboring different rickettsiae.

MATERIALS AND METHODS

Patients

Among a total of 42 patients admitted in the Infectious Diseases Department of the Municipal Hospital No. 10 of Khabarovsk city during the summer of 2003 with suspected tick-borne rickettsiosis, two patients conserved ticks that bit them. In one patient the tick was removed from the scrotum during medical examination. Both ticks were engorged and were identified as *Haemaphysalis* spp. More precise identification could not been done because of the deformation of the tick's body following engorgement. Both patients presented eschar at the place of the tick bite, although in Patient B it was small (5 mm in diameter) without significant inflammation. A skin biopsy of eschar was taken before antibiotic had been administered. Both patients recovered completely following doxycycline treatment.

Tick Sample Collection

Unfed, actively questing ticks were collected during June 2003 in the Khabarovsk area by dragging a standard 1 m^2 flannel flag over vegetation. Places for tick collection have been chosen based upon probabilities of infection acquisition. The collection sites are woodlands close to city resorts in which the major tree species were *Quercus mongolica* Fisch. ex Ledeb., *Populus tremula L., Alnus fruticosa Rupr., Fraxinus mandshurica Rupr., Juglans manshurica* Max. The undergrowth was predominantly ferns *(Pteridium aquilinum* (L.) Kuhn, *Dryopteris crassirhizoma* Nakai) and creepers (*Schizandra chinensis* Baill., *Vitis amurensis Rupr., Dioscorea nipponica* Makino). *Artemisia* spp., *Bidens tripartita* L., *Adenocaulon adhaerescens* Maxim., various *Poaceae*- and *Carexaceae*- covered cuttings, where ticks were especially abundant. In total 67 *H. japonica* douglasii (37 female and 30 male) ticks and 32 *H. concinnae* (21 female and 11 male) ticks were collected. They were identified according to the standard pictorial keys. All ticks were stored at −80°C and then transported in alcohol at ambient temperature.

DNA Extraction

DNeasy Tissue Kit (Qiagen, Hilden, Germany) was used to prepare DNA from ticks and skin biopsy samples. Ticks were washed in ethanol and dried with sterile fiber-free tissue. Then each tick was placed in a sterile 1.5 mL microcentrifuge tube and cut into minuscule particles with sterilized disposable scalpel razors. Then the cut ticks were mixed with 180 μL of tissue lysis buffer, supplemented with proteinase K and placed in a water bath at 55°C for 3 h. DNA was extracted following the manufacturer's instructions using sorbent-containing columns. DNA from skin samples was extracted immediately after biopsy as described above for ticks.

Oligonucleotide Primers, PCR, Sequencing, and Data Analysis

The primers used in this study (Eurogentec, Seraing, Belgium) and their characteristics are listed in TABLE 1.

We used the nested PCR method for primary detection of the rickettsial DNA in human samples and ticks. The *gltA* (citrate synthase) gene was chosen as the target for amplification because of its genus-specificity and conservativeness.

TABLE 1. List of the primers mentioned in the text

Primer name	Primer sequence 5′– 3′
CS2d	ATGACCAATGAAAATAATAAT
CSEndr	CTTATACTCTCTATGTACA
RpCS877	GGGGACCTGCTCACGGCGG
RpCS1258	ATTGCAAAAAGTACAGTGAACA
CS535r	GAATATTTATAAGACATTGC
CS409d	CCTATGGCTATTATGCTTGC

Primers CS2d and CSEndr[5] were used for the primary amplification. They amplify the full length of the *gltA* gene. Primers CS877f and CS1258r were used in the nested PCR assay. To amplify the full length of the *gltA* gene we used the products of the first PCR (with CS2D and CSEndr primers) as a template for the nested PCR, combining the primers CS2d with CS535r, CS409d with RpCS1258, and RpCS877 with CSendr.[5,10]

DNA sequencing reactions were done using fluorescent-labeled dideoxynucleotide technology (BigDye Terminator v3.0 Cycle Sequencing Ready Reaction Kit; Applied Biosystems, Foster, CA). PCR primers were used for sequencing as described in a previous section. Data were collected with an ABI prism 310 Genetic Analyzer capillary sequencer (Applied Biosystems). Sequences were edited and assembled using Genetix-Win Version 5.1.1 software (Genetic Information Processing Software, Japan). The *gltA* gene sequences used for comparison were obtained from the GenBank database, aligned, and then corrected manually to preserve codon alignment and conserved motifs. A total of 1,065 base pairs portions of *gltA* gene were used for constructing phylogenetic trees by UPGMA, neighbor-joining, minimum evolution, and maximum parsimony methods. They were calculated using the MEGA2 Version 2.1 software (available at http://www.megasoftware.net).[11] Evolutionary distances for the defined groups were calculated using the Kimura 2-parameter model. Internal node support was verified using the bootstrap method with 100 replicates.

DNA extraction, primary and nested PCR amplifications, and electrophoresis were performed in separate rooms. Precautions were taken to prevent contamination of samples, including usage of aerosol-resistant pipet tips. Quality control included negative and positive controls (DNA of *Rickettsia montanensis*) that were amplified simultaneously with samples tested.

Serological Studies

For the investigations sera collected from patients were dried on blotting papers as described previously[12] and transported to Marseilles where microimmunofluorescence testing[13] was performed using in-house-prepared antigens of *R. heilongjiangensis* (strain 054, ATCC VR-1524), *R. sibirica* (strain 246, ATCC VR-151), and *Rickettsia conorii* (strain Moroccan ATCC VR-141). Antigens were applied by pen point to 18-well microscope slides, dried for 30 min, and fixed. Appropriate positive- and negative-control sera were tested on each slide together with twofold dilutions of patient's sera made in 3% nonfat dry milk in phosphate-buffered saline (PBS). Slides were processed as described previously[14] using fluorescein isothiocyanate-conjugated goat anti-human γ chain and μ chain immunoglobulins (BioMérieux, Marcy l'Etoile, France). Endpoints for each antigen were the lowest concentrations of serum that definitely conferred fluorescence on bacteria.

RESULTS

We have amplified a complete gltA gene of *R. heilongjiangensis* from the skin biopsy of patient B. Clinical and serological data of both patients correspond to the previously described Far Eastern tick-borne rickettsiosis.[5] Sera collected from both patients showed seroconversion when tested with the antigen of *R. heilongjiangensis* with simultaneous increasing of IgG and IgM titers. Sera obtained during the first days of the disease (6th day for Patient A and 2nd day for patient B) were negative, but testing of convalescent sera revealed high titers of both IgG and IgM in two patients. The titers for the Patient A were 1/80 and 1/20 and for the patient B 1/320 and 1/320, respectively, for IgG and IgM. Clinical, epidemiological, and laboratory data of both patients are shown in TABLE 2.

We have amplified and sequenced the *gltA* gene DNA from both ticks obtained from these patients and found that the complete (full size) *gltA* gene sequences obtained from patient A and ticks obtained from patients A and B are completely identical to each other.

When we investigated ticks collected from wild vegetation at places of probable acquisition of rickettsial infection by the patients, we found that 9 (4 male and 5 female) of 32 (28.13%) collected *H. concinnae* ticks contained DNA (*gltA* gene) of *R. heilongjiangensis*. A total of 6 (all male) of 67 collected *H. japonica douglasii* ticks (8.96%) also contained rickettsial DNA, but 3 of these 6 rickettsiae were identified as *R. heilongjiangensis,* whereas the other 3 represented different species. One tick contained DNA of *Candidatus* Rickettsia tarasevichiae, and 2 others contained DNA of unknown rickettsial species, close to each other, provisionally called here *Candidatus* Rickettsia principis (with variants Hjd54 and Hjd61) because the ticks were collected close to the Prince Volkonsky village. The phylogenetic tree has been constructed by the Maximum Parsimony method (FIG. 1) in order to visualize the position of these rickettsiae among all other rickettsiae.

gltA sequences of both variants of *Candidatus* Rickettsia principis (Hjd54 and Hjd61) were deposited in the GenBank under the numbers AY578114 and

TABLE 2. Clinical and epidemiological data of two patients with proved *R. heilongjiangensis* infection

Feature or sign	Patient A (male)	Patient B (male)
Age (years)	34	73
Date of onset of clinical symptoms	June 14, 2003	June 25, 2003
Date of admission to hospital	June 19, 2003	June 26, 2003
Primary diagnosis at admission	Yersiniosis	Tick-borne rickettsiosis
Date of tick removal	June 20, 2003	June 24, 2003
Place of tick attachement	Scrotum	Left thigh
Suggested date of tick bite	June 9, 2003	June 22, 2003
Incubation period (days)	5	3
Temperature at admission	39°C	38.5°C
Chills	Yes	Yes
Malaise	Yes	Yes
Headache	Yes	Yes
Dizziness	Yes	No
Myalgias, arthralgias	Yes	Yes
Anorexia	Yes	Yes
Maculopapular rash	Yes	No
Eschar at the place of tick bite (diameter in cm)	No	Yes (0.5)
Regional to the eschar lymphadenopathy	No	Yes
Leukocyte count at admittance per mm^3	7700	8100
Leukocyte count after defervescence per mm^3	6800	10700
Electron spin resonance (ESR) at admittance	10	17
ESR after defervescence	15	13
Left shift (maximal, day of the disease)	25% (6)	20% (2)
Thrombocyte count at admittance per mm^3	160,000	270,000
Urinalysis	Normal	Proteinuria 0.033 g/L (2nd day)
AlAT activity (0–40 U/L)	52.5	38.7
AsAT activity (0–40 U/L)	24.9	13.2
Response to doxycycline, 100 mg × 2, 14 days	Good	Good

AY578115, respectively.

DISCUSSION

Here we described two new cases of acute tick-borne infection caused by *R. heilongjiangensis*. Clinically and epidemiologically suggested diagnoses were proved by serology and molecular biology by amplification and sequencing of bacterial DNA from the patients' specimens.

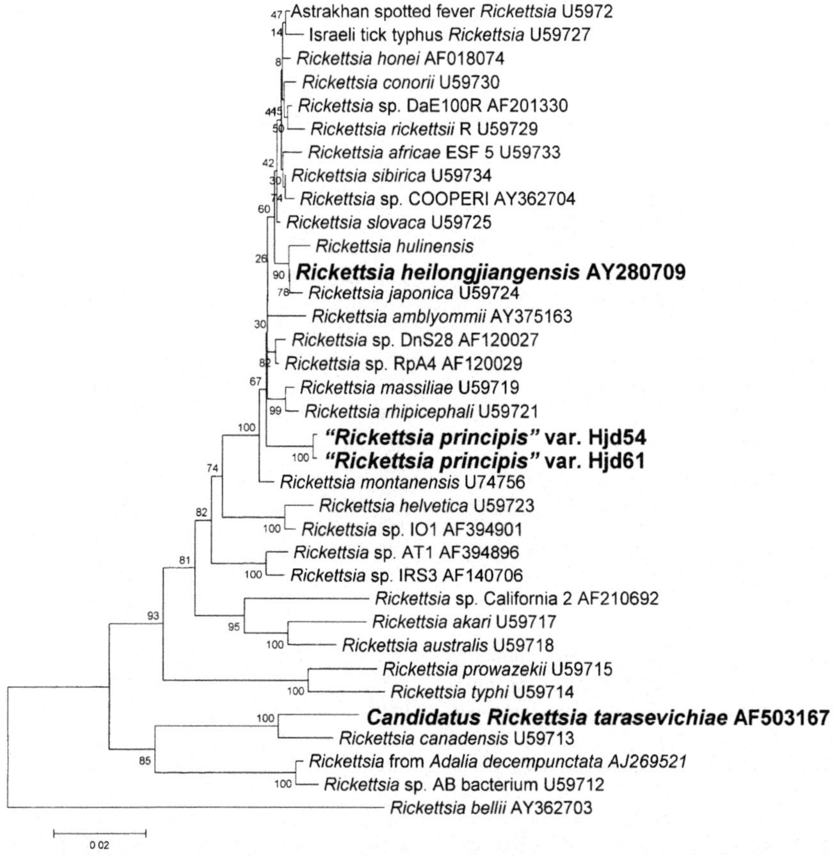

FIGURE 1. *gltA* gene–based phylogenetic tree showing the position of rickettsiae amplified from *Haemaphysalis* species ticks collected in the Russian Far East. Tree has been built by Minimum Evolution method with bootstrap probabilities of 100 (those >80 are shown) using MEGA 2.1 software.[11] GenBank accession numbers are shown after the species name.

Amplification of completely identical DNA of *R. heilongjiangensis* from the skin biopsy of an acutely ill patient and engorged tick removed from the patients several days before development of the disease (7–10 days before biopsy has been taken) suggests the direct transmission of the pathogenic rickettsia during tick blood sucking. *H. concinnae* and *H. japonica douglasii* are widespread in the Russian Far East.[2,7] Both species often bite humans.[7,15] A screening study of both species of ticks collected in the Khabarovsk area, where most of the patients probably acquired the infection, showed that 28% of *H. concinnae* ticks contain DNA of *R. heilongjiangensis*. So, a high incidence in the environment, a tendency to bite humans, and a high incidence of pathogenic rickettsia in ticks suggest *H. concinnae* to be the most important vector for tick-borne spotted fever caused by *R. heilongjiangensis* (Far Eastern tick-borne rickettsiosis).

H. japonica douglasii may also serve as a vector for the transmission of *R. heilongjiangensis,* but the rate of infection in ticks (4.48%) is significantly lower than in *H. concinnae* ticks. Unlike *H. concinnae, H. japonica douglasii* were shown to harbor different rickettsiae. DNA of "*Candidatus* Rickettsia tarasevichiae" has been previously amplified only from *Ixodes persulcatus* ticks.[16] Here we present the first evidence of the presence of this rickettsia in *Haemaphysalis* ticks. Two other *gltA* genes amplified from male ticks belong probably to the different strains of another new species of *Rickettsia*. Phylogenetic analysis showed that this rickettsiae form a distinct and well-defined (bootstrap probability 100 of 100) clade inside the spotted fever group of rickettsiae. The life cycle and pathogenicity in humans of these rickettsiae remains unknown.

Further epidemiological studies should be performed in order to assess the infection rates in other regions, the possibility of transmitting and harboring *R. heilongjiangensis* by other species of ticks, and to discover the natural reservoir of this infection.

ACKNOWLEDGMENTS

We are grateful to A. Yakovlev for worthy advice, G.S. Tomilka and V.I. Zholondz for assistance in sample collection, and J. Urquhart for thoroughly reviewing this manuscript.

REFERENCES

1. Mill, E.I. 1936. Tick-borne fever in Primorye. Far East. Med. J. **3:** 54–56.
2. Zdrodovskii, P.F. & E.M. Galinevich. 1960. The Rickettsial Diseases. Pergamon Press. London.
3. Zhang, J.Z., Y. Fan, Y.M. Wu, *et al.* 2000. Genetic classification of "Rickettsia heilongjiangii" and "Rickettsia hulinii," two Chinese spotted fever group rickettsiae. J. Clin. Microbiol. **38:** 3498–3501.
4. Fournier, P.E., J.S. Dumler, G. Greub, *et al.* 2003. Gene sequence-based criteria for identification of new rickettsia isolates and description of *Rickettsia heilongjiangensis* sp. nov. J. Clin. Microbiol. **41:** 5456–5465.
5. Mediannikov, O., Y. Sidelnikov, L. Ivanov, *et al.* 2004. Acute tick-borne rickettsiosis caused by *Rickettsia heilongjiangensis* in Russian Far East. Emerg. Infect. Dis. **10:** 810–817.
6. Belozerov, V.N. 1974. Experimental study of the seasonal adaptation of the ticks *Haemaphysalis concinna* Koch. from various regions of the Fast East. Med. Parazitol. (Mosk). **43:** 31–38.

7. FILIPPOVA, N.A. 1997. Ixodid ticks from subfamily *Ambliomminae. In* Fauna of Russia and Neighboring Countries: *Arachnidae.* S. Y. Tsalolikhin, Ed. Vol. 4, No. **5:** 28–49. Nauka. St. Peterburg.
8. KOLONIN, G.V., E.I. BOLOTIN, *et al.* 1981. Ecology of the ixodid tick, *Haemaphysalis japonica,* in the Maritime Territory. Med Parazitol (Mosk). **60:** 61–66.
9. SHPYNOV, S., P.-E. FOURNIER, N. RUDAKOV, *et al.* 2004. Detection of a rickettsia closely relate to *Rickettsia aeschlimannii*, "*Rickettsia heilongjiangensis,*" *Rickettsia* sp. RpA4, and *Ehrlichia muris* in ticks collected in Russia and Kazakhstan. J. Clin. Microbiol. **42:** 2221–2223.
10. ROUX, V., E. RYDKINA, *et al.* 1997. Citrate synthase gene comparison, a new tool for phylogenetic analysis, and its application for the rickettsiae. Int. J. Syst. Bacteriol. **47:** 252–261.
11. KUMAR, S., K. TAMURA & M. NEI. 2004. MEGA 3: Integrated software for Molecular Evalutionary Genetics: analysis and sequence alignment. Brief Bioinform. **5:** 150–163.
12. LA SCOLA, B. & D. RAOULT. 1997. Laboratory diagnosis of rickettsioses: current approaches to diagnoses of old and new rickettsial diseases. J. Clin. Microbiol. **35:** 2715–2727.
13. RAOULT, D. & V. ROUX. 1997. Rickettsioses as paradigms of new or emerging infectious diseases. Clin. Microbiol. Rev. **10:** 694–719.
14. FOURNIER, P.-E., M. JENSENIUS, H. LAFERL, *et al.* 2002. Kinetics of antibody responses in *Rickettsia africae* and *Rickettsia conorii* infections. Clin. Diagn. Lab. Immunol. **9:** 324–328.
15. ESTRADA-PENA, A. & F. JONGEJAN. 1999. Ticks feeding on humans: a review of records on human-biting *Ixodoidea* with special reference to pathogen transmission. Exp. Appl. Acarol. **23:** 685–715.
16. SHPYNOV, S., P.-E. FOURNIER, *et al.* 2003. "Candidatus Rickettsia tarasevichiae" in *Ixodes persulcatus* ticks collected in Russia. Ann. N. Y. Acad. Sci. **990:** 162–172.

Seroprevalence of *Anaplasma phagocytophilum* among Forestry Rangers in Northern and Northeastern Poland

JOANNA STAŃCZAK,[a] AND ANNA GRZESZCZUK[b]

[a]*Medical University of Gdańsk, Institute of Maritime and Tropical Medicine, 9 B Powstania Styczniowego Street, 81-519 Gdynia, Poland*

[b]*Medical University of Bialystok, Department of Infectious Diseases, 14 Zurawia Street, 15-540 Bialystok, Poland*

ABSTRACT: The aim of this study was to evaluate the seroprevalence of *A. phagocytophilum* in a group of workers from 13 forest management areas of northern and northeastern Poland. A total of 478 sera were tested by indirect immunofluorescence assay (IFA). Elevated IgG antibody titers were detected in 46 samples (9.6%). Of these, 34 (73.9%) had a titer of 1:64, 7 (15.2%) 1:128, and 5 (10.9%) 1:256. All seropositive persons disclaimed any clinical symptoms of HGA. The antibodies prevalent in persons with extreme outdoor activity and in indoor workers was comparable—9.6 % and 9.7%, respectively. Moreover, participants had no statisfically significant differences in the prevalence of antibodies related to sex, age, or years of employment.

KEYWORDS: *Anaplasma phagocytophilum*; forestry workers; seroprevalence; Poland

Ixodes (I.) ricinus ticks are responsible for the majority of tick bites in Poland and are known vectors of tick-borne encephalitis (TBE) viruses, *Borrelia burgdorferi* sensu lato and *Anaplasma phagocytophilum*.[1,2] The highest TBE and Lyme disease prevalence is recorded in the northeastern part of the country and recently the clinical cases of human granulocytic anaplasmosis (HGA) have also been diagnosed there.[3] Thus, the aim of our study was to evaluate the prevalence of antibodies to *A. phagocytophilum* among a high-risk group of forestry workers from this area.

Address for correspondence: Joanna Stańczak, Medical University of Gdańsk, Institute of Maritime and Tropical Medicine, 9 B Powstania Styczniowego street, 81-519 Gdynia, Poland. Voice: + 48 58 69 98 551; fax: +48-58-622-33-54.
e-mail: astan@amg.gda.pl

Ann. N.Y. Acad. Sci. 1078: 89–91 (2006). © 2006 New York Academy of Sciences.
doi: 10.1196/annals.1374.011

In 2003–2004, we analyzed a total of 478 sera collected from 388 males and 90 females (mean age 43 years) from 13 forest management areas of northern and northeastern Poland. The control group consisted of 50 patients suspected for toxoplasmosis. All sera were tested for for the presence of IgG and IgM antibodies to *A. phagocytophilum* by IFA (Focus Technologies, HGE IFA IgG/IgM, Cypress, CA, USA). Titers 1:64 and 1:20, respectively, were considered positive. The positive sera were then assayed at a twofold dilution. The selected positive sera were also tested for *B. burgdorferi* s.l. IgG antibodies by enzyme-linked immunosorbent assay (ELISA). All serum samples from the control group were negative. In the study group, the IgM antibodies to *A. phagocytophilum* were not detected while 9.6% ($n = 46$) of the sera was found to be IgG positive. Thirty-four (73.9%) of the positive samples had a titer of 1:64, seven (15.2%) had a titer of 1:128, and five (10.9%) had a titer of 1:256. The overall seropositivity encountered was two times lower than the seroprevalence (20.6%) detected in mid-eastern Poland[4] and slightly lower than that in southern Germany.[5] No statistically significant difference ($P = 0.85$) was observed in the prevalence of antibodies in persons with extreme outdoor activity (9.6%; $n = 39/406$) and in a group of office workers (9.7%; $n = 7/72$), working predominantly indoor and only occasionally professionally visiting forests. Interestingly, also in our previous work, we did not find an increased risk of *A. phagocytophilum* seropositivity among forestry rangers in comparison with the inhabitants living in the Bialowieża Primeval Forest, the area highly endemic for Lyme disease ($P > 0.05$; OR = 2.183; 95% CI = 0.26–18.49).[6] This may be caused by the exposure of humans to tick bites during recreation and mushroom or berry picking. Moreover, in the present study participants showed no statistically significant differences in the prevalence of antibodies related to gender: 9.3% males vs. 11.1% females ($P = 0.7$); age: 5.4% (groups of 21–30 years old; 4/74), 16.6% (groups ≥61 years old; 1/6) ($P = 0.14$); and years of employment: 0% (groups working < 1 year; 0/10), 16.7% (group working > 40 years; 1/6) ($P = 0.27$). Surprisingly, the highest seropositivity (14.6%; 13/89) was observed among workers who denied having tick bites, while the lowest seropositivity (6.1%; 6/98) was observed in persons who were frequently bitten. The latter, however, are used to checking their body for attached ticks, which may reduce the risk of *A. phagocytophilum* transmission, which requires a long attachment. Similar to our previous study,[6] the self-reported frequency of tick bites was not identified as increasing the risk of *A. phagocytophilum* seropositivity ($P = 0.2$). In 3 (8.1%) of 37 selected *A. phagocytophilum*-positive sera we also detected antibodies against *B. burgdorferi* s.l.

The results obtained constitute further evidence of the existence of natural foci of human granulocytic anaplasmosis in northern and northeastern Poland. Individuals with different exposures to tick bites show evidence of contact with the HGA agent, although none of the seropositive persons developed clinical manifestations of the disease. Infections seem to be asymptomatic or mild and unspecific. Thus, further studies on the pathogenicity of different *A. phagocytophilum* isolates seem to be required.

REFERENCES

1. STAŃCZAK, J. *et al.* 2002. Coinfection of *Ixodes ricinus* (Acari; Ixodidae) with the agents of Lyme borreliosis (LB) and human granulocytic ehrlichiosis (HGE). Int. J. Med. Microbiol. **291:** 198–201.
2. STAŃCZAK, J. *et al.* 2004. Ixodes ricinus as a vector of *Borrelia burgdorferi* sensu lato, *Anaplasma phagocytophilum* and *Babesia microti* in urban and suburban forests. Ann. Agric. Environ. Med. **11:** 109–114.
3. GRZESZCZUK, A. *et al.* 2006. Etiology of tick-borne febrile illnesses in adult residents of north-eastern Poland: report from a prospective study. Int. J. Med. Microbiol. 2006 Mar 8; e-pub ahead of print. PMID: 16530481.
4. TOMASIEWICZ, K. *et al.* 2004. The risk of exposure to *Anaplasma phagocytophilum* infection in mid-eastern Poland. Ann. Agric. Environ. Med. **11:** 261–264.
5. FINGERLE, V. *et al.* 1997. Human granulocytic ehrlichiosis in southern Germany: increased seroprevalence in high-risk group. J. Clin. Microbiol. **35:** 3244–3247.
6. GRZESZCZUK, A. *et al.* 2002. Serological and molecular evidence of human granulocytic ehrlichiosis focus in the Bialowieża Primeval Forest (Puszcza Bialowieska), northeastern Poland. Eur. J. Clin. Microbiol. Infect. Dis. **21:** 6–11.

Seroprevalence of Human Anaplasmosis in Slovene Forestry Workers

TEREZA ROJKO,[a] TINA URSIC,[b] TATJANA AVŠIČ-ŽUPANC,[b] MIROSLAV PETROVEC,[b] FRANC STRLE,[a] AND STANKA LOTRIC-FURLAN[a]

[a]*Department of Infectious Diseases, University Medical Centre Ljubljana, 1525 Ljubljana, Slovenia*

[b]*Institute of Microbiology and Immunology, Medical Faculty, 1525 Ljubljana, Slovenia*

ABSTRACT: The aim of the present study was to establish the prevalence and incidence of symptomatic and asymptomatic infection with *Anaplasma phagocytophilum* during the period of tick activity and to compare the risk of infection for forestry workers and indoor workers in Slovenia.

KEYWORDS: *Anaplasma phagocytophilum*; human anaplasmosis; forestry workers; seroprevalence

Human anaplasmosis (HA), formerly known as human granulocytic ehrlichiosis, is an emerging tick-borne zoonosis caused by *Anaplasma phagocytophilum (A. phagocytophilum)*. The first confirmed case of HA in Europe was described in Slovenia in 1997 and since then several cases of the disease have been reported.[1]

On account of the specific characteristics of their work, forestry workers are classified as persons with an increased risk for tick-borne diseases and 7.2–15.0% of forestry and other outdoor workers tested positive for IgG antibodies against *A. phagocytophilum* in different countries in Europe.[2]

The purpose of the present study was to establish the prevalence and incidence of symptomatic and asymptomatic infection with *A. phagocytophilum* during the period of tick activity and to compare the risk of infection with *A. phagocytophilum* for forestry workers and indoor workers in Slovenia.

The prospective study included 96 forestry workers and 53 indoor workers, residing in the same region as the forestry workers. All the study participants were examined twice in 2002; before the beginning of tick activity (March) and at the end of the period of tick activity (November). At each examination, principal demographic and epidemiological data were collected using a ques-

Address for correspondence: Tereza Rojko, Department of Infectious Diseases, University Medical Centre Ljubljana, Japljeva 2, 1525 Ljubljana, Slovenia. Voice: 386-1-522-21-10; fax: 386-1-522-24-56. e-mail: tereza.rojko@guest.arnes.si

Ann. N.Y. Acad. Sci. 1078: 92–94 (2006). © 2006 New York Academy of Sciences.
doi: 10.1196/annals.1374.012

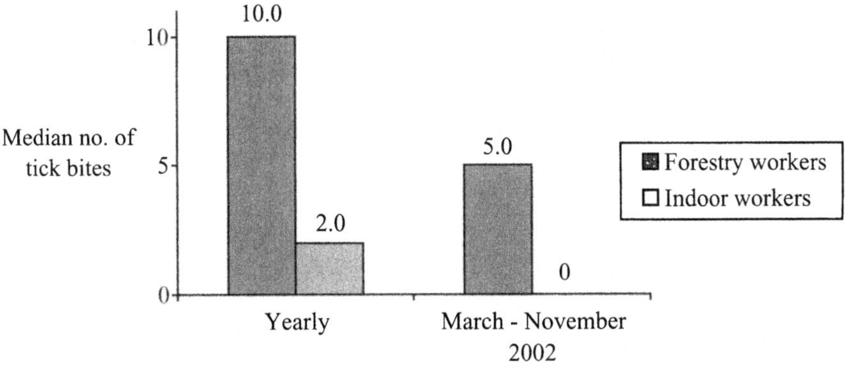

FIGURE 1. Tick exposure in the forestry workers group and in the indoor workers group.

tionnaire, and a blood sample was taken for serological analysis. The presence of IgG antibodies to *A. phagocytophilum* in the paired sera was determined with an indirect immunofluorescent assay (IFA). Antigen was prepared from a human promyelocytic cell line (HL60) infected with a tick-derived isolate of *A. phagocytophilum* (USG3). Antibody titers of 128 or higher were considered positive, titers of 32 and 64 reactive, whereas titers of < 32 were considered negative. Seroconversion was defined as a shift from a negative to a positive IgG result during the study period. Quantitative data were analyzed with the Kruskal–Wallis test while qualitative data were compared with the Yates corrected chi-square test or with the two-tailed Fisher's test. A P-value of <0.05 was considered to indicate statistical significance.

Forestry workers reported significantly more tick bites on average per year and during the study period than the indoor workers (FIG. 1). The seroprevalence rate to *A. phagocytophilum* of all the participants was 11.4% at the beginning of the study and 10.7% at the end of the study.

At the first examination, 8.3% of the forestry workers and 17% of the indoor workers tested positive, while 17.7% of the forestry workers and 13.2% of the indoor workers tested reactive for IgG to *A. phagocytophilum* ($P = 0.259$). At the second examination, 8.3% of the forestry workers and 15.1% of the indoor workers tested positive, while 24% of the forestry workers and 13.2% of the indoor workers tested reactive for IgG to *A. phagocytophilum* ($P = 0.172$).

There was no seroconversion in the study period. Four participants (2.7%), two from each group, experienced at least fourfold increase in antibody titers to *A. phagocytophilum,* but none of them had any clinical signs or symptoms of acute infection with *A. phagocytophilum*.

The seroprevalence rate of IgG antibodies to *A. phagocytophilum* among our study participants was comparable to the results of other European studies. Although the forestry workers stated significantly more tick bites on average per year as well as during the study period, there were more seropositive participants among the indoor workers, but the difference was not significant. The risk of acute infection with *A. phagocytophilum* during the study period was low for both groups.

REFERENCES

1. LOTRIC-FURLAN, S., M. PETROVEC, T. AVSIC-ZUPANC & F. STRLE. 2004. Comparison of patients fulfilling criteria for confirmed and probable human granulocytic ehrlichiosis. Scand. J. Infect. Dis. **36:** 817–822.
2. GRZESZCZUK, A., B. PUZANOWSKA, H. MIEGOĆ & D. PROKOPOWICZ. 2004. Incidence and prevalence of infection with *Anaplasma phagocytophilum*. Prospective study in healthy individuals exposed to ticks. Ann. Agric. Environ. Med. **11:** 155–157.

Molecular Epidemiology of Human and Bovine Anaplasmosis in Southern Europe

VICTORIA NARANJO,[a] FRANCISCO RUIZ-FONS,[a] URSULA HÖFLE,[a] ISABEL G. FERNÁNDEZ DE MERA,[a] DIEGO VILLANÚA,[a] CONSUELO ALMAZÁN,[b] ALESSANDRA TORINA,[c] SANTO CARACAPPA,[c] KATHERINE M. KOCAN,[b] CHRISTIAN GORTÁZAR,[a] AND JOSÉ DE LA FUENTE[a,b]

[a]*Instituto de Investigación en Recursos Cinegéticos IREC (CSIC-UCLM-JCCM), Ronda de Toledo s/n, 13005 Ciudad Real, Spain*

[b]*Department of Veterinary Pathobiology, Center for Veterinary Health Sciences, Oklahoma State University, Stillwater, Oklahoma 74078, USA*

[c]*Istituto Zooprofilattico Sperimentale della Sicilia, Via G. Marinuzzi No. 3, 90129 Palermo, Italy*

ABSTRACT: The genus *Anaplasma* (Rickettsiales: Anaplasmataceae) includes several pathogens such as *A. marginale* and *A. phagocytophilum* that have an impact on veterinary and human health. In this study, we characterized *A. marginale* and *A. phagocytophilum* infections in humans, wild and domestic animals, and ticks in southern Europe (particularly in south-central Spain and in Sicily) by means of serologic study, PCR, and sequence analysis of major surface proteins (*msp*) 1α and 4 and 16S rDNA. The results suggest that *A. marginale* infections in this region are maintained in cattle and deer, with ticks and tabanids serving as biological and mechanical vectors of the pathogen, respectively. Infections with *A. phagocytophilum* may occur in humans and are maintained in cattle, donkeys, deer, and birds and are most likely transmitted by several tick species with as yet an unknown role as reservoir hosts for other wild and domesticated mammals. The presence of concurrent infections in cattle and deer suggests that these pathogens may multiply in the same reservoir host and illustrates the complexity of the epidemiology of bovine and human anaplasmosis in this region.

KEYWORDS: *Anaplasma*; *A. marginale*; *A.* phagocytophilum; tick; deer

Address for correspondence: José de la Fuente, Department of Veterinary Pathobiology, Center for Veterinary Health Sciences, Oklahoma State University, Stillwater, OK 74078, USA. Voice: 405-744-0372; fax: 405-744-5275.
e-mail: jose_delafuente@yahoo.com; djose@cvm.okstate.edu

**Ann. N.Y. Acad. Sci. 1078: 95–99 (2006). © 2006 New York Academy of Sciences.
doi: 10.1196/annals.1374.013**

INTRODUCTION

Organisms in the genus *Anaplasma* are obligate intracellular pathogens that multiply in both vertebrate and invertebrate hosts.[1] The type species, *A. marginale*, causes bovine anaplasmosis and only infects ticks and ruminants, whereas *A. phagocytophilum*, the causative agent of human and animal granulocytic anaplasmosis, has a wide host range including ticks, ruminants, rodents, equines, birds, and humans.[1] Recent reports demonstrated that *A. marginale* and *A. phagocytophilum* coexist in geographic areas and that concurrent infections occur in ruminants and ticks.[2]

In this study, we characterized *A. marginale* and *A. phagocytophilum* infections in humans, wild and domestic animals, and ticks in southern Europe, and particularly in south-central Spain and in Sicily, Italy, by serology, polymerase chain reaction (PCR), and sequence analysis of major surface proteins (*msp*) 1α and 4 and 16S rDNA genotypes.[2–8]

MATERIALS AND METHODS

Samples

Samples were collected in the Community of Castilla–La Mancha in Spain and in Sicily, Italy.[4–8] Mammals tested included humans, cattle, donkeys, sheep, dogs, rodents, Iberian red deer, and European wild boar, as well as several species of birds and ticks.[4–8] (TABLE 1). Species of hematophagous Diptera were analyzed as mechanical vectors of *Anaplasma* spp. in Spain.[8]

TABLE 1. Evidences of *A. marginale* and *A. phagocytophilum* infections in vertebrate hosts and invertebrate vectors

	A. marginale		*A. phagocytophilum*	
Animal species	Serology	PCR	Serology	PCR
Human	ND	0/20 (0)	0/20 (0)	1/20 (5)
Cattle	231/305 (76)	126/317 (40)	24/33 (73)	20/157 (13)
Donkey	0/3 (0)	0/3 (0)	3/3 (100)	3/3 (100)
Sheep	6/6 (100)	0/8[a] (0)	2/5 (40)	0/8[a](0)
Iberian red deer	18/170 (11)	1/6 (17)	2/20 (10)	6/6 (100)
European wild boar	ND	0/18 (0)	ND	0/18 (0)
Rodent	ND	0/6 (0)	ND	0/6 (0)
Dog	ND	0/9 (0)	ND	0/9 (0)
Birds	ND	0/46 (0)	ND	10/46 (22)
Ticks	ND	47/270 (17)	ND	77/132 (58)
Diptera	ND	2/5 (40)	ND	0/5 (0)

NOTE: Data are presented as the number of positive samples/total number of samples analyzed (% seropositivity or % observed prevalence as determined by PCR and sequence analysis of *msp4* and/or 16S rDNA amplicons). ND, not determined.

[a]7 of 8 sheep samples were positive for *A. ovis*.[6]

Anaplasmosis Serologic Tests

The anaplasmosis competitive ELISA was performed using the *Anaplasma* Antibody Test Kit, cELISA from VMRD, Inc. (Pullman, WA) following the manufacturer's instructions.[4,6–8] This assay detects serum antibodies against the MSP5 protein of *Anaplasma* spp. in cattle. The immunofluorescence test for *A. phagocytophilum* was performed using IFA Antibody Test Kits from Fuller Laboratories (Fullerton, CA) and Focus Technologies (Cypress, CA) following the manufacturer's instructions.[6,8]

PCR

DNA was extracted from blood, spleen, tick, and Diptera samples and the *A. marginale msp1α* and the *Anaplasma* spp. *msp4* genes were amplified by PCR and sequenced as reported previously.[3–8] The *A. marginale* and *A. phagocytophilum* species-specific *msp4* PCRs (sensitivity of 5 copies *msp4*/ng DNA) were used for pathogen identification and phylogenetic analysis.[2–8] The 16S rRNA was used for genotyping *Anaplasma* strains.[8] The *msp4* coding region was completely sequenced.[2] Only the fragment containing the tandem repeats in the variable region of *msp1α* was sequenced.[2]

RESULTS AND DISCUSSION

Antibodies to *A. marginale* and *A. phagocytophilum* were detected in cattle and deer and in cattle, deer, sheep, and donkeys, respectively (TABLE 1). *A. marginale* DNA was detected in tabanids (*Atylotus loewianus, Tabanus nemoralis*), ticks (*Hyalomma m. marginatum, Rhipicephalus bursa, R. turanicus, Haemophysalis punctata, Dermacentor marginatus*), cattle, and deer, whereas *A. phagocytophilum* DNA was amplified from tick (*Hy. m. marginatum, R. bursa, D. marginatus, Ixodes* sp.), deer, cattle, donkey, birds (*Turdus merula, Fringilla coelebs, Passer domesticus, P. hipaniolensis, Emberiza cia, Lanius senator, Pica pica, Aegithalos caudatus*), and human samples (TABLE 1). Concurrent infections of the two *Anaplasma* species were found in ticks (*Hy. m. marginatum, R. bursa, D. marginatus*), cattle, and deer.

The analysis of *Anaplasma msp1α*, *msp4*, and 16S rDNA sequences[2–4,6,8] demonstrated that, as previously reported for New World strains, genetic variation among *A. marginale* strains occurred within cattle herds and geographic locations,[2,4,6–8] suggesting that multiple introductions of genetically diverse strains of the pathogen may have occurred in these geographic areas with further diversification by cattle movement.[2] Genetic diversity of *A. phagocytophilum* strains also occurred in this geographic region,[2,3,8] but ruminant and nonruminant strains of the pathogen were differentiated by the use of the *msp4* sequence.[3]

In summary, the results of this study suggest that *A. marginale* infections in this region are maintained in cattle and deer with ticks and tabanids serving as biological and mechanical vectors of the pathogen, respectively. Infections with *A. phagocytophilum* are maintained in cattle, donkeys, deer, and birds and are most likely transmitted by several tick species with as yet an unknown role as reservoir hosts for other wild and domesticated mammals such as sheep, European wild boar, rodents, and rabbits. Humans may become infected with *A. phagocytophilum*. The demonstration of concurrent infections in cattle and deer suggests that these pathogens may multiply in the same reservoir host.[8] However, although limited by the number of samples analyzed, the failure to identify identical *Anaplasma* genotypes in cattle and deer may reflect a certain degree of host tropism among *Anaplasma* genotypes.[8] Alternatively, infection-exclusion of some genotypes may occur as previously reported for *A. marginale* in cattle and ticks.[1] *A. phagocytophilum* strain genetic differences may be associated with variations in pathogenicity and host tropism,[3] although the exact relationship between these factors is presently unknown.

The results of this study illustrated the complexity of the epidemiology of bovine and human anaplasmosis in regions where *A. marginale* and *A. phagocytophilum* coexist and share common reservoir hosts and vectors. The increasing contact between wildlife, domestic animals, and human populations increases the risk of outbreaks of human and bovine anaplasmosis, as well as the difficulty of implementing surveillance and control measures.

ACKNOWLEDGMENTS

This research was supported by the Endowed Chair for Food Animal Research (K.M. Kocan), the Instituto de Ciencias de la Salud, Spain (ICS-JCCM) (Project 03052–00), and the Ministry of Health, Italy.

REFERENCES

1. KOCAN, K.M., J. DE LA FUENTE, E.F. BLOUIN, *et al.* 2004. *Anaplasma marginale* (Rickettsiales: Anaplasmataceae): recent advances in defining host-pathogen adaptations of a tick-borne rickettsia. Parasitology **129:** S285–S300.
2. DE LA FUENTE, J., A. LEW, H. LUTZ, *et al.* 2005. Genetic diversity of *Anaplasma* species major surface proteins and implications for anaplasmosis serodiagnosis and vaccine development. Anim. Health Res. Rev. **6:** 75–89.
3. DE LA FUENTE, J., R.B. MASSUNG, S.J. WONG, *et al.* 2005. Sequence analysis of the *msp4* gene of *Anaplasma phagocytophilum* strains. J. Clin. Microbiol. **43:** 1309–1317.
4. DE LA FUENTE, J., J. VICENTE, U. HÖFLE, *et al.* 2004. *Anaplasma* infection in free-ranging Iberian red deer in the region of Castilla-La Mancha, Spain. Vet. Microbiol. **100:** 163–173.

5. DE LA FUENTE, J., V. NARANJO, F. RUIZ-FONS, *et al.* 2004. Prevalence of tick-borne pathogens in ixodid ticks (Acari: Ixodidae) collected from European wild boar (*Sus scrofa*) and Iberian red deer (*Cervus elaphus hispanicus*) in central Spain. Eur. J. Wild. Res. **50:** 187–196.
6. DE LA FUENTE, J., A. TORINA, S. CARACAPPA, *et al.* 2005. Serologic and molecular characterization of *Anaplasma* species infection in farm animals and ticks from Sicily. Vet. Parasitol. **133:** 357–362.
7. DE LA FUENTE, J., A. TORINA, V. NARANJO, *et al.* 2005. Genetic diversity of *Anaplasma marginale* strains from cattle farms with different husbandry systems in the Province of Palermo, Sicily. J. Vet. Med. (B) **52:** 226–229.
8. DE LA FUENTE, J., V. NARANJO, F. RUIZ-FONS, *et al.* 2005. Potential vertebrate reservoir hosts and invertebrate vectors of *Anaplasma marginale* and *A. phagocytophilum* in central Spain. Vector Borne Zoonotic Dis. **5:** 390–401.

Human Exposure to *Anaplasma phagocytophilum* in Portugal

A.S. SANTOS,[a] F. BACELLAR,[a] AND J.S. DUMLER[b]

[a]*Centro de Estudos de Vectores e Doenças Infecciosas (CEVDI), Instituto Nacional de Saúde Dr. Ricardo Jorge, Águas de Moura, Portugal*

[b]*Division of Medical Microbiology, Department of Pathology, The Johns Hopkins University School of Medicine, Baltimore, Maryland, USA*

ABSTRACT: A retrospective study to detect *Anaplasma phagocytophilum* antibodies by indirect immunofluorescence (IFA) and Western blot (WB) assay was conducted in 367 potentially exposed patients from Portugal. The study included 26 patients with confirmed Lyme borreliosis (LB), 77 with suspected LB, 264 seronegative patients studied for possible tick-transmitted LB and boutonneuse fever (LB/BF) infection, and 96 healthy blood donors. Overall, patients with LB and suspected LB ($n = 2$ [7.7%] and $n = 6$ [7.8%], respectively) were more often seropositive ($n = 8$ [7.8%]; $P < 0.001$), whereas only 1 (0.4%; $P = 0.046$) patient in the LB/BF seronegative group had confirmed disease. This study is the first evidence of human exposure to *A. phagocytophilum* or an antigenically similar bacterium in Portugal, and suggests that LB patients are significantly more likely to contact *A. phagocytophilum*.

KEYWORDS: serologic study; *Anaplasma phagocytophilum* exposure; Lyme borreliosis; Portugal

INTRODUCTION

Human granulocytic anaplasmosis (HGA) is an emerging tick-borne illness caused by *Anaplasma phagocytophilum*. In Portugal, the bacterium is present in both *Ixodes ricinus* and *I. ventalloi* ticks,[1] but no data concerning the occurrence of HGA are available. Because *I. ricinus* frequently bite humans, we conducted a retrospective study of *A. phagocytophilum* antibodies in potentially exposed patients.

Address for correspondence: Ana Sofia Santos, Centro de Estudos de Vectores e Doenças Infecciosas, Instituto Nacional de Saúde Dr. Ricardo Jorge, Avenida da Liberdade 5, 2965-575 Águas de Moura, Portugal. Voice: +351-265-912222; fax: +351-265-912568; +351-265-912155.
e-mail: ana.santos@insa.min-saude.pt

**Ann. N.Y. Acad. Sci. 1078: 100–105 (2006). © 2006 New York Academy of Sciences.
doi: 10.1196/annals.1374.014**

MATERIAL AND METHODS

Sera from clinically ill individuals submitted to CEVDI's laboratory during 2002 for Lyme borreliosis (LB) and boutonneuse fever (BF) testing were used. The study included 26 patients with confirmed LB, 77 with suspected LB, 264 seronegative patients studied for possible tick-transmitted (LB/BF) infection, and 96 healthy blood donors. The first two cohorts included all confirmed and suspected LB cases with available sera.[2] The third cohort was a random sample comprising 25% of LB/BF seronegative patients detected monthly during the year. Institutional Review Board approval was obtained for these studies. Antibodies against *A. phagocytophilum* were detected by indirect immunofluorescence (IFA) and Western blot (WB) assays, as described previously.[3,4] *A. phagocytophilum* Webster and *Ehrlichia chaffeensis* Arkansas strains were used as antigen. *E. chaffeensis* cross-reactivity was evaluated in seropositive samples. The criteria for seropositivity were IFA titer $\geq$80 and WB reactivity with bands between 42–49 kDa for *A. phagocytophilum* and between 23–30 kDa for *E. chaffeensis*.[3,4] Confirmed *A. phagocytophilum* exposure was defined as a fourfold higher IFA titer to specific antigen than to *E. chaffeensis* or when IFA and WB were positive. Possible exposure was defined as a single IFA titer of $\geq$80 or less than fourfold titer difference between specific antigen and *E. chaffeensis*, and no specific WB bands. For statistical analysis, independence of proportions of reactive patient samples was compared between groups and the test population using the Chi-square test. P values of <0.05 were considered significant.

RESULTS

A total of 419 sera were obtained from 367 patients for serologic testing, not including 96 sera obtained from healthy blood donors. In 323 patients (88%) only one serum sample was available. Serologic results are summarized in TABLES 1 and 2 and in FIGURE 1. Confirmed and possible exposures to *A. phagocytophilum* were present in 1 (1%) and 3 (3%) of 96 blood donors, respectively. Patients with possible exposure were not more likely to have clinical findings ($P = 0.10$, χ^2 test); thus, only confirmed patients were considered infected. Overall, 2.2% (10/463) individuals had confirmed exposure. Patients with LB or suspected LB were significantly more likely to have *A. phagocytophilum* confirmed ($P \leq 0.002$). Of 367 studied, patients with LB ($n = 2$ [7.7%]) and suspected LB ($n = 6$ [7.8%]) together ($n = 8$, [7.8%]) were more often seropositive ($P < 0.001$), whereas only one (0.4%; $P = 0.046$) patient in the LB/BF seronegative group had confirmation. *E. chaffeensis* cross-reactions were detected only in three patients and only in the suspected LB group. No statistically significant differences in age, gender, and province were noted between the *A. phagocytophilum*-confirmed and -seronegative individuals; the

TABLE 1. Serologic testing results for *A. phagocytophilum* and *E. chaffeensis*

Study groups[a]	No. tested patients/ no. reactive	No. confirmed (%) / no. possible (%) exposure	
		A. phagocytophilum	*E. chaffeensis*
Patients with defined LB	26/5	2(7.7) / 3(11.5)	—
Patients with suspected LB	77/11	6 (7.8) / 2 (2.6)	— / 3(3.9)
Seronegative LB/BF patients	264/4	1(0.4) / 3(1.1)	—
Blood donors	96/4	1(1) / 3(3.1)	—

[a] LB = Lyme borreliosis; BF = Boutonneuse fever.

median age of confirmed patients was 42 years, 66.6% were women and 55.5% came from Estremadura province. However, among patients in Estremadura province, *A. phagocytophilum*-confirmed cases more often resided in Oeiras than in other locations ($P < 0.001$).

DISCUSSION

A. phagocytophilum is transmitted by *Ixodes* ticks and shares these vectors with other human infectious agents, such as *Borrelia burgdorferi* (etiologic agent of LB). This ecological characteristic of *A. phgocytophilum* allowed us to focus the study on patients with clinical and laboratory evidence of confirmed or suspected LB in order to detect potential HGA. HGA results in an acute, generally self-limited, nonspecific febrile illness, characterized by headache and myalgias, often accompanied by systemic symptoms and signs and hematological abnormalities such as thrombocytopenia, leukopenia, elevated serum hepatic transaminases, and/or elevated serum C-reactive protein concentrations. This nonspecific presentation is common with other tick-borne illnesses, justifying the inclusion of an additional patient group with clinical suspicion of LB or BF but lacking laboratory confirmation for these diagnoses. In this report we have found that significantly higher proportions of patients with suspected/confirmed LB had confirmed exposure to *A. phagocytophilum* than did the overall population. Our results corroborate previous findings, and show that the prevalence of *A. phagocytophilum* infection among Portuguese LB patients (7.8%) is in the range of 2–21% as described in other European countries.[5] However, the evaluation is partly impaired by the lack of analysis of paired sera that underestimates prevalence. Cultivation of Portuguese strains could also help to improve serologic tests. In conclusion, these data provide evidence that Portuguese *A. phagocytophilum* strains infect and evoke immune responses in tick-exposed humans, and that LB patients are significantly more likely to acquire *A. phagocytophilum* infection. The possibility of coinfection with both pathogens and comorbidity is worthy of clinical consideration.

TABLE 2. Epidemiological and clinical data from patients with confirmed exposure to *A. phagocytophilum*

							A. phagocytophilum serology	
Patient groups[a]		Gender[b]/ age (year)[c]	Origin of residence[d]	Clinical presentation/ laboratory findings[e]	Onset (days)	No. sera tested (time interval)[f]	IFA titer (Ig class)	WB result[g]
Defined LB	1	F/63	Elvas; Alto Alentejo	Febrile illness/—	—	2 (68 days)	80(IgG); 1280(IgG)	–
	2	F/57	Oeiras; Estremadura	—	—	1	640(IgG)	+
Suspected LB	3	M/U	Algarve	—	—	2 (13 days)	320(IgG)	+
	4	M/74	Oeiras; Estremadura	—	—	1	320(IgG)	+
	5	F/67	Lisboa; Estremadura	Vasculitis	—	1	320(IgM)	+
	6	F/9	Oeiras Estremadura	—	—	1	160(IgG)	+
	7	M/50	Oeiras; Estremadura	—	—	1	160(IgM)	+
	8	F/31	V F Xira; Ribatejo	—	—	3 (112 days)	80(IgG)	+
Seronegative LB/BF	9	F/19	F Foz; Beira Litoral	Febrile illness/ LT	8	1	80(IgM)	+

[a]LB = lyme borreliosis; BF = boutonneuse fever; [b]M = male; F = female; [c]U = unknown; [d]Patient locality and province; [e]LT = Elevated liver transaminases; [f]Average in days between consecutive sera samples; [g]Western blot result for polyvalent antibodies testing.

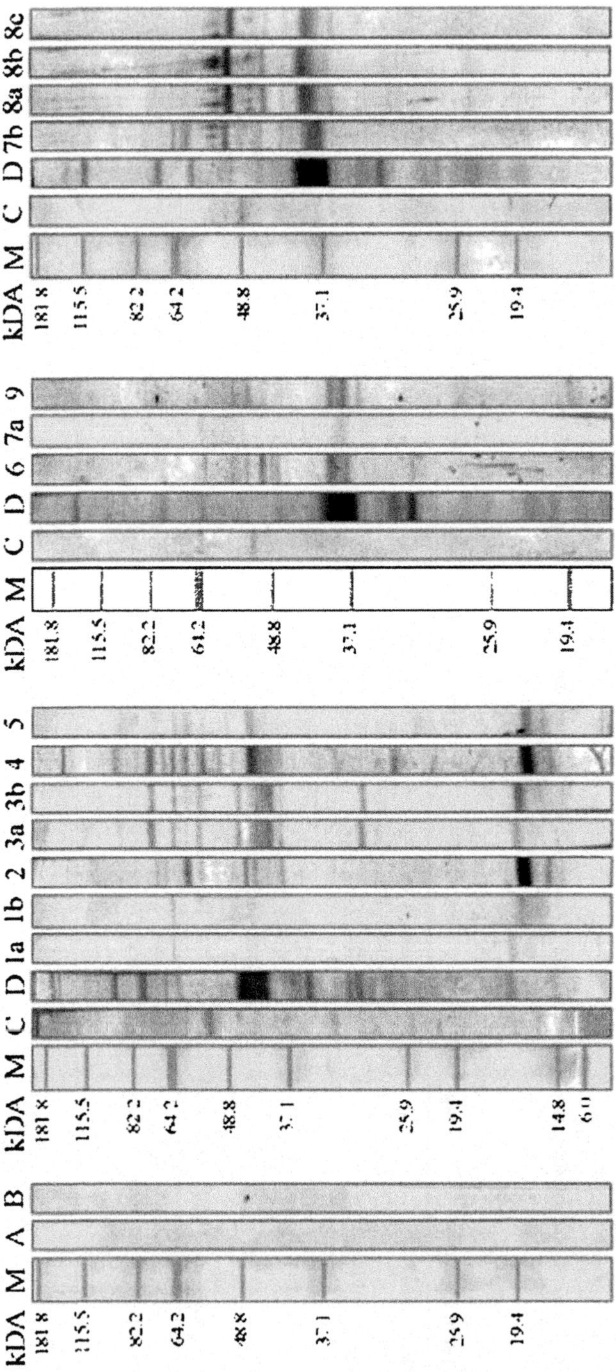

FIGURE 1. Western blot results for *A. phagocytophilum* IFA-positive patients.

ACKNOWLEDGMENTS

This research was partially supported by the Portuguese government through Fundação para a Ciência e a Tecnologia Grant BD/8610/2002 and through the National Institutes of Health Grant R01-AI41213 to JSD. A special thanks to Cevdi and Dumler's lab team.

REFERENCES

1. SANTOS, A.S. *et al.* 2004. Detection of *Anaplasma phagocytophilum* DNA in *Ixodes* ticks (Acari: Ixodidae) from Madeira Island and Setúbal District, Mainland Portugal. Emerg. Infect. Dis. **10:** 1643–1648.
2. LOPES DE CARVALHO, I., T. LUZ & M.S. NÚNCIO. 2004. 14 year follow-up study of Lyme borreliosis in Portugal. Presented at the third Congress of the European Society for Emerging Infections. Paris, France. October 17–20.
3 WONG, S.J., G.S. BRADY & J.S. DUMLER. 1997. Serologic responses to *Ehrlichia equi*, *Ehrlichia chaffeensis* and *Borrelia burgdorferi* in patients from New York State. J. Clin. Microbiol. **35:** 2198–2205.
4. UNVER, A. *et al.* 2001. Western blot analysis of sera reactive to human monocytic ehrlichiosis and human granulocyic ehrlichiosis agents. J. Clin. Microbiol. **39:** 3982–3986.
5. BLANCO, J.R. & J.A. OTEO. 2002. Human granulocytic ehrlichiosis in Europe. Clin. Microbiol. Infect. **8:** 763–772.

Human Granulocytic Anaplasmosis in Northeastern Italy

ANNA BELTRAME,[a] MAURIZIO RUSCIO,[b] ALESSANDRA ARZESE,[c] GIADA RORATO,[a] CAMILLA NEGRI,[a] ANGELA LONDERO,[a] MASSIMO CRAPIS,[a] LUIGIA SCUDELLER,[a] AND PIERLUIGI VIALE[a]

[a]*Clinic of Infectious Diseases, Department of Clinical and Morphological Research, School of Medicine, University of Udine, Udine 33100, Italy*

[b]*Microbiology Department, Hospital of S. Daniele, S. Daniele 33038, Italy*

[c]*Institute of Microbiology, University of Udine, Udine 33100, Italy*

ABSTRACT: Sporadic cases of human granulocytic anaplasmosis (HGA) have been reported in areas with a high prevalence of tick-borne diseases (TBDs) in Europe. We aimed at estimating the sero-prevalance of *A. phagocytophilum* and other TBDs in northeastern Italy in outpatients with a history of recent tick bite or suspected TBD. In the 1-year study, 79 patients were enrolled and 30 (38%) received a diagnosis of TBD: 24 (30%) with Lyme desease and 5 (6%) with HGE. Our findings indicate the presence of HGA in northernsterm Italy; so, since co-infection with Lyme disease appeared to be frequent, physicians assessing patients after a tick bite should consider HGA in the diagnosis.

KEYWORDS: human granulocytic anaplasmosis; tick-borne infection; seroprevalence; Italy

INTRODUCTION

Human granulocytic anaplasmosis (HGA) is a recently recognized tick-borne disease (TBD).[1] Sporadic cases of HGA have been reported in Europe, namely in areas at high prevalence of Lyme disease and TBE infection.[2,3] The Friuli Venezia Giulia (FVG) region in the northeastern part of Italy borders Slovenia and Austria, both of which are endemic areas for TBDs.[4,5] Since the clinical manifestation and the routine abnormal findings on laboratory tests of HGA are nonspecific, diagnosis relies on laboratory tests that can detect the etiologic agent (identification of morulas in the white cells or *Anaplasma phagocytophilum* DNA by polymerase chain reaction) as well as specific antibodies.[6] In the present study, we estimated the prevalence of antibodies to *A.*

Address for correspondence: Anna Beltrame, Department of Infectious Disease, Via Colugna No. 50, 33100 Udine, Italy. Voice: +39-0432-55-9355; fax: +39-0432-55-9360.
e-mail: anna.beltrame@med.uniud.it

Ann. N.Y. Acad. Sci. 1078: 106–109 (2006). © 2006 New York Academy of Sciences.
doi: 10.1196/annals.1374.015

phagocytophilum and other TBDs in a selected population of FVG residents coming to medical attention.

METHODS

From 2003 throughout 2004 we enrolled outpatients with a history of tick bite within the last 6 months and/or who presented to the Department of Infectious Diseases of Udine (FVG) with suspected tick-borne infection and all patients admitted to the hospital of Udine for aseptic meningitis. From each patient a paired (acute and convalescent) serum specimen and peripheral blood smears were collected. Serology for *A. phagocytophilum* (indirect immunofluorescence antibody [IFA] tests), TBE virus (enzyme immunoassay [EIA]) and *Borrelia burgdorferi* (enzyme-linked immunosorbent assay [ELISA]) were assessed in a central laboratory.[7] Peripheral blood smears were stained with Wright's stain and microscopic examination was done to reveal morulas in the cytoplasm of neutrophils. In the patients with aseptic meningitis, a cerebrospinal fluid (CSF) sample was sent to a central laboratory to test for antibodies to various meningoencephalitic agents, including tick-borne encephalitis (TBE) virus and *Borrelia burgdorferi* but not *A. phagocytophilum*. Cases of HGA were defined as confirmed (fever and seroconversion, or a fourfold change in serum antibody titer to *A. phagocytophilum,* and/or a positive PCR), probable (fever and acute and convalescent serum samples with unchanging IFA titers), or possible (serum samples with a titer of $\geq$1:128 of antibodies to *E. phagocytophila* at only the testing point).[6]

RESULTS

During the 1-year study, 79 patients were enrolled (70 outpatients, 9 inpatients) and 30 patients (38%) received a diagnosis of at least one TBD: 24 (30%) with Lyme borreliosis, 5 (6%) with TBE, and 5 (6%) with HGA. Of the latter, two were confirmed HGA cases (1 meningitis), one was a probable HGA case, and two patients had a possible HGA infection. All were outpatients. Three of the patients with HGA were coinfected with *B. burgdorferi* (TABLE 1). Only in the patient with aseptic meningitis were morulas detected in stained blood smears. Treatment with doxycycline, 100 mg twice daily for 10 days, was administered to all five patients, obtaining clinical improvement in all within 2 weeks.

CONCLUSION

HGA is an emerging zoonotic disease in the United States and Europe.[8] In Italy, the reported prevalence of *A. phagocytophilum* is 24.4% in *Ixodes ricinus*

TABLE 1. Clinical characteristics of HGA patients

Patient	Sex	Age (years)	History of tick bite	Area of occasional forest visit	Clinical presentation	IgG to HGA at diagnosis	IgM to HGA at diagnosis	Seroconversion	HGA diagnosis	Co-infection
Case 1	M	33	Yes	Trieste	Asymptomatic	128	0	No	Possible	No
Case 2	F	69	Yes	Pontebba	Asymptomatic	64	1280	Yes	Confirmed	Lyme disease
Case 3	F	41	Yes	Tarvisio	Fever, arthralgia, myalgia	256	0	No	Probable	Lyme disease
Case 4	F	48	Yes	Martignacco	Erythema migrans	128	0	No	Possible	Lyme disease
Case 5	F	17	No	Rivignano	Fever headaches Nucal rigidity Arthralgia, myalgia	128	80	Yes	Confirmed	No

ticks, while in humans it ranges between 1.5% (patients who live in *I. ricinus*–exposed areas) and 8.6% (tick-exposed individuals).[9,10] In FVG, several cases of tick-borne diseases are reported in the human population: Lyme borreliosis since 1986, and one case of meningo-encephalomyelitis TBE was recently reported.[11,12] In 1998, the first two confirmed cases of HGA were diagnosed in the region.[7] A seroprevalence study conducted 2 years later reported a low (0.6%) seroprevalence rate of HGA in forestry rangers.[13]

Although a low incidence of HGA has been reported in European countries, we found that 5 of 79 patients bitten by a tick in one year had HGA (6%) in our area. This prompts some considerations. First, HGA should always be included in the serologic testing of patients with a history of tick bite. Second, HGA should be considered in symptomatic patients even when Lyme disease is already confirmed, and HGA should be evoked in febrile EC since co-infection is frequent.

REFERENCES

1. CHEN, S.M. *et al.* 1994. Identification of a granulocytotropic *Ehrlichia* species as the agent of human disease. J. Clin. Microbiol. **32:** 589–595.
2. PETROVEC, M. *et al.* 1997. Human disease in Europe caused by granulocytic *Ehrlichia* species. J. Clin. Microbiol. **35:** 1556–1559.
3. BLANCO, J.R. & J.A. OTEO. 2002. Human granulocytic ehrlichiosis in Europe. Clin. Microbiol. Infect. **8:** 763–772.
4. LOTRIC-FURLAN, S. *et al.* 2001. Prospective assessment of the etiology of acute febrile illness after a tick bite in Slovenia. Clin. Infect. Dis. **33:** 503–510.
5. WALDER, G. *et al.* 2003. First documented case of human granulocytic ehrlichiosis in Austria. Wien. Klein. Wochenschr. **115:** 263–266.
6. BROUQUI, P. *et al.* 2004. Guidelines for the diagnosis of tick-borne bacterial diseases in Europe. Clin. Microbiol. Infect. **10:** 1108–1132.
7. RUSCIO, M. & M.CINCO. 2003. Human granulocytic ehrlichiosis in Italy: first report on two confirmed cases. Ann. N. Y. Acad. Sci. **990:** 350–352.
8. BAKKEN, J.S. & J.S. DUMLER. 2000. Human granulocytic ehrlichiosis. Clin. Infect. Dis. **31:** 554–560.
9. CINCO, M. *et al.* 1998. Detection of HGE agent-like *Ehrlichia* in *Ixodes ricinus* ticks in northern Italy by PCR. Wien. Klin. Wochenschr. **110:** 898–900.
10. NUTI, M. *et al.* 1998. *Ehrlichia* infection in Italy. Emerg. Infect. Dis. **4:** 663–665.
11. CICERONI, L. & S. CIARROCCHI. 1998. Lyme disease in Italy, 1983–1996. New Microbiol. **21:** 407–418.
12. BELTRAME, A. *et al.* 2005. Tick-Borne encephalitis in Friuli Venezia Giulia, north-eastern Italy. Infection **33:** 158–159.
13. CINCO, M. *et al.* 2004. Seroprevalence of tick-borne infections in forestry rangers from northeastern Italy. Clin. Microbiol. Infect. **10:** 1056–1061.

Human Infection with *Ehrlichia Canis* Accompanied by Clinical Signs in Venezuela

MIRIAM PEREZ,[a] MAIRIM BODOR,[b] CHUNBIN ZHANG,[c] QINGMIN XIONG,[c] AND YASUKO RIKIHISA[c]

[a]*Departamento de Medicina y Cirugía, Decanato de Ciencias Veterinarias, Universidad Centroccidental "Lisandro Alvarado," Barquisimeto Edo, Lara, Venezuela*

[b]*Servicio Emergencia Adultos, Hospital Central "Antonio María Pineda," Barquisimeto Edo, Lara, Venezuela.*

[c]*Department of Veterinary Biosciences, College of Veterinary Medicine, The Ohio State University, Columbus, Ohio 43210–1093, USA*

ABSTRACT: A total of 20 human patients with clinical signs compatible with human monocytic ehrlichiosis (HME), who were admitted to the emergency clinic in Lara State, Venezuela, were studied. Thirty percent (6/20) patients were positive for *Ehrlichia canis* 16S rRNA on gene-specific polymerase chain reaction (PCR). Compared with the U.S. strains, 16S rRNA gene sequences from all six patients had the same base mutation as the sequence of the *E. canis* Venezuelan human *Ehrlichia* (VHE) strain previously isolated from an asymptomatic human. This study is the first report of *E. canis* infection of human patients with clinical signs of HME.

KEYWORDS: 16s rRNA; *Ehrlichia canis*; human; infection; Venezuela

INTRODUCTION

Ehrlichia canis (*E. canis*), the agent of canine monocyte ehrlichiosis is an obligate intracellular Gram-negative bacterium, common in tropical and subtropical regions.[1] Since the discovery of *E. canis* in 1935 by Donatein and Lestoquard,[2] *E. canis* has been considered as a pathogen for the dog and other canids. In 1996, we reported the first human infection with *E. canis* and culture isolation of an *E. canis* strain from an apparently chronically infected asymptomatic human in Lara, Venezuela. We designated this strain Venezuelan human *Ehrlichia* (VHE).[3] *E. canis* is closely related to *Ehrlichia chaffeensis*

Address for correspondence: Yasuko Rikihisa, Department of Veterinary Biosciences, College of Veterinary Medicine, The Ohio State University, 1925 Coffey Road, Columbus, OH 43210-1093, USA. Voice: +1-614-292-9677; fax: +1-614-292-6473.
e-mail: rikihisa.1@osu.edu

Ann. N.Y. Acad. Sci. 1078: 110–117 (2006). © 2006 New York Academy of Sciences.
doi: 10.1196/annals.1374.016

that causes human monocytic ehrlichiosis (HME), and the clinical signs of canine monocytic ehrlichiosis caused by *E. canis* are very similar to those of HME.[1] Clinical signs of HME vary from asymptomatic infection to severe illness which requires hospitalization, or even death. Severe HME is characterized by fever, chill, headache, myalgia, anorexia, nausea or vomiting, and weight loss.[4] Thrombocytopenia, leukopenia, and liver enzyme abnormalities also are often reported. Meningitis, encephalitis, or encephalopathy also may occur as a result of infection, and *E. chaffeensis* may be detected in the cerebrospinal fluid. These clinical signs are similar to Rocky Mountain spotted fever, influenza, infectious mononucleosis, and dengue fever.[5,6] Patients with similar febrile illnesses of unknown etiology or patients suspected of having hemorrhagic dengue, infectious mononucleosis, viral and bacterial meningitis, and encephalitis are common in Venezuela.[6] Cases of potential human ehrlichiosis were previously reported in Venezuela: ehrlichial inclusion in a sick child who is seropositive to *E. chaffeensis* antigen and ehrlichial inclusions in platelets from HIV patients.[7,8] We postulate that some of these patients may be infected with *E. canis* VHE strain, which is treatable with early diagnosis. In the present study, we analyzed blood specimens from human patients with clinical signs suggestive of human ehrlichiosis for *E. canis* infection in Lara State, Venezuela.

MATERIALS AND METHODS

Patients who were admitted to the emergency care facility at the Central Hospital in Barquisimeto, Lara State, Venezuela, during spring of 2002 were screened for clinical signs suggestive of human ehrlichiosis (fever of >39°C, rash, headache, malaise, myalgia, arthralgia, and cytopenia). A total of 20 (45% female and 55% male) patients with ages ranging from 13 to 38 years old (median age, 24.5 years old) consented to having their blood tested for ehrlichiosis for the study. The study protocol was approved by the Institutional Review Board. Patients were asked to answer a questionnaire about possible contact with animals or tick bites. A total of 5–10 mL EDTA-anticoagulated blood specimens were collected from all 20 patients for analysis of hematological values, and *E. canis* infection using indirect fluorescent antibody test and polymerase chain reaction (PCR).[3] DNA was isolated from 200 μL of whole blood using a QIAamp blood kit (Qiagen Inc, Valencia, CA), according to the manufacturer's instructions. Nested PCR was performed to detect *E. canis* DNA with the primers ECC-ECB and HE3-ECA for 16S rRNA gene-specific primers.[9] Amplified PCR products were cloned. Each insert was sequenced by the dideoxy termination method using the ABI Prism BigDye terminator v3.0 cycle sequencing reaction kit and an ABI 310 or 3730 sequencer. Sequence alignments and analysis were performed using MegAlign, and MapDraw programs (DNAStar, Inc. Madison, WI).

RESULTS

Six (30%) of 20 specimens were *E.* canis 16S rRNA gene PCR positive showing a typical ~400 base pairs (bp) band after agarose gel electrophoresis (FIG. 1). *E. canis* Oklahoma strain DNA was used as positive control in the PCR. Base sequences of PCR products from six patients (GenBank No. DQ003032) had the identical sequence including the single base mutation at base position 199 as that of *E. canis* VHE strain (GenBank No. AF373612). As reported previously,[3,9] this base mutation is not found in the US strains.

Indirect fluorescent antibody tests of the acute phase serum specimens revealed 1: 80 titer against *E. canis* antigen in patient #27, but patients #2, #12, #15, and #18 were negative (<1:20). Serum samples were not available from patient #6. All six *E. canis*-PCR positive human patients had fever (>39°C) of 2–6 days of duration. In addition, five of six patients had headache and/or myalgia. Malaise, nausea, vomitting, arthralgia, rash, bone pain, diarrhea, and/or abdominal pain also were recognized in the five patients (TABLE 1). Hematological tests could be performed for five of the six PCR-positive patients. Three of five *E. canis*-infected patients exhibited hematological abnormalities. One patient showed leukopenia (4×10^9 leukocytes/L; normal range $5–10 \times 10^9$/L) and thrombocytopenia (78×10^9 platelets/L; normal range $150–400 \times 10^9$ </L), one patient had leukopenia (3.9×109 total leukocytes/L), and one patient had anemia (hemoglobin level 9.8 g/dL and hematocrit level 30%, normal range 12–17 g/dL and 35–46%, respectively) (TABLE 2). Clinical signs, hematological parameters, and the age distribution of 14 *E. canis*-PCR negative patients were very similar to those of the PCR-positive patients (data not shown).

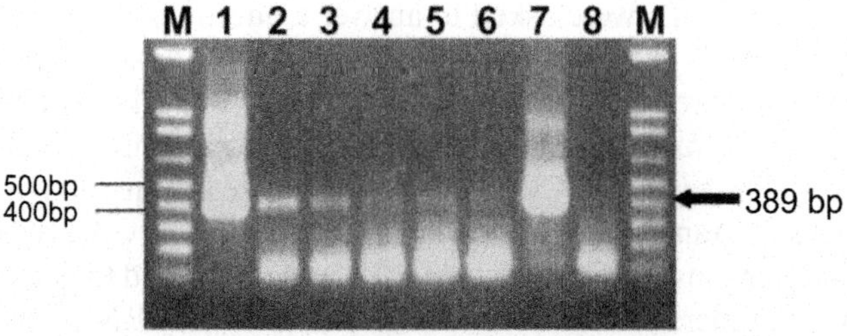

FIGURE 1. Agarose gel electrophoresis of *Ehrlichia canis* PCR products amplified from total DNA of patient blood specimens M: molecular Marker Lanes 1, 2, 3, 4, 5, and 6 represent patient 2, 12, 15, 16, 18 and 27, respectively. Lane 7: positive control (*E. canis* Oklahoma strain) Lane 8: negative control (Water) The arrow indicates the amplification of the 389 bp fragment of 16S rRNA gene of *E. canis*.

TABLE 1. Sex, age, and clinical signs of *E. canis*-infected patients

			Clinical signs									
Patient ID[a]	Sex	Age[b]	Fever	Headache	Myalgia	Malaise	Nausea	Vomiting	Bone Pain	Arthralgia	Cutaneous rash		Abdominal Pain
2	F[c]	16	X	X		X		X					
12	F	26	X	X			X		X				
15	M[d]	18	X		X					X	X	X	
16	M	38	X										
18	F	26	X		X							X	X
27	M	24	X	X	X			X					

[a]Patient identification; [b]Years old; [c]Female; [d]Male.

TABLE 2. Hematological findings of *E. canis*–infected patients

Patient ID[a]	Hemoglobin (g/dL)	Hematocrit (%)	Thrombocytes ($\times 10^9$/L)	Leukocytes ($\times 10^9$/L)
2	12.5	34	204	3.9
12	ND[b]	ND	ND	ND
15	14.3	42	78	4.0
16	15.0	46	152	7.6
18	9.8	30	255	7.9
27	14.2	42	202	7.0

[a]Patient identification; [b]Not determined.

DISCUSSION

Human ehrlichiosis and human anaplasmosis are zoonoses that are transmitted from infected animals to humans via infected tick bite. Recently, *Ehrlichia ewingii,* once thought to be an exclusive canine pathogen was discovered to cause human granulocytic ehrlichiosis, although the agent has not been isolated from patients.[10] In addition, *Anaplasma phagocytophilum,* previously known as an equine pathogen (*Ehrlichia equi*) and *Anaplasma phagocytophila* (a ruminant pathogen) are emerging human pathogens.[11,12] Considering actual culture isolation of infectious *E. canis* from the human without severe clinical signs,[3] the present study suggests *E. canis* can cause both asymptomatic and symptomatic human infection. Our observation shows that human infection with *E. canis* is not an isolated event in Lara, Venezuela. The PCR test is more appropriate for the present study, since an immunofluorescence assay (IFA) test is genus-specific, but cannot distinguish among *E. canis, E. ewingii,* and *E. chaffeensis.*[13] In addition acute-stage sera are often low titer or negative on IFA, and convalescent sera are not easily obtainable, because recovered patients rarely return to the hospital. PCR tests were independently performed in Venezuela and in the United States. Both laboratories had not been working with the VHE strain for numbers of years. All negative controls were negative. The Ohio laboratory also made sure that these are human blood specimens by sequencing human-specific DNA. Therefore, PCR cross-contamination is a remote possibility. Unlike *Ehrlichia chaffeensis,* genomic sequences of *E. canis* strains from various geographic regions appear to be highly conserved based on sequences of *p30* homologues and *virB9.*[14–16] Currently available *E. canis gltA* sequences from United States, Spain, and Italy (GenBank No. ay647155, ay615901, af304143) are also identical. Currently two *groEL* sequences are available: one is from the Florida strain and other is from the Jake strain (GenBank No. u96731, NZ˙AAEJ01000001) and they were identical. In the present study we have actually sequenced three additional *p30* paralogs of the VHE strain (*p30-2*, *p30-1*, *p30-17*, GenBank No. AY872188, AY872189, AY872190, respectively), and their sequences were identical as those of the Oklahoma strain. The results are in agreement with sequences of *p30* paralogs of the VHE strain by Yu *et al.*[17] Therefore, contrary to the general belief, so far, 16S rRNA gene is the best tool to distinguish various strains of *E. canis*.[9]

Brown dog tick *Riphicephalus sanguineus* is known to be primarily responsible for transmission of *E. canis* among dogs in North America.[18,19] *R. sanguineus* ticks have a world-wide distribution.[20] Human infection with *E. canis* has not been seriously investigated, because *R. sanguineus* rarely bites humans in North America.[21] However, human infestation with brown dog ticks has been reported in the Mediterranean region and in Central and South America.[20,22,23] In previous studies, we showed that dogs and *R. sanguineus* are infected with *E. canis* VHE strain in Venezuela, and suggested the potential transmission of *E. canis* VHE strain from dogs to humans.[9] In South America, *Ehrlichia*-PCR testing has been rarely performed. However, in Mexico, Argentina, Chile, Venezuela, and Brazil, human ehrlichial infections were reported based on serologic data using *E. chaffeensis* as antigen.[7,24–27] Because *E. chaffeensis* serologically cross-reacts with *E. canis* and *E. ewingii,*[13] it is possible that these patients are actually infected with the latter agents rather than with *E. chaffeensis*.

Venezuela is a tropical country with a large abundance of ticks and a high level of *E. canis* infection among dogs.[3,28] Three of the patients in this study had dogs at home, and two of the patients worked in rural areas (farmer or field soldier deployed to the wilderness). These results support the zoonotic potential of *E. canis*. None of these six *E. canis*-infected patients in the present study were immunosuppressed, or had underlying diseases. All of them were young, with age ranging from 16 to 38 years old (median age, 25 years old), which is in contrast to the high median age of HME patients in the United States (53 years old).[29]

Febrile episodes with unknown etiology are seen frequently in this tropical region; therefore, this is an important finding that provides strong justification for including *E. canis* infection among differential diagnosis for the febrile illnesses in this region of the world. Because the disease is treatable with antibiotics, especially when the therapy is initiated early, our findings could resolve the cause of a considerable proportion of febrile illnesses with unknown etiology in this region. Further research will hopefully define the geographic distribution and risk of human infection with *E. canis*.

ACKNOWLEDGMENTS

The study is partially funded by the Ohio State University Canine Research fund, and CDCHT-UCLA and FONACIT in Venezuela.

REFERENCES

1. RISTIC M. & C.J. HOLLAND. 1993. Canine ehrlichiosis. *In* Rickettsial and Chlamydial Diseases of Domestic Animals. Z. Woldehiwet & M. Ristic, Eds.: 169–186. Pergamon Press. Oxford, UK.

2. DONATEIN, A. & F. LESTOQUARD. 1935. Existence on algerie d'une rickettsia du chien. Bull. Soc. Pathol. Exot. **28:** 418–419.
3. PEREZ, M., Y. RIKIHISA & B. WEN. 1996. *Ehrlichia canis*-like agent isolated from a man in Venezuela: antigenic and genetic characterization. J. Clin. Microbiol. **34:** 2133–2139.
4. PADDOCK, C.D. & J.E. CHILDS. 2003. *Ehrlichia chaffeensis:* a prototypical emerging pathogen. Clin. Microbiol. Rev. **16:** 37–64.
5. SEXTON, D.J. & K.S. KAYE. 2002. Rocky Mountain spotted fever. Med. Clin. N. Am. **86:** 351–360.
6. COSTA DE LEONA, L., J. ESTEVEZ, F. MONSALVE DE CASTILLO, *et al.* 2004. Laboratory diagnosis of patients with exanthematic or febrile syndromes occurring in the Zulia State, Venezuela, during 1998 [in Spanish]. Rev. Med. Chil. **132:** 1078–1084.
7. ARRAGA-ALVARADO, C., M. MONTERO-OJEDA, A. BERNARDONI, *et al.* 1996. Human ehrlichiosis: report of the 1st case in Venezuela. Invest. Clin. **37:** 35–49.
8. DE TAMI I DEL, C., I.M. TAMI-MAURY. 2004. Morphologic identification of *Ehrlichia* sp. in the platelets of patients infected with the human immunodeficiency virus in Venezuela [in Spanish]. Rev. Panam. Salud. Publica. **16:** 345–349.
9. UNVER, A, M. PEREZ, N. ORELLANA, *et al.* 2001. Molecular and antigenic comparison of *Ehrlichia canis* from dogs, ticks, and a human in Venezuela. J. Clin. Microbiol. **39:** 2788–2793.
10. BULLER, R.S., M. ARENS, S.P. HMIEL, *et al.* 1999. *Ehrlichia ewingii*, a newly recognized agent of human ehrlichiosis. N. Engl. J. Med. **341:** 148–155.
11. GOODMAN, J.L., C. NELSON, C. VITALE, *et al.* 1996. Direct cultivation of the causative agent of human granulocytic ehrlichiosis. N. Engl. J. Med. **334:** 209–215.
12. DUMLER, J.S., A.F. BARBET, C.P.J. BEKKER, *et al.* 2001. Reorganization of genera in the families *Rickettsiaceae* and *Anaplasmataceae* in the order *Rickettsiales*; unification of some species of *Ehrlichia* with *Anaplasma*, *Cowdria* with *Ehrlichia*, and *Ehrlichia* with *Neorickettsia*; description of six new species combinations; and designation of *Ehrlichia equi* and "HGE agent" as subjective synonyms of *Ehrlichia phagocytophilum*. Int. J. Syst. Evol. Microbiol. **51:** 2145–2165.
13. RIKIHISA, Y., S.A. EWING & J.C. FOX. 1994. Western blot analysis of *Ehrlichia chaffeensis*, *E. canis* or *E. ewingii* infection of dogs and human. J. Clin. Microbiol. **32:** 2107–2112.
14. FELEK, S., R. GREENE & Y. RIKIHISA. 2003. Transcriptional analysis of *p30* major outer membrane protein genes of *Ehrlichia canis* in naturally infected ticks and sequence analysis of *p30-10* of *E. canis* from diverse geographic regions. J. Clin. Microbiol. **41:** 886–888.
15. FELEK, S., H. HUANG & Y. RIKIHISA. 2003. Sequence and expression analysis of virB9 of the type IV secretion system of *Ehrlichia canis* strains in ticks, dogs, and cultured cells. Infect. Immun. **71:** 6063–6067.
16. MCBRIDE, J.W., X.J. YU & D.H. WALKER. 2000. A conserved, transcriptionally active p28 multigene locus of *Ehrlichia canis*. Gene **254:** 245–252.
17. YU, X-J., J.W. MCBRIDE & D.H. WALKER.1999. Charcterizaton of the genus-common outer membrane proteins in *Ehrlichia*. *In* Rickettsiae and Rickettsial Diseases at the Turn of the Third Millenium, D. Raoult & P. Brouqui Eds.:103–107 Elsevior. Paris.

18. HUA, P., M. YUHAI, T. SHIDE, *et al.* 2000. Canine ehrlichiosis caused simultaneously by *Ehrlichia canis* and *Ehrlichia platys*. Microbiol. Immunol. **44:** 737–739.
19. LEWIS, G.E., M. RISTIC, R.D. SMITH, *et al.* 2000. The brown dog tick *Rhiphicephalus sanguineus* and the dog as experimental hosts of *Ehrlichia canis*. Am. J. Vet. Res. **38:** 1953–1955.
20. GODDARD, J. 1989. Focus of human parasitism by the brown dog tick, *Rhiphicephalus sanguineus* (Acari:Ixodidae). J. Med. Entomol. **26:** 628–629.
21. MURPHY, G.L., S.A. EWING, L.C. WHITWORTH, *et al.* 1998. A molecular and serologic survey of *Ehrlichia canis*, *E. chaffeensis*, and *E. ewingii* in dogs and ticks from Oklahoma. Vet. Parasitol. **79:** 325–339.
22. GULIELMONE, A.A., A.J.Y. MANGOLD & A.E. VIÑABAL. 1991. Ticks (Ixodidae) parasitizing humans in four provinces of Northwestern Argentina. Ann. Trop. Med. Parasitol. **85:** 539–542.
23. LIMA, V.L.C. DE., A.C. FIGUEIREDO, M.G. PIGNATTI y. M. MODOLO. 1995. Febre maculosa no municipio de Pedreira-estado de Sao Paulo-Brasil: relación entre ocorrencia de casos e parasitismo humano ixodídeos. Rev. Soc. Bras. Med. Trop. **28:** 135–137.
24. GONGORA-BIACHI, R.A., J. ZAVALA-VELAZQUEZ, C.J. CASTRO-SANSORES & P. GONZALEZ-MARTINEZ. 1999. First case of human ehrlichiosis in Mexico. Emerg. Infect. Dis. **5:** 481.
25. RIPOLL, C. M., C. E. REMONDEGUI, G. ORDONEZ, *et al.* 1999. Evidence of rickettsial spotted fever and ehrlichial infections in a subtropical territory of Jujuy, Argentina. Am. J. Trop. Med. Hyg. **61:** 350–354.
26. LOPEZ, J., M. RIVERA, J.C. CONCHA, *et al.* 2003. Serologic evidence for human ehrlichiosis in Chile [in Spanish]. Rev. Med. Chil. **131:** 67–70.
27. CALIC S.B., M.A. GALVAO, F. BACELLAR, *et al.* 2004. Human ehrlichioses in Brazil: first suspect cases. Braz. J. Infect. Dis. **8:** 259–262.
28. ARRAGA- ALVARADO, C. 1992. Ehrlichiosis canina en Maracaibo, Estado Zulia—Venezuela: reporte de 55 casos. Revista. Científica Universidad del Zulia. **2:** 41–52.
29. GARDNER, S.L., R.C. HOLMAN, J.W. KREBS, *et al.* 2003. National surveillance for human ehrlichiosis in the United States, 1997–2001, and proposed methods for evaluation of data quality. Ann. N.Y. Acad. Sci. **990:** 80–89.

Human Monocytic Ehrlichiosis and Human Granulocytic Anaplasmosis in the United States, 2001–2002

LINDA J. DEMMA, R.C. HOLMAN, J.H. McQUISTON, J.W. KREBS, AND D.L. SWERDLOW

Viral and Rickettsial Zoonoses Branch, Division of Viral and Rickettsial Diseases, Centers for Disease Control and Prevention, Atlanta, Georgia, USA

ABSTRACT: The epidemiologic features are described of cases of human monocytic ehrlichiosis and human granulocytic anaplasmosis in the United States.

KEYWORDS: ehrlichiosis; anaplasmosis; human monocytic ehrlichiosis; human granulocytic anaplasmosis; tick-borne diseases

INTRODUCTION

Human monocytic ehrlichiosis (HME) and human granulocytic anaplasmosis (HGA) are zoonotic tick-borne diseases. In the United States, HME is caused by *Ehrlichia chaffeensis* and HGA by *Anaplasma phagocytophilum*. This report describes incidence trends for these diseases and provides the first national surveillance summary that includes supplemental epidemiologic and clinical data.

METHODS

HME and HGA cases were analyzed to capture epidemiologic and clinical information for the year 2001–2002, the period in which these data were nationally notifiable.[1–4] The study used data provided by state health departments via the National Electronic Telecommunications System for Surveillance (NETSS), which provides national case counts on notifiable diseases,[4] and from supplemental standardized case report forms (CRFs), which were implemented for HME and HGA in 2001 by the Centers for Disease Control and Prevention (CDC).[2,3]

Address for correspondence: Linda J. Demma, Division of foodborne, Enteric, and Mycotic Diseases, National Center for Zoonotic, Vectorborne, and Enteric Diseases, CDC, Atlanta, Georgia 30333. Voice: 404-639-3343; fax: 404-639-3535.
e-mail: ldemma@cdc.gov

**Ann. N.Y. Acad. Sci. 1078: 118–119 (2006). © 2006 New York Academy of Sciences.
doi: 10.1196/annals.1374.017**

RESULTS

During the year 2001–2002, 32 states reported a total of 1,176 cases of HME and HGA to the CDC through NETSS. This represented 287 (32%) more cases than reported in 1999–2000.[5,6] The average annual incidences for HME and HGA were 0.6 and 1.4 cases per million population, respectively, and the incidence was highest among men >60 years of age. Cases of HME primarily occurred in the eastern and south-central regions, and HGA occurred in northeastern and upper midwestern regions, corresponding to the prevalence of tick vectors. A total of 883 cases of HME and HGA were reported through CRFs. We found that 42% of HME and 33% of HGA patients were hospitalized with life-threatening complications; 12% of HME and 6% of HGA cases had preexisting immunosuppressive conditions, and case fatality rates were 3.0% for HME and 0.34% for HGA.

CONCLUSIONS

We report an increase in HME and HGA incidence in the United States, a finding that may be related to enhanced awareness and reporting and to changes in human–vector interaction. Our data indicate that HME and HGA may be significant opportunistic infections and should be strongly considered by physicians when treating patients with a clinically compatible illness and possible tick exposure. Continued and improved surveillance activities will progressively reinforce our understanding and awareness of these zoonotic infections and guide control measures.

REFERENCES

1. DEMMA, L.J., *et al.* 2005. Epidemiology of human ehrlichiosis and anaplasmosis in the United States, 2001–2002. Am. J. Trap. Med. Hyg. **73:** 400–409.
2. CDC. 2000. Ehrlichiosis (HGE, HME, other or unspecified): 2000 case definition. CDC: Division of Public Health Surveillance.Atlanta, GA.
3. CDC, 1997. Case definitions for infectious conditions under public health surveillance. Morbidity and Mortality Weekly Report. CDC. Atlanta, GA, **RR10:** 46–47.
4. CDC, 1990. National Electronic Telecommunications System for Surveillance—United States. Morb. Mortal. Wkly. Rep. **40:** 502–503.
5. GARDNER, S.L., R.C. HOLMAN, J.W. KREBS, *et al.* 2003. National surveillance for the human ehrlichioses in the United States, 1997–2001, and proposed methods for evaluation of data quality. Ann. N.Y. Acad. Sci. **990:** 80–89.
6. MCQUISTON, J.H., C.D. PADDOCK, R.C. HOLMAN & J.E. CHILDS. 1999. The human ehrlichioses in the United States. Emerg. Infect. Dis. **5:** 635–642.

Anthropogenic Effects on Changing Q Fever Epidemiology in Russia

N.K. TOKAREVICH,[a] O.A. FREILYKHMAN,[a] N.M. TITOVA,[b] I.R. ZHELTAKOVA,[b] N.A. RIBAKOVA,[c] AND E.V. VOROBEYCHIKOV[d]

[a]*St. Petersburg Pasteur Institute, St. Petersburg, Russia*

[b]*Surveillance Centers, Leningrad, Russia*

[c]*Surveillance Centers, Vologda, Russia*

[d]*State Research Center of Highly Pure Biopreparations, St. Petersberg, Russia*

ABSTRACT: In the northwestern region of Russia (Leningrad province) cattle is proved to be the main source of *C. burnetii* infection in humans, both in menaced professionals and in formally nonmenaced groups. Liquidation of specialized cattle-breeding complexes (with their well-organized veterinary surveillance) and broadening of the circle of non-professionals that contact with agriculture or domestic animals infected with *C. burnetii* provide the prerequisites to Q fever spreading among various groups of population.

KEYWORDS: Q fever; *C. burnetii* infection; epidemiology; anthropogenic effects

INTRODUCTION

Production activity is a major factor that has an adverse influence on global environment and the health of the population. The recent deep social and economic transformation that has occurred in Russia makes it necessary to objectively assess its influence on the spreading of zoonotic infections, including Q fever, in humans and animals. This article aims to assess of anthropogenic effects on the prevalence and activity of agricultural Q fever foci and to show the peculiarities of infections epidemiology currently typical for Russia.

MATERIALS AND METHODS

From 1985 to 2004 in the northwestern region of Russia we studied blood sera of humans and livestock for the presence of antibodies to *Coxiella (C.)*

Address for correspondence: Nikolay K. Tokarevich, St. Petersburg Pasteur Institute, 14 Mira Str., St. Petersburg 197101, Russia. Voice: 812-2322136; fax: 812-2329217.
e-mail: zoonoses@mail.ru

**Ann. N.Y. Acad. Sci. 1078: 120–123 (2006). © 2006 New York Academy of Sciences.
doi: 10.1196/annals.1374.018**

burnetii; the sera samples were taken from various professional groups of healthy population (23,044), cattle (9,552), sheep (792), and domestic dogs (139). We studied those human and animal sera samples using standard techniques both of complement fixation and indirect fluorescent antibody tests.[1] Besides, we analyzed the published data on Q fever human morbidity in Russia and the results of *C. burnetii* seroprevalence research both in humans and in agricultural animals in some Russian regions over the same period. Statistical treatment of the results was conducted with the help of correlation and variance analysis.[2]

RESULTS

Livestock was found to be infected with *C. burnetii* in most districts of the northwestern region. The infection rate in animals depended on their maintenance conditions and varied within different time periods. Therefore, in the 1980s, when cattle were bred in large farming complexes without grazing in natural meadows, *C. burnetii* seroprevalence was much lower than in freely grazing animals (0.3 ± 0.1% and 3.3 ± 0.2%, respectively). Rearrangement of those large-scale collective farms and the creation of small private farms did not cause seroprevalence reduction in agricultural animals. By the end of the 1990s antibodies to *C. burnetii* were revealed in 2.2 ± 0.2% and 2.4 ± 0.3% of the cattle in Vologda and Leningrad provinces, respectively, and in Volgograd province the sheep infection rate was found to be 5.7 ± 0.8%.

Very similar *C. burnetii* seroprevalence rates were established for some other Russian regions: therefore, in the Southern Federal Region, antibodies to *C. burnetii* were revealed in 3.3 ± 1.2% of cattle in Volgograd province,[3] and in 3.2 ± 1.2% of cattle and 4.2 ± 1.5% of sheep in Rostov province.[4]

We analyzed the data on *C. burnetii* seroprevalence for various professional groups of the local population and for livestock in seven districts of Leningrad province. An explicit correlation is revealed between *C. burnetii* prevalence in humans (professionally at risk ("menaced") or not) and infection rate in local cattle with this pathogen ($r = 0.94$, $P = 0.0007$ and $r = 0.82$, $P = 0.0013$, respectively).

We found that cattle infected with *C. burnetii* were the major source of the pathogen infection in humans, both in menaced professionals (94.27%; $F = 104.5$, $P = 0.0001$), and those that were formally nonmenaced (84.12%; $F = 51.85$, $P = 0.0002$). Therefore, all other sources of *C. burnetii* infection contribute only 5.73% for menaced professionals and 15.88% for the remaining population.

In the 1980s and early 1990s in the northwestern region antibodies to *C. burnetii* were revealed more often in professionally menaced groups than in the remaining population (4.7 ± 0.8% and 3.1 ± 0.2%, respectively).

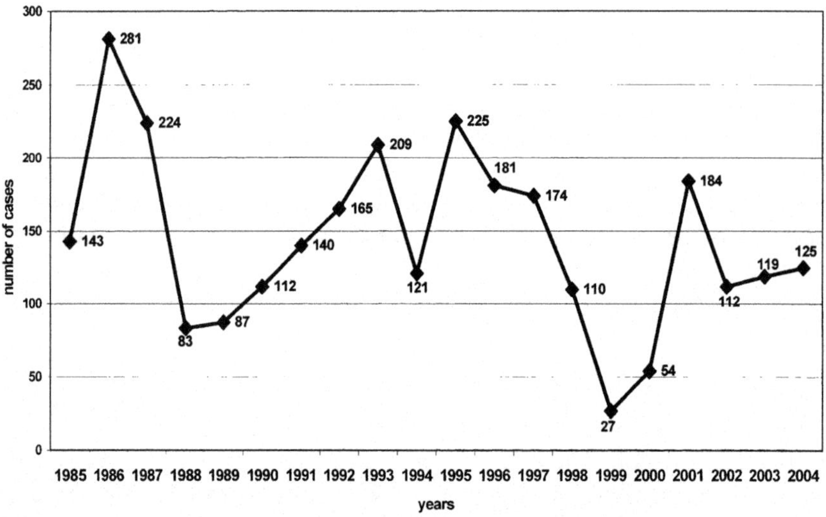

FIGURE 1. Q fever incidence in Russia (from 1985 to 2004).

However, in the next period the ratio was overturned and the figures changed to 3.2 ± 0.3% and 4.9 ± 0.9%, respectively.

It is probable that the tendency we found in the northwestern region is also valid for other Russian regions. For instance, within 10 years the frequency of the occurrence of antibodies to *C. burnetii* in the healthy population of Russia increased from 1.1 ± 0.1% (1993) and 1.8 ± 0.2% (1994) to as high as 16 ± 0.6% (2002) or 11.1 ± 0.6% (2003).

Within these 10 years in Russia there was a reduction in the number of livestock and consequently a reduction in the data evidence for the growth of *C. burnetii* infection rate in humans. Our opinion is that the paradox may be attributed to some degree to the liquidation of large specialized livestock farms with their well-established system of veterinary monitoring, and the creation of small private farms allowing agricultural animals to graze in Q fever-suspected territories, to the lack of veterinary monitoring of Q fever at those small farms, to uncontrolled migration of cattle from *C. burnetii*-infected farms, and to considerable growth in the number of nonprofessional contacts of population with agricultural animals. Furthermore, in recent years, there has been a dramatic rise in domestic dogs involved in *C. burnetii* circulation, particularly in cities. For example, in St. Petersburg *C. burnetii* antibodies were found in 2.6 ± 1.3% of the blood sera samples taken from sick dogs with diseases of unknown etiology.

Within the last 20 years, according to official Russian statistics, there were 2,876 patients with Q fever and the mean annual incidence was about 1 per million of population. The tendency for a decrease in the number of registered Q fever cases (see FIG. 1) does not reflect, in our opinion, the real prevalence, and results from its considerable hypodiagnostics. It is also evidenced by more frequent increase of antibodies to *C. burnetii* in the healthy population of Russia compared that of with the earlier period (10 years ago).

CONCLUSIONS

1. Rearrangement of agriculture enterprises and growth of the number of people that have nonprofessional contacts with agricultural or domestic animals infected with *C. burnetii* is responsible for Q fever prevalence in various groups of the population.
2. There is an obvious hypodiagnosis of Q fever in Russia.

REFERENCES

1. KRUTITSKAYA, L., N. TOKAREVICH, A. ZHEBRUN, *et al.* 1996. Autoimmune component in individuals during immune response to inactivated combined vaccine against Q fever. Acta Virol. **40:** 173–177.
2. MEDIK, V.A., M.S. TOKMACHEV & B.B. FISHMAN. 2000. Statistics in Medicine and Biology: Instructions in 2 Volumes M. Komarov Ed. **1:** 412. Theoretical Statistics Med. Russia.
3. TIHONOV, N.G.,T.P. PASHANINA, V.F. IGNATOVICH, *et al.* 1999. Q fever and spotted fever in Volgograd province [in Russian; summary in English]. Epidemiol. Infect. Dis. **4:** 12–15.
4. ZIKOVA, T.A. 2004. Epidemiological peculiarities of *C. burnetii* infection in Rostov province. Proceedings of International Conference on Modern Means for Immunodiagnostics, Immunoprophylaxis and Emergency Prevention of Actual Infections. P. Ogarkov & Y.U. Ivannikov Eds.: 61–62. St. Petersburg, Russia.

Human *Coxiella burnetii* Infections in Regions of Bosnia and Herzegovina, 2002

ZVIZDIĆ ŠUKRIJA, SADETA HAMZIĆ, DŽEVAD ČENGIĆ, EDINA BEŠLAGIĆ, NIHAD FEJZIĆ, DARKO ČOBANOV, JASMINKA MAGLAJLIĆ, SANDRA PUVAČIĆ, AND ZLATKO PUVAČIĆ

Medical Faculty University of Sarajevo, 71000 Sarajevo, Bosnia and Herzegovina

ABSTRACT: Acute infections in humans and animals caused by *Coxiella burnetii (C. burnetii)* are becoming an important medical problem for Bosnia and Herzegovina (B&H). From a clinical and epidemiological aspect, Q fever represents a complex medical problem, considering that one of the highest incidence rates of Q fever in Europe has been recorded during the last few years in B&H. The first case of this disease in B&H was described in 1950, by Muray *et al.*, and the first epidemic, with 16 infected individuals, was recorded the same year. Confirmed animal infections by *C. burnetii* in B&H were first reported in 1985 when, of all tested sheep, positive results were found in 12.4%. During 2001, 2.11% of tested sheep and goats were found to have a positive result, which was also confirmed by studies from the following years in particular regions of B&H. These studies suggest that endemic loci of infected animals are established in particular geographic regions in B&H, which is important to emphasize for better understanding of the sources and routes of *C. burnetii* transmission to the human population. This conclusion is based on the studies from 2000, when 2.17% of positive cattle, 1.85% of positive sheep, and 0.27% of positive goats were registered. During the same period, in B&H, in 6 different regions, 156 individuals with Q fever were registered as were 3 separate epidemics with 115 infected individuals. Official data on the number of detected animal *C. burnetii* infections during 2002 suggest that 10 positive cattle and 88 positive sheep or goats were registered. During 2003, 24 positive cattle, 29 positive goats, and 167 positive sheep were detected, while in 2004, 71 positive cattle, 4 positive goats, 37 positive sheep, and 72 positive animals from the sheep–goat group were registered. According to official reports from 2001, 19 individuals with Q fever were registered in B&H, while in 2002, the number of infected individuals increased to 250. In five cantons in B&H, 43 infected individuals were registered during 2002, while in Republika Srpska of B&H, 207 infected individuals in the region of Banja Luka were registered. From 1998 to 2003, 373 individuals with Q fever were reported in

Address for correspondence: Zvizdić Šukrija, Medical Faculty University of Sarajevo, Cekalusa 90, 71000 Sarajevo, Bosnia and Herzegovina. Voice/fax: +387-33-202-050.
e-mail: sukrijazvizdic@yahoo.com

**Ann. N.Y. Acad. Sci. 1078: 124–128 (2006). © 2006 New York Academy of Sciences.
doi: 10.1196/annals.1374.128**

B&H, whereof 265 individuals (71.04%) were infected during epidemics, and 108 (28.95%) sporadically. Q fever incidence rates in B&H were high during 1998 (5.68%ooo) and very high in 2000, with 115 individuals with an acute clinical form and an incidence rate of 6.95%ooo. The incubation time varied between 9 and 28 days.

KEYWORDS: *Coxiella burnetii*; human and animal infections; Q fever; Bosnia and Herzegovina, 2002

INTRODUCTION

Q fever is a zoonosis caused by rickettsia *Coxiella burnetii (C. burnetii)*. In wider areas of our region, this disease was first reported in the 1950s, while the first epidemic was reported in 1949 in Serbia. The first case of this disease in the federation of Bosnia and Herzegovina (B&H) was described in 1950 by Muray *et al.*, and the first epidemic, with 16 infected individuals, was recorded the same year. By 1991, individual cases of Q fever had been registered, both in animals and humans. Since 1998, an increased number of cases , as well as the presence of significant number of animal reactors, has been reported. Human infections were reported as sporadic cases or as smaller or larger epidemics. A significant number of infected individuals were registered in 1998 and 2000 in central Bosnia, in the region of Kakanj Municipality, with 86 and 56 infected individuals. Confirmed animal infections with *C. burnetii* in B&H were first reported in 1985, when positive results were found in 12.4%, of all tested sheep.[1] During the studies on the presence of this infection in domestic animals in 2001, Bajrović and Velić *et al.* detected its presence in 2.11% of tested sheep and goats, while Mačkić, testing the dogs for *C. burnetii* infection, detected their contamination in 3.22%.[2-4] These and subsequent studies indicate persistent presence of *C. burnetii*–infected animal loci in particular regions of B&H, which is important for better understanding of the sources and routes of *C. burnetii* transmission to the human population. This conclusion is based on studies from 2000, when 2.17% of positive cattle, 1.85% of positive sheep, and 0.27% of positive goats were registered. During the same period, in B&H, in six different regions, 156 individuals with Q fever were reported as were as 3 separate epidemics with 115 infected individuals. Official data on a number of detected animal *C. burnetii* infections during the year 2002, suggest that 10 positive cattle and 88 positive animals from the sheep–goat group were registered. In 2003, 24 positive cattle, 29 positive goats, and 167 positive sheep were detected, whereas in 2004, 71 positive cattle, 4 positive goats, 37 positive sheep, and 72 positive animals from the sheep–goat group were registered.[5]

In particular regions of B&H, endemic presence of *C. burnetii* was evident during the 1990s. This conclusion is based on the studies that suggest that, during 2000 in B&H, a positive serological *C. burnetii* diagnosis was reported in 2.17% of tested cattle, 1.85% of sheep, and 0.27% of goats. During the same period in B&H, in six different regions, 156 individuals with Q fever were as were 3 separate epidemics with 115 infected individuals. According to official reports from 2001, 19 individuals with Q fever were registered in B&H, while in 2002 the number of infected individuals increased to 250.[6] In the five cantons of Federation B&H, 43 infected individuals were reported during 2002, while, in Republika Srpska of B&H, 207 infected individuals on the region of Banja Luka were registered. During the period from 1998 to 2003, 373 individuals with Q fever were registered in B&H: of these persons, whereof265 (71.04%) were infected during epidemies, and 108 (28.95%) were infected sporadically. Q fever incidence rates in Federation B&H were high during 1998 (5.68%ooo), and very high in 2000, with 115 individuals with an acute clinical form and the incidence rate of 6.95%ooo. The incubation time varied between 9 and 28 days.

MATERIALS AND METHODS

During the period from January to December 2002, 113 sera samples from 76 individuals were tested. All tested individuals were suspected of having *C. burnetii* infections. In order to detect IgM and IgG antibodies to *C. burnetii* antigens, indirect immunofluorescence antibody Assay (IFA–Focus Technologies Cypress, CA) was used. The test is based on the "sandwich" method, and performed in two phases. The results were interpreted as follows:

- Q fever seropositive sera are those with detected presence of specific anti–*C. Burnetii* antibodies, in dilution of 1:64 and more, for IgM and IgG antibody class.
- In acutely infected individuals, specific anti-IgM antibodies to phase II *C. burnetii* antigen are present in the serum, in dilution of 1:64 and more, which means that they showed a seroconversion, or displayed quadruple increase of specific antibody titer. During the acute stage of disease, higher titers of IgM antibodies to phase II antigens are present, compared to the titer of IgM antibodies to phase I antigen.

RESULTS

During 2002, in five cantons of Federation B&H, 43 individuals with clinically and laboratory-confirmed Q fever diagnosis were reported. During the same period, in Republika Srpska of B&H, 207 infected individuals were registered; the overall number of infected individuals for 2002 in B&H was 250. The greatest numbers of infected individuals were registered in three cantons of Federation B&H. Nine individuals were registered in Una-Sana Canton, 8 in Zenica-Doboj Canton, and 24 in the city of Livno (Canton 10), during the only registered epidemic in 2002.

In the microbiology laboratory of the medical faculty of the University in Sarajevo in 2002, most of the analyzed and tested blood and serum samples were collected from Sarajevo, 29 or 38.1%, from Bihać, 14 or 18.4% of sera, and from Zenica, 9 or 11.7%. The testing confirmed a seropositivity in 8 (30.7%) individuals from Bihać, in 6 (23.0%) individuals from Sarajevo, and in 3 (11.4%) individuals from Zenica. This suggests that of all 76 tested individuals, *C. burnetii* infection was confirmed in 26 (34.2%), which means that they suffered from one of the possible clinical forms of Q fever. Of 26 seropositive individuals, 23 of them (88.5%) were males and 3 (11.5%) females. The youngest seropositive individual was 11 years old, and the oldest one was 59. The largest number of seropositive individuals belonged to the age group of 31–40 years. In this group of tested individuals, suspected Q fever was confirmed in 12 (46.15%) individuals who were found to have *C. burnetii* infection and seropositivity. Of 76 individuals, 113 sera were tested and analyzed. Of all tested sera, anti-*C. burnetii*-specific antibodies were detected in 50 (44.2%) sera. Two serum samples were tested in 31 individuals, the third serum was tested in 4 individuals, and four and five sera were tested in one individual. Of 26 seropositive individuals, the first tested sera were positive in 24 of them, and we confirmed a seroconversion in two individuals, analyzing their 2 serum samples. Of all 26 Q-positive individuals, an acute disease form was confirmed in 25 individuals (phase II IgM and IgG were higher than phase I), and serological results indicated reconvalescent form in 1 individual (phase I IgG antibody titer of 1:256 was equal to phase II IgG antibody titer, with the negative result for IgM antibodies). In the acute stage of disease, the lowest titer of positive IgM antibody result was 1:64, while the highest was 1:1024. The most frequent result of positive IgM antibody titer varied between 1:256 and 1:512. In individuals with confirmed seroconversion, phase II IgG antibody titer in the second serum was 1:128/256.

CONCLUSIONS

- On the basis of available literature and our own studies, we can conclude that Q fever has become an endemic disease in B&H.
- In the period before the 1990s, *C. burnetii* caused sporadic cases of human infections, and in the last several years it has been transformed to an endemic phenomenon.
- During 2002, Q fever occurred in sporadic forms on wider areas all over B&H.
- In B&H, 250 infected individuals were reported during 2002, of whom 43 (17.2%) were from the Federation and 207 (82.8%) in RS B&H.
- The cases of Q fever epidemies were registered in two regions of B&H (areas of Banja Luka and Livno).

REFERENCES

1. Nevjestić, A., L.J. Rukavina, M. Sabirović & V. Cvelić-Čabrilo. 1985. Ispitivanje rasprostranjenosti Q groznice kod ovaca u SR Bosni i Hercegovini. Vet Glasnik 885–889.
2. Bajrović, T. 2001. Učestalost pojedinih zaraznih bolesti u BiH. Seminar, Akutna epidemiološka situacija zaraznih bolesti u BiH.
3. Velić, R., T. Bajrović, L. Arapović, *et al.* 2001. Seroprevalenca Q-groznice kod preživara na širem području Bosne i Hercegovine tokom 2000. godine. Veterinarija **49:** 137–146.
4. Mačkić, S. 2002. Ispitivanje Q groznice kod pasa na području Federacije BiH. Magistarski rad, Veterinarski fakultet Sarajevo [magisterial project].
5. Izvješće, O. pojavi zaraznih bolesti u Bosni i Hercegovini, Ministarstvo vanjske trgovine iekonomskih odnosa Bosne i Hercegovine, Ured za veterinarstvo Bosne i Hercegovine, 2002, 2003, 2004.
6. Annual Communicable Diseases Report BiH for 2001, 2002. HQ/SFOR CJMED.

Gipuzkoa, Basque Country, Spain (1984–2004): A Hyperendemic Area of Q Fever

M. MONTES, G. CILLA, D. VICENTE, V. NIETO, M. ERCIBENGOA, AND E. PEREZ-TRALLERO

Microbiology Service, Hospital Donostia, San Sebastián, Spain

ABSTRACT: Overall 1,261 cases of Q fever were diagnosed between 1984 and 2004 in Gipuzkoa (Basque Country, Spain). Most (75.5%) of the cases ocurred in subjects 15–45 years of age. A total of 79.5% of the cases (n = 1003) ocurred between January and June. The annual incidence for acute Q fever in Gipuzkoa was 7.7, 15.8, 9.6, and 5.7 for the periods 1984989, 1990994, 1995999, and 2000004, respectively. In 94% of the cases IgM titer was $\geq$1/256. The most frequent clinical manifestation was pneumonia (79%). Only two cases of chronic Q fever were detected during the 21 years studied.

KEYWORDS: Q fever; serology; incidence; Spain

INTRODUCTION

Q fever is an endemic disease caused by *Coxiella burnetii* (*C. burnetii*), which has a worldwide distribution. *Coxiella* is an organism that is extremely resistant to external agents. The reservoirs of *C. burnetii* are mammals (mainly sheep), birds, and ticks. In humans, acute infection is usually asymptomatic, although it can sometimes lead to serious complications, and rarely produces chronic infection. The most common clinical manifestations of acute Q fever are febrile syndrome, atypical pneumonia, and hepatitis. The main presentation of chronic Q fever is endocarditis (patients with previous valvular heart disease). Diagnosis relies on serology, although molecular techniques are recently being developed.[1] This study analyzes the impact and characteristics of Q fever in Gipuzkoa over a 21-year period.

Address for correspondence: Milagrosa Montes , Phm. D., M.D., Microbiology Department, Donostia Hospital, P° Dr. Beguiristain s/n, 20014 San Sebastián, Gipuzkoa, Spain. Voice: +34-943-007046; fax: +34-943-007063.
e-mail: ludvirol@chdo.osakidetza.net

**Ann. N.Y. Acad. Sci. 1078: 129–132 (2006). © 2006 New York Academy of Sciences.
doi: 10.1196/annals.1374.020**

MATERIALS AND METHODS

Patients with Q fever diagnosed between 1984 and 2004 at the Microbiology Department of Donostia Hospital in Gipuzkoa were studied. Gipuzkoa is located in the northeastern area of the Basque Country in Spain, bordered by the Cantabric Sea and France to the north. Donostia Hospital is the public reference hospital for this geographical area. The serological diagnostic criteria used for acute Q fever (IFI, *C burnetii* phase II, Bio-Mèrieux, France) were: seroconversion of IgG and/or IgM (from negative to $\geq$1/128), a fourfold or greater increase in IgG titer in two paired sera, or IgM titer $\geq$1/256 in a single sample after removing rheumatic factor. The incidence of acute Q fever was calculated using official census data from 1991 and 2001 (total population 622,866–639,717).

RESULTS

Overall, 1,261 cases of Q fever were diagnosed (955 in men) ranging from 19 to 153 cases/year (FIG. 1). Most (75.5%) of the cases occurred in subjects aged 15–45 years and the disease was exceptional in subjects aged $<$15 years (n = 16) or $>$64 years (n = 57). A total of 79.5% of the cases (n = 1003) occurred between January and June (maximum May, with 264 cases in the 21 years of the study; minimum October, with 19 cases). Cases were distributed throughout the province. The annual incidence (cases/100,000 inhabitants) of

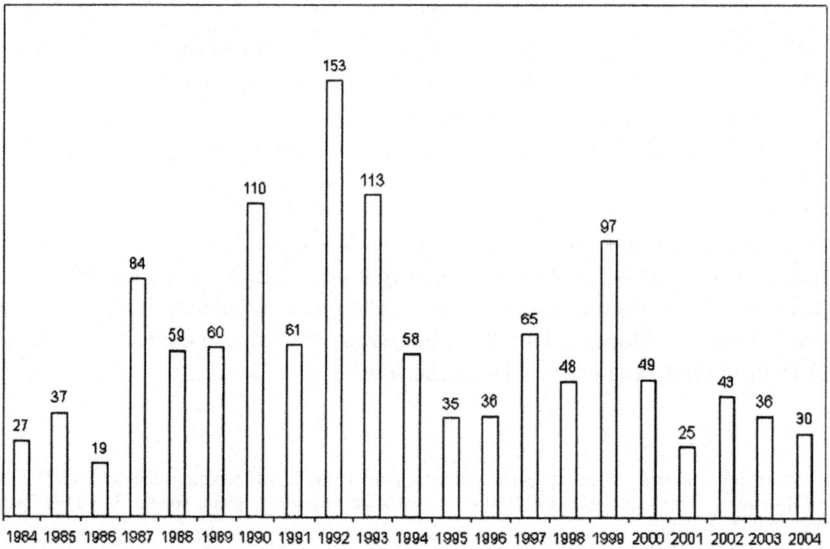

FIGURE 1. Cases of acute Q fever in Gipuzkoa.

acute Q fever in Gipuzkoa was 7.7, 15.8, 9.6, and 5.7 for the periods 1984–1989, 1990–1994, 1995–1999, and 2000–2004, respectively. Local outbreaks were detected especially in the 1990s. No outbreaks were observed after the year 2000. In 94% of the cases (1,188/1,261), IgM titer was $\geq$1/256. Among the 870 cases with a known diagnosis, the most frequent clinical manifestation was pneumonia (79%). Half of the patients were hospitalized (366/688 cases in which this datum was known) and 10 were admitted to the intensive care unit. Only two cases of chronic Q fever were detected during the 21 years studied.

DISCUSSION

Gipuzkoa is an endemic area of acute Q fever, in which 30–60 cases/year are serologically diagnosed in the absence of outbreaks. The disease occurs throughout Spain and its incidence is especially high in the Basque Country, where it is higher than in the rest of Europe and other continents.[2] Our data confirm this situation and are of special significance, considering that, in the absence of seroconversion or a fourfold rise in antibody titers, the minimum IgM titer that we considered for a diagnosis of acute Q fever was 1/256, which is more restrictive than the widely used titer $\geq$1/50 recommended by Tissot *et al.*[3] As in other geographical areas, most cases of Q fever occurred in young men and in the spring, coinciding with the lambing season, which explains the greater environmental contamination with *C. burnetii*. The most common form of presentation of acute Q fever in Gipuzkoa was atypical pneumonia (79% of cases with a known diagnosis). In other parts of Spain, such as Galicia and the Canary Islands, the most frequent form of presentation is fever with hepatitis,[4] while in Andalusia most patients present with prolonged febrile syndrome, and only one-third presents with pneumonia.[5] The low incidence of chronic Q fever in Gipuzkoa should be stressed. Only two cases of endocarditis were diagnosed during the 21 years studied, contrasting with reports from other parts of Spain,[6] and other countries, such as France[7] and England.[8] In conclusion, Gipuzkoa continues to be an endemic area of Q fever, where chronic Q fever is exceptional. However, a decreasing trend in the incidence of acute Q fever has been observed in the last few years.

REFERENCES

1. FOURNIER, P.E. & D. RAOULT. 2003. Comparison of PCR and serology assays for early diagnosis of acute Q fever. J. Clin. Microbiol. **41:** 5094–5098.
2. MAURIN, M. & D. RAOULT. 1999. Q fever. Clin. Microbiol. Rev. **12:** 518–553.
3. TISSOT DUPONT, H., X. THIRION & D. RAOULT. 1994. Q fever serology: cutoff determination for microimmunofluorescence. Clin. Diagn. Lab. Immunol. **1:** 189–196.

4. PASCUAL VELASCO, F. 1996. Distribution geogiáfica de la fiebre Q. España. *In* Fiebre Q, 1st ed. Junta de Castilla y Leon, Consejería Sanidad y Bienestar Social. Heraldo de Zamora: 120–127.
5. LÓPEZ, L.F., R. TORRONTERAS, J.D. ALCÁNTARA, *et al.* 1985. Fiebre Q como causa frecuente de síndrome febril no focalizado y utilidad de la inmunofluorescencia indirecta en su diagnóstico. Emferm. Infec. Microbiol. Clin. **3:** 279–280.
6. TELLEZ, A., P. ANDA, F. DE ORY, *et al.* 1992. Fiebre Q crónica: revisión de 58 casos. [Abstract 232]. Enferm. Infec. Microbiol. Clin. **10** (Suppl 2): s89.
7. BROUQUI, P., T.H. DUPONT, M. DRANCOURT, *et al.* 1993. Chronic Q fever: ninety-two cases from France, including 27 cases without endocarditis. Arch. Intern. Med. **153:** 642–648.
8. PALMER, S.R. & S.E.J. YOUNG. 1982. Q fever endocarditis in England and Wales, 1975–1981. Lancet **ii:** 1148–1149.

Serotesting of Human Q Fever Distribution in Bosnia and Herzegovina

S. HAMZIĆ, E. BEŠLAGIĆ, AND Š. ZVIZDIĆ

Medical Faculty, University of Sarajevo 71000 Sarajevo, Bosnia and Herzegovina

ABSTRACT: Q fever is a zoonotic disease with worldwide distribution. It occurs in different geographic regions and climate zones. From 1990 till the end of 1997, only three infected individuals were registered in Bosnia and Herzegovina, during the year 1991, with the incidence of 0.05% 000. From 1996 onward, there was a sudden aggravation of epizoological and epidemiological situation in particular regions of Bosnia and Herzegovina. We performed serotesting during the 4-year period from 2000 to 2003. We tested serum samples from 708 individuals from different regions of Bosnia and Herzegovina. Q fever was serologically diagnosed in 249 individuals. The overall seroprevalence was 35.2%. The acute disease form was confirmed in 79.9% of the whole seropositive sample. Most of the Q seropositive individuals were from Kakanj (17.3%), Mostar (15.3%), Sarajevo (12.5%), Bihać (9.6%), Zenica (9.2%), Gornji Vakuf (8.9%), Tešanj (4.4%), Visoko (2.8%), and Travnik (2.4%). The number and distribution of seropositive individuals suggests that Q fever is endemic in Bosnia and Herzegovina.

KEYWORDS: human Q fever; seroprevalence; distribution; Bosnia and Herzegovina

INTRODUCTION

Q fever is a zoonotic disease with worldwide distribution. It occurs in different geographic regions and climate zones. Since the signs and symptoms of Q fever are not disease-specific, it is difficult to determine the exact diagnosis without adequate laboratory testing. Many human infections are unapparent.

The aerosol route (inspiration of infected particles) is the primary route for human *Coxiella burnetii (C. burnetii)* infection.[1] *C. burnetii* is very resistant

Address for correspondence: S. Hamzić, Medical Faculty University of Sarajevo, Čekaluša 90, 71000 Sarajevo, Bosnia and Herzegovina.
e-mail: sadetahamzic@yahoo.com

Ann. N.Y. Acad. Sci. 1078: 133–136 (2006).
doi: 10.1196/annals.1374.022

to natural destruction and can persist for several weeks in areas where infected animals were present; the microorganism can also be transmitted by wind.[2,3] Q fever can occur in humans even without any apparent contact with animals. It usually occurs sporadically, and very frequently in the form of smaller or greater epidemics.

Q fever is primarily a professional risk for people in contact with domestic animals, such as cattle, sheep, and goats. Predisposing host factors are very important for occurrence of the disease.

In Bosnia and Herzegovina, E. S. Murray *et al.* described the first instance of Q fever in August 1950,[4] in Sokol village, with 16 infected individuals.[4,5] Since then, individual infection cases have been registered.

From 1990 till the end of 1997, only three infected individuals were registered in Bosnia and Herzegovina, during the year 1991. After that period, since 1998, the increased number of infected individuals has been registered in our country, where epidemic forms as well as sporadic cases of Q fever also occur.

Although the disease has been occurring in rather different geographic regions of Bosnia and Herzegovina, the region of central Bosnia was especially at risk, while one of the most endangered municipalities was Kakanj, which had the greatest number of registered disease cases.

MATERIALS AND METHODS

The study involved individuals whose sera were tested in the laboratory for specific diagnostic of human Q fever, in the Department for Microbiology of the Medical Faculty of the University of Sarajevo, during the period from July 2000 to December 2003. There were 708 individuals tested. Blood samples were collected by vein puncture, and isolated serum samples were kept at –20°C until the testing. The sera were tested by indirect immunofluorescence (IFA) for detection of specific human IgM and IgG, antibodies for *C. burnetii*. We used the commercial test kit, reagents for Q fever IFA IgM and IFA IgG, for *in vitro* diagnostic use from Focus Technologies (formerly MRL, Cypress, CA, USA).

RESULTS

During the 4-year follow-up period, of 708 individuals tested, we confirmed the presence of specific anti-*C. burnetii* antibodies in 249 (249/708; 35.2%) seropositive individuals (TABLE 1). The acute disease form was confirmed in 79.9% of whole seropositive sample.

The most Q-seropositive individuals were from Kakanj (17.3%), Mostar (15.3%), Sarajevo (12.5%), Bihać (9.6%), Zenica (9.2%), Gornji Vakuf (8.9%), Tešanj (4.4%), Visoko (2.8%), and Travnik (2.4%) (TABLE 2).

TABLE 1. Tested and Q-seropositive individuals in the period 2000–2003

		Positive	
Year	Tested	Number	%
2000	413	176	70.7
2001	127	24	9.6
2002	76	26	10.5
2003	92	23	9.2
Total:	708	249	100.0

Of 249 seropositive individuals, the highest Q fever seroprevalence was recorded in Zenica–Doboj Canton (88/249; 35.4%), Herzegovina–Neretva Canton (57/249; 22.9%), Sarajevo Canton (36/249; 14.5%), Una–Sana and Middle-Bosnia Canton with 29 seropositive individuals (29/249; 11.6%), Tuzla Canton (5/249; 2.0%), Bosnia–Podrinje Canton (3/249; 1.2%), Canton 10, Livno, and West-Herzegovina Canton (1/249; 0.4% of seropositive individuals) (TABLE 3).

CONCLUSIONS

- During our four-year study, in the period 2000–2003, Q fever was serologically confirmed in a wider area of Bosnia and Herzegovina.
- Q fever occurred in an epidemic form during 2000, and in sporadic forms during the years 2001, 2002, and 2003.
- The presence of anti–*C. burnetii*–specific human antibodies was confirmed in 249 (35.2%) individuals tested.

TABLE 2. Human Q fever distribution based on the city of residence with the most seropositive individuals

2000–2003		
	Positive	
City of residence	Number	%
Kakanj	43	17.3
Mostar	38	15.3
Sarajevo	31	12.5
Bihać	24	9.6
Zenica	23	9.2
Gornji Vakuf	22	8.9
Goranci	12	4.8
Tešanj	11	4.4
Visoko	7	2.8
Travnik	6	2.4

TABLE 3. Human Q fever incidence in the cantons

		Positive	
Canton	Individuals tested	Number	%
Una–Sana	50	29	11.6
Tuzla C	14	5	2.0
Zenica–Doboj	261	88	35.4
Bosnia–Podrinje	8	3	1.2
Middle Bisnia	55	29	11.6
Herzegovina–Neretva	143	57	22.9
West Herzegovina	2	1	0.4
Sarajevo Canton	160	36	14.5
Canton 10. Livno	15	1	0.4
Total	708	249	100.0

- The higher Q fever prevalence was in Zenica–Doboj Canton, with specific antibodies detected in 35.4% of tested individuals.
- The number and distribution of seropositive individuals suggest that Q fever is an endemic phenomenon in Bosnia and Herzegovina.

REFERENCES

1. MARRIE, T.J. 1990. Epidemiology of Q Fever *In* Q Fever. Vol. 1. The Disease. T.J. Marrie, Ed.: 49–70. CRC Press. Boca Raton, FL.
2. MARRIE, T.J. & D. RAOULT. 1997. Q fever—a review and issues for the next century. Int. J. Antimicrob. Agents. **8:** 145–161.
3. TISSOT DUPONT, H., S. TORRES, M. NEZRI & D. RAOULT. 1999. A hyperendemic focus of Q fever related to sheep and wind. Am. J. Epidemiol. **150:** 67–74.
4. GAON, J., S. BORJANOVIĆ, B. VUKOVIĆ, *et al.*1979. Special epidemiology of acute infective diseases. Svjetlost. Sarajevo, Bosnia.
5. PUVAČIĆ, Z. & A. ARNAUTOVIĆ. 1997.General and Special Epidemiology. Sarajevo.
6. Mc QUISTON, J.H., W.L. NICHOLSON, R. VELIĆ, *et al.* 2000. Investigation of Q Fever in Bosnia-Herzegovina 2000: An Example of International Cooperation. CDC.
7. HAMZIĆ, S. 2004. Sero-Epidemiological Studies on Q Fever among Populations in Areas at Risk. Doctoral dissertation, Medical Faculty, University of Sarajevo.

Ticks and Tick-Borne Rickettsiae Surveillance in Montesinho Natural Park, Portugal

M. SANTOS-SILVA,[a] R. SOUSA,[a] A.S. SANTOS,[a] D. LOPES,[b] E. QUEIJO,[c] A. DORETA,[b] L. VITORINO,[d] AND F. BACELLAR[a]

[a]*Centro de Estudos de Vectores e Doenças Infecciosas, Instituto Nacional de Saúde Dr. Ricardo Jorge, 2965 Águas de Moura, Portugal*

[b]*Câmara Municipal de Vinhais, Portugal*

[c]*Organização de Produtores Pecuários de Vinhais, Vinhais, Portugal*

[d]*Departamento de Biologia Vegetal/Centro de Genética e Biologia Molecular, Faculdade de Ciências, Universidade de Lisboa, Lisboa, Portugal*

ABSTRACT: This study constitutes the first contribution to the knowledge of tick dynamics and its implication in the epidemiology of rickettsial diseases in Montesinho Natural Park (MNP), Bragança district of Portugal. Of 76 ticks collected, 12 (15.8%) were *Dermacentor (D.) marginatus*, 36 (47.4%) *D. reticulatus*, and 28 (36.8%) *Rhipicephalus (R.) sanguineus*. Isolation assays were performed by shell-vial technique on 41 ticks. Israeli spotted fever strain was an isolate from *R. sanguineus*, and three isolates of *Rickettsia slovaca* were obtained from *D. reticulatus*. All 76 ticks were screened by PCR for *Rickettsia* sp., *Ehrlichia (E.) chaffeensis*, and *Anaplasma (A.) phagocytophilum*. *Rickettsia* RpA4 strain DNA was detected in 10 *D. marginatus* and 2 *D. reticulatus*, and Israeli spotted fever strain in 1 *R. sanguineus*. No *E. chaffeensis* or *A. phagocytophilum* infection was detected. New host records are provided for *D. reticulatus*. Also described for the first time in Portugal is the isolation of *R. slovaca* from *D. reticulatus* and the isolation of Israeli spotted fever strain from *R. sanguineus*. This confirms the association of the last rickettsiae strain with the same vector tick as previously described in Israel and Sicily.

KEYWORDS: ticks; *Dermacentor marginatus*; *Dermacentor reticulatus*; *Rhipicephalus sanguineus*; Rickettsiae; *Rickettsia slovaca*; Israeli spotted fever; Rpa4; Portugal

Address for correspondence: Margarida Santos-Silva, Center for Vectors and Infectious Diseases Research, National Institute of Health, Av. Liberdade 5, 2965-575 Aguas de Moura, Portugal. Voice: +351-265-912-222; fax: +351-265-912-568.
e-mail: m.santos.silva@insa.min-saude.pt

**Ann. N.Y. Acad. Sci. 1078: 137–142 (2006). © 2006 New York Academy of Sciences.
doi: 10.1196/annals.1374.023**

INTRODUCTION

Ticks are important parasites present in almost every geographic zone in the world. Several tick species are involved in the transmission of a wide range of pathogens with medical significance. In Portugal different rickettsial agents have been isolated and detected from different tick species in boutonneuse fever (BF), caused by *R. conorii*, the tick-borne disease with major implication on public health.[1] Bragança, a northeastern district of Portugal, presents the highest incidence rate of BF,[2] but no studies have been done on the prevalence of rickettsial agents in ticks from this region. As the importance of this geographical area in BF cases contrasts with the lack of information about ticks and tick-borne rickettsiae, the aim of the study was to assess tick species and pathogen interactions present in this area.

MATERIALS AND METHODS

Ticks were collected in Montesinho Natural Park (41°47N–42°N, 6°30′W-7°12′W), Bragança district, from September to November 2004 by dragging vegetation or directly removed from hosts. All specimens were identified by morphological characters using standard taxonomic keys[3,4] and kept alive until being processed for hemolymph test as previously described.[5] Suspected ticks were directly used to rickettsiae isolation by shell-vial technique.[6] Additionally, all ticks were processed individually for DNA extraction as described.[7] PCR approaches to detect DNA from *Rickettsia (R.)* spp., *Ehrlichia (E.) chaffeensis,* and *Anaplasma (A.) phagocytophilum* were performed using five primer sets. Rickettsiae DNA detection was achieved by amplification of *gltA* fragment gene (RpCs.877p–RpCs.1258n) and *ompA* fragment gene (Rr190.70p–Rr190.602n).[8] The amplification of *E. chaffeensis* DNA was assayed by a nested PCR using ECC-ECB and HE1-HE3 primers, which amplify a fragment of the 16S rRNA gene.[9] *A. phagocytophilum* DNA was amplified by msp465f and msp980r primers, derived from the highly conserved regions of major surface protein-2 (*msp2*) paralogous genes.[10] Positive PCR products were sequenced directly using an ABI automated sequencer (Applied Biosystems USA) according to the manufacturer's instructions. Sample amplifications were performed with the forward and reverse primers used for PCR identification. Sequences were identified using the BLAST software.[11]

RESULTS

From a total of 76 ticks collected in Montesinho Natural Park (MNP), 12 (15.8%) were *Dermacentor marginatus*, 36 (47.4%) *D. reticulatus,* and

TABLE 1. Number of ticks by species and stage collected in MNP and the detection assay results

Month	Origin	Site	*Dermacentor marginatus* (*n*)	*Dermacentor reticulatus* (*n*)	*Rhipicephalus sanguineus* (*n*)	*Rickettsial* detection/Tick instar		
						Anaplasma phagocytophilum	*Ehrlichia chaffeensis*	*Rickettsia*
September								
	Vegetation	Cerdeira	5M; 2F			–	–	RpA4/ 5M; 2F
		Trincheira	3F			–	–	RpA4/ 1F
		Gondosende		1M; 1F		–	–	–
	Canis familiaris	Paradinha			1F; 3 N	–	–	–
		Paramio		1M; 1F	3 N	–	–	–
		Fontes de			1N	–	–	–
		Transbaceiro				–	–	–
		Lagarelhos			18N	–	–	ISF/1N
		Tuízelo			2 N	–	–	–
October								
	Canis familiaris	Moimenta		1F		–	–	–
		Quintanilha	1F			–	–	RpA4/ 1F
	Canis lupus	Quintanilha		1F		–	–	–
November								
	Canis familiaris	Lagarelhos		1M; 3F				–
		Moimenta	1F	3 M				RpA4/ 1F; 2 M
		Paramio		1M				–
		Travanca		14 M; 7 F				–
		Sobreiró de Baixo		1 F				–
TOTAL			12	36	28			

M = male; F = female; N = nymph.

TABLE 2. *Rickettsia* isolation results

Origin	Site	Tick species	No positive ticks/ no tested	*Rickettsia* isolates
Canis familiaris	Moimenta	*D. marginatus*	0/1	
	Quintanilha		0/1	
Canis familiaris	Lagarelhos	*D. reticulatus*	1/2	*R. slovaca*
	Moimenta		0/3	
	Paramio		0/3	
	Sobreiró de Baixo		0/1	
	Travanca		2/15	*R. slovaca*
Canis lupus	Quintanilha	*D. reticulatus*	0/1	
Canis familiaris	Fontes de Transbaceiro	*R. sanguineus*	0/1	
	Lagarelhos		1/7	ISF
	Paramio		0/1	
	Paradinha		0/3	
	Tuízelo		0/2	
Total			4/41	

28 (36.8%) *Rhipicephalus sanguineus*. Seasonal distribution of ticks regarding origin, site of collection, and arthropod instars are provide in TABLE 1 and FIGURE 1. Four *Rickettsia* strains were isolated. Israeli spotted fever strain (ISF) was obtained from one *R. sanguineus* (1/14), after nymph molting, and three isolates of *Rickettsia slovaca*, from adult *D. reticulatus* (3/24) as shown in TABLE 2. Of the 76 ticks tested by PCR, 83.3% *D. marginatus* (10/12), 5.6% *D. reticulatus* (2/36), and 3.6% *R. sanguineus* (1/28) were positive for the presence of rickettsial DNA using *gltA* and *OmpA* primers. Rickettsiae detected in the *Dermacentor* ticks were identified as *Rickettsia* RpA4 (similarity values between 98% and 100% for OmpA gene sequence; GenBank, AF120022). ISF was detected in one *R. sanguineus* tick (similarity values 99% for *gltA* gene sequence; GenBank, U59727), the same specimen with positive isolate, referred to above. All *D. reticulatus* positive by shell-vial for *R. slovaca* were PCR-negative, which may be on account of PCR inhibition. No *E. chaffeensis* or *A. phagocytophilum* infection was detected on ticks (TABLE 1).

DISCUSSION

This is the first report of *D. marginatus* and *D. reticulatus* in MNP, although other studies have already documented the same species in other regions.[12] In Portugal *D. reticulatus* has a patchy distribution associated with continental climate inlands in the northeast region, with only one host known—*Canis lupus*. Here is also report the first association of *D. reticulatus* to *Canis familiaris*. In MNP the domestic dog appears to contribute not only to the maintenance of

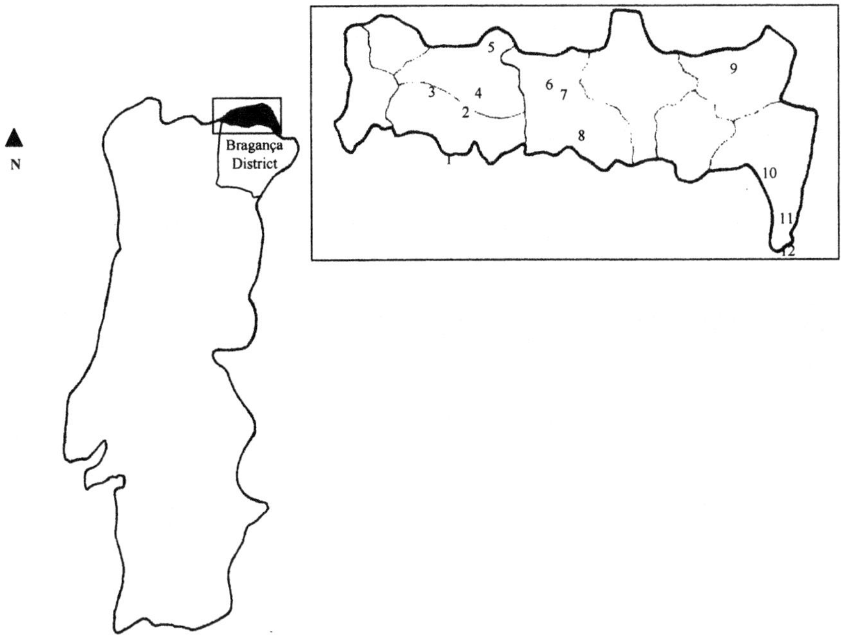

FIGURE 1. Montesinho Natural Park. Site of tick collection: Sobreiró de Baixo (1); Lagarelhos (2); Tuízelo (3); Travanca (4); Moimenta (5); Fontes de Transbaceiro (6); Paramio (7); Gondosende (8); Trincheira (9); Cerdeira (10); Quintanilha (11); Paradinha (12). Site of rickettsiae isolation/detection: Lagarelhos (2); Travanca (4); Moimenta (5); Trincheira (9); Cerdeira (10); Quintanilha (11).

R. sanguineus but represents an important host for *D. reticulatus*. This fact largely contrasts with all data recovered from other parts of the country, particularly in the south, where dogs are not involved in *D. reticulatus* cycle (Santos-Silva, unpublished data). This is the first isolation of *R. slovaca* from *D. reticulatus* and ISF strain from *R. sanguineus* in Portugal, confirming the association of this rickettsiae strain with the same vector tick as previously described in Israel and Sicily.[13] ISF isolate done after nymph molting reinforces the importance of transstadial transmission in rickettsial infectious maintenance. In Portugal *Rickettsia* RpA4 genotype has been already described in *D. marginatus* collected in the south (Vitorino *et al.* unpublished data). This study shows that *Rickettsia* RpA4 is also present in *D. reticulatus*, the same species in which it has been described for the first time in Russia.[14] The pathogenic role of this rickettsia has not yet been proven. In conclusion, the results obtained will be an important tool to develop strategies of control and prevention against ticks and tick-borne rickettsiae pathogens in Montesinho Natural Park.

ACKNOWLEDGMENTS

This research was supported by Fundação Calouste Gulbenkian (references 65286).We thank the Montesinho Natural Park team for their valuable assistance during this study.

REFERENCES

1. BACELLAR, F. 1999. Ticks and rickettsiae in Portugal. *In* Rickettsiae and Rickettsial Diseases. D. Raoult & P. Brouqui. Eds.: 103–109 Elsevier. Paris.
2. SOUSA, R. *et al.* 2003. Sobre a realidade epidemiologica da febre escaro-nodular em Portugal. Acta Medica **16:** 430–438.
3. DIAS, JATS AS CARRAÇAS (ACARINA-IXODOIDEA) DA PENÍNSULA IBÉRICA. Algumas Considerações Sobre A Sua Biogeografia E Relacionamento Com A Ixodofauna Afropaleártica E Afrotropical. 163 Instituto de Investigação Científica Tropical, Estudos Ensaios e Documentos, Lisboa.
4. WALKER, J. *et al.* 2000.The Genus *Rhipicephalus* (Acari: Ixodidae). A Guide to the Brown Ticks of the World. Cambridge University Press. Cambridge, UK.
5. BURGDORFER, W. 1970. Hemolymph test—a technique for detection of *Rickettsiae* in ticks. Am. J. Trop. Med. Hygiene **XIX:** 1010–1014.
6. PÉTER, O. *et al.* 1990. Isolation by a sensitive centrifugation cell culture system of 52 strains of spotted fever group rickettsiae from ticks collected in France. J. Clin. Microbiol. **XXVIII:** 1597–1599.
7. SANTOS, A.S. *et al.* 2004. Detection of *Anaplasma phagocytophilum* DNA in *Ixodes* ticks (Acari: Ixodidae) from Madeira Island and Setúbal District, Mainland Portugal. Emerg. Infect. Dis. **10:** 1643–1648.
8. REGNERY, R.L. *et al.* 1991. Genotypic identification of rickettsiae and estimation of intraspecies sequences divergence for portions of two rickettsial genes. J. Bacteriol. **173:** 1576–1589.
9. DAWSON, J.E. *et al.* 1996. Polymerase chain reaction evidence of *Ehrlichia chaffeensis*, an etiologic agent of human ehrlichiosis, in dogs from southeast Virginia. Am. J. Vet. Res. **57:** 1175–1179.
10. CASPERSEN, K. *et al.* 2002. Genetic variability and stability of *Anaplasma phagocytophila* msp2 (p44). Infect. Immun.**70:** 1230–1234.
11. ALTSCHUL, S.F. *et al.* 1990. Basic local alignment search tool. J. Mol. Biol. **215:** 403–410.
12. CAEIRO, V. 1999. General review of tick species present in Portugal. Parassitology **41:** 11–15.
13. GIAMMANCO, G.M. *et al.* 2003. Israeli spotted fever Rickettsia in Sicilian *Rhipicephalus sanguineus* ticks. Emerg. Infect. Dis. **9:** 892–893.
14. SHPYNOV, S. *et al.* 2004. Detection of a *Rickettsia* closely related to *Rickettsia aeschlimannii*, "*Rickettsia heilongjiangensis*", *Rickettsia* sp. strain RpA4, and *Ehrlichia muris* in ticks collected in Russia and Kazakhstan. J. Clin. Microbiol. **42:** 2221–2223.

Current Knowledge of Rickettsial Diseases in Italy

LORENZO CICERONI, ANTONELLA PINTO, SIMONETTA CIARROCCHI, AND ALESSANDRA CIERVO

Department of Infectious, Parasitic and Immune-Mediated Diseases, Istituto Superiore di Sanità, Viale Regina Elena, 299-00161 Rome, Italy

ABSTRACT: Rickettsial diseases continue to be the cause of serious health problems in Italy. From 1998 to 2002, 4,604 clinical cases were reported, with 33 deaths in the period from 1998 to 2001. Almost all the cases reported in Italy are cases of Mediterranean spotted fever (MSF). Other rickettsioses that have been historically documented are murine typhus and epidemic typhus. Since 1950, only sporadic cases of murine typhus have been reported, and Italy currently appears to be free of epidemic typhus. As in other European countries, imported cases of rickettsialpox, African tick-bite fever (ATBF), and scrub typhus have been reported. In 2004, three cases of a mild form of rickettsiosis were serologically attributed to *Rickettsia helvetica*.

KEYWORDS: rickettsial diseases; spotted fever group rickettsiosis; Mediterranean spotted fever; murine typhus; epidemic typhus; spotted fever group rickettsiae; *Rickettsia conorii*; Italy

In Italy, as in a number of countries, rickettsial diseases continue to be the cause of serious health problems. According to the most recent data from the Italian Ministry of Health, in the 5-year period from 1998 to 2002, 4,604 cases of rickettsiosis were reported,[1] with an average of 921 cases per year (1.6 per 100,000 population) (TABLE 1; FIG. 1) (though provisional data show that only 534 cases were reported in 2003). With regard to individual rickettsial diseases, no data on the number of cases are available from the Ministry of Health, nor are data available on those rickettsioses that have not been historically documented in Italy. In fact, in the Ministry's "Bolletino Epidemiologico" (Epidemiological Bulletin), the data on all rickettsioses, with the exception of epidemic typhus, are pooled.

Nonetheless, current knowledge indicates that nearly all the cases of rickettsiosis reported in Italy are cases of Mediterranean spotted fever (MSF) (also known as "boutonneuse fever," given the characteristic skin eruptions). MSF

Address for correspondence: Dr. Lorenzo Ciceroni, Department of Infectious, Parasitic and Immune-Mediated Diseases, Istituto Superiore di Sanità, Viale Regina Elena 299, 00161 Rome, Italy. Voice: +39-06-49902741; fax: +39-06-49387112; +39-06-49902934.
e-mail: ciceroni@iss.it

Ann. N.Y. Acad. Sci. 1078: 143–149 (2006). © 2006 New York Academy of Sciences.
doi: 10.1196/annals.1374.024

TABLE 1. Rickettsiosis in Italy, 1998–2003: regional distribution of clinical cases[a] (deaths[b])

	Year					
Region	1998 Clinical cases (deaths)	1999 Clinical cases (deaths)	2000 Clinical cases (deaths)	2001 Clinical cases (deaths)	2002 Clinical cases (deaths)	2003 Clinical cases (deaths)
Valle d'Aosta	1 (0)	0 (0)	1 (0)	0 (0)	0 (NA[c])	0[d] (NA)
Piedmont	13 (0)	15 (0)	10 (1)	9 (0)	7 (NA)	2[d] (NA)
Lombardy	22 (1)	5 (0)	4 (0)	13 (0)	8 (NA)	128[d] (NA)
Trentino-Alto Adige	3 (0)	0 (0)	0 (0)	0 (0)	1 (NA)	0[d] (NA)
Veneto	3 (0)	11 (0)	0 (0)	3 (0)	10 (NA)	2[d] (NA)
Friuli Venezia Giulia	0 (0)	0 (0)	0 (0)	1 (0)	1 (NA)	0[d] (NA)
Liguria	1 (0)	8 (0)	13 (0)	9 (0)	7 (NA)	4[d] (NA)
Emilia Romagna	9 (0)	10 (0)	11 (0)	11 (0)	10 (NA)	0[d] (NA)
Tuscany	11 (0)	8 (0)	8 (0)	5 (0)	7 (NA)	2[d] (NA)
Marche	10 (0)	9 (0)	2 (0)	4 (0)	4 (NA)	2[d] (NA)
Umbria	7 (0)	2 (0)	0 (0)	1 (0)	1 (NA)	2[d] (NA)
Lazio	74 (0)	151 (3)	116 (0)	103 (2)	84 (NA)	5[d] (NA)
Abruzzo	14 (0)	7 (1)	14 (0)	1 (0)	10 (NA)	3[d] (NA)
Campania	80 (1)	102 (2)	56 (0)	25 (0)	43 (NA)	37[d] (NA)
Molise	2 (0)	5 (0)	3 (0)	7 (1)	5 (NA)	1[d] (NA)
Apulia	2 (0)	7 (0)	2 (0)	4 (1)	1 (NA)	0[d] (NA)
Basilicata	4 (0)	2 (0)	0 (0)	0 (0)	0 (NA)	0[d] (NA)
Calabria	80 (2)	145 (1)	59 (0)	66 (0)	68 (NA)	30[d] (NA)
Sicily	450 (3)	629 (4)	393 (4)	397 (2)	498 (NA)	206[d] (NA)
Sardinia	127 (0)	147 (2)	107 (2)	80 (0)	125 (NA)	110[d] (NA)
Italy	913 (7)	1,263 (13)	799 (7)	739 (6)	890 (NA)	534[d] (NA)

[a]SOURCE: Ministry of Health. Bollettino Epidemiologico.
[b]SOURCE: Ufficio di Statistica-CNSPS (ISS) su dati ISTAT.
[c]Not available.
[d]Provisional data.

is caused by *Rickettsia (R.) conorii* and is transmitted by *Rhipicephalus sanguineus*. It was described as a new disease in Tunis in 1910. Although it is usually benign, complications can occur in the heart, pancreas, liver, and nervous system. The mortality rate is generally low (under 2.5%), although it is higher in persons suffering from other conditions (e.g., heart disease, diabetes mellitus, chronic liver disease, and glucose 6-phosphate dehydrogenase deficiency); it is also higher in cases that were diagnosed late or not adequately treated. Predisposition to the disease can be enhanced by alcohol abuse, smoking, and old age. In Italy, MSF was first identified in 1920 by Carducci in the Lazio region and then in 1927 by Ingrao in Sicily.[2] In the mid 1970s, an increase in MSF-related morbidity was reported in Italy,[3] as well as in other European

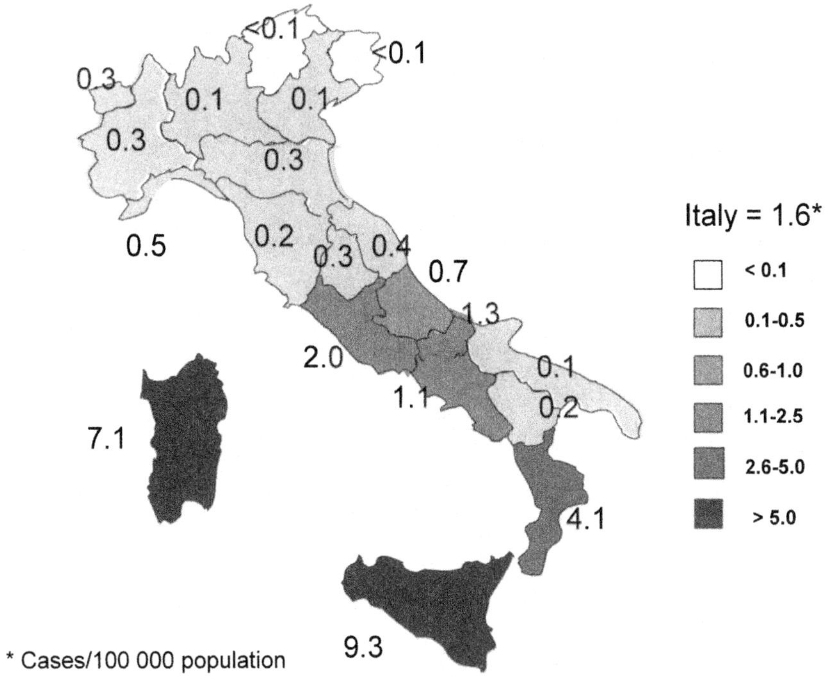

FIGURE 1. Incidence of rickettsiosis in Italy, 1998–2002.

countries (i.e., France, Spain, and Portugal). Specifically, the number of cases increased from about 90 in 1975 to about 340 in 1976 and to over 800 in 1979,[4] peaking at more than 1,500 cases in 1995[1,4] (FIG. 2). In the period from 1998 to 2002, as in previous years, most cases (91.3%) were reported in the regions of Sicily, Sardinia, Lazio, Calabria, and Campania[1] (Italy is divided into 20 regions) (TABLE 1). The cases in Sicily alone accounted for 51.4% of all clinical cases. Males represented 60.0% of cases, although they constitute only 48.5% of the population (TABLE 2). The highest incidence rates were found among the youngest (0–14 years) and the oldest age groups (over 65) (1.85 and 2.04 cases per 100,000 population, respectively) (TABLE 2). From 1998 to 2001, 33 deaths from rickettsiosis were reported (Elaboratrone statistica a cura dell'Ufficio di Statistica-CNSPS (ISS) su Dati ufficiali iSTAT di mortalità anni 1998–2001) (TABLES 1 and 2).

Another rickettsiosis that has been documented in Italy is murine typhus, also known as "endemic typhus." The etiologic agent of murine typhus is *Rickettsia typhi*, a rickettsia of the typhus group (TG), which is generally transmitted to the host through contamination of the skin, conjunctivae, or respiratory tract with the feces of the rat flea. In the 1940s murine typhus was the most widespread rickettsiosis, especially in Sicily during World War II, after the arrival of the Allied Forces.[5] After 1950 only sporadic cases were reported. However, several

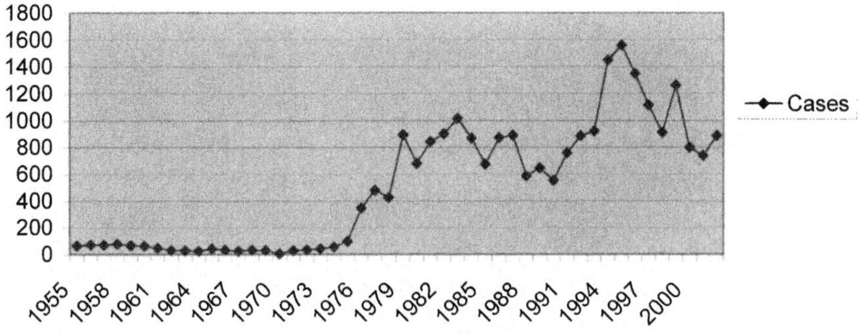

FIGURE 2. Cases of rickettsial diseases in Italy.

reports indicate that, at least until recently, ecological niches of *R. typhi* still exist, at least in Sicily; in particular, asymptomatic cases of murine typhus were reported in Sicily in the late 1980s.[6] A case of murine typhus was reported in Florence in 1991 in a person reportedly bitten by an unidentified insect during a trip to Sicily about 2 weeks before the onset of symptoms[7] and another case was reported in the Sicilian city of Catania in 1999 (G. Gulisano, personal communication).

A rickettsiosis with symptoms similar to those of murine typhus, yet with a more serious prognosis, is epidemic typhus, also known as "exanthematic typhus." The etiologic agent of epidemic typhus is *Rickettsia prowazekii*, a TG rickettsiosis transmitted by the human body louse. This disease has a long history in Italy: the first descriptions date back to 1546, when Fracastoro, in his text *De Contagione*, reported several epidemics between 1508 and 1530.[8] Other epidemic episodes were described in Sicily in the 17th century and in Apulia in 1902.[8] The last outbreak, which affected approximately 2,000 persons, 16–18% of whom died, was reported in Naples during World War II.[8] Since then, the incidence of epidemic typhus has sharply declined,[2] and according to official notifications, Italy appears to be free of this disease.[9] Nonetheless, the risk of infection with *R. prowazekii* is still present. In fact, since 1995, louse-borne diseases have shown a dramatic resurgence. In Russia and Burundi, outbreaks were reported in 1997, and in Peru an outbreak was reported in 1998.

In Italy, as in other European countries, imported cases of rickettsialpox,[10,11] African tick-bite fever (ATBF),[12,13] and scrub typhus have been reported.[14] Rickettsialpox is caused by *Rickettsia akari* and transmitted by *Allodermanyssus sanguineus* mites; it is endemic in urban areas of the eastern United States, and in Africa, Korea, Croatia, and the Ukraine. ATBF is caused by *Rickettsia africae*, a spotted fever group (SFG) rickettsia transmitted mainly by ticks of the *Amblyomma hebraeum* and *Amblyomma variegatum* spp. ATBF is found in sub-Saharan Africa; it typically occurs in clusters and has recently emerged as an imported disease in persons returning from Africa. Scrub typhus is caused

TABLE 2. Rickettsiosis in Italy, 1998–2003: age and sex distribution of clinical cases[a] (deaths[b])

	Year					
Region	1998 Clinical cases (deaths)	1999 Clinical cases (deaths)	2000 Clinical cases (deaths)	2001 Clinical cases (deaths)	2002 Clinical cases (deaths)	2003 Clinical cases (deaths)
Age (years)						
0–14	145 (0)	253 (1)	127 (0)	120 (0)	120 (NA[c])	65[d] (NA)
15–24	65 (0)	89 (0)	66 (0)	59 (0)	50 (NA)	35[d] (NA)
25–64	481 (2)	626 (4)	405 (3)	382 (2)	454 (NA)	314[d] (NA)
≥65	188 (5)	270 (8)	188 (4)	172 (4)	249 (NA)	112[d] (NA)
Unknown	34 (0)	25 (0)	13 (0)	6 (0)	17 (NA)	8[d] (NA)
Total	913 (7)	1,263 (13)	799 (7)	739 (6)	890 (NA)	534[d] (NA)
Sex						
Males	538 (6)	776 (6)	492 (3)	425 (4)	531 (NA)	354[d] (NA)
Females	374 (1)	486 (7)	307 (4)	314 (2)	359 (NA)	179[d] (NA)
Unknown	1 (0)	1 (0)	0 (0)	0 (0)	0 (NA)	1[d] (NA)
Total	913 (7)	1,263 (13)	799 (7)	739 (6)	890 (NA)	534[d] (NA)

[a]SOURCE: Ministry of Health: Bollettino Epidemiologico.
[b]SOURCE: Ufficio di Statistica-CNSPS (ISS) su dati ISTAT.
[c]Not available.
[d]Provisional data.

by *Orientia tsutsugamushi*, which is transmitted to humans and rodents by the bite of infected larval trombiculid mites; it is endemic in Japan, Korea, Southeast Asia, and Australia.

Until 2002, *R. conorii* was thought to be the only pathogenic SFG rickettsia in Italy. Since then, *R. helvetica* has been found in *Ixodes ricinus; Rickettsia slovaca* has been found in *Dermacentor marginatus* and *Haemaphysalis punctata*; Israeli spotted fever *Rickettsia* in *Rhipicephalus sanguineus*; and *R. africae* and *Rickettsia aeschlimannii* in *Hyalomma marginatum.*[15–18] In 2004, three cases of a mild form of rickettsiosis (without skin rash and characterized by fever, headache, and myalgia) were serologically attributed to *R. helvetica.*[19] The identification of a number of additional pathogenic tick-transmitted rickettsiae raises the question of whether SFG rickettsiae other than *R. conorii* are (or have been) involved in bacterial diseases in Italy. Given that the indirect immunofluorescence assay (IFA), performed with *R. conorii* and/or *R. typhi* as antigens, is the most commonly used serological test in Italy for diagnosing rickettsial disease, and that IFA does not identify the *Rickettsia* species causing the infection (because of problems of cross-reactivity), some of the cases serologically diagnosed as MSF may instead be attributable to other rickettsiae of the same group.

In conclusion, in Italy, rickettsial diseases constitute a public health problem that cannot be ignored, given both the number of clinical cases and the potentially serious consequences. Knowledge of the presence of rickettsiae in Italy can only be increased by more extensively applying gene amplification-based techniques and by using a multiple antigen IFA and/or cross-adsorption and Western blotting. In any case, clinicians need to be aware of the tick-borne diseases emerging in Italy.

REFERENCES

1. MINISTERO DELLA SALUTE. 1993–2003. Malattie Infettive. *In* Dati Epidemiologici. Bollettino, Anni (http://www.ministerosalute.it/promozione/malattie/bollettino).
2. GIUNCHI, G. 1961. Malattie da rickettsie. *In* Trattato Italiano di Medicina Interna. Parte quarta. Malattie Infettive e Parassitarie. P. Introzzi, Ed.: 1185–1240. Abruzzini Rome. Italy.
3. MANSUETO, S., G. TRINGALI & D.H. WALKER. 1986. Widespread, simultaneous increase in the incidence of spotted fever group rickettsioses. J. Infect. Dis. **154:** 539–540.
4. ISTITUTO NAZIONALE DI STATISTICA. 1958–1992. Malattie infettive e diffusive soggette a notifica obbligatoria. *In* Statistiche della Sanità. Anni 1955-1992. ISTAT, Roma.
5. MANSUETO, S. 1971. Sulla epidemiologia delle rickettsiosi dermotifose in Sicilia. Giornale di Malattie Infettive e Parassitarie **23:** 993–998.
6. VITALE, G., M. PSISTAKIS, R. LIBRIZZI, *et al.* 1988. Casi asintomatici di tifo murino in Sicilia. Giornale di Malattie Infettive e Parassitarie **40:** 904–909.
7. MORI, M. & F. PRIGNANO. 1991. Tifo murino con manifestazioni cutanee tipiche. Giornale Italiano di Dermatologia e Venereologia **126:** 445–447.
8. SCURO, L.A. 1961. Tifo esantematico. *In* Trattato Italiane di Medicina Interna. Parte Quarta. Malattie Infettive e Parassitarie. P. Introzzi, Ed.: 1193–1206. Abruzzini. Rome. Italy.
9. ISTITUTO NAZIONALE DI STATISTICA. 2004. Le notifiche di malattie infettive in Italia. Collana Informazioni. Anni 2000–2001. ISTAT, Roma.
10. RONGIOLETTI, F., L. MASSONE, C. GAMBINI, *et al.* 1991. Rickettsiosi varicelliforme ovvero, sui rischi della caccia. Giornale Italiano di Dermatologia e Venereologia **126:** 183–185.
11. LAZZARTONI, L., M. MERIGHI, F. SOLDANI, *et al.* 1996. Un caso d'importazione di rickettsiosi varicelliforme. Microbiologia Medica **11:** 521.
12. CARUSO, G., C. ZASIO, F. GUZZO, *et al.* 2002. Outbreak of African tick-bite fever in six Italian tourists returning from South Africa. Eur. J. Clin. Microbiol. Infect. Dis. **21:** 133–136.
13. RAOULT, D., P.E. FOURNIER, F. FENOLLAR, *et al. Rickettsia africae*, a tick-borne pathogen in travelers to sub-Saharan Africa. N. Engl. J. Med. **344:** 1504–1510.
14. BIANCHI, G.B., G. CROCE & U. CELLULARE. 1989. "Scrub typhus": un caso di importazione osservato a Milano. Boll. Ist. Sieroter. Milan. **68:** 1–4.
15. BENINATI, T., N. LO, H. NODA, *et al.* 2002. First detection of spotted fever group rickettsiae in *Ixodes ricinus* from Italy. Emerg. Infect. Dis. **8:** 983–986.
16. SANOGO, Y.O., P. PAROLA, S. SHPYNOV, *et al.* 2003. Genetic diversity of bacterial agents detected in ticks removed from asymptomatic patients in northeastern Italy. Ann. N. Y. Acad. Sci. **990:** 182–190.

17. GIAMMANCO, G.M., S. MANSUTO, P. AMMATUNA, *et al.* 2003. Israeli spotted fever *Rickettsia* in Sicilian *Rhipicephalus sanguineus* ticks. Emerg. Infect. Dis. **9:** 892–893.
18. BENINATI, T., C. GENCHI, C.S. CARACAPPA, *et al.* 2005. Rickettsiae in ixodid ticks, Sicily. Emerg. Infect. Dis. **11:** 509–511.
19. FOURNIER, P.E., C. ALLOMBERT, Y. SUPPUTAMONGKOL, *et al.* 2004. An eruptive fever associated with antibodies to *Rickettsia helvetica* in Europe and Thailand. J. Clin. Microbiol. **42:** 816–818.

No Serological Evidence for Rickettsial Diseases among Danish Elite Orienteerers

PETER SCHIELLERUP,[a] THOMAS DYHR,[b] JEAN MARC ROLAIN,[c] MARIANNE CHRISTENSEN,[a] RASMUS DAMSGAARD,[d] NIELS FISKER,[e] NIELS FROST ANDERSEN,[f] DIDIER RAOULT,[c] AND KAREN A. KROGFELT[a]

[a] *Unit of Gastrointestinal Infections, Department of Bacteriology, Mycology and Parasitology, Statens Serum Institut, DK-2300 Copenhagen S, Denmark*

[b] *Department of Neuroanaesthesia, Rigshospitalet, Copenhagen University Hospital, DK-2100 Copenhagen Ø, Denmark*

[c] *Unite des Rickettsies, CNRS UMR 6020, Faculté de Medecine, Université de la Mediterranée, 13385 Marseille, France*

[d] *Copenhagen Muscle Research Center, Rigshospitalet, DK-2100 Copenhagen Ø, Denmark*

[e] *Department of Clinical Immunology, Odense University Hospital, DK-5000 Odense C, Denmark*

[f] *Department of Haematology, Aarhus University Hospital, DK-8000 Aarhus C, Denmark*

ABSTRACT: A series of sudden unexpected cardiac deaths among Swedish elite orienteerers in the 1980s have resulted from the combination of infectious diseases and physical exercise. Studies in the late 1990s have pointed to *Chlamydia* and *Barontella*, which both had a high seroprevalence among Swedish elite orienteerers. We conducted a case–control study aimed to elucidate the serologic prevalence of rickettsial diseases among Danish elite orienteerers. Ticks are known as vectors for some rickettsial diseases. None of the orienteerers had a positive antibody titer against any of the tested Rickettsia despite a very high frequency of tick bites in this group.

KEYWORDS: rickettsiosis; elite sportsmen; elite orienteerers; case–control; serology; *C. burnetii*; *R. conorii*; *R. helvetica*; *R. felis*; *R. typhi*; *F. tularensis*

Address for correspondence: Peter Schiellerup, Unit of Gastrointestinal Infectious, Department of Bacteriology, Mycology and Parasitology, Statens Serum Institut, Artillerivej 5, DK-2300 Copenhagen S., Denmark. Voice: +45 3268 3176; fax: +45 3268 3085.
e-mail: pschiellerup@dadlnet.dk

Ann. N.Y. Acad. Sci. 1078: 150–153 (2006).
doi: 10.1196/annals.1374.025

INTRODUCTION

Elite sportsmen have been shown to have an impaired immune system after high-impact physical exercise.[1] In the 1980s, sudden unexpected cardiac deaths among Swedish elite orienteerers were investigated, and focus was brought on the combination of straining physical exercise and infectious diseases.[2–4] Elite orienteerers are highly exposed to ticks that are vectors for a wide range of infectious diseases. Since *Rickettsia* spp. have been shown to occur in Scandinavian ticks[5,6] it is likely that rickettsiosis occurs more often in this group of elite orienteerers.

We conducted a study aimed investigating whether elite orienteerers, who are massively exposed to ticks, are more susceptible to rickettsial diseases.

MATERIALS AND METHODS

Two-hundred and sixty-five persons were included in the study; 43 elite orienteerers, 63 elite sportsmen of indoor sports (badminton, handball, or basketball) and 159 blood donors. Elite sportsmen were defined as doing $\geq$15 hours of weekly physical exercise. The blood donors were matched for sex and age with the orienteerers. Each person delivered a blood sample during the winter of 2001–2002.[7] The participants had given their written, informed consent. The Danish Scientific Ethical Committee approved the study (KF 01–136/01).

The study involves no ethical conflicts or conflicts of interest on behalf of the authors. Blood samples were tested for antibodies against *Coxiella burnetii*, *Rickettsia conorii*, *R. helvetica*; *R. felis, R. typhi*, and *Francisella tularensis* by immunofluorescence microscopy. Samples were analyzed at the WHO Laboratory, Marseille.

RESULTS

A total of 8 (3.0%) participants, (3 blood donors and 5 elite handball players), 5 males and 3 females, had positive IgM antibody titers against *R. helvetica*; 7 of them (3 blood donors and 4 elite handball players) also had IgM antibody titers against *R. conorii*. None were positive for IgG antibodies. Of these, only one, a male handball player, reported a tick bite within the last year. Two donors had a positive *F. tularensis* antibody titer. All the controls were in the forest less than once a week. In the orienteerer group, 37 (86%) had noticed at least one tick bite within the last year, compared to 8 (5%) and 4 (6%) in the blood donor and elite sportsman groups, respectively. None of the participants had positive antibody titers against *R. typhi, R. felis,* or *C. burnetii* (TABLE 1).

TABLE 1. Exposures and seropositivity to tested bacteria in cases and controls

	Elite orienteerers ($n = 43$) no. pos./(%)	Blood donors ($n = 159$) no. pos./(%)	Elite sportsmen ($n = 63$) no. pos./(%)
Exposure			
Tick bite < 1 year	37 (86)	8 (5)	4 (6)
Danish forest 1–3 times/week	34 (79)	49 (31)	2 (3)
Swedish forest < 1 year	43 (100)	20 (13)	18 (29)
Foreign country < 6 months	43 (100)	64 (40)	42 (67)
Seropositivity			
R. helvetica	0	3 (2)	5 (8)
R. conorii	0	3 (2)	4 (6)
R. typhi	0	0	0
R. felis	0	0	0
C. burnetii	0	0	0
F. tularensis	0	2 (1)	0

CONCLUSION

This study was conducted to estimate the seropositivity against *C. burnetii*, *R. conorii*, *R. helvetica*, *R. felis*, *R. typhi*, and *F. tularensis* in a case–control study. The hypothesis was that elite orienteerers would have a higher prevalence of rickettsial antibodies, compared to the control groups, since orienteerers are heavily exposed to ticks, and since ticks have been shown to harbor *Rickettsia* spp.[5,6] Surprisingly, we found that the elite orienteerers were very rarely seropositive to the tested bacteria, compared to the control groups. Eighty-six percent of the orienteerers had noticed at least one tick bite within the last year, compared to 5% and 6% in the blood donor and elite sportsman groups, respectively. All, with no exception in the elite orienteerer group, had been running in a Swedish forest within the year prior to the blood sampling, compared to only 13% and 29% of the blood donor and elite sportsman groups, respectively. This underlines the results indicating that exposure and infection with *Rickettsia* might be from other sources or vectors than ticks. This finding is not explained by travel abroad to countries with endemic rickettsiosis.

The explanation could be that orienteerers are much more aware of tick bites and of the importance of immediate removal of the tick, and thereby less prone to rickettsial diseases and other infections. On the other hand, a major focus has been put on the awareness of tick-borne infections in the past 10–15 years, on the basis of the occurence of borreliosis, which is endemic in Scandinavia. Therefore, one would expect all participants in this study to be aware of ticks and remove them as fast as possible.

The true explanation for our findings could be a combination of very sensitive methods used in studies detecting *Rickettsia* spp. in ticks (usually PCR), in addition to the fact that orienteerers are much more aware of the need to use protective clothing after sudden unexpected deaths in Sweden from tick-borne disease, thereby reducing their risk of being infected. A Scandinavian study of three (e.g., Denmark, Sweden, and Norway) or more Nordic countries, including forest workers, elite orienteerers, and a control group, could be interesting and would possibly elucidate our findings.

ACKNOWLEDGMENTS

All participants in the study are thanked for their efforts. We thank Karl Kristian Terkelsen for establishing contact with the elite orienteerers and Dr. Per Wantzin for kindly recruiting blood donors. The study was partly financially supported by "Team Denmark."

REFERENCES

1. PEDERSEN B.K., T. ROHDE & M. ZACHO. 1996. Immunity in athletes. J. Sports Med. Phys. Fitness **36:** 236–245.
2. LARSSON E., L. WESSLEN, O. LINDQUIST, *et al.* 1999. Sudden unexpected cardiac deaths among young Swedish orienteerers: morphological changes in hearts and other organs. APMIS **107:** 325–336.
3. WESSLEN L., C. PAHLSON, O. LINDQUIST, *et al.* 1996. An increase in sudden unexpected cardiac deaths among young Swedish orienteerers during 1979–1992. Eur. Heart J. **17:** 902–910.
4. NILSSON K., O. LINDQUIST & C. PAHLSON. 1999. Association of *Rickettsia helvetica* with chronic perimyocarditis in sudden cardiac death. Lancet **354:** 1169–1173.
5. NIELSEN H., P.E. FOURNIER, I.S. PEDERSEN, *et al.* 2004. Serological and molecular evidence of *Rickettsia helvetica* in Denmark. Scand. J. Infect. Dis. **36:** 559–563.
6. NILSSON K., O. LINDQUIST, A.J. LIU, *et al.* 1999. *Rickettsia helvetica* in *Ixodes ricinus* ticks in Sweden. J. Clin. Microbiol. **37:** 400–403.
7. SCHIELLERUP P., T. DYHR, J.M. ROLAIN, *et al.* 2004. Low seroprevalence of *Bartonella* species in Danish elite orienteerers. Scand. J. Infect. Dis. **36:** 604–606.

Rocky Mountain Spotted Fever in the United States, 1997–2002

ALICE S. CHAPMAN, STACI M. MURPHY, LINDA J. DEMMA, ROBERT C. HOLMAN, AARON T. CURNS, JENNIFER H. McQUISTON, JOHN W. KREBS, AND DAVID L. SWERDLOW

Viral and Rickettsial Zoonoses Branch and Office of the Director, Division of Viral and Rickettsial Diseases, Centers for Disease Control and Prevention, Atlanta, Georgia 30033, USA

ABSTRACT: The increased incidence of Rocky Mountain spotted fever (RMSF) in 1997–2002 compared with previous years may be related to enhanced awareness and reporting of RMSF as well as changes in human–vector interaction. However, reports on RMSF mortality underscore the need for physician vigilance in considering a diagnosis of RMSF for febrile individuals potentially exposed to ticks and stress the importance of treating such persons regardless of the presence of a rash.

KEYWORDS: Rocky Mountain spotted fever; surveillance; epidemiology

INTRODUCTION

Rocky Mountain spotted fever (RMSF), caused by *Rickettsia rickettsii,* is a zoonotic tick-borne disease that may result in fatal illness if not treated promptly. This study summarizes the epidemiology of RMSF cases in the United States for 1997–2002.

METHODS

RMSF case data were analyzed using data for 1997–2002 from the National Electronic Telecommunications System for Surveillance, which provides national case counts on notifiable diseases, and from RMSF surveillance case report forms that capture demographic and epidemiologic information. These data are submitted to the Centers for Disease Control and Prevention by state health departments.

Address for correspondence: Alice S. Chapman, Centers for Disease Control and Prevention, 1600 Clifton Rd, Atlanta, GA 30033. Voice: 404-639-1521; fax: 404-639-2778.
e-mail: achapman@.cdc. gov

Ann. N.Y. Acad. Sci. 1078: 154–155 (2006). © 2006 New York Academy of Sciences.
doi: 10.1196/annals.1374.026

RESULTS

The average annual incidence of RMSF during 1997–2002 was 2.2 cases per million persons. The incidence for 2002 was 3.8 cases per million, the highest in 10 years (lowest incidence was 1.4 cases/million in 1998). The southeastern United States accounted for 59.4% of reported cases. RMSF cases occurred mostly among whites (89.6%) and males (58%). The incidence was lowest among children <5 and 10–19 years (1.6 cases/million persons) and highest among adults 60–69 years of age (3.1 cases/million persons). The overall case–fatality rate was 1.4%, with the highest rate (5%) among children <5 years of age.

CONCLUSIONS

The increased incidence of RMSF in 1997–2002 compared with previous years may be related to enhanced awareness and reporting of RMSF as well as to changes in human–vector interaction. The age-specific incidence of RMSF also differs from that previously reported, and studies are needed to determine if this change is a surveillance artifact or if it reflects changing epidemiology. The case fatality rate decreased since 1993–1996, and this may be on account of increased recognition of disease and subsequent treatment.[1] Increased fatalities among children <5 years could reflect increased severity of illness or delayed recognition of RMSF and associated treatment. Additionally, a reluctance to empirically treat very young children with doxycycline may persist despite published recommendations for such treatment.[2] Reports on RMSF mortality underscore the need for physician vigilance in considering a diagnosis of RMSF for febrile individuals potentially exposed to ticks, and stress the importance of treating such persons regardless of the presence of a rash.

REFERENCES

1. TREADWELL, T.A., R.C. HOLMAN, M.J. CLARKE, *et al.* 2000. Rocky Mountain spotted fever in the United States, 1993-1996. Am. J. Trop. Med. Hyg. **6:** 21–26.
2. HOLMAN, R.C., C.D. PADDOCK, A.T. CURNS, *et al.* 2001. Analysis of risk factors for fatal Rocky Mountain spotted fever: evidence for superiority of tetracyclines for therapy. J. Infect. Dis. **184:** 1437–1444.

Rickettsia felis in the Americas

M.A.M. GALVÃO,[a,b] J.E. ZAVALA-VELAZQUEZ,[c] J.E. ZAVALA-CASTRO,[c] C.L. MAFRA,[d] S.B. CALIC,[e] AND D.H. WALKER[b]

[a]*Escola de Nutrição, Universidade Federal de Ouro Preto, Ouro Preto, Minas Gerais, 35400-000, Brasil*

[b]*World Health Organization Collaborating Center for Tropical Diseases of University of Texas Medical Branch, Galveston, USA*

[c]*Universidad Autonóma de Yucatán, Mevida, Mexico*

[d]*Universidade Federal de Viçosa, Viçosa, Brasil*

[e]*Fundação Ezequiel Dias, Belo Horizonte, MG, Brasil*

ABSTRACT: The authors describe their work in the Americas in *Rickettsia felis* cases in humans and the presence of *Rickettsia felis* in vectors.

KEYWORDS: rickettsioses; *Rickettsia felis*

INTRODUCTION

Including the discovery and initial characterization of *Rickettsia felis* in 1992 by Azad *et al.*,[1] and the subsequent first description of a human case of infection by Schriefer *et al.* in 1994,[2] some descriptions of *R. felis* detection in vectors or human cases in the Americas, are demonstrated in FIGURE 1 and TABLE 1.[1–9]

In 2004 Galvão *et al.*[10] discussed the clinical aspects of the human cases described by Zavala-Velazquez *et al.* in Mexico, in the year 2000,[3] and by Raoult *et al.* in Brazil, in the year 2001.[4] Now we are reporting the occurrence of two more cases in Mexico and one case in Brazil, respectively. The distribution of signs and symptoms among all these cases is shown in TABLE 2.

DISCUSSION AND CONCLUSIONS

The clinical manifestations in these patients increase our knowledge of the illness caused by *R. felis* infection. Fever, myalgia, and arthralgia suggest

Address for correspondence: M.A.M. Galvão, Escola de Nutrição, Universidade Federal de Ouro Preto, Ouro Preto, Minas Gerais 35400-000, Brasil. Voice: 55+31+35591813; fax: 55+31+35591228. e-mail: magalvao@barroco.com.br

Ann. N.Y. Acad. Sci. 1078: 156–158 (2006). © 2006 New York Academy of Sciences.
doi: 10.1196/annals.1374.027

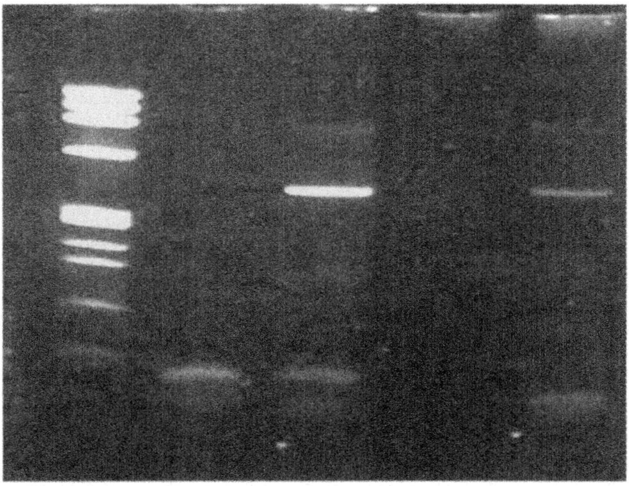

FIGURE 1. Detection of the *Rickettsia felis* 17-kDa gene by PCR in flea DNA from Brazil.

systemic acute-phase response, the exanthem reflects vasculitis, a common manifestation of rickettsial diseases, and the occurrence of gastrointestinal signs and symptoms (abdominal pain, nausea, vomiting, and diarrhea) indicate possible visceral involvement. Neurologic signs and symptoms suggest that we are facing a potentially severe new rickettsial disease. More studies are needed to determine the real clinical spectrum of symptoms present in *R. felis* rickettsiosis, as well as the epidemiology including the role of vectors, reservoirs, hosts, and the case–fatality ratio of this disease.

TABLE 1. ***Rickettsia felis*** **reported in the Americas**

Year	Authors	Vector/Host	Method	Country
1992	Azad *et al.*[1]	Fleas	PCR	USA
1994	Schriefer *et al.*[2]	Human	PCR	USA
2000	Zavala-Velazquez *et al.*[3]	Human	PCR/Serology	Mexico
2001	Raoult *et al.*[4]	Human	PCR/Serology	Brazil
2002	Oliveira & Galvão *et al.*[5]	Fleas	PCR	Brazil
2002	Boostrom *et al.*[6]	Fleas/Opossums	PCR/Serology	USA
2002	Zavala-Velazquez *et al.*[7]	Fleas	PCR	Mexico
2004	Blair, P.J. *et al.*[8]	Fleas	PCR	Peru
2005	Stevenson, H.L. *et al.*[9]	Fleas	PCR	USA

TABLE 2. Clinical signs and symptoms of *Rickettsia felis* infection in 3 cases from Brazil and 5 cases from Mexico

Signs/Symptoms	Percent (%)
Fever	100.0
Skin rash	100.0
Myalgia and arthralgia	75.0
Neurologic signs and symptoms	63.0
Gastrointestinal signs and symptoms	63.0

REFERENCES

1. AZAD, A.F., J.B. SACCI, JR., W.M. NELSON & G.F. DASH. 1992. Genetic characterization and transovarial transmission of a thyphus-like rickettsia found in cat fleas. Proc. Natl. Acad. Sci. USA **89:** 43–46.
2. SCHRIEFER, M.E., J.B. SACCI, JR. J.S. DUMLER & M.G. BULLEN.1994. AFAzad: identification of a novel rickettsial infection in a patient diagnosed with murine thyphus. J. Clin. Microbiol. **32:** 949–954.
3. ZAVALA-VELAZQUEZ, J.E., J.A. RUIZ-SOSA, R.A. SANCHEZ-ELIAS, *et al.* 2000. *Rickettsia felis* rickettsiosis in Yucatan. Lancet **356:** 1079–1080.
4. RAOULT, D., B. LA SCOLA, M. ENEA, *et al.* 2001. A flea-associated rickettsia pathogenic for humans.Emerg. Infect. Dis. **51:** 339–347.
5. OLIVEIRA, R.P., M.A.M. GALVÃO, CL MAFRA, *et al.* 2002. *Rickettsia felis* in *Ctenocephalides spp*. Fleas, Brazil. Emerg. Infect Dis. **8:** 317–319.
6. BOOSTROM, A., M.S. BEIER, J.A. MACALUSO, *et al.* 2002. Geographic association of *Rickettsia felis*-infected opossums with human murine typhus. Texas. Emerg. Infect. Dis. **8:** 549–554.
7. ZAVALA-VELAZQUEZ, J.E., J.E. ZAVALA-CASTRO, I. VADO-SOLIS, *et al.* 2002 Identification of *Ctenocephalides felis* fleas as host of *Rickettsia felis*, the agent of a spotted fever rickettsiosis in Yucatan, Mexico.Vector Borne Zoonotic Dis. **2:** 69–75.
8. BLAIR, P.J., J. JIANG, G.B. SCHOELER, *et al.* 2004. Characterization of spotted fever group rickettsiae in flea and tick specimens from northern Peru. J. Clin. Microbiol. **42:** 4961–4967.
9. STEVENSON, H.L., M.B. LABRUNA, J.A. MONTENIERI, *et al.* 2005.Detection of *Rickettsia felis* in a new world flea species, *Anomiopsyllus nudata* (Siphonaptera:ctenophthalmidae). J. Med. Entomol. **42:** 163–167.
10. GALVÃO, M.A.M., C.L. MAFRA, C.B. CHAMONE, *et al.* 2004. Clinical and laboratorial evidence of *Rickettsia felis* infections in Latin America. Rev. Bras. Med. Trop. **37:** 187–191.

Evidence of Infection in Humans with *Rickettsia typhi* and *Rickettsia felis* in Catalonia in the Northeast of Spain

MARÍA MERCEDES NOGUERAS,[a] NEUS CARDEÑOSA,[a] ISABEL SANFELIU,[b] TOMÁS MUÑOZ,[c] BERNAT FONT,[a] AND FERRAN SEGURA[a]

[a]*Infectious Diseases Program, Department of Internal Medicine, Corporació Parc Taulí, Sabadell, Barcelona, Spain*

[b]*UDIAT Diagnostic Center, Corporació Parc Taulí, Sabadell, Barcelona, Spain*

[c]*Pediatric Department, Corporació Parc Taulí, Sabadell, Barcelona, Spain*

ABSTRACT: **Murine typhus is a cause of fever of intermediate duration in the south of Spain, where antibodies against *Rickettsia typhi* and *Rickettsia felis* were observed in humans. This study presents the first report from the northeast of Spain. Human serum samples were tested by serological test. *R. typhi* and *R. felis* seroprevalences were 8.8% and 3.2%, respectively.**

KEYWORDS: ***Rickettsia typhi*; *Rickettsia felis*; murine typhus; epidemiology; seroprevalence; Spain; public health**

INTRODUCTION

Murine typhus (MT) is a cause of fever of intermediate duration in the south of Spain.[1] In this region, human infection by *Rickettsia typhi*, an MT etiological agent, as well as *Rickettsia felis* have been observed.[2] The latter may produce a clinical syndrome compatible with MT.[3] *R. felis* was identified in fleas from the south of Spain in 2002.[4] The objective of this study is to evaluate the presence of *R. typhi* and *R. felis* in the northeast of Spain.

METHODS

From September 1993 to January 1994, 217 sera from patients who presented at Sabadell Hospital were collected. The samples include sera from adults

Address for correspondence: María Mercedes Nogueras, Infectious Diseases Program, Department of Internal Medicine, Corporació Parc Taulí, Parc Taulí s/n; Sabadell 08208, Spain. Voice: 34-93-745-8252; fax: 34-93-716-0446.
e-mail: mnogueras@cspt.es

**Ann. N.Y. Acad. Sci. 1078: 159–161 (2006). © 2006 New York Academy of Sciences.
doi: 10.1196/annals.1374.028**

undergoing minor surgery and children cared for in the Pediatrics Emergency Service. Informed consent was obtained. Age, gender, and municipalities were surveyed. According to the number of inhabitants, municipalities were categorized into rural (R; <5,000), suburban (SU; 5,000–50,000), and urban (U; > 50,000) areas.

Human sera were evaluated by indirect immunofluorescence assay (IFA). This assay (MRL Diagnostics, Cypress, CA USA) was used to determine antibodies to *R. typhi*. *R. felis* antigen was obtained from the Unité de Rickettsies, France. Titers ≥ 1/40 were considered positive. Statistical analysis was performed using χ^2, Fisher exact test, and Student's *t*-test. A $P < 0.05$ was considered positive.

RESULTS

The mean age of the patients was 34.36 ± 23.20 years (0–91 years). One hundred and forty-four patients (66.4%) lived in urban areas, 59 (27.2%) lived in suburban areas, and four (6.5%) patients lived in rural areas. One hundred and eighteen patients (54.38%) were men and 99 (45.62%) were women.

Seroprevalence of *R. typhi* was 8.8% as 3.2% of samples were seropositive for *R. felis*. One sample presented antibodies to both rickettsial species at the same titer (1/40).

The mean age of subjects seropositive to *R. felis* tends to be higher (61 ± 13.14 years, 42–82 years) than those seropositive to *R. typhi* (41.92 ± 22.22 years, 1–78 years). There was a significant association between *R. felis* seropositivity and age ($P = 0.001$).

Neither *R. typhi* nor *R. felis* presented significant association between seropositive rates and living area. However, seroprevalence of both rickettsial species seem to be higher in the rural areas. In fact, *R. typhi* was present in 7.6%, 8.5%, and 21.4% in urban , suburban, and rural areas, respectively; *R. felis* seroprevalence was 3.5%, 1.7%, and 7.1% in urban, suburban, and rural areas.

DISCUSSION

This study represents a preliminary approach to *R. typhi* and *R. felis* infection in humans in the northeast of Spain. As in the south of Spain,[2] both rickettsial species seems to be present in our region. However, seropositive rates differ from those obtained in the south. In fact, *R. felis* seroprevalence in our region (3.2%) seems to be much lower than the 6.5% found in the south. On the contrary, *R. typhi* seroprevalence tends to be higher in the north (8.8% vs. 3.77%).

The only statistically significant association observed was that between *R. felis* seropositivity and age, probably because the opportunity for infection increases with time.

Until recently, *R. typhi* was associated with urban areas as well as with the presence of rats and their fleas. Nevertheless, the classic cycle seems to be replaced in some regions by the peridomestic animal cycle.[5] This fact may explain why higher rates were found in rural areas in our study.

In order to elucidate the *R. typhi* transmission cycle as well as to know the *R. felis* cycle, variables concerning contact with animals and occupation will be surveyed. Moreover, epidemiological studies in possible reservoirs and vectors such as cats, rodents, and their fleas should be conducted.

REFERENCES

1. BERNABEU-WITTEL, M., J. PACHON, A. ALARCON, *et al.* 1999. Murine typhus as common cause of fever of intermediate duration: a 17 year study in the South of Spain. Arch. Intern. Med. **159:** 872–876.
2. NOGUERAS, M. M., N. CARDEÑOSA, M. BERNABEU, *et al.* 2004. Evidence of infection in humans with *Rickettsia felis* and *Rickettsia typhi* in the South of Spain. *In*: Abstracts of the 44th ICAAC; Washington; October 30–November 2, 2004; Abstract L-997. American Society for Microbiology.
3. RAOULT, D., B. LA SCOLA, M. ENEA, *et al.* 2001.A flea-associated rickettsia pathogeneic for humans. Emerg. Infect. Dis. **7:** 73–81.
4. MARQUEZ, F. J., M. A. MUNIAIN, J. M. PEREZ & J. PACHON. 2002. Presence of *Rickettsia felis* in the cat fleas from southwestern Europe. Emerg. Infect. Dis. **8:** 89–91.
5. AZAD, A. F., S. RADULOVIC & J. A. HIGGINS. 1997. Flea-borne rickettsioses: ecologic considerations. Emerg. Infect. Dis. **3:** 319–327.

Boutonneuse Fever and Climate Variability

RITA DE SOUSA, TERESA LUZ, PAULO PARREIRA, MARGARIDA SANTOS-SILVA, AND FATIMA BACELLAR

CEVDI, National Institute of Health Dr. Ricardo Jorge, Águas de Moura, Portugal

ABSTRACT: Researchers have long appreciated the role of climate in vector-borne diseases, including the resurgence of boutonneuse fever (BF). Portugal usually is classified as having temperate Mediterranean climate. In this new century, in analyzing the data from the Meteorology Institute, this pattern has changed and an accentuated variability in climate is being observed. BF (*febre escaro nodular*) is endemic and high season is from late spring and summer. The brown dog tick *Rhipicephalus sanguineus.* is the vector and reservoir of *Rickettsia conorii complex* strains: *R. conorii* Malish and Israeli spotted fever strain. To assess the influence of climate change in BF seasonality our aim was to compare the human sera samples received at CEVDI–INSA for laboratory diagnosis of MSF for 5 months per year from October to February, ("off-season") from 2000 to 2005. Of 1,299 sera samples in persons with suspected clinical diagnosis of MSF, 45 (3.4%) were considered positive cases and the number of positive cases has doubled in the last 2 years. BF epidemiology clearly appears to be associated with climate change, especially with low precipitation values. Physicians should be aware of increasing off-season BF cases.

KEYWORDS: boutonneuse fever; *Rhipicephalus sanguineus* ticks; climate variability; Portugal

INTRODUCTION

In 1996 the Intergovernmental Panel on Climate Change from the United Nations stated that "climate change is likely to have wide-ranging and mostly adverse impacts on human health, with significant loss of life exacerbating and accelerating the tempo of contemporary infectious disease emergence, particularly for those diseases transmitted by an intermediate host, or vector."[1]

Address for correspondence: Fátima Bacellar, Centro de Estudos de Vectores e Doenças Infecciosas, Instituto Nacional de Saúde Dr. Ricardo Jorge, Avenida da Liberdade 5, 2965-575 Águas de Moura, Portugal. Voice: +351-265-912222; fax: +351-265-912568;+351-265-912155.
e-mail: fatima.bacellar@insa.min-saude.pt

Ann. N.Y. Acad. Sci. 1078: 162–169 (2006). © 2006 New York Academy of Sciences.
doi: 10.1196/annals.1374.029

Over the last decades several vector-borne diseases have emerged, expanded, or reemerged, for example, Lyme borreliosis, anaplasmosis/ehrlichiosis, malaria, and dengue. The resurgence of boutonneuse fever (BF) in Spain and Italy were examples of climate-related changes associated with the prevalence or distribution of pathogens and their vectors.[2]

The climate in Portugal is classified as temperate Mediterranean type. The normal value range (1961–1990) for average annual air temperature is 9°C to 23°C and precipitation varies from 600 mm in the lowlands south of River Tejo to 1500 mm in the mountainous north.[3]

In Portugal BF (Febre escaro nodular- ICD10: A77.1) is endemic and the most reported tick-borne disease having an annual incidence of $8.9/10^5$ inhabitants (1989–2003).[4]

Rickettsia conorii conorii and *Rickettsia conorii israeli* are the BF agents and their vector and reservoir is the brown dog tick.

The tick species *Rhipicephalus sanguineus* is adapted to higher temperatures with humidity being the limiting survival factor.[5] In the laboratory it can survive in extreme conditions (8°C–40°C/45% humidity) and its life cycle can be completed in less than 10 weeks.[6] *R. sanguineus* is generally an exophilic three-host tick, but usually, in domestic environments, the larvae, nymphs, and adults may be found on dogs all over the year. In the last few years variability in temperature and rainfall patterns is being observed, mainly generating warmer and drier winters.[7] The abundance and seasonal dynamics of ticks are influenced by climatic conditions and although the risk of transmission of rickettsiae is dependent on several parameters, such as prevalence of rickettsia-infected ticks and affinity of specific ticks for humans, climatic changes have influence on the abundance and tick activity and consequently on the incidence of BF. The aim of this work was to discern whether the known seasonality of BF is being changed by the influence of climate variability in Portugal.

MATERIALS AND METHODS

Meteorological Data Collection

Data on air temperature and precipitation were collected at the Institute of Meteorology (IM)<http://www.meteo.pt/>. Air temperature values from October to February during the 2000–2005 period are presented as average values combining data from all meteorological stations. Data on precipitation are presented for the October 2003 to February 2004 period individually for each meteorological station. This period was chosen as a representation of the differences observed from normal values.

Sample

One thousand, two hundred, and ninety-nine human samples of clinically suspected cases of BF were received at CEVDI for 5 months per year (October to February) during 2000–2005.

Case Definition

A patient seroconversion in two samples collected from October to February proved an "off-season" BF case, as well the positive isolation of rickettsiae from total blood. A positive case was defined also by a single sample with IgM > 1:32 in combination with IgG > 1:128 dilutions.[5]

Laboratory Tests

Immunofluorescence assay (IFA) was done as described elsewhere using an in-house *R. conorii* conorii antigen. The cutoff used in CEVDI is a titer of >128 for IgG and >32 for IgM. Suspected cases were considered with titers of IgM = 32 and IgG = 64 and a second sample was solicited. Isolation procedure was as described by Marrero and Raoult in 1989.[8] The isolates were identified by PCR and sequencing with *gtlA* and *rOmpA* primers pairs was described by Regnery *et al.* in 1991.[9]

RESULTS

Meteorological Data

The data obtained at IM show that an overall warmer and drier climate over the 19th and 20th centuries was observed in mainland Portugal (FIG. 1). From 2000 to 2003 the average annual air temperature was slightly above the average normal values (1969–1990). Monthly analysis from October 2003 to February 2004 showed that at the end of winter and early spring temperatures rose about 12°C over the normal mean values, but average minimum and maximum air temperatures were lower, without precipitation. Summer was rainy by contrast. In the period from October 2004 to February 2005 winter began with the highest temperatures ever registered, which were maintained until January 2005. Then the average daily minimum air temperature dropped to unusually negative values and the maximum air temperature did not exceed 15°C all over the territory. There was no rain from November to February. The average data are shown in FIGURE 2. The annual precipitation oscillated from the highest values in 2000–2001, higher than the normal values (the double in the inner northeast), to lower values in 2001–2002, peaking again in 2002–2003 (FIG. 3).

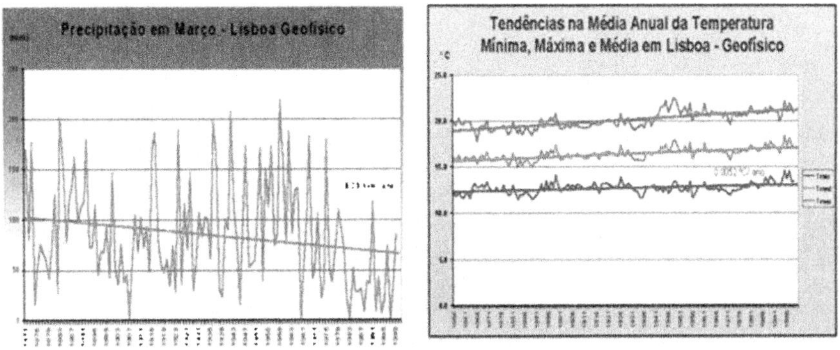

FIGURE 1. Precipitation and air temperature drifts in the 19 and 20th century. Data from IM, March, Lisbon.

Sample Data

Of the 1,299 total samples received in the "off-season" period, 45 (3,46%) were considered positive by case definition (FIG. 4). Of the 45 positive samples 18 (40%) were in males, of whom 58% were in the 60–65 year age group. All

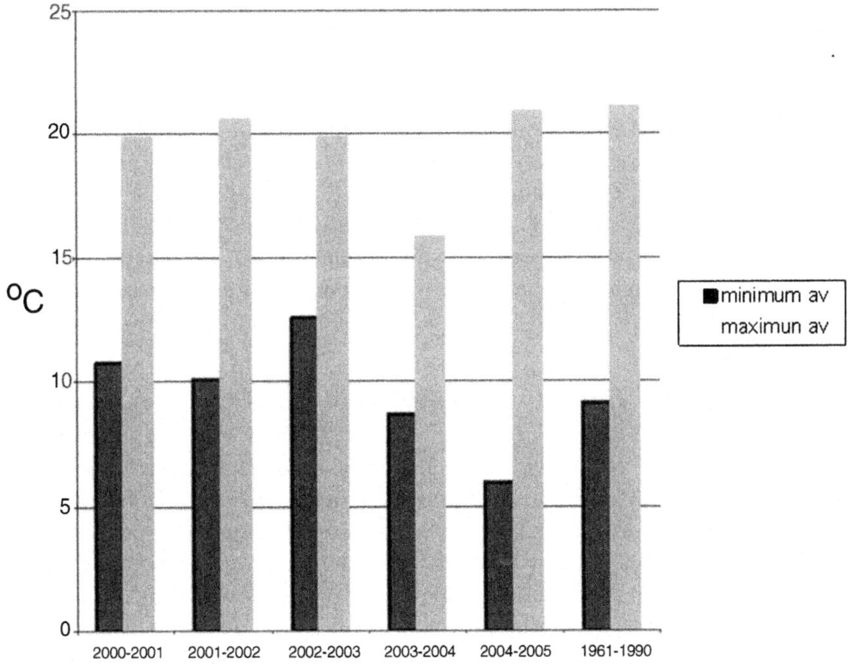

FIGURE 2. Total minimum and maximum air temperature average (°C) in Portugal (October 2000 to February 2005).

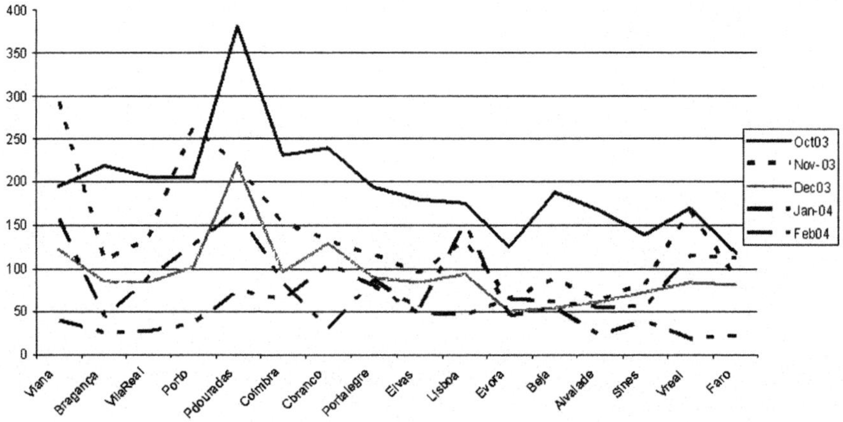

FIGURE 3. Average precipitation (mm) from the northern to southern regions of Portugal (October 2003 to February 2004).

patients were hospitalized, four of whom were admitted to the intensive care unit. In the case report forms the most mentioned symptom was arthralgia (14/45). Two patients had ticks attached. Seroconversion was observed in 51% (23/45) of the positive cases. In the first samples no antibodies were detected or were only at the cut-off level, and the second samples, received 2 to 4 weeks later, had antibody titers of IgM >32 to 1,024 and IgG>128 to 2,048. Half of the positive cases (22/45) had only one sample with IgM over 32 and IgG ranging from 128 to 2,048. The number of positive cases has doubled in the last 2 years. The number of seroconversion increased from one case in October 2000 to February 2001 to 10 cases in October 2003 to February 2004. One isolate was obtained from a total blood sample of a Caucasian male, 68 years old, living in Evora district, hospitalized in the end of January with a respiratory insufficiency syndrome. IFA serology in this sample was considered negative and a second sample was not obtained. The isolate was identified as *R. conorii conorii*.

DISCUSSION

It is a recognized fact that the average global surface temperature has already increased by about 0.6°C. The 1990s were the warmest decade for the northern hemisphere during the previous 1,000 years, and minimum temperatures have been increasing more rapidly than maximum temperatures, generating a concern manifested in several studies supported by the United Nations (United Nations Environment Programme [UNEP]). The precipitation in high latitudes (Northern Hemisphere) is increasing due to that fact, causing winters with shorter snow period.[10] Climate is an important determinant of the spatial and temporal distribution of vectors and pathogens. Therefore, change in

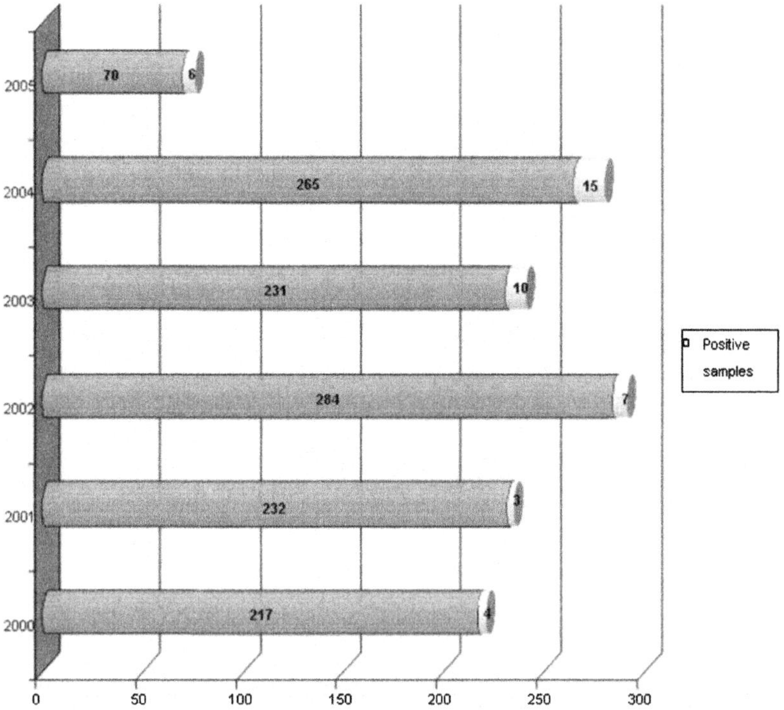

FIGURE 4. CEVDI samples received for BF laboratory diagnosis (October 2000 to February 2005).

climate and climate variability are having a profound impact on the ecology of vector populations, with an expected increase in the incidence and change in the seasonality of vector-borne diseases. On the other hand, maintaining an adaptive seasonality in life cycle is a basic ecological requisite for arthropods, for example, diapause, questing/quiescence behavior are altered physiological states that can be reversed by environmental cues (e.g., photoperiod, humidity level).[11] Ticks are known to be capable of surviving in patch microclimates and to top it off, in laboratory conditions, only changes in humidity appear to have a significant effect on questing nymphs and adults.[12] *R. sanguineus* is documented to be found all over the year in domestic environments, whose conditions of acclimatization allow the life cycle to be completed independently from the natural climate. Generalizing the climate variability observed in Portugal (formerly classified as a temperate Mediterranean type, with cold and heavily rainy winters) during the studied period, that is, warmer air temperature and lower precipitation in winters, one may suppose that even in natural environments *R. sanguineus* survival and broader activity periods are possible.Therefore, as mentioned, it is possible to have three generations in 1 year in the best environmental conditions. The autumn peak of nymphs observed in nature (and indirectly by the BF cases) is probably due to the newly emerged spring-fed larvae added to the nymphs or larvae of the previous year, inducing a multiplying effect in the numbers of questing ticks. Also, the absence of heavy precipitation permits that eggs stay viable, without being washed by rain, and thus hatching occurs in higher numbers also. All data combined, that is, isolation of rickettsiae from one blood sample and seroconversion in all cases presented here, resulted in an observed increase in "off season" BF cases.The serological test does not discriminate among rickettsioses and their results may include infections transmitted by vectors other than *R. sangineus,* such as fleas or other ticks. The results of this study allow the conclusion that BF is not characterized as a seasonal infection anymore and should be considered as the differential diagnosis of a flu-like syndrome in fall and winter. Health personnel and the general public will need to adapt their knowledge and habits to different climatic patterns and to different pathology patterns. Part of the challenge lies in promoting behavior, such as self-protection from tick bites even in winter, in order to prevent tick-borne diseases.

REFERENCES

1. Epstein, D. 1996. The climate of change: omens for the future. Perspectives in Health. PAHO Today.
2. Parola, P. & D. Raoult. 2004. Climate change and bacterial disease. Arch. Pediatr. **11:** 1018–1025.
3. Calheiros, J. & E. Casimiro. 2004."Potenciais Impactos De Alterações Climáticas Na Saúde Humana e Implicações para o Turismo em Portugal. *In* F.D. Santos *et al.*, Eds. Mudança climática em Portugal, cenários, Grádiva.
4. De Sousa, R. *et al.* 2003. Epidemiologic features of Mediterranean spotted fever in Portugal. Acta Med. Port. **16:** 429–436.
5. Bacellar, F., R. Sousa, A. Santos, *et al.* 2003. Boutonneuse fever in Portugal: 1995–2000. Data of a state laboratory. Eur. J. Epidemiol. **18:** 275–277.
6. Silva, M. & A. Filipe. 1998. "Ciclos biológicos de ixodídeos *(Ixodoidea: Ixodidae)* em condições de laboratório." Rev. Port. Ciên. Vet. **527:** 143–148.
7. Casimiro, E. *et al.* 2001. Climate change and human health impacts in Portugal: preliminary results. Proceedings of the 1st Meeting on Guidelines to Assess the Health Impacts of Climate Change, WHO/Health Canada/UNEP, Victoria, Canada (Feb 2001).
8. Marrero, M. & D. Raoult. 1989. Centrifugation-shell vial technique for rapid detection of Mediterranean spotted fever rickettsia in blood culture. Am. J. Trop. Med. Hyg. **40:** 197–199.
9. Regnery, R.L., C.L. Spruill & B.D. Plikaytis. 1991. Genotypic identification of rickettsiae and estimation of intraspecies sequence divergence for portions of two rickettsial genes. J. Bacteriol. **173:** 1576–1589.
10. Dore, M.H. 2005. Climate change and changes in global precipitation patterns: what do we know? Environ. Int. **31:** 1167–1181.
11. Perret, J.L., O. Rais & L. Gern. 2004. Influence of climate on the proportion of *Ixodes ricinus* nymphs and adults questing in a tick population. J. Med. Entomol. **41:** 361–365.

12. RANDOLPH, S.E. & K. STOREY. 1999. Impact of microclimate on immature tick-rodent host interactions (Acari: Ixodidae): implications for parasite transmission. J. Med. Entomol. **36:** 741–748.

Brazilian Spotted Fever: A Case Series from an Endemic Area in Southeastern Brazil

Epidemiological Aspects

RODRIGO N. ANGERAMI,[a] MARIÂNGELA R. RESENDE,[a] ADRIANA F.C. FELTRIN,[a] GIZELDA KATZ,[b] ELVIRA M. NASCIMENTO,[c] RAQUEL S.B. STUCCHI,[a] AND LUIZ J. SILVA[a]

[a]*Universidade Estadual de Campinas (UNICAMP), Campinas, São Paulo, Brazil*

[b]*Centro de Vigilância Epidemiológica "Alexandre Vranjac," São Paulo, Brazil*

[c]*Instituto Adolfo Lutz, São Paulo, Brazil*

ABSTRACT: **Brazilian spotted fever (BSF) is the most important tick-borne disease in Brazil and is caused by *Rickettsia rickettsii* and transmitted by the Ixodid tick Amblyomma cajennense, its main vector. We present epidemiologic aspects of a case series of patients admitted to the Hospital das Clínicas da UNICAMP from 1985 to 2003 with a confirmed diagnosis of BSF either by a fourfold rise in indirect immunofluorescence (IFA) titers of IgG antibodies reactive with *R. rickettsii* or isolation of *R. rickettsii* from blood or skin specimens. Seasonal variation of case occurrence seems to be associated with the life cycle of the tick. The recent reemergence of cases seems to be associated with the growing numbers of the capybara (*Hydrochaeris hydrochaeris*) and their expansion into urban areas.**

KEYWORDS: **Brazilian spotted fever; epidemiology; *Rickettsia rickettsii***

INTRODUCTION

Brazilian spotted fever (BSF) is the most important tick-borne disease in Brazil and is caused by *Rickettsia rickettsii* and transmitted by the Ixodid tick *Amblyomma cajennense*, its main vector.[1–3] This disease has been known in Brazil since 1929, when it was described as São Paulo exanthematic typhus in the state of São Paulo in southeastern Brazil.[4] During the period between

Address for correspondence: Rodrigo Nogueira Angerami, Rua da Urca, 354, San Conrado, Sousas, Campinas, SP Brazil, 13104-900. Voice: 55-19-3788-7451; fax: 55-19-3258-7930.
e-mail: rodrigoang@uol.com.br

**Ann. N.Y. Acad. Sci. 1078: 170–172 (2006). © 2006 New York Academy of Sciences.
doi: 10.1196/annals.1374.030**

1940 and the 1980s there was a marked drop in the number of reported cases of BSF in Brazil.[2,5]

Since the 1980s an apparent reemergence of the disease has been observed with an increase in the number of reported cases in the southeast of Brazil, especially in the states of São Paulo and Minas Gerais.[2,5]

A notifiable disease in Brazil since 2002, Brazilian spotted fever has occurred in the states of São Paulo, Minas Gerais, Espírito Santo, Rio de Janeiro, Bahia,[1,5] and more recently, Santa Catarina.

We describe the main epidemiological aspects of the patients with BSF admitted at Hospital das Clínicas da UNICAMP, a referral hospital in the region of Campinas[6] in São Paulo.

PATIENTS AND METHODS

This is a case series study based on a retrospective review of medical records and case notification files of patients admitted to the Hospital das Clínicas da UNICAMP from 1985 to 2003 with a confirmed diagnosis of BSF either by fourfold rise in indirect immunofluorescence assay (IFA) titers of IgG antibodies reactive with *R. rickettsii* or isolation of *R. rickettsii* from blood or skin specimens.

The confirmatory laboratory tests were performed at Instituto Adolfo Lutz, the state public health laboratory and regional referral center for rickettsial diseases.

RESULTS

Of the 23 patients that fulfilled the inclusion criteria, 17 (74%) were male and six (26%) were female. All the age groups were affected, but a higher incidence was observed among adults. Ages ranged from 5 to 66 years, with of median age of 31.2 years.

Despite the occurrence of cases throughout the year, the highest incidence was observed from May to October. The nearby municipality of Pedreira was the most important area of transmission, associated with the infection for seven (7/23, 30%) of the studied cases.

Both occupational and recreational activities in waterside areas, rivers, lakes or ponds, were the probable places of infection for 16 of 23 patients (69.5%). Eleven patients (47.8%) were probably infected during recreational activities, mostly fishing. Occupational activities with tick exposure were reported by 21.7% of patients from rural areas.

Tick exposure was reported by 82.6% (19/23) of the patients. The presence of capybaras (*Hydrochaeris hydrochaeris)* at the probable places of infection was reported by 43.4%; one patient referred to close contact with a healthy dog.

DISCUSSION

After approximately four decades of an unexplained epidemiologic silence, also observed in the United States,[7] we have been seeing a reemergence of BSF with an increasing number of cases annually in São Paulo state since 1985.[6] Of the 155 confirmed cases of BSF in this state between 1985 and 2004, 75.4% (117/155) occurred in the region of Campinas; 35 of them in the Municipality of Pedreira.[8] The epidemiologic importance of Pedreira was previously known[2] and is confirmed in our observations. The period of a higher number of cases reported by us was also observed previously and correlates with the tick life cycle.[2] The importance of capybaras as reservoirs for the tick vector in São Paulo state has been reported previously[3] and capybaras were referred by almost half of the patients in this series. Some domestic animals, however, such as dogs and horses, could be important reservoirs for ticks.[3] Another point of concern is the fact that, especially in the state of São Paulo, we have observed an urbanization of BSF and an apparent increase in tick exposure, and therefore of rickettsial infection, during recreational activities in all age groups. BSF is an important public health problem in the region and measures for its control should include public education on avoiding tick exposure in areas with increased risk. The capybara is a protected species and is gradually adapting to urban habitats. In the context of BSF control in this reported area the control of the capybara population should be considered a priority.

REFERENCES

1. SANGIONI, L.A., M.C. HORTA, M.C. VIANNA, *et al.* 2005. Rickettsial infection in animals and Brazilian Spotted Fever endemicity. Emerg. Inf. Dis. **11:** 265–270.
2. LEMOS, E.R.S., F.B.F ALVARENGA, M. CINTRA, *et al.* 2001. Spotted fever in Brazil: an epidemiological study and description of clinical cases in an endemic area in the state of São Paulo. Am. J. Trop. Med. Hyg. **65:** 329–334.
3. HORTA, M.C., M.B. LABRUNA, L.A. SANGIONI, *et al.* 2004. Prevalence of antibodies to spotted fever group rickettsiae in humans and domestic animals in a Brazilian Spotted Fever endemic area in the state of São Paulo, Brazil: serologic evidence for infection by *Rickettsia rickettsii* and other Spotted Fever Group rickettsia. Am. J. Trop. Med. Hyg. **71:** 93–97.
4. DIAS, E. & A.V. MARTINS. 1939. Spotted fever in Brazil: a summary. Am. J. Trop. Med. **19:** 103–108.
5. GALVÃO, M.A.M., J.S. DUMLER, C.L. MAFRA, *et al.* 2003. Fatal spotted fever rickettsiosis, Minas Gerais, Brazil. Emerg. Inf. Dis. **9:** 1402–1405.
6. LIMA, V.L.C., S.S.L. SOUZA, C.E. SOUZA, *et al.* 2003. Spotted fever in Campinas region, State of São Paulo, Brazil. Cad. Saúde Públ. **19:** 331–334.
7. THORNER, A.R., D.H.WALKER & W.A. PETRI. 1998. Rocky Mountain spotted fever. Clin. Inf. Dis. **27:** 1353–1360.
8. Centro de Vigilância Epidemiológica. Febre maculosa. Available at http://www.cve.saude.sp.gov.br/htm/Cve_fmaculosa.htm [accessed in July 30, 2005].

Prospective Evaluation of Rickettsioses in the Trakya (European) Region of Turkey and Atypic Presentations of *Rickettsia Conorii*

FIGEN KULOGLU,[a,b] JEAN MARC ROLAIN,[b] BAYRAM AYDOSLU,[a] FILIZ AKATA,[a] MURAT TUGRUL,[a] AND DIDIER RAOULT[b]

[a]*Trakya Universitesi Tıp Fakültesi, Klinik Bakteriyoloji ve Infeksiyon Hastalıkları AD, 22030, Edirne, Turkey*

[b]*Unité des rickettsies, IFR48, CNRS UMR 6020, Faculté de medeciné, Université de la Méditerranée, 13385 Marseille Cedex 05, France*

ABSTRACT: In 2004 between the months of May–November, 11 patients with spotted fever group (SFG) rickettsioses were admitted to the Trakya University Hospital in Edirne, Turkey. SFG rickettsioses were diagnosed clinically. Before treatment, punch biopsy from skin lesions, especially from the eschar, was performed. Serum specimens were tested by IFA using a panel of nine rickettsial antigens, including SFG rickettsiae and *R. typhi*. Western blotting and standard PCR were also performed. The average age of the 11 patients (4 male and 7 female) was 51 years. All the patients had high fever; 10 (91%) had maculopapular rash; 8 (73%) had rash in the palms or on the soles. Five patients had a unique eschar; two had double eschars (64%). Two patients presented with multiple organ failure and one of them died. All the patients had significant antibody titers against SFG rickettsiae. PCR experiments of skin biopsies were positive in six (60%) of 10 skin biopsy samples and DNA sequencing of the positive PCR products gave 100% homology with *Rickettsia conorii* Malish 7 for *opmA* and *gltA*. Trakya Region in an endemic area for rickettsioses. In this series, three patients presented with life-threatening diseases and one of them died. This patient was the first fatal case (2.8%). Atypic and serous life-threatening presentations of rickettsioses must be kept in mind for the differential diagnosis of febrile disease in Turkey.

KEYWORDS: Rickettsioses; Trakya Region; Turkey; *Rickettsia conorii*

Address for correspondence: Didier Raoult, Unité des Rickettsies, CNRS, UMR 6020, Faculté de Medecine, Université de la Méditerranée, 27 Boulevard Jean Moulin, 13385 Marseille Cedex 05, France. Voice: 33 4 91 32 43 75; fax: 33 4 91 38 77 72.
e-mail: didier.raoult@medecine.univ-mrs.fr

Ann. N.Y. Acad. Sci. 1078: 173–175 (2006). © 2006 New York Academy of Sciences.
doi: 10.1196/annals.1374.031

OBJECTIVES

Rickettsia conorri (*R. conorii*) infection designated as Mediterranean spotted fever has been identified in many Mediterranean locations including Turkey.[1,2] In 2004 between the months of May and November, 11 patients with spotted fever group (SFG) rickettsioses were admitted to the Trakya University Hospital, a tertiary care hospital with 860 beds in Edirne, Turkey.

MATERIALS AND METHODS

SFG rickettsioses were diagnosed clinically.[1–3] Before treatment, punch biopsy from skin lesions, especially from the eschar was performed.[1–3] In Edirne, serum specimens were tested by microimmunofluorescence assay (IFA) using commercially available antigens (*R. conorii* IFA IgG; Focus Technologies, Cypress, CA USA). In Marseille, serum specimens were retested by IFA using a panel of nine rickettsial antigens, including SFG rickettsiae (*R. conorii* subsp. *conorii, R. helvetica, R. slovaca, R. massiliae, R. aeschlimannii, R. sibirica mongolotimonae, R.conorii* subsp. *israelensis, R. felis, R. akari)* and *R. typhi*. Western blotting was also performed. DNA was extracted from skin biopsies using the QIAamp DNA Mini Kit (Qiagen, Hilden, Germany).[1–3] Standard polymerase chain reaction (PCR) was performed with primers suitable for hybridization within the conserved region of genes coding for outer membrane protein A (*ompA*) and citrate synthase (*gltA*).[1–3]

RESULTS

The average age of the 11 patients (4 male and 7 female) was 51 years. All the patients had high fever; 10 (91%) had maculopapular rash; 8 (73%) had rash in the palms or on the soles. Five patients had a unique, two had double eschars, so 64% of all the patients had eschar. Two patients who had a unique eschar, presented with dysfunction of organs and one of them died. Another patient without eschar presented with acute meningitis and seroconversion was determined in the cerebrospinal fluid by IFA. One patient had temporary hearing loss. In November, a 29-year-old male patient (farmer) presented with high fever without rash. All these patients had significant antibody titers against SFG rickettsiae. Western blotting was positive in three patients (27%). PCR experiments of skin biopsies were positive in 6 (60%) of 10 skin biopsy samples and DNA sequencing of the positive PCR products gave 100% homology with *Rickettsia conorii* subsp. *conorii* strain Malish for *ompA* (Genbank accession number AE008674) and *gltA* (Genbank accession number AE008677).

CONCLUSION

Trakya region is an endemic area for rickettsioses.[2] Especially older patients (two patients, 59 and 77 years of age) presented with life- threatening dysfunction of organs and one of them died. Since 2001, 35 patients were diagnosed as SFG rickettsioses in the Trakya region; this is the first fatal case (2.8%). Atypic and serious life-threatening presentations of rickettsioses must be kept in mind for the differential diagnosis of febrile diseases in the Turkey, in southern Europe. Although African tick-bite fever is the only identified rickettsioses with several inoculation eschars, two of our patients presented with double eschar and their serologic results were in accordance with *R. conorii*. The first case of 2005 is also a 68-year-old poor farmer with double eschar on his right leg and a tick dropped from his clothes. Although it relies on trained personnel and specific laboratory tools, the etiological diagnosis of tick-transmitted diseases is mandatory for comprehensive information on these infectious diseases.

REFERENCES

1. WALKER, D.H. & D. RAOULT. 2005. *Rickettsia rickettsii* and other spotted fever group Rickettsiae (Rocky Mountain Spotted Fever and Other Spotted Fevers). *In* Principles and Practice of Infectious Disease, 6th ed. G.L. Mandell, J.E. Bennett & R. Dolin, Eds.: 2287–2295. Elsevier, Churchill, Livingstone. Philadelphia, Pennsylvania.
2. KULOGLU, F., J.M. ROLAIN, P.E. FOURNIER, *et al.* 2004. First isolation of *Rickettsia conorii* from humans in the Trakya (European) region of Turkey. Eur. J. Clin. Microbiol. Infect. Dis. **23:** 609–614.
3. BROUQUI, P., F. BACELLAR, G. BARANTON, *et al.* 2004. Guidelines for the diagnosis of tick-borne bacterial diseases in Europe. Clin. Microbiol. Infect. **10:** 1108–1132.

Serologic Study of Rickettsioses among Acute Febrile Patients in Central Tunisia

N. KAABIA,[a] J.M. ROLAIN,[b] M. KHALIFA,[a] E. BEN JAZIA,[a] F. BAHRI,[a] D. RAOULT,[b] AND A. LETAÏEF[a]

[a]*Internal Medicine and Infectious Diseases Unit, CHU F, Hached Sousse Tunisia*

[b]*Unité des Rickettsies, CNRS UMR 6020, Faculté de Médecine Marseille-France*

ABSTRACT: Although Mediterranean spotted or "boutonneuse" fever (MSF) has been documented in central Tunisia, other spotted fever group rickettsioses (SFGR) and typhus group rickettsioses (TGR) have received little attention in our region. We sought to determine the role of rickettsioses, Q fever, ehrlichioses, and bartonelloses among patients with acute fever. The results of this study of 47 persons with acute fever of undetermined origin are reported in this paper. We concluded that SFGR, murine typhus, and acute Q fever are common causes of acute isolate fever in summer in central Tunisia and should be investigated systematically in patients with acute fever of unknown origin.

KEYWORDS: rickettsia; epidemiology; Q fever; bartonella; ehrlichia; Tunisia

INTRODUCTION

Rickettsiae are gram-negative and obligate intracellular bacteria belonging to the α group of *Proteobacteria*.[1] In the Mediterranean basin, especially in North Africa, there are few reports on the prevalence of human rickettsiosis, Q fever, and ehrlichiosis. In Tunisia, although Mediterranean spotted fever (MSF) is endemic, other rickettsial infections were reported among patients with acute febrile illness and the prevalence of antibodies against *Coxiella burnetii (C. burnetii)* has been reported to be as high as 26% in central Tunisia.[2,3,4] In addition, numerous other infectious agents are responsible for acute febrile illnesses (such as vector-borne bacterial and viral diseases, typhoid fever, and malaria). These causes vary by region, climatic conditions, and endemicity of some diseases.[5] In our region, we recently noted two outbreaks of West Nile virus (WNV) infection.[6,7] In this report we attempted to evaluate the prevalence

Address for correspondence: Amel Letaief, Service de Médecine Interne et Maladies Infectieuses, CHU F, Hached-Sousse 4000 Tunisia. Voice/fax: +216-73-21-11-83.
e-mail: amel.letaief@famso.rnu.tn

Ann. N.Y. Acad. Sci. 1078: 176–179 (2006). © 2006 New York Academy of Sciences.
doi: 10.1196/annals.1374.126

of antibodies against several SFG rickettsiae, *C. burnetii*, *Bartonella* spp., and *Ehrlichia* among patients with acute febrile illness of unknown origin.

MATERIALS AND METHODS

We included in this study all patients hospitalized for acute febrile illness between July 1 and October 31, 2004 in the Infectious Diseases Unit of the University Hospital in Sousse (central Tunisia). At admission, all patients had had three blood cultures, midstream urine, chest X ray, full blood count, an erythrocyte sedimentation rate (ESR), renal function, glycemia, sanguine ionogram, liver enzymes, and lumbar puncture if the patient had meningitic rigidity. Typhoid fever and WNV infection were excluded by laboratory tests. If no etiology was determined for this acute febrile illness, at least one serum sample was collected. Among the 47 patients, 69 blood samples were collected and stored at –20°C (27 had 1 sample, 18 had 2-week-interval paired samples, and 2 had 3 samples). Immunofluorescence assays were performed to detect antibodies against *C. burnetii, R. conorii, R. africae, R. massiliae, R. aeschlimannii, R. mongolitimonae, R. felis, R. typhi, R. prowazekii, B. quintana, B. henselae,* and human granulocytic ehrlichia. Titers of 1:128 of IgG antibodies with seroconversion in paired sera or titers of 1:32 of IgM antibodies (or both) against any species were considered as evidence of recent infection. For sera confirmed by IFA, Western immunoblotting (WB) procedures were performed as described elsewhere[8] with only *R. conorii*, *R. aeschlimannii*, *R. felis,* and *R. typhi* antigens. Seroconversion or a fourfold rise in antibody titers, or single high-titer phase II antibody ($\geq$ 200 IgG or $\geq$ 50 IgM) confirmed acute Q fever. Skin biopsy was made in one patient with acute febrile rash.

RESULTS

During the duration of the study, 47 patients (27 male and 20 female) were hospitalized for acute fever of undetermined origin. The mean age was 35.5 years (16–79 years) and 74.5% were from a rural region. Fifteen patients (32%) had contacts with domestic animals.

Four clinical patterns were observed: acute fever with lymphocytic meningitis (10 patients); isolated acute fever (19 patients); acute fever with rash (13 patients, associated with eschar in 4 cases); and acute fever with atypical pneumonia (5 patients). Normal blood cells or leukopenia and cytolysis were noted in 39 (83%) and 29 (62%) patients, respectively.

Serology allowed diagnosis of acute febrile illness for 31 patients (66%). Rickettsial infection was confirmed in 27 cases (57.5%) and acute Q fever in 4 cases (8.5%) (TABLE 1). Identification of the rickettsial agents performed using WB and cross-adsorption studies allowed diagnose of murine typhus, MSF, and 1 case of *R. felis* infection, whereas 8 cases remain undetermined. None of the sera tested showed antibodies against *Bartonella* spp. or HGE. In

TABLE 1. Clinical patterns and results of serologic testing

Clinical pattern	Isolate acute fever N (%)	Fever + meningitis N (%)	Fever + skin rash N (%)	Fever+ pneumonia N (%)	Total N (%)
Antibodies anti-					
R. typhi	8[a]	0	3	0	11 (23,5)
R. conorii	0	2	5	0	7 (15)
R. felis	0	0	1	0	1 (2)
Other SFGR	4	1	2	1	8 (17)
C. burnetii	2	0	0	2	4 (8)
Patients with positive serology (%)	14 (73.5)	3 (30)	11(84.5)	3 (60)	31 (66)
Negative serology	5	7	2	2	16 (34)
Total	19	10	13	5	47 (100)

[a]Two patients had serological evidence of concomitant or consecutive infection with murine typhus and acute Q fever.

the skin biopsy from the patient with rash, we were able to amplify *R. conorii*. Patients were treated with doxycycline or ciprofloxacin with a favorable outcome.

DISCUSSION

In this study, using well-documented and sophisticated serological methods,[9] we have confirmed the high frequency of rickettsial infections and acute Q fever among patients with febrile illness in the center of Tunisia. Interestingly, we report for the first time serological evidence of *R. felis* infection in North Africa. A recent study, performed in Algeria, has demonstrated the presence of *R. felis* in fleas.[10] Since *R. felis* has a worldwide distribution and infestation with these fleas is very common, our result was not surprising. In humans, few cases of infections with *R. felis* have been reported.[11,12]

Several cases remain undetermined in our series even after cross-adsorption studies, and skin biopsies are needed to better understand the epidemiology of rickettsial diseases in Tunisia. Finally, we confirm the absence of ehrlichiosis in Tunisia. We did not find serological evidence of *Bartonella* infection in our series. However, *B. quintana* infective endocarditis has recently been described in Tunisia.[13]

In conclusion, SFG rickettsiosis, murine typhus, and acute Q fever are the most common causes of acute isolate fever in summer in central Tunisia and should be investigated systematically in patients with fever of unknown origin.

REFERENCES

1. RAOULT, D. & J.G. OLSON. 1999. Emerging rickettsioses. *In* Emerging Infections 3. W.M. Scheld, W.A. Craig, J.M. Hughes, Eds.: 17–35. ASM Press. Washington DC.

2. LETAIEF, A.O., S. YACOUB, H.T. DUPONT, *et al.* 1995. Seroepidemiological survey of rickettsial infections among blood donors in central Tunisia. Trans. R. Soc. Trop. Med. Hyg. **89:** 266–268.
3. OMEZZINE-LETAIEF, A., H. TISSOT DUPONT, F. BAHRI, *et al.* 1997. Etude séroépidémiologique chez 300 malades fébriles hospitalisés dans un service de médecine et maladies infectieuses. Méd. Mal. Infect. RICAI **27:** 663–666.
4. OMEZZINE-LETAIEF, A., N. KAABIA, M. CHAKROUN, *et al.* 2005. Clinical and laboratory features of murine typhus in central Tunisia: a report of 7 cases. Int. J. Infect. Dis. **9:** 331–334.
5. NDIP, L.M., D.H. BOUYER, A.P. TRAVASSOS DA ROSA, *et al.* 2004. Acute spotted fever rickettsiosis among febrile patients, Cameroon. Emerg. Infect. Dis. **10:** 432–437.
6. TRIKI, H., S. MURRI, B. LE GUENNO, *et al.* 2001. West Nile viral meningoencephalitis in Tunisia. Med. Trop. **61:** 487–490.
7. NAOUFEL, K., M. KHALIFA, W. HACHFI, *et al.* 2004. West nile virus meningoencephalitis in central Tunisia: report of 13 case. Clin. Microbiol. **10**(S3): 662.
8. TEYSSEIRE, N. & D. RAOULT. 1992. Comparison of western immunoblotting and microimmuno-fluorescence for diagnosis of Mediterranean spotted fever. J. Clin. Microbiol. **30:** 455–460.
9. LA SCOLA, B. & D. RAOULT. 1997. Laboratory diagnosis of rickettsioses: current approaches to the diagnosis of old and new rickettsial diseases. J. Clin. Microbiol. **35:** 2715–2727.
10. BITAM, I., P. PAROLA, K. DITTMAR DE LA CRUZ, *et al.* 2006. First molecular detection of *Rickettsia felis* in fleas from Algeria. Am. J. Trop. Med. Hyg. **74:** 532–535.
11. RAOULT, D., B. LA SCOLA, M. ENEA, *et al.* 2001. A flea-associated rickettsia pathogenic for humans. Emerg. Infect. Dis. **7:** 73–81.
12. RICHTER, J., P.E. FOURNIER, J. PETRIDOU, *et al.* 2002. *Rickettsia felis* infection acquired in Europe and documented by polymerase chain reaction. Emerg. Infect. Dis. **8:** 207–208.
13. ZNAZEN, A., J.M. ROLAIN, N. HAMMAMI, *et al.* 2005. High prevalence of *Bartonella quintana* endocarditis in sfax (Tunisia). Am. J. Trop. Med. Hyg. **72:** 503–507.

Reemergence of Rickettsiosis in Oran, Algeria

NADJET MOUFFOK,[a] ANWAR BENABDELLAH,[a] HERVÉ RICHET,[b] JEAN MARC ROLAIN,[b] FATIHA RAZIK,[a] DJAMILA BELAMADANI,[a] SALIHA ABIDI,[a] RAMDANE BELLAL, FRÉDÉRIQUE GOURIET,[b] NORI MIDOUN,[a] PHILIPPE BROUQUI,[b] AND DIDIER RAOULT[b]

[a]*Service des maladies infectieuses et tropicales, Centre Hospitalier Universitaire d'Oran, 31 000 Oran, Algeria*

[b]*Unité des Rickettsies CNRS UMR 6020A, IFR48, Faculté de Médecine, Université de la Méditerranée, 13385 Marseille Cedex 05, France*

ABSTRACT: The presumptive cases of Mediterranean spotted fever have been identified in 1993 and since that time, its frequency has steadily increased. The prospective study, in summer 2004, was conducted in order to present the descriptive clinic and epidemiology, to identify more severe forms, the presence of the multiple eschars, and different rickettsial strains caused the disease in our region. In Oran, the cases were diagnosed clinically. In Marseille, serum specimens were tested by IFA using the panel of eight rickettsial antigen; Western blot and cross-adsorption studies were also performed in order to confirm the diagnosis. Ninety-three patients clinically diagnosed were recorded from July 3 to October 28, 2004. Eighty percent were male, the mean age was 44.3 years, 90% were exposed to dog and 32% reported tick bites. Clinical signs were as follow: presence of underlying disease (44%), sudden onset (78%), fever (100%), loss of weight (63%), conjunctivitis (43%), and a tache noire was noticed in 70%. Interestingly, two patients had two and three eschars, respectively. The rash was maculopapular (palm and sole) and purpuric in nine cases. Doxycycline was the most antibiotic (91%) with favourable outcome in 91% of the cases. Malignant form with death is reported for three patients (3.2%). Among the 93 patients, 104 serum from 65 patients were tested (serums of others patients were lost or ticket not found on tube. Sixty-three patients out of 65 had a positive serology by IFA with cross-reactive antibodies especially between *R. conorii*, *R. felis* and/or *R. typhi*. Two others negative serology were: one precocious serum and second from the patient, which presented symptoms of MSF and tested two serums, Western blot and cross-adsorption.

This work was presented at the poster session of the Fourth International Conference on Rickettsiae and Rickettsial Diseases in Logroño (La Rioja), Spain, June 2005.

Address for correspondence: Mouffok Nadjet, Centre Hospitalier Universitaire d'Oran, 72 boulevard Benzedjeb, 31 000 Oran, Algeria.
e-mail: nadjmouf_31@yahoo.fr

**Ann. N.Y. Acad. Sci. 1078: 180–184 (2006). © 2006 New York Academy of Sciences.
doi: 10.1196/annals.1374.033**

KEYWORDS: Rickettsiose; Mediterrean spotted fever; *R. conorii*; *R. felis*; *R. typhi*; multiple eschar

INTRODUCTION

Rickettsiae are gram-negative and obligate intracellular bacteria belonging to the alpha group of *Proteobacteria.*[1] Rickettsial diseases are distributed all over the world but Mediterranean spotted fever (MSF), a zoonosis caused by *Rickettsia conorii* subsp. *conorii* transmitted by the bites of the brown dog tick *Rhipicephalus sanguineus,* is the most common rickettsiosis occurring in the Mediterranean region.[2] Infections caused by *R. conorii* have been described in Algeria neighboring countries, such as Tunisia and Morocco, but have never been systematically studied in Algeria.[3,4] Therefore, a prospective study was performed in Oran (Algeria), the second largest city in Algeria, to assess the frequency of MSF and to describe the clinical and epidemiological characteristics of the disease.

MATERIALS AND METHODS

Patients

All patients seen at the infectious diseases (ID) consultation of the Oran Teaching Hospital with suspicion of MSF (high fever, maculopapular skin rash, headache, myalgia, arthralgia, and/or eschar) during the summer of 2004 were included in the study. Clinical and epidemiological data, laboratory results, treatments, and outcome were collected on a standardized questionnaire.

Methods

Sera collected from July 3 to October 28, 2004 from patients clinically suspected to have rickettsial infection were tested in a multiple-antigen immunofluorescence assay (IFA) as previously reported,[5] using nine SFG rickettsial antigens: *R. conorii* subsp. *conorii* strain seven, *R. africae*, *Rickettsia sibirica mongolotimonae*, *Rickettsia aeschlimannii*, *Rickettsia massiliae*, *R. helvetica*, *R. slovaca*, *R. conorii* supsp. *israelensis*, *Rickettsia felis*, and *R. typhi*. Titers of 1:128 in the case of IgG antibody with seroconversion in paired sera or titers of 1:32 in the case of IgM antibody (or both) against any species were considered as evidence of recent infection by *Rickettsia*. For identification at the species level, Western immunoblotting (WB) procedures were performed as described elsewhere.[6]

Data were entered and analyzed with EpiInfo version 6 (Centers for Disease Control and Prevention, Atlanta, Georgia, USA).

TABLE 1. Clinical signs, MSF, Oran, Algeria, 2004

Clinical Sign	N (%) ($N = 93$)
Sudden onset	73 (78.5%)
Altered state of health	12 (13%)
High fever	93 (100%)
Loss of weight ($\geq$ 2 kg)	57 (61%)
Asthenia	90 (97%)
Headache	78 (84%)
Maculopapular/nodular skin rash	67 (72%)
Purpuric skin rash	9 (9.7%)
Discrete eruption	17 (18%)
Tache noire	65 (70%)
Conjunctivitis	32 (34%)
Hemorragic conjunctivitis	2 (2%)
Epistaxis	7 (7.5%)
Splenomegaly	1
Hepatomegaly	4 (4%)
Adenopathy	9 (9.7%)
Cough	19 (20.4%)

RESULTS

Ninety-three patients seen at the Oran ID consultation for suspicion of MSF from July 3 to October 28, 2004 were included. Forty-three patients (46%) lived in the city of Oran and most patients (75 [81%]) were male. The mean age of the patients was 44.3 years (0–75 years) but females were significantly older than males (49.2 years $\pm$ 27 verses 43.2 $\pm$ 16, $P = 0.04$). Prior to the ID consultation, 65 patients (70%) had received antibiotics (mostly ampicillin, amoxicilline, and amoxicilline/clavulanate), 45 (48.4%) anti-inflammatory drugs, and 11 (12%) corticosteroids.

Eighty-five (91%) patients had been exposed to dogs, 30 (32%) reported tick bites, and the tick was still on 18 (19%) patients at the time of the consultation. Most contaminations (78, [84%]) probably took place at the patients' home.

Underlying diseases were present in 40 (43%) patients and included diabetes, cancer, valvulopathy, asthma, high blood pressure, history of myocardial infarction, and tobacco/alcohol consumption.

The interval of time between the first symptom and the ID consultation was 6.5 days (median 6 days). A maculopapular or purpuric rash was observed in 76 patients and an eschar was present in 65 patients (trunk or legs) (TABLE 1). Of interest, two patients had two and three eschars, respectively. Sixty-two patients (67%) were hospitalized.

Most patients (85 [91%]) were treated with doxycycline (200 mg/day for 3–6 days) other patients were treated with josamycine ($N = 5$), fluoroquinolones ($N = 3$), and chloramphenicol ($N = 1$).

Evolution was favorable for 90 patients (97%). A small desquamation was observed in 18 patients (19%), apyrexia was obtained after a mean of 2.6 days (1–6 days). Asthenia and myalgia persisted for few weeks to up to 6 months in 67 patients (72%). Complications occurred in 39 (42%) patients and included myocarditis (N = 5), pericarditis (N = 3), cardiac insufficiency (N = 2), renal insufficiency (N = 8), coma (N = 2), convulsion (N = 2), cerebellar syndrome (N = 3), lymphocytic meningitis (N = 4), polynevritis (N = 6), and diabetic acidoketosis (N = 4). Three patients died (3.2%). The first one presented neurological complications, the second had respiratory and renal insufficiency, and the third one, who had diabetes and severe purpura, died suddenly despite an initial favorable evolution.

Laboratory Results

Serum samples were available for 65 (70%) patients (104 serum samples) and antibodies against any of the rickettsial strains tested were detected for 63 patients. The two seronegative patients had serum samples collected early during the course of their disease. Final diagnosis of *R. conorii* subsp. *conorii* infection was obtained by WB and cross-adsorption studies for 53 patients. For two patients, serology was in favor of TG infection and for eight patients the rickettsial species remained undetermined.

DISCUSSION

R. conorii infections are the most common rickettsial infections in the Mediterranean region but no cases of MSF from Algeria have been published in the medical literature. In Oran, the first case of MSF has been clinically diagnosed for the first time in 1993 and since that time the number of cases has steadily increased to reach 134 in 2004. This is the reason leading to the performance of this study whose main objective was to better describe MSF in Algeria.

Our study shows, by using well-documented and sophisticated serological methods, that rickettsial infections are probably underestimated and underreported in Algeria. Apart from two cases of TG infection, the majority of our patients were infected by *R. conorii*. This finding is not surprising since the main clinical signs of MSF served as inclusion criteria. Because cross-reactive antibodies are usually extensive between *R. typhi* and *R. prowazekii*, our two cases with positive serology against TG rickettsia may have been either murine typhus or epidemic typhus as recently reported in Algeria.[7,8]

Our study also confirms the existence of severe forms of MSF characterized by the occurrence of complications in 42% of the patients and three deaths.[9] Another important finding is the presence of multiple eschars that are reported for the first time in this region. There is no clear explanation for this occurrence that could be on account of other rickettsial strains or to a change in the epidemiology/behavior of the *R. sanguineus* ticks. In conclusion, our study shows that SFG rickettsiosis and TG infections are common causes of acute fever in Algeria and, therefore, should be systematically investigated.

REFERENCES

1. RAOULT, D. & J.G. OLSON. 1999. Emerging rickettsiosis. *In* Emerging Infections 3. W.M. Scheld, W.A. Craig & J.M. Hughes, Eds.: 17–35. ASM Press. Washington DC.
2. RAOULT, D. & V. ROUX. 1997. Rickettsioses as paradigms of new or emerging infectious diseases. Clin. Microbiol. Rev. **10:** 694–719
3. LETAIEF, A.O., S. YACOUB, H.T. DUPONT, *et al.* 1995. Seroepidemiological survey of rickettsial infections among blood donors in central Tunisia. Trans. R. Soc. Trop. Med. Hyg. **89:** 266–268.
4. MESKINI, M., L. BEATI, A. BENSLIMANE & D. RAOULT. 1995. Seroepidemiology of rickettsial infections in Morocco. Eur. J. Epidemiol. **11:** 655–660.
5. KULOGLU, F., J.M. ROLAIN, P.E. FOURNIER, *et al.* 2004. First isolation of *Rickettsia conorii* from humans in the Trakya (European) region of Turkey. Eur. J. Clin. Microbiol. Infect. Dis. **23:** 609–614.
6. TEYSSEIRE, N. & D. RAOULT. 1992. Comparison of Western immunoblotting and microimmunofluorescence for diagnosis of Mediterranean spotted fever. J. Clin. Microbiol. **30:** 455–460.
7. MOKRANI, K., P.E. FOURNIER, M. DALICHAOUCHE, *et al.* 2004. Reemerging threat of epidemic typhus in Algeria. J. Clin. Microbiol. **42:** 3898–3900.
8. NIANG, M., P. BROUQUI & D. RAOULT. 1999. Epidemic typhus imported from Algeria. Emerg. Infect. Dis. **5:** 716–718.
9. RAOULT, D., P. ZUCHELLI, P.J. WEILLER, *et al.* 1986. Incidence, clinical observations and risk factors in the severe form of Mediterranean spotted fever among patients admitted to hospital in Marseilles 1983–1984. J. Infect. **12:** 111–116.

Geoinformational Mapping of Foci of Siberian Tick-Borne Rickettsiosis in Altai Krai

NADEZHDA YU.KUREPINA,[a] IRINA N. ROTANOVA,[b]
ANATOLY S. OBERT,[c] AND NIKOLAI V. RUDAKOV[d]

[a,b,c]*The Institute for Water and Environmental Problems, Siberian Branch of the Russian Academy of Sciences Barnaul, Russia*

[d]*The Omsk Research Institute of Natural Foci Infections, Omsk, Russia*

ABSTRACT: The high rate of Siberian tick-borne rickettsiosis morbidity in Altai Krai calls for research into its causation as well as public heath concern. We constructed a thematic map to assess the medical-geographical situation, reveal disease foci, and define the risk level in the geographical regions covered.

KEYWORDS: Siberian tick-borne rickettsiosis; land use confinement; medical-geographical efforts; geoinformation mapping; modelling of natural foci diseases

INTRODUCTION

The Institute for Water and Environmental Problems, of the Siberian Branch of the Russian Academy of Sciences has been investigating natural focal diseases including Siberian tick-borne rickettsiosis (STBR) with the use of geoinformation systems (GIS). The high rate of Siberian tick-borne rickettsiosis morbidity in Altai Krai calls for research into ist causation and planning for public health means of dealing with it. Mapping not only demonstrates the current epidemiological situation of the given territory, but also makes it possible to assess this situation. Moreover, the use of substantial amounts of data allows forecasting recommendations with a rather high level of reliability.

Construction of thematic maps on epidemiological situations is laborious and time-consuming. Computerization and advanced software products contribute greatly to the optimization of technological and technical processes of thematic map construction, and allow one to update the database on population morbidity, including other qualitative and quantitative information, and

Address for correspondence: Molodezhnaya St., Barnaul 656038, Altai Krai, Russia. Voice: 8-385-2-666-458; fax : 666-460/385-2-240-396.
e-mail: kurepina@iwep.ab.ru

Ann. N.Y. Acad. Sci. 1078: 185–187 (2006). © 2006 New York Academy of Sciences.
doi: 10.1196/annals.1374.034

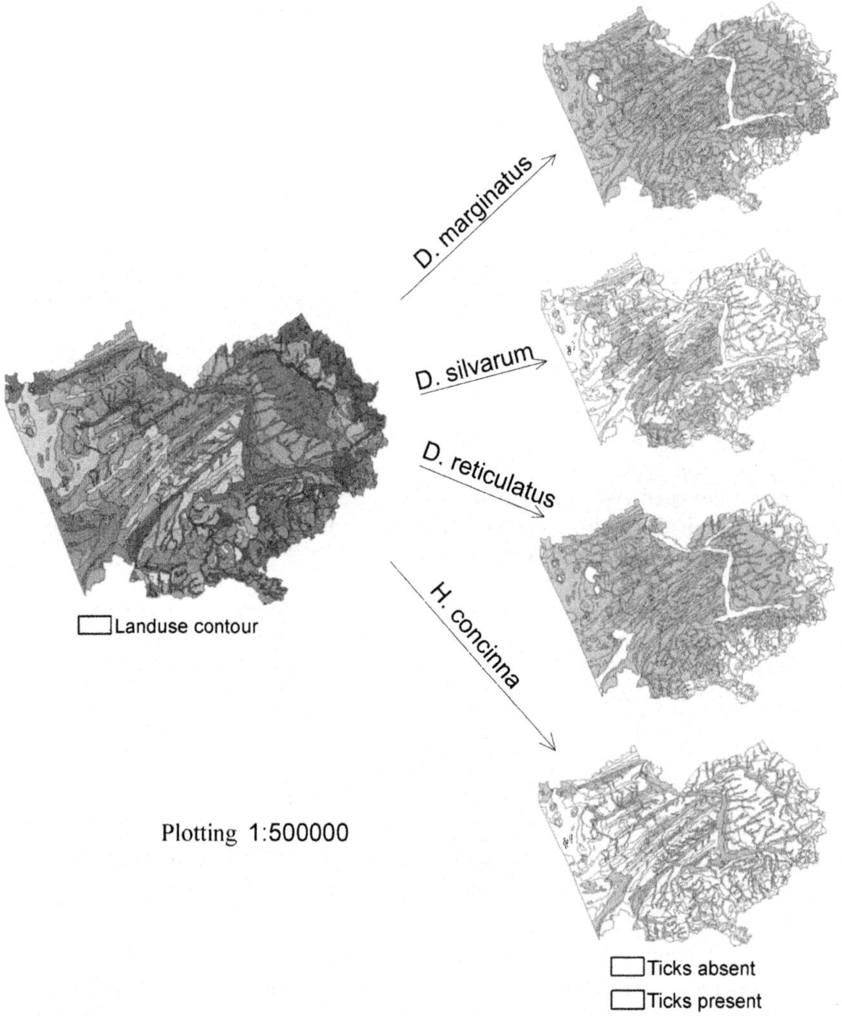

FIGURE 1. Landscape confinement of Siberian tick-borne rickettsiosis by its major carriers in Altai Krai territory.

to make thematic layers overlay, to integrate and visualize them more clearly. Our ability to plan preventive measures is improved by modeling scenarios of medical-geographical situations. Automation of major steps of research enables consideration of continuously changing abiotic, biotic, and athropogenic conditions that have effects on the medical-geographical and epidemiological situation.

The physical-geographical conditions of the Altai Krai territory contribute to the maintenance of natural factors that keep the STBR pathogen in circulation.

The main carriers of disease are the following tick species: *Dermacentor marginatus, D. silvarum, D. reticulates,* and *Haemaphysalis concinna*. Each of these tick species has its own habitat. Tick confinement to definite biotopes are analyzed by studying tick species composition. Since each landscape appears to have a specific set of natural prerequisites for diseases, tools for reduction of morbidity mapping is made on the landscape (FIG. 1). Most of research work was done in the ArcView GIS-environment. The step-by-step overlay analysis of thematic layers (locations where people were infected, distribution of main ticks' feeders, population density, land use, etc., as well as statistical data processing on morbidity for 20 years) made it possible to assess the medical-geographical situation, reveal disease foci, estimate the zones and ranges of medical-geographical problems, define the risk level, and reveal the most critical medical-geographical situations and their areas.

The merits of applying GIS technologies to medical-geographical investigations are in allowing more flexible and quick response mechanisms for monitoring an outbreak, as well as possibly making corrections of the prophylactic measures taken for any specific epidemiological situation.

ACKNOWLEDGMENTS

This research was supported by the Russian Humanitarian Research Foundation Grant 05-06-15026.

The Foci of Scrub Typhus and Strategies of Prevention in the Spring in Pingtan Island, Fujian Province

GUO HENGBIN, CAO MIN, TAO KAIHUA, AND TANG JIAQI

Huadong Research Institute of Medicine and Biotechnics, Nanjing (21002), China

ABSTRACT: This study investigates the foci of tsutsugamushi disease on Pingtan Island, China, with emphasis on cases of illness that occur in the spring. The investigation of the physical and medical geography in the endemic area and detection of 141 sera samples in the population of Pingtan Island is described. The serum samples were taken from 102 soldiers and 39 residents and were tested by the indirect immunofluorescence assay. An island resident's history of illness inquiry was also conducted, and the results are discussed. We were able to determine the most prevalent types of rats and mites from the rat bodies. We isolated the DNA from the spleen of 40 rats and from five groups of mites by polymerase chain reaction. *Orientia tsutsugamushi* was isolated by perctoneally injecting KM mice with the patient's untreated blood along with ground rat viscera (*R. losea*) and ground mites (*L. deliense*), respectively. The isolates were typed at the molecular level. This study will then present direct prevention strategies that emphasize personal hygiene, methods of personal protection, keeping living environments clean, and it will provide strategies to eliminate rats and mites. As a result of adopting these strategies, no case of scrub typhus occurred in this region in approximately 2003–2004.

KEYWORDS: tsutsugamushi disease in spring; IFA; *Orientia tsutsugamushi*; prevention strategies; scrub typhus; spring

Scrub typhus is a disease that is endemic in tropical and subtropical zones in Southeast Asia. In China the disease was previously found to be endemic in the southern regions. Since 1986, there have been epidemics of scrub typhus in northern China, in northern Jiangsu province, Sandong province, the city of Tianjin, and in Sanxi and Hebei provinces.

Since the 1990s, detection of the disease has been more prevalent, with cases being reported almost every year in basic units of the army. Moreover, cases in the spring that had not been reported previously were noted. Because the health

Address for correspondence: Guo Hengbin, Huadong Research Institute of Medicine and Biotechnics, Nanjing (21002), China. Voice: 0086-25-84412432; fax: 0086-25-84507094.
e-mail: hbguo2003@yahoo.com.cn

Ann. N.Y. Acad. Sci. 1078: 188–196 (2006). © 2006 New York Academy of Sciences.
doi: 10.1196/annals.1374.130

of the soldiers and the general population were threatened, it was necessary to investigate the epidemiology of scrub typhus in the southern regions of China and to develop prevention strategies.

MATERIALS AND METHODS

Geographic, Meteorological, and Vegetational Data

The data were obtained from the bureaus of meteorology, environmental protection, and forestry, respectively, in the form of field inquiries and surveys.

Serological Epidemiology and the History of the Illness

Serum samples from 141 individuals were collected from people living on Pingtan Island. The sera obtained from 102 soldiers and 39 residents were tested by indirect immunofluorescence. The criteria for positive were as follows: If more than 10 specific fluorogenic *O. tsutsugamushi (Ot)* particles were detected in one microscopic field of vision, the serum was considered positive. In addition, 367 residents were queried regarding the history of the illness.

Investigation of the Animal Host and Vector Mite

During the evening, rat cages were placed in different locations, including inside the house, dining room, haystack, vegetable plot, and ridge where rats are active. In the morning mites found on rats were collected. We then determined the most prevalent kind of rat and mite. The rate at which rats carried mites was also calculated. The liver, spleen, and kidney of the rats were then obtained under sterile conditions and were stored in 95% ethanol.

Isolation and Identification of Orientia Tsutsugamushi

Isolation of O. tsutsugamushi

Mice were injected in the peritoneum with whole human blood and viscera of rats or *Leptotrombidium deliense*, all of which were collected from the foci of the disease in Pingtan Island. The test results were scored as described in Ref. 2. The isolates were identified and analyzed by PCR and sequencing.

PCR Identification

The *O. tsutsugamushi* DNA was abstracted from the viscera of the infected rats by a routine method that used a template for PCR. Forward primer 5′-GGG GAT CCG GAT TTA GAG CAG AG-3′ was designed according to the 791–805 bp, 252–266 bp, 258–272 bp sequence of the 56-kDa protein gene of the *O. tsutsugamushi* standard strains Gilliam, karp, and kato, respectively. The backward primer 5′GGC GAA TTC AAA AAC TAG AAG TTA TAG CG-3′ was designed according to the 2114–2134 bp, 1584–1604 bp, 1645–1665 bp sequence of the 56-kDa protein gene of the *O. tsutsugamushi* standard strains Gilliam, karp, and kato, respectively. The PCR reaction system and the reactive condition were the same as described in Ref. 2. The PCR products were detected by 1% agarose gel electrophoresis.

Sequence Analysis

The gene fragments of the PCR products were cloned into vector PGEM-T, and the sequence was determined by the Shenyou Company in Shanghai and analyzed by soft DNA SYS .

Preventive Strategies

Propagating News About Prevention

Our institute made efforts to communicate knowledge about disease prevention to the army. Specialists from our institute produced a poster titled “How to prevent the disease of scrub typhus along the southeast coast.” In addition, once a year, a team of experts was sent to the army units.. Lectures were presented by the team, which included information and recommendations concerning personal hygiene and how to eradicate rats and mites before the epidemic season.

Personal Hygiene

During the height of the season for scrub typhus, when soldiers were marching, camping, or working in the field, they were instructed to tighten the cuff on their sleeves and trouser legs, and to tuck in their shirts. Soldiers were instructed not to sit or lie down on the grass and not to dry clothes or bedclothes by hanging them on the brush or on the branches.

Soldiers were also informed that dimethyl phthalate applied to the hands, neck, ears, and legs would provide protection for a period of 2 h. However, when exposed to water, protection ceases. When 70% dimethyl phthalate is combined with 30% dibutyl phthalate, protection could last up to 8 h, and protection may still be effective to some extent after the body is submerged in water three to four times. Clothes could be treated as follows: Apply 2–3% ambrocide on the cuffs of the trousers, sleeves, and the neckband to form a 10-cm wide protective belt. This should be done once, every three to five days. In addition, the clothes (including socks) should be soaked in a dibutyl phthalate emulsion of 670 mL per set/clothes. The solution used could still provide protection to clothing for up to four wash cycles. Tetracyclines could also be used as a preventive measure if conditions permit.

In addition, it was reported that mites typically do not bite until they have been on their host for at least 20–30 minutes and up to 1 h. Therefore, a regular change of clothes, bathing, and washing of body parts that have soft skin, such as armpits, waists, and the perineum reduced the chances of being bitten.

Cleanliness of the Living Environment

(1) *Inside the house*: Clean the house by throwing away trash, especially from dark and damp areas of the house. Spray insecticides, such as dichlorvos, inside and around the house and barracks. (**Warning:** people cannot stay in the house immediately after the insecticide is applied)

(2) *Camping in the field*: A region should be chosen that is dry, high, and sunny; campers should avoid the damp, dark, and low-lying land where rats might be. Overgrown weeds around the barracks or tent should be uprooted so that a 1-m perimeter belt is established. This area should be sprayed with insecticide. The uprooted weeds should be moved away from the barracks and buried deeply below the ground, burned, or sprayed with pesticide.

Preventive Strategies for Killing Rodents and Mites with Poison

(1) *Methods for wiping out the mites*: Remove overgrown weeds in camp or training sites, or along roads. Level the ground to expose it to more sunlight and lower the humidity, making it unsuitable for mite propagation. For areas where weeds cannot easily be removed, apply 1% dichlorvos 100–200 mL/m^2 every 7–10 days. Spread straw used to weave straw mattresses (3–5 cm thick) on the ground and spray once with 2% dichlorvos solution, 1000 mL/m^2. Turn the mattresses over and spray again. The mattresses should be dried in the sun and stored.

(2) Methods to prevent rodent infestation include regular cleaning of the inside and outside of buildings, keeping rooms clean, storing grain and vegetables at a height above 0.5 m and away from the wall, and filling holes and cracks in walls. The management of both food and human waste should be

FIGURE 1. Lush vegetation of the island.

intensified, and trash should be removed every day to decrease the habitat and the food resource for rodents.

Finally, the rat poison currently in use is a slow poison. It is made with warfarin, diphacinone, coumatetralyl, and bromadiolone. However, different types of poison should be alternated so that rats do not build resistance.

RESULTS

Foci of Scrub Typhus

Scrub typhus foci in south subtropical zones (FIGS. 1 and 2) are present. Summer is longer than winter, it is warm and humid; rarely is there any snow or frost. In the spring the temperature is lower than in the autumn. In this region, the average yearly temperature is 19.5°C. The coldest month (February)

FIGURE 2. Arid vegetation of the island.

averages 10.5°C. The hottest month (July) averages 27.8°C. Rain and high temperature exist in the same season. Yearly precipitation averages 119.3 mm, which is one of the lowest precipitation rates in the Fujian province. The average relative humidity is 81–84%, and rich vegetation is suitable for mite and rodent survival.

Seroepidemiology and History of Illness

We tested 141 sera samples: 102 soldiers' sera and 39 residents' sera by an indirect immunofluorescence assay. Forty-six of the 141 sera samples were positive, showing a positive rate of 32.6% (16/39). Thirty sera scored positive from soldiers (29.4%) and 16 sera were positive from residents (41%). After asking the 367 residents about the history of their illness, 36 reported having suffered the illness in the past; the incidence of illness was 17.2%.

Vector and Host

From April to May in 2004, we collected 2,432 trombidium and 40 gamasid mites from rat ears. *L. deliense* was found to be the dominant biological vector, numbering 2,300 (94.6%), of which 480 were collected in April. Besides these, 110 *L. intermedium* and 22 *L. yu* were also found.

Rodents are the main hosts for these mites. A total of 246 rats were captured, including 80 *Rattus losea,* 45 *R. flavipetus,* 5 *R. confucianus,* and 50 *R. norvegicus; and* 66 *Suncus murinus* in which *R. losea* made up 32.5% of the total collection. The carrying mite rate and carrying mite index of *R. losea*, *R. flevipectus*, *S. murinus*, *R. confucianus*, and *R. norvegicus* were 87.5%, 19.9%; 33.3%, 1.8%; 3.2%, 4.73%; 0.0%; and 10%, 9%, respectively. The carrying mite rate of the rodents was 45.1%.

Detection of Molecular Epidemiology

Forty rats and five groups of mites were tested by PCR. Four rats and two groups of mites scored positive at a specific 430 bp DNA band; FIG. 3).

Isolation and Identification of O. tsutsugamushi

Isolation of O. tsutsugamushi

Three *O. tsutsuugamushi* strains designated Pt1, Pt2, and Pt3 were isolated from the patient's untreated blood, the ground viscera of *R. losea*, and the ground mite, respectively. (No *O. tsusugamushi* strains were isolated from the inoculated *R. norvegicus* and *S. murinus).*

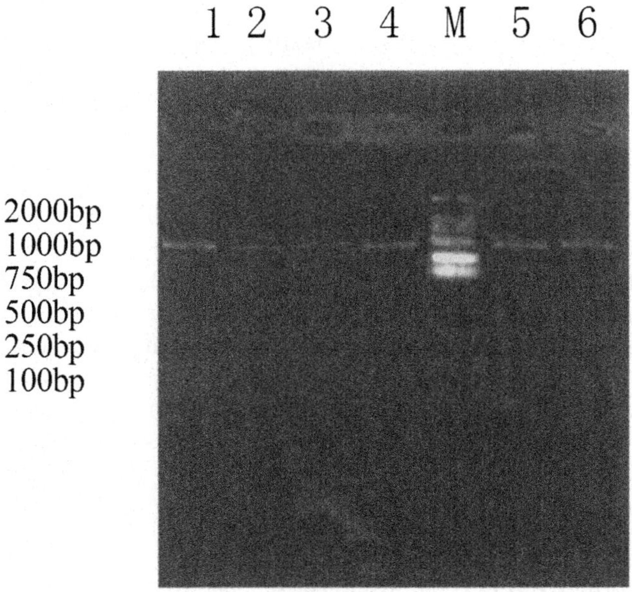

FIGURE 3. *Lanes* 1–4: *O. tsutsugamushi* DNA detected from the mice spleen; *lanes* 5–6: *O. tsutsugamushi* DNA detected from the vector mite.

Result of PCR

The DNA of three *O. tsutsugamushi* isolates was amplified using the primer P1 and P2 pair, with the result that 1350 bp specific bands of 56-kDa protein gene of *Ot were* displayed (FIGS. 4 and 5).

Result of Prevention

The above four preventive strategies that we adopted seemed to be effective for disease prevention within the troops and the hospital serving the troops as there were no new apparent cases of scrub typhus in the island in the troop garrison in 2000~2004, when the preventive measures were applied.

DISCUSSION

In recent years, cases of scrub typhus have emerged in Pington Island in the spring. This finding had not been reported before. We made an epidemiological investigation of scrub typhus in the spring, looking into the types of foci, landscape, vegetation, host, vector, and the endemic characteristics of the disease. This study verified the existence of foci of scrub typhus from many

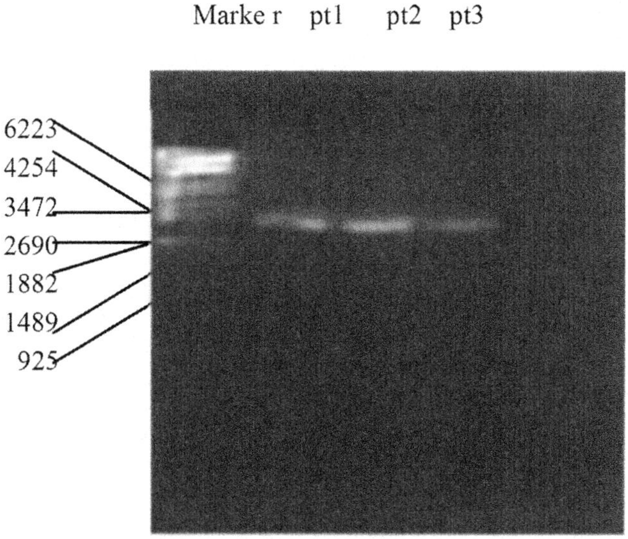

FIGURE 4. The gene fragments of 56 kDa protein of three *O. tsutsugamushi.* strains isolated from the Pingtan Islands.

areas, such as natural and medical geology, vector, host, seroepidemiology, and *Ot* isolates. The island is in the south subtropical zone; *R. losea* is the prevalent rat as the host, and *L. deliense* was found in the rat's body in early April. The mite's earlier appearance is consistent with the earlier appearance of scrub typhus cases in the spring. The results of seroepidemiological investigation showed that the positive rate of anti-*Ot* antibody was 32.6% (46/141). The anti-*Ot* antibody-positive rate of the residents (41%) was higher than that of army personnel (29.4%), which was a result of the basic sanitary unit's emphasis on disseminating the knowledge of illness prevention.

In this study of 22 sera of children younger than 14 years old, 9 were positive, which accounted for more than half of the total positive samples. This was due

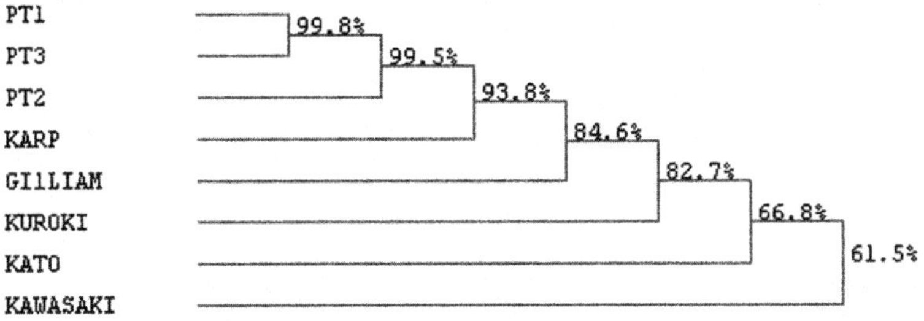

FIGURE 5. Comparison of 56 kDa protein gene sequences of *O. tsutsugamushi* from Pingtan Islands with reference strains.

to the fact that the grassland, which is also a prime mite habitat, was found to be a common play area for many children. This increased their risk of contracting the disease. In addition, we found that among army personnel who tested positive for anti-*Ot* antibody, 7 had suffered from the disease within the previous 3 years, and most of them were new recruits and therefore more susceptible to disease contraction. We also learned through informal inquiry that 63 of 365 residents had also suffered from the disease. This information revealed that the infection of *Ot* was especially prevalent in this area.

In this study *O. tsutsugamushi* strains, *R. losea* and *L. deliense*, respectively, were isolated from the patient. The investigation revealed that the dominant rat was *R. losea,* and *L. deliense* was the dominant mite. Investigation also revealed that the *Ot* infection rate was high. These findings are in accordance with results from research conducted by Yu Eengsu[5] in 1953. In short, through our comprehensive study on the epidemiology of scrub typhus, we developed relevant preventative strategies and concluded that these strategies were effective and could be used by troops.

REFERENCES

1. MIN, CAO, Guo HENGBIN, Li YUEXI, *et al.* 2003. The development of the method of detecting Ot DNA using gene chip technique. Chinese J. Zoonoses **19:** 19–22.
2. HENGBIN, GUO, Xu MANHUA, Yu MINGMING, *et al.* 1992. Studies on methods of the isolation of the attenuate Ot strains. Chinese J. Zoonoses **8:** 5–6.
3. WEI, Xi, Eds. Studies on the Medical Pathogen of Rickettsia. Scientific Technique Press. Shanghai. pp. 296–297.
4. YU ENGSU, CHEN XIANGREI,WU GUANGHUA, *et al.* 2000. Studies on Investigating The Scrub Typhus, pp. 50–93. Asia, The Medicine Press. Hong Kong.
5. FAN MINGYAN, Ed. 1988. Yu Engsutreatise collectanca. pp. 169–176. Chinese Journal of Epidemiology Press.

Detection and Identification of a Novel Spotted Fever Group Rickettsia in Western Australia

HELEN OWEN,[a] NATHAN UNSWORTH,[b] JOHN STENOS,[b]
IAN ROBERTSON,[a] PHILLIP CLARK,[a] AND STAN FENWICK[a]

[a]*School of Veterinary and Biomedical Sciences, Murdoch University, South Street, Murdoch, Western Australia 6150, Australia*

[b]*Australian Rickettsial Reference Laboratory, Douglas Hocking Research Institute, Geelong Hospital, Bellerine Street, Geelong, Victoria, Australia, 3220*

ABSTRACT: The extent to which rickettsiae are present in Western Australia (WA) is largely unknown. Recently there has been anecdotal evidence of a disease of unknown but possibly rickettsial origin occurring on Barrow Island, WA. Ticks were collected from people and screened using PCR. The rickettsial species was then cultured and its novelty and phylogenetic position examined. The infecting rickettsial species is divergent enough to be classified as a novel species. Sequence data suggest that the evolutionary route for Australian rickettsiae did not progress through a recent common ancestor. The pathogenic potential of the novel species is as yet unknown.

KEYWORDS: spotted fever group rickettsiae; Western Australia; Candidatus Rickettsia gravesii

Rickettsioses have long been recognized to occur in WA when murine typhus was first reported in 1927. Recently, a case of scrub typhus has been reported and preliminary epidemiological studies have detected spotted fever antibodies in human sera. While active research continues on *Rickettsia australis* and *R. honei*, rickettsiae isolated from the eastern states of Australia, there is a limited amount of current information on the nature of rickettsial presence in WA.[1–3]

Barrow Island, located 60 km off the northwest coast of WA, was targeted for sample collection after there was anecdotal evidence of a disease of unknown but possibly rickettsial origin occurring on the island. *Amblyomma triguttatum* ticks were removed from people on the island. DNA was extracted from the ticks using Chelex-100 resin (Bio-Rad Laboratories) and a protocol described

Address for correspondence: Helen Owen, School of Veterinary and Biomedical Sciences, Murdoch University, South Street, Murdoch, Western Australia, 6150. Voice: 08-9360-6379; fax: 08-9310-4144. e-mail: 19507648@student.murdoch.edu.au

Ann. N.Y. Acad. Sci. 1078: 197–199 (2006). © 2006 New York Academy of Sciences.
doi: 10.1196/annals.1374.038

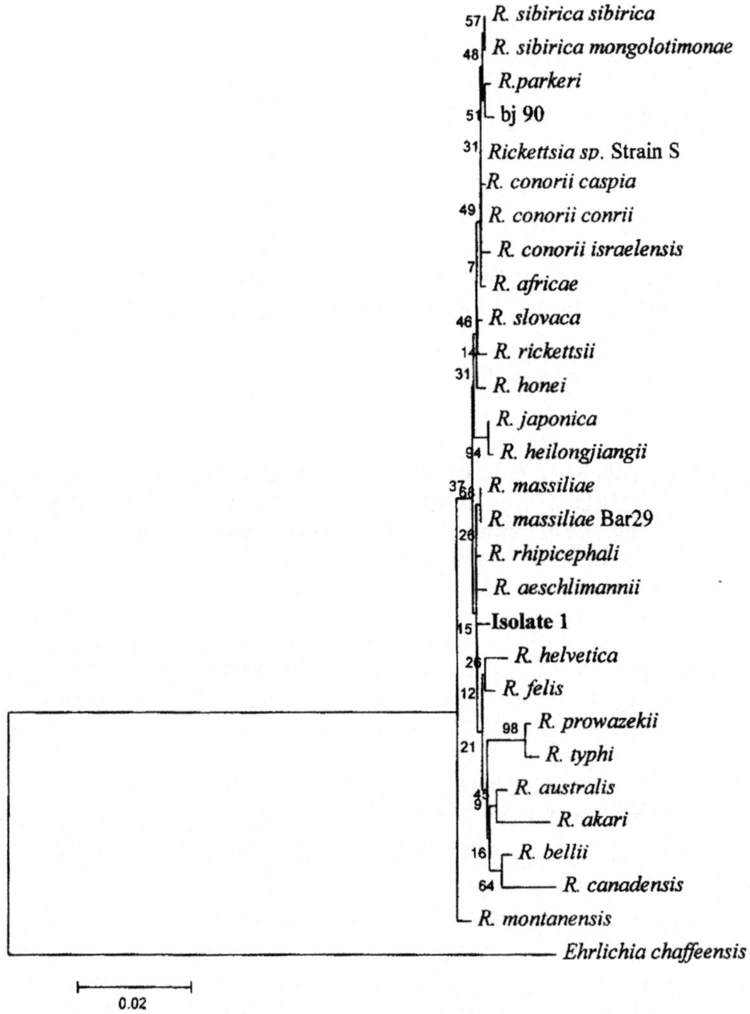

FIGURE 1. Phylogenetic tree of *Rickettsiae* constructed using the neighbor-joining method and based on 16S rRNA sequence comparison. Sequence alignment was performed using the multisequence alignment program CLUSTAL in the BISANCE software package. Phylogenetic and molecular evolutionary analyses were conducted using MEGA version 2.1.[6] The novel isolate described in this study is isolate one.

by Guttman *et al.*[4] Screening for the presence of rickettsiae was performed by PCR using *gltA* gene primers. The infecting rickettsial species was cultured in XTC_2 and Vero cell lines and its novelty and phylogenetic position examined by sequencing segments of the 16S rRNA, *gltA, ompA*, 17-kDa antigen, *ompB,* and *sca4*.

Fifteen of the 32 *Amblyomma triguttatum* ticks collected were PCR-positive. The infecting rickettsial species is novel according to the guidelines outlined by Fournier *et al.*[5] It clusters most consistently with the *R. massiliae* subgroup of the spotted fever group (FIG. 1), demonstrating 99.7%, 98.4%, 95.6%, 97.4%, 96.6%, and 99.2% nucleotide sequence similarity to members of the subgroup on the basis of its 16S rRNA, *gltA*, *ompA*, *ompB*, *sca4*, and 17-kDa antigen genes, respectively.

These sequence data suggest that the evolutionary route for Australian rickettsiae did not progress through a recent common ancestor, but is more likely the result of multiple independent introductions of rickettsiae. The next phase of the project will involve further characterization of the isolate and comparison with known pathogenic rickettsiae, investigation into whether a disease with symptoms consistent with a rickettsiosis occurs on the island, and, ultimately, an attempt to detect the isolate from a human sample. This will be done with the aim of determining the pathogenic potential of the isolate.

We propose the creation of one new species, *Candidatus* Rickettsia gravesii (graves.i.i. L. gen. n. *gravesii* after Stephen Graves, an Australian physician and microbiologist who is the founder and director of The Australian Rickettsial Reference Laboratory and has a long history of promoting and contributing to Australian rickettsial research).

REFERENCES

1. ODORICO, D., S. GRAVES, B. CURRIE, *et al.* 1998. New *Orientia tsutsugamushi* strain from scrub typhus in Australia. Emerg. Infect. Dis: 641–644.
2. GRAVES, S., L. WANG, Z. NACK & S. JONES. 1999. Rickettsia serosurvey in kimberly, Western Australia. Am. J. Trop. Med. Hygiene **60:** 786–789.
3. KILMINSTER, T. 1997. An investigation of typhus in Western Australia. B.Sc. thesis, Department of Microbiology, University of Western Australia, Nedlands, Western Australia.
4. GUTTMAN, D., P. WANG, I. WANG, *et al.* 1996. Multiple infections of *Ixodes scapularis* ticks by *Borrelia burgdorferi* as revealed by single-strand confirmation polymorphism analysis. J. Clin. Microbiol. **34:** 652–656.
5. KUMAR, S., K. TAMURA, I.B. JAKOBSEN & M. NEI. 2001. MEGA2: molecular evolutionary genetics analysis software. Arizona State University. Tempe, Arizona, USA.
6. FOURNIER, P., J. DUMLER, G. GREUB, *et al.* 2003. Gene sequence-based criteria for the identification of new *Rickettsia* isolates and description of *Rickettsia heilongjiangensis* sp. nov. J. Clin. Microbiol. **41:** 5456–5465.

Low Incidence of Tick-Borne Rickettsiosis in a Spanish Mediterranean Area

ANTONIO GUERRERO,[a] FLOR GIMENO,[a] JAVIER COLOMINA,[a] MERCÉ MOLINA,[a] JOSE ANTONIO OTEO,[b] AND MARIA CUENCA[a]

[a]*Hospital de La Ribera, Ctra Corbera Km 1, 46600 Alzira-Valencia, Spain*

[b]*Hospital de La Rioja, 26001 Logroño, Spain*

ABSTRACT: The aim of this study was to know the incidence of tick-borne rickettsial disease in a Mediterranean area. The incidence in 5 years for 100,000 inhabitants was 1.7 for tick-borne-lymphadenopathy (TIBOLA) and 0.4 for Mediterranean spotted fever (MSF). MSH incidence during the last few years has been lower than expected; in contrast TIBOLA seems to be an emerging disease. At the present time our data suggest a low tick-borne rickettsiosis incidence of MSF and a superior incidence of TIBOLA than MSF in the Spanish Mediterranean area.

KEYWORDS: incidence; tick-borne; rickettsiosis; Spain; Mediterranean

INTRODUCTION

In 1997 *Rickettsia slovaca* (*R. slovaca*) was associated to tick-borne-lymphadenophathy (TIBOLA). This disease is also called in Spain dermacentor-borne-necrosis-erythema-lymphadenopathy (DEBONEL).[1,2] Main tick-borne rickettsial diseases in Spain include Mediterranean spotted fever (MSF) and the emergent disease, TIBOLA/DEBONEL, but there are few regional epidemiologic data about them.[2] The aim of this study was to know the incidence of tick-borne rickettsial diseases in a Mediterranean area of 240,000 inhabitants (10th Public Health Area of Valencian Community).

METHODS

All patients referred to our center (Hospital de La Ribera) from January 1999 to January 2005 with a suspected rickettsial infection were included in

Address for correspondence: A. Guerrero, Hospital de La Ribera, Ctra Corbera Km 1, 46600 Alzira-Valencia, Spain. Voice: 96-245-82-62; fax: 96-245-81-51.
e-mail: aguerrero@hospital-ribera.com

**Ann. N.Y. Acad. Sci. 1078: 200–202 (2006). © 2006 New York Academy of Sciences.
doi: 10.1196/annals.1374.039**

the study. For each patient, epidemiologic and clinical data were collected by the consulting physician by use of a standardized questionnaire. For a microbiological diagnosis, IgG and IgM antibody estimation was performed by the use of immunofluorescence (IF) assay or polymerase chain reaction (PCR) technique. Clinical case definitions of disease included characteristic inoculation eschar, "tache noire," at the site of the tick bite.[3]

A field study was also made to detect rickettsial parasitic infection of tick. Seminested ompA PCR (using primers Rr190.70p and Rr190.701n for primary PCR, and Rr190.70p and Rr190.602n for the second run) of these specimens and subsequent sequence analysis was made for identified *R. slovaca* infection.[4]

RESULTS

One hundred ten serum samples were analyzed for MSF; 6 (5%) were IgG positive for *Rickettsia conorii* and only 1 (17%) of them presented typically clinical symptoms (diffuse rash, fever, and eschar). In four patients with TIBOLA (scalp location of the eschar and local lymphadenopathy) and exclusion of other etiology, the results of *R. slovaca* IF assay were negative in all of them. All patients had been bitten during the colder months of the year.

In the field study, 15 ticks were collected during 2003–2004. Three of them were identified as *Dermacentor marginatum*. PCR and sequence analysis of these specimens identified *R. slovaca* infection in one of them.

The incidence in 5 years for 100,000 habitants was 1.7 for TIBOLA and 0.4 for MSF.

DISCUSSION

Until recently the only rickettsiosis reported from Europe was MSF. *R. slovaca* has been detected in the European countries and a series of cases have been communicated from Hungary, France, and Spain.[5]

The incidence of MSF during the last few years has been lower than expected; in contrast TIBOLA seems to be an emerging disease. The rickettsioses have characteristic clinical features including fever, headache, maculopapular eruption, and eschar formation (primary lesion). An eschar or cutaneous necrosis caused by rickettsial vasculitis at the tick-bite site of inoculation, known as *tache noire* ("black spot"), is pathognomonic. However, it could be seen in all TIBOLA cases but of MSF in only half of the patients; for that reason the incidence, according to our inclusion criteria, can be infravalued.

Boutonneuse fever had demonstrated an increased incidence in Mediterranean countries, such as Spain, Italy, and Israel. However, at the present time our data suggest a low tick-borne rickettsiosis incidence of MSF and a superior incidence of TIBOLA than MSF in the Spanish Mediterranean area.

REFERENCES

1. Bernabeu-Wittel, M. & F. Segura-Porta. 2005. Enfermedades producidas por Rickettsia. Enferm Infecc. Microbiol. Clin.. **23:** 163–172.
2. Oteo, J.A. & V. Ibarra. 2002. DEBONEL (Dermacentor-borne-necrosis-erythema-lymphadenopathy). Una nueva enfermedad transmitida por garrapatas? Enferm. Infecc. Microbiol. Clin **20(2):** 51–52.
3. Brouqui, P., F. Bacellar, G. Baranton, *et al.* 2004. ESCMID Study Group on Coxiella, Anaplasma, Rickettsia and Bartonella; European Network for Surveillance of Tick-Borne Diseases. Guidelines for the diagnosis of tick-borne bacterial diseases in Europe. Clin. Microbiol. Infect. **10:** 1108–1132.
4. Regnery, R.L., C.L. Spruill & B.D. Plikaytis. 1991. Genotypic identification of rickettsiae and estimation of intraspecies sequence divergence for portions of two rickettsial genes. J. Bacteriol. **173:** 1576–1589.
5. Raoult, D., A. Lakos, F. Fenollar, *et al.* 2002. Spotless rickettsiosis caused by Rickettsia slovaca and associated with Dermatocentor ticks. Clin. Infect. Dis. **34:** 1331–1336.

Public Health Problem of Zoonoses with Emphasis on Q Fever

E. BEŠLAGIĆ, S. HAMZIĆ, O. BEŠLAGIĆ, AND Š. ZVIZDIĆ

Medical Faculty University of Sarajevo, 71000 Sarajevo, Bosnia and Herzegovina

ABSTRACT: Zoonoses are animal and human diseases. Q fever is primarily a zoonosis—an animal disease that can be transmitted to humans under certain conditions. Recent epidemiological studies suggest that Q fever should be considered as a public health problem in many countries where it is present, but unrecognizable due to inadequate disease controls. Through specific serological diagnosis of clinically suspected human Q fever cases, we are trying to determine a level of general *Coxiella burnetii (C. burnetii)* exposition among populations in different regions of Bosnia and Herzegovina. This would be a contribution in controlling the present and the future disease outbreaks, as well as its prevention, which is one of the prime objectives of public health. During the period from January to June 2004, in the Laboratory of the Department for Microbiology in the Medical Faculty of the University of Sarajevo, of 58 tested sera from 48 clinically suspected individuals, we confirmed the presence of specific anti-*C. burnetii* antibodies in 30 sera (51.7%), from 25 seropositive individuals (52.0%), by means of indirect immunofluorescent antibody (IFA) testing. Urgent steps must be taken in public education to help decrease the risk of *C. burnetii* infection among at-risk populations in regions of Bosnia and Herzegovina.

KEYWORDS: zoonoses; public health problem; Q fever; Bosnia and Herzegovina

INTRODUCTION

Zoonoses are animal and human diseases. Q fever is primarily a zoonosis—an animal disease that can be transmitted to humans under certain conditions. The causal agent is *Coxiella burnetii (C. burnetii)*. Its extraordinary ability to maintain pathogenesis in the external environment, as well as its resistance against drying and the effects of physical and chemical factors on one hand, and its confirmed aggressiveness and virulence of minimal infectious doses on the other hand, are increasing the risk of disease outbreak.[1] In nature, the

Address for correspondence: E. Bešlagić, Medical Faculty University of Sarajevo, Čekaluša 90, 71000 Sarajevo, Bosnia and Herzegovina. Voice/fax: +387-33-202-050.
e-mail: edinabeslagic@yahoo.com

Ann. N.Y. Acad. Sci. 1078: 203–205 (2006).
doi: 10.1196/annals.1374.094

causal *C. burnetii* is isolated from many wild animals. The most frequently infected domestic animals include cattle, sheep, and goats.[2]

Infection via *C. burnetii* aerosols can occur directly from birth fluids of infected animals, which can infect newborn animals, placenta, or wool.[3] Contaminated dust and aerosols can be transmitted even for a distance of a few kilometers by wind, during dry and windy weather.[4,5] Recent epidemiological studies suggest that Q fever should be considered as a public health problem in many countries where it is prevalent but unrecognizable on account of insufficient and inadequate disease controls. Many human infections are unapparent. Humans are often very susceptible to this disease, since very few microorganisms are necessary to cause infection.

MATERIALS AND METHODS

During the period from January to June 2004, 58 sera from 48 clinically suspected individuals were tested in the laboratory for specific diagnosis of human Q fever, in the Department for Microbiology of the Medical Faculty University of Sarajevo, where we used the indirect immunofluorescent antibody (IFA) test.

IFA test was performed as:

- IFA IgM Q fever test for detection of human IgM antibodies for *C. burnetii*
- IFA IgG Q fever test for detection of human IgG antibodies for *C. burnetii*

RESULTS

Of 48 clinically suspected individuals during the specified period, human Q fever was serologically confirmed in 25 persons (52.0%). Of 58 sera tested, specific anti-*C. burnetii* antibodies were detected in 30 sera (51.7%) (TABLE 1).

Of 58 tested sera from 48 individuals, anti-*C. burnetii* antibodies were detected in 30 sera. Specific IgM antibodies were found in 18 sera, and specific IgG antibodies in 28 sera. Specific anti-*C. burnetii* antibodies were not found in the sera from 27 individuals. Of 18 IgM positive sera, all 18 sera had the phase II higher than the phase I. In 27 IgG positive sera, the phase II was higher than the phase I, while we found the phase II to be equal to the phase I in 1 serum (TABLE 2).

TABLE 1. The sample tested during the period from January to June 2004

		Positive	
January–June 2004	Tested	Number	%
Individuals tested	48	25	52.0
Sera	58	30	51.7

TABLE 2. The values of phase I- and phase II-specific antibodies in positive sera

Sera	Phase II > Phase I	Phase II < Phase I	Phase II = Phase I
IgM-positive	18	–	–
IgG-positive	27	–	1

Seroconversions were reported in four cases. Specific anti-*C. burnetii* antibodies were found in the first 22 sera and 8 pairs of sera collected from tested individuals after 2 weeks. The results obtained show the acute Q fever phase in all individuals.

CONCLUSIONS

- Q fever is an important public health problem in Bosnia and Herzegovina.
- We serologically confirmed the acute form of Q fever in 52.0% of clinically suspected individuals during the first half of 2004, using IFA test.
- The presence of specific anti–*C. burnetii* antibodies in the sera from tested individuals confirmed the distribution of human Q fever in Bosnia and Herzegovina.
- Our studies suggest that the disease occurs among a wider group of population, which indicates the greater public health significance of this disease.
- Urgent steps should be taken to focus on public education, which would help decrease the risk of *C. burnetii* infection.
- Supervising and controlling disease outbreaks in already determined loci, detecting new loci through the research actions, and defining sources and routes of disease transmissions.

REFERENCES

1. RAOULT, D., H. TISSOT DUPONT, C. FOUCAULT, *et al.* 2000. Q fever 1985–1998. Clinical and epidemiologic features of 1383 infections. Medicine **79:** 109–123.
2. FISHBEIN, D.B. & D. RAOULT. 1992. A cluster of *Coxiella burnetii* infections associated with exposure to vaccinated goats and their unpasteurized dairy products. Am. J. Trop. Med. Hyg. **47:** 35–40.
3. TISSOT DUPONT, H., D. RAOULT, P. BROUQUI, *et al.* 1992. Epidemiologic features and clinical presentation of acute Q fever in hospitalized patients—323 French cases. Am. J. Med. **93:** 427–434.
4. MARRIE, T.J. & D. RAOULT 1997. Q fever—a review and issues for the next century. Int. J. Antimicrob. Agents **8:** 145–161.
5. TISSOT DUPONT, H., S. TORRES, M. NEZRI & D. RAOULT 1999. A hyperendemic focus of Q fever related to sheep and wind. Am. J. Epidemiol. **150:** 67–74.

Rickettsia slovaca Infection: DEBONEL/TIBOLA

V. IBARRA,[a] J.A. OTEO,[a] A. PORTILLO,[a] S. SANTIBÁÑEZ,[a] J.R. BLANCO,[a] L. METOLA,[a] J.M. EIROS,[b] L. PÉREZ-MARTÍNEZ,[a] AND M. SANZ[a]

[a]*Área de Enfermedades Infecciosas, Hospitales San Millán-San Pedro-de La Rioja, Logroño, Spain*

[b]*Centro Nacional de Microbiología, Instituto de Salud Carlos III. Madrid, Spain*

ABSTRACT: This study describes the epidemiological, clinical, and microbiological characteristics of a new tick-borne disease in Spain—*Dermacentor*-borne necrosis erythema lymphadenopathy (DEBONEL). The clinical presentations include an eschar at the site of the tick bite, surrounded by an erythema and painful regional lymphadenopathy. The disease appears during the colder months and its vector is *Dermacentor marginatus (D. marginatus)*. From January 1990 to December 2004, 54 patients presented at Hospital of La Rioja with these clinical and epidemiological data. The ratio of females to males was 32/22. The average age was 37 years. In all cases tick bites were located on the upper body (90% on the scalp). The median incubation period was 4.7 days. Signs and symptoms were mild in all cases. Only a small number of patients presented mild and nonspecific abnormalities in a complete blood cell count and mild elevation of erythrocyte sedimentation rates and C-protein reactive and liver enzyme levels. Serological evidence of acute rickettsiosis was observed in 19 patients (61%). In 29% sera tested by polymerase chain reactions (PCRs) were positive. The sequence obtained from a PCR product revealed 98% identity with *Rickettsia* sp. strains RpA4, DnS14, and DnS28. All ticks removed from patients were PCR-positive. Sequencing showed 8 of them identified as *R. slovaca* and 2 as *Rickettsia* sp. strains RpA4, DnS14, and DnS28.

KEYWORDS: *Rickettsia slovaca*; *Rickettsia* sp.; strain RpA4; strain DnS14; DnS28 strain; tick; *Dermacentor;* DEBONEL; TIBOLA

INTRODUCTION

Ticks have been described as vectors of human infectious diseases since the beginning of the 20th century. In the last few years, different emerging tick-borne diseases (TBDs) have been reported all over the world. The development

Address for correspondence: J. A. Oteo, Área de Enfermedades Infecciosas, Hospitales San Millán-San Pedro-de La Rioja, Avda. Viana No 1. 26001-Logroño (La Rioja), Spain. Voice: +34941297275; fax: +34941297267.
e-mail: jaoteo@riojasalud.es

**Ann. N.Y. Acad. Sci. 1078: 206–214 (2006). © 2006 New York Academy of Sciences.
doi: 10.1196/annals.1374.040**

of new techniques, including the shell vial assay for isolation of organisms and molecular methods for their characterization, has allowed the etiology of many of these diseases to be known.[1]

Since 1990, we have investigated TBDs in the Hospital of La Rioja. Lyme borreliosis (LB) and Mediterranean spotted fever (MSF) are the most prevalent TBDs in our area, although some cases of human anaplasmosis have been described.[2]

In November 1996 a 67-year-old man was evaluated for eschar on the thoracic region surrounded by an erythema (erythema migrans–like) after a tick bite. On examination he had several painful points of axillary lymphadenopathy. He kept the tick, which was identified as a female adult *Dermacentor marginatus (D. marginatus).* At this time, *D. marginatus* was only implicated as a vector of one case of LB.[3] No antibodies against *Borrelia burgdorferi, Rickettsia conorii, Anaplasma phagocytophilum,* or *Francisella tularensis* were found. A skin biopsy was performed in order to investigate the presence of *B. burgdorferi* by polymerase chain reaction (PCR) and isolation in BSK medium, but without success. The PCR from the tick was also negative for *B. burgdorferi.* The patient was treated with doxycycline and signs and symptoms were soon alleviated. *R. slovaca* was identified by PCR from the frozen tick few years later.[4] Since the description of this case, we revised all TBDs seen at our hospital up to that moment. In addition, we began to collect data and in same cases sera specimens from patients who presented with similar clinical findings.

MATERIALS AND METHODS

Patients

Patients' inclusion criteria were: (1) a crusted lesion or a point of necrosis (eschar) at the site of the tick attachment, surrounded by erythema and painful regional lymphadenopathy; and (2) tick bite by *D. marginatus* or a large tick during the period of maximum activity for *D. marginatus* (in our area from the end of October to the beginning of May).

The study was retrospective from January 1990 to December 2000 and prospective from January 2001 to December 2004.

Epidemiological and clinical data were obtained for patients studied prospectively. They were asked about the underlying diseases, date and site of the tick bite, method of removing the tick, geographical location and clinical features, evolution, and treatment. Routine laboratory tests (complete blood cell count, erythrocyte sedimentation rates, ASAT and ALAT, and C reactive protein) were performed. All these patients were reexamined 1, 4, 12, and 24 weeks after the initial visit. For the patients studied retrospectively these data were collected (when were available) from the clinical histories.

Patients diagnosed with erythema migrans (LB) and MSF during 2003–2004 were used as controls.

Ticks

D. marginatus was collected from people: 10 from patients and 4 from asymptomatic persons.

Microbiological Tests

From 31 patients an acute-phase serum (before starting the treatment) and late-phase sera samples (1st, 3rd, and 6th months) were taken. From 21 patients an EDTA-blood sample was obtained. In eight cases skin biopsy specimens of the border of the lesion were also obtained. All samples were centrifuged and frozen at $-70°$ C until examination. All *D. marginatus* collected from people were classified with regard to genus and species and kept in 70% alcohol before being processed.

Serological Tests

Acute and convalescent sera samples were tested by an indirect immunofluorescence assay (IFA) for the presence of IgG antibodies against *R. conorii* (BioMérieux, Lyon, France) and *Rickettsia slovaca* (antigen slides kindly donated by Dr. Fátima Bacellar, Centro de Estudos de Vectores e Doenças Infecciosas do Instituto Nacional de Saúde, Portugal). Seroconversion, or a fourfold increase in the sera obtained in the late phase, was considered evidence of recent infection by spotted fever group (SFG) *Rickettsia*.

Molecular Methods

The acute-phase sera, the EDTA blood samples, the skin biopsy samples, as well as all *D. marginatus* specimens removed from people were analyzed by PCR assays.

DNA was extracted using the QIAamp Tissue kit (Qiagen, Hilden, Germany) according to the manufacturer' recommendations. The presence of SFG *Rickettsia* in human samples was determined by PCR assays for *ompA*[5] and *gltA*[6] genes. All ticks were also screened for the presence of *B. burgdorferi, A. phagocytophilum,* and *F. tularensis* by the partial amplification of the 16S rDNA, *msp*2, and 17-kDa lipoprotein genes, respectively. Sequencing reactions were carried out at Universidad de Alcalá de Henares (Spain). Sequences obtained were compared with those available at Genbank using BLAST utility (National Center for Biotechnology Information <http://www.ncbi.nlm.nih.gov>).

RESULTS

Epidemiological and Clinical Data

From January 1990 to December 2004, 54 patients who presented at our hospital with TBDs had the inclusion criteria for the study. The study was retrospective for 20 patients and prospective for 34 patients. Thirty-two patients (59%) were female and 22 (41%) were male. The median age was 37 years (range 3–78). All patients were bitten during the colder months (from the end of October to the beginning of May), with a peak in November (12 cases) and April (10 cases). No was observed from June to September. All patients remembered that they had been bitten by a "very large" tick. Forty-seven patients (87%) removed the tick by hand, nearly always after previous administration of oil on the site of the tick attachment. Three of them kept the ticks. Seven patients were seen with the tick attached, which was removed by forceps. In all 10 cases the tick was identified as *D. marginatus.* All patients gave information about the geographical area where the tick had bitten them (96% in La Rioja Baja). The incubation period varied from 1 to 13 days (median 4.7). In 48 patients (90%) the tick bite was located on the scalp, while 3 patients were bitten on the arms and 2 on the thorax. In all cases, the tick bite was located on the upper portions of the body. The typical skin lesion at the site of the tick bite started as a crusted lesion. A necrotic eschar (0.5 to 2.5 cm in diameter), surrounded by erythema, appeared a few days later. In eight cases (15%) the erythema was higher than 5 cm (erythema migrans-like). In all patients whose tick bites were located on the scalp region, erythema was higher than 5 cm in diameter (33 cm in one patient). Multiple and painful regional lymphadenopathy was the other major feature, being present in all cases. Only four patients presented with the skin lesion few days before the onset of lymphadenopathy. Most patients with the skin lesion on the scalp region suffered from headache. Low-grade fever ($<38°C$) was observed in 18 patients (33%), and 2 patients (3.7%) had fever higher than 38.5°C. A macular rash, which did not affect either palms or soles, was observed in only one patient. Routine laboratory tests were normal in most cases. Five of 47 patients (10.6%) showed mild leukopenia and 2 of them (4.2%) had increased white cell counts. The lowest and the highest leukocyte counts were 2,800/mm^3 and 10,000/mm.^{3} Three patients (6.4%) suffered from thrombocytopenia. The lowest thrombocyte count was 100,000/mm^3. One patient showed anemia (Hb: 11 gr/dL), and anemia and thrombocytopenia were observed in one case. Nine patients (19%) had elevated erythrocyte sedimentation rates. C-reactiue protein was mildly elevated in 38% of cases (18 of 47). Eight of the 51 patients (16%) had an elevated ASAT and ALAT level (less than twice the upper limit of the normal values). In all these cases the laboratory tests performed 1 month later gave normal results.

All patients were treated with antibiotics. Doxycycline (100 mg/b.i.d. 14 days) was administered to 79% of patients. Ten were treated with azithromycin (10 mg/kg q.d. 5 days) and one child with josamycin (500 mg/b.i.d. 14 days). The treatment was changed in only one patient (doxycycline was substituted by azithromycin) on account of the development of nausea and diarrhea. In all cases learning of the signs and symptoms was observed. Evolution of disease was only available for 34 patients studied prospectively. Fever disappeared 48 h after the beginning of the treatment, and the painful lymphadenopathy remitted in 1 week for most patients. However, the evolution of the erythema and the eschar was slower, persisting both for 1 or 2 months. Sixty-six percent of patients fully recovered. The remaining ones developed persistent alopecia at the site of the tick bite (0.5–2 cm in diameter).

Serological Tests

IgG antibodies against *R. conorii* and *R. slovaca* were tested (IFA) in 31 patients. Evidence of recent infection by a *Rickettsia* SFG was observed in 19 patients (61%). In 18 of them (95%) seroconversion was detected. In one out of 19 patients (5%) we observed a fourfold increase in IgG antibodies. In 27 cases (87%) seroconversion or increase of the previous titers was detected during the first month. Four patients (13%) showed seroconversion in the third month.

PCR Assays

All EDTA blood samples and skin biopsies were PCR-negative. In 9 of 31 sera tested (29%) *ompA* PCRs were positive. Sequencing of PCR products was achieved in only one case. Comparison of the sequence with those deposited in Genbank revealed 98% identity with a fragment of *ompA* gene of *Rickettsia* sp. strains RpA4, DnS14, and DnS28.

None of the 14 *D. marginatus* ticks removed from people was infected with *A. phagocytophilum, B. burgdorferi,* and *F. tularensis.* All of them were infected by SFG *Rickettsia.* By sequencing procedure (*ompA* gene) *R. slovaca* (100% similarity) was identified in 8 of 10 *D. marginatus* ticks removed from patients. *Rickettsia* sp. strains RpA4, DnS14, and DnS28 (98% similarity) were identified in the two remaining ones, as well as in all *D. marginatus* removed from asymptomatic person.

DISCUSSION

This study describes the epidemiological, clinical, and laboratory findings for a new TBD in Spain. We first named the syndrome as *Dermacentor*-borne necrosis erythema lymphadenopathy (DEBONEL) because *D. marginatus* is the vector implicated in its transmission, and necrosis, erythema, and lymphadenopathy are the main clinical findings.[4,7]

All patients reported in this study were living in La Rioja, where LB and MSF are endemic.[2] To estimate the specificity of epidemiological and clinical data of DEBONEL, we have also studied prospectively all patients diagnosed with erythema migrans (early located LB) and MSF at our hospital during 2003–2004. During this period, 27 cases of LB, 24 cases of DEBONEL, and 7 cases of MSF were diagnosed. Therefore, DEBONEL was a prevalent TBD in La Rioja during this period (21% of cases).

The seasonal distribution of DEBONEL was significantly different ($P < 0.05$) from that in LB and MSF. DEBONEL occurred mainly during the colder months, whereas LB was diagnosed from late spring to early autumn, and MSF in summer. In addition, DEBONEL had a different geographical location. All cases were located in La Rioja Baja, in areas with low mountains (Wooded with Holm oaks) and with dry Indo-Mediterranean climate conditions.

In our study females were more frequently involved than males (female/male ratio: 1.45). On the contrary, LB and MSF were more frequent in males (female/male ratio: 0.5 and 0.16, respectively) ($P < 0.05$). The patients' median age was 37 years. Nine patients (16.6%) were children younger than 10 years. The average age of DEBONEL patients was significantly younger ($P < 0.05$) than the average age of LB (48 years) and MSF patients (55 years). In different Spanish studies, LB and MSF were also more frequent in males and their average age was higher.[8–11] The site of the tick bite in DEBONEL was different from that LB and MSF: Patients diagnosed with DEBONEL were usually bitten on the scalp region (90%), and always on the upper portions the body. However, the tick's attachment in LB and MSF can appear on any part on the body with predominance of the lower extremities[12] ($P < 0.05$). These epidemiological differences between the three TBDs may be explained by the differences in the characteristics and behaviors of the tick vector.

Adult *D. marginatus* ticks' greatest activity, in La Rioja, is observed from early autumn to winter,[13] which may explain why DEBONEL is more frequent during the colder months. The geographical distribution of the disease corresponds to the area of *D. marginatus* distribution in La Rioja.[13] On the other hand, this species seems to prefer to live on hairy animals. To get a host, it waits higher than other genera (A. Estrada, personal communication). These facts could explain why children and females are more frequently involved than males, why the tick bite is usually located on the scalp region, and the fact that bites outside the head are observed only in males, on hairy regions (arms, axillas, and thorax).

The clinical symptoms of DEBONEL are different from those of LB and MSF. Nevertheless, some clinical features are similar in the three TBDs. Erythema migrans (the hallmark of the early located LB) later expands, central clearing occurs frequently, and it has a big size and does not have a central necrosis. However, erythema of DEBONEL is usually small (only 15% were higher than 5 cm) and always has a central necrosis. General symptoms are rare and mild in both TBDs, except for headache, which is common in DEBONEL. Erythema and eschar at the inoculation site were significantly less common in patients with MSF than in patients with DEBONEL ($P < 0.05$). In addition, in MSF patients the eschar fully recovered and its size was smaller. On the contrary, some patients with DEBONEL, after healing of the eschar, developed a local alopecia. Regional lymphadenopathy was found less often in MSF than in DEBONEL ($P < 0.05$). The presence of fever and rash was also significantly less common in patients with DEBONEL than in patients with MSF ($P < 0.05$). MSF is usually a mild illness, but malignant forms have been reported in 5–12% of cases associated with 2.5% mortality.[10,11] In this new TBD signs and symptoms were mild in all cases and we have not observed a fatal evolution in any case.

Similar epidemiological and clinical cases have been reported in France and Hungary by Raoult and Lakos.[14–17] In Hungary the syndrome is known as tick-borne lymphadenopathy (TIBOLA).

In 1997 *R. slovaca* was identified by PCR in a skin biopsy from a French patient with similar clinical manifestations than those seen in La Rioja.[14] In 2003 *R. slovaca* was isolated from a skin biopsy, which demonstrates its implications in the etiology of the disease.[18]

In the present study we have attempted to identify the causative agent of this new TBD in La Rioja. Culture was attempted from 2 skin biopsies, in serum and blood collected on heparin samples, and from a PCR-positive tick removed from a patient in other laboratories (Centro de Estudos de Vectores e Doenças Infecciosas do Instituto Nacional de Saúde, Portugal and Unité des Rickettsies, Marsella, France) without success. Nevertheless, we have confirmed the SFG rickettsial infection by PCR in 29% of sera tested, and in all ticks (10) removed from patients. By means of a sequencing procedure *R. slovaca* (100% similarity) was identified in 8 of 10 *D. marginatus* ticks removed from patients. *Rickettsia* sp. strains RpA4, DnS14, and DnS28 (98% similarity) were identified in the two remaining ones, as well as in a serum sample from a patient.

Rickettsia sp. strains RpA4, DnS14, and DnS28 are closely related and branched with members of the *R. massiliae* group. These strains were first identified in *Dermacentor* sp. ticks collected in territories of the former Soviet Union.[19,20] The prevalence of *Rickettsia* sp. strains RpA4, DnS14, and DnS28 in *D. marginatus* collected from people and from several animal hosts in La Rioja is high (59.3%). Six of 14 (43%) *D. marginatus* ticks removed from people were infected by these strains. Moreover only 2 of 6 people (33%) developed disease. The remaining ones (4/6) did not develop any clinical manifestation.

The pathogenicity of *Rickettsia* sp. strains RpA4, DnS14, and DnS28 is still unknown and, up to date, these genotypes have not been implicated in human disease.

Our data suggest that *R. slovaca* is the main etiological agent of DEBONEL in La Rioja. Nevertheless, it is possible that other *Rickettsia* sp., such as strains RpA4, DnS14, and DnS28 may also be implicated in the etiology.

On the other hand, according to our results, DEBONEL may be a prevalent TBD in Spain and should be suspected following a tick bite. In addition, persons bitten by *D. marginatus* infected with *R. slovaca* had a higher risk of developing DEBONEL than did those bitten by *D. marginatus* infected with *Rickettsia* sp. strains RpA4, DnS14, and DnS28 (100% and 33%, respectively).

ACKNOWLEDGMENTS

This study was supported in part by grants from the Fondo de Investigación Sanitaria (FIS PI021810 and FIS G03/057), Ministerio de Sanidad y Consumo, Spain.

REFERENCES

1. PAROLA, P. & D. RAOULT. 2001. Ticks and tick-borne bacterial diseases in humans: an emerging infectious threat. Clin. Infect. Dis. **32:** 897–928.
2. OTEO, J.A. 2001. Tick-borne diseases in Spain. Clin. Microbiol. Infect. **7:** 31.
3. ANGELOV, L., P. DIMOVA & W. BERBENCOVA. 1996. Clinical and laboratory evidence of the importance of the tick *D. marginatus* as a vector of *B. burgdorferi* in some areas of sporadic Lyme disease in Bulgaria. Eur. J. Epidemiol. **12:** 449–506.
4. OTEO, J.A. & V. IBARRA. 2002. DEBONEL (*Dermacentor*-borne-necrosis-erythema-lymphadenopathy): una nueva enfermedad transmitida por garrapatas? Enferm. Infecc. Microbiol. Clin. **2:** 51–52.
5. FOURNIER, P.E., V. ROUX & D. RAOULT. 1998. Phylogenetic analysis of spotted fever group rickettsiae by study of the outer surface protein rOmpA. Int. J. Syst. Bacteriol. **48:** 839–849.
6. ROUX, V., E. RYDKINA, M. EREMEEVA, *et al.* 1997. Citrate synthase gene comparison, a new tool for phylogenetic analysis, and its application for the *Rickettsiae.* Int. J. Syst. Bacteriol. **47:** 252–261.
7. OTEO, J.A., V. IBARRA, J.R. BLANCO, *et al.* 2004. *Dermacentor*-borne necrosis erythema and lymphadenopathy: clinical and epidemiological features of a new tick-borne disease. Clin. Microbiol. Infect. **4:** 327–331.
8. OTEO, J.A. & V. MARTÍNEZ DE ARTOLA. 1995. Borreliosis de Lyme: aspectos epidemiológicos y etiopatogénicos. Enferm. Infecc. Microbiol. Clin. **13:** 550–555.
9. OTEO, J.A., J.R. BLANCO, V. MARTÍNEZ DE ARTOLA, *et al.* 2000. Eritema migratori (borreliosis de Lyme): características clínico epidemiológicas de 50 casos. Rev. Clin. Esp. **2:** 57–59.
10. JUFRESA, J., J. ALEGRE, J.M. SURINACH, *et al.* 1997. Clinical study of 86 patients with Mediterranean boutonneuse fever who were admitted in a university general hospital [in Spanish]. An. Med. Int. **14:** 328–331.
11. ANTÓN, E., B. FONT, I. SANFELIÚ, *et al.* 2003. Clinical and laboratory characteristics of 144 patients with Mediterranean Spotted Fever. Eur. J. Clin. Microbiol. Infect. Dis. **22:** 126–128.

12. OTEO, J.A., V. IBARRA, J.R. BLANCO, *et al.* 2003. Epidemiological and clinical differences among DEBONEL/TIBOLA and other tick-borne diseases in Spain. Ann. N. Y. Acad. Sci. **990:** 391–392.
13. OTEO, J.A., J.R. BLANCO, R. GRANDIVAL, *et al.* 1998. Ixodidiasis humana en La Rioja: estudio estacional y de especies. Presented at the I Reunión Nacional del Grupo de Rickettsias y Borrelias. Haro, La Rioja, Spain.
14. RAOULT, D., P.H. BERBIS, V. ROUX, *et al.* 1997. A new tick-transmitted disease due to *Rickettsia slovaca*. Lancet **350:** 112–113.
15. LAKOS, A. 1997. Tick-borne lymphadenopathy—a new rickettsial disease? Lancet **350:** 1006.
16. RAOULT, D., A. LAKOS, F. FENOLLAR, *et al.* 2002. A spotless rickettsiosis caused by *Rickettsia slovaca* and associated with *Dermacentor* ticks. Clin. Infect. Dis. **34:** 1331–1336.
17. LAKOS, A. 2002. Tick-borne lymphadenopathy (TIBOLA). Wien. Klin. Wochenschr. **114:** 648–654.
18. CAZORLA, C., M. ENEA, F. LUCHT, *et al.* 2003. First isolation of *Rickettsia slovaca* from a patient, France. Emerg. Infect. Dis. **9:** 135.
19. RYDKINA, E., V. ROUX, N. FETISOVA, *et al.* 1999. New *Rickettsiae* in ticks collected in territories of the former Soviet Union. Emerg. Infect. Dis. **5:** 811–814.
20. SPHYNOV, S., P. PAROLA, N. RUKADOV, *et al.* 2001. Detection and identification of spotted fever group *Rickettsiae* in *Dermacentor* ticks from Russia and Central Kazakhstan. Eur. J. Clin. Microbiol. Infect. Dis. **20:** 903–905.
21. OTEO, J.A., A. PORTILLO, S. SANTIBÁÑEZ, *et al.* 2006. Prevalence of spotted fever group *Rickettsia* (SFGR) species detected in ticks in La Rioja, Spain. Ann. N. Y. Acad. Sci. this volume.

Infective Endocarditis due to *Bartonella* spp. and *Coxiella burnetii*

Experience at a Cardiology Hospital in São Paulo, Brazil

RINALDO FOCACCIA SICILIANO,[a] TÂNIA MARA STRABELLI,[a] ROGÉRIO ZEIGLER,[a] CRISTHIENI RODRIGUES,[a] JUSSARA BIANCHI CASTELLI,[a] MAX GRINBERG,[a] SILVIA COLOMBO,[b] LUIZ JACINTHO DA SILVA,[b] ELVIRA MARIA MENDES DO NASCIMENTO,[b] FABIANA CRISTINA PEREIRA DOS SANTOS,[b] AND DAVID EVERSON UIP[a]

[a]*Heart Institute (InCor.HC.FMUSP) of University of São Paulo School of Medicine, Av. Dr. Enéias de Carvalho Aguiar, 44, Pinheiros, São Paulo, Brazil CEP 05403–000*

[b]*Adolfo Lutz Institute, São Paulo-Brazil, Av. Dr. Arnaldo, 355 Cerqueira Cesar, São Paulo, Brazil CEP 01246–902*

ABSTRACT: *Bartonella* spp. and *Coxiella burnetii* are recognized as causative agents of blood culture–negative endocarditis (BCNE) in humans and there are no studies of their occurrences in Brazil. The purpose of this study is to investigate *Bartonella* spp. and *C. burnetii* as a causative agent of culture-negative endocarditis patients at a cardiology hospital in São Paulo, Brazil. From January 2004 to December 2004 patients with a diagnosis of endocarditis at our Institute were identified and recorded prospectively. They were considered to have possible or definite endocarditis according to the modified Duke criteria. Those with blood culture–negative were tested serologically using the indirect immunofluorescent assay (IFA) for *Bartonella henselae, B. quintana,* and *C. burnetii.* IFA-IgG titers >800 for *Bartonella* spp. and *C. burnetii* were considered positive. A total of 61 patients with endocarditis diagnosis were evaluated, 17 (27%) were culture-negative. Two have had IgG titer greater than 800 (≥3,200) against *Bartonella* spp. and one against *C. burnetii* (phase I and II≥6,400). Those with *Bartonella*-induced endocarditis had a fatal disease. Necropsy showed calcifications and extensive destruction of the valve tissue, which is diffusely infiltrated with mononuclear inflammatory cells predominantly by foamy macrophages. The patient with *C. burnetii* endocarditis received specific antibiotic therapy. Reports of infective endocartitis due to *Bartonella* spp. and *C. burnetii* in Brazil reveal

Address for correspondence: Rinaldo Focaccia Siciliano, Rua Cardoso de Almeida 1006/apto13, Perdizes, São Paulo/Brazil CEP 05013-001. Voice: 55-11-38011724; fax: 55-11-30695349.
e-mail: rinaldo_focaccia@uol.com.br

Ann. N.Y. Acad. Sci. 1078: 215–222 (2006).
doi: 10.1196/annals.1374.123

the importance of investigating the infectious agents in culture-negative endocarditis.

KEYWORDS: endocarditis; blood culture–negative endocarditis; *Bartonella* spp.; *Coxiella burnetti*

INTRODUCTION

The main test for the etiological diagnosis of infective endocarditis is blood culture. However, previous studies have shown that up to 30% of the patients with diagnosed endocarditis may present negative blood cultures[1]. The main causes for these apparent negative blood cultures are bacteriological development inhibition due to use of antibiotics prior to blood sample collection or the fact that the causative microorganisms are hard to grow and be identified using conventional blood culture techniques. The empirical antibiotic treatment range used in these cases may not encompass all agents or may not be powerful enough for adequate treatment causing these patients to present high morbidity and mortality when compared with those who are culture-positive.[2,3] Recently, many publications in European countries have demonstrated a significant involvement of *Coxiella burnetti*, *Bartonella henselae*, and *B. quintana* in patients with blood culture–negative endocarditis (BCNE).[4,5] Although there is very little information in the medical literature about the occurrence of these agents as a cause of endocarditis in Latin America, we have decided to investigate *Bartonella* spp. and C. *burnetii* as causative agents of endocarditis at our Institute in São Paulo, Brazil.

METHODS

All episodes of endocarditis seen at the Heart Institute of the University of São Paulo School of Medicine between January 2004 and December 2004 were identified and recorded prospectively. The Heart Institute (InCor.HC.FMUSP) is a 460-bed referral university cardiology center. Patients were considered to have possible or definite endocarditis according to the modified Duke criteria.[6] All the cases considered as BCNE have had at least four negative blood cultures. Each 20 mL of blood was collected by means of an independent puncture in one aerobic and one anaerobic bottle and incubated for 30 days (BacT/Alert System-Organon Teknika Corp., Durham, NC). Histologic and microbiologic study of valvular material was performed when cardiac surgery was indicated. A standardized questionnaire was completed for each patient with BCNE. Recorded data included previous antibiotic therapy, valvar surgery, and presence of environmental exposure factors (homelessness, chronic alcoholism, presence of body lice, contact with farm animals, cats, and ingestion of raw

TABLE 1. Antibody titers against *C. trachomatis, C. psittacii, and C. pneumoniae* in patients with BCNE and positive serology for *Bartonella spp. or C. burnetti*

	IFA IgG		MIF IgG		
Case	*B. henselae/B. quintana.*	*C. burnetti*	*C. trachomatis*	*C. psittacii*	*C. pneumoniae*
1	≥3200	<400	32	128	64
2	≥3200	<400	64	64	<32
3	<400	≥6400	<32	<32	<32

milk). Patients under 18 years of age or with congenital cardiopathies were excluded.

Those with BCNE were tested by indirect immunofluorescent assay (IFA) for *B. henselae, B. quintana*, and *C. burnetii* (Nine Mile strain, Biological Products Branch of CDC, Atlanta, GA). Immunoglobulin G titers >800 to *Bartonella* spp. or *C. burnetii* were considered positive for endocarditis diagnosis.[7,8] When serology was positive, we also carried out IgG microimmunoflouorescence for *Chlamydia trachomatis, C. psittacii*, and *C. pneumoniae* to observe cross-reaction.

In our Institute, patients with BCNE were empirically treated with vancomycin plus gentamicin or penicillin plus oxacillin plus gentamicin, based on the high prevalence of *Streptococcus* spp., *S. aureus,* and *Enterococcus spp.* as the causative agent of endocarditis. When serology was positive, therapy was adjusted: for *Bartonella spp.* we used cephtriaxone plus gentamicin (>1 month) and for *C. burnetti* infection, doxycycline plus ciprofloxacin (>2 years).

RESULTS

A total of 61 patients with endocarditis were evaluated and 17 (27%) were blood culture–negative. Among these patients with BCNE there were 5 male and 12 female patients. The mean age was 42 years and 11 patients had prosthetic valvular endocarditis. Serology was performed in all 17 patients: 2 were positive for *B. henselae* and *B. quintana* with IgG titer ≥3,200, 1 was positive for *C. burnetti* with IgG titer ≥6,400 against *C. burnetii* (phase I and II), and all the others had titers inferior to 400 against *B. henselae, B. quintana*, and *C. burnetti.* In TABLE 1 we present the microimmunofluorescence for *Chlamydia* spp. of the patients with positive serology for *Bartonella* spp. or *C. burnetti.*

At admission the two patients with *Bartonella* endocarditis had fever and weight loss for more than 60 days. The echocardiography revealed little valvar vegetation and empiric antibiotics were started (penicillin + oxacillin + gentamicin). However, despite the treatment, they had rapid progression to a severe congestive heart failure due to valve dysfunction and died 7 and 10 days

TABLE 2. Clinical and epidemiological features of two patients with *Bartonella* endocarditis

Age (years)	Sex	Valve involved	Underlying valvular diseases	Epidemiology	Duration of symptoms	Clinical presentation	Outcome
66	M	Aortic	Valvular rheumatic disease	Domestic cat	90 days	Fever, weight loss, petechiae	Deceased
53	M	Aortic	Valvar prosthesis	Domestic cat (cat scratch on hand)	60 days	Fever, weight loss splenomegaly	Deceased

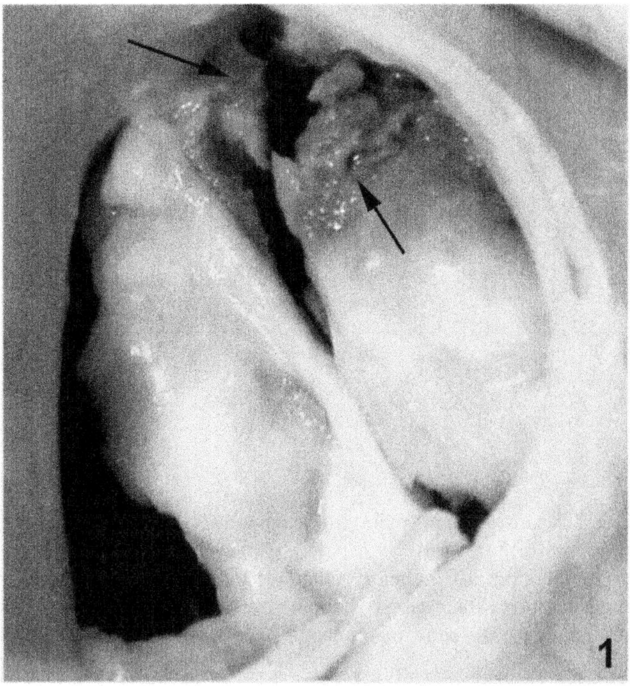

FIGURE 1. The gross feature of the aortic valve congenitally bicuspid with severe fibrosis and calcification. Note the presence of vegetation (*arrows*) with local destruction of the valve tissue.

after hospitalization, respectively. At the time of death there were no serology results. Other clinical features of these patients are presented in TABLE 2. The pathological findings at their necropsies confirmed the diagnosis of endocarditis (FIG. 1). In one of them, the lesion was found on a congenitally bicuspid aortic valve, and the condition was also complicated by the existence of a valve ring abscess. In both cases there was superficial fibrinous vegetation with few neutrophils and necrotic cellular debris. In both cases, the valvular stroma was fibrotic and reorganized, with multifocal calcification and inflammatory infiltrate of mononuclear cells, composed predominantly of foamy macrophages (FIG. 2). Special stains, including Brown-Hopps, Ziehl-Neelsen, and Grocott, showed no microorganisms in valvular tissue.

The patient with Q fever endocarditis was a 45-year-old man who had a history of valvular rheumatic disease and valve replacement in January 2004. He came to our service in October 2004 with history of weakness, fever, and weight loss for 60 days and, as he lived in a poor rural area, had contact with cattle and consumed raw milk daily. The echocardiography showed vegetation (1.7 cm) on aortic valvar prosthesis, so a course of empiric antibiotics was started (vancomycin + gentamicin). After 20 days the patient had no clinical improvement

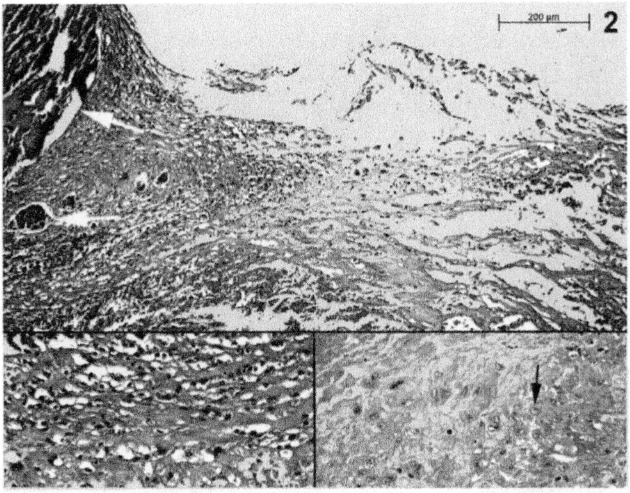

FIGURE 2. *Top*: histological findings of endocarditis on the bicuspid aortic valve. We can see calcification and extensive destruction of the valvular tissue, which is diffusely infiltrated by mononuclear inflammatory cells. *Bottom left*: superficial region of the valve with fibrin and necrotic cellular debris. The *bottom right* shows the valve stroma infiltrate composed predominantly of foamy macrophages (one pointed out by the *arrow*).

and, with the serological diagnosis of *Coxiella* infection, we changed the intravenous antibiotic treatment to oral doxycycline plus ciprofloxacin. The fever stopped after 5 days and the control echocardiography was negative to valvar vegetation after 2 months of therapy. Because the patient had mild valvar dysfunction, no replacement was performed. The patient is still under treatment and is asymptomatic.

DISCUSSION

C. burnetti infection usually manifests as a mild disease with spontaneous recovery. However, endocarditis is the most common presentation of the chronic disease. It has been shown in recent case series that *C. burnetti* or *Bartonella* spp. may cause 3–5% of all cases of infective endocarditis.[9,10] The species of *Bartonella* more often implicated in endocarditis are *B. quintana* and *B. henselae*, whereas other species are rarely observed.

The positive predictive value of the indirect IFA to *Bartonella* spp. and *C. burnetti* is high (95–100%)[7,11] but cross-reactions can occur between *Bartonella* spp. *C. burnetti*, and *Chlamydia spp.* We could not find cross-reactions between *Bartonella* species and *C. burnetti* in our positive cases. We also tested *C. trachomatis, C. psitacii*, and *C. pneumoniae,* but no significant cross-reaction was found.

The indirect IFA cannot reliably identify the causative species of *Bartonella* and our positive cases presented cross-reaction between *B. henselae and B. quintana*. Epidemiological evaluation may suggest that these patients may have been infected by *B. henselae* because both of them had a domestic cat and one of them presented a scar in the hand due to cat scratch that had occurred about 1 month before the onset of symptoms. *B. quintana* is transmitted by body lice and none of these patients presented a known epidemiological risk factor for this infection (e.g., homelessness or chronic alcoholism). The patient with Q fever endocarditis might have been exposed to *C. burnetti* because he lived in a rural area and mentioned that he had contact with cattle and had drunk untreated milk on a regular basis.

The two patients with endocarditis caused by *Bartonella* spp. progressed to death within 10 days after hospitalization due to an advanced state of deterioration of valvular and perivalvular tissue caused by endocarditis. Probably, the time taken to seek medical assistance (more than 60 days of fever) contributed to the severity of their disease.

The endocarditis diagnosis was based on Dukes criteria and on morphological pattern of the lesions with destructive and inflammatory alterations on the valves. The remodeling of the valvular tissue and the predominant inflammatory cellular composition of macrophages suggested the chronic character of the process. Our pathological findings were similar to those described by Lepidi *et al.*[12] studying 15 patients with confirmed *Bartonella* endocarditis.

In the present study, serology was able to identify the etiological agent in 3 of the 17 cases of BCNE (2 *Bartonella* spp. and 1 *C. burnetti*), which represents almost 5% (3/61) of all endocarditis diagnosed in our Institute. We performed a Mcdlinc databasc scarch and no cases of *Bartonella* spp. or *C. burnetti* endocarditis were reported in Brazil. However, different clinical presentations of *B. henselae* and *B. quintana* infection, such as bacillary angiomatosis, cat scratch, and trench fever disease, are well recognized in South America. Although we have information about the sporadic cases of human infection caused by acute *C. burnetti,* little information in the literature is available about the occurrence of Q fever in Brazil and it is possible that this is a human underdiagnosed and/or underreported disease. A misdiagnosis of endocarditis caused by those microorganisms and by the inappropriate antimicrobial chemotherapy may be responsible for relapses, increasing the morbidity and the mortality rate. In conclusion, the report of infective endocartitis due to *Bartonella* spp. and *C. burnetii* in Brazil reveals the importance of investigating those infectious agents in culture-negative endocarditis.

REFERENCES

1. BROUQUI, P. & D. RAOULT. 2001. Endocarditis due to rare and fastidious bacteria. Clin. Microbiol. Rev. **14:** 177–207.

2. ZAMORANO, J., J. SANZ, C. ALMEIDA, *et al.* 2003. Differences between endocarditis with true negative blood cultures and those with previous antibiotic treatment. J. Heart Valve Dis. **12:** 256–260.
3. HOUPIKIAN, P. & D. RAOULT. 2002. Diagnostic methods current best practices and guidelines for identification of difficult-to-culture pathogens in infective endocarditis. Infect. Dis. Clin. North Am. **16:** 377–392.
4. HOUPIKIAN, P. & D. RAOULT. 2005. Blood culture-negative endocarditis in a reference center: etiologic diagnosis of 348 cases. Medicine (Baltimore) **84:** 162–173.
5. LAMAS, C.C. & S.J. EYKYN. 2003. Blood culture negative endocarditis: analysis of 63 cases presenting over 25 years. Heart **89:** 258–262.
6. LI, J.S., D.J. SEXTON, N. MICK, *et al.* 2000. Proposed modifications to the Duke criteria for the diagnosis of infective endocarditis. Clin. Infect. Dis. **30:** 633–638.
7. FOURNIER, P.E., J.L. MAINARDI & D. RAOULT. 2002. Value of microimmunofluorescence for diagnosis and follow-up of *Bartonella* endocarditis. Clin. Diag. Lab. Immunol. **9:** 795–801.
8. MAURIN, M. & D. RAOULT. 1999. Q fever. Clin. Microbiol. Rev. **12:** 518–553.
9. RAOULT, D., H. TISSOT-DUPONT, C. FOUCAULT, *et al.* 2000. Q Fever 1985–1998: clinical and epidemiologic features of 1,383 infections. Medicine **79:** 109–123.
10. RAOULT, D., P.E. FOURNIER, M. DRANCOURT, *et al.* 1996. Diagnosis of 22 new cases of *Bartonella* endocarditis. Ann. Intern. Med. **125:** 646–652.
11. TISSOT-DUPONT, H., X. THIRION & D. RAOULT. 1994. Q fever serology: cutoff determination for microimmunofluorescence. Clin. Diagn. Lab. Immunol. **1:** 189-196.
12. LEPIDI, H., J.P. CASALTA, P.E. FOURNIER, *et al.* 2005. Quantitative histological examination of mechanical heart valves. Clin. Infect. Dis. **40:** 655–661.

Arthropod-Borne Diseases in Homeless

PHILIPPE BROUQUI[a] AND DIDIER RAOULT[b]

[a]*Maladies Infectieuses et Tropicales CHU Nord, AP-HM, 13915 Marseille, France*

[b]*Unité des rickettsies, CNRS UMR 6020, IFR 48, Faculté de Médecine, 13385 Marseille, France*

ABSTRACT: **Homeless people are particularly exposed to ectoparasite. The living conditions and the crowded shelters provide ideal conditions for the spread of lice, fleas, ticks, and mites. Body lice have long been recognized as human parasites and although typically prevalent in rural communities in upland areas of countries close to the equator, it is now increasingly encountered in developed countries especially in homeless people or inner city economically deprived population. Fleas are widespread but are not adapted to a specific host and may occasionally bite humans. Most common fleas that parasite humans are the cat, the rat, and the human fleas, *Ctenocephalides felis*, *Xenopsylla cheopis*, and *Pulex irritans*, respectively. Ticks belonging to the family *Ixodidae*, in particular, the genera *Dermacentor*, *Rhipicephalus*, and *Ixodes*, are frequent parasites in humans. *Sarcoptes scabiei var. hominis* is a mite (*Arachnida* class) responsible for scabies. It is an obligate parasite of human skin. The hematophagic-biting mite, *Liponyssoides sanguineus*, is a mite of the rat, mouse, and other domestic rodents but can also bite humans. Finally, the incidence of skin disease secondary to infestation with the human bedbug, *Cimex lectularius*, has increased recently. Bacteria, such as *Wolbacchia* spp. have been detected in bedbug. The threat posed by the ectoparasite in homeless is not the ectoparasite themselves but the associated infectious diseases that they may transmit to humans. Except for scabies all these ectoparasites are potential vectors for infectious agents. Three louse-borne diseases are known at this time. Trench fever caused by *Bartonella quintana* (*B. quintana*), epidemic typhus caused by *Rickettsia prowazekii*, and relapsing fever caused by the spirochete *Borrelia recurrentis*. Fleas transmit plague (*Xenopsylla cheopis and Pulex irritans*), murine typhus (*Xenopsylla cheopis*), flea-borne spotted rickettsiosis on account of the recently described species *Rickettsia felis* (*C. felis*), and occasionally cat scratch disease on account of *Bartonella henselae* (*C. felis*). The role of fleas as potential vector of *B. quintana* has recently been suggested. Among the hematophagic-biting mites, *L. sanguineus*, is responsible for the transmission of *Rickettsia akari*, the etiologic agent of rickettsialpox. Virtually, no data are available on tick-borne disease in**

Address for correspondence: Philippe Brouqui, Unité des rickettsies, CNRS UMR 6020, IFR 48, Faculté de médecine, 27 bd, J Moulin, 13385 Marseille, cedex 5, France.
e-mail: philippe.brouqui@medecine.univ-mrs.fr

Ann. N.Y. Acad. Sci. 1078: 223–235 (2006). © 2006 New York Academy of Sciences.
doi: 10.1196/annals.1374.041

this population. This article will deal with epidemiology, diagnosis, prevention, and treatment of these ectoparasite and the infectious diseases they transmit to the homeless people.

KEYWORDS: homeless; ticks; fleas; louse; scabies; eosinophils; *Bartonella quintana*; typhus; relapsing fever; *Borrelia recurrentis*

INTRODUCTION

Homelessness, defined as the absence of customary and regular access to a conventional dwelling or residence, is a growing social and public health problem in wealthy Western countries. The number of homeless living in the United States, the United Kingdom, and France has been estimated of at least 500,000, 120,000, and 400,000, respectively. In towns from industrialized countries the homeless population is estimated to be more than 1500 in Marseilles, France[1] and to more than 5500 individuals in Tokyo.[2] Homeless people are particularly exposed to ectoparasites.[3] The living conditions and the crowded shelters provide ideal conditions for the spread of lice, fleas, ticks, and mites. In a recently published study over 4 years, in which was included and followed 930 homeless people, lice were found in 22% and were associated with hypereosinophilia. *Bartonella quintana* (*B. quintana*) was isolated from blood culture in 50 patients (5.3%). The number of bacteremic patients increased from 3.4% to 8.4% over the 4 years of the study indicating that despite effective treatment for *B. quintana* bacteremia and the efforts made to delouse this population, *B. quintana* remains endemic.[4] In the same study, we detected a high seroprevalence to *Borrelia recurrentis*, *R. conorii*, and *R. prowazekii* antibodies, and 27 patients (3%) were infested with scabies showing the high degree of exposure to arthropod-borne diseases of these population.[5]

LOUSE-BORNE DISEASES

The Body Louse

Lice infestations have been prevalent among humans for thousands of years. The oldest eggs of *Pediculus* spp. have been found in the Judean desert and date from between 6900 to 6300 BC.[6] Although more than 3000 species of lice exist, the only three that affect humans are *Pediculus humanus capitis* (head lice), *Pediculus humanus humanus* (body lice), and *Phthirus pubis* (*P. pubis*). *P. humanus capitis* is known to affect all levels of society, *P. pubis* is transmitted sexually, and *P. humanus humanus* has been associated more frequently with conditions of lack of hygiene and extreme poverty. Recently, an independent lineage of head louse restricted to America that diverged 1 million years ago from the worldwide lice was described.[7] Infestation with *P. humanus humanus,* also known as "vagabond's disease," is an issue of concern among homeless.

Close body-to-body contact is strongly associated with transmission of lice. For that reason, infestation occurs more frequently in crowded environments, such as homeless shelters, refugee camps, and jails, especially when hygienic standards are lacking. Lice are extremely host specific and live in clothes, and feed exclusively on human several times a day.[8,9]

Body lice are defenseless, and their only natural enemy is their host.[10] The louse's life cycle begins as an egg, laid in the folds of clothing. As the body louse is highly susceptible to cold, the eggs are usually attached to inner clothing, close to the skin where the body temperature reaches 29–32°C. When seeking lice or eggs, the inner belts of underwear, trousers, or skirts are therefore the best places to look. Louse eggs are held in place by an adhesive produced by the mother's accessory gland.[11] When held at a constant temperature (i.e., when clothes are not removed) the eggs will hatch 6–9 days after being laid. The emerging louse immediately moves onto the skin to feed before returning to the clothing, where it remains until feeding again. A louse typically feeds five times a day. The growing louse molts three times, usually at days 3, 5, and 10 after hatching. After the final molt, the mature louse will typically live for another 20 days. Digestion of the blood meal is rapid. Erythrocytes are quickly hemolyzed and remain liquefied. At maturity, lice can mate immediately, and during the prolonged mating, both the male and the female will continue to feed throughout the process.[11] Females lay about eight eggs a day, and because they do not have a sperm storage organ, they must mate before laying eggs; thus, daily mating is critical. Population density is variable; usually only a few lice are observed on the same host, although we have observed people with more than 300 lice (Foucault C, unpublished data) . Theoretically, a pair of mating lice can generate 200 lice during their 1-month life span. A population can increase by as much as 11% per day, but this rate is rarely observed.[12] Although merely theoretical, this calculation shows how rapidly an outbreak of louse infestation could develop. Humidity is a critical factor for lice that are susceptible to rapid dehydration.[10] The optimal humidity for survival is in the range of 70–90%[11]; they cannot survive when this value falls below 40%. Conversely, under conditions of extremely high humidity, louse feces become sticky and can fatally adhere lice to clothing. The louse's only method of rehydration is to feed blood. The small diameter of the proboscis prevents the rapid uptake of blood; thus, frequent, small meals are necessary.[10] Temperature is also highly influential on the louse's physiology.

Laboratory lice prefer a temperature of between 29°C and 32°C.[11] In the wild, lice are able to maintain this temperature range by nestling in clothing. However, if a host becomes too hot because of fever or heavy exercise, infesting lice will leave him. Body lice die at 50°C, and this temperature is critical when washing clothes, as water or soap alone will not kill lice. Although eggs can survive at lower temperatures, their durability is limited to 16 days.

Three Louse-Borne Diseases

Three louse-borne diseases are known at this time. Trench fever, first described during World War I and caused by *B. quintana*, epidemic typhus caused by *R. prowazekii,* and relapsing fever caused by the spirochete *B. recurrentis.*[13] All diseases are associated with louse infestation and poverty. Louse-borne infections have recently reemerged in jails of Rwanda and in the refugee camps in Burundi,[14] in rural community in the Andes of Peru,[15] and in rural louse-infested population of Russia,[16] but this reemergence also occurred in large and modern cities of developed countries especially in the homeless population.

Trench Fever

Trench fever was the first clinical manifestation of infection on account of *Bartonella* species to be recognized. The name "trench fever" was chosen because the disease was associated with both Allied and German troops during World War I. The disease is characterized by a 5-day relapsing fever, with severe and persistent pain in the legs. It has been estimated that trench fever affected 1,000,000 people during World War I.[17] Epidemics of the disease were most frequently reported in Russia and on the eastern, central, and western European fronts during the two world wars of this century. The disease was supposedly imported from the Eastern front by German soldiers in 1914, and British troops were responsible for its spread to Mesopotamia.[17] After the war, the incidence of trench fever fell dramatically. During World War II, trench fever reemerged, and large-scale epidemics of the disease were again reported. Confusion is apparent in early articles on *B. quintana,* because it was named *Rickettsia quintana, Rickettsia weigli, Rickettsia da Rochalimae*, or *Rickettsia pediculi.*[18] A rickettsia-like organism, named *Rickettsia quintana,* was proposed as the etiologic agent of trench fever.[19,20] Vinson and Fuller reported the first successful axenic cultivation of the agent, which had been reclassified as *Rochalimaea quintana. R. quintana,* which has been subject to taxonomic reclassification, is now named *B. quintana.* The earliest studies for *B. quintana* were carried out on human volunteers[21] followed by macaques.[20] McNee *et al.* were the first to suggest that lice had a role in the transmission of trench fever.[22] The ubiquity of trench fever has been further demonstrated, with cases being reported in Japan, China, Mexico, and Burundi. Recent investigations have, however, led to the reemergence of *R. quintana* as an organism of medical importance.

Clinical manifestations of *B. quintana* infection are trench fever, chronic bacteremia, lymphadenopathy, endocarditis, and, in immunocompromised host, bacillary angiomatosis. Trench fever is the initial clinical presentation of *B. quintana* infection and represents the acute infection.[23] The incubation period is between 15 and 25 days. Clinical manifestations range from asymptomatic to severe, life-threatening illness. The presentation most often reported corresponds to a febrile illness of acute onset of a periodic nature accompanied by headache dizziness and pain in the legs. Headache is most often severe, at the front of the head and behind the eyes. Symptoms may therefore suggest meningitis. Pain may spread to legs and is often felt in the bones. Conjunctival congestion is frequently noticed. Fever is periodic, and the interval between attacks is usually between 4 and 8 days, with 5 days being the most commonly observed period. The term *quintan fever* refers to the recurring 5-day attacks. Each succeeding attack is usually less severe. Major polymorphonuclear leukocytosis is often observed.

Persistent and chronic bacteremia has long been recognized. Kostrzewski carried out a unique experiment by surveying louse feeders enrolled for typhus vaccine production.[19] Among the 100 participants exposed, all were sick and 10 were asymptomatic, of which 5 were chronically bacteremic. Among the symptomatic patients, two-thirds experienced several episodes of trench fever; these were not relapses, being several months apart. Asymptomatic carriers do not frequently have antibodies. In our experience chronic bacteremia can last up to 78 weeks.[24] Definitive link between chronic bacteremia and endocarditis, although likely, has not yet been proven.[25]

B. quintana endocarditis has first been reported in three non-HIV-infected, homeless men in France.[26] All three patients required valve replacements because of extensive valvular damage, and pathological investigation confirmed the diagnosis of endocarditis. *B. quintana* endocarditis is most often observed in homeless people with chronic alcoholism and exposure to body lice and in patients without previously known valvulopathy. *Bartonella* endocarditis is usually indolent and culture negative, and thus, diagnosis is often delayed, resulting in a mortality rate higher than that for some other forms of endocarditis.

B. quintana has also been reported as a cause of lymphadenopathy in cat owners.[27]

Bacillary angiomatosis is a vascular proliferative disease most often involving the skin, but it may involve other organs The disease was first described in HIV-infected patients[28] and organ transplant recipients,[29] but it can also rarely affect immunocompetent patients.[30] These may be caused by both *B. quintana* and *B. henselae* the agent of cat scratch disease. The clinical differential diagnosis includes pyogenic granuloma, hemangioma, subcutaneous tumors, and Kaposi's sarcoma. The skin lesions are very similar to those reported for verruga peruana, the chronic form of Carrion's disease.[31]

Evidence of *B. quintana* in homeless has been reported in France,[32–35] in the United States,[36] in Japan,[37] and in Russia.[38] In this population *B. quintana* causes trench fever,[39] chronic bacteremia,[40–42] endocarditis,[43,44] bacillary angiomatosis,[45] and undifferentiated fever in HIV-infected persons.[46] *B. quintana* endocarditis is characterized by the fact that it occurs in alcoholic lice-infested homeless without previous valvulopathy while *B. henselae* occurs in patient in contact with cat or cat fleas and with previous valvulopathy.[47]

Typhus

Epidemic typhus is a disease caused by *R. prowazekii*. The main reservoir is humans. It is a life-threatening, acute exanthematic febrile illness. The mortality rate varies from 2% to 40% for untreated cases. In self-resolving cases, the bacteria can persist for life in human, and under stressful conditions, recrudescence may occur as a milder Brill–Zinsser disease.[48] Because this clinical form is bacteremic, it can initiate an outbreak of epidemic typhus when body louse infestations are prevalent in the population. Outbreaks of epidemic typhus have always been associated with war, famine, refugee camps, cold weather, poverty, or lapses in public health. Recent outbreaks of epidemic typhus have been reported in Burundi and Russia and sporadic cases have been reported in Algeria.[49,50] In developed countries, similar poor living conditions predisposing to high prevalence of body lice infestation exist in homeless populations.[51] In a recent study among homeless living in Marseilles' shelters, we demonstrate that significant antibody titers to *R. prowazekii* were present in 0.75% of sera.[52] Moreover, in Marseilles, a sporadic case of imported typhus from Algeria in a homeless patient, a case of the recurrent bacteremic form of typhus—Brill–Zinsser–disease, and an acute autochthonous case of epidemic typhus have recently been reported.[53–55] Although no outbreaks of typhus have been notified yet in the homeless population, this disease is likely to reemerge in such situation.

Relapsing Fever

Relapsing fever is caused by the spirochete *Borrelia recurrentis* that is excreted in the feces.[56] The illness begins abruptly with chills, headache, and fever. Most of these symptoms, which are associated with myalgia, arthralgia, abdominal pains, anorexia, dry cough, and fatigue, are mild for the first few days. Fever ranges between 39.5°C and 40°C. A cough is frequently prominent and could be associated with both epistasis and hemoptysis. Neurological involvement is usual.[57] The most commonly reported neurological symptom is meningismus that is not generally severe unless associated with subarachnoid hemorrhage. Encephalitis and encephalopathy occur occasionally, manifesting as seizures, and somnolence. Physical signs may be observed, such as conjunctival injection or conjunctivitis, petechial skin rash on the trunk, splenomegaly that is often tender, and hepatomegaly. Jaundice is possible and is a diagnostic clue in louse-associated diseases. One of the complications of louse-borne relapsing fever is bleeding, purpura, and epistaxis being the more common findings. Other hemorrhagic phenomena include hemoptysis, hematemesis, hematuria, cerebral hemorrhages, bloody diarrhea, retinal hemorrhage, and splenic rupture. Clinical characteristics of relapsing fever are an initial febrile episode terminating in the crisis phenomenon, followed by an interval of apyrexia of variable length, which is followed by relapse, with return of fever and other clinical manifestations.[58] Periods of relapse are less severe and shorter than the first febrile attack, with each relapse being less severe. Occasionally, no relapses are observed. The duration of the primary febrile attack averages 5.5 days. The duration of afebrile intervals averages 9.25 days (range, 3–27 days). Most patients have only one relapse, although a few have two. The duration of relapse averages 1.9 days. Peak temperatures are lower during relapses. Finally, without treatment, the death rate varies from 10% to 40%; antibiotic therapy decreases it to 2–4%.[58] Few data are available in the literature concerning the occurrence of relapsing fever in the homeless population, whereas outbreaks are ongoing in Sudan and antibodies to *B. recurrentis* have been detected in rural Andean communities in Peru.[59] In Marseilles we found a significant higher seroprevalence of antibodies to *B. recurrentis* in this population. That seroprevalence increased in 2002 suggesting that an unnoticed small outbreak has occurred.[60]

Flea-Borne Diseases

Fleas are widespread but are not adapted to a specific host and may occasionally bite humans.[61] Most common fleas that parasite humans are the cat, the rat, and the human fleas, *Ctenocephalides felis*, *Xenopsylla cheopis,* and *Pulex irritans*, respectively. Fleas transmit plague (*Xenopsylla cheopis and Pulex irritans*), murine typhus (*Xenopsylla cheopis*), flea-borne spotted rickettsiosis on account of the recently described species *Rickettsia felis* (*Ctenocephalides felis, Pulex irritans*), and may be cat scratch disease on account of *Bartonella henselae* (*Ctenocephalides felis*). The role of fleas as potential vector of *B. quintana* has recently been suggested as it is found in *C. felis* and *Pulex irritans* (unpublished).[27,62] Antibodies to *Bartonella* spp. have been retrieved with a high prevalence in IDVA in inner city of Baltimore.[63] Murine typhus, a flea-associated infection, has not been found in both IVDU in New York[64] and homeless of Los Angeles.[65] *R. felis* infection has been reported to occur in southern Europe and the Americas in the rodent-exposed population[66,67] but no data are available in homeless. In our study on 930 homeless from Marseilles the seroprevalence of both *R. felis* and *R. typhi* were not significantly different from that of the control population.[68]

Tick-Borne Diseases

Ticks belonging to the family *Ixodidae*, in particular the genera *Dermacentor*, *Rhipicephalus,* and *Ixodes*, are frequent parasites in humans. Tick bite usually goes unnoticed. The tick may remain attached to the host without any local symptoms for several hours or days, the time usually necessary for disease transmission. In our areas, many dogs during summer time harbor *Rh.*

sanguineus that may bite humans occasionally.[69] Virtually, no data are available on tick-borne disease in this population. Although the seroprevalence of antibodies to the tick-borne (spotted fever group) rickettsia is not different between the homeless of Marseilles and the general population, we have reported severe cases of Mediterranean spotted fever in these patients, one of them being bitten by 22 ticks.[70]

Mite-Borne Diseases

The hematophagic-biting mite, *Liponyssoides sanguineus*, is a mite of the rat, mouse and other domestic rodents but can also bite humans.[71] *L. sanguineus*, is responsible for the transmission of *Rickettsia akari*, the etiologic agent of rickettsialpox. This disease has mainly been reported in the United States (New York), ex-USSR, Slovenia, Ukraine, the Republic of Korea, and the People's Republic of Korea. Recently, it has been shown that Rickettsial pox was persistent in New York City with 34 new cases diagnosed in 18 months survey.[72] It has been reported that 9% of IVDA of Harlem New York and 16 % of those in inner-city Baltimore had antibodies to *R. akari*.[73,74] In our population of homeless in Marseilles the seroprevalence of *R. akari* was 0.2% being not different to that of the controls.[75]

Sarcoptes scabiei var. hominis is also a mite (*Arachnida* class) responsible for scabies.[76] It is an obligate parasite of human skin. Human-to-human transmission usually occurs after prolonged skin contact, but scabies is a highly contagious disease. Scabies is a widespread disease occurring irrespective of race or socioeconomic conditions but epidemics occur frequently under unfavorable sanitation conditions. In the homeless scabies was reported with prevalence varying 3.8–56.5%.[77,78] Scabies are not potential vector for infectious agents.

The incidence of skin disease secondary to infestation with the human bedbug has recently increased in the United States, United Kingdom, and Canada.[79] Toronto Public Health documented complaints of bedbug infestations from 46 locations in 2003, most commonly apartments (63%), shelters (15%), and rooming houses (11%). Pest control operators in Toronto ($n = 34$) reported treating bedbug infestations at 847 locations in 2003, most commonly single-family dwellings (70%), apartments (18%), and shelters (8%). Bedbug infestations were reported at 20 (31%) of 65 homeless shelters. In one of the affected shelter, 4% of residents reported having bedbug bites.[80] Bacteria, such as *wolbacchia* have been detected in the bedbug *Cimex lectularius*.[81] Although no case of bedbug-borne infection has been reported yet, bedbug infestations can have an adverse effect on health and quality of life in the general population, particularly among homeless persons living in shelters.

Diagnosis of Arthropod-Borne Diseases

Except for relapsing fever in which blood smear may show the presence of the spirochete, diagnosis is usually confirmed by IFA. Serological cross-reaction occurs between closely related bacterial species and cross-adsorption assay with either IFA or Western blotting is mandatory to confirm the species involved.[82] This diagnostic is now available for *B. recurrentis* since this spirochete is cultivable *in vitro*.[83] Molecular detection and cultivation of the infectious agents are restricted to specialized laboratories.

Treatment and Prevention

Delousing

In the long term, control of lice has largely been a failure. Although, the simplest method for delousing is a complete change of clothing, this is not always either practical or even acceptable and other simple measures, such as washing, can be effective.[84] Powder dusting of the entire clothing with 10% DDT, 1% malathion, or 1% permethrin dust is another alternative.[85] There is no need to disinfect other belongings, with the exception of recently used blankets or clothes. Ivermectin, a macrocyclin lactone used for treatment of onchocerciasis, causes paralysis in many nematodes and arthropods. It has been used for the treatment of lice-infected swine and cattle, and been used successfully in the treatment of human head lice.[86] We have observed the dramatic effect of oral ivermectin in a cohort of homeless men from a shelter in Marseilles suggesting that ivermectin treatment may contribute to control body louse infestation among homeless people (unpublished data).

Treatment of Specific Infections

Most cases of trench fever, the acute form of *B. quintana* bacteremia, were reported prior to the antibiotic era. There were no fatal cases of trench fever, and the clinical manifestations lasted for 4–6 weeks before full recovery. However, successful treatment of some trench fever patients with tetracycline or chloramphenicol was reported after World War II, although these data remain anecdotal. Thus, it seems reasonable to prescribe doxycycline for such patients. Treatment of paucisymptomatic, persistent *B. quintana* bacteremia may be of importance for the prevention of endocarditis in these patients.[87] We recommend that patients with *B. quintana* bacteremia be treated with gentamycin (3 mg/kg of body weight i.v. once daily for 14 days), in combination with doxycycline (200 mg p.o. daily) for 28 days or 6 weeks in case of endocarditis.[88] The recommended treatment for epidemic typhus is doxycycline. A single dose of 200 mg has been reported to be as effective as the conventional therapy for epidemic typhus.[89,90] Doxycycline prophylaxis may be used in epidemic situations in adjunction to delousing measures. In patients with louse-borne relapsing fever a single dose of 100 mg of doxycycline is adequate treatment for most patients and is effective in clearing the spirochetes from blood smears.[91] Tick-borne rickettsioses should be treated with doxycycline 200 mg/day for 1–5 days.

Among arthropod-borne diseases, louse-borne infection is particularly worrying. Despite effective treatment of *B. quintana* bacteremia and the efforts made to delouse this population *B. quintana* remains endemic.[92] Moreover, we found hallmark of unnoticed outbreak of epidemic typhus and relapsing fever. The uncontrolled louse infestation in this population should alert the community of the possibility of severe reemerging louse-borne infection.

ACKNOWLEDGMENTS

This work was supported by "Conseil Général des Bouches du Rhône," "Programme hospitalier de recherche clinique 2000 and 2003 " du Ministère de la Santé, and by CNRS UMR 6020.

REFERENCES

1. SASAKI, T., M. KOBAYASHI & N. AGUI. 2002. Detection of Bartonella quintana from body lice (Anoplura: Pediculidae) infesting homeless people in Tokyo by molecular technique. J. Med. Entomol. **39:** 427–429.
2. RAOULT, D., C. FOUCAULT & P. BROUQUI. 2001. Infections in the homeless. Lancet Infect. Dis. **1:** 77–84.
3. BROUQUI, P., A. STEIN, H.T. DUPONT, *et al.* 2005. Ectoparasitism and vector-borne diseases in 930 homeless people from Marseilles. Medicine (Baltimore) **84:** 61–68.
4. CHOSIDOW, O. 2000. Scabies and pediculosis. Lancet **355:** 819–826.
5. REED, D.L., V.S. SMITH, S.L. HAMMOND, *et al.* 2004. Genetic analysis of lice supports direct contact between modern and archaic humans. PLoS Biol. **2:** e340.
6. BURGESS I.F. 1995. Human lice and their management. *In* Advances in Parasitology. J.R. Baker, R. Muller & D. Rollinson, Eds.: 271–342. Academic Press. London.
7. MAUNDER, J.W. 1983. The appreciation of lice. Proc. Roy. Inst. Gr. Brit. **55:** 1–31.
8. EVANS, F.C. & F.E. SMITH. 1952. The intrinsic rate of natural increase for the human louse *Pediculus humanus L.* Am. Nat. **86:** 299–310.
9. RAOULT, D. & V. ROUX. 1999. The body louse as a vector of reemerging human diseases. Clin. Infect. Dis. **29:** 888–911.
10. RAOULT, D., J.B. NDIHOKUBWAYO, H. TISSOT-DUPONT, *et al.* 1998. Outbreak of epidemic typhus associated with trench fever in Burundi [see comments]. Lancet **352:** 353–358.

11. RAOULT, D., R.J. BIRTLES, M. MONTOYA, *et al.* 1999. Survey of three bacterial louse-associated diseases among rural Andean communities in Peru: prevalence of epidemic typhus, trench fever, and relapsing fever. Clin. Infect. Dis. **29:** 434–436.
12. MAURIN, M. & D. RAOULT. 1996. Bartonella (Rochalimaea) quintana infections. Clin. Microbiol. Rev. **9:** 273–292.
13. KOSTRZEWSKI, J. 1949. The epidemiology of Trench fever. Bul de l'Académie polonaise des sciences et des lettres Classe de Médecine **7:** 233–263.
14. BYAM, W. & L.L. LLOYD. 1920. Trench fever : its epidemiology and endemiology. Proc. Roy. Soc. Med. **13:** 1–27.
15. VINSON, J.W., G. VARELA, & C. MOLINA-PASQUEL. 1969. Trench fever. 3. Induction of clinical disease in volunteers inoculated with Rickettsia quintana propagated on blood agar. Am. J. Trop. Med. Hyg. **18:** 713–722.
16. MCNEE, J.W., A. RENSHAW & E.H. BRUNT. 1916. "Trench fever" : a relapsing fever occurring with the British forces in France. Br. Med. J. **12:** 225–234.
17. FOUCAULT, C., J.M. ROLAIN, D. RAOULT & P. BROUQUI. 2004. Detection of Bartonella quintana by direct immunofluorescence examination of blood smears of a patient with acute trench fever. J. Clin. Microbiol. **42:** 4904–4906.
18. FOUCAULT, C., K. BARRAU, P. BROUQUI & D. RAOULT. 2002. *Bartonella quintana* Bacteremia among Homeless People. Clin. Infect. Dis. **35:** 684–689.
19. FOURNIER, P.E., H. LELIEVRE, S.J. EYKYN, *et al.* 2001. Epidemiologic and clinical characteristics of Bartonella quintana and Bartonella henselae endocarditis: a study of 48 patients. Medicine (Baltimore) **80:** 245–251.
20. DRANCOURT, M., J.L. MAINARDI, P. BROUQUI, *et al.* 1995. Bartonella (Rochalimaea) quintana endocarditis in three homeless men. N. Engl. J. Med. **332:** 419–423.
21. RAOULT, D., M. DRANCOURT, A. CARTA & J.A. GASTAUT. 1994. *Bartonella (Rochalimaea) quintana* isolation in patient with chronic adenopathy, lymphopenia, and a cat. Lancet **343:** 977.
22. STOLER, M.H., T.A. BONFIGLIO, R.T. STEIGBIGEL & M. PEREIRA. 1983. An atypical subcutaneous infection associated with acquired immune deficiency syndrome. Am. J. Clin. Pathol. **80:** 714–718.
23. KEMPER, C.A., C.M. LOMBARD, S.C. DERESINSKI & L.S. TOMPKINS. 1990. Visceral bacillary epithelioid angiomatosis: possible manifestations of disseminated cat scratch disease in the immunocompromised host : a report of two cases. Am. J. Med. **89:** 216–222.
24. TAPPERO, J.W., J.E. KOEHLER, T.G. BERGER, *et al.* 1993. Bacillary angiomatosis and bacillary splenitis in immunocompetent adults. Ann. Intern. Med. **118:** 363–365.
25. MAGUINA, C., P.J. GARCIA, E. GOTUZZO, *et al.* 2001. Bartonellosis (Carrion's Disease) in the modern era. Clin. Infect. Dis. **33:** 772–779.
26. BROUQUI, P., P. HOUPIKIAN, H.T. DUPONT, *et al.* 1996. Survey of the seroprevalence of Bartonella quintana in homeless people. Clin. Infect. Dis. **23:** 756–759.
27. BROUQUI, P., B. LASCOLA, V. ROUX & D. RAOULT. 1999. Chronic Bartonella quintana bacteremia in homeless patients. N. Engl. J. Med. **340:** 184–189.
28. GUIBAL, F., P. DE LA SALMONIERE, M. RYBOJAD, *et al.* 2001. High seroprevalence to Bartonella quintana in homeless patients with cutaneous parasitic infestations in downtown Paris. J. Am. Acad. Dermatol. **44:** 219–223.
29. JACKSON, L.A. & D.H. SPACH. 1996. Emergence of Bartonella quintana infection among homeless persons. Emerg. Infect. Dis. **2:** 141–144.
30. RYDKINA, E.B., V. ROUX, E.M. GAGUA, *et al.* 1999. Bartonella quintana in body lice collected from homeless persons in Russia. Emerg. Infect. Dis. **5:** 176–178.

31. ROLAIN, J.M., C. FOUCAULT, R. GUIEU, *et al.* 2002. Bartonella quintana in human erythrocytes. Lancet **360:** 226–228.
32. SPACH, D.H., A.S. KANTER, M.J. DOUGHERTY, *et al.* 1995. Bartonella (Rochalimaea) quintana bacteremia in inner-city patients with chronic alcoholism. N. Engl. J. Med. **332:** 424–428.
33. KOEHLER, J.E., F.D. QUINN, T.G. BERGER, *et al.* 1992. Isolation of Rochalimaea species from cutaneous and osseous lesions of bacillary angiomatosis. N. Engl. J. Med. **327:** 1625–1631.
34. KOEHLER, J.E., M.A. SANCHEZ, S. TYE, *et al.* 2003. Prevalence of Bartonella infection among human immunodeficiency virus-infected patients with fever. Clin. Infect. Dis. **37:** 559–566.
35. TARASEVICH, I., E. RYDKINA & D. RAOULT. 1998. Outbreak of epidemic typhus in Russia [letter; comment]. Lancet **352:** 1151.
36. NIANG, M., P. BROUQUI & D. RAOULT. 1999. Epidemic typhus imported from Algeria. Emerg. Infect. Dis. **5:** 716–718.
37. STEIN, A., R. PURGUS, M. OLMER & D. RAOULT. 1999. Brill-Zinsser disease in France [letter]. Lancet **353:** 1936.
38. BADIAGA, S., P. BROUQUI & D. RAOULT. 2005. Autochthonous epidemic typhus associated with *Bartonella quintana* bacteremia in a homeless person. Am. J. Trop. Med. Hyg. **72:** 638–639.
39. HOUHAMDI, L. & D. RAOULT. 2005. Excretion of living Borrelia recurrentis in feces of infected human body lice. J. Infect. Dis. **191:** 1898–1906.
40. CADAVID, D. & A.G. BARBOUR. 1998. Neuroborreliosis during relapsing fever: review of the clinical manifestations, pathology, and treatment of infections in humans and experimental animals. Clin. Infect. Dis. **26:** 151–164.
41. SOUTHERN, P.M.,JR. & J.P. SANFORD. 1969. Relapsing fever. A clinical and microbiological review. Medicine **48:** 129–149.
42. SOUSA, C.A. 1997. Fleas, flea allergy, and flea control: a review. Dermatol. Online J. **3:** 7.
43. ROLAIN, J.M., M. FRANC, B. DAVOUST & D. RAOULT. 2003. Molecular detection of Bartonella quintana, B. koehlerae, B. henselae, B. clarridgeiae, Rickettsia felis, and Wolbachia pipientis in cat fleas, France. Emerg. Infect. Dis. **9:** 338–342.
44. COMER, J.A., C. FLYNN, R.L. REGNERY, *et al.* 1996. Antibodies to Bartonella species in inner-city intravenous drug users in Baltimore, Md. Arch. Intern. Med. **156:** 2491–2495.
45. COMER, J.A., T. DIAZ, D. VLAHOV, *et al.* 2001. Evidence of rodent-associated Bartonella and Rickettsia infections among intravenous drug users from Central and East Harlem, New York City. Am. J. Trop. Med. Hyg. **65:** 855–860.
46. SMITH, H.M., R. REPORTER, M.P. ROOD, *et al.* 2002. Prevalence study of antibody to ratborne pathogens and other agents among patients using a free clinic in downtown Los Angeles. J. Infect. Dis. **186:** 1673–1676.
47. RICHTER, J., P.E. FOURNIER, J. PETRIDOU, *et al.* 2002. Rickettsia felis infection acquired in Europe and documented by polymerase chain reaction. Emerg. Infect. Dis. **8:** 207–208.
48. ZAVALA-VELAZQUEZ, J.E., J.A. RUIZ-SOSA, R.A. SANCHEZ-ELIAS, *et al.* 2000. Rickettsia felis rickettsiosis in Yucatan. Lancet **356:** 1079–1080.
49. PAROLA, P. & D. RAOULT. 2001. Ticks and tickborne bacterial diseases in humans: an emerging infectious threat. Clin. Infect. Dis. **32:** 897–928.
50. HUEBNER, R.J., W.L. JELLISON & C. POMERANTZ. 1946. Rickettsialpox—a newly recognized rickettsial disease. IV Isolation of a rickettsia apparently identical

with the causative agent of rickettsialpox from *Allodermanyssus sanguineus* a rodent mite. Public Health Rep. **61:** 1677–1682.

51. PADDOCK, C.D., S.R. ZAKI, T. KOSS, *et al.* 2003. Rickettsialpox in New York City: a persistent urban zoonosis. Ann. N. Y. Acad. Sci. **990:** 36–44.
52. COMER, J.A., T. TZIANABOS, C. FLYNN, *et al.* 1999. Serologic evidence of rickettsialpox (Rickettsia akari) infection among intravenous drug users in inner-city Baltimore, Maryland. Am. J. Trop. Med. Hyg. **60:** 894–898.
53. BADIAGA, S., A. MENARD, H. TISSOT-DUPONT, *et al.* Prevalence of skin infections in sheltered homeless of Marseilles, France. Eur. J. Dermatol. In press.
54. ARFI, C., L. DEHEN, E. BENASSAIA, *et al.* 1999. Dermatologic consultation in a precarious situation: a prospective medical and social study at the Hospital Saint-Louis in Paris. Ann. Dermatol.Venereol. **126:** 682–686.
55. TER POORTEN M.C. & N.S. PROSE. 2005. The return of the common bedbug. Pediatr. Dermatol. **22:** 183–187.
56. HWANG, S.W., T.J. SVOBODA, I.J. DE JONG, *et al.* 2005. Bed bug infestations in an urban environment. Emerg. Infect. Dis. **11:** 533–538.
57. RASGON, J.L. & T.W. SCOTT. 2004. Phylogenetic characterization of Wolbachia symbionts infecting Cimex lectularius L. and Oeciacus vicarius Horvath (Hemiptera: Cimicidae). J. Med. Entomol. **41:** 1175–1178.
58. LA SCOLA B. & D. RAOULT. 1997. Laboratory diagnosis of rickettsioses: current approaches to diagnosis of old and new rickettsial diseases. J. Clin. Microbiol. **35:** 2715–2727.
59. PORCELLA, S.F., S.J. RAFFEL, M.E. SCHRUMPF, *et al.* 2000. Serodiagnosis of Louse-Borne relapsing fever with glycerophosphodiester phosphodiesterase (GlpQ) from Borrelia recurrentis. J. Clin. Microbiol. **38:** 3561–3571.
60. BELL, T.A. 1998. Treatment of *Pediculus humanus* var. *capitis* infestation in Cowlitz County, Washington, with ivermectin and the LiceMeister comb. Pediatr. Infect. Dis. J. **17:** 923–924.
61. FOUCAULT, C., K. BARRAU, P. BROUQUI & D. RAOULT. 2002. *Bartonella quintana* Bacteremia among homeless people. Clin. Infect. Dis. **35:** 684–689.
62. ROLAIN, J.M., P. BROUQUI, J.E. KOEHLER, *et al.* 2004. Recommendations for treatment of human infections caused by Bartonella species. Antimicrob. Agents Chemother. **48:** 1921–1933.
63. HUYS, J., J. KAYHIGI, P. FREYENS & G.V. BERGHE. 1973. Single-dose treatment of epidemic typhus with doxycyline. Chemotherapy **18:** 314–317.
64. PERINE, P.L., D.W. KRAUSE, S. AWOKE & J.E. MCDADE. 1974. Single-dose doxycycline treatment of louse-borne relapsing fever and epidemic typhus. Lancet **2:** 742–744.

Clinical Diagnosis and Treatment of Human Granulocytotropic Anaplasmosis

JOHAN S. BAKKEN[a] AND J. STEPHEN DUMLER[b]

[a]*Department of Family Medicine, University of Minnesota at Duluth, School of Medicine, Duluth, and St. Luke's Infectious Disease Associates, St. Luke's Hospital, Duluth, Minnesota 55802, USA*

[b]*Division of Medical Microbiology, Department of Pathology, The Johns Hopkins University School of Medicine, Baltimore, Maryland 21205, USA*

ABSTRACT: Tick-borne rickettsiae in the genera *Ehrlichia* and *Anaplasma* are intracellular bacteria that infect wild and domestic mammals and, more recently, man. The increased desire of humans for recreational activities outdoors has increased the exposure to potential human pathogens that previously cycled almost exclusively within natural, nonhuman enzootic hosts. *Anaplasma phagocytophilum* causes an acute, nonspecific febrile illness of humans previously known as human granulocytotropic ehrlichiosis (HGE) and now called human granulocytotropic anaplasmosis (HGA). The first patient to have recognized HGA was hospitalized at St Mary's Hospital in Duluth, Minnesota, USA in 1990. However, the clinical and laboratory presentation of this infection remained undefined until 1994, when Bakken and collaborators published their experience with 12 patients who had HGA. By the end of December 2004, at least 2,871 cases of HGA had been reported from 13 U.S. states to the Centers for Disease Control and Prevention (CDC). A limited number of laboratory-confirmed cases have been reported from countries in Europe, including Austria, Italy, Latvia, the Netherlands, Norway, Poland, Slovenia, Spain, and Sweden. *Ixodes persulcatus*-complex ticks are the arthropod hosts for *Borrelia burgdorferi*, the agent of Lyme borreliosis, and are also the arthropod hosts for *A. phagocytophilum.* Most cases of HGA have been contracted in geographic regions that are endemic for Lyme borreliosis. Male patients outnumber female patients by a factor of 3 to 1 and as many as 75% of patients with HGA have had a tick bite prior to their illness. Seroepidemiologic studies have demonstrated that HGA for the most part is a mild or even asymptomatic illness. However, older individuals and patients who are immunocompromised by natural disease processes or medications may develop an acute, influenza-like illness characterized by high fever, rigors, generalized myalgias, and severe headache. Local skin reactions at the site of the tick bite have not been described, and nonspecific skin rashes have been reported only occasionally. Anaplasmosis is associated with variable but suggestive changes

Address for correspondence: Johan S. Bakken, M.D., Ph.D., St. Luke's Infectious Disease Associates, 1001 East First Street, Suite L201, Duluth, Minnesota 55802. Voice: 218-249-7990; fax: 218-249-7996.
e-mail: jbakken@slhduluth.com

**Ann. N.Y. Acad. Sci. 1078: 236–247 (2006). © 2006 New York Academy of Sciences.
doi: 10.1196/annals.1374.042**

in routine laboratory test parameters. Most patients develop transient reductions in total leukocyte and platelet concentrations. Relative granulocytosis accompanied by a left shift and lymphopenia during the first week of illness has been reported frequently. Serum hepatic transaminase concentrations usually increase two- to fourfold, and inflammatory markers, such as C-reactive protein and the erythrocyte sedimentation rate, rise during the acute phase. Abnormal laboratory findings may return toward normal range for patients who have been ill for more than 7 days, which may obfuscate the clinical decision making. Characteristic clusters of bacteria (morulae) are observed in the cytoplasm of peripheral blood granulocytes in 20% to 80% of infected patients during the acute phase of illness. The clinical diagnosis may be confirmed retrospectively by specific laboratory tests, which include positive polymerase chain reaction (PCR), identification of *A. phagocytophilum* in culture of acute-phase blood, or the detection of specific antibodies to *A. phagocytophilum* in convalescent serum. Virtually all patients have developed serum antibodies to *A. phagocytophilum* after completion of antibiotic therapy, and demonstration of seroconversion by indirect immunofluorescent antibody testing of acute-phase and convalescent-phase serum samples is currently the most sensitive and specific tool for laboratory confirmation of HGA. Treatment with doxycycline usually results in rapid improvement and cure. Most patients with HGA have made an uneventful recovery even without specific antibiotic therapy. However, delayed diagnosis in older and immunocompromised patients may place those individuals at risk for an adverse outcome, including death. Thus, prompt institution of antibiotic therapy is advocated for any patient who is suspected to have HGA and for all patients who have confirmed HGA.

KEYWORDS: human granulocytotropic anaplasmosis; tick-borne infections; zoonosis; doxycycline; rifampicin

INTRODUCTION

Human granulocytotropic anaplasmosis (HGA) is a tick-borne zoonotic infection that is caused by *Anaplasma phagocytophilum (A. phagocytophilum).*[1] HGA has become increasingly recognized in the United States of America and in several European countries. The increased desire to pursue outdoor recreational activities during the summer months has also amplified humans' potential exposure to pathogenic bacteria that are present in nonvertebrate bloodsucking enzootic hosts during a portion of their life cycle. Just like *Borrelia burgdorferi*, the agent of Lyme borreliosis, *A. phagocytophilum* cycles within hard-bodied ticks that are members of the *Ixodes persulcatus* complex. The tabular list of International Classification of Diseases (ICD-9) has categorized anaplasmosis under the subheading tick-borne rickettsioses and the numerical code for HGA is 082.49.[2]

Bakken *et al.* described an outbreak of human granulocytotropic ehrlichiosis (HGE) among older men from the upper midwest of the United States in 1994.[3]

Early genetic and serologic analysis of blood from infected patients indicated that the causative agent of HGE was closely related to the granulocytotropic veterinary ehrlichiae *Ehrlichia phagocytophila* and *Ehrlichia equi.*[4] A recent reorganization of the family *Anaplasmataceae* resulted in the unification of these granulocytotropic agents of animals and man into the genus *Anaplasma*, as *A. phagocytophilum* comb. nov.[1] *A. phagocytophilum* describes a granulocytotropic agent that causes infections in animals and humans and the clinical syndrome caused by this bacterium is now called HGA, which is synonymous with HGE.[1,5,6]

Epidemiology of HGA

Ixodes ticks are the established vectors for the agents that cause HGA, Lyme borreliosis, and human babesiosis (caused by the parasite *Babesia microti*).[7–9] *Ixodes scapularis* is distributed throughout the eastern and midwestern regions of the United States of America, while the endemic range for *Ixodes pacificus* is limited to the Pacific coast. Transovarial passage of *A. phagocytophilum* from adult female ticks to offspring larvae does not occur or does so only at very low frequency. Thus, progeny from the adult female tick are not infected and must acquire the infectious agent in a subsequent bloodmeal. However, the infectious cycle of *A. phagocytophilum* is maintained in nature primarily by ticks that feed on transiently or persistently infected reservoir hosts, potentially including small rodents, such as *Peromyscus leucopus* (the white-footed mouse), *Neotoma fuscipes* (the dusky-footed wood rat), *Cletrionomys glareolus* (the bank vole), and the white-tailed deer.[10–12] Humans become infected by accident, as they are likely to be dead-end hosts. The "grace period" for *A. phagocytophilum* transmission from ticks to mammal hosts has not been firmly established, but between one-thirds and two-thirds of mice acquired infection from ticks in the first 24 h after attachment.[13]

The Centers for Disease Control (CDC) had recorded more than 2,135 cases of HGA in the United States since 1994 and the end of 2002, the last year for complete data,[14] and more than 2,871 cases had been reported to the CDC by the end of 2004 (J. McQuiston, personal information). A limited number of HGA cases have also been reported from Austria, Italy, Latvia, the Netherlands, Norway, Poland, Spain, and Sweden, and in a recent publication Blanco and Oteo compared the clinical and laboratory features for 15 European patients with those reported for U.S. patients.[15] Seroepidemiologic investigations have shown that HGA for the most part is a mild or even asymptomatic infection[16–19] and most patients recover uneventfully in 1 to 2 weeks even in the absence of specific antibiotic therapy.[20] The estimated HGA case fatality rate is low (0.5% to 1%); however, it may be difficult to identify patients who are likely to develop serious or fatal disease.[8,21] Thus, prompt institution of active antibiotic therapy is advocated for all patients who have confirmed HGA and are symptomatic.[22]

Epidemiologic studies also provide evidence of high-risk regions and populations. From compilations of published literature, seroprevalence as high as 15.4% in Slovenia[23] and 14.9% in Wisconsin[17] have been reported among cross-sectional analyses. Studies of at-risk populations in New York State who are *B. burgdorferi*–seropositive have shown that 35.6% of these patients also seroreact with *A. phagocytophilum*, and the median *A. phagocytophilum* seroprevalence among patients considered at-risk (*B. burgdorferi*–seropositive status from occupational or residential exposure) was recently estimated at 9.6%.[16] Incidence rates as high as 52 to 58 cases/100,000 population have been demonstrated in some highly endemic regions of Connecticut[24] and Wisconsin.[17] Despite high seroprevalence rates, only a single case of transfusion-related HGA has been documented, underscoring the rarity of this event.[25]

Human infection by *A. phagocytophilum* results in an undifferentiated febrile illness with pancytopenia and biochemical evidence of hepatic injury.[26] The bacterium preferentially resides in the unusually harsh environment of neutrophils, a unique niche among bacteria. In limited histopathologic examinations of humans, normocellular to hypercellular bone marrows and focal hepatic inflammation with apoptoses, but few infected cells are observed.[26] Usually fewer than 1% of neutrophils are infected and pancytopenia cannot be explained by bacterial cytolysis. Thus, leukopenia, thrombocytopenia, and inflammatory hepatic foci must result from an alternate mechanism that leads to destruction or sequestration of these elements. Hemophagocytosis in mononuclear phagocyte organs suggests cytokine-mediated macrophage activation, an observation supported by abundant interferon (IFN)-γ in sera from animals and humans.[27,28] Unlike the situation for monocytotropic ehrlichiosis caused by *Ehrlichia chaffeensis*, granulomas are not identified in HGA, and CSF pleocytosis is distinctly rare.

Clinical Diagnosis

HGA is a clinical syndrome most commonly manifested by nonspecific fever, chills, headache, and myalgias (TABLE 1).[3,5,15,21,29] The symptoms and signs range from asymptomatic to fatal disease and there is a direct correlation between increased patient age and/or comorbid illnesses and the clinical severity of HGA.[30] Most symptomatic patients report exposure to ticks 1 to 2 weeks before the onset of symptoms. HGA can be severe, with nearly half of patients requiring hospitalization and up to 17% requiring admission to an intensive care unit.[21] Even though the case fatality rate is low, significant complications can occur and include a septic or toxic shock-like syndrome, respiratory insufficiency, invasive opportunistic infections with both viral and fungal agents, rhabdomyolysis, pancarditis, acute renal failure, hemorrhage, and neurologic diseases, such as brachial plexopathy and demyelinating polyneuropathy.[26,31–33]

TABLE 1. Published signs and symptoms (%) reported by patients who were treated for laboratory-confirmed HGA in the USA[5,29,54,58] and in Europe[15] contrasted with 207 cases treated at the Duluth Clinic, Duluth, Minnesota[21,54]

Prevalence of complaint	Symptom or sign	Published cases (%) $N = 4–243$	Duluth clinic cases (%) $N = 207$
Common	Fever	99–100	92
	Headache	61–93	70
	Myalgias	40–83	86
	Malaise	47–93	97
	Rigors	27–39	86
Less common	Anorexia	6	47
	Nausea	53	36
	Arthralgias	27–78	32
	Cough	20	22
Uncommon	Abdominal pain	20	4
	Confusion	No published data	17
	Rash	2–16	3[a]

[a]Erythema migrans in all cases.

Permutations of leukopenia, a left shift, thrombocytopenia, and hepatic transaminase elevations are present in the majority of patients and are suggestive clues to the diagnosis. Although both leukopenia and thrombocytopenia are present in many patients at presentation, these abnormalities usually normalize by the end of the second week. At least 20% and up to 80% of patients present with morulae in peripheral blood neutrophils that confirm the diagnosis.[5,34] Patients who have a clinical illness compatible with HGA should be considered for specific antibiotic treatment.[8,22,35]

The diagnosis can be confirmed by blood smear examination and PCR analysis during early infection, and by serologic testing in late infection or convalescence (TABLES 2 and 3). PCR amplification of *A. phagocytophilum* DNA from acute- phase blood[5,21,36] or isolation of *A. phagocytophilum* in HL-60 promyelocytic leukemia cell cultures inoculated with acute-phase blood[5,21,37] can confirm the diagnosis, but these test modalities are available in only a limited number of laboratories. Blood samples should be secured before the patient begins antibiotic treatment since therapy will rapidly reduce the detectable quantities of infected cells or bacterial DNA. Serologic testing using an indirect fluorescent antibody method with demonstration of fourfold change or seroconversion has been used most commonly to confirm HGA.[20,38,39] IgM tests are only reactive during the first 45 to 50 days after infection, and these tests are not more sensitive than those that detect IgG antibodies at the same time intervals.[40]

Treatment of HGA

In vitro investigations of a handful of clinical isolates have demonstrated that *Ehrlichia* and *Anaplasma* species are uniformly susceptible to the tetracycline

TABLE 2. Laboratory criteria for diagnosis of HGA[59]

Case definition	Laboratory test result for HGA
Probable infection	Morulae in peripheral blood smear neutrophils[a] *or* Single titer *A. phagocytophilum* serum IFA[b] ≥ 64 *or* Positive *A. phagocytophilum* PCR[c] of blood
Confirmed infection	IFA seroconversion or seroreversion[d] *or* Isolation of *A. phagocytophilum* from blood[e] *or* Morulae in peripheral blood smear neutrophils[a] *and* Single *A. phagocytophilum* serum IFA[b] ≥ 64 *or* Positive *A. phagocytophilum* PCR[c] of blood

[a]Light microscopy of Wright-stained peripheral acute-phase blood; Indirect immunofluorescent antibody test with *A. phagocytophilum*[b] antigen; Polymerase chain reaction with specific *A. phagocytophilum*[c] primers; Fourfold or greater change in serum antibody titer[d]; Isolation of *A. phagocytophilum*[e] in tissue culture inoculated with acute-phase blood.

antibiotics.[41–43] Doxycycline hyclate has traditionally been the agent of choice on account of favorable pharmacokinetic properties compared with other tetracycline derivatives. For the most part, HGA is a mild illness, but a relationship exists between more serious infections, including fatal outcome, and variables, such as advanced age, ongoing immunosuppressive therapy, chronic inflammatory illnesses, or underlying malignant diseases.[8,35,44–47] Because of the potential for serious or even fatal infection it is therefore recommended that all patients who have suspected or documented HGA should be treated with oral or intravenous doxycycline hyclate in the absence of specific contraindications to tetracycline drugs (TABLE 4). The recommended therapy for adults is doxycycline 100 mg given orally at 12-h intervals.[8,22,35,44] Children older than 8 years should also be treated with doxycycline given in divided doses with dosage adjusted to the patient's weight (4.4 mg/kg/24 h, maximum 100 mg per dose administered).[22,48] Doxycycline is the drug of choice for children who are seriously ill regardless of age.[48] Doxycycline therapy leads to clinical improvement in 24–48 h.[8,22,35,44,45,47] Thus, patients who fail to respond to treatment within this time frame should be reevaluated for alternative diagnoses and treatment.

TABLE 3. Relative sensitivity of diagnostic tests used for laboratory diagnosis confirmation of HGA[8,12,21,35,38,60,61]

Duration of illness (days)	Blood smear microscopy	HL-60 cell culture	PCR	Serologic test (IFA)
0–7	Medium	Medium	High	Low
8–14	Low	Low	Low	Medium
15–30			Low	High
31–60				High
>60				High

TABLE 4. Recommended adult and pediatric antibiotic treatment for HGA[5,8,22,48]

Antibiotic	Dose (adults)	Dose (children)	Duration (days)
Doxycycline hyclate	100 mg i.v.[a] or p.o.[b] q 12 h	2.2 mg/kg p.o. q 12 h[c]	5–14[d]
Tetracycline hydrochloride	500 mg p.o. q 6 h	25–50 mg/kg/day p.o. in 4 divided doses[c]	5–14[d]
Rifampin	300 mg p.o. q 12 h	10 mg/kg p.o. q 12 h	7 days

[a]Intravenous administration.
[b]Oral administration.
[c]Until fever has resolved and for three additional days.
[d]14 days recommended if coincubating *B. burgdorferi* infection is suspected.

The optimal duration of doxycycline therapy has not been established. In patients who have been treated for 7 to 10 days infections have resolved completely, and relapse or chronic infection has never been reported, even for those patients who were never treated with an active antibiotic. Reinfection, albeit rare, may occur, and was recently demonstrated in the case of a 65-year-old woman from Westchester County, New York State.[49] Adult patients who are considered at risk for coinfection with *B. burgdorferi* should continue doxycycline therapy for a full 14 days. A shorter course of doxycycline (5 to 7 days) has been advocated for pediatric-age-group patients because of the potential risk for adverse effects (dental staining) seen occasionally in young children.[22,48,50]

In vitro studies have shown that rifamycins also have excellent activity against *Ehrlichia* and *Anaplasma* species.[41–43,51] A small number of pediatric patients and pregnant women who had HGA were treated successfully with rifampin.[52–54] Thus, patients who have HGA and who are unsuited for tetracycline treatment because of a history of drug allergy or pregnancy, and children younger than 8 years of age who are not seriously ill should be considered for rifampin therapy. Studies with levofloxacin and trovafloxacin (no longer a registered medication) demonstrated some activity *in vitro*.[41–43] However, there is no published information available about the usefulness of these antibiotic drugs or other fluoroquinolones as clinical agents *in vivo*.

Approach to the Patient with a Nonspecific Febrile Illness

Patients who present with nonspecific fever after exposure to ticks should be evaluated by clinical examination and routine laboratory testing to determine whether the clinical illness is caused by a tick-borne infection. Reduced concentrations of leukocytes or platelets and increased band neutrophils in peripheral blood, and mild increases in serum hepatic transaminase concentrations warrant consideration for treatment with an antibiotic agent that includes *A. phagocytophilum* in the therapeutic spectrum. Specific laboratory tests

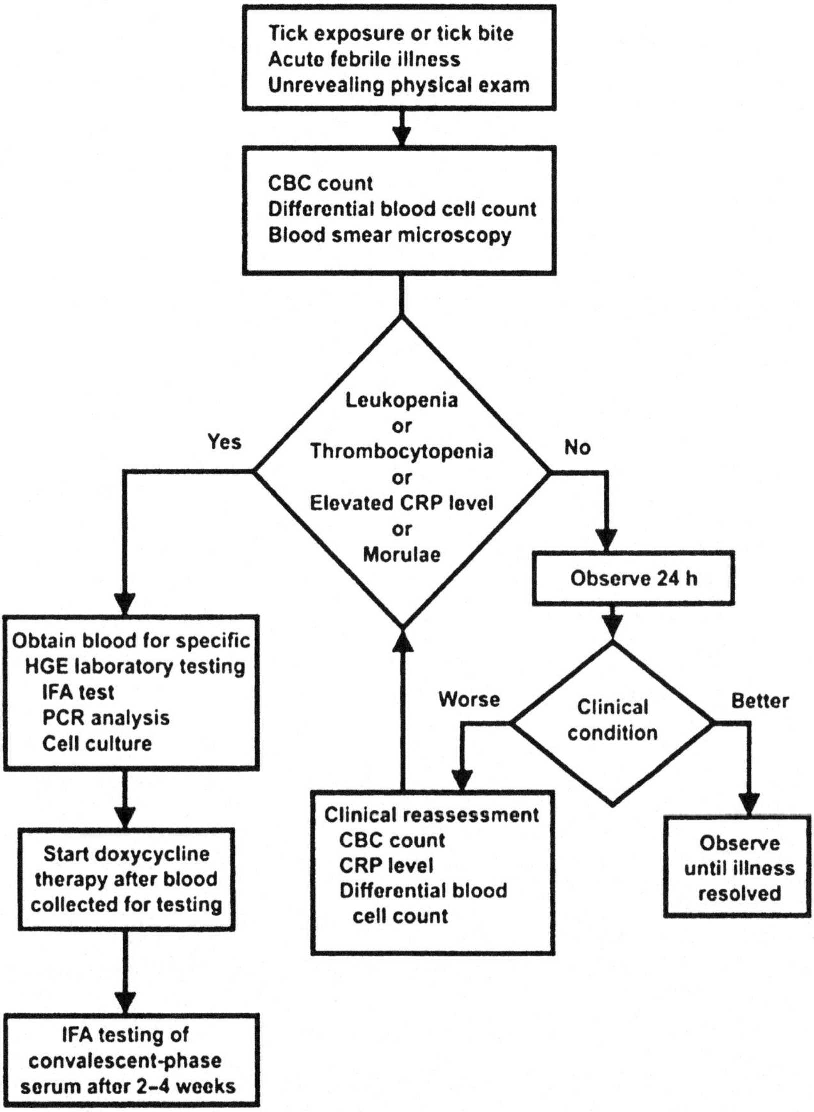

FIGURE 1. Suggested clinical, laboratory, and treatment approach to patients who present with a history of exposure to ticks and nonspecific fever.[12]

that confirm the diagnosis of anaplasmosis are generally not available in the acute care setting.[8,35,37,44,47,55,56] Thus, patients who are suspected of having HGA should begin empiric antibiotic treatment as soon as blood samples have been collected for confirmatory laboratory testing at a reference laboratory. Acute-phase blood samples should be paired with convalescent serum to detect seroconversion in those instances where blood smear microscopy, PCR, or cell culture testing were either unavailable or the results were inconclusive.

Patients suspected of having HGA may be managed in accordance with the laboratory testing and treatment algorithm shown in FIGURE 1. HGA is a reportable illness in many U.S. states and confirmed cases from the United States of America must be reported to the CDC or to the local state health department in the state where the diagnosis was made.

REFERENCES

1. DUMLER, J.S., A.F. BARBET, C.P. BEKKER, *et al.* 2001. Reorganization of genera in the families *Rickettsiaceae* and Anaplasmataceae in the order *Rickettsiales*: unification of some species of *Ehrlichia* with *Anaplasma*, *Cowdria* with *Ehrlichia* and *Ehrlichia* with Neorickettsia, descriptions of six new species combinations and designation of *Ehrlichia equi* and 'HGE agent' as subjective synonyms of *Ehrlichia phagocytophila*. Int. J. Syst. Evol. Microbiol. **51:** 2145–2165.
2. ANONYMOUS. 2002. International Classification of Diseases. The World Wide Web 9th Revision, Clinical Modification[6th]. PMIC, Los Angeles, CA. [electronic citation].
3. BAKKEN, J.S., J.S. DUMLER, S.M. CHEN, *et al.* 1994. Human granulocytic ehrlichiosis in the upper Midwest United States: a new species emerging? JAMA **272:** 212–218.
4. DUMLER, J.S., K.M. ASANOVICH, J.S. BAKKEN, *et al.* 1995. Serologic cross-reactions among *Ehrlichia equi*, *Ehrlichia phagocytophila*, and human granulocytic ehrlichia. J. Clin. Microbiol. **33:** 1098–1103.
5. AGUERO-ROSENFELD, M.E., H.W. HOROWITZ, G.P. WORMSER, *et al.* 1996. Human granulocytic ehrlichiosis: a case series from a medical center in New York State. Ann. Intern. Med. **125:** 904–908.
6. BAKKEN, J.S. & J.S. DUMLER. 1996. Human granulocytic ehrlichiosis in the United States. Infect. Med. **October:** 877–912.
7. ANDERSON, J.F. 2002. The natural history of ticks. Med. Clin. North Am. **86:** 205–218.
8. BAKKEN, J.S. & J.S. DUMLER. 2000. Human granulocytic ehrlichiosis. Clin. Infect. Dis. **31:** 554–560.
9. KRAUSE, P.J., K. MCKAY, C.A. THOMPSON, *et al.* 2002. Disease-specific diagnosis of coinfecting tickborne zoonoses: babesiosis, human granulocytic ehrlichiosis, and Lyme disease. Clin. Infect. Dis. **34:** 1184–1191.
10. CASTRO, M.B., W.L. NICHOLSON, V.L. KRAMER & J.E. CHILDS. 2001. Persistent infection in *Neotoma fuscipes* (Muridae: Sigmodontinae) with *Ehrlichia phagocytophila* sensu lato. Am. J. Trop. Med. Hyg. **65:** 261–267.
11. TELFORD, S.R., III, J.E. DAWSON, P. KATAVOLOS, *et al.* 1996. Perpetuation of the agent of human granulocytic ehrlichiosis in a deer tick-rodent cycle. Proc. Natl. Acad. Sci. USA **93:** 6209–6214.
12. BAKKEN, J.S. 2003. Human anaplasmosis: epidemiological, clinical, laboratory, and therapeutic aspects of a new emerging tick-borne infectious disease. Department of Microbiology, University of Tromsø, Norway.

13. KATAVOLOS, P., P.M. ARMSTRONG, J.E. DAWSON & S.R. TELFORD, III. 1998. Duration of tick attachment required for transmission of granulocytic ehrlichiosis. J. Infect. Dis. **177:** 1422–1425.
14. MCQUISTON, J.H., C.L. MCCALL & W.L. NICHOLSON. 2003. Ehrlichiosis and related infections. J. Am. Vet. Med. Assoc. **223:** 1750–1756.
15. BLANCO, J.R. & J.A. OTEO. 2002. Human granulocytic ehrlichiosis in Europe. Clin. Microbiol. Infect. **8:** 763–772.
16. AGUERO-ROSENFELD, M.E., L. DONNARUMMA, L. ZENTMAIER, *et al.* 2002. Seroprevalence of antibodies that react with *Anaplasma phagocytophila*, the agent of human granulocytic ehrlichiosis, in different populations in Westchester County, New York. J. Clin. Microbiol. **40:** 2612–2615.
17. BAKKEN, J.S., P. GOELLNER, M. VAN ETTEN, *et al.* 1998. Seroprevalence of human granulocytic ehrlichiosis among permanent residents of northwestern Wisconsin. Clin. Infect. Dis. **27:** 1491–1496.
18. HILTON, E., J. DEVOTI, J.L. BENACH, *et al.* 1999. Seroprevalence and seroconversion for tick-borne diseases in a high-risk population in the northeast United States. Am. J. Med. **106:** 404–409.
19. WOESSNER, R., B.C. GAERTNER, M.T. GRAUER, *et al.* 2001. Incidence and prevalence of infection with human granulocytic ehrlichiosis agent in Germany: a prospective study in young healthy subjects. Infection **29:** 271–273.
20. BAKKEN, J.S., I. HALLER, D. RIDDELL, *et al.* 2002. The serological response of patients infected with the agent of human granulocytic ehrlichiosis. Clin. Infect. Dis. **34:** 22–27.
21. BAKKEN, J.S., J. KRUETH, C. WILSON-NORDSKOG, *et al.* 1996. Clinical and laboratory characteristics of human granulocytic ehrlichiosis. JAMA **275:** 199–205.
22. BAKKEN, J.S. & J.S. DUMLER. 2002. *Ehrlichia* and *Anaplasma* species. *In* Antimicrobial Therapy and Vaccine. V. Yu, R. Weber, D. Raoult, Eds.: 875–882. Apple Trees Productions, LLC. New York.
23. CIZMAN, M., T. AVSIC-ZUPANC, M. PETROVEC, *et al.* 2000. Seroprevalence of ehrlichiosis, Lyme borreliosis and tick-borne encephalitis infections in children and young adults in Slovenia. Wien. Klin. Wochenschr. **112:** 842–845.
24. MCQUISTON, J.H., C.D. PADDOCK, R.C. HOLMAN & J.E. CHILDS. 1999. The human ehrlichioses in the United States. Emerg. Infect. Dis. **5:** 635–642.
25. LEIBY, D.A., A.P. CHUNG, R.G. CABLE, *et al.* 2002. Relationship between tick bites and the seroprevalence of *Babesia microti* and *Anaplasma phagocytophila* (previously *Ehrlichia* sp.) in blood donors. Transfusion **42:** 1585–1591.
26. LEPIDI, H., J.E. BUNNELL, M.E. MARTIN, *et al.* 2000. Comparative pathology, and immunohistology associated with clinical illness after *Ehrlichia phagocytophila*-group infections. Am. J. Trop. Med. Hyg. **62:** 29–37.
27. DUMLER, J.S., E.R. TRIGIANI, J.S. BAKKEN, *et al.* 2000. Serum cytokine responses during acute human granulocytic ehrlichiosis. Clin. Diagn. Lab. Immunol. **7:** 6–8.
28. MARTIN, M.E., K. CASPERSEN & J.S. DUMLER. 2001. Immunopathology and ehrlichial propagation are regulated by interferon-gamma and interleukin-10 in a murine model of human granulocytic ehrlichiosis. Am. J. Pathol. **158:** 1881–1888.
29. WALLACE, B.J., G. BRADY, D.M. ACKMAN, *et al.* 1998. Human granulocytic ehrlichiosis in New York. Arch. Intern. Med. **158:** 769–773.

30. BAKKEN, J.S., R.L. TILDEN, J.J. WALLS & J.S. DUMLER. 1999. Influence of occupation, preexisting illness, and chronic medication use on the severity of illness of human granulocytic ehrlichiosis. *In* Rickettsia and Rickettsial Diseases at the Turn of the Third Millenium. D. RAOULT & P. BROUQUI, Eds.: 195–198. Elsevier. Marseilles, France.
31. BAKKEN, J.S., S.A. ERLEMEYER, R.J. KANOFF, *et al.* 1998. Demyelinating polyneuropathy associated with human granulocytic ehrlichiosis. Clin. Infect. Dis. **27:** 1323–1324.
32. HOROWITZ, H.W., S.J. MARKS, M. WEINTRAUB & J.S. DUMLER. 1996. Brachial plexopathy associated with human granulocytic ehrlichiosis. Neurology **46:** 1026–1029.
33. JAHANGIR, A., C. KOLBERT, W. EDWARDS, *et al.* 1998. Fatal pancarditis associated with human granulocytic ehrlichiosis in a 44-year-old man. Clin. Infect. Dis. **27:** 1424–1427.
34. BAKKEN, J.S., M.E. AGUERO-ROSENFELD, R.L. TILDEN, *et al.* 2001. Serial measurements of hematologic counts during the active phase of human granulocytic ehrlichiosis. Clin. Infect. Dis. **32:** 862–870.
35. DUMLER, J.S. & D.H. WALKER. 2001. Tick-borne ehrlichioses. Lancet Inf. Dis. **1:** 21–28.
36. MASSUNG, R.F. & K.G. SLATER. 2003. Comparison of PCR assays for detection of the agent of human granulocytic ehrlichiosis, *Anaplasma phagocytophilum*. J. Clin. Microbiol. **41:** 717–722.
37. GOODMAN, J.L., C. NELSON, B. VITALE, *et al.* 1996. Direct cultivation of the causative agent of human granulocytic ehrlichiosis. N. Engl. J. Med. **334:** 209–215.
38. AGUERO-ROSENFELD, M.E. 2002. Diagnosis of human granulocytic ehrlichiosis: state of the art. Vect. Borne Zoon. Dis. **2:** 233–239.
39. BELONGIA, E.A., K.D. REED, P.D. MITCHELL, *et al.* 2001. Tickborne infections as a cause of nonspecific febrile illness in Wisconsin. Clin. Infect. Dis. **32:** 1434–1439.
40. WALLS, J.J., M. AGUERO-ROSENFELD, J.S. BAKKEN, *et al.* 1999. Inter- and intralaboratory comparison of *Ehrlichia equi* and human granulocytic ehrlichiosis (HGE) agent strains for serodiagnosis of HGE by the immunofluorescent-antibody test. J. Clin. Microbiol. **37:** 2968–2973.
41. HOROWITZ, H.W., T.C. HSIEH, M.E. AGUERO-ROSENFELD, *et al.* 2001. Antimicrobial susceptibility of *Ehrlichia phagocytophila*. Antimicrob. Agents Chemother. **45:** 786–788.
42. KLEIN, M.B., C.M. NELSON & J.L. GOODMAN. 1997. Antibiotic susceptibility of the newly cultivated agent of human granulocytic ehrlichiosis: promising activity of quinolones and rifamycins. Antimicrob. Agents Chemother. **41:** 76–79.
43. MAURIN, M., J.S. BAKKEN & J.S. DUMLER. 2003. Antibiotic susceptibilities of *Anaplasma (Ehrlichia) phagocytophilum* strains from various geographic areas in the United States. Antimicrob. Agents Chemother. **47:** 413–415.
44. FISHBEIN, D.B., J.E. DAWSON & L.E. ROBINSON. 1994. Human ehrlichiosis in the United States, 1985 to 1990. Ann. Intern. Med. **120:** 736–743.
45. OLANO, J.P. & D.H. WALKER. 2002. Human ehrlichioses. Med. Clin. North Am. **86:** 375–392.
46. PADDOCK, C.D., S.M. FOLK, G.M. SHORE, *et al.* 2001. Infections with *Ehrlichia chaffeensis* and *Ehrlichia ewingii* in persons coinfected with human immunodeficiency virus. Clin. Infect. Dis. **33:** 1586–1594.

47. Paddock, C.D. & J.E. Childs. 2003. *Ehrlichia chaffeensis*: a prototypical emerging pathogen. Clin. Microbiol. Rev. **16:** 37–64.
48. Anonymous. 2003. Ehrlichia infections (human ehrlichioses). *In* 2003 Report of the Committee of Infectious Diseases. L.K. Pickering, Ed.: 266–269. Red Book, American Academy of Pediatrics. Elk Grove Village, IL.
49. Horowitz, H.W., M. Aguero-Rosenfeld, J.S. Dumler, *et al.* 1998. Reinfection with the agent of human granulocytic ehrlichiosis. Ann. Intern. Med. **129:** 461–463.
50. Anonymous. 2004. Tetracyclines. *In* AHFS Drug Information. G.K. McEvoy, Ed.: 433–457. American Society of Health-System Pharmacists. Bethesda, MD.
51. Brouqui, P. & D. Raoult. 1992. *In vitro* antibiotic susceptibility of the newly recognized agent of ehrlichiosis in humans, *Ehrlichia chaffeensis*. Antimicrob. Agents Chemother. **36:** 2799–2803.
52. Buitrago, M.I., J.W. Ijdo, P. Rinaudo, *et al.* 1998. Human granulocytic ehrlichiosis during pregnancy treated successfully with rifampin. Clin. Infect. Dis. **27:** 213–215.
53. Elston, D.M. 1998. Perinatal transmission of human granulocytic ehrlichiosis. N. Engl. J. Med. **39:** 1941–1942.
54. Krause, P.J., C.L. Corrow & J.S. Bakken. 2003. Successful treatment of human granulocytic ehrlichiosis in children using rifampin. Pediatrics **112:** e252–e253.
55. Everett, E.D., K.A. Evans, R.B. Henry & G. McDonald. 1994. Human ehrlichiosis in adults after tick exposure. Diagnosis using polymerase chain reaction. Ann. Intern. Med. **120:** 730–735.
56. Olano, J.P., W. Hogrefe, B. Seaton & D.H. Walker. 2003. Clinical manifestations, epidemiology, and laboratory diagnosis of human monocytotropic ehrlichiosis in a commercial laboratory setting. Clin. Diagn. Lab. Immunol. **10:** 891–896.
57. Belongia, E.A., K.D. Reed, P.D. Mitchell, *et al.* 1999. Clinical and epidemiological features of early Lyme disease and human granulocytic ehrlichiosis in Wisconsin. Clin. Infect. Dis. **29:** 1472–1477.
58. Horowitz, H.W., M.E. Aguero-Rosenfeld, D.F. McKenna, *et al.* 1998. Clinical and laboratory spectrum of culture-proven human granulocytic ehrlichiosis: comparison with culture-negative cases. Clin. Infect. Dis. **27:** 1314–1317.
59. Walker, D.H., J.S. Bakken, P. Brouqui, *et al.* 2000. Diagnosing human ehrlichioses: current status and recommendations. Am. Soc. Microbiol. News **5:** 287–293.
60. Aguero-Rosenfeld, M.E., F. Kalantarpour, M. Baluch, *et al.* 2000. Serology of culture-confirmed cases of human granulocytic ehrlichiosis. J. Clin. Microbiol. **38:** 635–638.
61. Dumler, J.S. & J.S. Bakken. 1998. Human ehrlichioses: newly recognized infections transmitted by ticks. Annu. Rev. Med. **49:** 201–213.

Diagnosis of *Coxiella burnetii* Pericarditis Using a Systematic Prescription Kit in Case of Pericardial Effusion

P.Y. LEVY,[a] F. THUNY,[b] G. HABIB,[b] J.L. BONNET,[b] P. DJIANE,[b] AND D. RAOULT[a]

[a]*Unité des Rickettsies, CNRS UMR 6020; IFR48, Faculté de Médecine, Université de la Méditerranée, 27 Boulevard Jean Moulin, 13385 Marseille Cedex 05, France*

[b]*Département de Cardiologie; CHU Timone, Marseille, France*

ABSTRACT: *Coxiella burnetii*, regarded as a potential agent of pericarditis, wa found to be responsible for almost 5% of the cases of idiopathic pericardial effusion reported in this series. Diagnosis was aided by use of a systematic kit described in this paper.

KEYWORDS: *Coxiella burnetii*; pericarditis

INTRODUCTION

Detection and treatment of pericarditis remain a challenging problem and the ratio of cases of unknown etiology remains high, between 40% and 85%. Pericardial effusion may be caused by a wide variety of infectious including *C. burnetii*[1] or *B. quintana* [2] or noninfectious processes.[3–5] In an effort to reduce this ratio, we have previously developed,[6] for 204 patients hospitalized in Marseilles with pericardial effusion, a diagnostic strategy that mandated the systematic use of a battery of noninvasive tests for the diagnosis of benign pericardial effusion. This allowed reducing the number of cases of pericarditis classified as idiopathic when compared to an intuitive prescription of tests.[7,8] Q fever was the main reported etiology in our experience,[6] which is particularly interesting because it is a treatable disease. We decided then to study all patients hospitalized with pericardial effusion from 2003 to March 2005, which comprised 250 additional patients.

Address for correspondence: Unité des Rickettsies, CNRS UMR 6020; IFR48, Faculté de Médecine. Université de la Méditerranée. 27 Boulevard Jean Moulin, 13385 Marseille Cedex 05, France. Voice: 33-491-38-55-17; fax: 33-491-83-03-90.
e-mail: DidierRaoult@univ.mrs.fr

**Ann. N.Y. Acad. Sci. 1078: 248–251 (2006). © 2006 New York Academy of Sciences.
doi: 10.1196/annals.1374.127**

MATERIALS AND METHODS

Diagnostic procedure included: I/A questionnaire with the following items: underlying conditions (immunosuppression, collagen diseases); epidemiological factors (contacts with toxins, animals, ingestion of unpasteurized milk, and travel outside the living area within 6 months). II/A list of tests to be performed systematically was selected after a Medline search:

1. One aerobic and one anaerobic blood culture vial were used, in which the growth of nonfastidious bacteria could be assessed, and one sterile tubes/swab was used for viral detection in the pharynx.

2. Serum samples used for estimation of antibodies against specific infectious agents, antinuclear antibodies, and serum thyroid-stimulating hormone. When possible, blood cultures are drawn 1 h after the initial evaluation, viral cultures of throat on the third day, and serum samples for the detection of convalescent antibodies 2 to 3 weeks after onset.

The diagnosis was considered certain if neoplastic cells or positive culture were found in the effusion. In the absence of an invasive procedure, the diagnosis was considered certain if collagen disease or thyroid dysfunction was diagnosed with the kit tests, if there was a twofold rise in antibody titers, or when a titer above a specific cut-off value was obtained. In the absence of one of these findings, patients with known renal failure, collagen disease, thyroid dysfunction, or neoplastic disease were considered to have these conditions as the cause of their pericarditis.

RESULTS

From January 2003 to March 2005, the kit was prescribed for 362 patients and 162 were excluded because they have presented myocarditis or dry pericarditis (absence of effusion). Among 250 patients included, 76 (31%) had a final etiological diagnosis (TABLE 1). If we aggregate this series with the previously published one, a total of 454 cases of pericardial effusion was studied and 194 had a final diagnosis: 65 benefited from an etiological diagnosis (17.7%) by serologic evaluation of serum, and *C. burnetii* was diagnosed in 17 cases which represents 26.1% of the noninvasive diagnosis, 42% of the infectious etiology, and almost 5% of the whole group with idiopathic pericarditis. Acute Q fever was diagnosed in 16 cases and a chronic form in one case. Specific epidemiological factors were noticed in 12 cases. Seasonal variation was studied for the five main infectious diseases (*C. burnetii, M.pneumoniae, S. pneumoniae*, influenza virus, and enterovirus) diagnosed. Influenza was selectively diagnosed in winter and no other seasonal variation was noted.

TABLE 1. Etiological diagnosis of the pericardial effusion cases

Etiological diagnosis	Number of etiologies identified 1998–2002 [6]	2003–March 05 [PR]	Total
Total	204	250	454
Undiagnosed	76	174	250 (55%)
Coxiella burnetii	10	7	17 (3.7%)
Seasonal variation (spring, summer, fall, winter)			(5,3,4,5)
Bartonella species	1	0	1
L. pneumophila	1	0	1
Chlamydia spp. *Borellia, Brucella*	0	0	0
M. pneumoniae	4	1	5
Seasonal variation (spring, summer, fall, winter)			(2,0,3,0)
S.pneumoniae	1	2	3
Seasonal variation (spring, summer, fall, winter)			(1,0,0,2)
C. freundii, E.coli	1	1	2
Actinomyces	1		1
Tuberculosis	3	1	4
Toxoplasmosis	4[a]	0*	4[a]
Cytomegalovirus	4[a]	0*	4[a]
Parvovirus B19	0	1	1
Hepatitis C	4[a]	0*	4[a]
Influenza virus	1	3	4
Seasonal variation (spring, summer, fall, winter)			(0,0,0,4)
Adenovirus	1	0	1
Enterovirus	8	7	15
Seasonal variation (spring, summer, fall, winter)			(5,2,2,6)
Hypo/hyperthyroidism	20	12	32 (7%)
Positive antinuclear antibodies	19	3	22 (4.8%)
Rheumatoid arthritis	8	3	11 (2.4%)
Others autoimmune diseases	2	0	2
Neoplastic diseases	30	28	58 (12.8%)
Renal insufficiency	5	7	12 (2.6%)

[a]Modification of the criteria of diagnosis with the conclusion of the publication 6 (diagnoses based on positive IgM serological tests and classified as "possible," are now considered as undiagnosed).

DISCUSSION

Our study has been going on for 6 years, which limits studies on seasonal variations on the prevalence of infectious diseases. We made a definite diagnosis in 42% of all cases of pericardial effusion (including those highly clinically suspected) and 17.7% of these classified as idiopathic. (The inclusion of systematic testing for noninfectious agents in the kit resulted in the unexpected diagnosis of 17 cases of Q fever. Without the use of the kit, they would have been classified as idiopathic. *C. burnetii* was found to be the cause of almost 5% of our cases idiopathic pericardial effusion. In this study, Q fever was systematically sought because *C. burnetii* had recently been mentioned as an etiological agent of pericarditis.[1] As a result, during the 6 years of this study, 17 new cases of Q fever were diagnosed. Q fever has a worldwide distribution but very few cases of pericarditis related to it have been described because it is rarely searched for in those circumstances. In countries where the prevalence is the highest (Spain, United Kingdom, and France), the respective role of epidemiological specificity and of active research from specialized laboratories remained unclear.

CONCLUSIONS

Our kit is of a great interest for patient because it increases identification of a number of treatable diseases, such as Q fever. Specific treatment may shorten the evolution of the disease and avoid recurrences.

REFERENCES

1. LEVY, P.Y., P. CARRARA & D. RAOULT. 1999. *Coxiella burnetii* pericarditis: a report of 15 cases and review. Clin. Infect. Dis. **29:** 393–397.
2. LEVY, P.Y., P.E. FOURNIER, M. CARTA & D. RAOULT. 2003. Pericardial effusion due to *Bartonella quintana* in homeless man. J. Clin. Microbiol. **41:** 5291–5293.
3. TROUGHTON, R.W., C.R. ASHER & A.L. KLEIN. 2004. Pericarditis. Lancet **363:** 717–727.
4. LANGE, R.A. & L.D. HILLIS. 2004. Acute pericarditis. N. Engl. J. Med. **351:** 2195–2202.
5. MAISCH, B., P.M. SEFEROVIC, A.D. RISTIC & A.L. AND. 2004. Guidelines on the diagnosis and management of pericardial diseases. The taskforce on diagnosis and management of pericardial diseases of the European Society of Cardiology. Eur. Heart J. **25:** 587–610.
6. LEVY, P.Y., R. COREY, P. BERGER, *et al.* 2003. Etiologic diagnosis of 204 pericardial effusions. Medicine (Baltimore) **82:** 385–391.
7. LEVY, P.Y., J.P. MOATTI, V. GAUDUCHON, *et al.* 2005. Comparison of intuitive versus systematic strategies for etiological diagnosis of pericardial effusion. Scand. J. Infect. Dis. **37:** 216–220.
8. LEVY, P.Y., M. KAHN & D. RAOULT. 2005. Acute pericarditis. N. Engl. J. Med. **352:** 1154–1155.

Brazilian Spotted Fever: A Case Series from an Endemic Area in Southeastern Brazil

Clinical Aspects

RODRIGO N. ANGERAMI,[a] MARIÂNGELA R. RESENDE,[a] ADRIANA F.C. FELTRIN,[a] GIZELDA KATZ,[b] ELVIRA M. NASCIMENTO,[c] RAQUEL S.B. STUCCHI,[a] AND LUIZ J. SILVA[a]

[a]*Universidade Estadual de Campinas (UNICAMP), Campinas, São Paulo, Brazil*

[b]*Centro de Vigilância Epidemiológica "Alexandre Vranjac," São Paulo, Brazil*

[c]*Instituto Adolfo Lutz, São Paulo, Brazil*

ABSTRACT: This case series study is based on a retrospective review of medical records and case notification files of patients admitted to The Hospital das Clínicas da UNICAMP from 1985 to 2003 with a confirmed diagnosis of BSF either by fourfold rise in indirect immunofluorescence assay (IFA) titers of IgG antibodies reactive with *R. rickettsii* or isolation of *R. rickettsii* from blood or skin specimens. A median lethality of 41.9 % was observed between 1985 and 2004. The case-fatality ratio of 30 % in our study, lower than the overall São Paulo state ratio, could be explained by a higher index of suspicion and a larger experience in our hospital, a regional referral center for BSF. The presence of the classical triad of fever, rash, and headache as described in RMSF was observed in fever than half (35.2%) of our patients.

KEYWORDS: Brazilian spotted fever; *Rickettsia rickettsii*

INTRODUCTION

Brazilian spotted fever (BSF) is the most important tick-borne disease in Brazil. It has been known in Brazil since 1929 and is associated with a severe clinical course and a high case-fatality ratio.[1] Like Rocky Mountain spotted fever (RMSF) it is caused by *Rickettsia rickettsii.* Although sharing the same etiological agent and having the main clinical aspects closely resem-

Address for correspondence: Rodrigo Nogueira Angerami, Rua da Urca, 354, San Conrado, Sousas, Campinas, SP Brazil, 13104-900. Voice: 55-19-3788-7451; fax: 55-19-3258-7930.
e-mail: rodrigoang@uol.com.br

Ann. N.Y. Acad. Sci. 1078: 252–254 (2006). © 2006 New York Academy of Sciences.
doi: 10.1196/annals.1374.044

bling RMSF,[1] a more severe clinical course and a higher case-fatality ratio are associated with BSF.

In the state of São Paulo in southeastern Brazil, during 1985 to 2004, 155 confirmed cases of BSF were reported and the case-fatality ratio was 41.9%.[2,3] We describe the main clinical and laboratory aspects of a case series from the Hospital das Clínicas da UNICAMP, a referral hospital in the region of Campinas[6] in São Paulo.

PATIENTS AND METHODS

This is a case series study based on a retrospective review of medical records and case notification files of patients admitted to the Hospital das Clínicas da UNICAMP from 1985 to 2003 with a confirmed diagnosis of BSF either by fourfold rise in indirect immunofluorescence assay (IFA) titers of IgG antibodies reactive with *R. rickettsii* or isolation of *R. rickettsii* from blood or skin specimens.

The confirmatory laboratory tests were performed at Instituto Adolfo Lutz, the state public health laboratory and regional referral center for rickettsial diseases.

RESULTS

Of the 23 patients included, 13 (56.5%) had a diagnosis of BSF confirmed by isolation of *R. rickettsii* in blood or skin, while in eight (34.7%) the confirmation was by IFA. Two cases were confirmed by the clinical–epidemiologic correlation criteria.

Fever was observed in all patients. General symptoms including myalgia, headache, vomiting, and abdominal pain occurred in 80%, 66%, 42%, and 38% respectively. Exanthem, predominantly with maculopapular pattern, was observed in 52% of the patients. More severe clinical manifestations occurred in many patients and were associated with higher case fatality. Of these manifestations, icterus was present in 52%, central nervous system impairment in 43%, respiratory distress in 37%, and acute renal insufficiency in 35.3% of the patients. Different grades of hemorrhagic manifestations were seen in 69.5% of patients, disseminated petechiae and hemorrhagic cutaneous suffusions being the most frequent. There was one case of gangrene of the fingers. Thirty-three percent of the studied patients had hypotension and shock.

Thrombocytopenia and elevation of liver enzymes (AST and ALT) were present in 100% of the patients and were the most frequent laboratory changes. The estimated incubation period ranged from 1 to 21 days, with a median of 12 days. Case-fatality ratio was 30%.

DISCUSSION

Since its first description in 1929, BSF has been associated with increased lethality. In these first reports the case-fatality ratio ranged from 75% to 80%.[4] A median lethality of 41.9% was observed between 1985 and 2004. The case-fatality ratio of 30% in our study, lower than overall São Paulo state ratio, could be explained by a higher index of suspicion and a larger experience in our hospital, a regional referral center for BSF. The presence of the classical triad of fever, rash, and headache as described in RMSF was observed in less than half (35.2%) of our patients.[5,6] Icterus, neurological symptoms, respiratory distress, and acute renal insufficiency, present in our series, had been previously described in BSF[1] and RMSF and were associated with more severe clinical course and higher case-fatality.

BSF seems to be more severe than RMSF. However, it is possible to hypothesize that a delay in initiating treatment for BSF in most health services in Brazil, the absence of parenteral doxycycline in Brazil, or that *R. rickettsii* in Brazil is more virulent, could explain the higher lethality of BSF when compared with RMSF.

REFERENCES

1. Lemos, E.R.S., F.B.F. Alvarenga, M. Cintra, *et al.* 2001. Spotted fever in Brazil: an epidemiological study and description of clinical cases in an endemic area in the state of São Paulo. Am. J. Trop. Med. Hyg. **65:** 329–334.
2. Centro de Vigilância Epidemsiológica. Febre maculosa. Available at http://www.cve.saude.sp.gov.br/htm/Cve_fmaculosa.htm [accessed on July 30, 2005].
3. Lima, V.L.C., S.S.L. Souza, C.E. Souza, *et al.* 2003. Spotted fever in Campinas region, State of São Paulo, Brazil. Cad. Saúde Públ. **19:** 331–334.
4. Dias, E. & A.V. Martins. 1939. Spotted fever in Brazil: a summary. Am. J. Trop. Med. **19:** 103–108.
5. Masters, E.J., G.S. Olson, S.J. Weiner, *et al.* 2003. Rocky Mountain spotted fever: a clinician's dilemma. Arch. Intern. Med. **163:** 769–774.
6. Thorner, A.R., D.H. Walker & W.A. Petri. 1998. Rocky Mountain spotted fever. Clin. Inf. Dis. **27:** 1353–1360.

Revisiting Brazilian Spotted Fever Focus of Caratinga, Minas Gerais State, Brazil

M.A.M. GALVÃO,[a,b] L.D. CARDOSO,[a] C.L. MAFRA,[c] S.B. CALIC,[d] AND D.H. WALKER[b]

[a]*Escola de Nutrição, Universidade Federal de Ouro Preto, Ouro Preto, Minas Gerais, 35400-000, Brazil*

[b]*World Health Organization Collaborating Center for Tropical Diseases at the University of Texas Medical Branch, Galveston, Texas 77 555, USA*

[c]*Universidade Federal de Viçosa, Viçosa Brazil*

[d]*Fundação Ezequiel Dias, Belo Horizonte, MG, Brazil*

ABSTRACT: We revisited a Brazilian spotted fever focal area in Minas Gerais state, Brazil, in 2002, and performed a serologic survey in dogs and cats. The results of this survey are compared with the survey made 10 years before. The possible efficacy of vector control measures adopted in this area and the role of dogs and horses as sentinels of infection by *Rickettsia* are discussed.

KEYWORDS: rickettsioses; Brazilian spotted fever; serologic survey

INTRODUCTION

In October and November of 1992 an outbreak of Brazilian spotted fever occurred in the municipality of Caratinga, Minas Gerais state, Brazil, resulting in 12 deaths. In December 1992, a serologic survey was made showing the presence of antibodies with a titer of 1:64 or more to *Rickettsia rickettsii* by indirect antibody fluorescence (IFA) test in 11 of 44 dogs and 15 of 28 horses.[1]

METHODS

During 2002, 73 canine and 18 equine sera from different animals were collected in the same place where the outbreak occurred almost 10 years earlier.

Address correspondence to Escola de Nutrição, Universidade Federal de Ouro Preto, Ouro Preto, Minas Gerais. 35400-000, Brazil. Voice: 55 + 31 + 35591813; fax: 55 + 31 + 35591228.
e-mail:galvaomarcio@oi.com.br

Ann. N.Y. Acad. Sci. 1078: 255–256 (2006).
doi: 10.1196/annals.1374.045

(These animals were not the same animals investigated in the 1992 outbreak). Sera were separated by centrifugation and stored at –20°C until tested at Ouro Preto Federal University (UFOP), Brazil. In UFOP the sera were screened for the presence of *Rickettsia rickettsii* antibodies by IFA at a serum dilution of 1:64.[2] Sera that scored positive at 1:64 were titered to the end point.

RESULTS

Among the dogs none was seroreactive to *R. rickettsii*, while 3 horses (16%) had antibodies to *R. rickettsii* at a titer ≥ 1:64. A pool of ticks collected in the same area in 2002 showed by PCR the presence of rickettsia from a spotted fever group which, when sequenced, demonstrated similarity to *R. rickettsii* and *Rickettsia honei*.[3]

DISCUSSION AND CONCLUSIONS

We propose that the measures of vector control adopted by the Caratinga Public Health Service were responsible in part for the control of Brazilian spotted fever in this area. Even with the possibility of the presence of *R. rickettsii* by PCR in a pool of ticks collected in the same area in 2002, the reduction of the number of ticks achieved by the vector control measures accomplished the goal of reducing the number of human cases to a zero in the last ten years. This work shows the importance of dogs and horses as sentinels of this kind of infection.

REFERENCES

1. Galvão, M.A.M. 1996. Febre Maculosa em Minas Gerais: um estudo sobre a distribuição da doença no Estado e seu comportamento em área de foco periurbano. Ph.D. thesis, Tropical Medicine, Belo Horizonte, Faculdade de Medicina da UFMG.
2. CDC—Division of Viral and Rickettsial Diseases. Indirect fluorescent antibody technique for the detection of rickettsial antibodies: 11. National Center for Infection Diseases, CDC. Atlanta, Georgia, USA.
3. Cardoso, L.D., M.A.M. Galvão, R.N. Freitas, *et al.* 2006. Detection and characterization of *Rickettsia spp.* in Brazilian spotted fever focus of Caratinga, Minas Gerais state, Brazil. Cad. Saude Publica. (Reports in Public Health) **22:** 495–501.

Fatal Case of Brazilian Spotted Fever Confirmed by Immunohistochemical Staining and Sequencing Methods on Fixed Tissues

TATIANA ROZENTAL,[a] MARINA E. EREMEEVA,[b] CHRISTOPHER D. PADDOCK,[c] SHERIF R. ZAKI,[c] GREGORY A. DASCH,[b] AND ELBA R. S. LEMOS[a]

[a]*Laboratório de Hantaviroses e Rickettsioses, Departamento de Virologia, FIOCRUZ, Rio de Janeiro, Rio de Janeiro, Brazil*

[b]*Viral and Rickettsial Zoonoses Branch, Division of Viral and Rickettsial Diseases, National Center for Infectious Diseases, Centers for Disease Control and Prevention, Atlanta, Georgia 30333, USA*

[c]*Infectious Disease Pathology Activity, Division of Viral and Rickettsial Diseases, National Center for Infectious Diseases, Centers for Disease Control and Prevention, Atlanta, Georgia 30333, USA*

ABSTRACT: The authors describe the first characterization of *Rickettsia rickettsii* in a fatal case occurring in Rio de Janeiro State, Brazil.

KEYWORDS: Brazilian spotted fever; immunohistochemical technique; PCR; sequence

Brazilian spotted fever (BSF) is the most prevalent rickettsial disease in Brazil[1] and is caused by *Rickettsia rickettsii*. *Amblyomma cajennense* is the most important vector for *R. rickettsii*. The clinical presentation of BSF closely resembles that of Rocky Mountain spotted fever (RMSF).[1,2] BSF has been described in several regions of the States of Espírito Santo, Minas Gerais, Rio de Janeiro, São Paulo, as well as Bahia, Goiás, and Santa Catarina. Since 1997, cases of BSF have been occurring at Barra do Piraí Municipality, Rio de Janeiro State, where samples from the fatal case reported here were collected.

A 54-year-old black male presented with a history compatible with BSF manifested by respiratory and renal failure, which evolved to death on the seventh day of the disease because of lack of specific antibiotic therapy. This fatal

Address for correspondence: Tatiana Rozental, Lab. de Hantaviroses e Rickettsioses, Depto. Virologia, Pav. Rocha Lima 5° andar, FIOCRUZ. Av. Brasil, 4365, Manguinhos, Rio de Janeiro, RJ, Brazil. CEP: 21045-900. Voice: +55-21-2598-4545; fax: +55-21-2270-6397.
e-mail: rozental@ioc.fiocruz.br

Ann. N.Y. Acad. Sci. 1078: 257–259 (2006).
doi: 10.1196/annals.1374.046

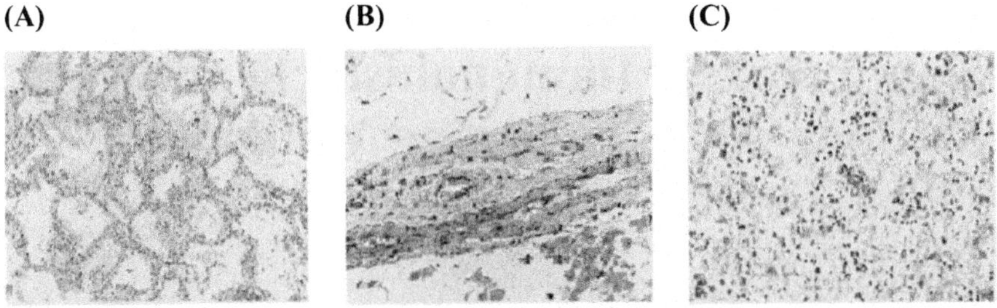

FIGURE 1. Fatal Brazilian spotted fever. (**A**) Lung tissue, hematoxylin, and eosin stain, 100×; (**B**) liver tissue, immunoalkaline phosphatase stain, 400×; (**C**) kidney tissue, immunoperoxidase stain, 400×.

case occurred at Barra do Piraí Municipality, located in the West region of Rio de Janeiro State, Brazil. A serum sample from the patient was evaluated for IgM and IgG antibodies reactive with *R. rickettsii* by using an indirect immunofluorescence assay as previously described[3] and was negative. The necropsied lung, liver, kidney, muscle, and spleen were formalin-fixed, paraffin-embedded, and the tissues were evaluated by using routine hematoxylin and eosin stain, which showed vasculitis, mainly in pulmonary tissue by (FIG. 1A). Indirect immunoalkaline phosphatase and immunoperoxidase staining were performed as described previously,[4,5] and showed spotted fever group rickettsiae in all tissues (FIG. 1B, C). DNA extraction was performed on deparaffinized 6-μm sections

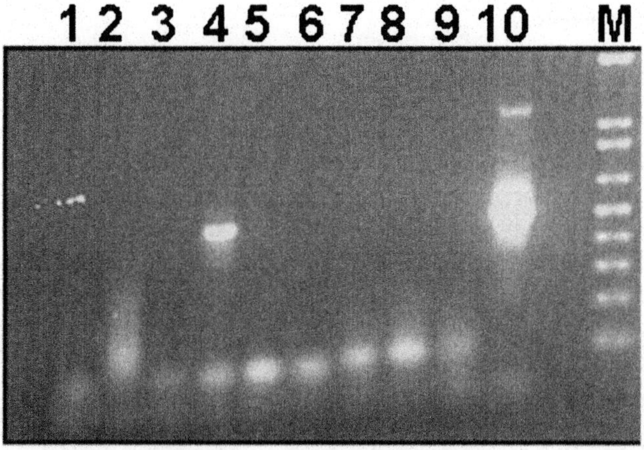

FIGURE 2. Semi-nested PCR of paraffin-embedded tissues using Rr190.70-701 primers followed by Rr190.70-602[7] primers of the rOmpA gene. The samples were loaded in the following order: lane 1, lung tissue [B2]; lane 2, liver tissue [C2]; lane 3, kidney tissue [D2]; lane 4, liver tissue [G2]; lane 5, muscle tissue [F2]; lane 6, kidney tissue [I2]; lane 7, spleen tissue [J2]; lane 8, negative control [no DNA template]; lane 10, PCR-positive control; lane M, DNA molecular weight marker XIV (Roche, Mannheim, Germany).

of each tissue using QIAamp DNA Mini Kit (Qiagen). PCR amplification was performed using Qiagen Master Mix reagents in a Gradient Master Cycle (Eppendorf, Westbury, NY, USA). An amplicon was reproducibly generated only from one of the two liver samples (G) following semi-nested PCR amplification of the rOmpA gene fragment (FIG. 2). The primers were made by CDC Core Facility, Atlanta, GA, USA. Sequence reactions were performed and analyzed as described previously.[6] The nucleotide sequence generated from the patient sample had 100% identity with the homologous sequence of the rOmpA fragment of *R. rickettsii* and was deposited in the NCBI GenBank under accession number DQ002505.

Although spotted fever cases confirmed by serology have been described previously in the state of Rio de Janeiro, this is the first time that an infection with *R. rickettsii* has been definitively identified there. More information on the epidemiology of BSF in Brazil will be obtained by molecular biology and isolation techniques with the ticks collected in the region.

REFERENCES

1. LEMOS, E.R.S., F.B.F. ALVARENGA, M.L. CINTRA, *et al.* 2001. Spotted fever in Brazil: a seroepidemiological study and description of clinical cases in an endemic area in the State of São Paulo. Am. J. Trop. Med. Hyg. **65:** 329–334.
2. WALKER, D.H. 1989. Rocky Mountain spotted fever: a disease in need of microbiological concern. Clin. Microbiol. Rev. **2:** 227–240.
3. PHILIP, R.N., E.A. CASPER, R.A. ORSEBEE, *et al.* 1976. Microimmunofluorescence test for the serological study of Rocky Mountain spotted fever and typhus. J. Clin. Microbiol. **3:** 51–61.
4. PADDOCK, C.D., P.W. GREER, T.L. FEREBEE, *et al.* 1999. Hidden mortality attributable to Rocky Mountain spotted fever: immunohistochemical detection of fatal, serologically unconfirmed disease. J. Infect. Dis. **179:** 1469–1476.
5. WHITE, W.L., J.D. PATRICK, L.R. MILLER. 1994. Evaluation of immunoperoxidase technique to detect *Rickettsia rickettsii* in fixed tissue sections. Am. J. Clin. Pathol. **101:** 747–752.
6. EREMEEVA, M.E., R.M. KLEMT, L.A. SANTUCCI-DOMOTOR, *et al.* 2003. Genetic analysis of isolates of *Rickettsia rickettsii* which differ in virulence. Ann. N. Y. Acad. Sci. **990:** 717–722.
7. EREMEEVA, M.E., X.-J. YU & D. RAOULT. 1994. Differentiation among spotted fever group rickettsiae species by analysis of restriction fragment length polymorphism of PCR-amplified DNA. J. Clin. Microbiol. **32:** 803–810.

Detection of *Rickettsia rickettsii* and *Rickettsia* sp. in Blood Clots in 24 Patients from Different Municipalities of the State of São Paulo, Brazil

FLÁVIA SOUSA GEHRKE,[a,c] ELVIRA MARIA MENDES DO NASCIMENTO,[a] ELIANA RODRIGUES DE SOUZA,[a,c] SILVIA COLOMBO,[a] LUIZ JACINTHO DA SILVA,[b] AND TERESINHA TIZU SATO SCHUMAKER[c]

[a]*Instituto Adolfo Lutz, São Paulo, São Paulo, Brazil*

[b]*Universidade Estadual de Campinas, São Paulo, Brazil*

[c]*Universidade de São Paulo, SP, Brazil*

ABSTRACT: The authors detected *Rickettsia* genus organisms using shell vial and polymerase chain reaction (PCR)/sequencing analysis in blood clots in patients suspected of having Brazilian spotted fever (BSF). DNA was detected using PCR with three sets of primers to access the *glt*A, *omp*A, and *omp*B genes. Sequence analysis was carried out using an automatic sequencer with Bioedit® software. Seventy-five percent of the culture samples were positive and all samples amplified rickettsial gene fragments. To date, 46% of the samples have been sequenced.

KEYWORDS: Brazilian spotted fever; *Rickettsia rickettsii*; *Rickettsia* sp.; *omp*B gene; São Paulo; Brazil

INTRODUCTION

In recent years, an upsurge of Brazilian spotted fever (BSF) has been recorded in the state of São Paulo as a result of forest destruction by human activities. Although several BSF epidemiological studies have been developed in this State, almost nothing is known about the *Rickettsia* genome isolated from humans. We report the results of tests performed in blood samples of patients from 13 counties of this endemic area.

Address for correspondence: Flávia Sousa Gehrke, Instituto de Ciências Biomédicas II, Av. Prof. Lineu Prestes, 1374, Cidade Universitária, 05508-900 São Paulo, Brazil. Voice: +55- 113091-7273; fax: +55 113091-7417.
e-mail: fgehrke@usp.br

**Ann. N.Y. Acad. Sci. 1078: 260–262 (2006). © 2006 New York Academy of Sciences.
doi: 10.1196/annals.1374.047**

MATERIALS AND METHODS

The protocol utilized was approved by the Ethical Committee on Human Experimentation of the Instituto de Ciências Biomédicas (Biomedical Sciences Institute/USP). The blood samples were collected from 24 patients from 8 counties of the Atibaia, Jaguari, and Camanducaia river basins, a BSF endemic area of the State of São Paulo, Brazil. The counties and the number of patients from each one (in parenthesis) were: Artur Nogueira (1), Campinas (5), Hortolândia (1), Louveira (3), Paulínia (1), Pedreira (1), Piracicaba (3), and Rio Claro (1). Five counties do not belong to the endemic area: Diadema (2), Mauá (1), São Bernardo (2), São Paulo (1), and Valinhos (2). During the course of the disease all patients presented with fever, malaise, myalgia, headache, abdominal pain, and maculopapular rash. In severe cases, resulting in death (11), patients also showed respiratory, neurological, gastric, and/or renal disorders. From the 24 patients only 17 sera samples were available for analysis by indirect immunofluorescence assay (IFA) for *Rickettsia rickettsii* antibody. The blood clots were inoculated in shell vials (VERO cells),[1] and positivity was confirmed using IFA prepared with *Rickettsia rickettsii*–positive human serum. Genomic DNA was extracted (phenol/chloroform),[2] and rickettsial DNA was detected using polymerase chain reaction (PCR)[3] with three sets of primers to access the citrate synthase gene-*glt*A,[4] 190-kDa surface protein gene-*omp*A,[5] and 120-kDa surface protein gene-*omp*B.[6] The amplified fragments were cloned[6] and sequence analysis was carried out using an automatic sequencer with Bioedit® software.

RESULTS AND DISCUSSION

Of the sera analyzed, 29.4% were positive by IFA. In blood clots, 75% of the culture samples analyzed were positive and 100% of the samples were amplified rickettsial gene fragments (gltA, ompA or ompB). To date, 46% of the samples have been sequenced and analysis of the ompB gene base sequences shows identity with the *Rickettsia rickettsii* (8 samples) or *Rickettsia* sp. (3 samples) ompB gene sequences, available in GenBank. The state of São Paulo, Brazil, harbors one of the most important foci of BSF. However, in this region, no genomic information about the disease agents in human beings of the *Rickettsia* species has been documented. Presently, the partial analysis of the rickettsial *omp*B gene detected in 11 patients shows identity with *R. rickettsii* or *Rickettsia* sp. Other analysis of the *Rickettsia* genes detected using PCR is in progress for species-specific identification. Adequate identification will allow us to have a clearer understanding of BSF epidemiology. Among the positive samples by shell vial, 46% are successfully maintained and can be used as antigen sources in the near future in serologic diagnosis in Brazilian Public Health Laboratories.

ACKNOWLEDGMENTS

This work was supported by FAPESP.

REFERENCES

1. NASCIMENTO, E.M.M., F.S. GEHRKE, R.A. MALDONADO, *et al.* 2005. Detection of Brazilian spotted fever infection by polymerase chain reaction in a patient from the state of São Paulo. Mem. Inst. Oswaldo Cruz **100:** 277–279.
2. TZIANABOS, T., E.B. ANDERSON & J.E. MACDADE. 1989. Detection of *Rickettsia rickettsii* DNA in clinical specimens by using polymerase chain reaction technology. J. Clin. Microb. **27:** 2866–2868.
3. REGNERY, R.L., C.L. SPRUIL & B.D. PLIKAYTIS. 1991. Genotypic identification of rickettsiae and estimation of intraspecies sequence divergence for portion of two rickettsial genes. J. Bacteriol. **173:** 1576–1589.
4. WOOD, D.O., L.R. WILLIAMSON, H. WINKLER, *et al.* 1987. Nucleotide sequence of the *Rickettsia prowazekii* citrate synthase gene. J. Bacteriol. **169:** 3564–3572.
5. EREMEEVA, M., Y. XUEJI & D. RAUOLT. 1994. Differentiation among spotted fever group rickettsiae species by analysis of restriction fragment length polymorphism of PCR-amplified DNA. J. Clin. Microbiol. **32:** 803–810.
6. SAMBROOK, J., E.F. FRITSCH & T. MANIATIS 1989. Molecular Cloning: A Laboratory Manual, 2nd ed. Cold Spring Harbor Laboratory Press. Cold Spring Harbor, New York.

Mediterranean Spotted Fever in Crete, Greece

Clinical and Therapeutic Data of 15 Consecutive Patients

A. GERMANAKIS,[b] A. PSAROULAKI,[a] A. GIKAS,[a] AND Y. TSELENTIS[a]

[a] *University of Crete, Faculty of Medicine, Laboratory of Clinical Bacteriology, Parasitology, Zoonoses and Geographical Medicine (WHO Collaborating Center for Research and Training in Mediterranean Zoonoses), Crete, Greece*

[b] *General Hospital of Sitia, Crete, Greece*

ABSTRACT: The clinical, epidemiological, and therapeutic aspects of 15 patients with Mediterranean spotted fever (MSF), admitted to the Internal Medicine Department of the General Hospital of Sitia (southeastern Crete, Greece) between December 2000 and July 2003, were studied. Diagnosis was made on the basis of clinical signs and symptoms and was confirmed by serology. Of the patients studied, 67% were men and 33% women, with a median age of 52 years (range of 23–76 years). Ten cases (67%) were diagnosed between May and July. Of all the patients, 93% had a history of contact with animals, mainly with sheep (11 patients, 73%), while 53% of them had a history of tick-bite (33%), or reported the presence of ticks in their environment (20%). The typical eschar lesion (*tache noir*) at the tick-bite site was present in 53% of the patients, while the rash was present in 87% of them. Laboratory findings included leukopenia (47%), thrombocytopenia (54%), elevation of transaminases (80%), hyponatremia (33%), and microscopic hematuria (80%). Four patients (27%) displayed pulmonary infiltrates on chest radiography. All patients were treated with doxycycline (200 mg daily) and recovered rapidly. Renal function deteriorated in one patient with chronic renal failure, but he recovered thereafter.

KEYWORDS: tick-borne diseases; Mediterranean spotted fever

Address for correspondence: A. Psaroulaki, University of Crete, Faculty of Medicine, Laboratory of Clinical Bacteriology, Parasitology, Zoonoses and Geographical Medicine (WHO Collaborating Center for Research and Training in Mediterranean Zoonoses), Crete, Greece. Voice: 003-30-28-10394743 ; fax: 003-028-10394740.

e-mail: annapsa@med.uoc.gr

Ann. N.Y. Acad. Sci. 1078: 263–269 (2006).
doi: 10.1196/annals.1374.049

INTRODUCTION

Spotted fever group (SFG) rickettsioses, caused by pathogenic SFG rickettsiae, are tick-borne diseases widely distributed throughout the world. Mediterranean spotted fever (MSF) is one of the oldest known rickettsioses. It occurs throughout the Mediterranean, including Italy, Spain, Portugal, Turkey, Cyprus, Palestine, Romania, Bulgaria, Tunisia, Algeria, Morocco, Libya, and Egypt. *Rickettsia conorii (R. conorii)*, which is responsible for MSF, is an obligate intracellular bacterium transmitted by ticks, especially the brown dog tick *Rhipicephalus sanguineus*, which is presumed to be the main vector and maybe reservoir of *R. conorii* in the Mediterranean area.[1]

In Greece, MSF was clinically described for the first time in the beginning of the 1930s. Serological investigations based on the Weil–Felix, complement fixation test and microagglutination test, had indicated the presence of *R.conorii* in human and animal populations in the past (until the 1980s), with high seroprevalence rates especially in rural areas.[2,3] Since the 1980s, seroepidemiological studies using IFAT and Western blotting have been undertaken in healthy populations all over Greece. In 1987, in a seroepidemiological survey conducted in southern Greece, Crete, *R. conorii* incidence rate was found to be 5,967/1,000.[4,5] In 1991, in a similar survey conducted in central Greece, Fokida, the seroprevalence of specific IgG antibodies to *R. conorii* was estimated at 58.3% (cut off $\geq$ 32) by using IFA, and at 45.3% by using Western blotting analysis.[6] *R. conorii* has been detected and isolated from ticks and human patients, using the shell-vial technique coupled with molecular methods.[7–9]

Unfortunately, rickettsioses are not nationally notified in Greece, and thus many cases are often either underdiagnosed or reported cases are often not well documented.

THE STUDY

Over a 3-year period (2000–2003), 15 cases of MSF were diagnosed in the Internal Medicine Department of the General Hospital of Sitia region (southeastern Crete, Greece) (FIG. 1). Clinical, epidemiological, laboratory, and therapeutic data of these cases are presented in this study.

MATERIALS AND METHODS

From December 2000 to July 2003, 15 cases of MSF were diagnosed in the General Hospital of Sitia. Clinical, epidemiological, laboratory, and therapeutic data of all patients were registered in a special protocol elaborated for the study. Diagnosis of the disease, on the basis of clinical symptoms and signs,

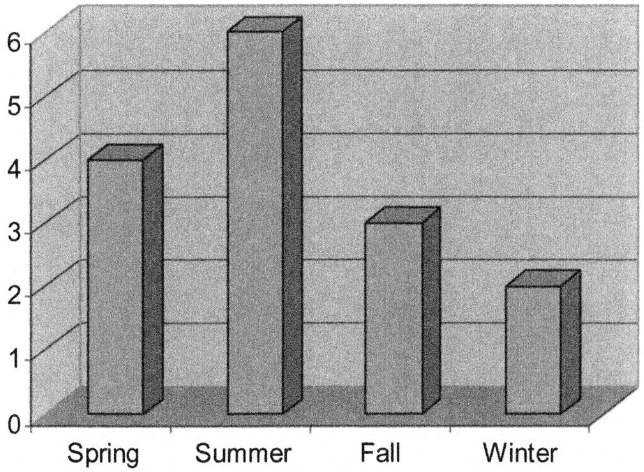

FIGURE 1. Distribution of cases of Mediterranean spotted fever by season.

was further confirmed by serology. Immunofluorescence assay test (IFAT) was performed in acute-phase and convalescent-phase serum samples of patients with suspected MSF. The presumptive clinical criteria for the diagnosis of MSF were the presence of fever, and/or headache, and/or skin rash, and/or an eschar at the tick bite site. A case was defined positive by either the presence of an IgM titer $\geq$ 1/100, or a fourfold or higher increase in IgG titers against *R. conorii* between two assays, or both findings. In the majority of patients, three serum samples were obtained: the first sample on admission, the second within a mean time of 2 weeks after admission, and the third, when possible, approximately 1 month later.

RESULTS ND DISCUSSION

Thirty-five serum samples from 15 patients with suspected MSF were tested by IFAT for the presence of IgG and IgM antibodies against *R. conorii* (Bio Merieux, Marcy L'Etoile, France). All patients had significant antibody titers against *R. conorii*, and the initial clinical diagnosis was laboratory-confirmed in all of them.

Clinical, Laboratory, and Epidemiological Data

Of the patients studied, 67% were men and 33% women, with a median age of 52 years (range of 23–76 years). The high male to female ratio (10/5) of these cases can be explained by the greater probability of men to have closer contact with animals, especially ticks (due to their agricultural occupation),

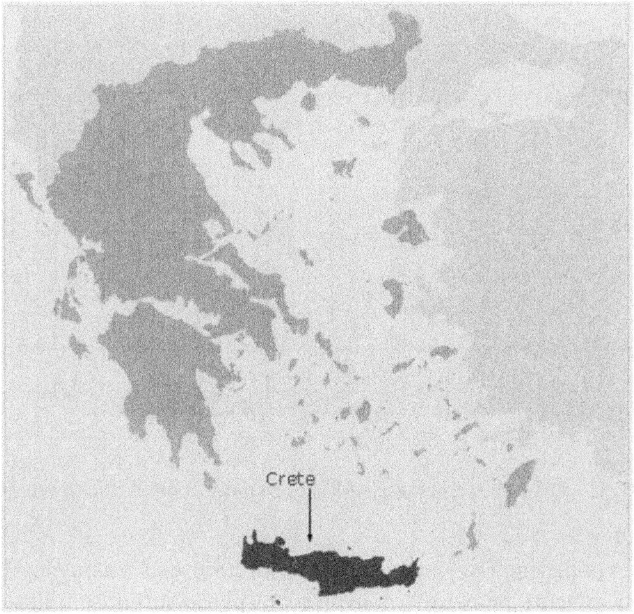

FIGURE 2. Geographic location of the island of Crete, Greece.

and therefore more likely to have exposure to *R. conorii.* Two cases were registered in 2000, 3 in 2001, 4 in 2002, and 7 in 2003. The seasonal distribution of MSF cases was observed to be in late spring and summer (FIG. 2). Of the cases studied, 67% (10/15) occurred from May to July, with a higher incidence of infection during July (5 cases). This is in agreement with the fact that the adult population of *R. sanguineus*, the main vector and reservoir of *R. conorii,* peaks in May, while larvae and nymphs are active during the summer months.[1] However, two cases occurred in winter (February, December). These winter cases are unusual for Mediterranean spotted fever. A common feature of these cases was the working environment of patients (either collecting olives or wild vegetation), and their exposure in places, where sheep and goats were grazing. The exposure to sheep and goats grazing does not mean exposure to *R. sanguineus* ticks, as they are not active during the winter. However, *R. sanguineus* ticks can survive inside human houses during the winter. One cannot also exclude that the patients may have been exposed and infected with other SFG rickettsiae transmitted by ticks active during the winter. Indeed, although IF is the reference method for the diagnosis of SFG rickettsioses, cross-reaction may occur within the genus *Rickettsia*. *R. massiliae, R. aeschlimanni*, and *R. sibirica mongolotimonae* are known to occur in Greece.

Of all, five of the patients (33%) had a history of tick bite, while three (20%) reported the presence of ticks in their environment.

Most of the patients (93%) had a history of contact with animals, while 11/15 patients (73%), reported contact with sheep or goats. These findings suggest that contact with ticks, especially ticks parasitizing on sheep and goats, constitutes the main risk factor, whereas other mammalian hosts (dogs, cats, and others) seem to play a secondary role in the transmission of the disease.

All patients had high fever (mean 39°C), and a maculopapular or purpuric rash was present in 87% of them. The most common clinical manifestations were fever (100%), rash (87%), malaise (87%), headache (80%), chills (67%), myalgia (60%), and conjunctivitis (53%). The typical eschar lesion (*tache noire*) at the tick bite site was present in 53% of the patients. Laboratory findings included leukopenia (47%), thrombocytopenia (54%), elevation of transaminases (high AST and ALT in 80% of the patients), and increased LDH level (87%), while hyponatremia and microscopic hematuria were observed in 33% and 80% of the cases, respectively. Four patients (27%) displayed pulmonary infiltrates on chest radiography. The main clinical and laboratory findings are shown in TABLE 1.

Therapeutic Data

The clinical effectiveness of antibiotic regimens administered in 15 patients was determined. Eight of 15 patients (53%) did not receive any antibiotic treatment before hospitalization. Among seven patients who had received antibiotics before admission, two were treated with doxycycline, one with macrolides, while four had received β-lactam compound. All patients were treated with doxycycline after their admission. Treatment with doxycycline

TABLE 1. Clinical symptoms and laboratory findings

	Percentage of patients[a,b]
Clinical Findings	
Fever	100%
Headache	80%
Chills	67%
Rash	87%
Malaise	87%
Myalgia	60%
Conjunctivitis	53%
Eschar lesion	53%
Pulmonary infiltrates	27%
Laboratory findings	
Leukopenia	47%
Thrombocytopenia	54%
Elevation of transaminases (AST and ALT)	80%
Hyponatremia	33%
Microscopic hematuria	80%

[a]The mean age for all patients was 52 years (a range of 23–76 years). [b] Male/female ratio: 10/5.

(200 mg daily) was started on the day of admission for all patients. The effectiveness of doxycycline was compared in terms of the duration of the fever. The duration of the treatment with doxycycline was 7–8 days for 10 patients (67%) and 9–14 days for 5 patients (33%). Fever subsided within 1 day for four patients (27%), 2 days for two (13%) patients, 3 days for eight (53%) patients, and 4 days for one patient (7%) after administration of doxycycline. The outcome was favorable in 14 patients (93%), and no relapse or complications were observed within 2 months, probably due to the fact that early treatment was initiated based on the availability of local epidemiological data. Renal function deteriorated in one patient with chronic renal failure, but he recovered thereafter.

CONCLUSION

MSF is endemic in Greece. The clinical symptoms of the disease are fever, headache, rash, myalgia and malaise, in combination with elevation of transaminases and thrombocytopenia. If epidemiological information is available, the disease can be easily suspected, and the initial diagnosis can be based primarily on clinical grounds and further confirmed by serology. Early administration of the appropriate treatment on clinical suspicion of rickettsial infection can diminish the complications, reduce the course of the disease, and contribute to a favorable outcome.

REFERENCES

1. RAOULT, D. & V. ROUX. 1997. Rickettsioses as paradigms of new or emerging infectious diseases. Clin. Microbiol. Rev. **10:** 694–719.
2. DOUKA-SEGDITSA, I., E. STOFOROS, M. MASTROYANNI-KORKOLOPOULOU & N.P. DRAGONAS. 1972. L'infection de l'homme par des rickettsies et neo-ickettsies en Greece.: depistage d'anticorps contre *R. prowazeki*, *R. mooseri*, *R. burnetii* et Neo-rickettsie. Q. Bull. Soc. Path Exot. Filial **65:** 46–50.
3. PATERAKI, E., Y. TSELENTIS, H. PAPAKYRIAKOU & I. BRUNEAU. 1970. Contribution a l'etude des rickettsioses en Greece. Arch. Inst. Pasteur Hellen. **16:** 23–29.
4. ANTONIOU, M., Y. TSELENTIS, T. BABALIS, *et al.* 1995. The seroprevalence of ten zoonoses in two villages of Crete, Greece. Eur. J. Epidemiol. **11:** 415–423.
5. ANTONIOU, M., I. ECONOMOU, X. WANG, *et al.* 2002. Fourteen-year seroepidemiological study of zoonoses in a Greek village. Am. J. Trop. Med. Hyg. **66:** 80–85.
6. BABALIS, T., H. TISSOT DUPONT, Y. TSELENTIS, *et al.* 1993. *Rickettsia conorii* in Greece: comparison of a microimmunofluorescence assay and western blotting for seroepidemiology. Am. J. Trop. Med. Hyg. **48:** 784–792.
7. BABALIS, T., Y. TSELENTIS, V. ROUX, *et al.* 1994. Isolation and identification of a rickettsial strain related to *R. massiliae* in Greek ticks. Am. J. Trop. Med. Hyg. **50:** 365–372.
8. PSAROULAKI, A., I. SPYRIDAKI, A. IOANNIDIS, *et al.* 2003. First isolation and identification of *Rickettsia conorii* from ticks collected in the region of Fokida in Central Greece. J. Clin. Microbiol. **41:** 3317–3319.

9. PSAROULAKI, A., A. GERMANAKIS, A. GIKAS, *et al.* 2005. First isolation and genotypic identification of *Rickettsia conorii* Malish 7 from a patient in Greece. Eur. J. Clin. Microbiol. Infect. Dis. **24:** 297–298.

Prevalence of *Rickettsia felis*-like and *Bartonella* Spp. in *Ctenocephalides felis* and *Ctenocephalides canis* from La Rioja (Northern Spain)

JOSÉ RAMÓN BLANCO, LAURA PÉREZ-MARTÍNEZ, MANUEL VALLEJO, SONIA SANTIBÁÑEZ, ARÁNZAZU PORTILLO, AND JOSÉ ANTONIO OTEO

Área de Enfermedades Infecciosas, Hospitales San Millán-San Pedro-de La Rioja, 26001–Logroño (La Rioja), Spain

ABSTRACT: Our aim was to determine the presence of *Rickettsia* spp. and *Bartonella* spp. in *Ctenocephalides felis* and *Ctenocephalides canis* from La Rioja (Spain). A total of 88 specimens were tested by polymerase chain reaction (PCR) using *gltA* and *ompB* genes as targets for *Rickettsia* spp., and 16S rRNA and *ribC* genes for *Bartonella* spp. *Rickettsia felis-like* (28.4%), *Bartonella clarridgeiae* (6.8%), and *Bartonella henselae* (3.4%) were detected in *Ctenocephalides* spp. Other *Bartonella* sp. different from *B. clarridgeiae* and *B. henselae* could also be present in fleas from La Rioja.

KEYWORDS: flea; *Ctenocephalides felis*; *Ctenocephalides canis*; *Rickettsia spp.*; *Rickettsia felis*; *Bartonella spp.*; *Bartonella henselae*; *Bartonella clarridgeiae*; polymerase chain reaction; Spain

INTRODUCTION

Fleas have a worldwide distribution and are vectors of several infectious diseases. Some of them, such as plague (*Yersinia pestis*) and murine typhus (*Rickettsia typhi*) have had a big effect on human history, causing death and population reduction.[1,2] In recent years, clinical observations joint with the application of molecular diagnosis techniques, such as polymerase chain reaction (PCR), have allowed to implicate "new pathogens" transmitted by fleas in different clinical pictures. In years 1994 and 2000, cases similar to murine typhus caused by *Rickettsia felis* (previously named ELB agent) and transmitted by the cat flea (*Ctenocephalides felis*), were detected in four patients

Address for correspondence: José A. Oteo, Área de Enfermedades Infecciosas, Hospitales San Millán-San Pedro-de La Rioja, Avda. Viana, N° 1, 26001–Logroño (La Rioja), Spain. Voice: 34-941 297275; fax: 34-941-297267.
e-mail: jaoteo@riojasalud.es

**Ann. N.Y. Acad. Sci. 1078: 270–274 (2006). © 2006 New York Academy of Sciences.
doi: 10.1196/annals.1374.050**

from Texas and Mexico by PCR.[3,4] Since then, several clinical cases have been published in the North and Central America and more recently, in Europe and Asia.[5–7] Furthermore, *C. felis* has been demonstrated to be the vector of *R. felis* and/or *Bartonella* spp.[5,8–17] Infections caused by different *Bartonella* species are associated with a variety of clinical manifestations.[18] For these reasons, in an effort to identify the possible etiologic agents and vectors for rickettsioses and/or bartonelloses in our area, *C. felis* and *Ctenocephalides canis* fleas were analyzed.

MATERIAL AND METHODS

From September to October 2003, a total of 88 fleas were surveyed to determine whether they harbored *Rickettsia* and/or *Bartonella*. Twenty-seven fleas were collected from 5 cats, and 61 specimens were removed from 15 dogs seen at veterinary practices in La Rioja (North of Spain). Fleas were identified using taxonomic keys and they were classified as *C. felis* ($n = 68$) and *C. canis* ($n = 20$). Samples were stored in 70% ethyl alcohol at room temperature and analyzed individually. They were washed for 5 min in sterile distilled water and crushed in sterile Eppendorf tubes with the tip of a sterile pipette. DNA was extracted by lysis with ammonium hydroxide.[19,20] The presence of *Rickettsia* spp. was determined by PCR assays for *gltA* and *ompB* genes.[12,21] *Bartonella* DNA was detected by amplification of a fragment of 16 rRNA gene.[22] Positive samples for *Bartonella* were confirmed in a second PCR run for *ribC* gene, using specific primers for *Bartonella quintana*, *Bartonella henselae,* and *Bartonella clarridgeiae.*[23] Blanks with water processed along with flea samples were included as controls. PCR products were purified and sequenced at Universidad de Alcalá de Henares (Spain). Sequences obtained were compared with those available at GenBank using BLAST utility (National Center for Biotechnology Information; available from: URL: http://www.ncbi.nlm.nih.gov).

RESULTS

PCR products of the expected size of rickettsial DNA for *gltA* and *ompB* genes were detected in 25 out of 88 studied fleas (28.4%). Fleas that harbored *Rickettsia* included 18 *C. felis* (collected on cats and dogs) and 7 *C. canis* (collected on dogs). In all cases, sequences of *gltA* rickettsial amplicons were 100% identical to one genotype closely related to *R. felis* (*Rickettsia* sp. TwKM03).

Bartonella DNA was detected in 12 *C. felis* and 1 *C. canis* when 16S rRNA gene was amplified by PCR (14.8%). Subsequently, PCR products of *ribC* gene from these 13 fleas were obtained and confirmed by sequencing. Six of them were found to carry *B. clarridgeiae* (46.1%), while *B. henselae* was detected in 3 out of 13 (23.1%). Primers for *B. clarridgeiae*, *B. henselae,* and *B. quintana* did not yield positive PCR results in the remaining 4 out of

13 specimens (30.8%). Furthermore, 4 out of 88 flea samples (4.5%) were co-infected with *R. felis*-like and *Bartonella* sp. No nucleic acids were amplified from the negative controls.

DISCUSSION

R. felis-like (28.4%), *B. clarridgeiae* (6.8%), and *B. henselae* (3.4%) were detected in *Ctenocephalides* spp. infesting peridomestic animals from the North of Spain.

R. felis is an emerging pathogen responsible for flea-borne spotted fever. Its presence (now detected in the North of Spain) was previously described in fleas from the South of our country,[12] among others (United States, Ethiopia, Brazil, France, United Kingdom, Thailand, and more recently, New Zealand and Uruguay).[2,5,7,11,13–17,24] Its pathogenic role has also been demonstrated in patients using serological assays or biological molecular techniques.[3–7] This is the first time that dog fleas (*C. canis*) were found to be infected with *R. felis*-like in Spain. The presence of *R. felis* in *C. canis* has been reported only in Perú, Thailand, and, very recently, in Brazil.[15,24,25]

B. henselae is the recognized agent of cat scratch disease (CSD). It is globally distributed and its transmission to cats by *C. felis* was demonstrated in 1996.[10] In our region, a high prevalence of antibodies against *B. henselae* was previously reported for cat owners,[26] and in this study *B. henselae*, among other species (at least, *B. clarridgeiae*) has been detected in cat fleas. In the last few years the presence of other *Bartonella* spp. different from *B. henselae* has been demonstrated in these vectors.[13,16] The detection of *B. clarridgeiae* could have clinical and epidemiological implications in our area. This *Bartonella* species has been suspected to be an agent of CSD, but its pathogenic role in humans has not been demonstrated yet.[27,28] Other *Bartonella* sp. different from *B. clarridgeiae* and *B. henselae* could be present in fleas from La Rioja. A molecular method for the identification of *Bartonella* species that may be useful to complete this research was presented at the Fourth International Conference on *Rickettsiae* and Rickettsial Diseases (Logroño, Spain, 2005).[29]

To our knowledge, this work provides the first evidence of flea-borne *Bartonella* spp. in Spain, as well as the first evidence of *R. felis*-like in fleas from La Rioja (North of Spain). Physicians should consider *Rickettsia* and *Bartonella* as potential causes of fever in cat or dog owners, as well as in patients bitten by fleas in our environment. Further studies will be necessary to know the epidemiological and clinical importance of these findings and their effects on public health.

ACKNOWLEDGMENTS

This study was supported in part by grants from the Fondo de Investigación Sanitaria (FIS G03/057), Ministerio de Sanidad y Consumo (Spain), and also from the Government of La Rioja (ANGI 2004/17; Plan Riojano I+D+I, 2003–2007).

REFERENCES

1. PERRY, R.D. & J.D. FETHERSTON. 1997. *Yersinia pestis*—etiologic agent of plague. Clin. Microbiol. Rev. **10:** 35–66.
2. AZAD, A.F., S. RADULOVIC, J.A. HIGGINS, *et al.* 1997. Flea-borne rickettsioses: ecologic considerations. Emerg. Infect. Dis. **3:** 319–327.
3. SCHRIEFER, M.E., J.B. SACCI, JR., J.S. DUMLER, *et al.* 1994. Identification of a novel rickettsial infection in a patient diagnosed with murine typhus. J. Clin. Microbiol. **32:** 949–954.
4. ZAVALA-VELÁZQUEZ, J.E., J.A. RUIZ-SOSA, R.A. SÁNCHEZ-ELÍAS, *et al.* 2000. *Rickettsia felis* rickettsiosis in Yucatan. Lancet **356:** 1079–1080.
5. RAOULT, D., B. LA SCOLA, M. ENEA, *et al.* 2001. A flea-associated *Rickettsia* pathogenic for humans. Emerg. Infect. Dis. **7:** 73–81.
6. RICHTER, J., P.E. FOURNIER, J. PETRIDOU, *et al.* 2002. *Rickettsia felis* infection acquired in Europe and documented by polymerase chain reaction. Emerg. Infect. Dis. **8:** 207–208.
7. PAROLA, P., R.S. MILLER, P. MC DANIEL, *et al.* 2003. Emerging rickettsioses of the Thai–Myanmar border. Emerg. Infect. Dis. **9:** 592–595.
8. ADAMS, J.R., E.T. SCHMIDTMANN & A. AZAD. 1990. Infection of colonized cat fleas *Ctenocephalides felis* (Bouche), with a *Rickettsia*-like microorganisms. Am. J. Trop. Med. Hyg. **43:** 400–409.
9. HIGGINS, J.A., S. RADULOVIC, M.E. SCHRIEFER, *et al.* 1996. *Rickettsia felis:* a new species of pathogenic R*ickettsia* isolated from cat fleas. J. Clin. Microbiol. **34:** 671–674.
10. CHOMEL, B.B., R.W. KASTEN, K. FLOYD-HAWKINS, *et al.* 1996. Experimental transmission of *Bartonella henselae* by the cat flea. J. Clin. Microbiol. **34:** 1952–1956.
11. OLIVEIRA, R.P., M.A.M. GALVAO, C.L. MAFRA, *et al.* 2002. *Rickettsia felis* in *Ctenocephalides* spp. fleas, Brazil. Emerg. Infect. Dis. **8:** 317–319.
12. MÁRQUEZ, F.J., M.A. MUNIAIN, J.M. PÉREZ, *et al.* 2002. Presence of *Rickettsia felis* in the cat flea from South-Western Europe. Emerg. Infect. Dis. **8:** 89–91.
13. ROLAIN, J.M., M. FRANC, B. DAVOUST, *et al.* 2003. Molecular detection of *Bartonella quintana*, *B. koehlerae*, *B. henselae*, *B. clarridgeiae*, *Rickettsia felis*, and *Wolbachia pipientis* in cat fleas, France. Emerg. Infect. Dis. **9:** 338–342.
14. KENNY, M.J., R.J. BIRTLES, M.J. DAY, *et al.* 2003. *Rickettsia felis* in the United Kingdom. Emerg. Infect. Dis. **9:** 1023–1024.
15. PAROLA, P., O.Y. SANOGO, K. LERDTHUSNEE, *et al.* 2003. Identification of *Rickettsia* spp. and *Bartonella* spp. in fleas from the Thai-Myanmar border. Ann. N. Y. Acad. Sci. **990:** 173–181.
16. KELLY, P.J., N. MEADS, A. THEOBALD, *et al.* 2004. *Rickettsia felis, B. henselae* and *B. clarridgeiae*, New Zealand. Emerg. Infect. Dis. **10:** 967–968.
17. VENZAL, J.M., L. PÉREZ-MARTÍNEZ, M.L. FÉLIX, *et al.* 2006. Prevalence of *Rickettsia felis* in *Ctenocephalides felis* and *Ctenocephalides canis* from Uruguay. Ann. N. Y. Acad. Sci. This volume.
18. BLANCO, J.R. & D. RAOULT. 2005. Diseases produced by *Bartonella*. Enferm. Infecc. Microbiol. Clin. **23:** 313–319.

19. GUY, E.C. & G. STANEK. 1991. Detection of *Borrelia burgdorferi* in patients with Lyme disease by the polymerase chain reaction. J. Clin. Pathol. **44:** 610–611.
20. RIJPKEMA, S., D. GOLUBIC, M. MOLKENBOER, *et al.* 1996. Identification of four genomic groups of *Borrelia burgdorferi* sensu lato in *Ixodes ricinus* ticks collected in Lyme borreliosis endemic region of northern Croatia. Exp. Appl. Acarol. **20:** 23–30.
21. ROUX, V., E. RYDKINA, M. EREMEEVA, *et al.* 1997. Citrate synthase gene comparison, a new tool for phylogenetic analysis, and its application for the rickettsiae. Int. J. Syst. Bact. **47:** 252–261.
22. BERGMANS, A.M., J.W. GROOTHEDDE, J. F. SCHELLEKENS, *et al.* 1995. Etiology of cat scratch disease: comparison of polymerase chain reaction detection of *Bartonella* (formerly *Rochalimaea*) and *Afipia felis* DNA with serology and skin tests. J. Infect. Dis. **171:** 916–923.
23. BERESWILL, S., S. HINKELMANN, M. KIST, *et al.* 1999. Molecular analysis of riboflavin synthesis genes in *Bartonella henselae* and use of the *ribC* gene for differentiation of *Bartonella* species by PCR. J. Clin. Microbiol. **37:** 3159–3166.
24. HORTA, M., D.P. CHIEBAO, P.M. OHARA, *et al.* 2005. Prevalence of *Rickettsia felis* in the fleas *Ctenocephalides felis felis* and *Ctenocephalides canis* from two Indian communities in Sao Paolo municipality, Brazil. Presented at the Fourth International Conference on Rickettsiae and Rickettsial Diseases. Logroño, Spain, June 18.
25. PACHAS, P.E., C. MORON, A. HOYOS, *et al.* 2001. *Rickettsia felis* identified in *Ctenocephalides canis* fleas from Peruvian Andes. Presented at the ASR/*Bartonella* Joint Conference. Big Sky, Montana, USA.
26. BLANCO, J.R., J.A. OTEO, V. MARTÍNEZ DE ARTOLA, *et al.* 1998. Seroepidemiology of *Bartonella henselae* infection in a risk group. Rev. Clin. Esp. **198:** 805–809.
27. KORDICK, D.L., E.J. HYLYARD, T. L. HADFIELD, *et al.* 1997. *Bartonella clarridgeiae*, a newly recognized zoonotic pathogen causing inoculation papules, fever, and lymphadenopathy (cat scratch disease). J. Clin. Microbiol. **35:** 1813–1818.
28. MARGILETH, A.M. & D. F. BAEHREN. 1998. Chest-wall abscess due to cat-scratch disease (CSD) in an adult with antibodies to *Bartonella clarridgeiae*: case report and review of the thoracopulmonary manifestations of CSD. Clin. Infect. Dis. **27:** 353–357.
29. GARCÍA-ESTEBAN, C., R. ESCUDERO, J.F. BARANDIKA, *et al.* 2005. A molecular method for the identification of *Bartonella* species in clinical and environmental samples. Presented at the Fourth International Conference on Rickettsiae and Rickettsial Diseases. Logroño, Spain, June 18.

Prediction of Habitat Suitability for Ticks

AGUSTÍN ESTRADA-PEÑA

Department of Parasitology, Veterinary Faculty, Miguel Servet, 177, 50013 Zaragoza, Spain

ABSTRACT: This article offers a personal perspective on the efforts to model the habitat suitability for ticks. A serious effort is being made to provide maps at an adequate resolution that can plot the probability of the presence and even the abundance of tick parasites of humans and that can be used to develop control programs for species with economic interest. Some of the methods being currently evaluated are described together with a summary of the techniques using remotely sensed information to capture the abiotic features of the habitat. Some examples, developed using different methods, are provided for the Mediterranean region and part of the U.S., together with an overview of the performance of these models.

KEYWORDS: mapping; habitat suitability; ecological niche; models

INTRODUCTION

Ticks are a serious pest of domestic animals, and a threat for human health. Efforts continue at global, regional, and local scales to understand the major factors governing tick distribution, as a necessary step in the development of sound, ecologically based, control measures. The need for models in tick research arises from the problems associated with the use of acaricides in Australia, because of the development of resistance by the cattle tick. Further efforts were applied to model some seasonal activity patterns of *Ixodes ricinus* in Europe. With the arrival of the computer age and the use of robust statistical methods to test the feasibility of the model's output, we have at hand a new set of tools devoted to understanding the effect of environmental factors on the tick's ability to colonize a site. As well suitable implementations are available to provide animal and human health authorities with information about the areas of risk for and the seasonal dynamics if ticks. In the last few years, models built around the main features of the life cycle of ticks have been produced, which not

Address for correspondence: Agustín Estrada-Peña, Department of Parasitology, Veterinary Faculty, Miguel Servet, 177, 50013 Zaragoza, Spain. Voice: 34-976-761-558; fax: 34-976-761-612.
e-mail: aestrada@unizar.es

Ann. N.Y. Acad. Sci. 1078: 275–284 (2006).
doi: 10.1196/annals.1374.052

only provided information about the main demographic processes of ticks at a given site, but also allowed new insights into the sometimes complex regulation of the tick life cycle. We then realized that these kinds of models could be used as a powerful tool to predict the risk for humans in defined areas or habitats. While some powerful tick models exist that are able to predict the habitat suitability or the abundance of ticks at defined regions, we are still far from predicting the risk for the tick-borne diseases. These diseases are commonly the result of complex transmission patterns, the presence and abundance of important reservoir hosts, and, perhaps most important, the dynamics and behavior of humans in the vicinity of tick foci. This article is not a review of the different models built around the ticks and tick-borne diseases, but a personal insight into the main developments in this field, and how these models could help to protect humans by providing information about the "hot spots" of ticks.

Background

At geographical scales beyond experimentation, empirical models provide one of the only ways to develop and test hypotheses about ecological features affecting tick distribution.[1] It is a central premise of biogeography that climate exerts a dominant control over the natural distribution of species. Like many other arthropods, ticks are sensitive to climate features, these climate variables promoting, enhancing, or even stopping different and more or less critical parts of their life cycle.

Models exploring the relationships between species occurrence and sets of predictor variables produce two kinds of useful outputs: (1) estimates of the probability that a species of tick might occur at given unrecorded locations and (2) an area's suitability for the species. Probabilities of occurrence could be interpreted as estimates of the probability that the species might find a suitable habitat in a given area. A number of modeling strategies for predicting the impacts of climate on tick distribution have been developed. These have focused on the identification of the "bioclimate envelope" (alternatively termed *climate space*) either through methods that correlate current species distribution with climate variables or through an understanding of the physiological response to the climate. The bioclimate envelope modeling approach has its foundations in the ecological niche theory. Hutchinson[2] defined the fundamental ecological niche as comprising those environmental conditions within which a species can survive and grow. Hutchinson proposed that the fundamental niche would completely define the ecological properties of a species: a conceptual space whose axes include all of the environmental variables affecting that species. Bioclimate envelopes can be defined as constituting the *climatic component of the fundamental ecological niche*.

The validity of the bioclimatic envelope approach has been questioned by pointing to many other factors that play an important part in determining tick species distribution and their dynamics over time such as hosts, landscape patterns of vegetation, the management of the flocks, or even the competition between tick species for resources (hosts in this case). Cumming[3] proposed that continental-scale distributions of ticks are principally determined by climate, and it is therefore suggested that many species distributions can in fact be considered to be in equilibrium with the current climate at the macroscale. Genetic adaptation of species is rarely considered, being range shifts frequently seen as the expected response to the climate. It is usually expected that evolutionary change occurs only on long time scales and that the tolerance range of a species remains the same as it shifts its geographical range.[4] However, studies have shown that climate-induced range shifts can involve not only migration into newly suitable areas, but also selection against phenotypes that are poorly adapted to local conditions.[5] Thomas *et al.*[6] demonstrated the importance of rapid evolutionary change in the distribution range of plants. The potential impact of rapid evolutionary change means that the climatic tolerances of a tick may alter from the original founders, making the fundamental niche unstable over time.

At high resolution, vegetation, landscape pattern, and connectivity between patches of suitable vegetation and climate are the major determinants of tick distribution and abundance. Ticks spent most of their life cycles in sheltered microhabitats in the ground, and only a small part feeding on hosts. Acquisition and transmission of pathogens are therefore separated by long periods of free-living existence during which the tick develops from one stage to the next. As with most terrestrial arthropods, the development rates of the ticks are mediated by temperature, while their survival rates are mediated by water losses. A general knowledge of microclimatic requirements for tick survival and development, in terms of temperature and moisture conditions, obtained from laboratory experiments under controlled conditions or long-term field work can be applied to the development of models aimed at describing the different patterns of life cycles under varied field conditions.

Methods

There are many modeling techniques available to explore the correlation between occurrence of a species and sets of predictor variables and there is a possibility that techniques may differ in their ability to summarize useful relationships between response and predictor variables. If the goal of a model is to predict occurrence of species outside their known range, interest could be focused on methods that optimize the overall fit. Although some studies have compared the performance of different modeling techniques for some plant and animal taxa, none has, to my knowledge, yet investigated systematically how these methods can be applied to different tick species. Modeling approaches on the basis of similarities between datapoints only use presence data (ignoring absences) to create the environmental envelopes of the species. The Gower similarity approach assigns each cell in the output layer an average multivariate distance, termed the *Gower metric*, between the cell and the closest presence cell in the training set.[7] A related approach, using presence data alone, is based on ordination of data in a multivariate space of environmental variables (ENFA[8]). Extracted factors are uncorrelated and have biological significance. The marginality factor describes how far the species optimum is from the mean environmental profile of the studied area. The tolerance factor describes how specialized the species is by reference to the available range of environments in the studied area.

Other approaches to modeling the tick environmental envelope, like neural networks, generalized linear or additive models, or spatial interpolators require the simultaneous use of a set of presences and a set of absences. A set of absences is very difficult to obtain in the modeling of tick distributions, as most reports include only the data for the presence of the tick. One method based on genetic algorithms, called GARP,[9] uses a pseudo-absence set, derived from the sites where no presence has been recorded.

Commonly, modeling methods use a training set and a test set. The training set is a subgroup of data from which the model is tuned up. Then the model is tested against the "test set," and the performance of the model in predicting distribution is tested, using several indexes (see below). The evaluation of performance of the model requires the derivation of matrices of confusion that identified true positive (a), false positive (b), false negative (c), and true negative (d) cases predicted by the model. The number of false negatives is particularly useful because it measures the number of residuals, or amount of unexplained variation of data. From the values in the matrix we calculate sensitivity (percentage of true positives correctly predicted: $a/a(a+c)$) and specificity (percentage of true negatives correctly predicted: $d/(b+d)$). A method to evaluate the model performance involves ROC curves and has been already used in logistic regression methods applied to habitat prediction for ticks.[3] The curve is obtained by plotting sensitivity versus (number 1-especificity) for varying probability thresholds. Good model performance is characterized by a curve that maximizes sensitivity for low values of (number 1 specificity). High performance models are indicated by large areas under the ROC curves (i.e., large areas under the curve [AUC]). Usually AUC values of 0.7–0.9 indicate useful applications, and values of >0.9 indicate high accuracy.[10] While the AUC measure is considered useful for comparing the performance of presence–absence models in a threshold-independent way, truly predictive modeling requires some probability at which to accept the presence of the target organism. The ROC procedure offers a way of identifying an optimum probability threshold by simply reading the point on the curve at which the sum of sensitivity and specificity is maximized.[11]

Most important for tick studies on habitat suitability is the accurate selection, not only of the method to be used, but of the variables to be used in this computation. If climate is used to build the ecological space of a given tick species, variables with biological impact on the tick life cycle should be selected, and those with small importance on the survival or colonization ability of the tick should be avoided. Most commonly, available software can produce different sets of models and, then, based on measurements of model performance, the adequate set of climate variables is selected. From here, it is inferred that modeling methodologies selected for the prediction of tick habitat should be able to explain the biological meaning of the conclusions about the outline of the species distribution. Methods like GARP are "black boxes" that cannot provide an insight about the biological and ecological processes behind the obtained results.

Integration of Methods to Produce Habitat Suitability Maps

It is proposed herein that identifying the suitable climate space for a tick species through the use of bioclimate envelope models should be the first step in a broader modeling framework, consisting in a hierarchy of factors operating at different scales. Thus, at a continental scale, climate can be considered the dominant factor, while at more local scales factors like topography and land-cover types become increasingly important. Further down the hierarchy, if conditions of higher levels are satisfied, factors including biotic interactions and microclimate may become significant. Thus, the distribution of a tick may be primarily defined by climatic tolerances if the data resolution is 50 km^2, whereas, as the resolution is downscaled to around 5 km,^{2} land-cover type may become the dominant control over species presence. Similarly, as thc rcsolution is downscaled to less than 1 km,^{2} biotic interactions may become important. Early attempts to predict tick habitat suitability commonly used direct or interpolated climate databases. These input parameters may be prone to error because of interpolation routines. Now, satellite imagery is routinely available, and several environmental features with ecological meaning may be extracted from these source data. Remote sensing is defined as "the measurement of some property of an object of interest by a sensor that is not in direct physical contact with the object." Remotely sensed information was used in the pioneering work by Hugh-Jones *et al.*[12] to identify habitats for the tick *Amblyomma variegatum* in the Caribbean, where it is a serious pest of domestic animals. Other than for economically sounding tick species, such as *Rhipicephalus appendiculatus,*[13] much effort has been devoted to the development and application of models driven by remotely sensed variables to map the distribution of human-threatening ticks, like *Ixodes ricinus* in Europe and *I. scapularis* in USA. These two species are prominent vectors to humans of several pathogens, and knowledge about their distribution features, habitat preferences, and landscape impact on tick population structure is of capital importance in the built-up of solid frameworks providing information about risk by ticks.

An important source of problems is the incorporation of all these resolutions into a solid and useful framework. Our studies (under development) are investigating the ecological niche of the most prominent tick species in the Mediterranean region, as predicted under actual climate conditions and under conditions of forecasted climate change. These studies are being carried out using two artificial neural networks (ANN) that operate at two different scales to describe the ecological niche of these ticks. The first scale is a set of 19 climate variables as observed at a resolution of 5 km in the Mediterranean region. The first ANN evaluates the records of each tick species in the region and produces a map of climate suitability, while the second ANN manages the vegetation at 1-km resolution and produces the final output. The maps for the different species obtained by this procedure have high AUC values (between 0.75 and 0.85), suggesting a good agreement with the observed distribution of the ticks, and also display a high resolution indispensable for adequate control measures. FIGURE 1 shows the climate environmental niche for some of these tick species. Studies are in progress to provide decisionmakers with a robust and statistical definition of the ecological niche of these species as a preliminary step in the understanding of the processes expected to be produced under conditions of climate change and tick habitat variation.

I. scapularis is one of the most important ticks concerning human health in the United States and Canada, vector of the Lyme disease spirochete and several other pathogens. Considering the threat of this tick for human health, it is important to develop a robust approach to understand the effects of the climate and vegetation on its distribution. The purpose here is to provide human-health authorities with accurate maps about the distribution of the vector, together with an estimation of changes in density according to landscape and climate changes. A preliminary approach was developed using remotely sensed data for the United States, and a predictive map of the habitat suitability for the tick in the continent was built up[14] at a resolution of 8 km. Thereafter, the effects of climate trends in the last few years in the U.S. were modeled for that tick, and a clear tendency toward the increase of habitat suitability was noted in some regions of northeast and northern states.[15] However, it has been demonstrated that the abundance of that tick is closely related not only to climate factors, but also to the landscape configuration. Furthermore, in a field study about the abundance of nymphal *I. ricinus* in individual patches of vegetation distributed into a heterogeneous habitat, it has been demonstrated that the tick is more abundant in those patches close to the "connectivity lines" between patches, as modeled for host movements among the patches.[16] This is an effect derived from graph theory, where every patch of vegetation is considered as a "place for ticks" forming part of a metapopulation. This metapopulation is built up with the specimens found in the different patches of vegetation and that are being interchanged through the host use of these patches. This effect has also been modeled[17] and checked against data from field captures in a larger area. Results are strongly supportive of this hypothesis and show

(A)

(B)

(C)

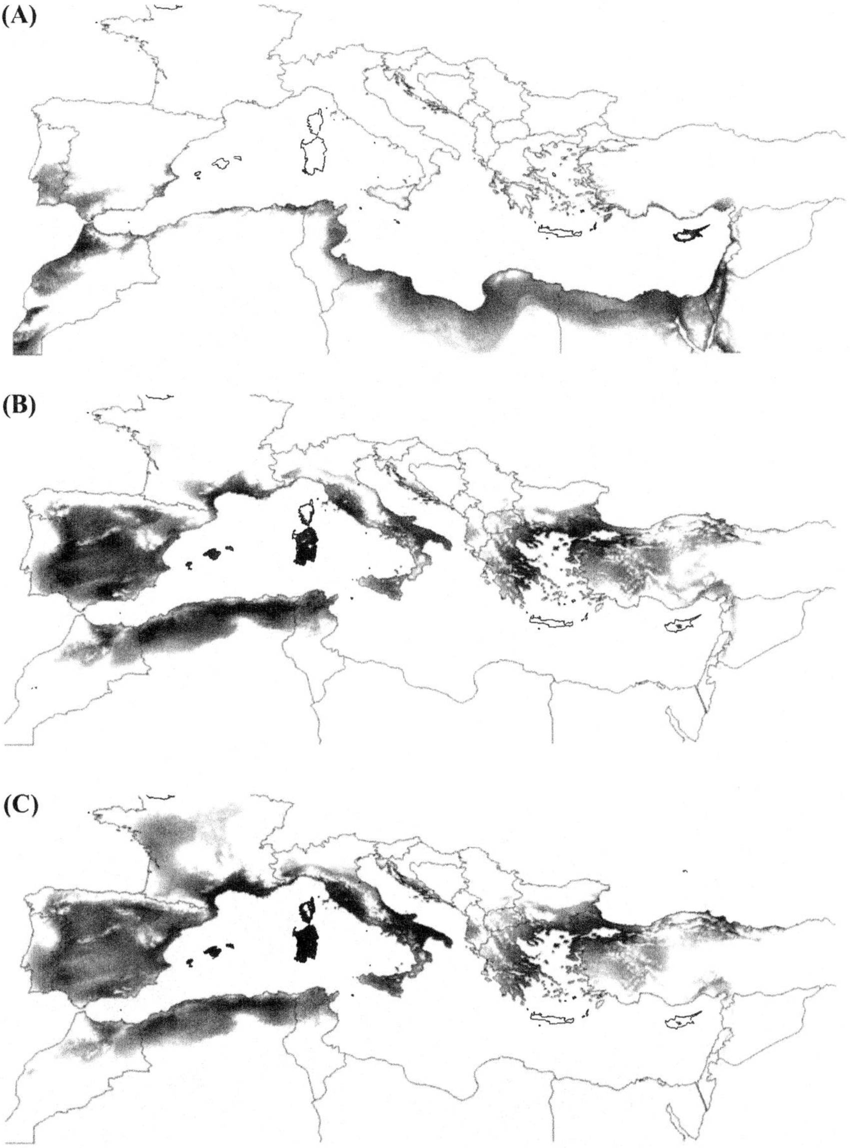

FIGURE 1. Some examples of the predicted distribution in the Mediterranean region of three economically prominent tick species. The gray scale shows different values of habitat suitability, darker being higher. A: *Boophilus annulatus*; B: *Rhipicephalus bursa*; C: *Rhipicephalus turanicus*. The last species is also a well-known parasite of humans. The AUC values of these models (indicating their predicting performance) are 0.88, 0.79, and 0.81, respectively.

(A)

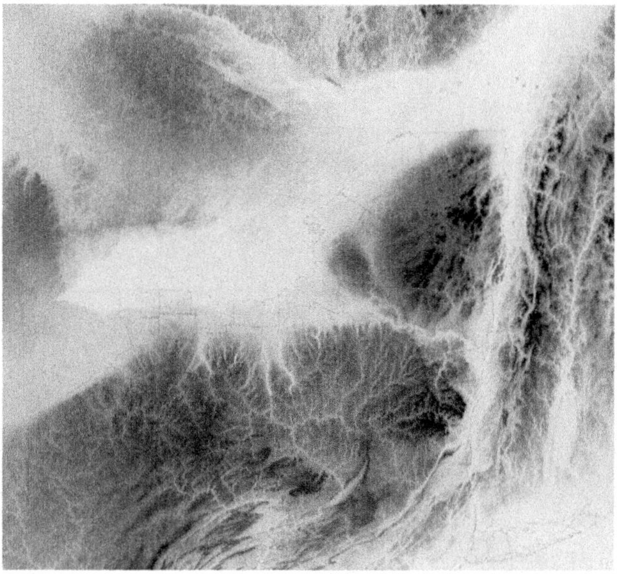

(B)

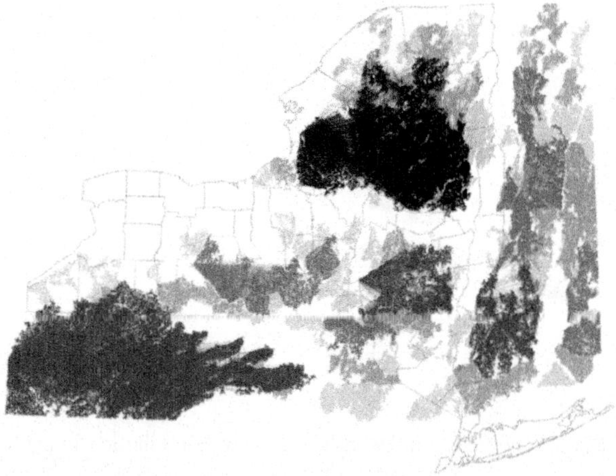

FIGURE 2. The modeling of the abundance of *Ixodes scapularis* in the United States. The figure shows the state of New York. (**A**) a general picture of the State, with the borders of the counties and an elevation model (darker is higher altitude). (**B**) the calculated recruitment of *I. scapularis* in the state, according to climate variables, patches of vegetation and host abundance (darker is higher). The recruitment values have been demonstrated to be highly correlated with the actual abundance of the tick.

that the well-connected patches are key stepping stones in the maintenance and survival of the tick metapopulation, and that tick abundance can be correlated by simple models using climate variables and landscape composition and topology. Within this framework, and modeling both the climate (macroscale) and the vegetation and hosts (microscale), a map of tick density is being built for every individual vegetation patch of every county in the United States. Climate is based on the results obtained by the PRISM model, which provides data at a resolution of 1 km for the co-terminous United States, while hosts' presence/absence data are being obtained from maps available through the GAP project (http://gap.uidaho.edu). Data about vegetation are obtained from the repository of digital maps available at http://edcww.cr.usgs.gov (LULC data). FIGURE 2 shows the details for the tick density values obtained from the predictive modeling as applied to the state of New York. AUC values for the predictive macroscale values are around 0.9, while studies of the correlations between predicted tick density and actual tick values provided a R^2 value of 0.81.

This methodology is a step ahead in the efforts for modeling tick-borne diseases. We are now able to predict not only the habitat suitability for ticks, but also the density of the different stages, thereby supplying human-health authorities with a new tool to provide advice about the risk to pick up a tick.

REFERENCES

1. ORMEROD, S.J. & A.R. WATKINSON 2000. Large-scale ecology and hydrology: an introductory perspective from the editors of Journal of Applied Ecology. J. App. Ecol. **37**(Suppl. 1): 1–5.
2. HUTCHINSON, G.E. 1957. Concluding remarks. Cold Spring Harbor Symp. Quant. Biol. **22:** 415–427.
3. CUMMING, G.S. 2002. Comparing climate and vegetation as limiting factors for species ranges of African ticks. Ecology **83:** 255–268.
4. PEARSON, R.G. & T.P. DAWSON. 2003. Predicting the impacts of climate change on the distribution of species: are bioclimate envelope models useful? Glob. Ecol. Biogeog. **12:** 361–371.
5. DAVIS, M.B. & R.G. SHAW 2001. Range shifts and adaptive responses to quaternary climate change. Science **292:** 673–679.
6. THOMAS, C.D., E.J. BODSWORTH, R.J. WILSON, *et al.* 2001. Ecological and evolutionary processes at expanding range margins. Nature **411:** 577–581.
7. CARPENTER, G., A.N. GILLISON & J. WINTER 1993. DOMAIN: a flexible modelling procedure for mapping potential distributions of plants and animals. Biodivers. Conserv. **2:** 667–680.
8. PERRIN, N. 1984. Contribution à l'écologie du genre Cepaea (Gastropoda): approche descriptive et expérimentale de l'habitat et de la niche écologique. Ph.D. thesis. University of Lausanne, Lausanne, Switzerland.
9. ANDERSON, R.P., D. LEW & A.T. PETERSON 2003. Evaluating predictive models of species' distributions: criteria for selecting optimal models. Ecol. Model. **162:** 211–232.

10. SWETS, J.A. 1988. Measuring the accuracy of diagnostic systems. Science **240:** 1285–1293.
11. ZWEIG, M.H. & G. CAMPBELL 1993. Receiver-operating characteristics (ROC) plots: a fundamental evaluation tool in clinical medicine. Clin. Chem. **39:** 1272–1276.
12. HUGH-JONES, M.E., N. BARRÉ, G. NELSON, *et al.* 1989. Remote recognition of *Amblyomma mariegatum* habitats in Guadeloupe using Landsat-TM imagery. Acta Vet. Scand. Suppl. **84:** 259–261.
13. RANDOLPH, S.E., & D.J. ROGERS 1997. A generic population model for the African tick *Rhipicephalus appendiculatus*. Parasitology **115:** 265–279.
14. ESTRADA-PEÑA, A. 1998. Geostatistics and remote sensing as predictive tools of tick distribution: a cokriging system to estimate *Ixodes scapularis* (Acari: Ixodidae) habitat suitability in the United States and Canada from advanced very high resolution radiometer satellite imagery. J. Med. Entomol. **35:**989–995.
15. ESTRADA-PEÑA, A. 2002. Increasing habitat suitability in the United States for the tick that transmits Lyme disease: a remote sensing approach. Environ. Health Perspect. **110:** 635–640.
16. ESTRADA-PEÑA, A. 2003. The relationships between habitat topology, critical scales of connectivity and abundance of *Ixodes ricinus* in a heterogeneous landscape in northern Spain. Ecography **26:** 661–673.
17. ESTRADA-PEÑA, A. 2004. Effects of habitat suitability and landscape patterns on tick (Acarina) metapopulation processes. Landscape Ecol. **20:** 529–541.

Natural Infection, Transovarial Transmission, and Transstadial Survival of *Rickettsia bellii* in the Tick *Ixodes loricatus* (Acari: Ixodidae) from Brazil

MAURICIO C. HORTA,[a] ADRIANO PINTER,[a]
TERESINHA T. S. SCHUMAKER,[b] AND MARCELO B. LABRUNA[a]

[a] *Department of Preventive Veterinary Medicine and Animal Health, Faculty of Veterinary Medicine, University of São Paulo, São Paulo, Brazil*

[b] *Department of Parasitology, Institute of Biomedical Sciences, University of São Paulo, São Paulo, Brazil*

ABSTRACT: An *Ixodes loricatus* engorged female, infected with *Rickettsia bellii,* was collected from an opossum (*Didelphis aurita*) in Mogi das Cruzes, São Paulo State, Brazil. Two consecutive laboratory tick generations (F_1 and F_2) reared from this single engorged female were evaluated for *Rickettsia* infection by polymerase chain reaction (PCR) targeting specific *Rickettsia* genes. Immature ticks fed on naïve Wistar rats (*Rattus norvegicus*) and adult ticks fed on opossum (*D. aurita*), both free of ticks and rickettsial infection. PCR performed on individual ticks from the F_1 (20 larvae, 10 nymphs, and 10 adults) and the F_2 (30 larvae, 30 nymphs, and 15 adults) yielded expected bands compatible with *Rickettsia.* All the PCR products that were sequenced, targeting *gltA* gene, resulted in sequences identical to each other and 99.7% (349/350) similar to the corresponding sequence of *R. bellii* in GenBank. The *R. bellii* infection on ticks from the second laboratory generation (F_2) was confirmed by other PCR protocols and successful isolation of *R. bellii* in cell culture. We report for the first time a *Rickettsia* species infecting *I. loricatus*, and the first report of *R. bellii* in the tick genus *Ixodes.* We conclude that there was an efficient transovarial transmission and transstadial survival of this *Rickettsia* species in the tick *I. loricatus.* Our results suggest that *R. bellii* might be maintained in nature solely by transovarial transmission and transstadial survival in ticks (no amplifier vertebrate host is needed), since there has been no direct or indirect evidence of infection of vertebrate hosts by *R. bellii.*

KEYWORDS: *Rickettsia bellii*; *Ixodes loricatus*; transovarial transmission; transstadial survival

Address for correspondence: Maurício C. Horta, Faculdade de Medicina Veterinária e Zootecnia, Universidade de São Paulo, Av. Prof. Dr. Orlando Marques de Paiva, 87, São Paulo, 05508-270, Brazil.Voice: 55-11-3091-7701; fax: 55-011-3091-7928.
e-mail: maurivet@yahoo.com

Ann. N.Y. Acad. Sci. 1078: 285–290 (2006). © 2006 New York Academy of Sciences.
doi: 10.1196/annals.1374.053

INTRODUCTION

Rickettsia bellii has been reported in 15 tick species belonging to the genera *Amblyomma* (7 species), *Dermacentor* (5), *Haemaphysalis* (1) *Argas* (1), and *Ornithodoros* (1), from different parts of Brazil and United States.[1–3] Among the genus *Rickettsia, R. bellii* is indeed the species with the greatest number of tick records. Although infection rates by *R. bellii* in tick populations are usually high (up to 40%),[2] there has been no direct or indirect evidence of animal or human infection by *R. bellii.* This scenario suggests that *R. bellii* is very well adapted to ticks, possibly being maintained in tick populations without the participation of amplifier vertebrate hosts.

The present study evaluated the infection, transovarial transmission, and transstadial survival of *R. bellii* in *Ixodes loricatus* ticks. Until the present study, this tick species had never been reported to be infected by any *Rickettsia* species.

MATERIALS AND METHODS

In January 2004, an *Ixodes loricatus* engorged female was collected from an opossum (*Didelphis aurita*) in Mogi das Cruzes (23°38′29.5″ S, 46°11′05.5″ W), São Paulo State, Brazil. The female tick was brought to the laboratory and held in an incubator at 25 ± 1°C and 95% relative humidity (RH) for egg oviposition. After oviposition, two drops of the hemolymph were collected from the female legs, one being tested by the hemolymph test[4] using Gimenez staining,[5] and the other used for DNA extraction[6] by PCR targeting a 401-bp fragment of the rickettsial *gltA* gene, using primers CS-78 and CS-323.[2]

Tick First Generation (F_1)

Larvae hatched from eggs oviposited by the female cited above were used to start the first laboratory tick generation. Infestations by larvae or nymphs were performed on naive Wistar rats (*Rattus norvegicus*) (2 rats for each tick stage) and infestation by adult ticks was performed on a *D. aurita,* as previously described.[7] The *R. norvegicus* were obtained from a laboratory animal house free of any ectoparasites. The *D. aurita* was trapped in the University of São Paulo campus and was free of ticks. Before infesting this opossum with the F_1 adult ticks, we collected a blood sample and tested it by the PCR cited above, while the blood serum was tested by immunofluorescence assay (IFA) using *Rickettsia rickettsii* and *R. bellii* antigens.[8]

A total of 200 larvae, 30 nymphs, or 10 adult pairs were infested on each individual host. Engorged ticks recovered from infested hosts were held in an incubator at 25 ± 1°C and 95% RH for molting or egg oviposition. From this laboratory first generation of *I. loricatus* (20 unfed larvae, 10 unfed nymphs, and 10 unfed adults) were individually examined for rickettsial infection by the PCR technique described above. PCR products from 6 larvae, 6 nymphs, and 6 adults were sequenced[6] and compared with NCBI Nucleotide BLAST searches.

Tick Second Generation (F_2)

Three F_1 engorged females were used to produce larvae that composed the second laboratory tick generation. From this step, each female offspring was evaluated separately, being composed of three groups of F_2 ticks. Larvae and nymphs of each group infested *R. norvegicus* as described above. From each of the three groups of F_2 ticks, 10 unfed larvae, 10 unfed nymphs, and 5 unfed adults were individually examined by the PCR technique described above. PCR products from 1 larva, 1 nymph, and 1 adult from each of the three groups of F_2 ticks were sequenced[6] and compared with NCBI Nucleotide BLAST searches.

One larva, 1 nymph, and 1 adult from each of the three groups of F_2 ticks were tested by two other PCR protocols, one targeting a 837-bp fragment of the *gltA* gene of *R. bellii,* using primers CS-239 and CS-1069,[2] and the other targeting a 549-bp fragment of the 17-kDa protein gene of *R. bellii,* using primers 17k-5 and 17k-3.[2] PCR products of the expected size were sequenced[6] and compared with NCBI Nucleotide BLAST searches. In all PCR reactions of the present study, *Rickettsia parkeri*–infected tick DNA and water were used as positive and negative controls, respectively.[9]

Isolation of Rickettsia *in Cell Culture*

Isolation of *Rickettsia* in cell culture was attempted on eight F_2 adult ticks. For this purpose, tick homogenate of each individual tick was inoculated onto two shell vials as previously described,[2,10] with the exception that we used the C6/36 cell line (from *Aedes albopictus*)[11] to seed the shell vials. Inoculated shell vials were incubated at 28°C. Infection of the inoculated shell vials was evaluated every 3 days by Gimenez staining. Isolation was considered successful if typical *Rickettsia*-like organisms were visualized by Gimenez staining.[5] Isolation was confirmed by performing a PCR and DNA sequencing on infected cells from two shell vials, using primers CS-78 and CS-323, as described above.

RESULTS

The *I. loricatus* engorged female that originated the laboratory tick colony was demonstrated to contain typical *Rickettsia*-like organisms inside hemocytes by the hemolymph test. The PCR performed on its hemolymph yielded an expected band, which after being sequenced was 99.7% (349/350), similar to the corresponding sequences of a Brazilian and a North American *R. bellii* isolate in GenBank (AY362703, U59716), and 99.4% (348/350) similar to a second Brazilian isolate of *R. bellii* (AY375161).

F_1 and F_2 Ticks

All the PCRs using primers CS-78 and CS-323 performed on the individual ticks (larvae, nymphs, and adults) from the two laboratory generations (F_1 and F_2 ticks) yielded expected bands compatible with *Rickettsia.* All the PCR products that were sequenced resulted in sequences identical to each other and 99.7% (349/350) similar to the corresponding sequences of a Brazilian and a North American *R. bellii* isolate in GenBank (AY362703, U59716), and 99.4% (348/350) similar to a second Brazilian isolate of *R. bellii* (AY375161).

PCRs using primers CS-239 and CS-1069 performed on 3 larvae, 3 nymphs, and 3 adult F_2 ticks yielded expected bands for all tested samples. The sequenced fragments were identical to each other. As its 5′ end overlapped the 3′ end of the fragment determined by the primers CS-78 and CS-323, both sequences were grouped to form a single partial sequence of 1,109 bp of the rickettsial *gltA* gene. This fragment was 99.8% (1,107/1,109) similar to the corresponding sequences of a North American *R. bellii* isolate in GenBank (U59716), and 99.7% (1,106/1,109) similar to two Brazilian isolates (AY362703, AY375161).

PCRs using primers 17k-5 and 17k-3 performed on 3 larvae, 3 nymphs, and 3 adult F_2 ticks yielded expected bands for all tested samples. The sequenced fragments were identical to each other and 100% (484/484) similar to the corresponding sequences of a Brazilian *R. bellii* isolate in GenBank (AY362702), and 99.6% (482/484) similar to a North American isolate of *R. bellii* (AF445380).

PCR performed on the blood of the opossum used to feed F_1 adult ticks was negative and its blood serum did not react with *R. rickettsii* or *R. bellii* antigens.

Isolation of Rickettsia *in Cell Culture*

Typical *Rickettsia*-like organisms were isolated in the cell culture from four of the eight F_2 adult ticks, as demonstrated by Gimenez staining on C3/36 cells from the shell vials. PCR performed on infected cells from two shell vials confirmed the isolation of *R. bellii,* as demonstrated by the DNA sequence of a partial sequence of the *gltA* gene, which was 99.7% (349/350) similar to the corresponding sequences of a Brazilian and a North American *R. bellii* isolate in GenBank (AY362703, U59716), and 99.4% (348/350) similar to a second Brazilian isolate of *R. bellii* (AY375161).

DISCUSSION

The present study reports for the first time a *Rickettsia* species infecting *I. loricatus*, and the first report of *R. bellii* in the tick genus *Ixodes*. Our *I. loricatus* laboratory colony started with a single naturally infected engorged female. Immature ticks fed on naive Wistar rats, which were provided by a laboratory animal house that was free of ectoparasites. Adult ticks were fed on an opossum with negative PCR and serology for *Rickettsia* infection. Thus, we consider insignificant the chances of our ticks becoming infected by rickettsiae through the parasitism on rats or opossums during the study. As 100% of the tested ticks from the first and the second generations were shown to be infected by *R. bellii,* we conclude that there was an efficient transovarial transmission and transstadial survival of this *Rickettsia* species in the tick *I. loricatus.*

Unfortunately, we did not compare the feeding performance and survival rates of the *R. bellii*–infected ticks with uninfected ticks, and therefore we could not evaluate whether infection by *R. bellii* had any deleterious effect on *I. loricatus* ticks. Even though our results suggest that *R. bellii* might be maintained in nature solely by transovarial transmission and transstadial survival in ticks (no amplifier vertebrate host is needed), there has been no direct or indirect evidence of infection of vertebrate hosts by *R. bellii.*

ACKNOWLEDGMENTS

We thank Edison L. Durigon (University of São Paulo) for providing the C3/36 cell line, and IBAMA for providing the opossum capture and rearing license. This work was supported by Fundação de Amparo a Pesquisa do Estado de São Paulo (FAPESP Grants 02/10759-0 to M.C.H., 03/04728-7 to T.T.S.S., and 03/13872-4 to M.B.L.).

REFERENCES

1. PHILIP, R.N., E.A. CASPER, R.L. ANACKER, *et al.* 1983. *Rickettsia bellii* sp. nov.: a tick-borne Rickettsiae, widely distributed in the United States, that is distinct from the spotted fever and typhus biogroups. Int. J. Syst. Bacteriol. **33:** 94–106.
2. LABRUNA, M.B., T. WHITWORTH, M.C. HORTA, *et al.* 2004b. *Rickettsia* species infecting *Amblyomma cooperi* ticks from an area in the state of São Paulo, Brazil, where Brazilian spotted fever is endemic. J. Clin. Microbiol. **42:** 90–98.
3. LABRUNA, M.B., T. WHITWORTH, D.H. BOUYER, *et al.* 2004. *Rickettsia bellii* and *Rickettsia amblyommii* in *Amblyomma* ticks from the state of Rondônia, Western Amazon, Brazil. J. Med. Entomol. **41:** 1073–1081.

4. BURGDORFER, W. 1970. The hemolymph test. Am. J. Trop. Med. Hyg. **19:** 1010–1014.
5. GIMÉMEZ, D.F. 1964. Staining rickettsiae in yolk-sac cultures. Stain Technology **39:** 135–140.
6. HORTA, M.C., A. PINTER, A. CORTEZ, *et al.* 2005. *Rickettsia felis* (Rickettsiales: Rickettsiaceae) in *Ctenocephalides felis felis* (Siphonaptera: Pulicidae) in the State of São Paulo, Brazil. Arq. Bras. Med. Vet. Zootec. **57:** 321–325.
7. SCHUMAKER, T.T.S., M.B. LABRUNA, I.S. ABEL, *et al.* 2000. Life cycle of *Ixodes (Ixodes) loricatus* (Acari: Ixodidae) under laboratory conditions. J. Med. Entomol. **37:** 714–720.
8. HORTA, M.C., M.B. LABRUNA, L.A. SANGIONI, *et al.* 2004. Prevalence of antibodies to spotted fever group rickettsiae in humans and domestic animals in a Brazilian spotted fever endemic area in the state of São Paulo, Brazil: serological evidence for infection by *Rickettsia rickettsii* and another spotted fever group rickettsia. Am. J. Trop. Med. Hyg. **71:** 93–97.
9. SANGIONI, L.A., M.C. HORTA, M.C.B. VIANNA, *et al.* 2005. Rickettsial infection in animals and Brazilian spotted fever endemicity. Emerg. Infect. Dis. **11:** 265–269.
10. KELLY, J.P., D. RAOULT & P.R. MASON. 1991. Isolation of spotted fever group rickettsiae from triturated ticks using a modification of the centrifugation vial technique. Trans. R. Soc. Trop. Med. Hyg. **85:** 397–398.
11. IGARASHI, A. 1978. Isolation of a Singh's *Aedes albopictus* cell clone sensitive to dengue and chikungunya viruses. J. Gen. Virol. **40:** 531–544.

Prevalence of Bacterial Agents in *Ixodes persulcatus* Ticks from the Vologda Province of Russia

MARINA E. EREMEEVA,[a] ALICE OLIVEIRA,[a]
JENNILEE B. ROBINSON,[a] NINA RIBAKOVA,[b]
NIKOLAY K. TOKAREVICH,[c] AND GREGORY A. DASCH[a]

[a] *Viral and Rickettsial Zoonoses Branch, Centers for Disease Control and Prevention, Atlanta, Georgia, USA*

[b] *Center for Epidemiological Surveillance, Vologda, Russia*

[c] *Pasteur Institute of Epidemiology and Microbiology, St. Petersburg, Russia*

ABSTRACT: The prevalence of rickettsiae, ehrlichiae, and the rickettsia-like endosymbiont called Montezuma relative to that of *Borrelia* was determined in questing *Ixodes persulcatus* (*I. persulcatus)* ticks collected in 2002–2003 from Vologda Province, Russia. *Ehrlichia muris, Anaplasma phagocytophilum,* Montezuma, and new spotted fever group rickettsiae were detected by polymerase chain reaction (PCR) for the first time in this area. The rickettsiae were all *Candidatus* Rickettsia tarasevichiae, the furthest west this organism has been detected. After *Borrelia,* Montezuma was the agent most frequently detected; it may be present throughout the distribution of *I. persulcatus* in Russia. Ehrlichiae and rickettsiae frequently share the same tick host with *Borrelia burgdorferi sensu lato* so cotransmission and mixed infections in vertebrate hosts, including humans, may occur.

KEYWORDS: *Ixodes persulcatus*; *Ehrlichia muris*; *Candidatus* Rickettsia tarasevichiae; *Anaplasma*; *Borrelia burgdoreri*; Vologda; coinfection

INTRODUCTION

The Vologda Province of Russia is located in northwest Russia east of St. Petersburg, at the western end of the distribution of *Ixodes persulcatus (I. persulcatus)* taiga ticks. The active tick questing season lasts from the second week of April through the middle of May. While May–June is characterized by

Address for correspondence: Marina E. Eremeeva, Viral and Rickettsial Zoonoses Branch, Mail Stop G-13, National Center for Infectious Diseases, Centers for Disease Control and Prevention, 1600 Clifton Road NE, Atlanta, GA 30333. Voice: 404-639-4612; fax: 404-639-4436.
e-mail: MEremeeva@cdc.gov

The findings and conclusions in this report are those of the author(s) and do not necessarily represent the views of the funding agency.

Ann. N.Y. Acad. Sci. 1078: 291–298 (2006). © 2006 New York Academy of Sciences.
doi: 10.1196/annals.1374.054

a relatively low level of tick activity, single ticks can be found until September. This area is highly endemic for the agents of Lyme disease and tick-borne encephalitis and their prevalence correlates with high densities of ticks (150 to 500 ticks per km^2) as assessed by flagging.[1] In previous studies in Russia *I. persulcatus* was also infected with rickettsiae, ehrlichiae, and the rickettsia-like endosymbiont Montezuma.[2–12] In this study we investigated the prevalence of these bacterial agents in *I. persulcatus* ticks and compared those findings to the prevalence of *Borrelia burgdorferi sensu lato*.

MATERIALS AND METHODS

Tick Collection, Identification, and DNA Preparation

Questing ticks were collected from vegetation by flagging during the spring and summer months from several sites in Vologda Province in 2002 and 2003. All ticks were identified as *I. persulcatus* using standard taxonomic keys. Ticks were surface-disinfected using a series of washes with 10% bleach for 1–2 min, 70% ethanol for 3–5 min, and three times with distilled water; excess water was removed using filter paper. The ticks were frozen in liquid nitrogen, and crushed into powder using Kontes pestles (Kimble–Kontes, Fisher). The powder was resuspended in lysis buffer supplemented to 1 mg/mL of proteinase K (Qiagen, Valencia, CA, USA) and incubated overnight at 56°C. DNA from ticks collected in 2002 was extracted using a DNeasy Extraction Kit (Qiagen), while DNA from ticks collected in 2003 was extracted using a Promega Wizard DNA Extraction Kit (Promega, Madison, WI, USA). DNA was stored at 4°C prior to further analysis.

Polymerase Chain Reaction (PCR) Detection Methods

Tick DNA was tested using several PCR-based methods. Single-step PCR assays detected the fragments of the citrate synthase gene (*glt*A) of *Rickettsia*[13] or the 16S rRNA gene of the rickettsia-like endosymbiont Montezuma,[14] using Taq MasterMix (Qiagen). The original PCR cycling conditions and primer sequences were modified to increase specificity of each test and will be reported elsewhere. Amplicons of the correct size were detected by electrophoresis in 1% agarose gel supplemented with 0.5 μg/mL of ethidium bromide. 16S ribosomal DNA of ehrlichiae was detected with a quantitative PCR assay[15] using SYBR Green PCR Reagent kit (PE Applied Biosystems, Foster City, CA, USA) and the *i*-Cycler with real time PCR detection system (Bio-Rad, Hercules, CA, USA). Automatic acquisition and subsequent data analyses were performed using the *i*-Cycler software. A TaqMan multiplex assay was used to test the tick DNAs simultaneously for the presence of *Anaplasma phagocytophilum (A. phagocytophilum)* and *B. burgdorferi sensu lato*.[16] The PCR reaction was performed using the Brilliant Quantitative PCR Core Reagent Kit with SureStart TM Taq polymerase (Stratagene, La Jolla, CA, USA). The reactions were run on the 7900HT TaqMan Instrument (Applied BioSystems). Quantitative assays were run in duplicate; the results were considered positive if both replicates were positive and produced statistically indistinguishable Ct values. Statistical analysis of results was done using the chi-square test.

DNA Sequencing

Sequence reactions were performed using the ABI PRISM BigDye^{TM} Terminator Cycle 3.1 Sequencing kit as recommended by the manufacturer (Applied BioSystems). The sequenced products were purified with Qiagen Dye Removal Kit (Qiagen) and run on an Applied Biosystems 3100 Sequencer, and data were analyzed with Nucleic Acid Sequence Analyzer (Applied Biosystems), CAP sequence assembler (www.infobiogen.fr), and ClustalW multiple sequence alignment software (http://clustalw.genome.jp/, Kyoto University Bioinformatics Center, Kyoto, Japan). Sequence identities were detected using NCBI BLAST search engine.

RESULTS AND DISCUSSION

Detection of Ehrlichial Agents by Using a SYBR Green PCR Assay

The SYBR Green 1 Double-Stranded Binding Dye specifically binds to an amplified fragment of the *rrs* of ehrlichiae (146–154 bp in size). The specific primers are located in the 5′-region of the *rrs* of *Ehrluchia chaffeensis (E. chaffeensis)*, and they were originally described as specific for this species.[15] However, multiple sequence alignment of this region from *E. chaffeensis* and other ehrlichiae indicates that *rrs* from other closely related organisms, including *E. muris* and *A. phagocytophilum* can be amplified and detected with these primers because of significant nucleotide sequence homology in primer annealing sites (FIG. 1). However, the homologous fragments in these organisms have important internal sequence differences that allow the identification of the specific organism being detected. If sequencing is not available, ehrlichial agents can be differentiated on the basis of their unique *Alu*I restriction patterns using this *rrs* fragment (FIG. 1).

DNA of ehrlichiae was detected in 14 of 84 ticks (16.7%) collected in 2002 (TABLE 1). Positive ticks were found in each of three collection sites investigated. The species of ehrlichiae was as follows: Nine ticks were infected with *E. muris* and one tick was infected with *E. chaffeensis*-like organism.

```
AF260591   1 TAACACATGCAAGTCGAACGGACAATTACCTATAGCCTTTTTGGTTACAGGTA-GTTGTT[21]GCATAGGAATCTACCTAGTAGTATGGAATAGCCATTAGAAATGATGGGTAATACTGTATAATCCCTGCGGGGG
AB028319   1 .....................................................-......[21].........................................................................
AB013009   1 ......................T.GC....C.....T.....A.C..T....TC.C.A..[21].........................................................................
U15527     1 ......................T.GC....C.....T.....A.C..T....TC.C.A..[21].........................................................................
AB013008   1 ......................T.GC....C.....T.....A.C..T....TC.C.A..[21].........................................................................
AB024928   1 ...............................................T.....-......[21].........................................................................
AF416764   1 ............................G.T....A......-.....T.AA..-A....[21]..G......................................................................
AB074459   1 ....................A.T....GTT.....--.C....AC..T.AA..-A.....[21]..G.............................T............A...........................
AB074460   1 ....................A.TTG-..A..G...---...--AC..T..T.--.CA...[21]............G...................TG..........CA........C..................
AY055469   1 ......................TT...CT.....G.T.GC.AT------...G...A...[21].........................................................................
consensus  1 TAACACATGCAAGTCGAACGgAcaatTacctaTAgccttTTtggtTAcAggTa-gttgTT[21]GCaTAGGAATCTaCCTAGTAGTATGGAATAGCcaTTAGAAATGAtgGGTAATACtGTATAATCCCTGCGGGGG
```

FIGURE 1. Nucleotide sequence alignment of a portion of *rrs* gene of *Anaplasmataceae* used for detection by SYBR Green PCR assay. Primer locations are shown in bold and are underlined. The *rrs* sequences from NCBI GenBank: *Ehrlichia* sp. HI-2000, AF260591; *Ehrlichia* sp. Anan, AB028319; *E. muris* strain NA-1, AB013009; *E. muris* strain AS145, U15527; *E. muris* strain I268, AB013008; *Ehrlichia* sp. HF565, AB024928; *E. chaffeensis* strain Arkansas, AF416764; *Ehrlichia* sp. TS37, AB074459; *Candidatus* N. mikurensis strain IS58, AB074460; *A. phagocytophilum* strain USG3, AY055469. The position of *Alu*I restriction sites (AGCT) is underlined. No *Alu*I restriction sites are present in the *E. chaffeensis* amplicon.

TABLE 1. Prevalence of tick-borne bacteria in *I. persulcatus* from Vologda Province

Organism and gene target detected	Year of tick collection*					
	2002			2003		
	Male	Female	Total	Male	Female	Total
B. burgdorferi (rrl)	10/32	16/52	26/84	14/64	7/14	21/78
A. phagocytophilum (msp2)	6/32	8/52	14/84	0/64	0/14	0/78
Ehrlichia and Anaplasma (rrs)	7/32	7/52	14/84	0/64	0/14	0/78
Rickettsia (gHA)	1/32	0/52	1/84	8/64	3/14	11/78
Rickettsia-like endosymbiont Montezuma (rrs)	1/32	27/52	28/84	3/64	10/14	13/78

* The number of positive ticks (in numerator) and total number of ticks examined (in denominator).

The species identity of ehrlichial organisms associated with four other ticks was not determined on account of the very low DNA copy numbers detected per tick. No ehrlichial DNAs were found in the 2003 tick samples.

Detection of A. phagocytophilum *and* B. burgdorferi *by Multiplex PCR*

The ABI TaqMan assay detected the presence of *A. phagocytophilum msp*2 gene in 14 of 84 ticks (16.7%) collected in 2002 from a single collection site; three of these tick DNAs tested positive in the SYBR Green assay mentioned earlier. No positive ticks were found among the 2003 tick samples. Although this TaqMan assay is highly specific and sensitive,[16] sequence confirmation cannot be done directly on this product because of the small size of the fragment detected (77 bp). Attempts to amplify a 546-bp fragment of *rrs* of *A. phagocytophilum* using a nested PCR assay[17] yielded no PCR product. However, the DNA of *B. burgdorferi* was detected in 31% of ticks collected in 2002 and 26.9% of ticks collected in 2003. Five (5.9%) of 84 ticks collected in 2002 were positive both for *B. burgdorferi* and *A. phagocytophilum*.

Detection of Rickettsiae and Montezuma Endosymbiont by Direct PCR

The 382-bp amplicon of *glt*A was detected in 14.1% of 2003 tick samples. Only one tick sample was positive among ticks collected in 2002. All *glt*A nucleotide sequences from the ticks were identical to there of *Candidatus* Rickettsia tarasevichiae (NCBI AF503167).[9] The DNA of the rickettsia-like endosymbiont Montezuma (NCBI AF493952,[6]) was primarily detected in association with female ticks (51.9% and 42.9% in 2002 and 2003, respectively). The markedly lower prevalence of Montezuma agent DNA in adult male *I. persulcatus* (1 out of 32 ticks tested) suggests that it is maintained transovarially like *Rickettsia* and may be lost in males during gonadal differentiation.

TABLE 2. Prevalence of mixed infections in *I. persulcatus* ticks from Vologda Province

	Year of tick collection	
Organism detected	2002	2003
B. burgdorferi + ehrlichiae	5.9%	0
B. burgdorferi + *Rickettsia*	0	1.3%
B. burgdorferi + rickettsia-like endosymbiont Montezuma	4.8%	3.8%
Ehrlichiae + rickettsia-like endosymbiont Montezuma	7.1%	0
Rickettsia + rickettsia-like endosymbiont Montezuma	0	3.8%
B. burgdorferi + ehrlichiae + rickettsia-like endosymbiont Montezuma	3.6%	0
B. burgdorferi + *Rickettsia* + rickettsia-like endosymbiont Montezuma	0	1.3%

DISCUSSION

We detected for the first time the presence of *E. muris* and *A. phagocytophilum* in *I. persulcatus* ticks from Vologda Province of Russia. We also first detected infection in Vologda ticks with *Candidatus* R. tarasevichiae and the rickettsia-like endosymbiont Montezuma, thus greatly extending the western limits where these organisms have been detected in *I. persulcatus.*[6,9] The prevalence of ehrlichial infection established for *I. persulcatus* ticks during this study is comparable to that found in other endemic regions of Russia situated both west and east of Vologda Province. In particular, in Baltic regions, 9.4% of *I. persulcatus* are infected with *E. muris* and 0.6% are positive for *A. phagocytophilum*, and 53.6% of ticks harbor *Borrelia.*[2] *E. muris* was detected in 14% of ticks from the Perm Province of the southern part of the Ural region,[8] while surveillance studies conducted in Western Siberia detected *E. muris* in 3.3% to 5.3% of ticks, *Anaplasma* in 2.1% to 13.3%, and *B. burgdorferi sensu lato* in 38% of ticks.[7,10]

In other analyses for *Borrelia* performed on a cohort of ticks collected in the same area and time, 64.7% of infected ticks harbored *B. afzelii,* 19.7% *B. garinii*, and 15.6% of infected ticks harbored both spirochetes. In our cohort, mixed infections with two or more microorganisms were found in 27% and 10% ticks collected during 2002 and 2003 irrespective of combinations (TABLE 2). These data indicate that there is a possibility of coinfection with different organisms when these ticks feed on humans and this exposure may require complex treatment regimens. The disease potential of *E. muris, Candidatus* R. tarasevichiae, and the rickettsia-like symbiont Montezuma in humans are, as yet, poorly defined. Awareness of possible human cases of these diseases is important during the summer and spring months in tick-infested areas of northwest Russia much as in the other areas where *I. persulcatus* circulates.[5,18,19]

This is important because if left untreated, some of these tick-borne diseases may be debilitating or even fatal.

ACKNOWLEDGMENTS

We are thankful to Robert Massung for his advice with primer selection for nested PCR, Kim Slater for her assistance with ABI 3100 Sequencer, and Shwante Rogers, Elizabeth Bosserman, and Maria Zambrano for their assistance with DNA sequencing. Alice Oliveira was supported by the James A. Ferguson Emerging Infectious Diseases Fellowship Program, National Center for Infectious Diseases, Atlanta, GA.

REFERENCES

1. Rybakova, N.A., V.V. Sochnev, E.N. Agievich, *et al.* 2000. Epidemiological characteristics of natural zoonoses in Vologda region [in Russian]. Epidemiol. Infect. Dis. **4:** 8–11.
2. Alekseev, A., H. Dubinina, I. Van De Pol & L. Schouls. 2001. Identification of *Ehrlichia* spp. and *Borrelia burgdorferi* in *Ixodes* ticks in the Baltic regions of Russia. J. Clin. Microbiol. **39:** 2237–2242.
3. Beklemishev, A., A. Dobrotvorsky, A. Piterina, *et al.* 2003. Detection and typing of *Borrelia burgdorferi* sensu lato genospecies in *Ixodes persulcatus* ticks in West Siberia, Russia. FEMS Microbiol. Lett. **227:** 157–161.
4. Korenberg, E.I. 1999. Ehrlichioses–a new infectious pathology problem for Russia [in Russian]. Med. Parazitol. I **4:** 10–16.
5. Korenberg, E., L. Gorban, Y. Kovalevskii, *et al.* 2001. Risk for human tick-borne encephalitis, borrelioses, and double infection in the pre-Ural region of Russia. Emerg. Infect. Dis. **7:** 459–462.
6. Mediannikov, O.Yu., L.I. Ivanov, M. Nishikawa, *et al.* 2004. Microorganism "Montezuma" of the order Rickettsiales: the potential causative agent of tick-borne disease in the Far East of Russia [in Russian]. Zh. Mikrobiol. Epidemiol. Immunobiol. **1:** 7–13.
7. Morozova, O.V., A.K. Dobrotvorsky, N.N. Livanova, *et al.* 2002. PCR detection of *Borrelia burgdorferi* sensu lato, tick-borne encephalitis virus, and the human granulocytic ehrlichiosis agent in *Ixodes persulcatus* ticks from Western Siberia, Russia. J. Clin. Microbiol. **40:** 3802–3804.
8. Ravyn, M.D., E.I. Korenberg, J.A. Oeding, *et al.* 1999. Monocytic *Ehrlichia* in *Ixodes persulcatus* ticks from Perm, Russia. Lancet **353:** 722–723.
9. Shpynov, S.N., P.-E. Fournier, N.V. Rudakov & D. Raoult. 2003. "Candidatus Rickettsia tarasevichiae" in *Ixodes persulcatus* ticks collected in Russia. Ann. N.Y. Acad. Sci. **990:** 162–172.
10. Shpynov, S.N., N.V. Rudakov, V.K. Yastrebov, *et al.* 2004. New evidence for the detection of *Ehrlichia* and *Anaplasma* in Ixodes ticks in Russia and Khazakhstan. Med. Parazitol. (Moscow) **2:** 10–14.
11. Shpynov, A., P.-E. Fournier, N. Rudakov, *et al.* 2004. Detection of a rickettsia closely related to *Rickettsia aeschlimannii*, "*Rickettsia heilongjiangensis,*" *Rickettsia* sp. strain RpA4, and *Ehrlichia muris* in ticks collected in Russia and Kazakhstan. J. Clin. Microbiol. **42:** 2221–2223.

12. SIDELNIKOV, YU. N., O.YU. MEDIANNIKOV, L.I. IVANOV & N.I. ZDANOVSKAYA. 2003. First case of human granulocytic ehrlichiosis from the Far East of Russian Federation [in Russian]. Klin. Med. **2:** 67–68.
13. REGNERY, R.L., C.L. SPRUILL & B.D. PLIKAYTIS. 1991. Genotypic identification of rickettsiae and estimation of intraspecies sequence divergence for portions of two rickettsial genes. J. Bacteriol. **173:** 1576–1589.
14. ROBINSON, J.B., N.K. TOKAREVICH, M.E. EREMEEVA & G.A. DASCH. 2003. Detection and molecular characterization of rickettsiae present in *Ixodes persulcatus* Schulze from the Vologda region of Russia. Entomological Society of America. Cincinnati, OH, October 26–29, 2003. Abstract 0475.
15. LI, J.S., E. YAGER, M. REILLY, *et al.* 2001. Outer membrane protein-specific monoclonal antibodies protect SCID mice from fatal infection by the obligate intracellular bacterial pathogen *Ehrlichia chaffeensis*. J. Immunol. **166:** 1855–1862.
16. COURTNEY, J.W., L.M. KOSTELNIK, N.S. ZEIDNER & R.F. MASSUNG. 2004. Multiplex real-time PCR for detection of *Anaplasma phagocytophilum* and *Borrelia burgdorferi*. J. Clin. Microbiol. **42:** 3164–3168.
17. MASSUNG, R.F., K. SLATER, J.H. OWENS, *et al.* 1998. Nested PCR assay for detection of granulocytic ehrlichiae. J. Clin. Microbiol. **36:** 1090–1095.
18. SEMENOV, A.V., A.N. ALEXEEV, E.V. DUBININA, *et al.* 2001. Detection of genotypical heterogeneity of *Ixodes persulcatus* Schulze (Acari: Ixodidae) population in the north-west region of Russia and specific features of the distribution of tick-borne pathogens of Lyme disease and ehrlichia infections in different genotypes. Med. Parazitol. [in Russian] **3:** 11–15.
19. VOROBYEVA, N.N., E.I. KORENBERG & Y.V. GRYGORYAN. 2002. Diagnosis of tick-borne diseases in the endemic region of Russia. Wien. Klin. Wochenschr. **114:** 610–612.

Ecology and Molecular Epidemiology of Tick-Borne Rickettsioses and Anaplasmoses with Natural Foci in Russia and Kazakhstan

NIKOLAY RUDAKOV,[a] STANISLAV SHPYNOV,[a]
PIERRE-EDOUARD FOURNIER,[b] AND DIDIER RAOULT[b]

[a]*Omsk Academy of Medicine and Omsk Institute of Natural Foci Infections, Omsk, Russia*

[b]*Université de la Mediterranée, Marseille, France*

ABSTRACT: During our more than 20 years of monitoring, we have used epidemiological, field, and experimental methods for characterization of natural foci of tick-borne rickettsioses in Russia. The main results were obtained through genetic methods (PCR sequence) at the Université de la Mediterranée (Marseille, France). We describe considerable heterogeneity of tick-borne α_1-proteobacteria: 16 microorganisms the of the order Rickettsiales were detected in Russia and Kazakhstan. *R. sibirica*–caused North Asiatic tick-borne rickettsiosis is the main tick-borne rickettsiosis in Russia, with wide distribution in Siberia and the Russian Far East and high epidemic activity of natural foci of different landscape types. Our results show circulation of different pathogenic rickettsiae in the same endemic territories. In the Far East region, *R. sibirica* subsp. *R. sibirica*, *R. sibirica* subsp. *BJ*-90, and *R. heilongjiangensis* were detected; in the Altay and Krasnojarsk regions, *R. sibirica* subsp. *R. sibirica* and *R. heilongjiangensis*; and in the Kurgan district of West Siberia, *R. sibirica* subsp. *R. sibirica* and *R. slovaca*. The roles of more than 15 new genotypes of α_1-proteobacteria in infectious disease in Russia and Kazakhstan are in need of further study.

KEYWORDS: ecology; epidemiology; rickettsiae; rickettsioses; Russia

Three rickettsioses of the spotted fever group (SFG) are recognized in Russia. Two of them are officially registered: (1) *R. sibirica* "Tick-borne rickettsiosis" (TBR) or North Asiatic tick-borne rickettsiosis (NATR) and (2) Astrakhan spotted fever (*R. conorii* complex), which is found mainly in the Astrakhan region.[1] The third tick-borne rickettsiosis, caused by *R. heilongjiangensis*,

Address for correspondence: Nikolay Rudakov, Omsk Research Institute of Natural Foci Infections, 644080, prospect Mira, 7 Omsk, Russia. Voice: 007(3812)65-04-88; fax: 007(3812)65-14-77.
e-mail: rickettsia@mail.ru

**Ann. N.Y. Acad. Sci. 1078: 299–304 (2006). © 2006 New York Academy of Sciences.
doi: 10.1196/annals.1374.009**

was detected in the Chabarovsk region of the Russian Far East.[2] The clinical characteristics of this rickettsiosis are NATR-like.

Not only these three pathogenic rickettsiae had the potential to cause recognized rickettsioses in Russia: *R. sibirica*, Astrakhan spotted fever rickettsia, and *R. heilongjiangensis* were also detected in ticks.

Other pathogenic rickettsiae were detected in the North Asiatic foci, and these include *R. slovaca* (*D. marginatus*), *R. sibirica* subsp.*BJ-90* (*D. silvarum*), *R. heilongjiangensis* (*H. concinna, D. nuttallii*), *R. helvetica* (*Ixodes persulcatus*), and *R. aeschlimannii* (*Haemophisalis punctata*). Our observations over 20 years showed no increase in the proportion of ticks containing pathogenic SFG rickettsiae in the different landscape types of the NATR foci. Different SFG rickettsiae with unknown pathogenicity were detected in ticks in Russia and Kazakhstan; these were *R. tarasevichiae*, *R.* sp. *RpA4*, *R.* sp. *DnS14*, *R.* sp. *DnS28*.

More than 61,000 TBR cases were detected in Russia by 1936. Active foci of NATR are spreading widely in the Asiatic part of Russia (comprising 18 administrative territories in Siberia and the Far East), North and East Kazakhstan, North China, and Mongolia. Varying epidemic activity of foci over different time periods were consistent with natural cycles. In recent years, Siberian tick typhus (STT) infections are re-emerging in Russia, as evidenced by a tenfold increase in morbidity since 1979. This highly significant increase in NATR morbidity was registered in the 1980s and 1990s in West Siberia (Altai region mainly) and the Russian Far East,[3,4] where a high morbidity level has been maintained. The major indexes of morbidity in West Siberia were detected in the Altay region (more than 50 per 100,000), in East Siberia, that is, in the Chakassia, Tuva, and Ust-Ordyn areas of Buratya (21–55 per 100,000), in the Far East, specifically in the Chabarovsk and Primorye regions (more than 10 per 100,000). The maximum number of deaths (3460 cases) occurred in 2001, at which time 64.7% of the cases were registered in West Siberia, 21.8% in East Siberia, and 13.5% in the Russian Far East. More than 80% of the infections are detected in the Altay and Krasnojarsk regions, but about 50% of the deaths occur in the Altay region of West Siberia alone.

The nosoarea of STT is correlated with the distribution of *R. sibirica* which extends from the Kurgan region on the west to the Primorje region of the Far East. Strains of *R. sibirica* were isolated and genetically identified from different species of tick vectors (*Dermacentor nuttallii, D. marginatus, D. silvarum, Haemophysalis concinna, Ixodes persulcatus*) in different parts of the NATR nosoarea in West and East Siberia and the Russian Far East, as well as from blood samples from patients diagnosed with NATR. The main hosts of *R. sibirica* are ticks of the genera *Dermacentor* (*D. nuttallii, D. marginatus, D. silvarum*) and *Haemaphysalis* (*H. concinna*). Strains of *R. sibirica* dominate among pathogenic strains isolated on guinea pig models. NATR's *Dermacentor* steppe and forest/steppe foci are populated with three subtypes of the major vector species (*D. nuttallii, D. marginatus, D. silvarum*), which are the most widely spread among the vector species and are of the highest epidemic importance. Morbidity indexes are 50–200 per 100,000 in the mountainous and forest/steppe foci favored by *D. nuttallii*. The distribution of *Dermacentor* depends mainly on climatic conditions (temperature and humidity), which in turn have a zonal character.

The main epidemiological characteristics of NATR are a seasonal peak of morbidity, a correlation between bites of specific tick vectors and endemic territories, a correlation of morbidity with age and occupation based on the intensity of contact of various professional and social groups of the regional human population with the landscape types where the natural foci of infection were found.

Seasonal increases in NATR morbidity were dependent on the specific biological cycles of the tick vectors and the climatic conditions of the endemic territories. A significant increase in morbidity in spring and early summer with a peak in May and a low rate in the autumn are specific for steppe and forest/steppe foci with *Dermacentor* ticks as vectors (main NATR foci in West and East Siberia, the Russian Far East). Territorial seasonal differences may be connected with level of snow cover and temperature at snow thaw. The greatest number of early cases of NATR are registered in mountain/steppe foci in the Altai region, which is inhabited by *Dermacentor nuttallii*. The autumn infections were determined to be the result of new generations of *Dermacentor imago*. Some of summer NATR cases may be caused by bites of infected nymphs of *D. nutallii* and *H. concinna*.

The seasonal activity of *H. concinna imago* differed from that of *Dermacentor* ticks in that it resulted in a summer peak of diseases, with morbidity occurring mainly among adults whose occupations brought them in contact with forested areas. The relict distribution of *H. concinna* in postglacial Eurasia is reflected in the "spotted" mapping of the territories of these ticks in different parts of the NATR foci in Altai, Krasnojarsk, Kemerovo, and other territories of Siberia, mainly in the Russian Far East. *H. concinna* ticks occupy bushes and boggy forests, often in association with *D. silvarum* (in its preferred steppe-forest sites) and *Ixodes persulcatus* (in its preferred forest sites). *R. heilongjiangensis* was detected in *H. concinna* ticks in several parts of the NATR nosoarea: in ticks from Russian Far East (Primorije region) and in the most epidemiologically active NATR foci in the Altay (*H. concinna*) and Krasnojarsk (*H. concinna* and *D. nuttallii*) regions.[5,6] The main seasonal differences in morbidity (a peak in July) are detected in the southern districts of the Chabarovsk region, where *H. concinna* are dominant. *R. heilongjiangensis* was identified through genotyping in *H. concinna* ticks and biomaterials from patients in these territories.[2]

R. heilongjiangensis is not a unique pathogenic rickettsia for the vector *H. concinna* in Siberia and the Russian Far East. Strains of *R. sibirica* subsp. *sibirica* were isolated from *H. concinna* ticks in the Altai (Siberia) and Primorije (Far East) regions. So, *R. heilongjiangensis* TBR is not the only Far East TBR.

R. slovaca was identified in *D. marginatus* ticks in the Voronezh and Stavronol regions of the European part of Russia. One strain of *R. slovaca* was isolated in our laboratory from *D. marginatus* ticks in the Kurgan district of West Siberia (nosoarea of NATR) by M.S. Shaiman in 1969 and was identified as *R. slovaca* 34 years later by S. Shpynov *et al.* in Marseille.[7] Cases of TIBOLA were not diagnosed in Russia until recently.

R. aeschlimannii was identified by genotyping in *H. punctata* taken from the Alma-Ata district of Kazakhstan. In the most recent periods, cases of "NATR" were registered in the same territories.[6] A rickettsial strain closely related to *R. helvetica* was detected in taiga ticks in the Omsk district.

Strain Primorije32/84 *R. sibirica* subsp. *BJ-90* was isolated by T. Reshetnikova in our laboratory from *D. silvarum* ticks in 1990, six years before the Chinese strains were identified. Our results show circulation of different pathogenic rickettsiae in the same NATR endemic territories. In the Far East region *R. sibirica* subsp. *R. sibirica*, *R. sibirica* subsp. *BJ*-90 and *R. heilongjiangensis* were detected; in the Altay and Krasnojarsk regions, *R. sibirica* subsp. *R. sibirica* and *R. heilongjiangensis*; and in the Kurgan district of West Siberia, *R. sibirica* subsp. *R. sibirica*, and *R. slovaca*.

A new rickettsial species—*R. tarasevichiae*, named in honor of the well-known Russian rickettsiologist academician I.V. Tarasevich—was widely detected in *Ixodes persulcatus* in Russia.[8] Three new rickettsiae closely related to *R. massiliae* (*R.* sp. *RpA4, R.* sp. *DnS14, R.* sp. *DnS28*), first detected by Rydkina *et al.*[9] in the Astrachan and Altay regions were revealed in ticks of the genus *Dermacentor*, mainly in foci of the STT and STT-free territories of Russia and Kazakhstan, but the pathogenicity of these new genotypes is so far unknown. Strains of these genotypes were isolated in our laboratory by I. Samoilenko *et al.*[10] A new genotype (86% homology of *omp*A gene with *R.* sp. *AT-1*) was identified in *I. persulcatus* in the Altay and Krasnojarsk regions.

Analysis of the distribution of SFG rickettsiae show their close ecological connection with particular species of vectors. *R. slovaca* and *R.* sp. *RpA4* coexist in the tick species *D. marginatus* and *D. reticulatus* (the western part of geographic range for *Dermacentor* in Eurasia), and *R.* sp. *DnS14* and *R. sibirica* coexist in *D. nuttallii* and *D. silvarum* (eastern part of the geographic range). *D. marginatus* and *D. reticulatus* are both vector and reservoir for *R. sibirica* in regions with the greatest number of species of the genus *Dermacentor* (the southern part of Western Siberia). *D. silvarum* is known not only as a vector for *R. sibirica*, but also for *R.* sp. *DnS14*, which has been found by genotyping in Buryatiya and *R. heilongjiangii* in the Altay, Krasnoyarsk, and Primoije regions of Russia. And only the SFG genotype of *R.* sp. *DnS28* may now be bound ecologically with a single species of ticks, *D. nuttallii.*

New data about distribution of Anaplasmataceae was compiled in Russia. The morbidity of HGE was detected by means of serologic methods (IFA, ELISA) in the Altay and Novosibirsk regions. *Ehrlichia muris* was detected in the Asiatic part of Russia,[6] The Schotti variant was detected in the Omsk region in *I. persulcatus*. *Anaplasma phagocytophila* was identified in the same species of ticks in the Altay and Primorije regions; *A. bovis* was identified in *H. concinna* in Primorije.

CONCLUSIONS

Heterogeneity and coexistence of different species of Rickettsiales in the same ticks and territories may influence the epidemic activity of natural foci and the registration of morbidity. The close similarity of the clinical and epidemiological manifestations of tick-borne infections caused by Rickettsiales means that identification of the infective agent and differentiation from the many other possible species is a difficult task.

R. sibirica NATR is the main TBR in Russia, with a wide distribution in Siberia and the Russian Far East and high epidemic activity of natural foci within different landscape types. *R. heilongjiangensis* TBR is not "Far East" TBR because: (i) *R. heilongjiangensis* was detected in classic NATR foci in the Altai and Krasnojarsk regions of Siberia with high levels of NATR morbidity; (ii) *R. sibirica* subsp. *sibirica* and *R. sibirica* subsp. *BJ-90* strains were isolated in the Russian Far East; (iii) seasonal cycles and age and occupational patterns of morbidity in TBR foci are dependent on the biological activity of different species of vectors; *R. sibirica* and *R. heilongjiangensis* were detected in *H. concinna* both in Siberia and the Russian Far East. We speculate that there is a possibiluty that NATRs are hyperdiagnosed because of the presence of infections caused by other pathogenic Rickettsiales, mainly *R. heilongjiangensis* and *Anaplasma phagocytophila*.

The roles of more than 15 new genotypes of α_1-proteobacteria in infectious disease and verification of diagnoses of tick-borne infections (including NATR and anaplasmoses) in Russia and Kazakhstan are in need of further study.

REFERENCES

1. Tarasevich, I.V., V.A. Makarova, N.F. Fetisova, *et al.* 1991. Studies of a "new" rickettsiosis "Astraktan" spotted fever. Eur. J. Epidemiol. **7:** 294–298.
2. Mediannikov O., Y. Sidelnikov, E. Ivanov, *et al.* 2004. Acute tick-borne rickettsiosis caused by *Rickettsia heilongjiangensis* in Russian Far East. Emerg. Infect. Dis. **10:** 810–817.
3. Rudakov, N.V. 1996. Tick- borne rickettsiosis in Russia (epidemiology and current conditions of natural foci). *In* Rickettsiae and Rickettsial Diseases: proceedings of the Vth International Symposium. Bratislava: VEDA, pp. 216–219.
4. Rudakov, N.V., I.V. Samoilenko, V.V. Yakimenko, *et al.* 1999. The re-emerging of Siberian tick typhus: field and experimental observations. *In* D. Raoult & P. Brougui, Eds.: 269–273. Rickettsiae and Rickettsial Diseases at the Turn of the Third Millenium. Elsevier, Paris.

5. SHPYNOV S., N. RUDAKOV, V. JASTREBOV, *et al.* 2003. First detection of *Rickettsia heilongjangensis* in *Haemaphisalis concinna* ticks in Russia. ZNISO, Moscow. **12:** 16–20.
6. SHPYNOV S., P.-E. FOURNIER, N. RUDAKOV, *et al.* 2004. Detection of Rickettsia closely related to *Rickettsia aeschlimannii*, "*Rickettsia heilongjiangensis*," Rickettsia sp. Strain RpA4, and *Ehrlichia muris* in ticks collected in Russia and Kazakhstan. J. Clin. Microbiol. **42:** 2221–2223.
7. SHPYNOV S., N. RUDAKOV, I. SAMOYLENKO, *et al.* 2004. Genetic identification of rickettsiae of the tick-borne spotted fever group, isolated in the foci of tick-borne rickettsiosis. J. Microbiol. Moscow. **5:** 43–48.
8. SHPYNOV, S., P.-E. FOURNIER, N. RUDAKOV & D. RAOULT 2003. "*Candidatus Rickettsia tarasevichiae*" in *Ixodes persulcatus* ticks collected in Russia. Rickettsiology: present and future directions. Ann. N.Y. Acad. Sci. **990:** 162–172.
9. RYDKINA, E., V. ROUX, N. RUDAKOV, *et al.* 1999. New Rickettsiae in ticks collected in territories of the former Soviet Union. Emerg. Infect. Dis. **5:** 811–814.
10. SAMOYLENKO, I.E., L.V. KUMPAN, S.N. SHPYNOV, *et al.* 2005. Methods of isolation and cultivation of rickettsiae of "new genotypes" from nosoarea of the North Asian tick typhus in Siberia. Fourth International Conference on Rickettsiae and Rickettsial Diseases [book of abstracts]. Logrono (La Rioja), Spain.

Prevalence of *Rickettsia felis* in *Ctenocephalides felis* and *Ctenocephalides canis* from Uruguay

JOSÉ M. VENZAL,[a] LAURA PÉREZ-MARTÍNEZ,[b] MARIA L. FÉLIX,[a] ARÁNZAZU PORTILLO,[b] JOSÉ R. BLANCO,[b] AND JOSÉ A. OTEO[b]

[a]*Departamento de Parasitología Veterinaria, Facultad de Veterinaria, Universidad de la República, Montevideo, Uruguay*

[b]*Área de Enfermedades Infecciosas, Hospitales San Millán-San Pedro-de La Rioja, Logroño (La Rioja), Spain*

ABSTRACT: Our aim was to determine the presence of *Rickettsia* spp. in 66 fleas from Uruguay. Rickettsial DNA was amplified using *gltA* and *ompB* PCR primers. *Rickettsia* spp. were found in 41% of the fleas (25 *Ctenocephalides felis* and 2 *Ctenocephlides canis*). Sequences resulted in the identification of *Rickettsia felis* and four genotypes closely related to this species (*Rickettsia* sp. TwKM03, California 2, Hf187, and RF2125). The presence of *R. felis* in fleas from Uruguay in was demonstrated. This is the second species of *Rickettsia* identified in Uruguay in the past 2 years using molecular approaches, and it is helping to clarify the etiology of rickettsial diseases in the region.

KEYWORDS: *Rickettsia felis*; *Ctenocephalides felis*; *Ctenocephalides canis*; Uruguay

INTRODUCTION

Rickettsial diseases are zoonoses caused by bacteria grouped in the order Rickettsiales. The classification within *Rickettsiales* is continually being modified as new data become available, particularly those based on molecular phylogenetic studies.[1] Among the principal human rickettsial diseases are those caused by pathogens in the genus *Rickettsia*, recognized as the spotted fever group (SFG) and the typhus group rickettsiae. In addition to the well-described pathogens of the genus *Rickettsia*, multiple species have been isolated from the invertebrate vectors and have been considered "nonpathogenic" *Rickettsiae*. However, with the development of improved diagnostic tools, several new

Address for correspondence: José Manuel Venzal, Departamento de Parasitología Veterinaria, Facultad de Veterinaria, Universidad de La República, Av. Alberto Lasplaces 1550, 11600 Montevideo, Uruguay. Fax: +598-2-628-0130.
e-mail: dpvuru@hotmail.com

**Ann. N.Y. Acad. Sci. 1078: 305–308 (2006). © 2006 New York Academy of Sciences.
doi: 10.1196/annals.1374.056**

species of *Rickettsia* have been demonstrated to be pathogens of animals and humans. In the past 20 years, nine new species or subspecies of tick-borne spotted fever rickettsiae have been identified as emerging pathogens throughout the world.[1]

In Uruguay, the initial cases of rickettsioses were identified in 1990, and reported as being caused by *Rickettsia conorii*, the agent used as the antigen in microimmunofluorescence testing.[2] In 2001, new cases of rickettsial diseases were confirmed and *Amblyomma triste* was implicated as the likely vector.[3] Conti-Díaz later confirmed *A. triste* as the vector, and proposed that the pathogen could be a species of *Rickettsia* different from *R. conorii.*[4] In 2004, *Rickettsia parkeri* was detected in *A. triste* from Uruguay, which suggested that this species could be responsible for at least some of the human rickettsioses in Uruguay.[5] Clearly, the full complement of rickettsial pathogens present and being transmitted within Uruguay has not been defined.

In addition to ticks, fleas should be considered as potential vectors of rickettsiae within Uruguay. Fleas have a worldwide distribution and in the last few years they have been associated with emerging human infections, including flea-borne spotted fever due to *R. felis.* The presence of *R. felis* has been demonstrated in fleas from numerous countries and flea-borne spotted fever is an emerging human disease with confirmed cases in Texas, Mexico, France, Brazil, Germany, and Thailand.[6–10] The aim of this study was to search for the presence of *Rickettsia* spp. in *Ctenocephalides felis* and *Ctenocephalides canis* from Uruguay.

MATERIAL AND METHODS

From April to September 2004, a total of 66 fleas (62 *C. felis* and 4 *C. canis*) were collected from 15 cats and dogs in the county of Montevideo (Uruguay). All fleas were analyzed individually. DNA was extracted by lysis with ammonium hydroxide.[11,12] *gltA* and *ompB* PCR assays for the detection of *Rickettsia* spp. were carried out as previously reported.[13,14] Amplicon products were subsequently sequenced at Universidad de Alcalá de Henares (Madrid) to establish species strain identification. Sequences were compared with those available at GenBank using the BLAST utility.[15]

RESULTS AND DISCUSSION

Rickettsia spp. was identified in 41% collected fleas (25 *C. felis* and 2 *C. canis*). Sequences of the rickettsial amplicon results were 100% identical to *R. felis* and to four genotypes closely related to this species: *Rickettsia* sp. strains TwKM03, California 2, Hf187, and RF2125. Curiously, *Rickettsia* sp. strains California 2 and RF2125 have been previously detected in fleas

(GenBank accession nos. AF210695 and AF516333, respectively). The presence of *R. felis* in *C. canis* has been reported only in Perú, Thailand, and very recently, in Spain and Brazil.[16–19] This study demonstrates the presence of *R. felis* in fleas from Uruguay. The high prevalence suggests the likelihood of exposure of both animals and humans. These bacteria should be considered as potential causes of fever of intermediate duration with no diagnostic orientation in patients from this country.

ACKNOWLEDGMENTS

We are grateful to Dr. Guy Palmer for critical reading of the manuscript.

This study was supported in part by grants from the Fondo de Investigación Sanitaria (FIS G03/057), Ministerio de Sanidad y Consumo, (Spain) and also from the Government of La Rioja (ANGI 2004/17; Plan Riojano I+D+I, 2003-2007).

REFERENCES

1. Parola, P., B. Davoust & D. Raoult. 2005. Tick- and flea-borne rickettsial emerging zoonoses. Vet. Res. **36:** 469–492.
2. Conti Díaz, I.A., I. Rubio, R.E. Somma Moreira, *et al.* 1990. Rickettsiosis cutáneo-ganglionar por *Rickettsia conorii* en el Uruguay. Rev. Inst. Med. Trop. (Sao Paulo) **22:** 313–318.
3. Conti Díaz, I.A. 2001. Rickettsiosis por *Rickettsia conorii* (fiebre botonosa del Mediterráneo o fiebre de Marsella). Estado actual en Uruguay. Rev. Med. Uruguay **17:** 119–124.
4. Conti-Díaz, I.A. 2003. Rickettsiosis caused by *Rickettsia conorii* in Uruguay. Ann. N. Y. Acad. Sci. **990:** 264–266.
5. Venzal, J.M., A. Portillo, A. Estrada-Peña, *et al.* 2004. *Rickettsia parkeri* in *Amblyomma triste* from Uruguay. Emerg. Infect. Dis. **10:** 1493–1495.
6. Schriefer, M.E., J.B. Sacci Jr., J.S. Dumler, *et al.* 1994. Identification of a novel rickettsial infection in a patient diagnosed with murine typhus. J. Clin. Microbiol. **32:** 949–954.
7. Zavala-Velázquez, J.E., J. A. Ruiz-Sosa, R.A. Sánchez-Elías, *et al.* 2000. *Rickettsia felis* rickettsiosis in Yucatan. Lancet **356:** 1079–1080.
8. Raoult, D., B. La Scola, M. Enea, *et al.* 2001. A flea-associated *Rickettsia* pathogenic for humans. Emerg. Infect. Dis. **7:** 73–81.
9. Richter, J., P.E. Fournier, J. Petridou, *et al.* 2002. *Rickettsia felis* infection acquired in Europe and documented by polymerase chain reaction. Emerg. Infect. Dis. **8:** 207–208.
10. Parola, P., R.S. Miller, P. Mc Daniel, *et al.* 2003. Emerging rickettsioses of the Thai-Myanmar border. Emerg. Infect. Dis. **9:** 592–595.
11. Guy, E.C. & G. Stanek. 1991. Detection of *Borrelia burgdorferi* in patients with Lyme disease by the polymerase chain reaction. J. Clin. Pathol. **44:** 610–611.

12. RIJPKEMA, S., D. GOLUBIC, M. MOLKENBOER, *et al.* 1996. Identification of four genomic groups of *Borrelia burgdorferi* sensu lato in *Ixodes ricinus* ticks collected in Lyme borreliosis endemic region of northern Croatia. Exp. Appl. Acarol. **20:** 23–30.
13. ROUX, V., E. RYDKINA, M. EREMEEVA, *et al.* 1997. Citrate synthase gene comparison, a new tool for phylogenetic analysis, and its application for the rickettsiae. Int. J. Syst. Bact. **47:** 252–261.
14. MÁRQUEZ, F. J., M. A. MUNIAIN, J.M. PÉREZ, *et al.* 2002. Presence of *Rickettsia felis* in the cat flea from South-Western Europe. Emerg. Infect. Dis. **8:** 89–91.
15. http://www.ncbi.nlm.nih.gov
16. PACHAS, P.E., C. MORON C.A. HOYOS, *et al.* 2001. *Rickettsia felis* identified in *Ctenocephalides canis* fleas from Peruvian Andes. Presented at the ASR/*Bartonella* Joint Conference. Big Sky, Montana, USA.
17. PAROLA, P., O.Y. SANOGO, K. LERDTHUSNEE, *et al.* 2003. Identification of *Rickettsia* spp. and *Bartonella* spp. in fleas from the Thai-Myanmar border. Ann. N. Y. Acad. Sci. **990:** 173–181.
18. BLANCO, J.R., L. PÉREZ-MARTÍNEZ, M. VALLEJO, *et al.* 2005. Prevalence of *Rickettsia felis*-like and *Bartonella* spp. in *Ctenocephalides felis* and *Ctenocephalides canis* from La Rioja (Northern Spain). Ann. N. Y. Acad. Sci. This volume.
19. HORTA, M., D.P. CHIEBAO, P.M. OHARA, *et al.* 2005. Prevalence of *Rickettsia felis* in the fleas *Ctenocephalides felis felis* and *Ctenocephalides canis* from two Indian communities in Sao Paolo municipality, Brazil. Presented at the Fourth International Conference on Rickettsiae and Rickettsial Diseases. Logroño, Spain, June 18.

Highly Variable Year-to-Year Prevalence of *Anaplasma phagocytophilum* in *Ixodes ricinus* Ticks in Northeastern Poland: A 4-Year Follow-up

ANNA GRZESZCZUK[a] AND JOANNA STAŃCZAK[b]

[a]*Medical University of Białystok, Department of Infectious Diseases, 15-540 Białystok, Poland*

[b]*Medical University of Gdańsk, Institute of Maritime and Tropical Medicine, 81-519 Gdynia, Poland*

ABSTRACT: *Anaplasma phagocytophilum* is transmitted mainly by *Ixodes ricinus* ticks in Europe. We followed-up *A. phagocytophilum* infection rate in *I. ricinus* in three selected collection sites in northeastern Poland during a four-year period. Overall infection rate was 14.1% (208/1474) with highest infection rate among females (36.8% verses males 8.2% and nymphs 0.9%). We noted a very big year-to-year variation of infection prevalence in each collection cite every year reflecting changeable granulocytic anaplasmosis risk for humans and animals.

KEYWORDS: *Anaplasma phagocytophilum*; *Ixodes ricinus*; host-seeking ticks; Poland

Northeastern Poland is an endemic region for tick-borne diseases, where Lyme disease and tick-borne encephalitis morbidity is 5 to 10 times higher than the one for the whole country (48.2 verses 10/100,000 and 9.4 verses 0.69/100,000, respectively, 1994).[1] The current study aimed to evaluate the infection rate of *Anaplasma phagocytophilum* in *Ixodes ricinus* in springs, and to compare the prevalence of *A. phagocytophilum* in particular collection sites during a 4-year follow-up.

We studied 1474 host-seeking *I. ricinus* (449 females, 461 males, and 910 nymphs) collected by flagging lower vegetation in three different study sites in the Bialowieża Primeval Forest and the Knyszyn Primeval Forest from 2000 to 2004. Polymerase chain reaction (PCR) was performed with the primers EHR 521 and EHR 747. The identity of selected ($n = 40$) amplicons was confirmed by independent DNA sequencing.

Address for correspondence: Anna Grzeszczuk, Medical University of Białystok, Department of Infectious Diseases, 14 Zurawia str., 15-540 Białystok, Poland. Voice: 48-85-7-409 480; fax 48-85-7-416-421.

e-mail: oliwa@amb.edu.pl

Ann. N.Y. Acad. Sci. 1078: 309–311 (2006). © 2006 New York Academy of Sciences.
doi: 10.1196/annals.1374.057

TABLE 1. Prevalence of *A. phagocytophilum* in questing *I. ricinus* ticks collected from different study sites in northeastern Poland. A 4-year follow-up

		No positive *I. ricinus* / No tested (% positive)				
Collection site	Year	Female	Male	Adults	Nymphs	Total
Bialowieża	2000	70/153 (45.6)[1]*	14/137(8.9)*	84/290 (29.0)[o]	0/30[o]*	84/320 (26.3)
Palace Park	2001	41/111 (36.9)[1]*	6/114 (5.3)*	47/225 (20.9)[o]	0/45[o]*	47/270 (17.4)
	2002	nt	nt	nt	nt	nt
	2003	-	1/1(nc)	1/1(nc)	-	1/1 (nc)
	2004	0/14[1]	1/20 (5.0)	1/34 (2.9)	0 /12	1/46 (2.2)
Subtotal		**111/278 (39.9)**	**22/272 (8.0)**	**133/550 (24.2)**	**0/87 (0.0)**	**133/637 (20.9)**
Królowy Most	2001	9/20 (45.0)[1]*	1/22 (4.5)[1]*	10/42 (23.8)[o]	0/21[o]*	10/63 (15.9)
	2002	3/4 (nc)[1]	5/7 (nc)[1]	8/11 (72.2)[o]	0/5[o]*	8/18 (50.0)
	2003	0/9[1]	0/10[1]	0/19	1/21(4.8)	1/40 (2.5)
	2004	22/30 (73.3)[1]*	4/30 (13.3)[1]*	26/60 (43.3)[o]	1/59(1.7)[o]*	27/119(22.7)
Subtotal		**34/63 (54.0)**	**10/69 (14.5)**	**44/132 (33.3)**	**2/106 (1.9)**	**46/238 (19.3)**
Supraśl	2001	6/18 (33.3)[1]*	0/21[1]*	6/39 (15.4)[o]	1/50 (2.0)[o]*	7/89 (7.9)
- Pólko	2002	8/13 (61.5)[1]	4/20 (20)[1]	12/33 (36.4)[o]	1/20 (5.0)[o]*	13/53 (24.5)
	2003	4/61 (6.6)[1]	2/58 (3.5)[1]	6/119 (5.0)[o]	1/246 (0.4)[o]*	7/365 (1.9)
	2004	2/16 (12.5)[1]*	0/21[1]*	2/37 (5.4)[o]	0/55	2/92 (2.2)
Subtotal		**20/108 (18.5)**	**6/120 (5.0)**	**26/228 (11.4)**	**3/371 (0.8)**	**29/599 (4.8)**
Total		**165/449* (36.8)**	**38/461* (8.2)**	**203/910[o] (22.3)**	**5/564[o]* (0.9)**	**208/1447 (14.1)**

Nt: not tested; nc: not calculated (No < 10).
*difference statistically significant between females and males; or females and nymphs (χ^2 or Fisher's exact test).
[1]difference statistically significant between years within one sex in particular collection site (χ^2or Fisher's exact test).
[o]difference statistically significant between adults and nymphs (χ^2 or Fisher's exact test).

To our knowledge, this is the first reported 4-year extensive survey across multiple recreational areas in Poland. DNA of *A. phagocytophilum* was detected in a total of 208 (14.1%) out of 1474 ticks investigated (TABLE 1). The overall prevalence was significantly lower in nymphs (0.8%; 5/564) than in adults (25.2%; 203/910), and in males (8.2%; 38/461) than in females (36.7%; 165/449). Such differences were observed every year. Prevalence among locations ranged from 4.8% to 20.9%. We noted significant year-to-year infection rate changes in all localities ranging, for example, from 6.6% up to maximum 73.3% in females and from 0% up to 20% in males. The overall *A. phagocytophilum* prevalence revealed in the present study (14.1%) is in agreement with the infection rates noted in northern (14%) and mideastern Poland (13.1%),[2,3] and is about twice as high as in eastern Slovakia (8.3%).[4] However, all works corroborate the highest *A. phagocytophilum* prevalence in host-seeking females.

An American 3-year follow-up study demonstrated a decline from 9.5% *I. scapularis* ticks infected with *A. phagocytophilum* in 1995 to 0.5% and 0% in consecutive years in the southern Coastal Maine,[5] which was in contrast with pretty stable year-to-year *Borrelia burgdorferi* infection rates. The obtained results point out a variability in the prevalence of *A. phagocytophilum* infection in ticks, providing evidence of a considerably different risk of acquiring *A. phagocytophilum* infection after a tick bite.

REFERENCES

1. GRZESZCZUK A. *et al.* 2005. Etiology of tick-borne febrile illnesses in adult residents of north-eastern Poland. Report from a prospective clinical study. Int. J. Med. Microbiol. **296**(Suppl 1): 242–249.
2. STAŃCZAK, J. *et al.* 2004. *Ixodes ricinus* as a vector of *Borrelia burgdorferi* sensu lato, *Anaplasma phagocytophilum* and *Babesia microti* in urban and suburban forests. Ann. Agric. Environ. Med. **11:** 109–114.
3. TOMASIEWICZ K. *et al.* 2004. The risk of exposure to *Anaplasma phagocytophilum* infection in mid-eastern Poland. Ann. Agric. Environ. Med. **11:** 261–264.
4. DERDÁKOVÁ M. *et al.* 2003. Molecular evidence for *Anaplasma phagocytophilum* and *Borrelia burgdorferi* in *Ixodes ricinus* ticks from eastern Slovakia. Ann. Agric. Environ. Med. **10:** 269–271.
5. HOLMAN M.S. *et al.* 2004. *Anaplasma phagocytophilum*, Babesia *microti* and *Borrelia burgdorferi* in *Ixodes scapularis*, Southern Coastal Main. Emerg. Infect. Dis. **10:** 744–746.

Detection of *Anaplasma phagocytophilum*, *Coxiella burnetii*, *Rickettsia* spp., and *Borrelia burgdorferi* s. l. in Ticks, and Wild-Living Animals in Western and Middle Slovakia

KATARÍNA SMETANOVÁ, KATARÍNA SCHWARZOVÁ, AND ELENA KOCIANOVÁ

Institute of Virology, Slovak Academy of Sciences, Slovak Republic

ABSTRACT: In this study, three tick species (*Ixodes ricinus*, *Dermacentor marginatus*, and *D. reticulatus*), small terrestrial mammals, and game were examined by PCR for the presence of tick-borne pathogens *Anaplasma phagocytophilum*, *Coxiella burnetii*, *Rickettsia* spp., and *Borrelia burgdorferi* Sensu lato.

KEYWORDS: ticks; wild-living animals; *Anaplasma*; *Coxiella*; *Rickettsia*; *Borrelia*; tick-borne pathogens

INTRODUCTION

Tick-borne diseases represent a significant worldwidely distributed health problem. Ticks, especially *Ixodes ricinus* are known to transmit various viral, bacterial, and protozoal pathogens in Europe.[1–6] Besides the most severe tick-borne diseases, such as Lyme borreliosis and tick encephalitis, other febrile tick-transmissible illnesses are supposed to occur in Slovakia. The aim of this study was to detect *Anaplasma phagocytophilum*, *Coxiella burnetii*, *Rickettsia* spp., and *Borrelia burgdorferi* s. l. in ticks and wild-living animals.

MATERIALS AND METHODS

Ticks were collected by flagging vegetation in suburban forest and pasture habitats in 28 locales in western and middle Slovakia in 2003 and 2004. Some ticks were obtained from small domestic animals. Ticks were determined as

Address for correspondence: Katarína Smetanová, Institute of Virology, Slovak Academy of Sciences, Dúbravská cesta 9, Bratislava 842 45. Voice: 01421259302430; fax: 01421254774284.
e-mail: viruksam@savba.sk

Ann. N.Y. Acad. Sci. 1078: 312–315 (2006). © 2006 New York Academy of Sciences.
doi: 10.1196/annals.1374.058

to species and maintained at 4°C until processed. Small terrestrial mammals were captured using live traps baited with oat flakes for 4 days. Approximately 50 traps were set for each of the three selected locales in middle Slovakia, where positive ticks had been detected. Parts of spleen kept in 70% ethanol were used for the research. Spleen sample from game—18 wild boars (*Sus scrofa*), 3 red deer (*Cervus elaphus*), and 2 roe deer (*Capreolus capreolus*) obtained during the hunting season 2004/2005—were also included in the study.

Ticks were disinfected in 70% ethanol, rinsed with distilled water, and cut in sterile tubes prior to the processing. DNA from ticks and spleen was isolated using DNeasy Tissue kit (Qiagen, Germany) according to the manufacturer's recommendations PCR with specific primer sets RpCS.877p and RpCS.1258n, Ehr521 and Ehr790, CBCOE and CBCOS were used for detection of *Rickettsia* spp., *Anaplasma phagocytophilum,* and *Coxiella burnetii*, respectively.[7–9] Randomly selected DNA samples from *I. ricinus* collected on localities with high incidence of Lyme borreliosis and suspected natural focci of *B. burgdorferi* s. l. were tested for the presence of *B. burgdorferi* s.s., *B. garinii,* and *B. afzelii* using genospecies specific primer sets, GI, GII, and GIII.[10] PCR products were resolved on 1–2% agarose electrophoresis gel and visualized under UV light.

RESULTS

Three different species of ticks were found (*I. ricinus*, *Dermacentor marginatus,* and *D. reticulatus*), and *I. ricinus* was dominant. Adults of 271 *I. ricinus*, 16 *D. marginatus,* and 9 *D. reticulatus* ticks from 23 localities in middle Slovakia were analyzed. *I. ricinus* ticks (56) from five localities in western Slovakia were also analyzed. Sixty-seven trapped mice were identified as *Apodemus flavicollis* (38), *Clethrionomys glareolus* (23), *A. sylvaticus* (3), and *Microtus arvalis* (3).

I. ricinus ticks from middle Slovakia were positive for *Rickettsia* spp. (6.3%), *A. phagocytophilum* (4.4%), and *C. burnetii* (0.4%). From randomly selected 53 *I. ricinus* ticks tested for the presence of *B. burgdorferi* genospecies, 2 were found to be infected with *B. garinii* and 2 with *B. afzelii*, whereas 1 tick was coinfected with all the three detected genospecies. *Rickettsia* spp. was detected in *D. marginatus* (5 positive ticks from 16 analyzed) as well as in *D. reticulatus* (1/9) ticks.

I. ricinus ticks from western Slovakia were positive for *Rickettsia* spp. (3/56), *B. burgdorferi* s. l. (3/56), and to *C. burnetii* (1/56).

Six percent positivity to studied pathogens was detected in spleen samples of small terrestrial mammals. One *M. arvalis* male was positive to *Rickettsia* spp., one *A. flavicollis* male to *C. burnetii,* and two *A. flavicollis*, one male and one female, to *A. phagocytophilum*. Three spleen samples of game (1 wild boar, 1 roe deer, and 1 red deer) were positive to *A. phagocytophilum*. The red deer was also coinfected with *C. burnetii*.

CONCLUSION

Besides severe tick-borne diseases as Lyme borreliosis and tick encephalitis, other febrile tick-transmissible illnesses are assumed to occur in Slovakia. Monitoring and mapping of natural foci, where ticks are abundant, are extremely important and helpful for early diagnosis of Lyme disease and other suspected febrile tick-borne illnesses. Special attention should be given to residents and visitors to places with potential natural foci.

ACKNOWLEDGMENT

This work was supported by VEGA grant No. 2/4045, Grant Agency of Ministry of Education and Slovak Academy of Sciences.

REFERENCES

1. NILSSON, K. *et al.* 1999. *Rickettsia helvetica* in *Ixodes ricinus* ticks in Sweden. J. Clin. Microbiol. **37:** 400–403.
2. VON LOEWENICH, F.D. *et al.*·2003. Human granulocytic ehrlichiosis in Germany: evidence from serological studies, tick analyses, and a case of equine ehrlichiosis. Ann. N. Y. Acad. Sci. **990:** 116–117.
3. SIXL, W. *et al.* 2003. Investigation of *Anaplasma phagocytophila* infections in *Ixodes ricinus* and *Dermacentor reticulatus* ticks in Austria. Ann. N. Y. Acad. Sci. **990:** 94–97.
4. ŘEHÁČEK, J. 1984. *Rickettsia slovaca*, the organism and its ecology. Acta Sci. Nat. Acad. Sci. Bohemostov Brno **18:** 1–50.
5. SEKEYOVA, Z. *et al.* 2000. Characterization of a new spotted fever group rickettsia detected in *Ixodes ricinus* (Acari: Ixodidae) collected in Slovakia. J. Med. Entomol. **37:** 707–713.
6. GERN, L. *et al.* 1999. Genetic diversity of *Borrelia burgdorferi* sensu lato isolates obtained from *Ixodes ricinus* ticks collected in Slovakia. Eur. J. Epidemiol. **15:** 665–669.
7. REGNERY, R.L. *et al.* 1991. Genotypic identification of rickettsiae and estimation of intraspecies sequence divergence for portions of two rickettsial genes. J. Bacteriol. **173:** 1576–1589.
8. KOLBERT, C. 1996. Detection of the agent of human granulocytic ehrlichiosis by PCR. *In* PCR Protocols for Emerging Infectious Diseases. D. Persing, Ed.: 106–111. ASM Press. Washington, D.C.
9. HENDRIX, L.R. *et al.* 1993. Cloning and sequencing of *Coxiella burnetii* outer membrane protein gene com1. Infect. Immun. **61:** 470–477.

10. DEMAERSCHALCK, I. *et al.* 1995. Simultaneous presence of different *Borrelia burgdorferi* genospecies in biological fluids of Lyme disease patients. J. Clin. Microbiol. **33:** 602–608.

Prevalence of *Anaplasma phagocytophilum*, *Rickettsia* sp. and *Borrelia burgdorferi* sensu lato DNA in Questing *Ixodes ricinus* Ticks from France

LÉNAÏG HALOS,[a] GWENAËL VOURC'H,[b] VIOLAINE COTTE,[b] PATRICK GASQUI,[b] JACQUES BARNOUIN,[b] HENRI-JEAN BOULOUS,[a] AND MURIEL VAYSSIER-TAUSSAT[a]

[a]*UMR 956 BIPAR, Ecole Nationale Vétérinaire d'Alfort, 94700 Maisons-Alfort, France*

[b]*Unité d'Epidémiologie Animale, INRA, 63122 Saint-Genès Champanelle, France*

ABSTRACT: A total of 4701 *Ixodes ricinus*, collected during the summer of 2003, were analyzed for three pathogens. DNA was detected from the three pathogens. Co-detection of more than one pathogen was observed.

KEYWORDS: *Ixodes ricinus*; tick; *Borrelia burgdorferi* sensu lato; *Anaplasma phagocytophilum*; *Rickettsia* sp.; prevalence

Ixodes ricinus ticks can transmit multiple pathogenic bacteria responsible for diseases in animals and humans. In Europe, *Borrelia burgdorferi* sensu lato (sl), *Anaplasma phagocytophilum*, and spotted fever group (SFG) *Rickettsia* sp. are considered as potentially emerging pathogens.[1,2] The first step for tick-borne diseases' risk assessment in a geographical area is the detection of the pathogens in the infected vector in natural conditions. Our objective was to study ticks infection rates for *B. burgdorferi* sl, *A. phagocytophilum*, and SFG *Rickettsia* sp., in pastures and woods of a selected area in France.

A total of 4701 *I. ricinus* ticks, including 102 females, 123 males, and 4476 nymphs, were collected during spring 2003 by drag sampling on vegetation.[3] Collection sites were 61 pastures and the 61 nearest woods, randomly selected in the area of Combrailles, Puy-de-Dôme, France. This region has endemic Lyme disease, SFG rickettsioses serological evidence in humans,[4] and reported tick-borne diseases in cattle. DNA was extracted, using the method described by Halos *et al.*,[5] from individual adults and from nymphs pooled

Address for correspondence: Muriel Vayssier-Taussat, UMR 956 BIPAR, Ecole Nationale Vétérinaire d'Alfort, Maisons-Alfort, France. Voice: 33-1-43-96-71-51; fax: 33-1-43-96-73-32.
e-mail: mvayssier@vet-alfort.fr

Ann. N.Y. Acad. Sci. 1078: 316–319 (2006).
doi: 10.1196/annals.1374.059

by 5 (56 pools), 10 (220 pools), or 50 (20 pools) within the same sampling site providing 521 DNA samples in total. We detected DNA from the three pathogens using specific polymerase chain reaction (PCR) amplification of a 550-bp fragment of 16s rDNA from *A. phagocytophilum*,[6] a 350-bp fragment of 16s rDNA from *B. burgdorferi* sl, [7] and a 380-bp fragment of the *glt*A gene from *Rickettsia* sp.[8] For each targeted pathogen, 10–20 positive PCR products were sequenced (Qiagen, Hilden, Germany). We present here preliminary descriptive analysis based on crude observed PCR results in individual and pooled samples. Trends were tested by chi-square on Excel software. $P < 0.05$ was regarded as significant.

Among the 521 DNA extracts, 95 (18%) carried *B. burgdorferi* sl DNA, 81 (16%) carried *Rickettsia* sp. DNA, and 79 (15%) carried *A. phagocytophilum* DNA. Eighteen *B. burgdorferi* sl sequences were 100% related to *B. garinii* or *B. afzelii* (accession number = DQ087518). Sequences were not discriminant enough to allow distinction at the genospecies level. Twelve *Rickettsia* sp. sequences were 100% related to *R. helvetica* (accession number = DQ087520) which has been associated with severe diseases in the past few years.[9] One *Rickettsia* sequence (accession number = DQ087521) was related to a SFG *Rickettsia* sp. of unknown pathogenicity. Ten *A. phagocytophilum* sequences were 95–100% related to *A. phagocytophilum* pathogenic strains (accession number = DQ087519).

Among adult ticks, sexual difference was observed only for *B. burgdorferi* sl, for which female ticks were significantly more infected than male ticks ($P = 0.03$). This can be related to a difference of host preference between immature males and immature females and is consistent with results of a previous study.[10]

For *B. burgdorferi* sl, ticks collected in woods carried more DNA than ticks collected on pastures (respectively, 69% and 31% of PCR positive samples) ($P < 0.0005$). For *A. phagocytophilum,* ticks collected on pastures carried more DNA than ticks collected in wood (respectively, 63% and 37% of PCR positives samples) ($P = 0.03$). These differences could be related to different reservoir hosts for the two bacteria. *B. burgdorferi* sl hosts are known to be small wood rodents while reservoir hosts of *A. phagocytophilum* remain unknown even if roe deer are highly suspected.[11] However, our results do not support the roe deer hypothesis but rather are in favor of a host living in herbal sites. Such difference between pastures and woods habitats for *A. phagocytophilum* has already been shown[12] and corroborates the fact that domestic herbivores such as horses, cows, and sheep display clinical granulocytic anaplasmoses.[13] Therefore, the currently admitted similarity in the life cycles of *A. phagocytophilum* and *B. burgdorferi* is probably not relevant.[14]

Co-detection of more than one pathogen occurs in 24.5% of nymph pools, in 13% of females, but none for males. This result suggests a geographical co-segregation of the agents rather than a co-carriage by individual ticks.

Nevertheless, it underlines the potential risk of co-transmission among people or animals that are usually bitten by several ticks.

Nymphal individual infection rates will be statistically estimated based on pools of PCR results in order to evaluate the effect of stage on pathogens' carriage and to confirm the trends observed on crude results. GIS data, and environmental, vegetation and climatic factors, are currently under investigation.

Our results show that the three bacteria have similar prevalence in the Puy-de-Dôme but seem to have different life cycles. Pastures as well as woods appear to be a risk area for humans and animals. There is a need for parallel investigation on human and animal occurrences of the diseases caused by *B. burgdorferi* sl, SFG *Rickettsia* sp., and *A. phagocytophilum* in the geographical area of study.

REFERENCES

1. PAROLA, P. 2004. Tick-borne rickettsial diseases: emerging risks in Europe (review). Comp. Immunol. Microbiol. Infect. Dis. **27:** 297–304.
2. RAOULT, D. & V. ROUX. 1997. Rickettsioses as paradigms of new or emerging infectious diseases. Clin. Microbiol. Rev. **10:** 694–719.
3. VASSALLO, M., B. PICHON, J. CABARET, *et al.* 2000. Methodology for sampling questing nymphs of *Ixodes ricinus* (Acari: *Ixodidae*), the principal vector of Lyme disease. Europe. Entomol. Soc. Am. **37:** 335–339.
4. PAROLA, P., L. BEATI, M. CAMBON & D. RAOULT. 1998. First isolation of *Rickettsia helvetica* from *Ixodes ricinus* ticks in France. Eur. J. Clin. Microbiol. Infect. Dis. **17:** 95–100.
5. HALOS, L., T. JAMAL, L. VIAL, *et al.* 2004. Determination of an efficient and reliable method for DNA extraction from ticks. Vet. Res. **35:** 709–713.
6. MASSUNG, R.F., K. SLATER, J.H. OWENS, *et al.* 1998. Nested PCR assay for detection of granulocytic *Ehrlichiae*. J. Clin. Microbiol. **36:** 1090–1095.
7. MARCONI, R.T. & C.F. GARON. 1992. Development of polymerase chain reaction primers sets for diagnosis of Lyme disease and for species-specific identification of Lyme disease isolates by 16S rRNA signature nucleotide analysis. J. Clin. Microbiol. **30:** 2830–2834.
8. REGNERY, R.L., C.L. SPRUILL & B.D. PLIKAYTIS. 1991. Genotypic identification of *Rickettsiae* and estimation of intraspecies sequence divergence for portions of two rickettsial genes. J. Bacteriol. **173:** 1576–1589.
9. NILSSON, K., O. LINDQUIST & C. PAHLSON. 1999. Association of *Rickettsia helvetica* with chronic perimyocarditis in sudden cardiac death. Lancet **354:** 1169–1173.
10. DE MEEÛS, T., L. BEATI, C. DELAYE, *et al.* 2002. Sex-biased genetic structure in the vector of Lyme disease, *Ixodes ricinus*. Evolution Int. J. Org. Evolution **56:** 1802–1807.
11. LIZ, J.S., J.W. SUMNER, K. PFISTER & M. BROSSARD. 2002. PCR detection and serological evidence of granulocytic ehrlichial infection in Roe Deer (*Capreolus capreolus*) and Chamois (*Rupicapra rupicapra*). J. Clin. Microbiol. **40:** 892–897.

12. Bown, K.J., M. Begon, M. Bennett, *et al.* 2003. Seasonal dynamics of *Anaplasma phagocytophila* in a rodent-tick (*Ixodes trianguliceps*) system, United Kingdom. Emerg. Infect. Dis. **9:** 63–70.
13. Blanco, J.R. & J.A. Oteo. 2002. Human granulocytic ehrlichiosis in Europe [review]. Clin. Microbiol. Infect. **8:** 763–772.
14. Levin, M.L., F. Des Vignes & D. Fish. 1999. Disparity in the natural cycles of *Borrelia burgdorferi* and the agent of human granulocytic ehrlichiosis. Emerg. Infect. Dis. **5:** 204–208.

Prevalence of Spotted Fever Group *Rickettsia* Species Detected in Ticks in La Rioja, Spain

J.A. OTEO,[a] A. PORTILLO,[a] S. SANTIBÁÑEZ,[a] L. PÉREZ-MARTÍNEZ,[a] J.R. BLANCO,[a] S. JIMÉNEZ,[b] V. IBARRA,[a] A. PÉREZ-PALACIOS,[b] AND M. SANZ[a]

[a]*Área de Enfermedades Infecciosas, Hospitales San Millán-San Pedro-de La Rioja, Avda de Viana 1, 26001 Logroño, La Rioja, Spain*

[b]*Consejería de Salud. Gobierno de La Rioja. Logroño, La Rioja, Spain*

Abstract: Our objective was to learn the prevalence of spotted fever group (SFG) *Rickettsia* detected in ticks in La Rioja, in the north of Spain. From 2001 to 2005, 496 ticks representing 7 tick species were analysed at the Hospital de La Rioja. Ticks were removed from humans with or without rickettsial syndrome ($n = 59$) or collected from mammals ($n = 371$) or from vegetation by dragging ($n = 66$). The presence of SFG *Rickettsia* in these ticks was investigated by semi-nested PCR (*ompA* gene) and sequencing. A phylogenetic tree using Clustal method (neighbor-joining) was constructed with these data. Only 3 of 170 *Hyalomma marginatum* ticks carried SFG Rickettsia. Sequencing analysis demonstrated the presence of *Rickettsia aeschlimannii* (1.8%). Furthermore, *Rickettsia massiliae* and BAR29 were found in 3 of 120 *Rhipicephalus sanguineus* specimens (2.5%). In contrast, 81 of 83 tested *Dermacentor marginatus* ticks were PCR-positive (97%). *Rickettsia slovaca* (40.6%) and *Rickettsia* sp. strains RpA4, DnS14, DnS28 and JL-02 (59.3%) were found within this tick species. No SFG *Rickettsia* was detected using *ompA* primers when *Ixodes ricinus, Rhipicephalus bursa, Rhipicephalus turanicus, Rhipicephalus eversti eversti, Hyalomma detritum scupense and Rhipicephalus* sp. were analyzed. We detected 17.5% of ticks associated with different SFG *Rickettsia*: *R. aeschlimannii*, *R. massiliae*, BAR29, *R. slovaca* and *Rickettsia* sp. strains RpA4, DnS14, DnS28 and JL-02. Their presence has to be taken into account since most of them have been recognized as human pathogens.

Keywords: spotted fever group *Rickettsia*; ticks; *Rickettsia aeschlimannii*; *Rickettsia massiliae*; BAR-20; *Rickettsia slovaca*; Spain; *Rickettsia* sp. strains RpA4, DnS14, DnS28, and JL-02

Address for correspondence: J. A. Oteo, Área de Enfermedades Infecciosas, Hospitales San Millán-San Pedro-de La Rioja, Avda de Viana n° 1, 26001 Logroño, La Rioja, Spain. Voice: +34 941 297275; fax: +34 941 297267.
e-mail: jaoteo@riojasalud.es

Ann. N.Y. Acad. Sci. 1078: 320–323 (2006).

doi: 10.1196/annals.1374.060

INTRODUCTION

Spotted fever group rickettsioses (SFGR) are recognized as important emerging tick-borne human infections. Tick-borne diseases including SFG rickettsioses occur among residents living in La Rioja (in the north of Spain).[1] Recently, new molecular methods have enabled the development of sensitive and fast tools for the detection and identification of tick-borne pathogens. Accordingly, in an effort to identify the possible etiologic agents for SFG rickettsioses affecting humans in our area, as well as their prevalence, ticks were analyzed for evidence of rickettsial infections.

MATERIALS AND METHODS

A total of 496 ticks, representing 7 tick species, were collected in La Rioja (Spain) during 2001–2005 and analyzed at the "Patógenos Especiales Laboratory" from the Hospital de La Rioja. Ticks were removed from humans (n = 59) and from various animal hosts during vaccination programs and hunting seasons (n = 371), and they were also collected on vegetation by dragging (n = 66) in distinct areas throughout the year (TABLE 1). Each tick was disinfected by immersion in alcohol 70%, then washed in sterile water and dried on sterile paper under a laminar flow hood. DNA was crushed from individual ticks using QIAamp Tissue Kit (Qiagen Inc, Hilden, Germany) following the manufacturer's protocol. The presence of SFGR in these ticks was investigated by regular and semi-nested PCR assays (*gltA* and *ompA* genes, respectively). A negative control with distilled water instead of tick DNA template in the PCR master mixture and a positive control (DNA from *Rickettsia conorii*) were included in each test. The PCR products were purified and DNA sequencing was performed. A phylogenetic analysis was inferred from the comparison of *ompA* sequences using the Clustal method (neighbor-joining) (FIG. 1).

TABLE 1. Tick species from different hosts studied by molecular methods

	Hosts[a]						
Tick species	Humans	Cows	Dogs	Wild boar	Deer	Sheep	Vegetation
H. marginatum	5 (2)	110 (1)	0	0	0	0	55 (0)
R. sanguineus	7 (0)	0	113 (0)	0	0	0	0
I. ricinus	22 (0)	90 (0)	0	0	0	0	0
D. marginatus	14 (14)	0	0	51 (51)	4 (4)	4 (3)	11 (10)
R. bursa	6 (0)	0	0	0	0	0	0
R. turanicus	4 (0)	0	0	0	0	0	0
H. detritum scupense	1 (0)	0	0	0	0	0	0

[a]No. of ticks found to be infected with SFGR are indicated in brackets.

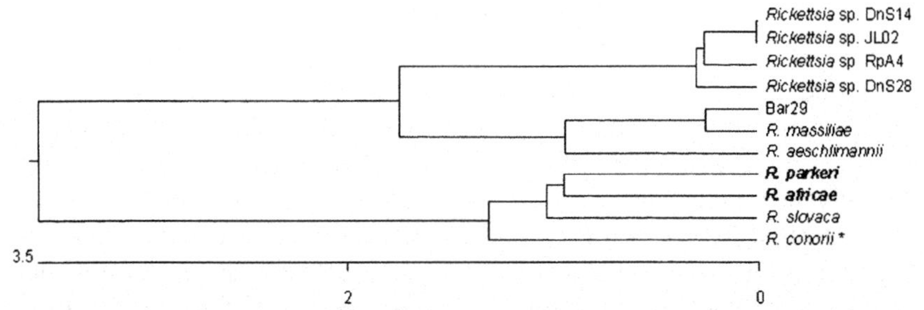

FIGURE 1: *ompA* gene-based phylogenetic tree of SFGR found in La Rioja (Spain). SFGR indicated in bold (*R. parkeri* and *R. africae*) have been found in samples received from Uruguay and Africa, respectively. **R. conorii* was detected from a blood sample of a patient (data not shown). Units at the bottom of the phylogenetic tree indicate the percentage of nucleotide substitutions.

RESULTS AND DISCUSSION

Only 3 of 170 *Hyalomma marginatum* ticks carried SFGR. Sequencing analysis demonstrated the presence of *Rickettsia aeschlimannii* (1.8%). Furthermore, *Rickettsia massiliae* and BAR29 were found in 1 and 2 of 120 *Rhipicephalus sanguineus* specimens, respectively (2.5%). In contrast, 81 of 83 tested *Dermacentor marginatus* ticks were PCR-positive for SFGR (97%). *Rickettsia slovaca* (40.7%) and *Rickettsia* sp. strains RpA4, DnS14, DnS28, and JL-02 (59.3%) were found within this tick species. No SFGR were detected for the remaining tick species (TABLE 1). *R. conorii* was not identified in ticks from our region, but we have molecular evidence of human infection caused by this agent (data not shown). Ticks of our study were not found to carry *Rickettsia helvetica*, although its presence in *Ixodes* ricinus has been reported in Spain.[2] In summary, 17.5% ticks were associated with different SFGR: *R. aeschlimannii*, *R. massiliae*, BAR29, *R. slovaca,* and *Rickettsia* sp. strains RpA4, DnS14, DnS28, and JL-02. Their presence has to be taken into account since most of them have been recognized as human pathogens.[3]

ACKNOWLEDGMENTS

This study was partly supported by a grant from the Fondo de Investigación Sanitaria (G03/057), Ministerio de Sanidad y Consumo, Spain.

REFERENCES

1. OTEO, J.A. & V. IBARRA. 2002. DEBONEL (*Dermacentor*-borne-necrosis-erythema-lymphadenopathy). A new tick-borne disease? Enferm. Infecc. Microbiol. Clin. **20:** 51–52.

2. FERNÁNDEZ-SOTO, P., R. PÉREZ-SÁNCHEZ, A. ENCINAS-GRANDES, *et al.* 2004. Detection and identification of *Rickettsia helvetica* and *Rickettsia* sp. IRS3/IRS4 in *Ixodes ricinus* ticks found on humans in Spain. Eur. J. Clin. Microbiol. Infect. Dis. **23:** 648–649.
3. RAOULT, D., P.E. FOURNIER, P. ABBOUD, *et al.* 2002. First documented human *Rickettsia aeschlimannii* infection. Emerg. Infect. Dis. **8**: 748–749.

Prevalence of *Rickettsia slovaca* in *Dermacentor marginatus* Ticks Removed from Wild Boar (*Sus scrofa*) in Northeastern Spain

A. ORTUÑO,[a] M. QUESADA,[b] S. LÓPEZ,[c] J. MIRET,[b] N. CARDEÑOSA,[b] J. CASTELLÀ,[a] E. ANTON,[b] AND F. SEGURA[b]

[a]*Parasitologia i Malalties Parasitàries, Department Sanitat i Anatomia Animal, Facultat de Veterinària, Universitat Autònoma de Barcelona, 08193 Bellaterra, Barcelona, Spain*

[b]*Corporació Parc Taulí, Hospital de Sabadell, Sabadell, Barcelona, Spain*

[c]*Federació Catalana de Caça, Representació Territorial de Lleida 25002, Spain*

ABSTRACT: *Rickettsia slovaca*, the causative agent of TIBOLA, is transmitted by *Dermacentor* ticks. *Dermacentor marginatus* is the most widely species distributed in northeasler Spain, and the wild boar constitutes the main host. *D. marginatus* ticks were collected from hunterkilled wild boar and were tested by PCR/RFLP. Rickettsial DNA–positive ticks were sequenced using the *ompA* PCR primers. The prevalence of *R. slovaca* in *D. marginatus* ticks was 17.7%. Other spotted fever group rickettsiae were detected in ticks, but these were not definitely identified.

KEYWORDS: *Rickettsia slovaca*; *Dermacentor marginatus* ticks; Sus scrofa; Spain

INTRODUCTION

In the epidemiology of tick-borne diseases, there is a wild cycle that permits the circulation of the pathogen in nature and a domestic cycle that is developed under the influence of human activity and domestic animals; both cycles are connected by ticks.

Rickettsia slovaca (R. slovaca) is the causative agent of TIBOLA (tick-borne lymphadenopathy), one of the newly emerging tick-borne diseases.[1] Dermacentor ticks act as vectors. In our area, the most widely distributed species

Address for correspondence: Dr. Anna Ortuño, Parasitologia i Malalties Parasitàries, Department of Sanitat i Anatomia Animal, Facultat de Veterinària, Universitat Autònoma de Barcelona, 08193 Bellaterra, Barcelona, Spain. Voice: 34-3-581-10-49; fax: 34-3-581-20-06.
e-mail: Ana.Ortuno@uab.es

**Ann. N.Y. Acad. Sci. 1078: 324–327 (2006). © 2006 New York Academy of Sciences.
doi: 10.1196/annals.1374.061**

is *Dermacentor marginatus (D. marginatus)*, a tick whose main hosts are ungulates, such as European wild boar, *Sus scrofa,* and it is highly prevalent in forests and pastures during the winter season. Nevertheless, the epidemiology of *R. slovaca* is not entirely clarified, especially with respect to the wild cycle. The aim of this study was to determine the prevalence of *R. slovaca* in *D. marginatus* ticks that were removed from European wild boar during the hunting season winter 2004.

MATERIALS AND METHODS

Sixty-nine ticks were removed from 24 hunter-killed wild boar during January and February 2004 in different areas of Catalonia (northeastern Spain). Ticks were held in a humidity chamber before freezing until they were processed. These ticks were identified following taxonomic keys.

Forty-five ticks were analyzed by using polymerase chain reaction restriction fragment length polymorphism (PCR/RFLP). The appropriate sample number of ticks was determined using the Epi-Info Computer Program for an estimated prevalence of 20% with a confidence level of 95% and a precision of 7%. First, ticks were disinfected in 70% alcohol, blotted and dried for a few minutes, transferred to Eppendorf tubes, and triturated prior to DNA preparation. DNA was extracted by using a QIAamp® DNA Mini Kit (Qiagen GmbH, Germany), as recommended by the manufacturer. PCR amplification was performed by using oligonucleotide primer pairs Rp CS877p and Rp CS.1258n generated from the citrate syntase gene of *R. prowazekii*, Rr 190.70p and Rr 190.602n generated from the 190 kDa antigen gene of *R. rickettsii,* and BG1-21 and BG2-20 generated from the 120-kDa antigen gene of *R. rickettsii.* PCRs were carried out using a Perkin–Elmer 9,600 thermocycler, according to the protocol described by Regnery *et al.,*[2] and the amplified products were analyzed on a 1.8% agarose gel (Bio-Rad, Hurcules, CA) in 0.5 × Tris-borate-EDTA (TBE) buffer. As positive control, purified DNA from *R. slovaca* that was kindly provided by Unité des Rickettsies, Marseille (France) was used. As a negative control for each tick sample, distilled water as in the tick samples was included. Amplified products were digested with *Alu*I, *Rsa*I, and *Pst*I restriction endonucleases, according to the protocol described by Eremeeva *et al.*[3] Electrophoretic separation was performed in a gel consisting of 1.8% agarose in TBE buffer.

The DNA fragments were visualized by ethidium bromide staining, and fragment sizes were compared with the sizes from a DNA Molecular Weight Marker VI (Boehringer Mannheim, GmbH, Germany).

All PCR-positive samples were sequenced. The sequences were obtained with forward sequencing primer of *omp*A, with the ABI 373 A Stretch sequencing system. Sequences were identified by comparison using the BLAST (Basic Local Aligment Search Tool) program.

TABLE 1. **Results obtained by RFLP and sequencing**

Wild boar	Tick	PCR	RFLP profiles	Sequencing	Similarity
J-0304–07	*D.m.*, male	+	*R. slovaca*	*R. slovaca*	100%
J-0304–10	*D.m,* male	+	*R. slovaca*	*R. slovaca*	100%
J-0304–12	*D.m,* male	+	*R. slovaca*	*R. slovaca*	99%
J-0304–14	*D.m*, male	+	*R. slovaca*	*R. slovaca*	100%
J-0304–19	*D.m.*, female	+	Doubtful	*R. slovaca*	100%
J-0304–21	*D.m*, male	+	*R. slovaca*	*R. slovaca*	99%
J-0304–42	*D.m*, male	+	*Rickettsia* spp.	*R. slovaca*	99%
J-0304–43	*D.m*, female	+	*Rickettsia* spp.	*R. slovaca*	99%

RESULTS

All ticks were identified as adults of *D. marginatus*. Rickettsial DNA was detected in 31 of the 45 ticks analyzed (68%). Results obtained by RFLP and sequencing are presented in TABLE 1.

Eight ticks were infected by *R. slovaca* (100% or 99% similarity with *R. slovaca* for the *omp*A gene). Prevalence of *R. slovaca* in *D. marginatus* ticks was 17.7%. Eight game pigs were parasitized by *R. slovaca*-infected ticks.

Twenty-three ticks were infected by other rickettsiae. The degree of *omp*A similarity with JL02, Rp A4, and Dn S14 in 13 ticks ranged from 88% to 97%, whereas 9 ticks exhibited *omp*A similarity with *R. slovaca* that ranged from 78% to 89%. In one tick the degree of *omp*A similarity was 82% with *R. parkeri*.

DISCUSSION

The prevalence of infection with *R. slovaca* in *D. marginatus* ticks in northeastern Spain was very similar to that obtained by Sanogo *et al*., who detected a prevalence of 15.7% in France.[4] In southern Croatia, the prevalence observed was higher (36.8%).[5] These results revealed that *R. slovaca* is widely distributed in *D. marginatus* ticks in our area. Outdoor activities, especially all of these related to the introduction of human being and domestic animals into tick-pathogen-wildlife cycles, such as hunting, could be considered an important risk factor. Nevertheless, the epidemiology of *R. slovaca* is still not clarified, especially with respect to wildlife cycle; until now, it is not known what role the wild boar plays in the epidemiology of this rickettsiosis.

On the other hand, our results showed further evidence of the diversity of *Rickettsia* spp. in ticks. Recently, new rickettsiae have been identified in ticks. *Rickettsia* sp. strain RpA4 was first detected in *Rh. pumilio* in the Soviet Union and DnS14 genotypes were detected in *D. nutallii* in Siberia.[6] In Spain, amplicons were obtained from ticks that showed high similarities to *Rickettsia* sp. JL02, *Rickettsia* sp. RpA4, *Rickettsia* sp. DnS14, and *Rickettsia* sp. DnS28,[7] and *Rickettsia* DnS28 was identified in *D. marginatus* ticks removed from wild boar.[8]

Further studies are necessary in order to clarify the epidemiology of *R. slovaca* and the diversity of rickettsiae in ticks.

REFERENCES

1. PAROLA, P. 2004. Tick-borne rickettsial diseases: emerging risks in Europe. Comp. Immunol. Microbiol. Infect. Dis. **27:** 297–304.
2. REGNERY, R.L., C.L. SPRUILL & B.D. PLIKAYTIS. 1991. Genotypic identification of rickettsiae and estimation of intraspecies sequence divergence for portions of two rickettsial genes. J. Bacteriol. **173:** 1576–1589.
3. EREMEEVA, M., X. YU & D. RAOULT. 1994. Differentiation among spotted fever group rickettsiae species by analysis of restriction fragment length polymorphism of PCR-amplified DNA. J. Clin. Microbiol. **32:** 803–810.
4. SANOGO, Y.O., B. DAVOUST, P. PAROLA, *et al.* 2003. Prevalence of *Rickettsia* spp. in *Dermacentor marginatus* ticks removed from game pigs (*Sus scrofa*) in southern France. Ann. N.Y. Acad. Sci. **990:** 191–195.
5. PUNDA-POLIC, V., M. PETROVEC, T. TRILAR, *et al.* 2002. Detection and identification of spotted fever group rickettsia in ticks collected in southern Croatia. Exp. App. Acarol. **28:** 169–176.
6. RYDKINA, E., V. ROUX, N. RUDAKOV, *et al.* 1999. New rickettsiae in ticks collected in territories of the former Soviet Union. Emerg. Infect. Dis. **5:** 811–814.
7. OTEO, J.A., V. IBARRA, J.R. BLANCO, *et al.* 2004. *Dermacentor-barne* necrosis erythema and lymphadenopathy: clinical and epidemiological features of a new tick-borne disease. Clin. Microbiol. Infect. **10:** 327–331.
8. DE LA FUENTE, J., V. NARANJO, F. RUIZ-FONS, *et al.* 2004. Prevalence of tick-borne pathogens in ixodid ticks collected from European wild boar (*Sus scrofa*) and Iberian red deer (*Cervus elaphus hispanicus*) in central Spain. Eur. J. Wild. Res. **50:** 187–196.

Prevalence Data of *Rickettsia slovaca* and Other SFG Rickettsiae Species in *Dermacentor marginatus* in the Southeastern Iberian Peninsula

F.J. MÁRQUEZ,[a] A. ROJAS,[b] V. IBARRA,[c] A. CANTERO,[d]
J. ROJAS,[b] J.A. OTEO,[c] AND M.A. MUNIAIN[e]

[a]*Departamento Biología Animal, Biología Vegetal y Ecología, Universidad de Jaén, Paraje Las Lagunillas s/n, 23071-Jaén, Spain*

[b]*Vircell S.L., Santa Fe, Granada, Spain*

[c]*Área de Enfermedades Infecciosas, Hospital de la Rioja, Logroño, Spain*

[d]*Servicio de Medicina Interna, H. San Juan de la Cruz, Úbeda, Jaén, Spain*

[e]*Sección de Enfermedades Infecciosas, H.U. Virgen Macarena, Sevilla, Spain*

ABSTRACT: In southern Spain, *Dermacentor marginatus* ticks can be infected with several genospecies of spotted fever Group (SFG) *Rickettsia*. We developed a nested polymerase chain reaction assay by using a species-specific probe targeting the *ompA* gene to detect and differentiate between the two groups of rickettsiae previously described in *D. marginatus*. SFG rickettsia has been detected in 85.15% of ticks studied (26.7% of positives have been to *R. slovaca*, the causative agent of TIBOLA-DEBONEL, and 73.3% to SFG rickettsia closely related to strains RpA4–JL-02-DnS14–DnS28).

KEYWORDS: *Rickettsia slovaca*; *Rickettsia* sp.; *Dermacentor marginatus*; Spain

INTRODUCTION

Dermacentor marginatus (D. marginatus) is believed to play the role of a natural reservoir of different *Rickettsia* species along its distribution area (western and central Europe).[1,2] Immature ticks usually feed on small mammals and birds, whereas adult ticks mainly feed on large mammals and frequently on humans.

Address for correspondence: Dr. Francisco J. Márguez, Departamento de Biología Animal, Biología Vegetal y Ecología, Universidad de Jaén, Paraje Las Lagunillas s/n, 23071-Jaén, Spain.
e-mail: jmarquez@ujaen.es

Ann. N.Y. Acad. Sci. 1078: 328–330 (2006).
doi: 10.1196/annals.1374.062

METHODS

DNA was extracted from 101 individual *D. marginatus* ticks (46 males, 55 females) captured between 2001–2004 in natural parks of Jaén and Granada (Andalusia, Spain) using the Nucleo Spin kit (Macherey-Nagel, Germany). Specific rickettsial sequences were detected by using multiplex nested PCR with primers that amplify a portion of *ompA* gene. Two PCR reaction mixes were made with $MgCl_2$ concentrations of 2.0 mM and 1.8 mM respectively. The first PCR reaction has an annealing temperature of 52°C for 30 cycles. Second PCR reaction has an annealing temperature of 66°C for 35 cycles. Primers used in the first amplification round were Rr190.70p (5′- ATG GCG AAT ATT TCT CCA AAA) and Rr190.620n (5′- AGT GCA GCA TTC GCT CCC CCT) (Regnery *et al.*,1991), with a product size of 532 pb. The second amplification round differentiates *R. slovaca* (forward primer RslovF–5′- ATG CAG CAT TTA GTG ATA ATG TT and reverse RslovR–5′- CCG TAA TAA TAC GAT CGG CTG) (estimate product size 162 pb) from other SFG rickettsia previously detected in *D. marginatus*[2–4] (forward primer RJDF-5′- TGC AGC ATT TAA TGA TCT TGC; reverse primer RJDR–5′-AAC AAG TTA CCT CCC GTC AC) with an estimated product size of 239 pb. Subsequent direct sequencing of first PCR amplified products was performed on selected samples in order to provide an objective and precise identification of rickettsiae species found in each tick.

RESULTS

SFG rickettsia was detected in 86 *D. marginatus* ticks (85.15%). *Rickettsia slovaca* has been detected by amplification and confirmed by sequence in 27 (26.7%) adult *D. marginatus* ticks. Amplification for the other two genospecies of SFG rickettsia closely related to strains RpA4–JL-02–DnS14–DnS28, previously described,[3] has been successful in 74 (73.3%) of positive specimens. Double infections have not been detected.

CONCLUSIONS

R. slovaca has a large geographic distribution range in Europe, and has been found in these ticks in many European countries, where *D. marginatus* ticks have been screened for it. The prevalence of infection by *R. slovaca* in *D. marginatus* ticks registered in Andalusia (26.7%) is similar to that obtained by other investigators in France, (15.7%–36%), Switzerland (30.3%–45.4%), and Croatia (36.8%), but higher than the value obtained in Hungary (21.1%), Portugal (5.5%) or Russia (1.8%–3.3%) (see Refs. 2 and 4). *Rickettsia slovaca,* causative agent of TIBOLA–DEBONEL,[5] has been considered nonpathogenic for the past 30 years until the first human clinical case was reported and this rickettsia has been isolated from humans.[6]

The other SFG rickettsia genospecies, of unknown pathogenicity and sympatric with *R. slovaca*, could be included in the *R. rhipicephali–R. massiliae* genogroup. Observed prevalence for this rickettsia is higher than that previously registered in other *D. marginatus* populations (1.4% in France).[4] Absence of double infections in these preliminary data should be interpreted in light of the fact that *D. marginatus* is unable to maintain simultaneously two different species of *Rickettsia.*

ACKNOWLEDGMENTS

This study was supported by a grant from the Fondo de Investigación Sanitaria (FIS-05/PI041521).

REFERENCES

1. MÁRQUEZ, F.J., V. IBARRA, J.A. OTEO, *et al.* 2003. Which spotted fever group rickettsia are present in *Dermacentor marginatus* ticks in Spain? Ann. N. Y. Acad. Sci. **990:** 141–142.
2. SANOGO, Y.O., B. DAVOUST, P. PAROLA, *et al.* 2003. Prevalence of *Rickettsia* spp. in *Dermacentor marginatus* ticks removed from game pigs (*Sus scrofa*) in southern France. Ann. N. Y. Acad. Sci. **990:** 191–195.
3. RYDKINA, E., V. ROUX, N. RUDAKOV, *et al.* 1999. New rickettsiae in ticks collected in territories of the former Soviet Union. Emerg. Infect. Dis. **5:** 811–814.
4. PUNDA-POLIC, V., M. PETROVEC, T. TRILAR, *et al.* 2002. Detection and identification of spotted fever group rickettsiae in ticks collected in southern Croatia. Exp. Appl. Acarol. **28:** 169–176.
5. OTEO, J.A., V. IBARRA, J.R. BLANCO, *et al.* 2004. *Dermacentor*-borne necrosis erythema and lymphadenopathy: clinical and epidemiological features of a new tick-borne disease. Clin. Microbiol. Infect. **10:** 327–331.
6. CAZORLA, C., M. ENEA, F. LUCHT, *et al.* 2003. First isolation of *Rickettsia slovaca* from a patient, France. Emerg. Infect. Dis. **9:** 135.

Spotted Fever Group Rickettsiae in Ticks Feeding on Humans in Northwestern Spain

Is *Rickettsia conorii* Vanishing?

PEDRO FERNÁNDEZ-SOTO,[a] RICARDO PÉREZ-SÁNCHEZ,[b] RUFINO ÁLAMO-SANZ,[c] AND ANTONIO ENCINAS-GRANDES[a]

[a]*Laboratorio de Parasitología, Facultad de Farmacia, Universidad de Salamanca, 37007 Salamanca, Spain*

[b]*Departamento de Patología Animal, IRNA (CSIC), 37007 Salamanca, Spain*

[c]*Dirección General de Salud Pública, Consejería de Sanidad, Junta de Castilla y León, 47011 Valladolid, Spain*

ABSTRACT: **During a 7-year study, we identified and analyzed by PCR 4,049 ticks removed from 3,685 asymptomatic patients in Castilla y León (northwestern Spain). A total of 320 ticks (belonging to 10 species) were PCR-positive for rickettsiae. Comparison of amplicon sequences in databases enabled us to identify eigth different spotted fever group (SFG) rickettsiae: *Rickettsia slovaca*, *Rickettsia* sp. IRS3/IRS4, *R. massiliae*/Bar29, *R. aeschlimannii*, *Rickettsia* sp. RpA4/DnS14, *R. helvetica*, *Rickettsia* sp. DmS1, and *R. conorii*. Although Mediterranean spotted fever (MSF) is an endemic disease in Castilla y León, *R. conorii* was found in only one *Rhipicephalus sanguineus* tick, whereas other pathogenic SFG rickettsiae were much more prevalent in the same area. Our data suggest that in Castilla y León, many MSF or MSF-like cases attributed to *R. conorii* could have been actually caused by other SFG rickettsiae present in ticks biting people in this region of Spain.**

KEYWORDS: **Mediterranean spotted fever; *Rickettsia conorii*; *R. slovaca*; *R. aeschlimannii*; Castilla y León; Spain**

Castilla y León is the largest regional community in the northwestern part of Spain and the largest regional area of Europe. Traditionally, the Mediterranean spotted fever (MSF) caused by *R. conorii* was thought to be the only prevailing rickettsial disease in this community and the main sources of knowledge are

Address for correspondence: Dr. Pedro Fernández-Soto, Laboratorio de Parasitología, Facultad de Farmacia, Universidad de Salamanca, Avenida Campo Charro s/n. 37007, Salamanca, Spain. Voice: +34-923-294535; fax: +34-923-294515.
e-mail: pfsoto@usal.es

**Ann. N.Y. Acad. Sci. 1078: 331–333 (2006). © 2006 New York Academy of Sciences.
doi: 10.1196/annals.1374.063**

numerous seroepidemiological studies carried out on humans and animals. To date, other rickettsioses have been poorly investigated in this region and it should be useful to expand those investigations, including the molecular identification (and ideally, isolation) of the rickettsiae.

The purpose of this study is to report the molecular identification of spotted fever group (SFG) rickettsiae in ticks removed from human subjects in a 7-year study carried out in Castilla y León, and, additionally, to highlight the near absence of *R. conorii* in ticks from this part of Spain in spite of the frequent MSF cases reported in that area for years.

From 1997 to 2003, 4,049 ticks removed from 3,685 asymptomatic patients assisted for tick bites in the hospitals and health care centers of Castilla y León were sent to our laboratory for identification and analysis. Rickettsial DNA was detected by PCR using primers RpCS.877p-RpCS.1258n to target a 380–397-bp *glt*A fragment and Rr190.70p-Rr190.701n to target a 629–632-bp *omp*A fragment, as previously described.[1] The amplicons obtained were purified, sequenced, and compared in GenBank for identification.

The 4,049 ticks identified belonged to 14 ixodid and 1 argasid species. Of the 3,685 patients, 91% were parasitized by a single tick and 9% by two or more. A total of 320 ticks (belonging to 10 species) were PCR-positive for rickettsiae, since in them the *glt*A, the *omp*A, or both amplicons were amplified, representing a global infection rate of 7.90%. For unknown reasons, in 72 of the 320 rickettsiae-positive ticks (22.5%) the sequencing of the amplicons failed and, consequently, it was not possible to identify the *Rickettsia* species involved. In the remaining 248 positive ticks (77.5%), comparison of amplicon sequences in the database enabled us to identify, by order of decreasing abundance, *R. slovaca* in 59 ticks (58 *D. marginatus* and 1 *D. reticulatus*), *Rickettsia* sp. IRS3/IRS4 in 51 ticks (50 *I. ricinus* and 1 *D. marginatus*), *R. massiliae*/Bar29 in 48 ticks (37 *R. turanicus*, 6 *R. sanguineus*, 4 *I. Ricinus,* and 1 *R. pusillus*), *R. aeschlimannii* in 42 ticks (26 *H. marginatum*, 7 *R. bursa*, 3 *I. ricinus*, 3 *R. turanicus*, 2 *R. sanguineus,* and 1 *H. punctata*), *Rickettsia* sp. RpA4/DnS14 in 35 ticks (25 *D. marginatus*, 9 *D. reticulatus,* and 1 *I. ricinus*), *R. helvetica* in 8 ticks (8 *I. ricinus*), *Rickettsia* sp. DmS1 in 4 ticks (4 *D. marginatus*) and, finally, *R. conorii* in 1 tick (1 *R. sanguineus*).

During 1997 to 2003, 353 MSF cases reported in Castilla y León were attributed to *R. conorii* according to the clinical signs, serological results, and the previous epidemiological data.[2] However, in the current study, *R. conorii* was found in only one *R. sanguineus* tick. The scarcity of *R. conorii* is not only unexpected, but also very important from an epidemiological point of view. Recently, we reported the presence of *R. aeschlimannii* in six tick species that frequently feed on humans in this community, raising the suspicion that many cases of MSF in this region of Spain could have been due to *R. aeschlimannii.*[1] The results presented here show that several pathogenic rickettsiae are much more prevalent than *R. conorii* in ticks biting people in Castilla y León. This observation suggests that, for years in our community, many MSF or

MSF-like cases attributed to *R. conorii*—on the basis of clinical, serological, and epidemiological data—could have been actually caused by other rickettsiae, such as *R. aeschlimannii* and *R. slovaca* and, perhaps, by other SFG rickettsiae among those recently identified, whose pathogenicity is not yet undoubtedly proven. The demonstration of this hypothesis requires further genotypic and phenotypic characterization—and ensuing comparison—of both the human and the tick rickettsial isolates.

ACKNOWLEDGMENTS

This work was supported by Consejería de Sanidad, Junta de Castilla y León, and Ministerio de Sanidad y Consumo, Fondo de Investigación Sanitaria (FIS), Red Temática de Investigación Cooperativa EBATRAG-G03/057.

REFERENCES

1. FERNÁNDEZ-SOTO P., A. ENCINAS-GRANDES & R. PÉREZ-SÁNCHEZ. 2003. *Rickettsia aeschlimannii* in Spain: molecular evidence in *Hyalomma marginatum* and five other tick species that feed on humans. Emerg. Infect. Dis. **9**: 889–890.
2. BOLETINES EPIDEMIOLÓGICOS DE CASTILLA Y LEÓN. 1997–2003. Dirección General de Salud Pública y Consumo. Consejería de Sanidad. Valladolid, Spain.

A Rickettsial Mixed Infection in a *Dermacentor Variabilis* Tick from Ohio

JENNIFER R. CARMICHAEL[a] AND PAUL A. FUERST[b]

[a]*Department of Molecular Genetics, The Ohio State University, Columbus, Ohio 43210, USA*

[b]*Department of Evolution, Ecology, and Organismal Biology, The Ohio State University, Columbus, Ohio 43210, USA*

ABSTRACT: We present the first report of superinfection in a *Dermacentor variabilis* tick from nature. The single tick, collected in Ohio, was found infected with *Rickettsia belli*, *R. nontanensis*, and *R. rickettsii*.

KEYWORDS: *Dermacentor variabilis*; *Rickettsia bellii*; *Rickettsia montanensis*; *Rickettsia rickettsii*; superinfection; interference

OBJECTIVE

Many arthropod species are vectors for rickettsial agents of human disease. Some rickettsial species have evolved a stable means of transovarial maintenance within their arthropod hosts, whereas other rickettsiae are acquired when the arthropod vector feeds on infected vertebrates (horizontal transmission). It is widely accepted that an initial infection of an arthropod by one rickettsial species prevents the acquisition and transmission of a secondary rickettsial form.[1–5] The cause of this phenomenon, referred to as interference, is unknown. During the 2003 screen of rickettsial-infected *Dermacentor variabilis* ticks from Ohio collected by the Ohio Department of Health, an isolate was found to be infected with multiple rickettsial forms, *Rickettsia bellii*, *Rickettsia montanensis*, and *Rickettsia rickettsii*. The tick was hemolymph-positive for *Rickettsia* sp. and positive for a direct fluorescent antibody test specific to *R. rickettsii*. Subsequent analyses were performed to verify the multiple infection.

Address for correspondence: Paul A. Fuerst, Department of Evolution, Ecology, and Organismal Biology, The Ohio State University, 318 W. 12th Avenue, Columbus, OH 43210. Voice: 614-292-6403; fax: 614-292-2030.
e-mail: fuerst.1@osu.edu

Ann. N.Y. Acad. Sci. 1078: 334–337 (2006). © 2006 New York Academy of Sciences.
doi: 10.1196/annals.1374.064

MATERIALS AND METHODS

Species of *Rickettsia* were identified by a semi-nested polymerase chain reaction (PCR) assay of the rickettsial 17 kDa surface antigen gene. Genomic DNA was extracted from both the tick and any associated bacteria. Semi-nested PCR was done in a Whatman Biometra thermal cycler (Biometra Biomedicine Analytik, Goettigen, Germany) with paired genus-specific oligionucleotide primers (17 kDa-5′ GCTTTACAAAATTCTAAAAACCATATA; 17kDa-3′ CTTGCCATTGTCCRTCAGGTTG; and 17kDa-3′nest TCACGGCAATATTGACC), designed based upon work done in our laboratory.[6] The resulting amplicon was sequenced to determine the rickettsial source. Further, the gene product was cloned to quantify the proportion of 17 kDa sequences of each species. Finally, a multiplex PCR was performed using species-specific primers, one primer set specific for *R. bellii*, and a second specific to *R. rickettsii* and *R. montanensis*. Assays were repeated in triplicate to confirm results.

RESULTS

The primary nucleotide sequence of the 17 kDa gene amplification product was inconsistent with a single species. Dual electropherogram peaks were seen at numerous base positions. Further analysis of the electropherogram suggested that the superimposed sequences corresponded to a combination of *R. bellii* and *R. montanensis*. Although not conclusive, results were also consistent with the additional presence of *R. rickettsii*. Subsequent analysis of cloned sequences showed specific sequences for *R. bellii*, *R. montanensis*, or *R. rickettsii*. The latter two species are distinguished by four nucleotide differences in the region sequenced, whereas *R. bellii* shows a number of nucleotide differences from the two Spotted Fever group forms. Finally, multiplex PCR analysis resulted in the amplification of two products, one product specific in size to *R. bellii* and the other containing a product consistent with a mixture of *R. montanensis* and *R. rickettsii*. Repeated analyses were consistent with the results reported here.

CONCLUSIONS

These results indicate the occurrence of a tick naturally superinfected with three different rickettsial forms, *R. bellii*, *R. montanensis,* and *R. rickettsii*. This represents the first molecular confirmation of multiple infection by rickettsiae of an arthropod in nature. Semi-nested PCR, a highly sensitive assay, in concert with vector cloning of specific portions of the 17 kDa surface antigen gene, could specifically identify each rickettsial species present in our tick isolate. In addition, the repeatability of the assays, as well as a multiplex PCR approach, further supports our conclusions. Thus, these data provide strong support for the presence of three rickettsial species in the tick isolate. In interpreting this case, we believe it is most likely that *R. bellii* was acquired transovarially. It is a very common species obtained from tick isolates in Ohio, representing up to 80% of isolates from *Dermacentor* ticks.[7–10] The less common *R. montanensis* may also have been acquired either transovarially, or more likely, from a feeding event. The latter mode of transmission may also explain the presence of *R. rickettsii,* which is relatively uncommon among the rickettsial flora in Ohio. Horizontal transmission is very important for this species' maintenance, and helps to explain the low frequency of *R. rickettsii* in nature ($<1\%$).[4,11,12] Interference, believed to occur in rickettsiae, seems to be specific to the prevention of transovarial transmission of two species. It may not exclude the possibility of a host acquiring a multiple infection, which has been observed in some laboratory studies.[2,13] Therefore, multiple infected vectors may exist in nature, and may serve a vital (if only intermediate) role for the maintenance of these rickettsiae. Highly sensitive assays, such as those we utilized in our study, identified *R. bellii*, *R. montanensis*, and *R. rickettsii* in a single tick isolate and can be very important in future screening approaches in nature.

ACKNOWLEDGMENTS

We wish to thank ODH-VBDU for their cooperation in providing ticks.

REFERENCES

1. Azad, A.F. & C.B. Beard. 1998. Rickettsial pathogens and their arthropod vectors. Emerg. Infect. Dis. **4:** 179–186.
2. Burgdorfer, W., S.F. Hayes & A.J. Mavros, 1981. Non-pathogenic rickettsiae in *Dermacentor andersoni*: a limiting factor for the distribution of *Rickettsia rickettsii*. *In* Rickettsiae and Rickettsial Diseases. W. Burgdorfer & R.L. Anacker, Eds.: 585–594. Academic Press. New York.
3. Macaluso, K.R., D.E. Sonenshine, S.M. Ceraul & A.F. Azad. 2002. Rickettsial infection in *Dermacentor variabilis* (Acari: Ixodidae) inhibits transovarial transmission of a second *Rickettsia*. J. Med. Entomol. **39:** 809–813.
4. McDade, J.E. & V.F. Newhouse. 1986. Natural history of *Rickettsia rickettsii*. Annu. Rev. Microbiol. **40:** 287–309.
5. Price, W.H. 1953. Interference phenomenon in animal infections with Rickettsiae of Rocky Mountain spotted fever. Proc. Soc. Exp. Biol. Med. **82:** 180–184.
6. Stothard, D.S. 1995. The evolutionary history of the genus *Rickettsia* as inferred from 16S and 23S ribosomal RNA genes and the 17 Kilodalton cell surface antigen gene. Ph.D. Thesis, The Ohio State University, Ohio. 214 pp.
7. Gordon, J.C., S.W. Gordon, E. Peterson & R.N. Philip. 1984. Epidemiology of Rocky Mountain spotted fever in Ohio, 1981: serologic evaluation of canines

and rickettsial isolation from ticks associated with human case exposure sites. Am. J. Trop. Med. Hyg. **33:** 1026–1031.

8. LINNEMANN, C.C.J., A.E. SCHAEFFER, W. BURGDORFER, *et al.* 1980. Rocky Mountain spotted fever in Clermont County, Ohio. II. Distribution of population and infected ticks in an endemic area. Am. J. Epidemiol. **111:** 31–36.
9. PRETZMAN, C., N. DAUGHERTY, K. POETTER & D. RALPH. 1990. The distribution and dynamics of *Rickettsia* in the tick population of Ohio. Ann. N. Y. Acad. Sci. **590:** 227–236.
10. RAOULT, D. & V. ROUX. 1997. Rickettsia as paradigms of new or emerging infectious diseases. Clin. Microbiol. Rev. **10:** 694–719.
11. BURGDORFER, W. 1988. Ecological and epidemiological considerations of Rocky Mountain spotted fever and scrub typhus. *In* Biology of Rickettsial Diseases, Vol. 1. D.H. Walker, Ed.: pp. 33–50 CRC Press. Boca Raton, FL.
12. HACKSTADT, T. 1996. The biology of Rickettsiae. Infect. Agent Dis. **5:** 127–143.
13. NODEN, B.H., S. RADULOVIC, J.A. HIGGINS & A.F. AZAD, 1998. Molecular identification of *Rickettsia typhi* and *R. felis* in co-infected *Ctenocephalides felis* (Siphonaptera: Pulicidae). J. Med. Entomol. **35:** 410–414.

Rocky Mountain Spotted Fever in Arizona: Documentation of Heavy Environmental Infestations of *Rhipicephalus sanguineus* at an Endemic Site

WILLIAM L. NICHOLSON,[a] CHRISTOPHER D. PADDOCK,[b] LINDA DEMMA,[a] MARC TRAEGER,[c] BRIAN JOHNSON,[c] JEFFREY DICKSON,[c] JENNIFER McQUISTON,[a] AND DAVID SWERDLOW[a]

[a]*Viral and Rickettsial Zoonoses Branch, Centers for Disease Control and Prevention, Atlanta, Georgia 30333, USA*

[b]*Infectious Disease Pathology Activity, Centers for Disease Control and Prevention, Atlanta, Georgia 30333, USA*

[c]*Indian Health Service, Public Health Service, Whiteriver, Arizona 85941, USA*

ABSTRACT: A recent epidemiologic investigation identified 16 cases and 2 deaths from Rocky Mountain spotted fever (RMSF) in two eastern Arizona communities. Prevalence studies were conducted by collecting free-living ticks (Acari: Ixodidae) from the home sites of RMSF patients and from other home sites within the community. Dry ice traps and flagging confirmed heavy infestations at many of the home sites. Only *Rhipicephalus sanguineus* ticks were identified and all developmental stages were detected. It is evident that under certain circumstances, this species does transmit *Rickettsia rickettsii* to humans and deserves reconsideration as a vector in other geographic areas.

KEYWORDS: Acari; Ixodidae; *Rickettsia rickettsii;* RMSF; tick survey

INTRODUCTION

Rocky Mountain spotted fever (RMSF) occurs over a wide geographical area in the United States, but most cases are reported from the eastern and

The paper was presented at the Fourth International Conference on Rickettsiae and Rickettsial Diseases, Logrono (La Rioja), Spain, 2005.
Address for correspondence: Dr. William L. Nicholson, Viral and Rickettsial Zoonoses Branch, Centers for Disease Control and Prevention, Mail Stop G-13, 1600 Clifton Road, Atlanta, GA 30333. Voice: 404-639-1095; fax: 404-639-4436.
e-mail: wan6@cdc.gov

**Ann. N.Y. Acad. Sci. 1078: 338–341 (2006). © 2006 New York Academy of Sciences.
doi: 10.1196/annals.1374.065**

central states. Arizona has reported cases of RMSF since at least 1912,[1] but has consistently identified only one or two cases in any year. Since 1981, only three cases have been reported from the state. The normal tick vectors, *Dermacentor andersoni* and *Dermacentor variabilis* are not widely distributed in the state, which may account for the low prevalence.

In 2003, the Centers for Disease Control and Prevention (CDC) and the Indian Health Service (IHS) identified a case of RMSF in a small child residing in the rural eastern White Mountain area of Arizona.[2] The case was confirmed by laboratory testing of the serum sediment by polymerase chain reaction (PCR) and nucleotide sequencing to determine the species. This led to an intensive epidemiologic study that is still under way today (2005). We were able to identify 13 additional cases and another fatality in this relatively small population (ca. 12,148 people). Thus, the incidence in this population was nearly 300 times greater than that anywhere else in the United States (1800 cases per million versus 5.6 cases per million nationally for children under 19 years old). As a part of the CDC epidemiologic study, tick surveys were conducted at selected homes. The initial efforts identified only the brown dog tick, *Rhipicephalus sanguineus*, in collections made around the homesites. Molecular proof of natural *Rickettsia rickettsii* infection and documentation of nymphal ticks feeding on children indicated that additional ecologic studies were needed. In late August and early September, 2004, tick collections were conducted by CDC entomologists and IHS environmental health officers. Our search focused on the peridomestic environment of case houses and houses near those sites. We also made a concerted effort to sample more broadly so that we might detect additional tick species that might be present in this area.

METHODS

Dry ice traps were placed at case patient ($n = 8$) and control ($n = 9$) homes. The traps consisted of pieces of dry ice placed into a ventilated plastic container. The container was placed in the center of a flannel cloth (ca. 1 m^2). The traps were allowed to operate for approximately 1 h to allow ticks to be attracted to the traps. After the allotted time, the traps were collected and the ticks removed for counting and laboratory analysis. Collections from flags and hand-picking from various substrates were also employed to supplement the collections with additional brown dog ticks and to search for additional tick species in more sylvatic environments ($n = 2$ areas).

RESULTS AND DISCUSSION

The results of our collections are summarized in TABLE 1. All ticks were identified as *Rhipicephalus sanguineus*, which supported our earlier tick

TABLE 1. Summary of *Rhipicephalus sanguineus* ticks collected from home sites in the White Mountain region of eastern Arizona, August–September 2004

Stage	Engorgement status	Number collected
Adults	Flat	214
Adults	Engorged	2
Nymphs	Flat	1,064
Nymphs	Engorged	325
Larvae	Flat	4,514
Larvae	Engorged	10
Total		6,124

collections in that only this species was identified in the area on dogs or on flags. All developmental stages were collected in the peridomestic area around houses, indicating active infestations. We found no significant difference in the prevalence of ticks at case or control houses. All subdivisions tested had some level of tick infestation. Absolute numbers of ticks also did not correlate with the household having clinical cases of RMSF. However, at one house where a fatality had occurred in 2004, we found extremely high infestations of the stucco wall and ceiling voids. Several non-case home sites were also found to be heavily infested. Questing, molting, and engorged ticks were located in various sites including the bare soil surface under houses, within voids of concrete piers, between siding and slabs of the houses, between crevices of fiberglass brick veneer panels, within seams of discarded mattresses, under discarded furniture, and interspersed in broom bristles. One home site yielded few ticks until we examined the structural grooves of a single plastic milk crate and found them filled with scores of ticks.

The dry ice trapping method provided an inexpensive and convenient method of surveillance for ticks in the peridomestic environment. Use of flannel allowed for lightweight and foldable collection surfaces. This made it convenient to remove the ice container, fold the cloth, and place it into a ziplock bag for transport. The bag could then be opened in the laboratory and the ticks removed and enumerated later. Dry ice trapping can be affected by weather, particularly wind and rain. Carbon dioxide is 2.9 times as heavy as air and radiates out from the source in all directions. Prevailing winds can modify the odor plume so that some directions of travel will be biased. In very windy conditions, no distinct plume can develop and the traps are not effective. In our short study, the traps performed well and provided the much needed numbers of ticks. While dry ice traps are operating, tick collectors may then focus their efforts on flagging or dragging additional areas, examining animals for ticks, and providing health education. Thus, this method of surveillance was employed on a much wider home site survey of the communities in the area (data to be reported later).

Clearly the association of people and their dogs with a single species in high numbers led us to investigate the brown dog tick as the likely vector in this area. Many years ago, this tick had been implicated in transmission of *Rickettsia rickettsii* in Mexico.[3,4] Experimental transmission through two complete life cycles (6 stages) was clearly shown in 1933.[5] Both transovarial and transstadial transmission were demonstrated. Since that time, the relatively low frequency of feeding on humans by this tick has led to the idea that this tick plays no significant role in RMSF transmission. Our studies of this outbreak have modified that viewpoint. This species should receive additional consideration in other geographic areas.

REFERENCES

1. McClintic, T.B. 1912. Investigations of and tick eradications in Rocky Mountain spotted fever. Public Health Rep. **27:** 732–760.
2. Demma, L.J., M.S. Traeger, W.L. Nicholson, *et al.* 2005. Rocky Mountain spotted fever from an unexpected tick vector in Arizona. N. Engl. J. Med. **353:** 587–594.
3. Bustamante, M.E. & G. Varela IV. 1947. Estudios de fiebre manchada en Mexico. Papel del *Rhipicephalus sanguineus* en la transmission de la fiebre manchade en la Republica Mexicana. Rev. Inst. Salubr. Enferm. Trop. **8:** 139–141.
4. Bustamante, M.E., G. Varela & C.O. Mariotte. 1946. Estudios de fiebre manchada en Mexico. Fiebre Manchada en la Laguna. Rev. Inst. Salubr. Enferm. Trop. **7:** 39–49.
5. Parker, R.R., C.B. Philip & W.L. Jellison. 1933. Rocky Mountain spotted fever: potentialities of tick transmission in relation to geographic occurrence in the United States. Am. J. Trop. Med. **8:** 341–379.

An Outbreak of Rocky Mountain Spotted Fever Associated with a Novel Tick Vector, *Rhipicephalus sanguineus*, in Arizona, 2004

Preliminary Report

LINDA J. DEMMA, M. EREMEEVA, W. L. NICHOLSON, M. TRAEGER, D. BLAU, C. PADDOCK, M. LEVIN, G. DASCH, J. CHEEK, D. SWERDLOW, AND J. McQUISTON

Centers for Disease Control and Prevention, Atlanta, Georgia, USA
Indian Health Service, Public Health Service, Whiteriver, Arizona, USA

ABSTRACT: This study describes preliminary results of an investigation of RMSF in Arizona associated with the brown dog tick, *Rhipicephalus sanguineus*. High numbers of dogs and heavy infestations of ticks created a situation leading to human disease.

KEYWORDS: RMSF; epidemiology; tick; *Rickettsia rickettsii*

BACKGROUND

Rocky Mountain spotted fever (RMSF), caused by *Rickettsia rickettsii,* is a zoonotic tick-borne disease that may result in fatal illness. RMSF is rare in Arizona on account of a climate unsupportive of the typical U.S. tick vectors, *Dermacentor* spp. During 2004, 13 cases of RMSF, including one death, were identified in eastern Arizona. We conducted environmental investigations and applied molecular tools to identify the source of the outbreak.[1]

METHODS

Partially engorged ticks were collected from patients' dogs, and flat (nonengorged) adult ticks were collected from the peridomestic environment around

Address for correspondence: Dr. William L. Nicholson, Viral and Rickettsial Zoonoses Branch, Centers for Disease Control and Prevention, Mail Stop G-13, 1600 Clifton Road, Atlanta, Georgia, USA 30333. Voice: 404-639-1095; fax: 404-639-4436.
e-mail: wan6@cdc.gov

Ann. N.Y. Acad. Sci. 1078: 342–343 (2006). © 2006 New York Academy of Sciences.
doi: 10.1196/annals.1374.066

the patients' homes. Polymerase chain reaction, DNA sequence analysis, and culture isolation were used to confirm the presence of *R. rickettsii*.

RESULTS

All patients reported exposure to tick-infested dogs, and 3 (23%) of 13 patients described a prior tick bite. All collected ticks were identified as *Rhipicephalus sanguineus* (brown dog tick); no *Dermacentor* spp. ticks were found. *R. rickettsii* DNA was detected in flat ticks and in ticks found on dogs. Cultures of *R. rickettsii* were established from flat and partially engorged ticks.

CONCLUSIONS

The identification of *R. rickettsii* in *Rh. sanguineus* ticks collected from patients' homes implicates this tick as a vector of RMSF, and provides the first evidence that this tick may be associated with RMSF in the United States. Transmission appears to be caused by high densities of *Rh. sanguineus* in peridomestic settings and by frequent dog–tick–human interactions. Ongoing preventive efforts include extensive tick control, community education, and physician awareness. The broad distribution of this common tick in the United States raises concerns over its potential to transmit *R. rickettsii* in similar settings.

REFERENCE

1. DEMMA, L.J., M.S. TRAEGER W.L. NICHOLSON, *et al.* 2005. Rocky Mountain spotted fever from an unexpected tick vector in Arizona. N. Engl. J. Med. **353:** 587–594.

Incidence and Distribution Pattern of *Rickettsia felis* in Peridomestic Fleas from Andalusia, Southeast Spain

F.J. MÁRQUEZ,[a] M.A. MUNIAIN,[b] J.J. RODRÍGUEZ-LIEBANA,[a] M.D. DEL TORO,[b] M. BERNABEU-WITTEL,[c] AND A.J. PACHÓN[c]

[a]*Departamento Biología Animal, Vegetal y Ecología, Universidad de Jaén, Campus Las Lagunillas s/n, 23071 Jaén, Spain*

[b]*Sección de Enfermedades Infecciosas. H.U. Virgen Macarena, Sevilla, Spain*

[c]*Servicio de Enfermedades Infecciosas, H.U. Virgen del Rocio, Sevilla, Spain*

ABSTRACT: The presence of *Rickettsia felis* was investigated in three species of pulicid fleas (*Ctenocephalides felis, Ctenocephalides canis* and *Pulex irritans*) collected in 38 locales in Andalusia (Spain) over the period 1999–2004. Amplification of a fragment of *OmpB* gene was positive in 54.17 % of lots of *Ct. felis*. The identity of the PCR bands was confirmed as *R. felis* by sequence data obtained directly from the PCR amplicon. No rickettsia was found in *Ct. canis* nor *P. irritans*.

KEYWORDS: *Rickettsia felis*; flea; *Ctenocephalides* sp.; *Pulex irritans*; Spain

INTRODUCTION

Rickettsia felis (R. felis), formerly thought to be an ELB agent, is an SFG Rickettsia that was detected in 1990 when tissues from the cat flea, *Ctenocephalides felis (Ct. felis)*, were examined under electron microscopy. Once detected, several antigenic and molecular studies concerning this rickettsia were developed. *R. felis* is maintained in the cat flea by transovarian transmission. Infection in fleas and humans has been described in North and South America, Europe, Africa, and Australasia.[1]

MATERIALS AND METHODS

Specimens of three species of flea from the Pulicidae family, commonly found on domestic dogs and cats, were collected from 38 sites in Andalusia

Address for correspondence: Dr. Francisco J. Márguez, Departamento de Biología Animal, Biología Vegetal y Ecología, Universidad de Jaén, Parage Las Lagunillas s/n, 23071-Jaén, Spain.
e-mail: jmarquez@ujaen.es

Ann. N.Y. Acad. Sci. 1078: 344–346 (2006).
doi: 10.1196/annals.1374.067

(Spain) over a 6-year period (1999–2004). Collected fleas were fixed in 70 % ethyl alcohol and stored at 4°C until they were processed. Taxonomic determination was made using current taxonomic keys. DNA extraction, amplification, and sequencing were accomplished over 72 lots of *Ctenocephalides felis felis* (Bouché, 1835) (56 male, 196 female, flea average per lot: 3.5), 8 lots of *Ct. canis* (Curtis, 1876) (11 male, 15 female, flea average per lot: 3.25), and 12 lots of *Pulex irritans* (*P. irritans*) (Linnaeus, 1758) (21 male, 10 female, flea average per lot: 2.58) collected on dogs (77 lots) and cats (15 lots).

DNA was extracted from the flea lots, ranging from 1 to 11 specimens of the same species per lot, using the NucleoSpin DNA tissue kit (Machery-Nagel, Germany). Elution of DNA was made in 100 μL of TE buffer (1 mM Tris HCl, 0.1 mM EDTA). Extraction blanks consisting of water processed along with flea samples were also included as control.

A Biometra DNA thermalcycler (Gottingen, Germany) was used for all PCR amplification. Three microliters of each DNA extraction was added to 27 μL of master mixture for each reaction. Final reagent concentration was 0.2 μM for each primer, 200 μM for each deoxynucleotide triphosphate (Promega Corp., Madison, WI, USA), 2 U of Biotaq™ polymerase (BioLine, London, UK), and 1× Bioline buffer. The following thermocycler parameters were used with the primer pairs for 120-kDa outer-membrane protein (*ompB*) (rfompbf: 5′–GAC AAT TAA TAT CGG TGA CGG, and rfompbr: 5′-TGC ATC AGC ATT ACC GCT TGC): 96°C (90 sec), followed by 35 cycles of 94°C (30 sec), 50°C (30 sec), and 72°C (45 sec), followed by an extension period (72°C, 7 min).[2]

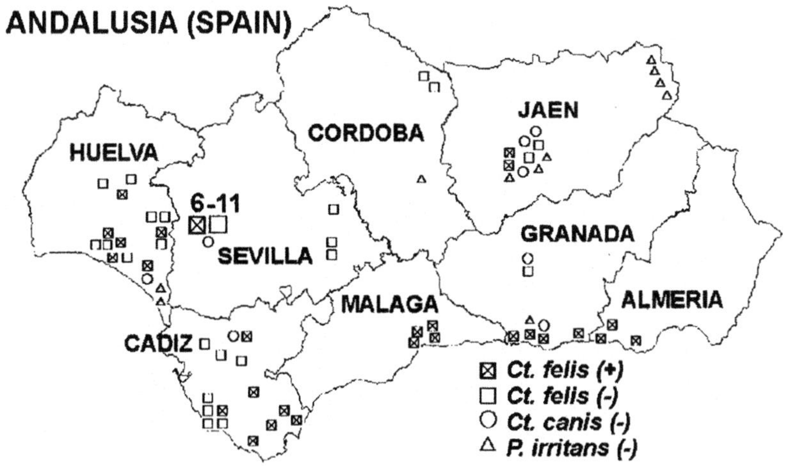

FIGURE 1. Distribution of fleas from the Pulicidae family tested for the presence of *R. felis* in Andalusia.

RESULTS

Amplification was positive for *R. felis* in 39 lots of *Ct. felis* (54.17%) (FIG. 1). No rickettsiae were found in the other anthropic fleas (*P. irritans* and *C. canis*) considered in this study. The PCR showed a high sensitivity, and species identification was confirmed by sequence.

The sequences obtained were compared with sequences from other *Rickettsia* species in GenBank using the BLAST utility (NCBI, Bethesda, MD, USA) and were identical to the previously reported sequence for *R. felis*. The fragment amplified for *ompB* corresponds to positions 599 to 1,259 in GenBank accession AF210695. All internal controls were negative.

CONCLUSIONS

Our findings show molecular evidence of *R. felis* infection in *Ct. felis*, which suggests that this pathogen is well established in Andalusia. In consequence, companion cats and dogs infested with *Ct. felis*, as well as people in contact with their pets, are exposed to pathogenic *R. felis*.

ACKNOWLEDGMENTS

This work was supported in part by a grant from the Fondo de Investigación Sanitaria (FIS-05/PI041521).

REFERENCES

1. PAROLA, P., B. DAVOUST & D. RAOULT. 2005. Tick- and flea-borne rickettsial emerging zoonoses. Vet. Res. **36**: 469–492.
2. MÁRQUEZ, F.J., M.A. MUNIAIN, J.M. PÉREZ-JIMÉNEZ, *et al.* 2002. Presence of *Rickettsia felis* in the cat flea from SW Europe. Emerg. Infec. Dis. **8**: 89–91.

Molecular Identification of *Rickettsia felis*-like Bacteria in *Haemaphysalis sulcata* Ticks Collected from Domestic Animals in Southern Croatia

DARJA DUH,[a] VOLGA PUNDA-POLIĆ,[b] TOMI TRILAR,[c]
MIROSLAV PETROVEC,[a] NIKOLA BRADARIĆ,[b]
AND TATJANA AVŠIČ-ŽUPANC[a]

[a]*Institute of Microbiology and Immunology, Ljubljana, Slovenia*

[b]*University Hospital and School of Medicine Split, Split, Croatia*

[c]*Slovenian Museum of Natural History, Ljubljana, Slovenia*

ABSTRACT: *Haemaphysalis sulcata* ticks collected from sheep and goats in southern Croatia were found infected with rickettsiae. Molecular analysis of the complete *gltA* gene and portion of 17 kDa and *ompB* genes revealed the presence of *Rickettsia felis*-like bacteria in up to 26% of tested ticks.

KEYWORDS: *Rickettsia*; *Haemaphysalis sulcata* ticks; molecular analysis; southern Croatia

INTRODUCTION

Ticks of the genus *Haemaphysalis* were rarely associated with rickettsiae. However, several recent studies done mostly in Asia and Russia revealed the presence of rickettsiae in *Hemaphysalis* ticks based on molecular research.[1,2] All bacteria described in these studies are well-known human pathogens previously detected in different species of ticks or fleas, namely, *Rickettsia rickettsii, R. japonica, R. sibirica*, and *R. felis*.[1-3] Description of *R. felis* in ticks is unusual because this bacterium is primarily vectored by fleas.[4] Yet, by phylogenetic analysis *R. felis* was placed in a spotted fever group (SFG) of rickettsiae, which includes species mostly transmitted by ticks.[4,5]

Herein we declare that *H. sulcata* ticks collected in southern Croatia are rather highly infected with rickettsiae most similar to *R. felis*. Two rickettsial

Address for correspondence: Dr. Darja Duh, Institute of Microbiology and Immunology, Zaloška 4, 1000 Ljubljana, Slovenia. Voice: +38615437455; fax: +38615437401.
e-mail: darja.duh@mf.uni-lj.si

**Ann. N.Y. Acad. Sci. 1078: 347–351 (2006). © 2006 New York Academy of Sciences.
doi: 10.1196/annals.1374.068**

diseases are commonly recognized in southern Croatia—Mediterranean spotted fever caused by *R. conorii* and rickettsial pox caused by *R. akari.*[6] However, this is the first report of the presence of *R. felis*-like bacteria in Croatia.

MATERIAL AND METHODS

In autumn 2000, spring 2001, and spring 2002, a total of 795 adult *H. sulcata* ticks were collected from domestic animals (sheep, goats) in southern Croatia. Ticks were washed in 70% ethanol and sterile water and homogenized with a pestle. DNA was extracted from 101 individual ticks using the QIAamp DNA Mini Kit (Qiagen, GmbH, Hilden, Germany). The efficiency of DNA extraction was confirmed by polymerase chain reaction (PCR) assay, which amplifies the mitochondrial 16S rRNA gene of tick origin.[7]

Each tick was tested for the presence of rickettsiae by using PCR and sequence analysis of complete *gltA* gene and a portion of 17 kDa, *ompA*, and *ompB* genes.[3,4,8] Sequencing on both strands was carried out in an automated sequencer using BigDye Terminator Cycle Sequencing Ready Reaction Kit (PE Applied Biosystems, Foster City, CA). Sequences obtained were analyzed with computer programs of the Lasergene 1999 software package (Dnastar, Madison, WI) and submitted to GenBank for determination of the accession numbers.

RESULTS

H. sulcata ticks were found to be infected with rickettsiae with the infection rate varying from 19.6 to 26% depending on when the ticks were collected (TABLE 1). With an exception of *ompA* gene, which could not be amplified, the nucleotide sequence determination of all other genes allowed us to identify the detected rickettsiae. The rickettsiae in ticks were identical among each other and most similar to *R. felis* (*Rickettsia* sp. California-2) in 97.5%, 95.2%, and 98% regarding the complete *gltA*, partial 17 kDa, and *ompB* genes, respectively. When the nucleotide sequence of complete *gltA* gene was translated, the *Rickettsia* from ticks showed even higher similarity to *R. felis* (98.8%), indicating that some of the nucleotide differences were due to wobbles at third or second codon position. Phylogenetic tree inferred with the PAUP program Phylogenetic Analysis Using Parsimony using distance analysis (K2P model), neighbor-joining tree evaluation, and 1,000 bootstraps confirmed clustering of *Rickettsia* from Croatian *H. sulcata* ticks with *R. felis* within the SFG (FIG. 1).

DISCUSSION

Ixodid ticks are the most important vectors and reservoirs of SFG rickettsiae, although at least two species within the SFG are associated with

TABLE 1. Collection of *H. sulcata* ticks in southern Croatia and determination of infection with rickettsiae by amplification of complete *gltA* gene and a portion of 17 kDa, *ompA*, and *ompB* genes

Month and year of collection	Number of Collected *H. sulcata*	Number of Tested *H. sulcata*	Number of Positive *H. sulcata*[a]
October 2000	111	51	10 (19.6%)
May 2001	4	4	1 (25%)
March 2002	680	46	12 (26%)
Overall	795	101	23 (22.8%)

[a]Ticks were considered positive if amplicons of complete *gltA*, partial 17 kDa, and *ompB* genes could be obtained.

mites and fleas. Ticks of the genera *Rhipicephalus, Dermacentor, Ixodes,* and *Amblyoma* are well known for transmitting different rickettsiae including human rickettsial pathogens.[5] However, the genus *Haemaphysalis* has been until recently of no interest regarding rickettsioses. Several reasons for this are: (1) *Haemaphysalis* ticks rarely bite humans and were therefore of little significance in the epidemiology of human pathogens; (2) the vectors

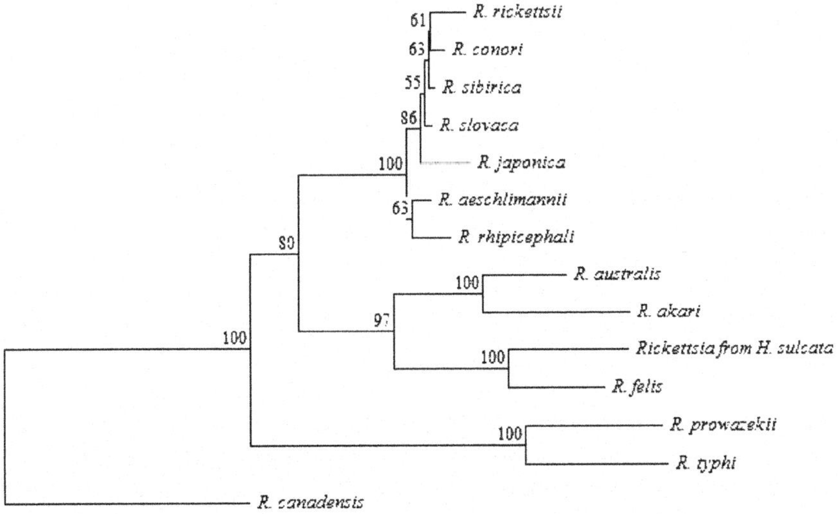

FIGURE 1. Phylogenetic relationships of representative rickettsiae deposited in the GenBank and detected in this study inferred from multiple sequence alignment of complete *gltA* gene. Accession numbers of rickettsiae: *R. rickettsii* (U59729), *R. conori* (U59730), *R. sibirica* (U59734), *R. slovaca* (U59725), *R. japonica* (U59724), *R. aeschlimannii* (U59722), *R. rhipicephali* (U59721), *R. australis* (U59718), *R. akari* (U59717), *Rickettsia* from *H. sulcata* (DQ081187), *R. felis* (AF210692), *R. prowazekii* (U59715), *R. typhi* (U59714), and *R. canadensis* (U59713). The number on each branch shows the percentage of occurrence in 1,000 bootstrap replicates.

for human rickettsial pathogens were already described and hence *Haemaphysalis* ticks were not tested for their presence; and (3) only four species of *Haemaphysalis* ticks are present in American continents and only one has been associated with rickettsiae, namely, *R. rickettsii* and *R. canada.*[9] With the description of *R. japonica* and several other rickettsial pathogens in *Haemaphysalis* ticks, their role as vectors and reservoirs of rickettsiae was established.[1,2]

H. sulcata are widely distributed in the Mediterranean region and are mostly found on livestock. We have collected them from sheep and goats in autumn and spring from 2000 to 2002 in southern Croatia. Their seasonal occurrence is between October and March and therefore they were almost absent in May 2001 (TABLE 1). Among 101 tested *H. sulcata* ticks, 19.6–26% of ticks were found infected with rickettsiae as determined by molecular research. Nucleotide sequence analysis of the complete *gltA*, partial 17 kDa, and *ompB* genes revealed that *Rickettsia* species in these ticks was most similar to *Rickettsia* sp. California-2, which was renamed as *R. felis* (Raoult, personal communication).

R. felis is a human pathogen causing flea-borne spotted fever reported in the United States, Mexico, Brazil, Germany, and France.[10] Although it is primarily transmitted by cat flea *Ctenocephalides felis,* it was detected in four *H. flava* ticks in Japan.[1,4] The authors were not certain whether the *Rickettsia* temporarily propagated in the tick from the animal infected by the cat flea or the tick inherited them by ovarial transmission.[1] In our study, *R. felis*-like bacteria from *H. sulcata* collected in southern Croatia were continuously detected in this tick species with high infection rate. Thus it is unlikely that the ticks accidentally acquired *Rickettsia* during the blood meal. *C. felis* fleas are semiobligate parasites on their preferred dog and cat hosts. In the absence of mentioned hosts, the fleas can be found on different animals including small mammals, which can also harbor *H. sulcata* ticks. However, only larvae and nymphs feed on small mammals and because we have detected the *R. felis*-like bacteria in adult ticks, we assumed that they probably acquired the bacteria by transovarial and/or trans-stadial transmission.

Southern Croatia is an endemic region for Mediterranean spotted fever.[6] In the case report of *R. felis* human infection in Europe, the authors described that the patient's symptoms resembled those of Mediterranean spotted fever.[11] They tested the patient for antibodies to *R. conorii* and found elevated titers. When molecular techniques were applied, they identified *R. felis* as the causative agent. Given the resemblance of spotted fever caused by *R. conorii* and *R. felis* and the high infection rate of *H. sulcata* ticks with *R. felis*-like bacteria, the human infection with this bacterium should be taken into consideration in southern Croatia. Identification and isolation of *R. felis*-like bacteria in a human patient would further warrant renaming this bacterium as proposed here: *Rickettsia kastelanii.*

REFERENCES

1. ISHIKURA, M., S. ANDO, Y. SHINAGAWA, *et al.* 2003. Phylogenetic analysis of spotted fever group rickettsiae based on gltA, 17-kDa, and rOmpA genes amplified by nested PCR from ticks in Japan. Microbiol. Immunol. **47:** 823–832.
2. LEE, J.H., H.S. PARK, K.D. JUNG, *et al.* 2003. Identification of the spotted fever group rickettsiae detected from *Haemaphysalis longicornis* in Korea. Microbiol. Immunol. **47:** 301–304.
3. ROUX, V., E. RYDKINA, M. EREMEEVA, *et al.* 1997. Citrate synthase gene comparison, a new tool for phylogenetic analysis, and its application for the rickettsiae. Int. J. Syst. Bacteriol. **47:** 252–261.
4. RAOULT, D., B. LA SCOLA, M. ENEA, *et al.* 2001. A flea-associated *Rickettsia* pathogenic for humans. Emerg. Infect. Dis. **7:** 73–81.
5. AZAD, A.F. & C.B. BEARD. 1998. Rickettsial pathogens and their arthropod vectors. Emerg. Infect. Dis. **4:** 179–186.
6. PUNDA, V., I. MILAS, N. BRADARIC, *et al.* 1984. Mediteranska pjegava groznica u Jugoslaviji. Lijec. Vjesn. **106:** 286–288.
7. BLACK, W.C. & J. PIESMAN. 1994. Phylogeny of hard- and soft-tick taxa (Acari: Ixodida) based on mitochondrial 16S rDNA sequences. Proc. Natl. Acad. Sci. USA **91:** 10034–10038.
8. HIGGINS, J.A. & A.F. AZAD. 1995. Use of polymerase chain reaction to detect bacteria in arthropods: a review. J. Med. Entomol. **32:** 213–222.
9. LABRUNA, M.B., L.M. CAMARGO, E.P. CAMARGO, *et al.* 2005. Detection of a spotted fever group Rickettsia in the tick *Haemaphysalis juxtakochi* in Rondonia. Brazil Vet. Parasitol. **127:** 169–174.
10. KENNY, M.J., R.J. BIRTLES, M.J. DAY, *et al.* 2003. *Rickettsia felis* in the United Kingdom. Emerg. Infect. Dis. **9:** 1023–1024.
11. RICHTER, J., P.E. FOURNIER, J. PETRIDOU, *et al.* 2002. *Rickettsia felis* infection acquired in Europe and documented by polymerase chain reaction. Emerg. Infect. Dis. **8:** 207–208.

Expression of rOmpA and rOmpB Protein in *Rickettsia massiliae* during the *Rhipicephalus turanicus* Life Cycle

MOTOHIKO OGAWA, KOTARO MATSUMOTO, PAROLA PHILIPPE, DIDIER RAOULT, AND PHILIPPE BROUQUI

Unité des Rickettsies, CNRS UMR 6020, IFR 48, Faculté de Médecine, Université de la Méditerranée, Marseille, France

ABSTRACT: *Rhipicephalus turanicus* tick colony infected in the laboratory with *Rickettsia massiliae* showed that the rickettsia is transovarially and transdatially tramsmitted. The expression of rOmpB did not change with temperature or the stages of the tick life cycle. In contrast, rOmpA was less expressed during the larval stage.

KEYWORDS: *Rhipicephalus turanicus*; *Rickettsia massiliae*; rOmpA; rOmpB; larvae

INTRODUCTION

Rickettsiae are obligate intracellular Gram-negative bacteria that belong to the alpha subdivision of *Proteobacteria*. The spotted fever group (SFG) rickettsia contains a large variety of strains. Some are human pathogens causing tick-borne eruptive fevers, whereas others have only been isolated from arthropods. *Rickettsia massiliae* was first isolated from *Rhipicephalus (R.) turanicus* ticks in the southern part of France.[1] *R. turanicus* is wildly distributed in this area and mainly bites cattle, sheep, goats, and occasionally humans and dogs. Human cases have been detected but not reported yet (D. Raoult, unpublished data). Recently, we succeeded in establishing a laboratory colony of *R. turanicus* infected with *R. massiliae.*[2] *R. massiliae* is transovarially transmitted to the progeny. Rates of transovarial and transtadial transmission are nearly 100%.[2] These results suggest that in this model *R. massiliae* has a symbiotic relationship with its host tick. *R. peacockii* is reported as an endosymbiont of *Dermacentor andersonii* and is closely related to the virulent *R. rickettsii.*[3] *R. peacockii* does not express rOmpA.[4] It has been suggested that rOmpA would be an important aspect of virulence of some rickettsia.[5] *R. rickettsii* is

Address for correspondence: Didier Raoult, Unité des Rickettsies, CNRS UMR 6020, IFR 48, Faculté de Médecine, Université de la Méditerranée, Marseille, France.
e-mail: didier.raoult@medecine.univ-mrs.fr

**Ann. N.Y. Acad. Sci. 1078: 352–356 (2006). © 2006 New York Academy of Sciences.
doi: 10.1196/annals.1374.069**

reactivated when the temperature increased in ticks and its protein profile is influenced by temperature, but not by host cell species.[6] We wonder herein whether the expression of rOmpA and rOmpB in *R. massiliae* varied depending upon the tick cycle and whether this was influenced by temperature and feeding.

Ticks and Preparation of Tick Specimens

R. turanicus ticks infected with *R. massiliae* are maintained in the laboratory at 25°C in high humidity.[2] For tick feeding, New Zealand white rabbits are used. For evaluation at different temperatures during the life cycle of the ticks, the ticks at each stage and the eggs were kept at room (25°C), high (37°C), and low (4°C) temperatures for 6 days before the assay. The ticks were washed with 70% ethanol and rinsed in distilled water, and then crushed in phosphate-buffered saline (PBS) at pH 7.2. Eggs were directly crushed in PBS. The suspension of tick cells was deposed on multi-well glass plates and used in the assay.

Immunofluorescence Assay

rOmpA and rOmpB were detected by double-stain immunofluorescence assay.[7] In this assay, a convalescent serum from a guinea pig experimentally infected with *R. massiliae* was used followed by FITC-conjugated anti-guinea pig IgG-whole molecule (Sigma Chemical Co. *Steinheim*, Germany). Monoclonal antibodies against rOmpA and rOmpB previously raised in the laboratory[8] were used and followed by anti-mouse Texas red-conjugated antibody (Jackson ImmunoResearch Europe Ltd. Co., Cambridgeshire, UK). *R. massiliae* grown in Vero cells at 32°C was used as control. To evaluate the expression of rOmpA and rOmpB, rickettsiae were counted on serial photos taken with a confocal microscope, TCS-4D (Leica Microsystems, Wetzlar, Germany) (FIG. 1).

Statistical Analysis

Differences in number of rickettsia expressing the proteins were tested by the chi-square test. A difference was considered to be statistically significant when the P value was less than 0.05.

Whatever the temperatures and the stages were, no significant changes were observed in the expression of rOmpB (FIG. 2). The percentage of rickettsia expressing rOmpA varied significantly from 83.5% to 94.2% depending upon the stage of the ticks but not with the temperature. The lower percentage of rickettsia expressing rOmpA was found in larva when they were not engorged (TABLE 1). A return to baseline expression was observed in larva when tested

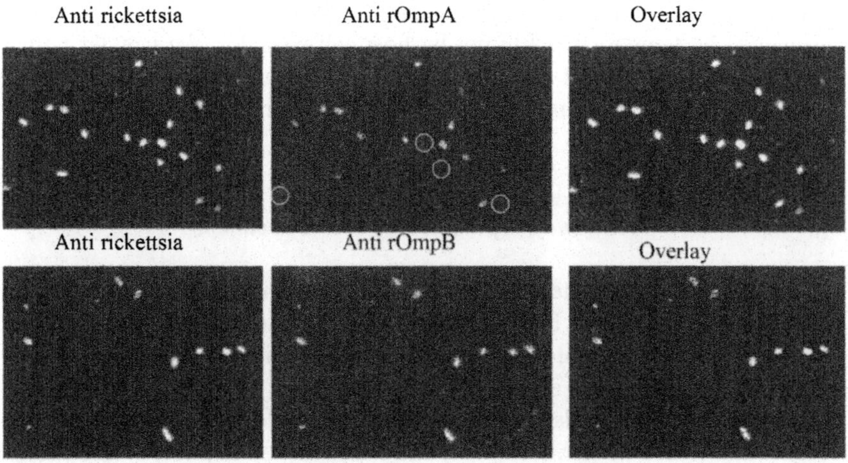

FIGURE 1. rOmpA and rOmpB expression of *Rickettsia massiliae* in *Rhipicephalus turanicus* larva ticks: double-stain immunofluorescence with anti-rOmpA (FITC) and rOmpB (Texas red) monoclonal antibodies. Pictures were taken with a confocal microscope (Leica Microsystems, Wetzlar, Germany). Magnification × 400.

after engorgement and may be associated with the reactivation phenomenon (FIG. 2). Specificity of monoclonal antibodies against both rOmpA and rOmpB were confirmed in Western blotting (data not shown). Our results showed that level of rOmpA expression varied through the life cycle of the ticks, whereas level of rOmpB expression was stable. This result suggests that rOmpA might be influenced by environmental changes of tick cells throughout the tick's life

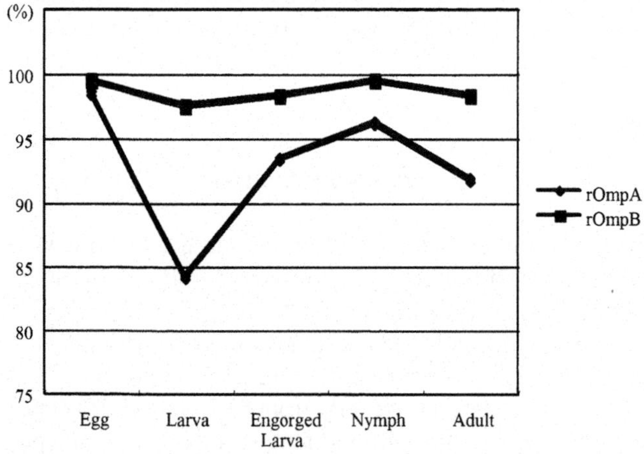

FIGURE 2. Percentage of rickettsia expressing rOmpA and rOmpB during the life cycle of *Rhipicephalus turanicus.*

TABLE 1. Expression of rOmpA and rOmpB of *Rickettsia massiliae* during a life cycle of tick and at different temperatures

	Normal (25°C)[a]		High (37°C)		Low (4°C)	
	rOmpA	rOmpB	rOmpA	rOmpB	rOmpA	rOmpB
Egg	438/444 (98.6)[b,c]	458/460 (99.6)	675/717 (94.1)[e]	590/598 (98.7)	640/658 (97.3)	730/735 (99.3)
Larva	684/815 (83.9)[d]	726/766 (94.8)	664/795 (83.5)	650/671 (96.9)	1007/1191 (84.6)	794/828 (95.9)
Engorged larva	489/523 (93.5)[c]	441/448 (98.4)	NT	NT	NT	NT
Nymph	573/595 (96.3)	692/695 (99.6)	473/477 (99.2)	597/602 (99.2)	360/368 (97.8)	476/483 (98.6)
Adult	487/530 (91.9)[c]	599/609 (98.4)	425/451 (94.2)	437/446 (98.0)	400/430 (93.0)	510/515 (99.0)

[a]Ticks were kept at normal (25°C), high (37°C), and low (4°C) temperatures for 6 days before the assay. Crashed whole body of ticks was used as antigen in immunofluorescence assay.

[b]rOmpA-or rOmpB-positive rickettsial particle/all rickettsial particle (%).

[c]There is a statistically significant difference between rOmpA and rOmpB expression at each stage of tick life cycle.

[d]There is a statistically significant difference of rOmpA expression among the different stages of ticks.

[e]There is a statistically significant difference of rOmpA expression among the ticks at different temperatures.

cycle. Variation in the expression of major surface membrane proteins during tick life cycle has been reported with *Borrelia burgdorferii*. In this model OspC is absent and OspA present on the bacteria in unfed ticks and the OspA level decreased while the OspC level increased during tick feeding and after, suggesting that OspC would be a transmission-specific outer membrane protein.[9] Temperature is well known to influence gene expression in bacteria as do many other environmental factors, such as osmolarity, pH, and growth phase.[9] This has been reported for tick-associated bacteria, such as *B. burgdorferii*, but also for *Rickettsia*.[10] During the tick cycle in our system neither rOmpA nor rOmpB were influenced by temperature.

In conclusion, rOmpA is expressed in the tick *R massiliae* all along its life cycle, but is slightly less expressed in the resting time of the larva stage.

REFERENCES

1. BEATI, L. & L. RAOULT. 1993. *Rickettsia massiliae* sp.nov., a new spotted fever group rickettsia. Int. J. Syst. Bact. **43:** 839–840.
2. MATSUMOTO, K., P. BROUQUI, D. RAOULT & P. PAROLA. Transmission of *Rickettsia massiliae* Bar29 in *Rhipicephalus turanicus* (Acari: Ixodidae). Med. Vet. Entomol. In press.
3. SIMSER, J., A.T. PALMER, U.G. MUNDERLOH & T.J. KURTTI. 2001. Isolation of a spotted fever group rickettsia, *Rickettsia peacockii*, in a Rocky Mountain wood tick, *Dermacentor andersoni*, cell line. Appl. Environ. Microbiol. **67:** 546–552.
4. BALDRIDGE, G.D., N.Y. BURKHARDT, J.A. SIMSER, *et al.* 2004. Sequence and expression analysis of the ompA gene of *Rickettsia peacockii*, an endosymbiont of the Rocky Mountain wood tick, *Dermacentor andersoni*. Appl. Environ. Microbiol. **70:** 6628–6636.
5. LI, H. & D.H. WALKER. 1998. RompA is a critical protein for the adhesion of *Rickettsia rickettsii* to host cells. Microbial. Pathogenesis **24:** 289–298.
6. POLICASTRO, P.F., U.G. MUNDERLOH, E.R. FISCHER & T. HACKSTADT. 1997. *Rickettsia rickettsii* growth and temperature-inducible protein expression in embryonic tick cell lines. J. Med. Microbiol. **46:** 839–845.
7. HOUHAMDI, L., P.E. FOURNIER, R. FANG, *et al.* 2002. An experimental model of human body louse infection with *Rickettsia prowazekii*. J. Infect. Dis. **186:** 1639–1646.
8. XU, W. & D. RAOULT. 1997. Production of monoclonal antibodies against *Rickettsia massiliae* and their use in antigenic and epidemiological studies. J. Clin. Microbiol. **35:** 1715–1721.
9. KONKEL, M.E. & K. TILLY. 2000. Temperature-regulated expression of bacterial virulence genes. Microbes Infect. **2:** 157–166.
10. HAYES, S.F. & W. BURGDORFER. 1982. Reactivation of *Rickettsia rickettsii* in *Dermacentor andersoni* ticks: an ultrastructural analysis. Infect. Immun. **37:** 779–785.

Lice Infestation and Lice Control Remedies in the Ukraine

I. KURHANOVA

L'viv Research Institute of Epidemiology and Hygiene of Ministry of Health of Ukraine, L'viv, Ukraine

ABSTRACT: A permanent decrease was seen in the prevalence of lice infestation among population of the Ukraine from 1990–2004. The prevalence of lice infestation among children under 14 years of age was 6–27 times more than that in adults. The highest figures were among children 7–14 years old. During all of the observation period there were changing tendencies relative to the groups. The greatest number of cases of infection with lice were noted in the months when the control inspections of children in education institutions were performed.

KEYWORDS: lice infestation; pediculicides

A permanent decrease of lice infestation among people during 1990–2004 in the Ukraine was reported (FIG. 1).

The prevalence of lice infestation among children by to 14 years of age 14 were 6–27 times higher than those among adults in different years of this period. The highest figures were among children 7–14 years old. During this time the proportion of infested children decreased from 85.00% to 52.89% and the proportion of infested adults increased from 15.0% to 47.115% (FIG. 2) in the infected population of the Ukraine.

During this observation period different tendencies in the change of the intensive figures in different regions of Ukraine were noticed.

Head lice infestation constantly dominated in all the population in Ukraine and its level was observed with the current increase in the level of body and mixed lice. Among children under 14 head lice infestation was 98.62%, and among adults 89.41% (FIG. 3).

The highest prevalence was in September and other months when the control inspections of children in schools and other education institutions were performed. The prevalence of body and mixed lice infestation in the cold period of year in all groups of the population is shown in FIGURE 4.

Address for correspondence: I. Kurhanova, L'viv Research Institute of Epidemiology and Hygiene, Zelena St., 12 L'viv, Ukraine. Fax: 0322-76-30-67.
e-mail: irynka@liTech.lviv.ua; Rickettsia@list.ru

Ann. N.Y. Acad. Sci. 1078: 357–360 (2006). © 2006 New York Academy of Sciences.
doi: 10.1196/annals.1374.070

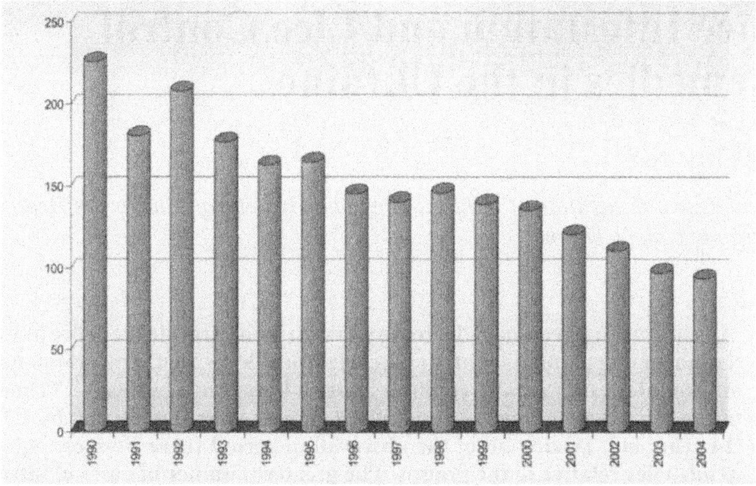

FIGURE 1. Infestation of lice among the population of the Ukraine from 1990–2004.

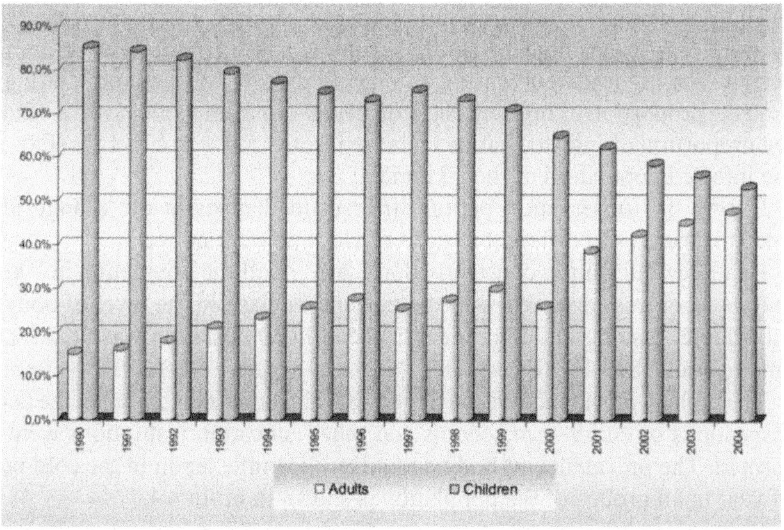

FIGURE 2. Percentage of infestation of lice in adults and children under 14 years of age from 1990–2004.

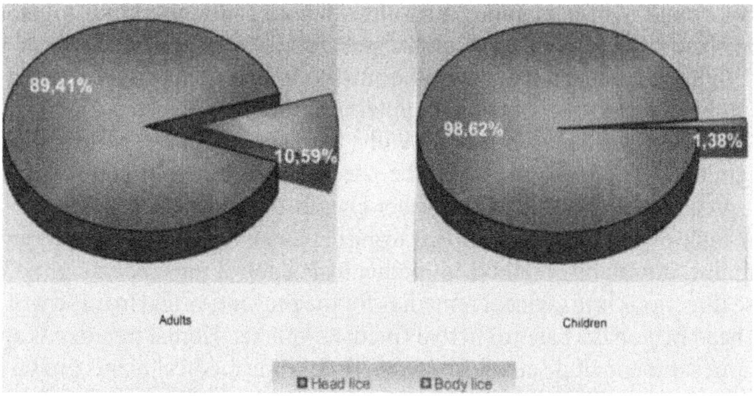

FIGURE 3. Percentage of head and body lice among adults and children up to 14 years of age in 1999–2004.

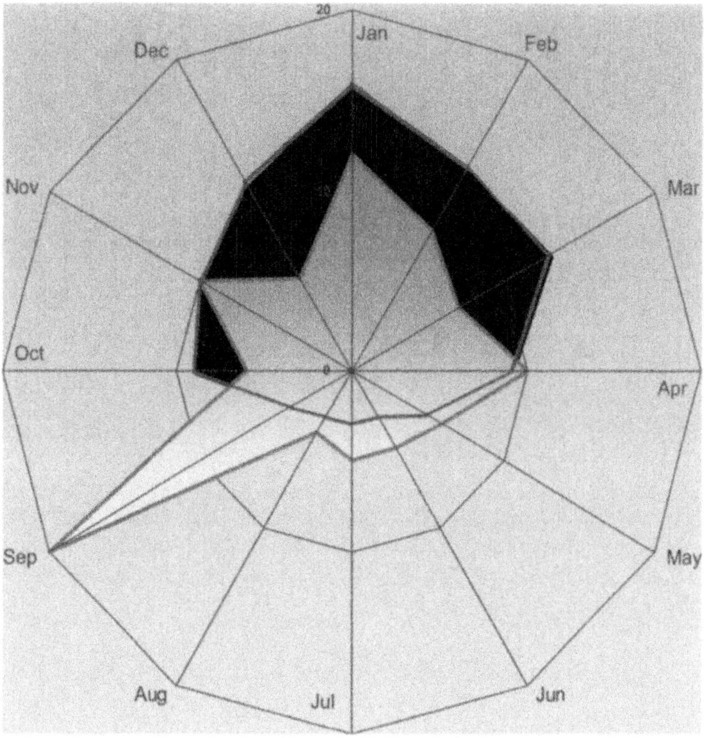

FIGURE 4. Prevalence of body and mixed lice in cold months in all groups of the population in 1999–2004.

The complex of lice control measures includes different kinds of pediculicides. The direct application of the pediculicides is conducted by medical staff in health, educational, and other institutions or by persons at home. The pediculicides are sold overs the-counter through pharmacies.

The pediculicides are made with benzyl benzoate, permethrin, and malathion. None of them should be used to treat children under 3 years of age, pregnant women, and some other classes of persons.

In light of the resistance of lice to insecticides, which are broadly used in medicine, alternative methods of pediculosis control must be streight. One of these directions is to concoct remedies for the prevention and treatment of body and head lice on the basis of herbal (medical) plants. Herbal treatments against lice infestation include ethanol extracts from some medical plants grown in the Ukraine, such as *Padus racemosa, Tanacetum vulgare, Euphorbia cyparissias, Scabiosa arvensis, Viola tricolor,* and *Coronaria flos-cuculi*. We have devised methods of manufacturing new prescriptions of lotion, and have studied the effect on lice of a laboratory population of *Pediculus humanus*—the result was a death rate of 100% of lice. A series of pediculicide lotions made of medical plants, which are used in folk scientific medicine, has been produced, and are patented in the Ukraine.

Prevalence of *Rickettsia felis* in the Fleas *Ctenocephalides felis felis* and *Ctenocephalides canis* from Two Indian Villages in São Paulo Municipality, Brazil

MAURICIO C. HORTA,[a] DANIELA P. CHIEBAO,[a]
DANIELE B. DE SOUZA,[b] FERNANDO FERREIRA,[a]
SÔNIA R. PINHEIRO,[a] MACELO B. LABRUNA,[a]
AND TERESINHA T.S. SCHUMAKER[b]

[a]*Department of Preventive Veterinary Medicine and Animal Health, Faculty of Veterinary Medicine, University of São Paulo, São Paulo, Brazil*

[b]*Department of Parasitology, Institute of Biomedical Sciences, University of São Paulo, São Paulo, Brazil*

ABSTRACT: We evaluated the presence of *Rickettsia* infection among fleas collected on domestic dogs in two Guarani Indian communities in the suburban area of São Paulo Municipality, Brazil. A total of 114 *Ctenocephalides felis felis* and 47 *Ctenocephalides canis* were collected from 40 dogs. A total of 41 *C. felis felis* (36.0%) and 9 *C. canis* (19.1%) fleas yielded expected bands by PCR, which were all shown by DNA sequencing to be indentical to the corresponding sequence of a fragment of the *Rickettsia felis gltA* gene deposited in GenBank. The overall prevalence of *R. felis* was 31.0% (49/161).

KEYWORDS: *Rickettsia felis*; *Ctenocephalides felis felis*; *Ctenocephalides canis*; flea; dog; Brazil

INTRODUCTION

The University of São Paulo has developed a multidisciplinary project in two Indian villages in São Paulo Municipality, Brazil. One of the project's aims is to investigate the occurrence of zoonoses among Indian residents. Human inhabitants and domestic dogs share the same habitat in the villages, increasing the risk of the acquisition of several zoonoses. *Ctenocephalides felis felis* and *Ctenocephalisdes canis* are the most important flea species of

Address for correspondence: Maurício C. Horta, Faculdade de Medicina Veterinária e Zootecnia, Universidade de São Paulo, Av. Prof. Dr. Orlando Marques de Paiva, 87, São Paulo, Brazil, 05508-270. Voice: 55-11-3091-7701; fax: 55-011-3091-7928.
e-mail: maurivet@yahoo.com

Ann. N.Y. Acad. Sci. 1078: 361–363 (2006).
doi: 10.1196/annals.1374.071

pet animals.[1] Besides the dermatological implications, these ectoparasites can transmit several disease agents to animals and humans.[1,2]

Rickettsia felis (the etiological agent of flea-borne rickettsiosis) has been previously reported in *C. felis felis* fleas from Brazil.[3] Although *R. felis* has been reported in *C. felis felis* worldwidely, its presence in the cosmopolitan flea *C. canis* has been reported only in Peru and Thailand.[4,5] In the current study, we evaluated the presence of *Rickettsia* in fleas collected on domestic dogs in Indian villages located in São Paulo Municipality, Brazil.

MATERIALS AND METHODS

This study was performed in two Indian villages located in the suburban area of the São Paulo municipality: Morro da Saudade village ($\approx$26.3 ha) with a current population of 695 Indians and Krukutu village ($\approx$25.9 ha) with a population of 145 Indians. From August to November 2004, 161 fleas (89 from Morro da Saudade and 72 from Krukutu) were collected from 40 resident dogs, which represented $\approx$28% of the entire canine population of the two villages. Fleas were disinfected in iodine alcohol, and placed in a freezer at -80 °C after being taxonomically identified.[2] The DNA of each flea was individually extracted[3] and tested by PCR targeting a 401-bp fragment of the rickettsial *gltA* gene.[6] Twenty-five *gltA*-positive fleas were also tested by a PCR targeting an 862-bp fragment of the rickettsial *ompB* gene.[6] PCR products were sequenced and compared with NCBI (National Center for Biotechnology Information) Nucleotide BLAST (Basic Local Alignment Search Tool) searches.

RESULTS AND DISCUSSION

The 161 collected fleas were identified as 114 *C. felis felis* (72 from Morro da Saudade and 42 from Krukutu) and 47 *C. canis* (17 from Morro da Saudade and 30 from Krukutu). A total of 41 *C. felis felis* (36.0%) and 9 *C. canis* (19.1%) fleas yielded expected bands by PCR targeting a fragment of the *gltA* gene, all of which were demonstrated by DNA sequencing to be 100% identical to each other and to the corresponding sequence of *R. felis* deposited in GenBank (AF210692). Twenty-five of the 50 *R. felis-gltA* positive fleas were also tested by a second PCR targeting an 862 fragment of the *ompB* rickettsial gene, yielding expected bands.

R. felis was detected in 29 (40.3%) and 12 (28.6%) *C. felis felis* fleas from Morro da Saudade and Krukutu villages, respectively. This rickettsia was also detected in one (5.9%) and eight (26.7%) *C. canis* fleas from Morro da Saudade and Krukutu villages, respectively.

In Brazil, *R. felis* has been detected in *C. felis felis*[3] and in patients showing clinical signals of spotted fever.[7] Thus, rickettisiosis due to *R. felis* may be taken into account in the investigation on zoonoses in the Indian villages. This study represents the first report of a natural infection of *R. felis* in *C. canis* in Brazil.

ACKNOWLEDGMENTS

This work was supported by Fundação de Amparo a Pesquisa do Estado de São Paulo (FAPESP grants 02/10759-0 to M.C.H., 03/13872-4 to M.B.L., 03/04728-7 to T.T.S.S).

REFERENCES

1. KRÂMER, F. & N. MENCKE. 2001. Flea Biology and Control: The Biology of the Cat Flea, Control and Prevention with Imidacloprid in Small Animals. Springer. Heidelberg.
2. LINARDI, P.M. & L.R. GUIMARÃES. 2000. Sifonápteros do Brasil. Museu de Zoologia USP/FAPESP. São Paulo.
3. HORTA, M.C., A. PINTER, A. CORTEZ, *et al.* 2005. *Rickettsia felis* (Rickettsiales: Rickettsiaceae) in *Ctenocephalides felis felis* (Siphonaptera: Pulicidae) in the State of São Paulo, Brazil. Arq. Bras. Med. Vet. Zootec. **57:** 321–325.
4. BLAIR, P.J., J. JIANG, G.B. SCHOELER, *et al.* 2004. Characterization of spotted fever group rickettsiae in flea and tick specimens from northern Peru. J. Clin. Microbiol. **42:** 4961–4967.
5. PAROLA, P., O.Y. SANOGO, K. LERDTHUSNEE, *et al.* 2003. Identification of *Rickettsia* spp. and *Bartonella* spp. in fleas from the Thai-Myanmar border. Ann. N. Y. Acad. Sci. **990:** 173–181.
6. LABRUNA, M.B., T. WHITWORTH, D.H. BOYER, *et al.* 2004. *Rickettsia bellii* and *Rickettsia amblyommii* in *Amblyomma* ticks from the State of Rondônia, Western Amazon, Brazil. J. Med. Entomol. **41:** 1073–1081.
7. RAOULT, D., B. LA SCOLA, M. ENEA, *et al.* 2001. A flea-associated *Rickettsia* pathogenic for humans. Emerg. Infect. Dis. **7:** 2–11.

Population Survey of Egyptian Arthropods for Rickettsial Agents

AMANDA D. LOFTIS,[a] WILL K. REEVES,[a] DANIEL E. SZUMLAS,[b] MAGDA M. ABBASSY,[b] IBRAHIM M. HELMY,[b] JOHN R. MORIARITY,[a] AND GREGORY A. DASCH[a]

[a] *Viral and Rickettsial Zoonoses Branch, Centers for Disease Control and Prevention, Atlanta, Georgia, USA*

[b] *Vector Biology Research Program, United States Naval Medical Research Unit No. 3, Cairo, Egypt*

ABSTRACT: Between June 2002 and July 2003, 987 fleas, representing four species, and 1019 ticks, representing one argasid and eight ixodid species, were collected from Egyptian animals. These arthropods were tested for rickettsial agents using polymerase chain reaction. DNAs from *Anaplasma* and *Ehrlichia* spp. were detected in 13 ticks. Previously undescribed *Bartonella* spp. were detected in 21 fleas. *Coxiella burnetii* was detected in two fleas and 20 ticks. *Rickettsia typhi* was detected in 27 fleas from 10 cities. Spotted fever group rickettsiae were detected in both fleas and ticks and included *Rickettsia aeschlimanii* and an unnamed *Rickettsia* sp.

KEYWORDS: *Rickettsia; Bartonella; Coxiella;* Siphonaptera; Acari; Ixodidae; Argasidae

Serologic surveys in Egypt have documented the exposure of humans, rodents, and domestic animals to rickettsial pathogens, including *Bartonella* spp., *Coxiella burnetii*, *Ehrlichia canis*, and typhus and spotted fever group *Rickettsia* spp.[1–4] Our objective was to determine the presence, geographic distribution, and species of rickettsial agents in ectoparasitic arthropods from Egypt.

Fleas and ticks were collected from wild and domestic animals from 17 cities in Egypt (FIG. 1) between July 2002 and July 2003. The collection location, date, host animal, ectoparasite species, and life stage and gender (adult arthropods) were recorded for each ectoparasite. A total of 987 fleas were collected and tested from 221 animals. Nine hundred (91.2%) of these fleas were

Address for correspondence: Amanda D. Loftis, CDC, 1600 Clifton Road NE, MS G-13, Atlanta, GA 30333.Voice: 404-639-4610; fax: 404-639-4436.
e-mail: aol2@cdc.gov

Ann. N.Y. Acad. Sci. 1078: 364–367 (2006).
doi: 10.1196/annals.1374.072

FIGURE 1. Map of Egypt showing arthropod collection sites (stars), July 2002–July 2003, overlaid on a political map of governorate and country boundaries.

Xenopsylla cheopis, and all but one of the *X. cheopis* were collected from Norway rats (*Rattus norvegicus*) ($n = 466$), or black rats (*Rattus rattus*) ($n = 433$). The remaining 87 fleas were collected from rats (*Rattus* spp.), a domestic goat (*Capra hircus*), a house mouse (*Mus musculus*), least weasels (*Mustela nivalis*), and Rueppel's foxes (*Vulpes rueppelli*), and included 38 *Ctenocephalides felis*, 37 *Leptopsylla segnis*, and 12 *Echinophaga gallinacea*. A total of 1,019 ticks were collected and tested from 109 animals. Ticks were collected primarily from cattle ($n = 610$), camels (145), and buffalo (94), and were identified as *Argas persicus* (29 from chicken and rabbit pens), *Rhipicephalus* (*Boophilus*) *annulatus* (625), *H. anatolicum anatolicum* (10), *H. anatolicum excavatum* (55), *H. dromedarii* (174), *H. impeltatum* (2), *H. marginatum rufipes* (3), *H.* spp. nymphs (55), *R. sanguineus* (49), and *R. turanicus* (17).

DNA was extracted from arthropods using a liquid-handling robot and 96-well plate-based DNA extraction kit (Promega BioSciences, Wizard SV96, San Luis Obispo, CA). Real-time polymerase chain reaction (PCR) was used to test arthropod DNAs for bacterial agents, including *Anaplasma*/*Ehrlichia* spp. (16S rDNA), *Bartonella* spp. (*glt*A gene), *Coxiella burnetii* (IS1111 transposable element), and *Rickettsia* spp. (17kD antigenic gene). Conventional PCR assays were used to amplify longer gene fragments from positive fleas and ticks, and these amplicons were sequenced to identify the bacterial agents.

DNA from *Anaplasma marginale* was detected in two ticks collected from cattle, and an agent with 80% homology of the citrate synthase gene to "*Anaplasma platys*" was detected in one tick. *Ehrlichia* spp. were detected in 10 ticks; eight ticks contained an *Ehrlichia* sp. with 99% homology of a 16S rDNA fragment to *Ehrlichia canis* and "*Ehrlichia ovina*," and two ticks contained an uncultured *Ehrlichia* sp. that is similar or identical to an agent reported from ticks from Asia and Africa. *A. marginale* is of veterinary and economic significance; the pathogenicity of the other agents is unknown.

DNA from *Bartonella* spp. was detected in 21 fleas (20 *X. cheopis* and 1 *L. segnis*). Two distinct 16S-23S ITS and two *gro*EL gene sequences were obtained from these fleas and appear to represent three distinct species of *Bartonella*. The pathogenicity of these previously undescribed agents is unknown, but at least one agent shows phylogenetic similarity to human pathogens. No *Bartonella* were detected in tick DNA extracts.

DNA from *C. burnetii* was detected in two fleas (one *X. cheopis* and one *C. felis*) and 20 ticks (16 *A. persicus*, three *Hyalomma* spp. adults, and one *R. sanguineus*), using a real-time PCR assay specific for the IS1111 transposable element, and confirmation using traditional PCR assays was possible in one flea and six ticks.

R. typhi, the agent of murine typhus, was detected in 2.9% (27/900) of *X. cheopis* and in 2.7% (1/37) of *L. segnis*. These fleas are experimental vectors of murine typhus. Typhus-positive fleas were collected from rats from 10 cities in the Nile Delta, Suez Canal area, and the coast of the Red Sea. These data document the continued presence of *R. typhi* throughout Egypt and are consistent with published seroprevalence data. *R. aeschlimanii* was detected in five adult *Hyalomma* spp. *R. conorii* was not detected in any ticks. An unnamed spotted fever group *Rickettsia* sp. was detected in all 12 *E. gallinacea*; this agent has been described from *Ctenocephalides* spp. in the United States and Thailand. Two additional, unidentified spotted fever group *Rickettsia* spp. were detected in fleas. The presence of these rickettsial agents in fleas and ticks might contribute to the presence of antibodies against *Rickettsia* spp. in the Egyptian population.

NOTE: The findings and conclusions in this report are those of the authors and do not necessarily represent the views of the United States Department of the Navy, U.S. Department of Defense, U.S. Department of Health and Human Services, or the United States Government.

REFERENCES

1. BOTROS, B.A. *et al.* 1995. Canine ehrlichiosis in Egypt: sero-epidemiological survey. Onderstepoort J. Vet. Res. **62:** 41–43.

2. CHILDS, J.E. *et al.* 1995. Prevalence of antibodies to *Rochalimaea* species (cat-scratch disease agent) in cats. Vet. Rec. **136:** 519–520.
3. CORWIN, A. *et al.* 1993. Community-based prevalence profile of arboviral, rickettsial, and Hantaan-like viral antibody in the Nile River Delta of Egypt. Am. J. Trop. Med. Hyg. **48:** 776–783.
4. REYNOLDS, M.G. *et al.* 2004. Serologic evidence for exposure to spotted fever and typhus group rickettsioses among persons with acute febrile illness in Egypt. Presented at the Fourth International Conference on Emerging Infectious Diseases. Atlanta, GA, USA, Feb 29–March 3.

First Molecular Detection of *R. conorii*, *R. aeschlimannii*, and *R. massiliae* in Ticks from Algeria

I. BITAM,[a,b] P. PAROLA,[b,c] K. MATSUMOTO,[b] J. M. ROLAIN,[b] B. BAZIZ,[d] S. C. BOUBIDI,[a] Z. HARRAT,[a] M. BELKAID,[a] AND DIDIER RAOULT[b]

[a]*Unité d'Entomologie Médicale, Institut Pasteur d'Algérie, Algiers, Algeria*

[b]*Unité des Rickettsies CNRS UMR 6020, IFR48, Faculté de Médecine, Marseille, France*

[c]*Laboratoire de Parasitologie et Mycologie, INSERM U399, IFR 48, Faculté de Médecine, Marseille, France*

[d]*Institut National Agronomique, El Harrach, Alger 16036, Algeria*

ABSTRACT: Ticks collected in Northern Algeria between May 2001 and November 2003 were tested by PCR for the presence of *Rickettsia* spp. DNA using primer amplifying *gltA* and *OmpA* genes. Three different spotted fever group rickettsias were amplified from these ticks: *R. Conorii* subsp. *P. conorii* strain Malish in *Rhipicephalus sanguineus*, *R. aeschlimannii* in *Hyalomma marginatum*, and *R. massiliae* in *Rhipicephalus turanicus*. Our results confirm the presence of *R. conorii* in ticks in Algeria and provide the first detection of *R. aeschlimannii* and *R. massiliae* in Algeria.

KEYWORDS: ticks; Algeria; *Rickettsia conorii*; *Rickettsia aeschlimannii*; *Rickettsia massiliae*; *Hyalomma*; *Rhipicephalus*

Tick-borne spotted fever group (SFG) rickettsioses are caused by obligate intracellular Gram-negative bacteria belonging to the genus *Rickettsia* within the order *Rickettsiales*. Although some rickettsias have been known as human pathogens since the beginning of the century, tick-borne SFG rickettsioses are also recognized as important emerging zoonoses worldwide.[1] In Algeria, tick-borne SFG rickettsioses have been poorly studied. Clinicians often claim to have admitted patients with spotted fever and these cases are usually considered as Mediterranean spotted fever (MSF) due to *Rickettsia conorii* subsp. *conorii*.[2] This disease is endemic in the Mediterranean area, where it is transmitted by the brown dog *Rhipicephalus sanguineus*. Although MSF is known

Address for correspondence: Didier Raoult, Unité des Rickettsies, CNRS UMR 6020, IFR 48, WHO Collaborative Center for Rickettsial Reference and Research, Faculté de Médecine, 27 Bd Jean Moulin, 13385 Marseille Cedex 5, France. Voice: (33) 4-91-32-43-75; fax: (33) 4-91-83-03-90.
e-mail: didier.Raoult@medecine.univ-mrs.fr.

**Ann. N.Y. Acad. Sci. 1078: 368–372 (2006). © 2006 New York Academy of Sciences.
doi: 10.1196/annals.1374.073**

to occur in Algeria, cases are poorly documented, usually solely by general SFG serology.[3,4] The aim of the present work was to survey ticks collected in Algeria and to detect rickettsial DNA by means of polymerase chain reaction (PCR).

MATERIALS AND METHODS

Collection and Identification of Ticks

A total of 78 ticks were collected from animals from various sites in Algeria between May 2001 and November 2003. All ticks were collected in Northern Algeria. In May–June 2001, 18 *Hyalomma marginatum* were collected from cattle and sheep at Ain Ennoussour (36°17′'N, 4°29′E). In June 2001, 3 *R. sanguineus* group ticks were collected from goats at El Tarf (36°46′N, 8°18′E). In February 2002, one *H. marginatum* was collected from cattle at Jijel (36°49′N, 5°45′E). In June 2002, 33 *Rh. bursa* were detected in cattle at Imeghdacen (36°48′N, 4°29′E). In May 2003, two *Rh. turanicus* were collected from cattle at Tizi Ouzou (36°43′N, 4°03′E). In August 2003, five *H. detritum* were collected from cattle at Tafaraoui (35°29′N, 0°32′O). Finally, in November 16 *Rh. sanguineus* group ticks were collected from hedgehogs in Algiers (36°46′N, 3°02′E).). All ticks were adults attached on mammals, when they were collected during diverse fieldwork in Algeria. Ticks were determined using taxonomic keys by one of us (I.B.). They were kept in alcohol until 2004, when they were tested by PCR for the presence of rickettsial DNA. Ticks of the *R. sanguineus* group, which were found to be positive for rickettsial DNA, were identified to the species level (as *R. sanguineus* or *R. turanicus*) by sequencing a portion of the 12S rRNA gene, as previously described.[5]

Polymerase Chain Reaction and Sequencing

Ticks were rinsed with distilled water, dried on sterile filter paper, and then crushed individually in sterile Eppendorf tubes. DNA was extracted by using the QIAamp Tissue Kit (QIAGEN, Hilden, Germany) according to the manufacturer's instructions. Rickettsial DNA was detected by PCR using primers RpCS.877p and RpCS.1258n (Eurogentec, Seraing, Belgium), which amplify a 396 base-pair fragment of the citrate synthase gene (*gltA*) of *Rickettsia* as previously described.[6,7] Additionally, a 629–632 base-pair fragment of the *ompA* gene was amplified using Rr190.70p and Rr190.701n.[7] Negative controls consisted of DNA extracted (in the same biosafety cabinet) from live uninfected lice from colonies of our laboratories (1 louse for 10 tested ticks). A positive control (*R. montanensis*) was included in each test. In order to identify the detected *Rickettsia* sp., PCR products were purified and DNA sequencing was performed as previously described.[6,7] All obtained sequences were assembled, edited, and compared to those available in GenBank as previously described.[8]

RESULTS

Using the *gltA* primers, PCR products of rickettsial DNA were detected in 21 ticks and from the positive controls. No nucleic acids were amplified from the negative controls. All ticks were also shown to be positive on the second PCR screening using *OmpA* primers. By sequencing the *gltA* and *OmpA* amplified fragments, three sequences of the expected size were, respectively, identified. These sequences were shown to correspond to three spotted fever group rickettsias. *R. conorii* subsp. *conorii* strain Malish (100% similarity with GenBank accession number AE008677 and U1028 for *gltA* and *OmpA*, respectively) was detected in a *R. sanguineus* specimen collected on a hedgehog. *R. aeschlimannii* (99.6% and 98.7% similarity with GenBank accession number AY259084 and AY259083 for *gltA* and *OmpA*, respectively) was identified in 11 *H. m. marginatum* and 1 *H. detritum detritum*, collected on cattle and sheep. Finally, *R. massiliae* (100% and 98.2% similarity with GenBank accession number RSU59720 and RMU43799 for *gltA* and *OmpA*, respectively) was detected in 4 *R. turanicus* (2 collected on cattle, 1 collected on goats, 1 collected on hedgehog) and 4 specimens of *R. sanguineus* (collected on a hedgehog).

DISCUSSION

Although *R. sanguineus* is the well-known vector of *R. conorii* subsp. *conorii*, the agent of MSF, this is the first time to our knowledge that this rickettsia is detected by molecular methods in ticks from Algeria. After an asymptomatic incubation for 6 days, MSF is abrupt and typical cases present with high fever, flu-like symptoms, and a black eschar at the tick bite site. Although *R. sanguineus* is well adapted to urban environments, it is relatively host-specific and rarely feeds on people unless its preferred host (the domestic dog) is not available. The probability for people to be bitten by several infected ticks is low and the inoculation eschar is usually unique in MSF.[1] One to several days following the onset of fever, a generalized maculopapular rash develops that often involves the palms and soles. The disease may be fatal as severe forms including major neurological manifestations and multiorgan involvement may occur in 5–6% of the cases.[1]

This work provides also the first detection of *R. aeschlimannii* in Algeria. This SFG rickettsia is an emerging pathogen that has been associated with ticks, particularly *H. m. marginatum* and *H. m. rufipes* ticks in Southern Europe and Africa, and *R. appendiculatus* in South Africa. To date, only two cases of *R. aeschlimannii* infection have been described, including one in a patient bitten in Morocco and one in South Africa.[1] The clinical signs of *R. aeschlimannii* infection seem to be similar to those of MSF. However, differences in the behaviors of the ticks and tick–rickettsia relationships might lead to clinical particularities. *Hyalomma* ticks, particularly *H. m. marginatum* are well known to bite people in Algeria (more than *Rh. sanguineus*). As demonstrated at sites in southern Europe, *H. m. marginatum* seem to be highly infected by *R. aeschlimannii* at least in some areas.[9] Thus, cases clinically identified in Algeria as MSF could be in fact due to *R. aeschlimannii*, particularly if patients present with several eschars.

Finally, the detection of *R. massiliae* in ticks in Algeria needs more attention. It has been associated with *R. sanguineus* and *R. turanicus* in Europe and sub-Saharan Africa.[5] Its pathogenicity for humans is unknown. However, *R. massiliae* has been detected in tick saliva,[5] suggesting that the bacterium could be transmitted through the tick bite. Furthermore, among 15 patients of MSF in Catalonia, Spain, eight sera reacted at high titers with only *R. conorii* and *R. massiliae* antigens, and the titers against *R. massiliae* were clearly higher than for *R. conorii*. Finally, isolation of *Rickettsia massiliae* from a patient has been recently reported in Sicily.[10]

Clinicians in Algeria or those elsewhere who may see patients returning from this country, have now to be aware not only of possible *R. conorii* subsp. *conorii* but also *R. aeschlimannii* infection in patients presenting signs of spotted fever and/or an eschar. However, further studies are needed to better describe the epidemiology and clinical features of spotted fever group rickettsioses in Algeria. Indeed, *R. felis*, the agent of the so-called flea-borne spotted fever, has also been recently detected in the country.[8] Although it may be misdiagnosed as and treated similarly to the tick-borne SFG rickettsioses, the epidemiological aspects of tick- or flea-transmitted SFG rickettsioses regarding the risk of exposure as well as preventive aspects are different and need consideration.

ACKNOWLEDGMENTS

We are grateful to Dr. Zine and Dr. Kemiha from the Veterinary Inspection at the Jijel department.

REFERENCES

1. PAROLA, P., C. PADDOCK & D. RAOULT. 2005. Tick-borne rickettsioses around the world: emerging diseases challenging old concepts. Clin. Microbiol. Rev. Submitted for publication.
2. ZHU, Y., P.E. FOURNIER, M. EREMEEVA & D. RAOULT. 2005. Proposal to create subspecies of *Rickettsia conorii* based on multi-locus sequence typing and an emended description of *Rickettsia conorii*. BMC Microbiol. **5:** 1–11.
3. DUPONT, H.T., P. BROUQUI, B. FAUGERE & D. RAOULT. 1995. Prevalence of antibodies to *Coxiella burnetti, Rickettsia conorii*, and *Rickettsia typhi* in seven African countries. Clin. Infect. Dis. **21:** 1126–1133.
4. BERNARD, J.G., J. BERENI & J. HAINAULT. 1963. Present status of the rickettsioses in Algeria [in French]. Bull. Soc. Pathol. Exot. Filiales. **56:** 620–628.

5. MATSUMOTO, K., P. BROUQUI, D. RAOULT & P. PAROLA. 2005. Transmission of *Rickettsia massiliae* Bar29 in the tick, *Rhipicephalus turanicus*. Med. Vet. Entomol. In press.
6. ROLAIN, J.M., M. FRANC, B. DAVOUST & D. RAOULT. 2003. Molecular detection of *Bartonella quintana, B. koehlerae, B. henselae, B. clarridgeiae, Rickettsia felis*, and *Wolbachia pipientis* in cat fleas, France. Emerg. Infect. Dis. **9:** 228–342.
7. PAROLA, P., Y.O. SANOGO, K. LERDTHUSNEE, *et al.* 2003. Identification of *Rickettsia spp.* and *Bartonella spp.* in fleas from the Thai-Myanmar border. Ann. N. Y. Acad. Sci. **990:** 173–181.
8. BITAM, I., P. PAROLA, B. BAZIZ, *et al.* 2005. First molecular detection of *Rickettsia felis* in fleas from Algeria. Am. J. Trop. Med. Hyg. In press.
9. MATSUMOTO, K., P. PAROLA, P. BROUQUI & D. RAOULT. 2004. *Rickettsia aeschlimannii* in *Hyalomma* ticks from Corsica. Eur. J Clin. Microbiol. Infect. Dis. **23:** 732–734.
10. VITALE, G., S. MANSUETO, J.M. ROLAIN & D. ROUALT. 2006. *Rickettsia massiliae* human isolation. Emerg. Inf. Dis. **12:** 174–175.

Ornithodoros moubata, a Soft Tick Vector for *Rickettsia* in East Africa?

SALLY J. CUTLER,[a] PAUL BROWNING,[b] AND JULIE C. SCOTT[a]

[a]*Veterinary Laboratories Agency, Addlestone, Surrey, UK*

[b]*Royal Holloway University of London, Egham, Surrey, UK*

ABSTRACT: *Omithodoros moubata* complex (Argasidae) ticks collected from human dwellings in central Tanzania were found to carry a novel rickettsial species that clustered among the spotted fever group. Atlhough no evidence of human infection was evident, these ticks feed primarily on man, thus providing opportunity for zoonotic infection.

KEYWORDS: *Rickettsia*; *Omithodoros moubata*; *Soft ticks*; *Argasidae*

BACKGROUND

In countries such as Tanzania arthropod-borne disease is common, with much of the population living in close proximity with disease vectors. Indeed, surveys of traditional Tanzanian "Tembe" dwellings show up to 88% of these to be infested with *Ornithodoros moubata* complex ticks.[1] These ticks are vectors for *Borrelia duttonii*, the cause of tick-borne relapsing fever endemic in this region. Rickettsiae like *Borrelia* are arthropod-borne, but are typically associated with transmission by a variety of hard tick (*Ixodid* species).

MATERIALS AND METHODS

Ticks

Collections were made from traditional dwellings in four different villages in the Dodoma Rural District, Central Tanzania. Dead ticks were imported under licence (AHZ/2074A/2001/13) and heat-inactivated (>80°C for 30 min) prior to proteinase K digestion and DNA extraction. DNA extracts were prepared from 204 samples collected from 81 huts.

Address for correspondence: Dr. S. G. Cutler, Bacterial Zoonoses, Department of Statutory and Exotic Bacterial Diseases, Veterinary Laboratories Agency, Woodham Lane, New Haw, Addlestone, Surrey KT15 3NB, UK. Voice: +44 (0) 1932-357807: fax: +44 (0) 1932-357423.
e-mail: s.cutler@vla.defra.gsi.gov.uk

**Ann. N.Y. Acad. Sci. 1078: 373–377 (2006). © 2006 New York Academy of Sciences.
doi: 10.1196/annals.1374.074**

Blood Samples

Blood samples (finger-prick) were collected from villagers ($n = 488$) and total DNA was extracted. These were tested blind to assess exposure of the population for rickettsial infection.

Rats

Rats ($n = 17$) were captured a sacrificed, ears, blood, and tail tips were collected, and total DNA was extracted as described above.

Chickens

Chickens belonging to sampled householders were purchased ($n = 11$), sacrificed, and samples of blood, wing tissue, and wattle were collected, and total DNA was extracted as above.

PCR

Two DNA targets were used for amplification of rickettsial DNA: 17-kDa antigen and an internal portion of the citrate synthase gene (*gltA*). Ticks were initially screened using a multiplex assay for *Bartonella, Yersinia pestis,* and *Rickettsia*, previously published for fleas.[2] Reactive samples upon initial screening were retested using individual primer pairs and amplicons subjected to sequence analysis. Ticks, chicken, and rat samples were also tested using amplification of the *gltA* gene (primers Rp877 and CS1273r) to confirm the presence of *Rickettsia* and provide sequence for phylogenetic analysis.[3–5]

RESULTS

Four of 204 extracted tick samples produced products using the 17-kDa PCR (1.9%). Sequence analysis confirmed this to be partial sequence of rickettsial 17-kDa antigen, aligning among *R. australis, R. akari,* and *R. felis* (FIG. 1). Ten of the 204 ticks produced amplicons using the *glt*A gene target (4.9%). Citrate synthase amplicons clustered with those of *R. helvetica* and *R. cooleyi* (FIG. 2). Both assays produced rickettsial amplicons from ticks from one dwelling (IK32a), while sample from two other houses were only positive in one assay alone (IM15 and MK37 using 17kDa and *glt*A, respectively). Surprisingly, some of the amplicons using citrate synthase primers were not rickettisal in origin, but instead belonged to *Acinetobacter* spp. which showed total homology over the primer-binding region. Only the 17-kDa assay was

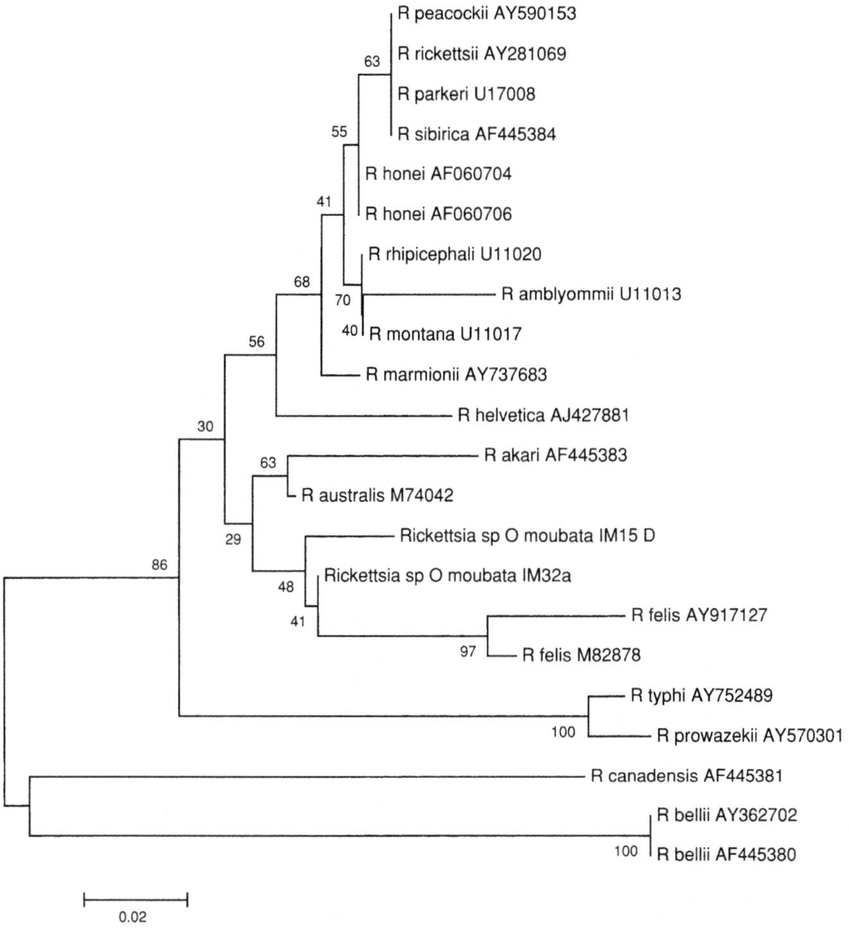

FIGURE 1. Phylogenetic neighbor-joining tree of 17-kDa antigen sequences illustrating the similarity of the *Ornithodoros*-associated *Rickettsia* with other rickettsiae (bootstrap value 250; accession numbers DQ092217 and DQ092218).

used to screen rats and chickens, all of which were negative. Human blood samples were similarly negative.

DISCUSSION

Despite the diverse vectorial capability of rickettsiae and successful laboratory infections of other tick species, including soft-bodied *Argasid* ticks, reports of naturally infected *Ornithodoros* ticks are limited.[6,7] We describe naturally infected *O. moubata* complex ticks collected from traditional African human dwellings in Tanzania, where ticks feed primarily on human occupants.

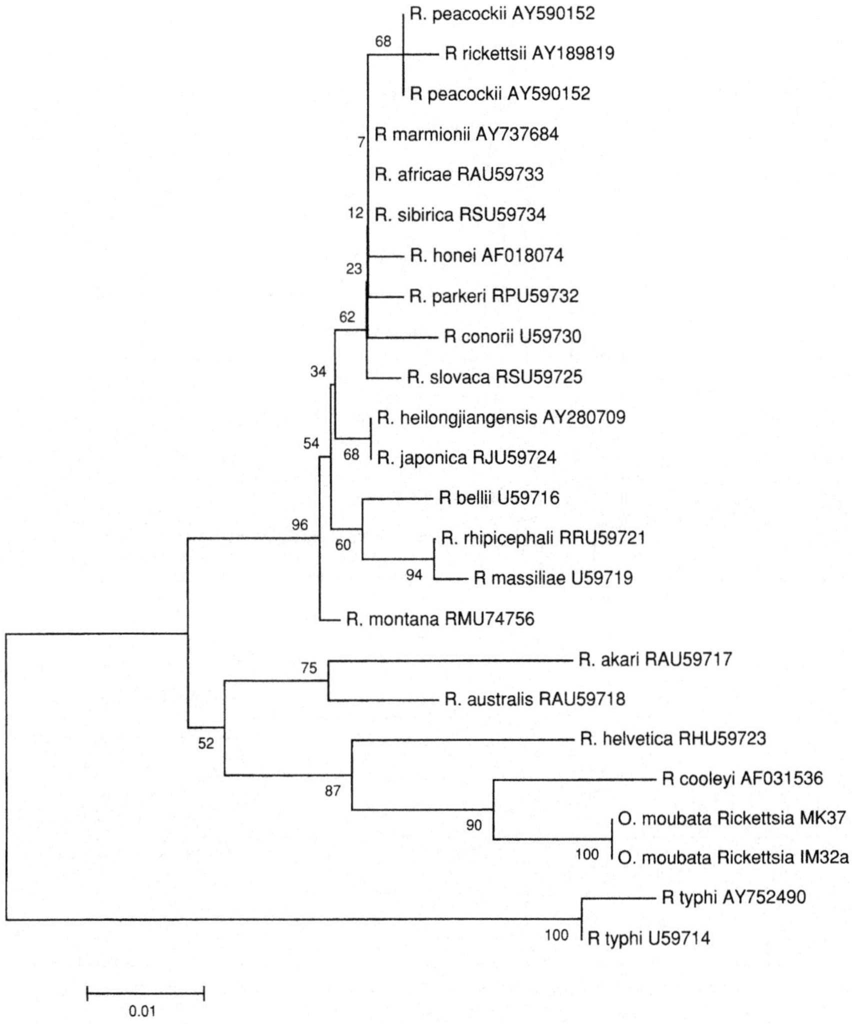

FIGURE 2. Phylogenetic neighbor joining tree of *gltA* (citrate synthase) sequences illustrating the similarity of the *Ornithodoros*-associated *Rickettsia* with other *Rickettsia* (bootstrap value 250; accession numbers DQ092215 and DQ092216).

Despite this human exposure, no evidence of rickettsial DNA was found in the blood samples taken from villagers where infected ticks were detected. Given the poor success of demonstrating DNA from clinical cases of rickettsiosis, this negative finding is not a surprise. Similarly, attempts to identify potential alternative reservoirs commonly found in traditional households (chickens and rats) were unsuccessful in establishing the natural reservoir for this rickettsial species, although this may reflect the small sample size tested. Although rickettsial disease is no stranger in sub-Saharan Africa, the rickettsial species found in *Ornithodoros* ticks appeared to be distinct. Whether this species can infect man remains to be established.

Two different genetic targets were used and sequence analysis confirmed the presence of *Rickettsia* spp. Primers used for amplification of citrate synthase were not specific, producing an amplicon from *Acinetobacter* spp., a ubiquitous soil microbe, but also found in both ticks and lice. However, this target is useful for phylogenetic analysis of rickettsiae. Phylogenetic trees using either *glt*A or 17-kDa gene portions revealed this rickettsial species detected in our study to be a unique cluster among the spotted fever group.

ACKNOWLEDGMENTS

We thank Dr. Alison Talbert and associates, who assisted in the collection of field samples, and the Royal Society for funding.

REFERENCES

1. TALBERT, A., A. NYANGE & F. MOLTENI. 1998. Spraying tick-infested houses with lambda-cyhalothrin reduces the incidence of tick-borne relapsing fever in children under five years old. Trans. R. Soc. Trop. Med. Hyg. **92:** 251–253.
2. STEVENSON, H., Y. BAI, M. KOSOY, *et al.* 2003. Detection of novel *Bartonella* strains and *Yersinia pestis* in prairie dogs and their fleas (Siphonaptera: Ceratophyllidae and Pulicidae) using multiplex polymerase chain reaction. J. Med. Entomol. **40:** 329–337.
3. BROUQUI, P., A. STEIN, H. DUPONT, *et al.* 2005. Ectoparasitism and vector-borne diseases in 930 homeless people from Marseilles. Medicine (Baltimore) **84:** 61–68.
4. ROUX, V. & D. RAOULT. 1999. Body lice as tools for diagnosis and surveillance of reemerging diseases. J. Clin. Microbiol. **37:** 596–599.
5. ROUX, V., E. RYDKINA, M. EREMEEVA, *et al.* 1997. Citrate synthase gene comparison, a new tool for phylogenetic analysis, and its application for the rickettsiae. Int. J. Syst. Bacteriol. **47:** 252–261.
6. REHACEK, J., J. URVOLGYI & E. KOVACOVA. 1977. Massive occurrence of rickettsiae of the spotted fever group in fowl tampan, *Argas persicus*, in the Armenian S.S.R. Acta Virol. **21:** 431–438.
7. PHILIP, R.N.C., E.A. ANACKER, R.L. CORY, *et al.* 1983. *Rickettsia bellii* sp. nov.: a tick-borne Rickettsia, widely distributed in the United States, that is distinct from the spotted fever and typhus biogroups. Int. J. Syst. Bacteriol. **33:** 94–106.

Detection of Members of the Genera *Rickettsia*, *Anaplasma*, and *Ehrlichia* in Ticks Collected in the Asiatic Part of Russia

STANISLAV SHPYNOV,[a] PIERRE-EDOUARD FOURNIER,[a] NIKOLAY RUDAKOV,[b] IRINA TARASEVICH,[c] AND DIDIER RAOULT[a]

[a]*Unité des Rickettsies CNRS UMR6020, IFR48, Faculté de Médecine, Université de la Méditerranée, 27 Boulevard Jean Moulin, 13385 Marseille Cedex 05, France*

[b]*Omsk Research Institute of Natural Foci Infections, 7, Prospect Mira, Omsk 644080, Russia*

[c]*The Gamaleya Research Institute of Epidemiology and Microbiology, 18, Gamaleya St., Moscow 123098, Russia*

ABSTRACT: A total of 395 adult ixodid ticks from three genera (*Dermacentor*, *Haemaphysalis*, and *Ixodes*) collected from the Urals to the Far East of Russia were tested by PCR and sequencing for the presence of spotted fever rickettsiae, anaplasmae, and ehrlichiae. Four, pathogens recognized in humans were detected in ticks: *Rickettsia sibirica*, *R. heilongjiangensis*, *R. helvetica*, and *Anaplasma phagocytophilum*. In addition, rickettsiae and ehrlichiae of unknown pathogenicity were detected, including *Rickettsia* sp. RpA4, *Rickettsia* sp. DnS14, *Rickettsia* sp. DnS28, "*Candidatus* R. tarasevichiae," a rickettsia closely related to *R. helvetica*, *A. bovis*, *Ehrlichia muris*, "*Ehrlichia*-like" "Schotti variant," and bacterium "Montezuma." Our findings indicated the distribution of rickettsiae and ehrlichiae in hard ticks in Russia.

KEYWORDS: ixodid ticks; rickettsiae; anaplasmae; ehrlichiae; Russia

INTRODUCTION

The main vectors of *R. sibirica* are ticks from the genera *Dermacentor* and *Haemaphysalis concinna* (*H. concinna*).[1] Astrakhan fever rickettsia is transmitted by *Rhipicephalus pumilio* (*R. pumilio*).[2] The recently described

Address for correspondence: Pierre-Edouard Fournier, Unité des Rickettsies CNRS UMR6020, IFR48, Faculté de Médecine, Université de la Méditerranée, 27 Boulevard Jean Moulin, 13385 Marseille Cedex 05, France. Voice: 33-491-38-55-17; fax: 33-491-38-77-72.
e-mail: Pierre-Edouard.Fournier@medecine.univ-mrs.fr

Ann. N.Y. Acad. Sci. 1078: 378–383 (2006). © 2006 New York Academy of Sciences.
doi: 10.1196/annals.1374.075

Rickettsia heilongjiangensis (*R. heilongjiangensis*) was detected in *H. concinna* ticks in Krasnoyarsk region, Siberia. Recently, *R. slovaca* was found in Russia in *Dermacentor marginatus* (*D. marginatus*). In addition to pathogenic rickettsiae, several new rickettsiae of unknown pathogenicity have recently been described in ixodid ticks in Russia.[3] In the Altay republic, *Rickettsia* sp. DnS14 and *Rickettsia* sp. DnS28 were found in *D. nuttalli*; in Astrakhan, *Rickettsia* sp. RpA4 was found in *R. pumilio*; in Burjatia, *Rickettsia* sp. DnS14 was found in *D. silvarum*; in the Voronezh, Omsk, and Novosibirsk regions of Russia, *Rickettsia* sp. RpA4 was found in *D. reticulatus*; and "*Candidatus* Rickettsia tarasevichiae" was detected in *Ixodes persulcatus* (*I. persulcatus*) ticks collected in various territories of Russia.[4] To date, no case of human ehrlichiosis has been reported in Russia, but *Anaplasma* and *Ehrlichia* sp. have been detected in *Ixodes* ticks in different regions of Russia.[5] However, few studies have been conducted to date and the knowledge on tick-borne infections in Russia is limited.

MATERIALS AND METHODS

From April to June 2002, 395 hard ticks from three genera (*Dermacentor, Haemaphysalis,* and *Ixodes*) were collected in Russia. Ticks were collected on the vegetation by the flagging technique in forest-steppe zone of Russia, from the Urals to the Far East. For DNA extraction, the MagNA Pure LC DNA isolation kit 2 was used according to the manufacturer's instructions (Roche Diagnostics, Basel, Switzerland). Rickettsial DNA was amplified and sequenced a 5′-fragment of 590 bp from the *ompA* gene using the primers 190-70 and 190-701, and the complete *gltA* gene using the two primer pairs CS1d-CS535r and CS409d-RP1258n. Ehrlichial DNA was detected using the 16EHRD and 16EHRR primers, which amplify a 345-bp fragment of the 16S rRNA gene of all known ehrlichiae and anaplasmae. PCR reactions and sequencing were carried out as previously reported.[4]

RESULTS

D. nuttalli ticks collected in the Krasnoyarsk area were infected with *R. sibirica*, *Rickettsia*. sp. DnS28, DnS14, and *R. heilongjiangensis* (TABLE 1). The latter species was also detected in *H. concinna* ticks in the Altay, Krasnoyarsk, and Prymorye areas. *Rickettsia* sp. RpA4 was identified in *D. marginatus* ticks collected in the Orenburg and Altay territories, in *D. reticulatus* ticks from the Orenburg, Chelyabinsk, Novosibirsk, and Altay areas, in *H. concinna* ticks from Altay, and *I. persulcatus* ticks from the Altay and Prymorye areas. *Rickettsia*. sp. DnS14 was detected in *D. marginatus* ticks from the Orenburg and Altay territories, in *D. nuttalli* from the Krasnoyarsk area, and

TABLE 1. Source and identification of bacteria described in this study

Region	Tick species	Number of positive ticks/total examined (%)	Gene	GenBank accession numbers	Identification
Orenburg area	*D. marginatus*	5/20 (25)	*ompA/gltA*	AF120022/AF120029	*Rickettsia* sp. *RpA4*
		3/20 (15)	*ompA/gltA ompA/gltA*	AF120021/AF120028	*Rickettsia* sp. *DnS14*
	D. reticulatus	3/12 (25)		AF120022/AF120029	*Rickettsia* sp. *RpA4*
Cheliabinsk area	*D. reticulatus*	7/32 (21.9)	*ompA/gltA*	AF120022/AF120029	*Rickettsia* sp. RpA4
Omsk area	*I. persulcatus*	32/53 (60.4)	*gltA*	AF503167	≪*Candidatus* Rickettsia tarasevichiae≫
		1/53 (1.9)	*gltA*	U59723	*Rickettsia helvetica*
		21/53 (39.6)	16S	AF493952	Bacterium “Montezuma”
		4/53 (7.5)	16S	U15527	*Ehrlichia muris*
		2/53 (3.8)	16S	AF104680	*Ehrlichia-like* sp. “Schotti variant”
Novosibirsk area	*D. reticulatus*	6/6 (100)	*ompA/gltA*	AF120022/AF120029	*Rickettsia* sp. RpA4
	D. marginatus	0/34	*ompA/gltA*		
	I. persulcatus	1/2 (50)	*gltA*	AF503167	≪*Candidatus* Rickettsia tarasevichiae≫
		1/2 (50)	16S	AF493952	Bacterium “Montezuma”
Altay area	*D. marginatus*	9/27 (33.3)	*ompA/gltA*	AF120022/AF120029	*Rickettsia* sp. *RpA4*
		3/27 (11.1)	*ompA/gltA*	AF120021/AF120028	*Rickettsia* sp. *DnS14*
	D. reticulatus	17/41 (41.5)	*ompA/gltA*	AF120022/AF120029	*Rickettsia* sp. *RpA4*
	I. persulcatus	3/8 (37.5)	*gltA*	AF503167	≪*Candidatus* Rickettsia tarasevichiae≫
		3/8 (37.5)	16S	AF493952	Bacterium “Montezuma”
		1/8 (12.5)	16S	M73224	*Anaplasma phagocytophila*
		3/8 (37.5)	16S	U15527	*Ehrlichia muris*
		1/8 (12.5)	*ompA/gltA*	AF120022/AF120029	*Rickettsia* sp. *RpA4*
	H. concinna	6/20 (30)	*ompA/gltA*	AF120022/AF120029	*Rickettsia* sp. *RpA4*
		4/20 (20)	*ompA/gltA*	AH012829/Y285776	≪*Rickettsia heilongjiangensis*≫

Continued.

TABLE 1. Continued.

Region	Tick species	Number of positive ticks/total examined (%)	Gene	GenBank accession numbers	Identification
Krasnoyarsk area	*D. nuttalli*	4/33 (12.1)	*ompA/gltA ompA/gltA*	U43807/U59734	*Rickettsia sibirica*
		1/33 (3)	*ompA/gltA*	AF120021/AF120028	*Rickettsia* sp. *DnS14*
		7/33 (21.2)	*ompA/gltA*	AF120018/AF120027	*Rickettsia* sp. *DnS28*
		1/33 (3)	*ompA/gltA*	AH012829/AY285776	≪*Rickettsia heilongjiangensis*≫
	H. concinna	1/18 (5.5)		AH012829/AY285776	≪*Rickettsia heilongjiangensis*≫
Prymorye area	*D. silvarum*	23/30 (76.7)	*ompA/gltA gltA*	AF120021/AF120028	*Rickettsia* sp. *DnS14*
	I. persulcatus	25/44 (56.8)	16S	AF503167	≪*Candidatus* Rickettsia tarasevichiae≫
		5/44 (11.4)	16S	AF493952	Bacterium “Montezuma”
		2/44 (4.5)	*ompA/gltA*	M73224	*A. phagocytophilum*
		2/44 (4.5)	*ompA/gltA*	AF120022/AF120029	*Rickettsia* sp. *RpA4*
	H. concinna	2/15 (13.3)	16S	AH012829/AY285776	≪*Rickettsia heilongjiangensis*≫
		2/15 (13.3)		U03775	*Anaplasma bovis*

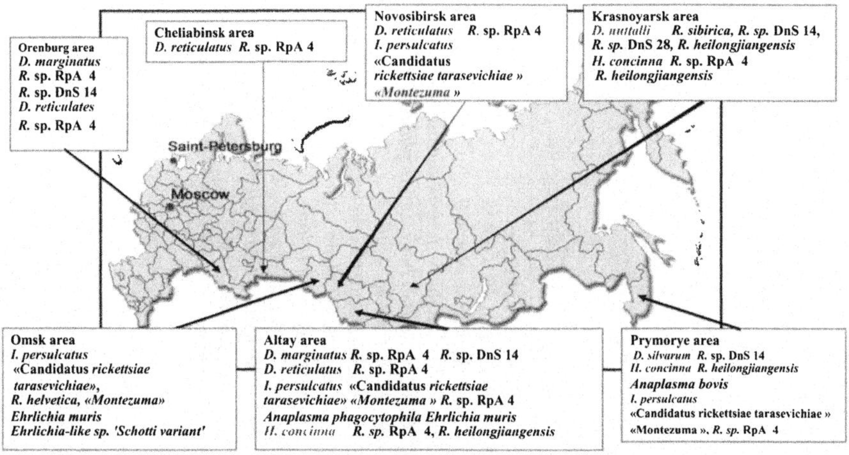

FIGURE 1. Map of Russia detailing the distribution of *Rickettsia*, *Anaplasma,* and *Ehrlichia* species found in hard ticks in Russia.

in *D. silvarum* from the Prymorye area. *Rickettsia* sp. DnS28 has been identified in *D. nuttalli* ticks collected in the Krasnoyarsk region. "*Candidatus* Rickettsia tarasevichiae" was detected in *I. persulcatus* ticks collected in the Omsk, Novosibirsk, Altay, and Prymorye areas. *R. helvetica* was discovered in *I. persulcatus* ticks in the Omsk area. *A. phagocytophila* was found in *I. persulcatus* ticks in the Altay and Prymorye areas. *A. bovis* was identified in *H. concinna* ticks from the Prymorye areas. *E. muris* was identified in *I. persulcatus* ticks from the Omsk and Altay areas. "*Ehrlichia*-like" "Schotti variant" was found in *I. persulcatus* from the Omsk area. Finally, bacterium "Montezuma" was found in *I. persulcatus* ticks from the Omsk, Novosibirsk, Altay, and Prymorye areas.

DISCUSSION

Our study demonstrated that 14 representatives of the order *Rickettsiales* are currently known to be present in Russia. When segregated by vector, these bacteria can be clustered into three groups (TABLE 1, FIG. 1). The first group (including *Rickettsia* species only) is ecologically linked with ticks of the genera *Dermacentor* and *Rhipicephalus* from the family *Rhipicephalinae*. This group includes *R. sibirica* sensu stricto, *R. conorii* subsp. *caspia,*[6] *R. slovaca*, *Rickettsia* sp. RpA4, *Rickettsia* sp. DnS14, and *Rickettsia* sp. DnS28. The second group includes bacteria detected in *I. persulcatus* ticks. This group comprises representatives of the genera *Rickettsia*, *Ehrlichia*, and *Anaplasma*, that is, *A. phagocytophilum*, *E. muris*, "*Ehrlichia*-like" " Schotti variant," *R. helvetica*, "*Candidatus* Rickettsia tarasevichiae," and bacterium "Montezuma." The third group only consists of *R. heilongjiangensis* and *A. bovis*, found in *H. concinna*. Among these bacteria, six are recognized human pathogens, including *R. sibirica* sensu stricto, *R. conorii* subsp. *caspia*, *R. slovaca*, *R. heilongjiangensis*, *R. helvetica*, and *A. phagocytophilum*. The distribution areas of these different tick species overlap, thus explaining the presence of a variety of species of *Rickettsia*, *Ehrlichia,* and *Anaplasma* in each territory.

In conclusion, our study is an important contribution to the knowledge of distribution of rickettsiae and ehrlichiae in Russia. Clinicians should be aware that several pathogenic tick-borne bacteria are endemic in Russia.

REFERENCES

1. ZDRODOVSKII, P.F. & H.M. GOLINEVICH. 1960. North Asian tick-borne rickettsiosis or tick-borne typhus fever. *In* The Rickettsial Diseases. P.F. Zdrodovskii & H.M. Golinevich, Eds.: 311–332. Pergamon Press. New York.
2. TARASEVICH, I.V., V.A. MAKAROVA, N.F. FETISOVA, *et al.* 1991. Studies of a "new" rickettsiosis, "Astrakhan" spotted fever. Eur. J. Epidemiol. **7:** 294–298.
3. RYDKINA, E., V. ROUX, N. FETISOVA, *et al.* 1999. New rickettsiae in ticks collected in territories of the former Soviet Union. Emerg. Infect. Dis. **5:** 811–814.
4. SHPYNOV, S., P.E. FOURNIER, N. RUDAKOV & D. RAOULT. 2003. "*Candidatus* Rickettsia tarasevichiae" in *Ixodes persulcatus* ticks collected in Russia. Ann. N.Y. Acad. Sci. **990:** 162–172.
5. RAVYN, M.D., E.I. KORENBERG, J.A. OEDING, *et al.* 1999. Monocytic *Ehrlichia* in *Ixodes persulcatus* ticks from Perm, Russia. Lancet **353:** 722–723.
6. ZHU, Y., P.E. FOURNIER, M. EREMEEVA & D. RAOULT. 2005. Proposal to create subspecies of *Rickettsia conorii* based on multi-locus sequence typing and an emended description of *Rickettsia conorii*. BMC Microbiol. **5:** 11.

Characterization of *Dermacentor variabilis* Molecules Associated with Rickettsial Infection

KEVIN R. MACALUSO,[a] ALBERT MULENGA,[b] JASON A. SIMSER,[c] AND ABDU F. AZAD[c]

[a]*Department of Pathobiological Sciences, School of Veterinary Medicine, Louisiana State University, Skip Bertman Drive, SVM-3213, Baton Rouge, Louisiana 70803, USA*

[b]*Department of Entomology, Texas A & M University, 2475 Minie Belle Heep Building, College Station, Texas 77843-2475, USA*

[c]*Department of Microbiology and Immunology, School of Medicine, University of Maryland at Baltimore, 655 W. Baltimore Street, BRB13-009, Baltimore, Maryland 21201, USA*

ABSTRACT: To ultimately define the virulence factors of rickettsiae, an understanding of the biology of the organism is essential. Comprehension of the pathogen–human interaction is critical to the development of control measures; and, in the case of vector-borne diseases, the role of the vector in maintaining and transmitting pathogens to vertebrate hosts is crucial to ultimate control. Recent studies have identified tick molecules that are likely involved in the tick–rickettsiae interchange, including tick response to infection and possible molecules exploited by rickettsiae during transmission events. We have further characterized several tick-derived molecules, including a histamine release factor, serine proteases, and lysozymes.

KEYWORDS: *Rickettsia*; transovarial transmission; tick; adhesion

INTRODUCTION

Arthropods play an essential role in the transmission of several emerging and reemerging infectious disease–causing agents including rickettsial organisms. The importance of arthropods as reservoirs and vectors of rickettsiae has been examined in regard to the influence of the host cell environment on rickettsiae,[1] as well as the impact of the rickettsiae on the arthropod host.[2–5] We

Address for correspondence: Kevin R. Macaluso, Department of Pathobiological Sciences, School of Veterinary Medicine, Louisiana State University, Skip Bertman Drive, SVM-3213, Baton Rouge, LA 70803. Voice: 225-578-9677; fax: 225-578-9701.
e-mail: kmacaluso@vetmed.lsu.edu

**Ann. N.Y. Acad. Sci. 1078: –388 (2006). © 2006 New York Academy of Sciences.
doi: 10.1196/annals.1374.076**

are investigating the transmission dynamics for both typhus group (TG) and spotted fever group (SFG) rickettsiae in order to gain further understanding of the ecology of rickettsial diseases. Of particular interest is the maintenance and transmission of SFG rickettsiae by ticks.

The relationships between different *Rickettsia* spp. and ixodid ticks have evolved by which ticks serve as both the vector and reservoir of SFG rickettsiae.[6] Rickettsiae systematically infect all organs of the tick, including the organs essential for both horizontal and vertical transmission, that is, salivary glands, midgut, and ovary tissues.[7] The salivary glands are critical for active transmission of the bacteria to a vertebrate host during tick bloodmeal acquisition. The role of the midgut in rickettsial infection is vital to successful rickettsial transmission, as the rickettsiae must overcome an adverse environment and establish infection.[7] Midgut infection may also contribute to dissemination of rickettsiae via both regurgitation and tick feces. However, the role of regurgitation in rickettsial transmission to the mammalian hosts remains controversial. The infection of the ovaries is important because it allows for the rickettsiae to be maintained in the developing egg, via invasion of developing tick oocytes, in a process termed *transovarial transmission*. During the tick life cycle, the rickettsial infection is subsequently passed along to other stages that is, larval, nymphal, and adult via transstadial transmission. This intimate relationship between the bacteria and vector has resulted in a mechanism of maintenance that ensures rickettsial survival even in the absence of a transmission event through a vertebrate host.

One focus of our research has been to delineate the molecular mechanisms of tick–rickettsiae interactions. We have used molecular and PCR-based techniques, including expression library screening,[8] differential display,[9] subtractive hybridization,[10] and expression sequence and homolog cloning.[11,12] This multifaceted approach to assess rickettsial infection in ticks has proven successful for the unique identification of clones, while demonstrating a common trend in upregulation of tick molecules in response to bacterial infection. Our studies have led to the identification of several tick-derived molecules[11–14] that are suspected to facilitate tick infection and rickettsial transmission (TABLE 1). Among the molecules already characterized are histamine release factor, serine proteases, and lysozymes.

A putative histamine release factor (DVHRF) in the American dog tick, *Dermacentor variabilis (D. variabilis),* was originally identified by subtractive hybridization[10] and has been functionally characterized.[14] Identification of DVHRF in ticks was counterintuitive. Studies conducted during the 1980s provided ample evidence to indicate that histamine at the tick-feeding site is detrimental to tick physiology. In recent years, studies have demonstrated that ticks block the effects of histamine by secreting histamine-binding proteins into the feeding site (reviewed in Ref.14). The full-length DVHRF cDNA is 945 base pair(bp) with a 522-bp open reading frame that encodes a 20-kDa (173 amino acid) polypeptide. Sequence analysis identified two HRF signature

TABLE 1. Activity and predicted function of novel tick genes identified from uninfected and *Rickettsia*-infected *D. variabilis* using molecular techniques

Predicted Function	Putative Identification	Method of Discovery	Expression during Rickettsial Infection
Adhesion or invasion	Mucin-like protein	SH/DD	−−−
	Clathrin adaptor protein	SH	+++
	Tetraspanin	SH	+++
	signal transducer and activator of transcription-1/3 protein inhibitor	SH	−−−
	ATPase/clathrin-coated	DD	+++
	Catenin	DD	+++
Tick immune and stress response	Ferritin	SH/EST	+++
	α-dehydrogenase reductase	SH	+++
	Glutathione-S transferase	SH/EST	+++
	Nucleosome assembly protein	SH	+++
	Cyclin A2 protein	SH	+++
	ATPase Cu^{2+} transporting	DD	+++
	Tubulin α-chain	DD	+++
	Defensin	HC	−−−
	Prophenoloxidase-activating factor	EST	+++
Tick–host interactions	α-2 macroglobulin	SH	+++
	Salivary glue precursor	DD	+++
	IgE-dependent histamine release factor	SH	+++
	ENA vasodilator	DD	+++
	Calreticulin	HC	−−−
Unknown	Probable elongation factor	SH	+++
	Similarity to *Drosophila melanogaster* CG17525	SH	+++
	Glycine-rich protein	DD	+++

SH = subtractive hybridization PCR; DD = differential display PCR; HC = homolog cloning; EST = RACE EST (full-length) sequencing of full-length *D. variabilis* hemocyte mRNA.
Sequences have been deposited into GenBank.

amino acid sequences to be conserved in DVHRF, indicating close structural similarity between DVHRF and other characterized HRF homologs. Northern and Western blotting analyses of partially fed and unfed ticks indicate that neither DVHRF transcriptional nor translational regulation is influenced by the tick-feeding activity. Like its counterparts from the mammalian system, tick DVHRF is expressed in various tissues, as assessed by both Northern and Western blotting analyses. Furthermore, functional characterization of an *Escherichia coli*–expressed recombinant DVHRF (rDVHRF) induced histamine secretion from a rat basophilic leukemic cell line in a dose-dependent manner. Mouse polyclonal antibodies to rDVHRF strongly reacted to a ~20-kDa DVHRF protein band on experimentally induced *D. variabilis* saliva protein blots, suggesting that this protein was potentially injected into the host during feeding. Our results suggest the existence of a versatile tick-derived control mechanism for levels of histamine at tick-feeding sites. Efforts to dissect the role of molecules, such as DVHRF, in rickettsial transmission by ticks are under way.

Understanding the mechanisms of rickettsial survival in the arthropod host is also important in understanding the rickettsial disease epidemiology. The successful outcome of rickettsial infection in vector ticks leaves us with many unanswered questions as to how rickettsiae bypass host immune responses and cope with the ticks' antimicrobial activities. Studies to define rickettsial components that facilitate their survival in the tick vector are now ongoing. Similarly, an appreciation of the biological factors facilitating vector-derived susceptibility to rickettsial infection is understated. Using expressed sequence tags, libraries, and homolog cloning strategies, we identified several putative antimicrobial molecules and provided characterization for two classes of molecules, serine protease and lysozyme, both in *Dermacentor* ticks. The serine protease closely resembled a factor D–like antimicrobial molecule originally described in the Japanese horseshoe crab, *Tachypleus tridentatus*. Although not examined in a *Rickettsia*- tick model system, upregulation of the factor D–like molecule was observed when *D. variabilis* organisms were challenged with other bacteria.[12] Additionally, an ortholog was identified in a *Dermacentor andersoni* (*D. andersoni*) cell line, indicating cross-species conservation. The identification of a c-type lysozyme in *D. variabilis* hemocytes was also linked to the immune response of the tick to *E. coli* challenge.[11] Similar to that described for the factor-D molecule, a lysozyme ortholog was also identified in the *D. andersoni* cell line. Studies are under way to examine the mechanism(s) by which rickettsiae avoid these immune molecules.

The identification and expression analysis of immune-like and stress response molecules, as well as host immunomodulatory molecules, suggest that ticks are actively responding to rickettsial infection and are not simply serving as transmission vehicles. This response obviously requires SFG rickettsiae to adapt in response to the host cell environment, as observed for other rickettsiae.[15–18] The mechanisms of SFG rickettsial infection in the tick host and the dynamics of rickettsial biology within the tick are currently under examination.

REFERENCES

1. POLICASTRO, P.F., U.G. MUNDERLOH, E.R. FISCHER & T. HACKSTADT. 1997. *Rickettsia rickettsii* growth and temperature-inducible protein expression in embryonic tick cell lines. J. Med. Microbiol. **46:** 839–845.

2. BURGDORFER, W. & L.P. BRINTON. 1975. Mechanisms of transovarial infection of spotted fever rickettsiae in ticks. Ann. N. Y. Acad. Sci **266:** 61–72.
3. MACALUSO, K.R., D.E. SONENSHINE, S.M. CERAUL & A.F. AZAD. 2001. Infection and transovarial transmission of rickettsiae in *Dermacentor variabilis* ticks acquired by artificial feeding Vector. Borne. Zoonotic. Dis. **1:** 45–53.
4. MACALUSO, K.R., D.E. SONENSHINE, S.M. CERAUL & A.F. AZAD. 2002. Rickettsial infection in *Dermacentor variabilis* (Acari: Ixodidae) inhibits transovarial transmission of a second *Rickettsia*. J. Med. Entomol. **39:** 809–813.
5. NIEBYLSKI, M.L., M.G. PEACOCK & T.G. SCHWAN. 1999. Lethal effect of *Rickettsia rickettsii* on its tick vector (*Dermacentor andersoni*). Appl. Environ. Microbiol. **65:** 773–778.
6. AZAD, A.F. & C.B. BEARD. 1998. Rickettsial pathogens and their arthropod vectors. Emerg. Infect. Dis. **4:** 179–186.
7. MUNDERLOH, U.G. & T.J. KURTTI. 1995. Cellular and molecular interrelationships between ticks and prokaryotic tick-borne pathogens. Annu. Rev. Entomol. **40:** 221–243.
8. MACALUSO, K.R., A. MULENGA, J.A. SIMSER & A.F. AZAD. 2003. Interactions between rickettsiae and *Dermacentor variabilis* ticks: analysis of gene expression. Ann. N. Y. Acad. Sci. **990:** 568–572.
9. MACALUSO, K.R., A. MULENGA, J.A. SIMSER & A.F. AZAD. 2003. Differential expression of genes in uninfected and *Rickettsia*-infected *Dermacentor variabilis* ticks as assessed by differential-display PCR. Infect. Immun. **71:** 6165–6170.
10. MULENGA, A., K.R. MACALUSO, J.A. SIMSER & A.F. AZAD. 2003. Dynamics of *Rickettsia*-tick interactions: identification and characterization of differentially expressed mRNAs in uninfected and infected *Dermacentor variabilis*. Insect Mol. Biol. **12:** 185–193.
11. SIMSER, J.A., K.R. MACALUSO, A. MULENGA & A.F. AZAD. 2004. Immune-responsive lysozymes from hemocytes of the American dog tick, *Dermacentor variabilis* and an embryonic cell line of the Rocky Mountain wood tick, *D. andersoni*. Insect Biochem. Mol. Biol. **34:** 1235–1246.
12. SIMSER, J.A., A. MULENGA, K.R. MACALUSO & A.F. AZAD. 2004. An immune responsive factor D-like serine proteinase homologue identified from the American dog tick, *Dermacentor variabilis*. Insect Mol. Biol. **13:** 25–35.
13. MULENGA, A., J.A. SIMSER, K.R. MACALUSO & A.F. AZAD. 2004. Stress and transcriptional regulation of tick ferritin HC. Insect Mol. Biol. **13:** 423–433.
14. MULENGA, A., K.R. MACALUSO, J.A. SIMSER & A.F. AZAD. 2003. The American dog tick, *Dermacentor variabilis*, encodes a functional histamine release factor homolog. Insect Biochem. Mol. Biol. **33:** 911–919.
15. JAURON, S.D., C.M. NELSON, V. FINGERLE, *et al.* 2001. Host cell-specific expression of a p44 epitope by the human granulocytic ehrlichiosis agent. J. Infect. Dis. **184:** 1445–1450.
16. SINGU, V., H. LIU, C. CHENG & R.R. GANTA. 2005. *Ehrlichia chaffeensis* expresses macrophage- and tick cell-specific 28-kilodalton outer membrane proteins. Infect. Immun. **73:** 79–87.
17. LOHR, C.V., K.A. BRAYTON, V. SHKAP, *et al.* 2002. Expression of *Anaplasma marginale* major surface protein 2 operon-associated proteins during mammalian and arthropod infection. Infect. Immun. **70:** 6005–6012.
18. GARCIA-GARCIA, J.C., F.J. DE LA, E.F. BLOUIN, *et al.* 2004. Differential expression of the msp1alpha gene of *Anaplasma marginale* occurs in bovine erythrocytes and tick cells. Vet. Microbiol. **98:** 261–272.

Ticks, Tick-Borne Rickettsiae, and *Coxiella burnetii* in the Greek Island of Cephalonia

A. PSAROULAKI, D. RAGIADAKOU, G. KOURIS, B. PAPADOPOULOS, B. CHANIOTIS, AND Y. TSELENTIS

University of Crete, Faculty of Medicine, Laboratory of Clinical Bacteriology, Parasitology, Zoonoses and Geographical Medicine (WHO Collaborating Center for Research and Training in Mediterranean Zoonoses), Crete, Greece

ABSTRACT: **Domestic animals are the hosts of several tick species and the reservoirs of some tick-borne pathogens; hence, they play an important role in the circulation of these arthropods and their pathogens in nature. They may act as vectors, but, also, as reservoirs of spotted fever group (SFG) rickettsiae, which are the causative agents of SFG rickettsioses. Q fever is a worldwide zoonosis caused by *Coxiella burnetii (C. burnetii)*, which can be isolated from ticks. A total of 1,848 ticks (954 female, 853 male, and 41 nymph) were collected from dogs, goats, sheep, cattle, and horses in 32 different localities of the Greek island of Cephalonia. *Rhipicephalus (Rh.) bursa, Rh. turanicus, Rh. sanguineus, Dermacentor marginatus (D. marginatus), Ixodes gibbosus (I. gibbosus), Haemaphysalis (Ha.) punctata, Ha. sulcata, Hyalomma (Hy.) anatolicum excavatum* and *Hy. marginatum marginatum* were the species identified. *C. burnetii* and four different SFG rickettsiae, including *Rickettsia (R.) conorii, R. massiliae, R. rhipicephali, and R. aeschlimannii* were detected using molecular methods. Double infection with *R. massiliae* and *C. burnetii* was found in one of the positive ticks.**

KEYWORDS: **ticks; *Rickettsia conorii*; *Rickettsia massiliae*; *Rickettsia rhipicephali*; *Rickettsia aeschlimannii*; *Coxiella burnetii*; Greece**

INTRODUCTION

Ticks are blood-sucking arthropods, which are ectoparasites of domestic and wild animals. They play an important role as vectors of microorganisms and are able to transmit a greater variety of pathogenic microorganisms than any other arthropod group. Because of climatic changes toward global warming,

Address for correspondence: A. Psaroulaki, Laboratory of Clinical Bacteriology, University of Crete, Crete, Greece. Voice: +003 02810394743; fax: +003 02810394740
e-mail: annapsa@med.uoc.gr

Ann. N.Y. Acad. Sci. 1078: 389–399 (2006). © 2006 New York Academy of Sciences.
doi: 10.1196/annals.1374.077

tick species may adapt to new area and might be considered as epidemiological markers for a number of infectious agents transmitted by them.

They may act as vectors, but also as reservoirs of spotted fever group (SFG) rickettsiae, which are the causative agents of SFG rickettsioses. Members of the SFG rickettsiae are usually associated with ixodid ticks, which transfer them to vertebrates via salivary secretions and among themselves both transtadially and transovarially.[1] Currently, at least four species of tick-borne rickettsiae—*R. conorii, R. massiliae, R. rhipicephali*, and *R. sibirica mongolotimonae*—exist in Greece. All of them have been detected in ticks, while *R. conorii* and *R. mongolotimonae* have also been detected in humans.[2–4]

C. burnetii, which is the causative agent of Q fever, is an obligate intracellular bacterium living in the phagolysosomes of the host cell. Ticks can transmit *C. burnetii* both transtadially and transovarially to their progeny, thereby, acting as a reservoir of this pathogen. *C. burnetii* has been found in several tick species, but the role of ticks in the transmission of the pathogen in human is probably minimal. However, they may be important in the dissemination of this pathogen in the environment because of the high concentration of *C. burnetii* in tick feces.[5] Q fever is endemic in Greece, and *C. burnetii* has been isolated from patients.[6,7]

Cephalonia, with an area of some 900 km^2, is the largest of the Ionian Islands and the fifth largest in all of Greece. The mild Mediterranean climate of the island and the presence of the large numbers of domestic productive and wild animals offer conditions that favor the existence of several tick species. Its population, estimated to be 35,000, is doubled during the summer months with the influx of foreign and domestic tourists. It is a mountainous island, with the mountain Ainos rising 1,626 m above sea level, and six more mountains with peaks over 1,000 m. The economy of the island is traditionally based on agriculture and livestock rearing. Nearly 75% of the total areas are pastureland, favoring the raising of sheep and goats. Although cases of rickettsiosis and Q fever are reported in the island of Cephalonia, the tick fauna was never investigated. The purpose of the study was (i) to identify the tick species present in the island and the association with their hosts and (ii) to detect rickettsiae and *C. burnetii* in the collected ticks.

MATERIALS AND METHODS

Tick Collection

During the years 1998 (April–July) and 1999 (May–June and October), a total of 1,848 ticks (952 female, 853 male, and 41 nymph) were collected from domestic animals (goats, sheep, cows, horses, and dogs) from 32 different localities around the island. Ticks from livestock were collected directly from hosts either by the authors or in some cases by shepherds with or without our presence. As regards dogs, all except two were domestic or stray animals living in villages or towns and not shepherd dogs. The ticks were immediately placed in vials with 70% ethanol, properly labeled, and were later identified in the laboratory using existing taxonomic keys.[9]

Detection of Rickettsiae and **C. burnetii**

Of the ticks collected, 852 were selected and were examined using molecular methods in order to detect rickettsiae and *C. burnetii*. Ticks were washed for 10 min in iodized alcohol, rinsed with distilled water, and dried on sterile paper. Consequently, tick samples were placed in 50 μL of 10 mM Tris·HCl (pH 8.0), heated at 90°C for 10 min, crushed with a sterile plastic homogenizer, and treated with 10 μL of proteinase K (10 U/mL) at 50°C for 3 h. In order to detect SFG rickettsiae, polymerase chain reaction (PCR) was performed using the primer pair Rp CS.877p/Rp CS.1258n that amplifies a fragment of the rickettsial *glt*A gene.[10,11] In the *glt*A-positive samples, a second PCR was performed using the primer pair Rr190.70p/Rr190.602n that amplifies a fragment of the rickettsial *omp*A gene.[10,12] The sequences obtained during this study were compared for similarity with those of other rickettsiae, lodged in GenBank using the BLAST program (http://www.ncbi.nlm.nih.gov), as well as with the data of previous reports.[11,12]

In order to detect *C. burnetii*, a "nested" PCR assay was performed using the primers Hfrag1/Hfrag2 in the first PCR and the primers HF1/HF2 for the "nested" PCR.[7,13]

The "genomic" primers CB1/CB2 were, also, used in a second PCR assay and the CB1/CB2 PCR products were digested using the enzymes TaqI and AluI.[7,13] The patterns obtained by restriction fragment length polymorphism (RFLP) analysis were compared to those obtained with the reference strains (Nine Mile and Q212).

RESULTS

Tick Species

A total of 1,848 ticks (954 female, 853 male, and 41 nymph) were collected from 5 host species in 32 localities (from goats in 15 localities, sheep in 14 localities, cows in 4 localities, horses in 4 localities, and dogs in 7 localities) (TABLE 1). Ticks of the genus *Rhipicephalus* were the most abundant. *Rh. turanicus* and *Rh. bursa* parasitized the livestock in 11 and 18 localities, respectively (FIG. 1). *Rh. sanguineus* was found in 7 localities, exclusively on dogs (FIG. 1). A specimen of *Rh. sanguineus*, not included in TABLE 1, was found on a human.

TABLE 1. Tick species by host collected in Cephalonia

	Number of ticks by host																
	Goats		Sheep			Cattle		Horses		Dogs			Total number of ticks				
Tick species	Female	Male	Female	Male	Nymph	Female	Male	Female	Male	Female	Male	Nymph	Female	Male	Nymph	Total	%
Rh. bursa	137	95	34	26		15	15		2				186	138		324	17.53%
Rh. turanicus	147	131	131	72		13	16	34	16	2	2		327	237		564	30.52%
Rh. sanguineus										155	104	29	155	104	29	288	15.58%
D. marginatus	107	129	84	70									191	199		390	21.10%
I. gibbosus	9	15	2			4		1	1				16	16		32	1.73%
Ha. punctata	3	3	4	7	12								7	10	12	29	1.57%
Ha. sulcata	8	3											8	3		11	0.60%
Hy. a. excavatum	1	3	10	14		10	27	6	9				27	53		80	4.33%
Hy. m. marginatum	5		12	46		10	35	7	10	3	2		37	93		130	7.03%
	417	379	277	235		52	93	48	38	160	108	29	954	853	41	1,848	

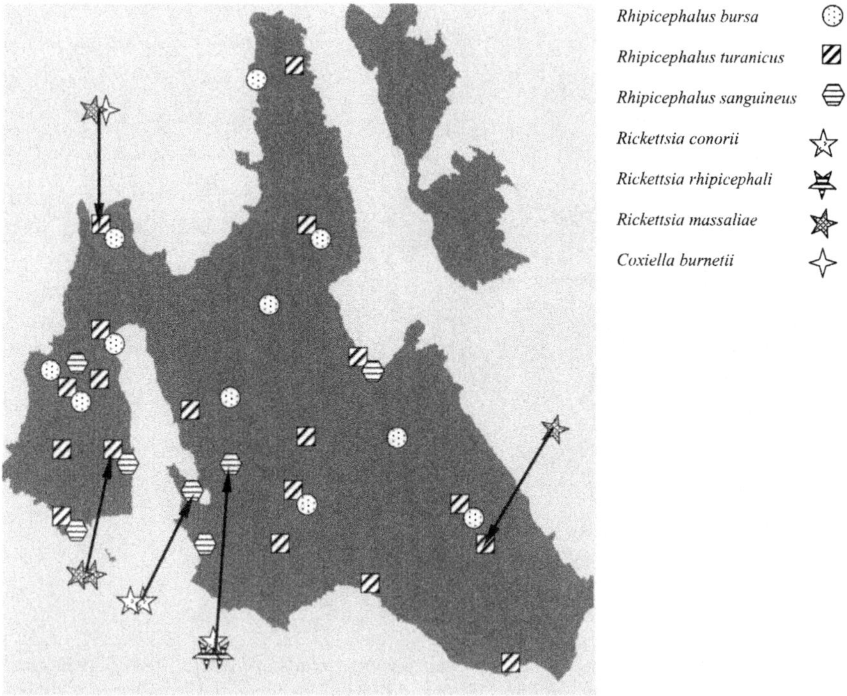

FIGURE 1. Distribution of *Rhipicephalus* species and of *Rickettsia* spp. and *C. burnetii* found in these tick species in Cephalonia.

Hyalomma ticks were present on several hosts in different parts of the island. *Hy. anatolicum excavatum* was collected from goats, sheep, cattle, and horses in 5 localities, while *Hy. marginatum marginatum* from all the 5 host species in 11 localities (FIG. 2).

Dermacentor marginatus (D. marginatus), *Haemaphysalis* (*Ha.) punctata,* and *Ha. sulcata* ticks were found only on small ruminants. *D. marginatus*, the most abundant among them, was found in 6 localities, while *Ha. punctata* and *Ha. sulcata* in 3 and 2 localities, respectively. Few specimens of *Ixodes gibbosus* (*I. gibbosus)* were collected from 3 localities (FIG. 2). All ticks collected from dogs belonged to the species *Rh. sanguineus*.

Rickettsial Infections

Fifteen tick specimens collected from 8 localities were found to be infected with *Rickettsiae* and/or *Coxiella* species. The following tick species were found to be infected with SFG rickettsiae: *Rh. sanguineus, Rh. turanicus, Hy. anatolicum excavatum. C. burnetii* was detected in *D. marginatus*, *Hy. marginatum marginatum,* and *Rh. turanicus* ticks.

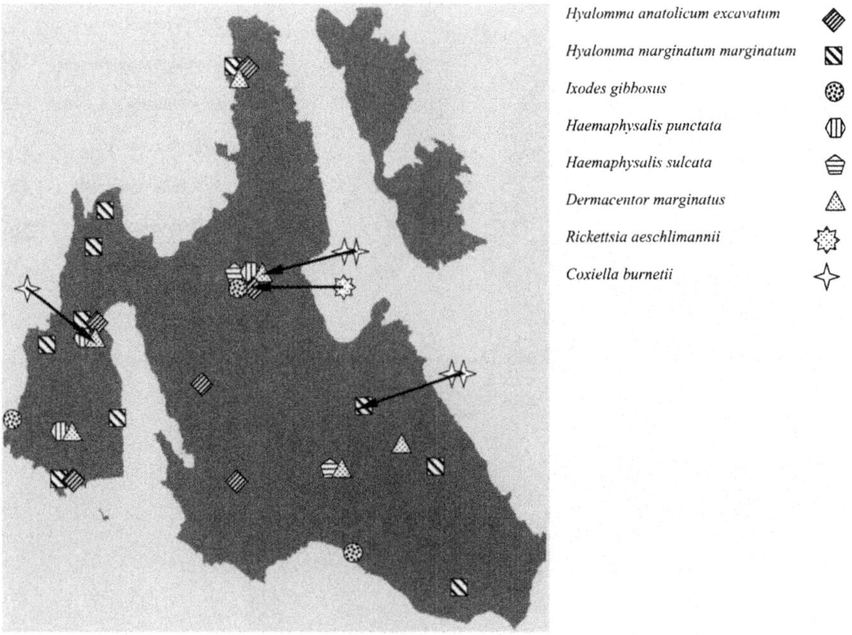

FIGURE 2. Distribution of *Hyalomma, Ixodes, Haemaphysalis, and Dermacentor* species and of *R. aeschlimannii* and *C. burnetii* found in these tick species in Cephalonia.

R. conorii was found in three specimens of *Rh. sanguineus* (a nymph, a female, and a nymph) collected from dogs in two villages. In one of these two villages, two other *Rh. sanguineus* ticks collected from dogs harbored *R. rhipicephali*. Four *Rh. turanicus* ticks, collected from cattle, sheep, and goats in three villages, were found to be infected with *R. massiliae*. One of them had a mixed infection with *C. burnetii* and *R. massaliae*. *C. burnetii* was also found in three *D. marginatus* ticks collected from goats in two villages, and in two specimens of *Hy. marginatum marginatum* collected from sheep in one village. A *Hy. anatolicum excavatum* tick, parasitizing on sheep, was found to be infected with *R. aeschlimannii*.

DISCUSSION

Rickettsiae of SFG are tick-borne microorganisms with effective transovarial and transstadial transmission. The main hosts are ticks (*Dermacentor*, *Rhipicephalus*, *Haemophysalis*, *Ixodes,* and *Amlyomma*). There are few publications concerning the tick fauna of Greece.[9,14] In a study that was conducted in Northern Greece during 1983–1986, 11,610 ticks, belonging to 7 genera (18 tick species and subspecies), were collected from domestic animals (cattle, sheep, goats, and dogs) in 64 localities throughout Macedonia (northern Greece).[9,14] In the aforementioned study, *Rh. bursa*, *Hy. marginatum marginatum,* as well as *Boophilus annulatus* were found in all bioclimatic zones. *Rh. turanicus, Rh. sanguineus, I. Gibbosus,* and *Hy. anatolicum excavatum* were essentially represented in the mesoMediterranean bioclimatic zone, while *I. ricinus, D. marginatus,* and *Haemaphysalis* ticks (*Ha. inermis, Ha. punctata, Ha. sulcata, Ha. parva*) were found frequently in the biotopes of the attenuated mesoMediterranean and subMediterranean bioclimats.

In the present study, ticks belonging to 5 genera (9 tick species and subspecies) were collected from domestic animals (cattle, sheep, goats, horses, and dogs) in 32 localities of the island of Cephalonia. *Rh. bursa, Rh. turanicus, Rh. sanguineus, D. marginatus, I. gibbosus, Ha. punctata, Ha. sulcata, Hy. anatolicum excavatum,* and *Hy. marginatum marginatum* were the species identified.

Ticks of the genus *Rhipicephalus* were the most abundant. This fact was expected to be in concordance with past studies since the island belongs to the mesoMediterranean bioclimatic zone to which these ticks are well adapted. Additionally, the fact that all ticks collected from dogs belonged to the species *Rh. sanguineus* can be justified since almost all dogs examined were domestic or stray ones, and the study took place mainly during spring and early summer periods of activity of these ticks. *Rh. sanguineus* is a tick species that can feed in all stages on dogs and does not require another host species to complete its life cycle. Ticks of the *Rh. sanguineus* species complex may be vector of various pathogens including *Rickettsia conorii* (the etiological agent of the Mediterranean spotted fever (MSF)), Astrakhan fever rickettsia, Bar29.[15,16]

Hyalomma ticks were widely distributed in different parts of the island in a wide range of hosts (goats, sheep, cattle, and horses). *Hy. anatolicum excavatum* ticks were collected from 5 localities, while *Hy. marginatum marginatum* ticks were found in 11 localities, parasitizing all five host species (FIG. 2). The widespread distribution of *Hy. marginatum marginatum* in Cephalonia confirms the results of previous studies conducted in northern Greece.[9,14] This is of epidemiological importance since *Hy. marginatum marginatum* may be a reservoir as well as a vector of *R. aeschlimannii*. This tick species is endemic in southern Europe, and readily bites people using an attack strategy. These ticks emerge from their habitat and run toward potential hosts when these appear nearby.[17] Birds constitute the main hosts of larvae and nymphs of *Hy. marginatum marginatum*, and play an important role in the dissemination of this tick species.

D. marginatus, *Ha. punctata,* and *Ha. sulcata* ticks were found only on small ruminants. Few specimens of *I. gibbosus* were collected from three localities (FIG. 2). The absence of ticks of the genus *Dermacentor*, *Haemaphysalis,* and *Boophilus* in cattle and horses could be explained from the low number of these hosts examined during autumn and winter, the periods of activity of these ticks in Greece. *D. marginatus* ticks are well known to bite people in Europe, and are vectors of *R. slovaca*, the causative agent of tick-borne lymphadenopathy (TIBOLA).[10,25,33]

The following tick species were found to be infected with SFG rickettsiae: *Rh. sanguineus, Rh. turanicus, Hy. anatolicum excavatum.* The rickettsial species detected in ticks during this study were: *R. massiliae, R. conorii, R. rhipicephali,* and *R. aeschlimannii.* Among these, *R. conorii, R. massiliae,* and *R. rhipicephali* have been detected in previous studies in ixodidae ticks in Greece,[2,18] while *R. aeschlimannii* was detected for the first time.

R. conorii is the most well-known SFG rickettsia in Greece. This bacterium, transmitted by *Rh. sanguineus*, causes "boutonneuse" or MSF, an endemic disease in Greece. *R. conorii* has been isolated from ticks and patients.[2,4] Seroepidemiological surveys in different areas of Greece have revealed an important prevalence of antibodies against *R. conorii*, especially among the rural population.[6,19,20] Until recently, MSF was thought to be the only rickettsial disease prevalent in Greece, but in recent studies it has been demonstrated that a second tick-transmitted rickettsial disease occurs in Greece, due to a rickettsial strain related to "*R. sibirica mongolotimonae.*" This rickettsia was simultaneously detected in a hospitalized patient and in a *Hy. anatolicum excavatum* tick.[3]

R. massiliae, an SFG rickettsia of unknown pathogenicity, which had been isolated for the first time from a *Rh. turanicus* tick in France,[21] was found in four specimens belonging to the same tick species in Cephalonia. A strain of *R. massiliae* (named GS) had earlier been isolated from a *Rh. sanguineus* tick in central Greece.[18]

R. aeschlimannii was detected in a *Hy. anatolicum excavatum* tick, removed from a sheep. This is the first detection of *R. aeschlimannii* in Greece and, to the best of our knowledge, the first detection of *R. aeschlimannii* in *Hy. anatolicum excavatum* ticks. *Hy. marginatum ticks* are regarded to be a reservoir as well as a vector of *R. aeschlimannii*. This rickettsia was first characterized as a new SFG rickettsia following its isolation from *Hy. marginatum marginatum* ticks in Morocco[22] it has, also, been detected in *Hy. marginatum rufipes* in Niger and Mali. The first human infection due to *R. aeschlimannii* was reported in 2002, in a patient returning from Morocco to France.[23] A second case was identified in a patient in South Africa.[24] In Europe, *R. aeschlimannii* has been detected in *Hy. marginatum marginatum* ticks in Croatia[25] in 6 tick species in Spain, including *Hy. marginatum marginatum*[26] and in *Hy. marginatum rufipes* and *Hy. marginatum marginatum* ticks in Corsica.[17] The results of these studies as well as our results indicate that *R. aeschlimannii* is widespread in southern Europe.

C. burnetii is an obligate intracellular bacterium living in the phagolysosomes of the host cell, which causes Q fever, a worldwide zoonosis. It most commonly spreads by means of inhalation or ingestion but it can, also, be transmitted by ticks with a bite or feces.[27] *C. burnetii* has been found in several tick species, including sheep-ticks of the genus *Dermacentor* in

Germany,[28] *Ha. longicornis* ticks in Korea,[29] and *Rhipicephalus* ticks in Southern Switzerland.[15] *C. burnetii* has been isolated from *Rh. sanguineus* and *Hyalloma* spp. ticks in Cyprus.[7] *D. marginatus* has been considered to be the main vector of coxiella in some areas in the southern part of central Slovakia.[30] In the southern part of central Slovakia and in northern Hungary, *D. marginatus* was the tick species with the highest positivity for *C. burnetii* using the PCR–RFLP method.[31]

In our study *C. burnetii* was detected in *D. marginatus*, *Hy. marginatum marginatum,* and *Rh. turanicus* ticks collected from domestic animals. This was the first detection of *C. burnetii* in ticks in Greece. Additionally, a mixed infection of *R. massiliae* and *C. burnetii* was demonstrated. Coexistence of different tick micro-organisms (rickettsiae, borreliae, ehrlichiae, tick-borne encephalitis complex viruses, etc.) has been reported. To our knowledge, this is the first evidence of dual infections with a SFG *Rickettsia* and *Coxiella* in ticks. Reports of people with both Q fever and other tick-borne diseases are rare. In a recent study, 5 of the 6 patients suffering from Q fever were, also, infected with SFG rickettsiae. Of these 6 cases, 3 were probably due to a concomitant infection after a tick bite.[32] As Q fever is endemic in Greece, the role of the ticks in the epidemiology of Q fever needs to be further investigated.

The results of this study extend the knowledge of the geographic distribution of SFG rickettsiae and their tick vectors. Except for *Rhipicephalus* spp. ticks, *Hyalomma* spp. ticks seem to be of great importance in the epidemiology of SFG rickettsiae.

REFERENCES

1. RAOULT, D. & V. ROUX. 1997. Rickettsioses as paradigms of new or emerging infectious diseases. Clin. Microbiol. Rev. **10:** 694–719.
2. PSAROULAKI, A., I. SPYRIDAKI, A. IOANNIDIS, *et al.* 2003. First isolation and identification of Rickettsia conorii from ticks collected in the region of Fokida in Central Greece. J. Clin. Microbiol. **41:** 3317–3319.
3. PSAROULAKI, A., A. GERMANAKIS, A. GIKAS, *et al.* 2005. Simultaneous detection of "Rickettsia mongolotimonae" in a patient and in a tick in Greece. J. Clin. Microbiol. **43:** 3558–3559.
4. PSAROULAKI, A., A. GERMANAKIS, A. GIKAS, *et al.* 2005. First isolation and genotypic identification of Rickettsia conorii Malish 7 from a patient in Greece. Eur. J. Clin. Microbiol. Infect. Dis. **24:** 297–298.
5. PAROLA, P. & D. RAOULT. 2001. Tick-borne bacterial diseases emerging in Europe. Clin. Microbiol. Infect. **7:** 80–83.
6. ANTONIOU, M., I. ECONOMOU, X. WANG, *et al.* 2002. A fourteen-year seroepidemiological study of zoonoses in a Greek village. Am. J. Trop. Med. Hyg. **66:** 80–85.
7. SPYRIDAKI, I., A. PSAROULAKI, F. LOUKAIDES, *et al.* 2002. Isolation of *Coxiella burnetii* by a centrifugation shell-vial assay from ticks collected in Cyprus: detection

by nested polymerase chain reaction (PCR) and by PCR-restriction fragment length polymorphism analyses. Am. J. Trop. Med. Hyg. **66:** 86–90.
8. Papadopoulos, B. 1990. Les tiques des animaux domestiques et les hématozoaires qu'elles transmettent en Macédoine (Grèce). Ph.D. Thesis, University of Neuchâtel, 1–248.
9. Brouqui, P., F. Bacellar, G. Baranton, *et al.* ESCMID Study Group on Coxiella, Anaplasma, Rickettsia and Bartonella; European Network for Surveillance of Tick-Borne Diseases. 2004. Guidelines for the diagnosis of tick-borne bacterial diseases in Europe. Clin. Microbiol. Infect. **10:** 1108–1132.
10. Roux, V., E. Rydkina, M. Eremeeva & D. Raoult. 1997. Citrate synthase gene comparison, a new tool for phylogenetic analysis, and its application for the rickettsiae. Int. J. Syst. Bacteriol. **47:** 252–261.
11. Fournier, P.E., V. Roux & D. Raoult. 1998. Phylogenetic analysis of spotted fever group rickettsiae by study of the outer surface protein rOmpA. Int. J. Syst. Bacteriol. **48:** 839–849.
12. Spyridaki, I., A. Psaroulaki, F. Loukaides, *et al.*2002. Isolation of *Coxiella burnetii* by a centrifugation shell-vial assay from ticks collected in Cyprus: detection by nested polymerase chain reaction (PCR) and by PCR-restriction fragment length polymorphism analyses. Am. J. Trop. Med. Hyg. **66:** 86–90.
13. Papadopoulos, B., P.-C. Morel & A. Aeschlimann. 1996. Ticks of domestic animals in the Macedonia region Greece. Vet. Parasitol. **66:** 25–40.
14. Bernasconi, M.V., S. Casati, O. Peter & J.C. Piffaretti. 2002. *Rhipicephalus* ticks infected with *Rickettsia* and *Coxiella* in Southern Switzerland (Canton Ticino). Infect. Genet. Evol. **2:** 111–120.
15. Cardenosa, N., V. Roux, B. Font, *et al.*2000. Short report: isolation and identification of two spotted fever group rickettsial strains from patients in Catalonia, Spain. Am. J. Trop. Med. Hyg. **62:** 142–144.
16. Matsumoto, K., P. Parola, P. Brouqui & D. Raoult. 2004. *Rickettsia aeschlimannii* in Hyalomma ticks from Corsica. Eur. J. Clin. Microbiol. Infect. Dis. **23:** 732–734.
17. Babalis, T., Y. Tselentis, V. Roux, *et al.* 1994. Isolation and identification of a *Rickettsial strain* related to Rickettsia massiliae in Greek ticks. Am. J. Trop. Med. Hyg. **50:** 365–372.
18. Babalis, T., H.T. Dupont, Y. Tselentis, *et al.* 1993. *Rickettsia conorii* in Greece: comparison of a microimmunofluorescence assay and western blotting for seroepidemiology. Am. J. Trop. Med. Hyg. **48:** 784–792.
19. Alexiou-Daniel, S., K. Manika, M. Arvanitidou & A. Antoniadis. 2002. Prevalence of *Rickettsia conorii* and *Rickettsia typhi* infections in the population of northern Greece. Am. J. Trop. Med. Hyg. **66:** 76–79.
20. Beati, L. & D. Raoult. 1993. *Rickettsia massiliae* sp. nov., a new spotted fever group rickettsia. Int. J. Syst. Bacteriol. **43:** 839–840.
21. Beati, L., M. Meskini, B. Thiers & D. Raoult. 1997. *Rickettsia aeschlimannii* sp. nov., a new spotted fever group rickettsia associated with *Hyalomma marginatum* ticks. Int. J. Syst. Bacteriol. **47:** 548–554.
22. Raoult, D., P.E. Fournier, P. Abboud & F. Caron. 2002. First documented human *Rickettsia aeschlimannii* infection. Emerg. Infect. Dis. **8:** 748–749.
23. Pretorius, A.M. & R.J. Birtles. 2002. *Rickettsia aeschlimannii*: a new pathogenic spotted fever group rickettsia, South Africa. Emerg. Infect. Dis. **8:** 874.

24. PUNDA-POLIC, V., M. PETROVEC, T. TRILAR, *et al.* 2002. Detection and identification of spotted fever group rickettsiae in ticks collected in southern Croatia. Exp. Appl. Acarol. **28:** 169–176.
25. FERNÁNDEZ-SOTO, P., A. ENCINAS-GRANDES & R. PÉREZ-SÁNCHEZ. 2003. *Rickettsia aeschlimannii* in Spain: molecular evidence in *Hyalomma marginatum* and five other tick species that feed on humans. Emerg. Infect. Dis. **9:** 889–890.
26. MAURIN, M. & D. RAOULT. 1999. Q fever. Clin. Microbiol. Rev. **12:** 518–553.
27. STING, R., N. BREITLING, R. OEHME & P. KIMMIG. 2004. The occurrence of *Coxiella burnetii* in sheep and ticks of the genus *Dermacentor* in Baden-Wuerttemberg. Dtsch. Tierarztl. Wochenschr. **111:** 390–394.
28. LEE, J.H., H.S. PARK, W.J. JANG, *et al.* 2004. Identification of the *Coxiella* sp. detected from *Haemaphysalis longicornis* ticks in Korea. Microbiol. Immunol. **48:** 125–130.
29. REHACEK, J. 1987. Epidemiology and significance of Q fever in Czechoslovakia. Zentralbl. Bakteriol. Mikrobiol. Hyg. [A]. **267:** 16–19.
30. SPITALSKA, E. & E. KOCIANOVA. 2003. Detection of *Coxiella burnetii* in ticks collected in Slovakia and Hungary. Eur. J. Epidemiol. **18:** 263–266.
31. ROLAIN, J.M., F. GOURIET, P. BROUQUI, *et al.* 2005. Concomitant or consecutive infection with *Coxiella burnetii* and tickborne diseases. Clin. Infect. Dis. **40:** 82–88.
32. OTEO, J.A., V. IBARRA, J.R. BLANCO, *et al.* 2004. *Dermacentor*-borne necrosis erythema and lymphadenopathy: clinical and epidemiological features of a new tick-borne disease. Clin. Microbiol. Infect. **10:** 327–331.

Molecular Characterization of *Rickettsia rickettsii* Infecting Dogs and People in North Carolina

LINDA KIDD,[a] BARBARA HEGARTY,[a] DANIEL SEXTON,[b] AND EDWARD BREITSCHWERDT[a]

[a]*North Carolina State University College of Veterinary Medicine, Raleigh, North Carolina, USA*

[b]*Duke University Medical Center, Durham, North Carolina, USA*

ABSTRACT: Rocky Mountain spotted fever (RMST) is an important cause of morbidity and mortality in people and dogs in the United States. Disease manifestations are strikingly similar in both species, and illness in dogs can precede illness in people. *R. rickettsii* has been identified as a Select Agent by the CDC as a Category C priority pathogen by the National Institute of Allergic and Infectious Diseases because it is amenable to use as a bioterror agent. The clinical and temporal relationship of naturally occurring diseases in dogs and people suggests that dogs could serve as sentinels for natural infection and bioterrorist attacks using this organism. Recognizing genetic modifications in naturally occurring disease agents in order to distinguish them from intentionally released agents are priorities put forth by the NIAID. To determine whether the rickettsiae naturally infecting dogs is the same as those that infect persons in a given geographical region, we characterized rickettsial isolates obtained from three dogs and two persons diagnosed with RMSF in North Carolina. Portions of three genes (*ompA, rrs*, and *gltA*) amplified by PCR were cloned and sequenced or directly sequenced. Reactions were run in duplicate in forward and reverse directions. Gene sequences were aligned with known sequences deposited in GenBank and with each other. Sequences of the 5′ region of the ompA gene were 100% homologous with a tick strain (Bitterroot) of *R. rickettsii* for all five isolates. Sequences of the *rrs* gene were 99.89.9% homologous with a tick strain (Sawtooth) of *R. rickettsii*. *rrs* gene sequences from one dog and the two persons was identical. Sequences of one dog isolate differed from these by one base pair. Sequences from another dog isolate differed by two base pairs. Sequences of the *gltA* gene are pending. This confirms on a molecular level that *R. rickettsii* causing naturally occurring RMSF in dogs in North Carolina is highly homologous to *R. rickettsii* that causes the disease in people in the same region. Sequence data will be deposited in GenBank,

Address for correspondence: Linda Kidd, TheScripps Research Institute, Department of Immunology, 10550 N. Torrey Pines Road, La Jolla, CA 92037. Voice: 919-539-2315.
e-mail: lkidd@scripps.edu

Ann. N.Y. Acad. Sci. 1078: 400–409 (2006).
doi: 10.1196/annals.1374.079

thereby providing genetic information regarding naturally occurring *R. rickettsii*.

KEYWORDS: Rocky Mountain spotted fever; *Rickettsia rickettsii*; dog; gene sequence

BACKGROUND AND PURPOSE

Diseases caused by organisms in the genus *Rickettsia* are heralded as paradigms of emerging infectious disease.[1,2] In addition to the clinical features of rickettsial disease, diagnosis typically uses methods such as serological testing that do not differentiate among species, so the species of the infecting *Rickettsia* is generally presumed on the basis of typical disease manifestations and the geographic location in which the tick bite was encountered. Advances in molecular techniques have contributed to the recognition and characterization of classic and novel rickettsial species as disease-causing agents in expanding geographic locales.[1–3]

Characterizing species of *Rickettsia* in a given region has important implications for local residents, travelers, and even global biodefense. Rocky Mountain spotted fever (RMSF) is an important cause of morbidity and mortality in people and dogs in the United States.[4–6] North Carolina has one of the highest reported incidences of the disease in people.[5,6] Disease manifestations are strikingly similar in people and in dogs.[1,2,5,7–9] Illness in dogs can precede illness in people in the same household.[4,10–13] *Rickettsia rickettsii* (*R. rickettsii*) has been identified as a select agent by the Centers for Disease Control and Prevention, and as a category C priority pathogen by the National Institute of Allergic and Infectious Disease (NIAID) because of concerns that it is amenable to use as a biological weapon.[14] The clinical and temporal relationship of the naturally occurring disease in dogs and people suggests that dogs can serve as sentinels for natural infection and perhaps bioterrorist attacks using this organism. Recognizing genetic modifications in naturally occurring disease agents in order to distinguish them from intentionally released agents are priorities recently put forth by the NIAID. In addition, molecular characterization of *R. rickettsii* isolated from dogs has not been reported.

The purpose of this study was to characterize *Rickettsia* isolates obtained from naturally infected dogs and persons in North Carolina by gene sequencing in order to determine whether *Rickettsia* species infecting dogs are the same as those that infect people in a given geographical region, and to provide sequence information regarding this important infectious agent.

MATERIALS AND METHODS

Rickettsia were cultured from the blood of 3 dogs and 2 people[14] presenting with clinical findings consistent with RMSF using Vero or DH82 cells as previously described.[15] DNA was extracted using the Qiaamp® DNA Blood Mini Kit following the manufacturer's instructions (bacterial culture and suspension and tissue protocol). An extraction control was included in the procedure. PCR was performed in order to amplify portions of three genes [*gltA, rrs, and ompA* (5′region)] using primers shown in TABLE 1. Negative (water) controls were included in all assays, and extraction controls were included in the *rrs* and *ompA* PCRs. Both the 5′ and 3′ portions of the *gltA* gene were amplified using a modification of a previously described protocol.[16] The reaction mixture contained forward and reverse primers at a final concentration of 1μM, 25 μL of PCR Master Mix (Promega, Madison, WI), 1–2 μL of template DNA and H_20 to a final reaction volume of 50 μL. Thermocycler conditions included a denaturation step for 2 min at 95°C, then 40 cycles of denaturation (95°C for 30 sec), annealing (45°C for 30 sec), and extension (72°C for 1 min), and a final extension at 92°C for 5 min. The *rrs* gene was amplified using the 8F and 1492R primers (TABLE 1) under the following conditions for all but one isolate. For each reaction, 27.5 μL of H_20, 6 μL of 25 mM $MgCl_2$ (3 mM final concentration), 5 μL 10 × buffer, and 0.75 μL AmpliTaqGold® (Applied, Mannheim, Germany) were incubated with 1 μL of a 1:10 dilution of DNase I (Roche Diagnostics) at room temperature for 30 minutes, and then 95°C for 15 min. This was then added to 7 μL of a mixture of 2 mM of each dNTP (final concentration of 280 μM), 1 μL of 25 pmol/μL (final concentration of 0.5uM) of each primer, and 1–2 μL DNA template to make a final reaction volume of 50 μL. The DNase treatment was adapted from a previously described protocol, and used to overcome nonspecific amplification of residual bacterial DNA in the AmpliTaqGold® polymerase, and/or the water.[18] Thermocycler conditions included a denaturation step at 94°C for 5 min, then 40 cycles of denaturation (94°C for 45 sec), annealing (55°C for 30 sec), and extension (72°C for 1 min), and a final extension of 72°C for 10 min. For one isolate, *Rickettsia*-specific *rrs* primers (20F and 1455R, TABLE 1) were used under the following conditions to amplify 16S rDNA. Twenty-five

TABLE 1. Primers used in amplification of genes

Primer pair	Gene	Sequence	Reference
5′gltA	*gltA*	5′ ATTAGGGTTATACTTACTG3′	
Cs890R		5′GCTTTAGCTACATATTTAGG3′	16
877p	*gltA*	5′GGGGACCTGCTCACGGCGG3′	16
1258n		5′ATTGCTAAAAGTACAGTGAACA3′	16
8F	*rrs*	5′ AGAGTTTGATCCTGGCTCAG3′	17[a]
1492R		5′-GGTTACCTTGTTACGACTT-3′	17[a]
20F	*rrs*	5′TGGCTCAGAACGAACGCTATC3′	
1455R		5′ TTTACCGTGGTTGGCTGCCT3′	
70F	*ompA*	5′ATGGCGAATATTTCTCCAAAA3′	19
701R		5′GTTCCGTTAATGGCAGCATCT3′	20

[a]Modified from original reference.

microliters of PCR Master Mix (Promega) forward and reverse primers at a final concentration of 1uM, 2 μL of DNA template and water were combined to make a final reaction volume of 50 μL. Thermocycler conditions consisted of an initial denaturation step of 94°C for 5 min, and 40 cycles of denaturation at 95°C for 45 sec, annealing at 58°C for 45 sec, extension at 72°C for 1.5 min, and a final extension of 72°C for 5 min. The 5′ region of the *ompA*, gene was amplified using a method adapted from a previously described protocol.[20] A reaction mixture containing primers at a final primer concentration of 0.3 uM, 5 μL of 10× AccuPrime *Pfx* reaction mix, 1–2 μL of DNA template, 1 μL of AccuPrime™ *Pfx* DNA polymerase (2.5 U/μL), and H_2O to a final reaction volume of 50 μL. Thermocycler conditions consisted of an initial denaturation step at 95°C for 2 min, then 35 cycles of denaturation (95°C for 15 sec), annealing (50°C for 30 sec), and extension (68°C for 1 min) with a final extension of 72°C for 5 min. PCR products were purified using the Qiaquick® PCR purification kit or the Qiaquick® gel extraction kit according to manufacturer's instructions. Products were cloned and sequenced, or subjected to direct sequencing. Cloning reactions were performed using the vector PGEM T Easy System I® or the TOPO TA Cloning® kit according to manufacturer's instructions Sequencing reactions were performed on the LICOR 4200 DNA sequencer according to manufacturer's instructions or through Davis Sequencing (www.davissequencing.com). Reactions were performed in duplicate, on separate clones or PCR reactions, in forward and reverse directions. Consensus sequences were created for each gene for each isolate. Regions of local sequence similarity were initially assessed using the Basic Local Alignment Search Tool (BLAST) for sequences deposited in GenBank (http://www.ncbi.nlm.nih.gov/BLAST). Because sequences deposited in GenBank for a given gene can differ in length, individual base pair variations that exist among members of the same species may not be accounted for when assessing homology using BLAST. To account for these differences, consensus sequences were created if more than one sequence was available in GenBank for a given species of *Rickettsia*. GenBank sequences for members of the genus *Rickettsia* with a species designation were downloaded and aligned using the CLUSTAL W multiple alignment tool (Bioedit version 5.0.6). Partial sequences were included only if a complete sequence for the same species was available in another deposit. *Rickettsia* species and GenBank accession numbers used in the creation of the consensus sequences are shown in TABLE 2. To make comparisons, sequences were aligned and edited to the same length for each gene for each isolate. The GenBank sequences were similarly aligned and trimmed so that sequences were the same length as the isolates. Consensus sequences for the isolates and the GenBank *Rickettsia* consensus sequences were then aligned simultaneously using CLUSTAL W multiple alignment tool (Bioedit version 5.0.6). Differences were enumerated manually, and also evaluated using a sequence identity matrix program (Bioedit Version 5.0.6).

RESULTS

PCR water controls and DNA extraction controls were negative. Contiguous sequences resulting from amplification of the two portions the *gltA* gene resulted in amplicons ranging from 1181 to 1196 base pairs in length, corresponding to base pair −26 to 1156 in relation to the *gltA* gene of *R. conorii* (GenBank accession number AE008677). *Rickettsia* species in GenBank with available sequence in the −26 to +21 region of the *gltA* gene are limited to *R. conorii*, *R. typhi,* and *R. prowazekii*. Therefore, using the BLAST program, sequence homology for the amplified *gltA* gene for dog and human isolates described in this study matched most closely to *R. conorii* (99.1–99.3%). Known sequences for the *R. rickettsii gltA* gene begin at base pair 21, therefore sequence homology estimates in relation to *R. rickettsii* could only be made on sequences trimmed as described in the MATERIALS AND METHODS section. The length of the trimmed sequences and the percentage homology to the *R. rickettsii* consensus sequence is shown in TABLE 3. The trimmed *gltA* consensus sequences for the dog and human isolates had a calculated sequence identity of 0.996–0.998 to *R. rickettsii*. The isolates of dog origin were identical to each other, while the isolates of human origin shared 0.997 identity. The sequence identity for dog isolates compared to human isolates ranged from 0.997 to 0.998. The sequence identity between the dog isolates and the *R. rickettsii* consensus sequence was not equal to 1.00 because of base pair variability at two sites in the *R. rickettsii* consensus sequence. When this variation was taken into account and homology was calculated manually, the dog isolate *gltA* sequences were 100% homologous with *R. rickettsii*, and homology of the human isolates to *R. rickettsii* ranged between 99.8% and 99.9% (TABLE 3). The isolate sequences were most homologous to the *R. rickettsii* consensus sequence. Sequences from *R. sibirica* (0.992–0.994 sequence identity) and *R. parkeri* (0.992–0.993 sequence identity) were the next most similar to the *gltA* sequences for the dog and human isolates. When compared to consensus sequences for known species of *Rickettsia*, *R. rickettsii* was most homologous to *R. parkeri*, *R. conorii,* and *R. slovaca* (0.992 sequence identity). *R. parkeri* and *R. siberica* shared the highest identity for all sequences compared (0.999 sequence identity).

Sequences resulting from the amplification of the 16S rRNA gene for four of the isolates using the 8F and 1492R eubacterial primers ranged from 1270 to 1438 base pairs in length, corresponding to positions 11 to 1455 of the *R. rickettsii* (GenBank accession number L36217) *rrs* gene. Using these primers resulted in amplification of a 16S DNA sequence from a *Mycoplasma* species in addition to *Rickettsia* for two dog isolates. Preliminary analysis showed 16S sequences matched most closely with *M. orale* (GenBank accession number AY796060.1, strain NC10112) and *M. timone* (GenBank accession number AY050170.1). *Rickettsia*-specific 16SDNA primers were designed in order to efficiently amplify rickettsial 16S DNA from one of the *Mycoplasma*-

TABLE 2. Rickettsia species and accession numbers used for sequence comparison

	GenBank accession number for each gene		
	gltA	*rrs*	*ompA*
Israeli tick typhus	U59727		AY197564, AY197563, AY197566, U43797
R. aeschlimannii	AY259084	U74757	
R. africae	U59733	L36098	
R. akari	U59717	L36099, U12958	
R. amblyomma(mii)		U11012	
R. astrakhan			U43791
R. australis	U59718	L36101, U17644, U12459	
R. bellii	U59716, AY362703, 375161	L36103, U11014	
R. canada	U20241, U59713	AF503168, U15162, L36104	
R. conorii	U59730	U12460, L36105, L36107	AY346453, U43794, U43806, U45244, U46918
R. heilongjiangii	AF178034, AY285776	AF178037	AY280711, AF179362
R. helvetica	AJ427878, U59723		
R. honei	U59726, AF022817		AF018075
R. japonica	U59724	L26213	
R. limoni		AF322442	
R. marmionii		AY737685	
R. massiliae	U59719	L36106, L36214	U43793, U43799
R. montana(ensis)	U74756	L36215, U11016	U43801
R. parkerii	U59732	L36673, U12461	
R. peacockii			AY319292, AH013412, AY357765, AY357766, AH0134413
R. prowazekii	AY730679, AY570300, M17149	M21789	
R. rhipicephali	U59721	L36216, U11019	U43803
R. rickettsii	AY189819, U59729	AY573599, L36217, M21293, 411021	AY319293, U43804
R siberica	U59734	D38628, U12462	U43807
R. slovaca	AY129301, U59725, AF484070	L36221	U43808
R. typhi	U20245, U59714	L36221, U12463,	
Thai tick typhus			U43809

TABLE 3. Homology of isolate consensus sequences to consensus sequences of *R. rickettsii*

Isolate No.	Origin	*gltA*	*rrs*[a]	*ompA*
1989CO1	Dog	100% (1141/1141)	99.8% (1223/1225)	100% (509/509)
1994CO2	Dog	100% (1141/1141)	99.9% (1224/1225)	100% (509/509)
1991CO3	Dog	100% (1141/1141)	99.9% (1224/1225)	100% (509/509)
1995HO1	Human	99.8% (1139/1141)	99.9% (1224/1225)	100% (509/509)
1995HO2	Human	99.9% (1140/1141)	99.9% (1224/1225)	100% (509/509)

[a]Final length includes gaps created in the alignment of all species.

containing isolates (TABLE 1). This resulted in an amplicon of 1221 base pairs in length corresponding to positions 69 to 1290. Using the BLAST program, all *Rickettsia* 16S DNA amplicons from the isolates matched most closely with *R. rickettsii* (strain Sawtooth GenBank, accession number RRU11021). Homology was 99.9% for all isolates. Consensus sequences for the *rrs* gene created from each isolate were 99.8–99.9% homologous with a consensus sequence created for the *rrs* gene sequences for *R. rickettsii*. (TABLE 2). Consensus *rrs* sequences from dog and human isolates were almost identical to each other (sequence identity 0.998–1.0 between dog and human isolates). The sequence identity with the *R. rickettsii* consensus sequence for one dog isolate was 0.998, while the rest of the dog and human isolates had a sequence identity of 0.999 with *R. rickettsii*. The species with the next closest sequence match to the dog and human isolates were *R. conorii* and *R. slovaca*, (sequence identity ranging from 0.995–0.996). The species with the most homologous gene sequences (other than the isolates and *R. rickettsii*) were *R. heilonjiangii* to *R. japonica,* and *R. conorii* to *R. astrakhan* (sequence identity 0.999).

Sequences resulting from the amplification of the 5′ region of the *ompA* gene ranged from 516–590 base pairs in length corresponding to base pair 91 to 680 of the *ompA* (190 kd) gene for *R. rickettsii* (GenBank accession number M31227). These sequences were 100% homologous to *R. rickettsii* strain Bitterroot (GenBank accession numbers RRU43804 and M31227). Trimmed sequences of the 5′ region of the *ompA* gene for the isolates were 100% homologous to each other (TABLE 2). These were also 100% identical to strain Bitterroot of *R. rickettsii* (GenBank accession number RRU43804). When consensus sequences were used for comparison, the identity between isolates and GenBank sequences was highest for the isolates with regard to each other (sequence identity of 1.00), and the consensus sequence for *R. rickettsii* (sequence identity 0.994). The identity between isolates and the consensus sequence of *R. rickettsii* was not 1.00 because of two variable base pairs in the *R. rickettsii* consensus sequence. *R. slovaca* and *R. siberica* (sequence homology of 0.958 and 0.962) represented the next most homologous species to the isolates. The *Rickettsia* species most homologous to *R. rickettsii* were also

R. slovaca and *R. sibirica* (sequence identity 0.958 for both). Using these consensus sequences, the species that shared the most identity for *ompA* (other than *R. rickettsii* and the isolates described in this study) were *R. astrakhan* and *R. sibirica* (sequence identity of 0.976).

DISCUSSION

Sequences from the genes *gltA, rrs*, and the 5′ region of *ompA* have been used, along with those of *gene D* and *ompB*, in establishing gene sequence-based criteria for the identification of new *Rickettsia* species. Although there is a high degree of homology in the *rrs* and *gltA* genes for members of this genus, the 5′ region of the *ompA* gene is highly variable and can be used to help assign an isolate to a particular species.[2,21] The results in this study identified a high degree of homology among the dog and human isolates, and between these isolates and *R. rickettsii*. The BLAST analysis program cannot always adequately demonstrate homology because several sequences that vary in length are available for each *Rickettsia* gene. Therefore assigning homology based on the number of matching bases using BLAST is not adequate. In this study, a consensus of GenBank sequences were generated and used for comparison between the dog and human isolates. Two variable base pairs occurred in the *R. rickettsii* consensus sequences for the *gltA* and *ompA* genes due to corresponding base pair differences in the available GenBank sequences. Base pairs from the isolates matched one of the base pairs from a GenBank *R. rickettsii* sequence in all instances, and thus their homology to *R. rickettsii* was higher when calculated manually than when calculated using the sequence identity matrix program (Bioedit version 5.1.0). Additionally, the primer used to amplify the 5′ region of the *gltA* gene (TABLE 1) was designed using an *R. conorii* sequence, because sequence for this region of the *gltA* gene for *R. rickettsii* has not been described previously. Amplificons from the 5′ region of the *gltA* gene of the isolates was 46 base pairs longer than most *Rickettsia gltA* sequences deposited in GenBank and therefore trimming was necessary to determine homology to other known species. Regardless of the method used to assess homology, all gene sequences for both the dog and the human isolates were more similar to *R. rickettsii* than any other available species consensus sequence, and differences were not large enough to assign the isolates to a new species, according to established criteria.[21]

Mycoplasma species DNA was amplified from two isolates when universal *rrs* (16S rDNA) eubacterial primers were used. This was presumed to be a cell line contaminant, as *Mycoplasma* was found in two additional isolates using the same cell line to culture blood from one dog and one human patient not included in this study. Preliminary sequence analysis showed highest degrees of homology with *M. orale* (GenBank accession number AY796060.1 NC10112) and *M. timone* (GenBank accession number AY050170.1). *Mycoplasma* contamination is a common problem in cell lines.[22] However, coinfection with *Mycoplasma* and *Rickettsia* in these dogs and human patients cannot be ruled out as the original uninoculated cell lines were not available for analysis.

These results confirm on a molecular level that *R. rickettsii* isolated from dogs with clinical manifestations consistent with RMSF in North Carolina is highly homologous to *R. rickettsii* that causes the disease in people in the same region. These are the first sequences reported for *R. rickettsii* isolated from dogs. Complete (untrimmed) sequence data for the isolates have been deposited in GenBank (accession numbers DQ150680 through DQ150694), thereby providing genetic information regarding naturally occurring *R. rickettsii.*

ACKNOWLEDGMENTS

This work was supported by the American Kennel Club Grant number 2610 and Merial Limited, Duluth, GA.

REFERENCES

1. RIZZO, M. *et al.* 2003. Rickettsial disease: classical and modern aspects. New Microbiol. **27:** 87–103.
2. RAOULT, D. & V. ROUX. 1997. Rickettsioses as paradigms of new or emerging infectious diseases. Clin. Microbiol. Rev. **10:** 694–719.
3. LA SCOLA, B. & D. RAOULT. 1997. Laboratory diagnosis of rickettsioses: current approaches to diagnosis of old and new rickettsial diseases. J. Clin. Microbiol. **35:** 2715–2727.
4. WARNER, R.D. & W.W. MARSH. 2002. Rocky Mountain spotted fever. J. Am. Vet. Med. Assoc. **221:** 1413–1417.
5. http://www.cdc.gov/ncidod/dvid/rmsf/organism.htm.
6. PADDOCK, C.D. *et al.*, 2002. Assessing the magnitude of fatal Rocky Mountain spotted fever in the United States: comparison of two national data sources. Am. J. Trop. Med. Hyg. **67:** 349–354.
7. GASSER, A.M., A.J. BIRKENHEUER & E.B. BREITSCHWERDT. 2001. Canine Rocky Mountain spotted fever: a retrospective study of 30 cases. J. Am. Anim. Hosp. Assoc. **37:** 41–48.
8. KEENAN, K.P. *et al.* 1977. Pathogenesis of infection with *Rickettsia rickettsii* in the dog: a disease model for Rocky Mountain spotted fever. J. Infect. Dis. **135:** 911–917.
9. KEENAN, K.P. *et al.* 1977. Studies on the pathogenesis of *Rickettsia rickettsii* in the dog: clinical and clinicopathologic changes of experimental infection. Am. J. Vet. Res. **38:** 851–856.
10. PADDOCK, C.D. *et al.* 2002. Short report: concurrent Rocky Mountain spotted fever in a dog and its owner. Am. J. Trop. Med. Hyg. **66:** 197–199.

11. ELCHOS, B.N. & J. GODDARD. 2003. Implications of presumptive fatal Rocky Mountain spotted fever in two dogs and their owner. J. Am. Vet. Med. Assoc. **223:** 1450–1452, 1433.
12. GORDON, J.C. *et al.* 1983. Rocky Mountain spotted fever in dogs associated with human patients in Ohio. J. Infect. Dis. **148:** 1123.
13. CARPENTER, C.F. *et al* 1995. Incidence of ehrlichial and rickettsial infection in patients with unexplained fever and recent history of tick bite in central North Carolina. J. Infect. Dis. **180:** 900–903.
14. AZAD, A.F. & S. RADULOVIC. 2003. Pathogenic rickettsiae as bioterrorism agents. Ann. N. Y. Acad. Sci. **990:** 734–738.
15. BREITSCHWERDT, E.B. *et al.* 1991. Efficacy of chloramphenicol, enrofloxacin, and tetracycline for treatment of experimental Rocky Mountain spotted fever in dogs. Antimicrob. Agents Chem. **35:** 2375–2381.
16. ROUX, V. *et al.* 1997. Citrate synthase gene comparison, a new tool for phylogenetic analysis and its application for the rickettsiae. Int. J. Syst. Bact. **47:** 252–261.
17. WILSON, K.H., R.B. BLITCHINGTON & R.C. GREENE. 1990. Amplification of bacterial 16s ribosomal DNA with polymerase chain reaction. J. Clin. Microbiol. **28:** 1942–1946.
18. HEININGER, A. *et al.* 2003. DNase pre-treatment of master mix reagents improves the validity of universal 16s rRNA gene PCR results. J. Clin. Microbiol. 1763–1765.
19. REGENERY, R., L.C.L. SPRUILL & B.D. PLIKAYTIS. 1991. Genotypic identification of rickettsiae and estimation of intraspecies sequence divergence for portions of two rickettsial genes. J. Bacteriol. **173:** 1576–1589.
20. FORNIER, P.E., V. ROUX & D. RAOULT. 1998. Phylogenetic analysis of spotted fever group rickettsiae by study of the outer surface protein *rOmpA*. Int. J. Syst. Bacteriol. **48:** 839–849.
21. FORNIER, P.E. *et al.* 2003. Gene sequence based criteria for identification of new rickettsia isolates and description of *Rickettsia heilonjiangensis* sp. nov. J. Clin. Microbiol. **41:** 5456–5465.
22. ROTTEM, S. & M.F. BARIELE. 1993. Beware of mycoplasmas. Trends Biotechnol. **11:** 143–151.

Bartonella Infection in Domestic Cats and Wild Felids

BRUNO B. CHOMEL,[a] RICKIE W. KASTEN,[a] JENNIFER B. HENN,[a] AND SOPHIE MOLIA[a,b]

[a]*Department of Population Health and Reproduction, School of Veterinary Medicine, University of California, Davis, California 95616, USA*

[b]*Centre de Coopération Internationale en Recherche Agronomique pour le Développement, Département d'Élevage et Médecine Vétérinaire, Domaine de Duclos, Prise d'eau, 97170 Petit Bourg, Guadeloupe, French West Indies*

ABSTRACT: *Bartonella* are vector-borne, fastidious Gram-negative bacteria causing persistent bacteremia in their reservoir hosts. Felids represent a major reservoir for several *Bartonella* species. Domestic cats are the main reservoir of *B. henselae*, the agent of cat-scratch disease. Prevalence of infection is highest in warm and humid climates that are optimal for the survival of cat fleas, as fleas are essential for the transmission of the infection. Flea feces are the likely infectious substrate. Prevalence of *B. henselae* genotypes among cat populations varies worldwide. Genotype Houston I is more prevalent in the Far East and genotype Marseille is dominant in western Europe, Australia, and the western United States. Cats are usually asymptomatic, but uveitis, endocarditis, neurological signs, fever, necrotic lesions at the inoculation site, lymphadenopathy, and reproductive disorders have been reported in naturally or experimentally infected cats. Domestic cats are also the reservoir of *B. clarridgeiae* and co-infection has been demonstrated. *B. koehlerae* has been isolated from domestic cats, and was identified in cat fleas and associated with a human endocarditis case. *B. bovis* was isolated from a few cats in the United States and *B. quintana* DNA was recently identified in a cat tooth. *Bartonella* spp. have also been isolated from free-ranging and captive wild felids from North America and Africa. Whereas, *B. henselae* was identified in African lions and a cheetah, some strains specific to these wild cats have also been identified, leading to the concept of a *B. henselae* group including various subspecies, as previously described for *B. vinsonii*.

KEYWORDS: *Bartonella*; domestic cats; wild cats; wild felids

Address for correspondence: Bruno B. Chomel, D.V.M., Ph.D., Department of Population Health and Reproduction, School of Veterinary Medicine, University of California, Davis, CA 95616. Voice: 1-530-752-8112; fax: 530-752-2377.
e-mail: bbchomel@ucdavis.edu

**Ann. N.Y. Acad. Sci. 1078: 410–415 (2006). © 2006 New York Academy of Sciences.
doi: 10.1196/annals.1374.080**

Bartonella species are fastidious Gram-negative bacteria that infect mainly mammalian erythrocytes and endothelial cells and cause long-lasting bacteremia in their reservoir hosts.[1] *Bartonella* are usually vector-borne bacteria and the vector can vary depending on the bacterial species involved and the hosts and reservoirs. Domestic cats are not only the principal reservoir for *B. henselae*, the main agent of cat scratch disease (CSD), but also for *B. clarridgeiae*, which has been suspected in a few cases of CSD,[1] and *B. koehlerae*, which has been recently reported to be the cause of human endocarditis.[2] The role of domestic cats in the epidemiology of *B. bovis* (formerly *B. weissii*) has not been clearly established, as only a limited number of isolates have been obtained from cats in Illinois and Utah.[3] Interestingly, *B. quintana* which has no known animal reservoir was recently identified by DNA extraction from the tooth of a cat.[4] Fleas play a major role in the transmission of *B. henselae* among cats[5] and are likely to be involved in transmission of other feline *Bartonella*.[6] However, other potential vectors such as ticks and biting flies have been recently identified to harbor *Bartonella* DNA, including *B. henselae*.[7,8]

Cats infected with *B. henselae, B. clarridgeiae*, and even *B. koehlerae* are usually bacteremic for weeks to months, and some cats have been reported to be bacteremic for more than a year.[9,10] Cats infected with *B. henselae* and *B. clarridgeiae* are more likely to have bacteremia relapses than cats infected with *B. koehlerae*.[10] Young cats ($\leq$1 year) are more likely than older cats to be bacteremic[11] and stray cats are more likely to be bacteremic than pet cats, as shown in different studies conducted in France (TABLE 1).[12–15] Prevalence of infection varies considerably among cat populations (stray or pets) with an increasing gradient from cold climates (0% in Norway) to warm and humid climates (68% in the Philippines) (TABLE 2).[16–21] At least two genotypes have been identified and designated as Houston-1 (type I) and Marseille (previously bacillary angiomatosis–trench fever [BATF]).[1,9] The respective prevalence of these two genotypes varies considerably among cat populations from different geographical areas (TABLES 1 and 2). *B. henselae* type Marseille is the dominant genotype in cat populations from the western United States, western Europe, and Australia, whereas type Houston-1 is dominant in Asia.[9] However, within a given country, the prevalence of these genotypes may vary. In France, Marseille type was the most common genotype in cats from the Nancy and Paris areas, whereas type Houston I was the main genotype in cats from Lyon or Marseille (TABLE 1). A few Australian and European studies have reported that most human cases of CSD were caused by *B. henselae* type Houston-1 despite the fact that type Marseille was found to be the dominant genotype in the cat population, suggesting that type Houston-1 strains could be more virulent in humans.[9]

Naturally infected cats are usually healthy carriers of the bacterium.[1,9] However, cases of uveitis and rare cases of endocarditis have been molecularly

TABLE 1. ***Bartonella henselae*** **(B. h.) and** ***B. clarridgeiae (B. c.)*** **bacteremia prevalences in stray and pet domestic cats in France**

		B. henselae						
		Type I	Type II	Types I/II		*B. h./B. c.*		
Origin and location	*N*	*N* (%)	*N* (%)	*N* (%)	*B. clarridgeiae*	*N* (%)	Total (%)	Reference
Stray								
Nancy	94	17 (34)	18 (36)	0 (0)	15 (30)	0 (0)	50 (53)	12
Marseille	61	15 (40)	7 (18)	0 (0)	16 (42)	0 (0)	38 (62)	13
Pets								
Paris	436	11 (15)	36 (50)	2 (2.8)	15 (21)	8 (11)	72 (16.5)	14
Lyon	99	6 (75)	0 (0)	0 (0)	2 (25)	0 (0)	8 (8.1)	15

associated with infection caused by *B. henselae*.[9] Seropositive cats were more likely to have kidney disease and urinary tract infections, stomatitis, and lymphadenopathy.[1] However, in experimentally infected cats, fever, lymphadenopathy, mild neurological signs, and reproductive disorders have been reported.[1]

The presence of cat fleas (*Ctenocephalides felis*) is essential for the maintenance of the infection within the cat population.[5] It has been shown that *B. henselae* can multiply in the digestive system of the cat flea and survive at least 3 days in the flea feces.[22,23] Experimentally, only cats inoculated with flea feces became bacteremic, whereas those on which fleas were deposited in retention boxes or which were fed fleas did not become bacteremic.[24] Therefore, the main source of infection appears to be the flea feces that are inoculated by contaminated cat claws.

A few studies have been conducted to determine whether captive or free-ranging wild felids can be infected with various *Bartonella* species. When *B. henselae* was used as antigen in an immunofluorescence test (IFA), *Bartonella* antibodies were detected in 17% (9/54) of the cats of the genus *Panthera* (lions, tigers, leopards, and jaguars), 18% of 11 cheetahs (*Acinonyx jubatus*), and 47% (24/49) of various small wild cats of the genus *Felis* kept in various

TABLE 2. ***Bartonella henselae*** **(*B. h.*) and** ***B. clarridgeiae (B. c.)*** **bacteremia prevalences in domestic cats from various climatic zones**

		B. henselae		*B. clarridgeiae*				
Location	*N*	Type I	Type II	Types I/II		*B. h./B. c.*	Total (%)	Reference
Norway	100	0	0	0	0	0	0 (0)	16
Sweden	91	0	2	0	0	0	2 (2.2)	17
Japan	690	43	1	0	5	1	50 (7.2)	18
Italy (North)	769	29	85	26	0	0	140 (18)	19
Jamaica	62	5	7	0	0	0	12 (19)	20
Philippines	31	13	0	0	6	4	19 (61)	21

zoological parks in California.[25] Similarly, *B. henselae* antibodies were detected in more than half (53%) of the 62 free-ranging bobcats (*Lynx rufus*) tested from California and from 35% of 74 free-ranging pumas (*Felis concolor*) from California as well as from 18% of 28 free-ranging Florida panthers (*F. concolor coryi*) and 28% of 7 pumas from Texas.[25,26] In a larger study of 479 samples collected from pumas and 91 samples collected from bobcats in North America, Central America, and South America, the overall prevalence of *B. henselae* antibodies was 19.4% in pumas and 23.1% in bobcats. In the United States, pumas from the southwestern states were more likely to be seropositive for *B. henselae* (prevalence ratio [PR] = 2.82, 95% confidence interval [CI] = 1.55, 5.11) than pumas from the Northwest and Mountain states. Similarly, adults were more likely to be *B. henselae*–seropositive (PR = 1.77, 95% CI = 1.07, 2.93) and to have higher antibody titers ($P = 0.026$) than juveniles and kittens. *B. henselae* antibody prevalence was 22.4% (19/85) in bobcats from the United States and 33.3% (2/6) in the Mexican bobcats.[27] Furthermore, *Bartonella* have been isolated from pumas and bobcats in California (Chomel *et al.*, International Conference on Emerging Infectious Diseases, Atlanta, 1998) and shown to be infective for domestic cats.[28] Similarly, a few serosurveys and isolation of *Bartonella* species have been performed on African wild felids from Africa. *B. henselae* genotype II was isolated from an African cheetah from Zimbabwe[29] and from a lion from South Africa.[30] In the latest study, seroprevalence for *B. henselae* antibodies based on an enzyme-linked immunoabsorbent assay (ELISA) test was 29% (18/62) in lions from three game ranches in the Free State Province of South Africa.[30]

In another study, blood and/or serum samples were collected from a convenience sample of 113 lions and 74 cheetahs captured in Africa between 1982 and 2002.[31] Whole blood available from 58 of the lions and 17 of the cheetahs was cultured for evidence of *Bartonella* spp., and whole blood from 54 of the 58 lions and 73 of the 74 cheetahs was tested for the presence of *Bartonella* DNA by TaqMan polymerase chain reaction (PCR). Three (5.2%) of the 58 lions and 1 (5.9%) of the 17 cheetahs were bacteremic. Two lions were infected with *B. henselae* determined by PCR/RFLP (restriction fragment length polymorphism) of the citrate synthase gene. The third lion and the cheetah were infected with previously unidentified *Bartonella* strains. Interestingly, 23% of the cheetahs and 3.7% of the lions tested by TaqMan PCR were positive for *Bartonella* spp. Serum samples from the 113 lions and 74 cheetahs were tested for the presence of IFA antibodies against *B. henselae* and antibody prevalence was 17% for the lions and 31% for the cheetahs.

Whereas, *B. henselae* was identified in African lions and a cheetah, some strains specific to these wild cats have also been identified, leading to the concept of a *B. henselae* group likely to include various subspecies, as previously described for *B. vinsonii*.

REFERENCES

1. Chomel, B.B., H.J. Boulouis & E.B. Breitschwerdt. 2004. Cat scratch disease and other zoonotic *Bartonella* infections. J. Am. Vet. Med. Assoc. **224:** 1270–1279.
2. Avidor, B., M. Graidy, G. Efrat, *et al.* 2004. *Bartonella koehlerae*, a new cat-associated agent of culture-negative human endocarditis. J. Clin. Microbiol. **42:** 3462–3468.
3. Regnery, R.L., N. Marano, P. Jameson, *et al.* 2000. A fourth *Bartonella* species, *Bartonella weissii*, species nova, isolated from domestic cats [abstract 4]. *In* Proceedings of the 15th Meeting of the American Society of Rickettsiology, p. 15.
4. La, V.D., L. Tran-Hung, G. Aboudharam, *et al.* 2005. *Bartonella quintana* in domestic cat. Emerg. Infect. Dis. **11:** 1287–1289.
5. Chomel, B.B., R.W. Kasten, K. Floyd-Hawkins, *et al.* 1996. Experimental transmission of *Bartonella henselae* by the cat flea. J. Clin. Microbiol. **34:** 1952–1956.
6. Rolain, J.M., M. Franc, B. Davoust & D. Raoult. 2003. Molecular detection of *Bartonella quintana, B. koehlerae, B. henselae, B. clarridgeiae, Rickettsia felis*, and *Wolbachia pipientis* in cat fleas, France. Emerg. Infect. Dis. **9:** 338–342.
7. Sanogo, Y.O., Z. Zeaiter, G. Caruso, *et al.* 2003. *Bartonella henselae* in *Ixodes ricinus* ticks (Acari: Ixodida) removed from humans, Belluno province, Italy. Emerg. Infect. Dis. **9:** 329–332.
8. Chung, C.Y., R.W. Kasten, S.M. Paff, *et al.* 2004. *Bartonella* spp. DNA associated with biting flies from California. Emerg. Infect. Dis. **10:** 1311–1313.
9. Boulouis, H.J., C.C. Chang, J.B. Henn, *et al.* 2005. Factors associated with the rapid emergence of zoonotic *Bartonella* infections. Vet. Res. **36:** 383–410.
10. Yamamoto, K., B.B. Chomel, R.W. Kasten, *et al.* 2002. Experimental infection of domestic cats with *Bartonella koehlerae* and comparison of protein and DNA profiles with those of other *Bartonella* species infecting felines. J. Clin. Microbiol. **40:** 466–474.
11. Chomel, B.B., R.C. Abbott, R.W. Kasten, *et al.* 1995. *Bartonella henselae* prevalence in domestic cats in California: risk factors and association between bacteremia and antibody titers. J. Clin. Microbiol. **33:** 2445–2450.
12. Heller, R., M. Artois, V. Xemar, *et al.* 1997. Prevalence of *Bartonella henselae* and *Bartonella clarridgeiae* in stray cats. J. Clin. Microbiol. **35:** 1327–1331.
13. La Scola, B., B. Davoust, M. Boni & D. Raoult. 2002. Lack of correlation between *Bartonella* DNA detection within fleas, serological results, and results of blood culture in a *Bartonella*-infected stray cat population. Clin. Microbiol. Infect. **8:** 345–351.
14. Gurfield, A.N., H.J. Boulouis, B.B. Chomel, *et al.* 2001. Epidemiology of *Bartonella* infection in domestic cats in France. Vet. Microbiol. **80:** 185–198.
15. Rolain, J.M., C. Locatelli, L. Chabanne, *et al.* 2004. Prevalence of *Bartonella clarridgeiae* and *Bartonella henselae* in domestic cats from France and detection of the organisms in erythrocytes by immunofluorescence. Clin. Diagn. Lab. Immunol. **11:** 423–425.
16. Bergh, K., L. Bevanger, I. Hanssen & K. Loseth. 2002. Low prevalence of *Bartonella henselae* infections in Norwegian domestic and feral cats. APMIS **110:** 309–314.

17. ENGVALL, E.O., B. BRANDSTROM, C. FERMER, *et al.* 2003. Prevalence of *Bartonella henselae* in young, healthy cats in Sweden. Vet. Rec. **152:** 366–369.
18. MARUYAMA. S, Y. NAKAMURA, H. KABEYA, *et al.* 2000. Prevalence of *Bartonella henselae, Bartonella clarridgeiae* and the 16S rRNA gene types of *Bartonella henselae* among pet cats in Japan. J. Vet. Med. Sci. **62:** 273–279.
19. FABBI, M., L. DEGIULI, M. TRANQUILLO, *et al.* 2004. Prevalence of *Bartonella henselae* in Italian stray cats: evaluation of serology to assess the risk of transmission of *Bartonella* to humans. J. Clin. Microbiol. 2004 **42:** 264–268.
20. MESSAM, L.L., R.W. KASTEN, M.J. RITCHIE & B.B. CHOMEL 2005. *Bartonella henselae* and domestic cats, Jamaica. Emerg. Infect. Dis. **11:** 1146–1147.
21. CHOMEL, B.B., E.T. CARLOS, R.W. KASTEN, *et al.* 1999. *Bartonella henselae* and *Bartonella clarridgeiae* infection in domestic cats from The Philippines. Am. J. Trop. Med. Hyg. **60:** 593–597.
22. HIGGINS, J.A., S. RADULOVIC, D.C. JAWORSKI & A.F. AZAD.1996. Acquisition of the cat scratch disease agent *Bartonella henselae* by cat fleas (Siphonaptera: Pulicidae). J. Med. Entomol. **33:** 490–495.
23. FINKELSTEIN, J.L., T.P. BROWN, K.L. O'REILLY, *et al.* 2002. Studies on the growth of *Bartonella henselae* in the cat flea (Siphonaptera: Pulicidae). J. Med. Entomol. **39:** 915–919.
24. FOIL, L., E. ANDRESS, R.L. FREELAND, *et al.* 1998. Experimental infection of domestic cats with *Bartonella henselae* by inoculation of *Ctenocephalides felis* (Siphonaptera: Pulicidae) feces. J. Med. Entomol. **35:** 625–628.
25. YAMAMOTO, K., B.B. CHOMEL, L.J. LOWENSTINE, *et al.* 1998. *Bartonella henselae* antibody prevalence in free-ranging and captive wild felids from California. J. Wildl. Dis. **34:** 56–63.
26. ROTSTEIN, D.S., S.K. TAYLOR, J. BRADLEY & E.B. BREITSCHWERDT. 2000. Prevalence of *Bartonella henselae* antibody in Florida panthers. J. Wildl. Dis. **36:** 157–160.
27. CHOMEL, B.B., Y. KIKUCHI, J.S. MARTENSON, *et al.* 2004. Seroprevalence of *Bartonella* infection in American free-ranging and captive pumas (*Felis concolor*) and bobcats (*Lynx rufus*). Vet. Res. **35:** 233–241.
28. YAMAMOTO, K., B.B. CHOMEL, R.W. KASTEN, *et al.* 1998. Homologous protection but lack of heterologous-protection by various species and types of *Bartonella* in specific pathogen-free cats. Vet. Immunol. Immunopathol. **65:** 191–204.
29. KELLY, P.J., J.J. ROONEY, E.L. MARSTON, *et al.* 1998. *Bartonella henselae* isolated from cats in Zimbabwe. Lancet **351:** 1706.
30. PRETORIUS, A.M., J.M. KUYL, D.R. ISHERWOOD & R.J. BIRTLES. 2004. *Bartonella henselae* in African lion, South Africa. Emerg. Infect. Dis. **10:** 2257–2258.
31. MOLIA, S., B.B. CHOMEL, R.W. KASTEN, *et al.* 2004. Prevalence of *Bartonella* infection in wild African lions (*Panthera leo*) and cheetahs (*Acinonyx jubatus*). Vet. Microbiol. **100:** 31–41.

Anaplasmosis

Focusing on Host–Vector–Pathogen Interactions for Vaccine Development

JOSÉ DE LA FUENTE,[a,c] PATRICIA AYOUBI,[b] EDMOUR F. BLOUIN,[a] CONSUELO ALMAZÁN,[a] VICTORIA NARANJO,[c] AND KATHERINE M. KOCAN[a]

[a]*Department of Veterinary Pathobiology, Center for Veterinary Health Sciences, Oklahoma State University, Stillwater, Oklahoma 74078, USA*

[b]*Microarray Core Facility, Department of Biochemistry and Molecular Biology, Oklahoma State University, Stillwater, Oklahoma 74078, USA*

[c]*Instituto de Investigación en Recursos Cinegéticos IREC (CSIC-UCLM-JCCM), Ronda de Toledo s/n, 13005 Ciudad Real, Spain*

ABSTRACT: *Anaplasma marginale* and *A. phagocytophylum* are intracellular rickettsiae that cause bovine anaplasmosis and human granulocytic anaplasmosis, respectively. The ultimate vaccine for the control of anaplasmosis would be one that reduces infection and transmission of the pathogen by ticks. Effective vaccines for control of anaplasmosis are not available despite attempts using different approaches, such as attenuated strains, infected erythrocyte and tick cell-derived purified antigens, and recombinant pathogen and tick-derived proteins. Three lines of functional analyses were conducted by our laboratory to characterize host–tick–*Anaplasma* interactions to discover potential vaccine candidate antigens to control tick infestations and the infection and transmission of *Anaplasma* spp.: (1) characterization of *A. marginale* adhesins involved in infection and transmission of the pathogen, (2) global expression analysis of genes differentially expressed in HL-60 human promyelocytic cells in response to infection with *A. phagocytophilum*, and (3) identification and characterization of tick-protective antigens by expression library immunization (ELI) and analysis of expressed sequence tags (EST) in a mouse model of tick infestations and by RNA interference in ticks. These experiments have resulted in the characterization of the *A. marginale* MSP1a as an adhesin for bovine erythrocytes and tick cells, providing support for its use as candidate vaccine antigen for the control of bovine . Microarray analysis of genes differentially expressed in human cells infected with *A. phagocytophilum* identified key molecules involved in

Address for correspondence: José de la Fuente, Department of Veterinary Pathobiology, Center for Veterinary Health Sciences, Oklahoma State University, Stillwater, OK 74078, USA. Voice: 405-744-0372; Fax: 405-744-5275.

e-mail: jose_delafuente@yahoo.com [ar] djose@cvm.okstate.edu.

Ann. N.Y. Acad. Sci. 1078: 416–423 (2006). © 2006 New York Academy of Sciences.
doi: 10.1196/annals.1374.081

pathogen infection and multiplication. The screening for tick-protective antigens resulted in vaccine candidates reducing tick infestation, molting, and oviposition and affecting *Anaplasma* infection levels in ticks.

KEYWORDS: **tick; *Anaplasma marginale*; *Anaplasma phagocytophilum*; RNA interference; vaccine**

INTRODUCTION

Organisms in the genus *Anaplasma* are obligate intracellular pathogens that multiply in both vertebrate and invertebrate hosts.[1] The type species, *A. marginale*, causes bovine anaplasmosis and only infects ticks and ruminants, while *A. phagocytophilum*, the causative agent of human and animal granulocytic anaplasmosis, has a wide host range including ticks and several species of birds and mammals, including humans.[1,2]

The effective control of anaplasmosis requires the control of infection in the mammalian host and the reduction of the infection and transmission of the pathogen by the tick vector. Several approaches have been used to develop vaccines for the control of anaplasmosis, particularly of bovine anaplasmosis.[3,4] *A. centrale,* generally less pathogenic than *A. marginale*, isolated by Sir Arnold Theiler in the early 1900s, is the most widely used vaccine for the control of bovine anaplasmosis.[3,4] Others have attempted the isolation of attenuated *A. marginale* strains with limited success.[3,4] The partial purification of *A. marginale* antigens from infected bovine erythrocytes was used to develop a commercial vaccine for the control of bovine anaplasmosis in the United States, but the results were variable.[3,4] The recent application of the tick cell culture system using *Ixodes scapularis* IDE8 cells for the multiplication of *A. marginale* and *A. phagocytophilum* provides the possibility to purify *Anaplasma* antigens from a more controlled and safe source compared to infected mammalian cells.[3,4] Tick cell culture-derived *A. marginale* antigens have been assayed in vaccine formulations for the control of bovine anaplasmosis with results similar to those obtained with bovine erythrocyte-derived antigens.[3,4] The cloning, expression, and characterization of recombinant *Anaplasma* major surface proteins (MSPs) have permitted the evaluation of these proteins in immunization pen trials for the control of bovine anaplasmosis.[3,4] These experiments have demonstrated that antibody and cell immune responses are required for effective control of anaplasmosis and that multiple antigens may be required for an effective vaccine.[4,5] However, to date, the use of partially purified antigenic preparations and recombinant proteins has resulted in only partial control of infection, as evidenced by a reduction of the clinical signs of bovine anaplasmosis.[3,5] Furthermore, little has been done toward the identification of antigens that could potentially reduce the infection and transmission of the pathogen by ticks, an important goal of vaccines for the control of anaplasmosis.[3]

Functional genomics or systems biology constitute an alternative approach for discovery of pathogen and vector-derived protective antigens for the control of anaplasmosis. The development of high-throughput screening and global expression analysis technologies together with bioinformatic tools and the sequence of *A. marginale* and *A. phagocytophilum* genomes provide the basis for this approach. As discussed herein, the application of this approach for the characterization of key molecules involved in host–tick–pathogen interactions will likely lead to the development of effective vaccines for the control of anaplasmosis.

MATERIALS AND METHODS

The majority of the materials and methods of the series of experiments considered in this analysis were described previously.[6–16] Three lines of experiments were conducted to characterize host–tick–pathogen interactions for discovery of candidate vaccine antigens for the control of tick infestations and prevention of the infection and transmission of *Anaplasma* spp. by ticks: (1) the characterization of *A. marginale* adhesins involved in infection and transmission of the pathogen,[6–11] (2) the global expression analysis of genes differentially expressed in HL-60 human promyelocytic cells in response to infection with *A. phagocytophilum,*[12] and (3) the identification and characterization of tick-protective antigens by expression library immunization (ELI) and analysis of expressed sequence tags (ESTs) in a mouse model of tick infestations and by RNA interference (RNAi) in ticks.[13–16]

RESULTS AND DISCUSSION

A. marginale MSP1a is Involved in and Confers Partial Protection to Infection and Transmission of the Pathogen

Characterization of the interaction of *A. marginale* with tick cells and bovine erythrocytes resulted in the identification of MSP1a as a key component in this process. The *A. marginale* MSP1a is encoded by the single-copy gene, *msp1*α, which differs in molecular weight among geographic strains of the pathogen because of a variable number of tandem repeats.[4] MSP1a forms a heterodimer with MSP1b and was shown to be an adhesin for bovine erythrocytes and tick cells.[6,17] The adhesion domain of MSP1a was identified on the extracellular N-terminal region of the protein that contains the repeated peptides.[7] MSP1a was also shown to be involved in the infection and transmission of *A. marginale* strains by *Dermacentor* spp.[8] Expression of MSP1a was found to be downregulated in tick cells, as compared with organisms derived from bovine erythrocytes.[9] In addition, both native and recombinant MSP1a were found to be glycosylated, a feature of the repeated N-terminal peptides, which appears to contribute to the adhesive properties and thus the function of the protein.[10]

MSP1a induces strong T cell responses[18,19] and contains conserved B cell epitopes within the repeated peptides that are recognized by immunized and protected cattle.[20] Furthermore, antibodies produced in rabbits against the whole recombinant MSP1a or a synthetic repeated peptide significantly reduced *A. marginale* infection of cultured tick cells.[7,21]

The biological function of MSP1a in infection and transmission of *A. marginale*, together with its immunological properties, suggested that MSP1a may be a good candidate for inclusion in vaccines for the control of bovine anaplasmosis. Experiments conducted to evaluate the protective capacity of recombinant MSP1a alone or in combination with whole *A. marginale* antigens from infected cultured IDE8 tick cells demonstrated that a preferential antibody response to MSP1a correlates with lower percentage reductions in packed cell volume (PCV) and thus reduced clinical disease in cattle.[9]

Although transmission of *A. marginale* was not blocked in ticks fed on immunized cattle, antibodies to recombinant MSP1a reduced pathogen infection of tick cells *in vitro*[21] as well as infection levels in *D. variabilis*.[11]

As suggested by work done on other rickettsial MSPs,[22] the functional and structural characteristics of MSP1a N-terminal region could be used to develop algorithms for *in silico* search of available genome sequences and biological throughput screening of polypeptides with potential value for diagnosis and vaccine development in *Anaplasma* spp. and other rickettsiae.

A. phagocytophilum Modulates the Expression of Genes Involved in the Regulation of Mechanisms used by the Pathogen for Infection and Multiplication in the Host Cells

Another approach for the identification of putative protective antigens is the characterization of the response of host cells to *Anaplasma* infection. These studies are done by looking at the differential gene expression in infected mammalian and tick cells to compare gene expression profiles in both cell types and select candidate genes for further characterization and the development of therapeutic and protective interventions.

Initial experiments were conducted using microarrays of synthetic polynucleotides of 21,329 human genes to identify genes that are differentially expressed in HL-60 human promyelocytic cells in response to infection with *A. phagocytophilum.*[12] Six genes (leukocyte immunoglobulin-like receptor, spinocerebellar ataxia type 8 [SCA8], KIAA0833 protein calmodulin-binding transcription activator 1, molybdenum cofactor sulfurase [HMCS], and two genes of unknown function) were upregulated greater than 30-fold in the *A. phagocytophilum*-infected cells, while expression of downregulated genes most often did not change more than 6-fold.[12]

The genes differentially regulated in *A. phagocytophilum* infected cells were those essential for cellular mechanisms including growth and differentiation, cell transport, signaling and communication, and protective response against infection, some of which are most likely necessary for infection and multiplication of *A. phagocytophilum* in host cells.[12] Experiments are in progress in our laboratory to expand these studies to the characterization of differential gene expression in *Anaplasma*-infected tick cells. The identification of similar genes regulated by *A. phagocytophilum* infection in mammalian and tick cells may provide new targets for vaccine development.

Protective Antigen 4D8 is Involved in the Control of Tick Growth and Development and Affects the Infection of Ticks by A. marginale

A complementary approach to anaplasmosis vaccine development is to devise methods for the control of tick populations, thus providing the possibility of combining pathogen- and tick-derived antigens in vaccine formulations designed to target tick control and reduction of tick-borne pathogen transmission.[22] The tick-protective antigen 4D8 was discovered by ELI and analysis of ESTs in a mouse model of tick infestations for identification of cDNAs protective against *I. scapularis.*[13,14] The gene and protein sequences were found to be conserved in invertebrates and vertebrates and the gene was expressed in all tick developmental stages and in adult tissues of several tick species, suggesting a conserved function for 4D8.[13,15] Immunization with recombinant *I. scapularis* 4D8 controlled larval, nymphal, and adult tick infestations, supporting the development of tick vaccines based on 4D8.[13–15]

Tick vaccines control tick populations and could affect the transmission of tick-borne pathogens by decreasing the vector capacity of ticks and interfering with the development of pathogens.[23] The infection with two different isolates of *A. marginale* was evaluated in guts and salivary glands of *D. variabilis* male ticks after RNAi following injection of 4D8 dsRNA and feeding on rickettsemic calves. The results were similar for both isolates and evidenced a decrease in salivary gland infection levels of dsRNA-injected ticks when compared to control mock-injected ticks by quantitative *msp4* PCR.[24]

CONCLUSIONS

The ideal vaccine for anaplasmosis would be one that induces protective immunity and prevents infection and transmission of the pathogen by ticks. Current vaccines do not prevent infection and persistently infected cattle are a major reservoir of *A. marginale*, thus serving as a source of infection for mechanical transmission and biological transmission by ticks. Although transmission-blocking antigens have not been identified from the tick vector or the pathogen, the results discussed herein using functional genomic approaches suggest that the combination of pathogen adhesins with tick antigens involved in the control of tick developmental processes associated with pathogen survival and multiplication will likely contribute to the development of vaccines that induce protective immunity to control pathogen infection and tick infestations and reduce transmission of pathogens by ticks.

Future research directions will focus on the characterization of gene expression profiles in tick cells following pathogen infection *in vitro* and *in vivo*, the identification and characterization of new *Anaplasma*- and tick candidate protective antigens, and the combination of antigen preparations for the control of pathogen infection, tick infestations, and the transmission of *Anaplasma* spp. for an effective control of anaplasmosis.

ACKNOWLEDGMENTS

This research was supported by the Oklahoma Agricultural Experiment Station Project 1669, the Endowed Chair for Food Animal Research (K.M. Kocan) and the Instituto de Ciencias de la Salud, Spain (ICS-JCCM) (project 03052-00).

REFERENCES

1. KOCAN, K.M., J. DE LA FUENTE, E.F. BLOUIN, *et al.* 2004. *Anaplasma marginale* (Rickettsiales: Anaplasmataceae): recent advances in defining host-pathogen adaptations of a tick-borne rickettsia. Parasitology **129:** S285–S300.
2. DE LA FUENTE, J., R.B. MASSUNG, S.J. WONG, *et al.* 2005. Sequence analysis of the *msp4* gene of *Anaplasma phagocytophilum* strains. J. Clin. Microbiol. **43:** 1309–1317.
3. KOCAN, K.M., J. DE LA FUENTE, A.A. GUGLIELMONE & R.D. MELÉNDEZ. 2003. Antigens and alternatives for control of *Anaplasma marginale* infection in cattle. Clin. Microbiol. Rev. **16:** 698–712.
4. DE LA FUENTE, J., A. LEW, H. LUTZ, *et al.* 2005. Genetic diversity of *Anaplasma* species major surface proteins and implications for anaplasmosis serodiagnosis and vaccine development. Anim. Health Res. Rev. **6:** 75–89.
5. PALMER, G.H., F.R. RURANGIRWA, K.M. KOCAN & W.C. BROWN. 1999. Molecular basis for vaccine development against the ehrlichial pathogen *Anaplasma marginale*. Parasitol. Today **15:** 253–300.
6. DE LA FUENTE, J., J.C. GARCÍA-GARCÍA, E.F. BLOUIN & K.M. KOCAN. 2001. Differential adhesion of major surface proteins 1a and 1b of the ehrlichial cattle pathogen *Anaplasma marginale* to bovine erythrocytes and tick cells. Int. J. Parasitol. **31:** 145–153.
7. DE LA FUENTE, J., J.C. GARCIA-GARCIA, E.F. BLOUIN & K.M. KOCAN. 2003. Characterization of the functional domain of major surface protein 1a involved in adhesion of the rickettsia *Anaplasma marginale* to host cells. Vet. Microbiol. **91:** 265–283.

8. DE LA FUENTE, J., J.C. GARCIA-GARCIA, E.F. BLOUIN, *et al.* 2001. Major surface protein 1a effects tick infection and transmission of *Anaplasma marginale.* Int. J. Parasitol. **31:** 1705–1714.
9. GARCIA-GARCIA, J.C., J. DE LA FUENTE, E.F. BLOUIN, *et al.* 2004. Differential expression of the *msp1α* gene of *Anaplasma marginale* occurs in bovine erythrocytes and tick cells. Vet. Microbiol. **98:** 261–272.
10. GARCIA-GARCIA, J.C., J. DE LA FUENTE, G. BELL, *et al.* 2004. Glycosylation of *Anaplasma marginale* major surface protein 1a and its putative role in adhesion to tick cells. Infect. Immun. **72:** 3022–3030.
11. DE LA FUENTE, J., K.M. KOCAN, J.C. GARCIA-GARCIA, *et al.* 2003. Immunization against *Anaplasma marginale* major surface protein 1a reduces infectivity for ticks. J. Appl. Res. Vet. Med. **1:** 285–292.
12. DE LA FUENTE, J., P. AYOUBI, E.F. BLOUIN, *et al.* 2005. Gene expression profiling of human promyelocytic cells in response to infection with *Anaplasma phagocytophilum.* Cell. Microbiol. **7:** 549–559.
13. ALMAZÁN, C., K.M. KOCAN, D.K. BERGMAN, *et al.* 2003. Identification of protective antigens for the control of *Ixodes scapularis* infestations using cDNA expression library immunization. Vaccine **21:** 1492–1501.
14. ALMAZÁN, C., K.M. KOCAN, D.K. BERGMAN, *et al.* 2003. Characterization of genes transcribed in an *Ixodes scapularis* cell line that were identified by expression library immunization and analysis of expressed sequence tags. Gene Ther. Mol. Biol. **7:** 43–59.
15. ALMAZÁN, C., U. BLAS-MACHADO, K.M. KOCAN, *et al.* 2005. Characterization of three *Ixodes scapularis* cDNAs protective against tick infestations. Vaccine **23:** 4403–4416.
16. DE LA FUENTE, J., C. ALMAZÁN, E.F. BLOUIN, *et al.* 2005. RNA interference screening in ticks for identification of protective antigens. Parasitol. Res. **96:** 137–141.
17. MCGAREY, D.J., A.F. BARBET, G.H. PALMER, *et al.* 1994. Putative adhesins of *Anaplasma marginale:* major surface polypeptides 1a and 1b. Infect. Immun. **62:** 4594–4601.
18. BROWN, W.C., G.H. PALMER, H.A. LEWIN & T.C. MCGUIRE. 2001b. $CD4^+$ T lymphocytes from calves immunized with *Anaplasma marginale* major surface protein 1 (MSP1), a heteromeric complex of MSP1a and MSP1b, preferentially recognize the MSP1a carboxyl terminus that is conserved among strains. Infect. Immun. **69:** 6853–6862.
19. BROWN, W.C., T.C. MCGUIRE, W. MWANGI, *et al.* 2002. Major histocompatibility complex class II DR-restricted memory CD4(+) T lymphocytes recognize conserved immunodominant epitopes of *Anaplasma marginale* major surface protein 1a. Infect. Immun. **70:** 5521–5532.
20. GARCIA-GARCIA, J.C., J. DE LA FUENTE, K.M. KOCAN, *et al.* 2004. Mapping of B-cell epitopes in the N-terminal repeated peptides of *Anaplasma marginale* major surface protein 1a and characterization of the humoral immune response of cattle immunized with recombinant and whole organism antigens. Vet. Immun. Immunopathol. **98:** 137–151.
21. BLOUIN, E.F., J.T. SALIKI, J. DE LA FUENTE, *et al.* 2003. Antibodies to *Anaplasma marginale* major surface proteins 1a and 1b inhibit infectivity for cultured tick cells. Vet. Parasitol. **111:** 247–260.
22. DE LA FUENTE, J., J.C. GARCIA-GARCIA, A.F. BARBET, *et al.* 2004. Adhesion of outer membrane proteins containing tandem repeats of *Anaplasma* and *Ehrlichia*

species (Rickettsiales: Anaplasmataceae) to tick cells. Vet. Microbiol. **98:** 313–322.
23. DE LA FUENTE, J. & K.M. KOCAN. 2003. Advances in the identification and characterization of protective antigens for development of recombinant vaccines against tick infestations. Expert Rev. Vaccines **2:** 583–593.
24. DE LA FUENTE, J. & K. KOCAN. 2006. Strategies for development of vaccines for control of ixodid tick species. Parasite Immunol. **28:** 275–283.

Evaluation of *E. ruminantium* Genes in DBA/2 Mice as Potential DNA Vaccine Candidates for Control of Heartwater

BIGBOY H. SIMBI,[a] MICHAEL V. BOWIE,[b] TRAVIS C. McGUIRE,[c] ANTHONY F. BARBET,[b] AND SUMAN M. MAHAN[a,b]

[a]*University of Florida/USAID/SADC/ Heartwater Research Project, Central Veterinary Laboratory, P.O. Box CY 551, Causeway, Harare, Zimbabwe*

[b]*Department of Pathobiology, College of Veterinary Medicine, University of Florida, Gainesville, FL 32610, USA*

[c]*Department of Veterinary Microbiology and Pathology, College of Veterinary Medicine, Washington State University, Pullman, WA 99164, USA*

ABSTRACT: Heartwater caused by the rickettsia *Ehrlichia ruminantium* (*E. ruminantium*) is an acute and fatal tick-borne disease of domestic and some wild ruminants. A user-friendly vaccine does not exist. We selected and tested nine genes of *E. ruminantium* for protection against challenge in a DBA/2 mouse model, in order to identify candidate genes for incorporation into a recombinant vaccine. Of the nine DNA vaccine constructs tested, four DNA constructs 14HWORF1/VR1012, 14HWORF2/VR1012, 27HWORF1/VR1012, and HSP58/VR1012 were not protective and were excluded from the study. The remaining five DNA constructs—MAP2/ VR1012, 1HWORF3/ VR1012, 4HWORF1/ VR1012, 18HWORF1/ VR1012, and 3GDORF3/ VR1012—offered partial protection against lethal challenge demonstrated by reduced mortalities compared to control groups. Protection was augmented when DNA primed mice were boosted with a respective homologous recombinant protein. Protection in these five groups was associated with the induction of cell-mediated or T helper 1 (Th1) type of immune responses characterized by the production of large amounts of interferon-gamma and interleukin-2 in *in vitro* proliferation assays using *E. ruminantium* antigens for stimulation. These responses were enhanced when the DNA-vaccinated DBA/2 mice were boosted with specific homologous recombinant protein vaccination. In a preliminary follow-up study, protection conferred by DNA vaccination with individual gene constructs was not enhanced when the protective constructs were administered in combination (including the *map-1* gene of *E. ruminantium*). Further evaluation of these and other untested DNA constructs is necessary to optimize their

Address for correspondence: Suman M. Mahan, Department of Pathobiology, College of Veterinary Medicine, University of Florida, Gainesville, Fl 32610, USA. Voice: +1- 352-392-4700; fax: +1-352-392-9704.
e-mail: MahanS@mail.vetmed.ufl.edu

**Ann. N.Y. Acad. Sci. 1078: 424–437 (2006). © 2006 New York Academy of Sciences.
doi: 10.1196/annals.1374.131**

expression *in vivo* in the presence of molecular adjuvants, such as the IFN-gamma gene, GM-CSF gene, IL-12 gene, and CpG motifs to fully evaluate their protective value.

KEYWORDS: **DNA vaccination; *E. ruminantium*; heartwater; cell-mediated immunity**

INTRODUCTION

Heartwater is a disease of domestic and wild ruminants caused by *Ehrlichia ruminantium (E. ruminantium)*, a rickettsial agent transmitted by ticks of the *Amblyomma* species.[1] The disease is endemic in sub-Saharan Africa and on three Caribbean islands.[2] Control of the disease has been hampered by the absence of safe and effective vaccines. Cell-mediated immune responses against this intracellular organism play a key role in protection.[3–6] This has been demonstrated in cattle where recombinant *E. ruminantium* antigens (MAP-1 and MAP-2) and autologous-infected endothelial cells or monocytes activate subpopulations of $CD4^+$ and gamma delta T cells. These cells secrete interferon-gamma (IFN-γ) and other cytokines characteristic of cell-mediated immunity, which can mediate and cause intracellular killing of *E. ruminantium.*[7–9] A DBA/2 and a C57Bl/6 mouse model have also been used to study mechanisms of immunity to *E. ruminantium* infection.[3,5,6,10–12] A protective immune response in mice immunized by infection with *E. ruminantium* and treatment with antibiotics was mediated by T cells and was associated with a polarized T helper 1 (Th1)-type immune response. Infection studies in C57BL/6 knockout mice highlighted the importance of $CD4^+$ T cells.[5,6] The first DNA vaccine for heartwater using the *map-1* gene encoding the immunodominant MAP-1 protein of the Crystals Springs strain of *E. ruminantium* was evaluated in DBA/2 mice.[10] The *map-1* gene administered as a DNA vaccine induced cell-mediated protective immune responses and partially protected DBA/2 mice against homologous lethal challenge resulting in a survival rate of 13% to 27%. However, boosting DNA vaccinated (primed) mice with homologous recombinant MAP-1 protein augmented the immune responses and increased levels of protection to 67% survival.[11] Analysis of the immune responses indicated that although *map-1* DNA immunization induced both humoral and cellular responses, protection was associated with induction of a Th1-type response with production of high levels of IFN-γ and IL-2,[11] which were augmented after the recombinant protein boost. In the present study, the DBA/2 mouse model of infection was used to screen for additional protective *E. ruminantium* genes by DNA vaccination followed by boosting with a recombinant protein. The protective immune response induced by each of the selected genes was also analyzed.

MATERIALS AND METHODS

Source of Mice Used

DBA/2 mice were obtained from Taconic Quality Laboratory Animal Services for Research (Germantown, NY, USA). Some of the mice used were bred in our laboratory at the Central Veterinary Laboratory, Harare, Zimbabwe. All trials were conducted using 8- to 16-week-old female mice. As well as immunized and challenged animals, two mice were normally included in each group for determination of cytokine and lymphocyte proliferation responses.

Origin of DNA Constructs

Thirty-four positive *Escherichia coli* (*E. coli)* recombinant clones that reacted strongly with immune sheep sera were obtained from a genomic DNA library of *E. ruminantium.*[13] Of these, clones 1HW, 4HW, 14HW, 18HW, 27HW, and 3GD were selected for evaluation as DNA vaccine candidates on the basis of their homology with other known genes designated as outer membrane proteins in GenBank or based on homology with genes that encode protective antigens in other infectious agents.[13] The respective genes were amplified by polymerase chain reaction (PCR) and cloned into a DNA vaccine vector VR1012 (Vical Inc., San Diego, CA, USA) as described previously[10,11] resulting in the following plasmid constructs with the given open reading frames (ORFs): 1HWORF3/VR1012, 4HWORF1/VR1012, 14HWORF1/VR1012, 14HWORF2/VR1012, 18HWORF1/VR1012, 27HWORF1/VR1012, and 3GDORF3/VR1012. The plasmid MAP2/VR1012 was constructed based on the sequences of the previously cloned *map-2* gene[14] while the plasmid HSP58/VR1012 was created by PCR amplification of the 58-kDa heat-shock protein[15] followed by cloning into the VR1012 plasmid vector.

Vaccination of DBA/2 Mice with DNA Vaccines and Recombinant Proteins

The vaccination procedure followed was as described previously[10] unless specified otherwise. In brief, the mice were primed by injection with 50 μL of 0.5% bupivacaine (Sigma, St. Louis, MO, USA) into each of the hind limb muscles. Three days later, the muscles were injected with 50 μg of individual DNA plasmid vaccine constructs in 50 μL of sterile PBS to give a total inoculating dose of 100 μg of plasmid DNA. Control mice were injected with plasmid VR-1012 vector DNA only. After 2 weeks, the same immunization procedure was repeated and after a further 2 weeks, the mice were bled for post-DNA vaccination sera. The mice were then either challenged with live *E. ruminantium* organisms or prepared for further vaccination with recombinant proteins or β-propiolactone inactivated, cell culture-derived *E. ruminantium* organisms as described below. Recombinant proteins were made as described previously.[13] After the second DNA vaccine inoculation, the respective mice were injected subcutaneously with 100 μg of a homologous recombinant protein (MAP1, MAP2, 1HWORF3, 4HWORF1, 18HWORF1, and 3GDORF3) emulsified in 10 μg of Quil A adjuvant. Recombinant bacterial alkaline phosphatase (rBAP) was used as a control protein. The mice were bled for sera 2 weeks after a second protein boost for use in antibody determinations by recombinant antigen-based ELISA as described previously.[16]

Vaccination of DBA/2 Mice with Combined Six DNA Genes and Six Recombinant Proteins

Six groups of 20 DBA/2 mice were used in each experimental group. In three of the groups, six DNA constructs (MAP1/VR1012, MAP2/VR1012, 1HWORF3/VR1012, 18HWORF1/VR1012, 4HWORF1/VR1012, and 3GDORF3/VR1012) were pooled to give a total of 600 μg DNA and inoculated into the mice 3 days post-bupivacaine injection as described after. After two inoculations of the combo DNA vaccines at an interval of 2 weeks, 100 μg each of a combo of six recombinant proteins (MAP1, TMMAP2, 1HWORF3, 18HWORF1, 4HWORF3, and 3GDORF3), totaling 600 μg was given as a booster inoculation to one group. In the second group, 600 μg of rBAP was given as control irrelevant protein and in the third group β-propiolactone-inactivated cell culture-derived *E. ruminantium* organisms[17] were given as a booster. For the remaining three groups, one group received the combo 6 recombinant protein + Quil A, while another group only received β-propiolactone- inactivated cell culture-derived *E. ruminantium* organisms in Quil A. The last group was the control group that received VR1012 DNA vaccination followed by rBAP boost in Quil A.

Challenging Mice with Live E. ruminantium *Organisms*

Two weeks after the last vaccination, the mice were challenged intravenously through the tail vein with a predetermined lethal dose of the *E. ruminantium* Highway strain (from Zimbabwe) frozen stabilate (0.1 mL) as described previously.[5] This was the strain from which the genes had been cloned, with the exception of 3GDORF3, which was cloned from a genomic DNA library of the Gardel strain (from Guadeloupe). The two mice in each group set aside for the evaluation of cytokines and lymphocyte proliferation responses were not challenged. The mice were observed for signs of sickness, such as ruffled coats and sluggish movement. The number of mice that died and the days to death were recorded until all the sick mice recovered completely. Protection by DNA vaccination in mice was evaluated by the survival rate and prolonged life after challenge with a lethal dose of *E. ruminantium.*

Recombinant Antigen ELISAs

Antibody responses to *E. ruminantium*-immunizing antigens resulting from DNA vaccination or in combination with recombinant protein vaccination were detected by a specific indirect ELISA as described previously.[16,18] The plates were coated with 100 μL of 3 μg/mL (0.3 μg/well) recombinant protein antigen, washed and incubated with 100 μL/well of 1:100 dilution of mouse sera in 1% powdered milk (w/v) in 0.1% Tween 20 PBS (pH 7.2). A 1:2,000 dilution of horse HRP-conjugated rabbit anti-mouse IgG second-step antibody (Kirkegaard and Perry Laboratories, Gaithersburg, MD, USA) was added and color development occurred after addition of a peroxidase substrate. The optical density was measured using a dual wavelength of 405 nm and 492 nm and the cut-off value was calculated as 2 × the Optical Density (OD) of the negative control serum expressed as a percentage of the OD of the positive control serum.[17,18]

Lymphocyte Proliferation and Cytokine Assays

The two mice set aside for lymphocyte and cytokine proliferation assays were sacrificed by cervical dislocation. Spleen cells were isolated from each mouse and a total of 2.5×10^5 cells were dispensed into each well in duplicate in 96-well microtiter plates (Corning Inc., Acton, MA, USA) and incubated at 37°C in a humidified atmosphere of 5% CO_2/95% air with 10 μg of specific recombinant protein- or cell culture-derived whole *E. ruminantium* antigens. As control, cells were incubated with no added antigen or with rBAP. On the third day, 0.5 μCi of tritiated thymidine (Amersham, UK) was added and incubated further for 24 h. The cells were harvested and thymidine incorporation was measured as an indicator of lymphocyte proliferation using a liquid scintillation counter (Beckman Instruments, Fullerton, CA, USA). Supernatants were collected on the third day and tested in duplicate for the cytokines IL-2, IL-4, IL-5, IL-6, IL-10, and IFN-γ using ELISAs as described by the manufacturer (Endogen Inc., Rockford, IL, USA). Cytokines present in each sample were then measured using curves generated from known standards.

Statistical Analysis

Mean days to death between vaccinated groups and control groups were compared using the Student's *t*-test while Fisher's exact test was used to determine the level of significance in protection between the various DNA and recombinant protein vaccination groups and control groups.

RESULTS

Clinical Response of Mice to E. ruminantium *Challenge*

The testing of several *E. ruminantium* genes as DNA vaccines was accomplished in two experiments. In experiment 1, six gene constructs were tested, namely, MAP2/VR1012, 18HWORF1/VR1012, 3GDORF3/VR1012, 1HWORF3/VR1012, 14HWORF1/VR1012, and 14HWORF2/VR1012. In experiment 2, an additional three DNA constructs were tested: 4HWORF1/VR1012, 27HWORF1/VR1012, and HSP58/VR1012. In each experiment a VR1012 plasmid control group was included. In all the experiments, the DBA/2 mice showed signs of sickness 10 days after challenge with live *E. ruminantium* organisms. The mobility of the mice was reduced with increasing sickness. In the DNA vaccination- only group, the mice died between 13 days and 18 days after challenge. However, when recombinant antigens were used to boost DNA vaccination singly or as a combo, the mice lived longer with deaths recorded between 13 days and 24 days after challenge. After day 24, no further deaths were recorded and all the sick mice had completely recovered. The protective DNA constructs as determined by increased survival rates and prolonged survival time periods were further tested by combining DNA vaccination with recombinant antigen boosting.

Survival Responses of DNA Vaccinated Mice to Challenge

Survival was prolonged in the groups vaccinated individually with the four DNA vaccine constructs—MAP2/VR1012, 1HWORF3/VR1012, 18HWORF1/VR1012, and 3GDORF3/VR1012— compared to VR1012 plasmid control ($P < 0.05$) (experiment 1, TABLE 1). In experiment 2, the 4HWORF1/VR1012 DNA vaccine group survived for a longer time period, but this, was not statistically significantly different from the control. The survival periods and survival rates of mice vaccinated with the DNA constructs 14HWORF1/VR1012, 14HWORF2/VR1012, 27HWORF1/VR1012, and HSP58/VR1012 were also not significantly different from the controls (TABLE 1). The five gene constructs that showed a significant number of survivors and/or longer survival periods were retested using 30 DBA/2 mice per group. There were more survivors in the vaccinated groups with the highest observed in the group 18HWORF1/VR1012 (6/29 survivors, $P = 0.05$) with a longer survival period (16.3 ± 3.2 days, $P = 0.03$) than mice from the VR1012 control group (TABLE 1). Mice from the group 4HWORF1/VR1012 had a longer

TABLE 1. Screening of nine *E. ruminantium* genes for protection by DNA vaccination

	Average days to death		Survivors			
DNA Construct	Vaccine	Control	Vaccine		Control	
Experiment 1						
MAP2/VR1012	15.8 ± 0.8[a]	14.0 ± 1.3	2/11	18%	1/11	9%
18HWORF1/VR1012	15.5 ± 1.2[a]	14.0 ± 1.3	2/13	15%	1/11	9%
3GDORF3/VR1012	16.1 ± 1.5[a]	14.0 ± 1.3	3/13	23%	1/11	9%
1HWORF3/VR1012	15.3 ± 1.3[a]	14.0 ± 1.3	2/13	15%	1/11	9%
14HWORF1/VR1012	14.6 ± 1.4	14.0 ± 1.3	1/13	8%	1/11	9%
14HWORF2/VR1012	13.9 ± 1.6	14.0 ± 1.3	0/13	0%	1/11	9%
Experiment 2						
4HWORF1/VR1012	16.4 ± 1.6	15.8 ± 1.2	3/14	21%	0/13	0%
27HWORF1/VR1012	14.3 ± 0.9	15.8 ± 1.2	0/14	0%	0/13	0%
HSP58/VR1012	14.7 ± 0.9	15.8 ± 1.2	1/13	8%	0/13	0%

[a]Statistical significant difference ($P < 0.05$, Student's t-test) in survival time, that is, days to death when compared to the VR1012 immunized control mice.

survival period (17.7 ± 3.8, $P = 0.001$), but did not have a significantly higher survival rate (1/30, $P > 0.9$) than the control group.

Boosting DNA Vaccinated Mice with Recombinant Protein Augmented Protection against E. ruminantium *Challenge*

The 18HWORF1/VR1012, 4HWORF1/VR1012, MAP2/VR1012, 1HWORF3/VR1012, and 3GDORF3/VR1012 that exhibited partial protection in the initial screening experiments (experiment 1 or 2, TABLE 1) were selected for further evaluation for protective efficacy in a prime boost regimen. Deaths in vaccinated mice were recorded starting at 12 days post challenge and stabilized after day 22. When DNA vaccination with a single construct was followed by a homologous recombinant protein boost, three constructs 18HWORF1/VR1012 (10/30, $P = 0.02$); MAP2/VR1012 (9/28, $P = 0.04$); and 4HWORF1/VR1012 (10/30, $P = 0.05$) gave significantly higher survival rates than the control nonrecombinant VR1012 plasmid (TABLE 2). TABLE 2 shows either 0, 1, or 2 mice surviving in the control groups. Although not statistically significant, higher numbers of mice survived challenge in the remaining two DNA vaccine groups compared to the controls. These results show that protection as measured by survival rate was augmented by use of the prime boost method of immunization.

Combo DNA and Recombinant Protein Vaccination against E. ruminantium *Challenge*

The five recombinant plasmid DNA constructs that showed partial protection singly by DNA vaccination only (TABLE 1) and when augmented by

TABLE 2. Survival rate of DBA/2 mice after DNA vaccination augmented by boosting with respective homologous recombinant proteins

DNA construct and recombinant protein	Survivors				*P* value
	Vaccine		Control		
MAP2/VR1012	9/28[a]	32%	0/29	0%	0.004
18HWORF1/VR1012	10/30[a]	33%	1/27	4%	0.02
3GDORF3/VR1012	7/26	27%	1/27	4%	0.06
1HWORF3/VR1012	8/29	28%	2/30	7%	0.09
4HWORF1/VR1012	10/30[a]	33%	2/30	7%	0.05

[a]Statistical significance by Fisher's exact two-tailed test ($P < 0.05$).

vaccination with homologous recombinant proteins (TABLE 2) were combined with a MAP1/VR1012 gene construct[11] and tested as a 6-DNA combo vaccine. DBA/2 mice primed with the 6-DNA combo vaccination and boosted with a nonspecific protein rBAP were not protected against *E. ruminantium* challenge (0/18 survivors), a result similar to that of the group vaccinated with nonrecombinant plasmid control VR1012 (0/20 survivors), and naïve unvaccinated control mice (0/10 survivors) (TABLE 3). However, when DBA/2 mice primed with 6-combo DNA inoculation were boosted with inactivated *E. ruminantium* organisms, the survival rate against lethal challenge was 10/20 (50%) compared to 5/18 (28%) in mice that received a booster vaccination of 6 respective combo recombinant proteins (TABLE 3). Vaccination with either inactivated *E. ruminantium* organisms or 6- combo recombinant proteins without DNA vaccine priming yielded significantly higher survival rates of 11/20 (55%) and 6/19 (32%), respectively, against challenge than did the VR1012 plasmid control 0/20 (0%) and naïve controls 0/10 (0%). There was no significant difference in the survival rates of the mouse groups vaccinated with inactivated *E. ruminantium* organisms alone (55%) or as a booster to DNA combo priming (50%). However, boosting with inactivated organisms protected more mice than boosting with the 6- combo recombinant proteins (TABLE 3). The DBA/2 mice primed with the 6- combo DNA and boosted with 6 recombinant proteins had a significantly higher survival rate (28%) than that obtained when boosting with an irrelevant protein (0%) (TABLE 3). In general, in this trial there was no evidence for augmentation of protection by any of the prime-boost vaccination regimens.

Antibody Responses in Vaccinated Mice

Antibodies were detected in the mice after DNA vaccination with the highest numbers of seropositive mice found in the 3GDORF3/VR1012, 21/30 (70%) group, followed by MAP2/VR1012, 14/30 (47%), 18HWORF1/VR1012,

TABLE 3. Survival rates of DNA combo, recombinant protein, and inactivated organism vaccination trials

Vaccination regimen		Number of mice surviving/ total	% Survival	Significance Fishers' exact (*P* value)
DNA vaccine	Boosting antigen			
None	Inactivated whole *E. ruminantium* organisms	11/20	55	Yes $P = 0001$
6- combo DNAs	Inactivated whole *E. ruminantium* organisms	10/20	50	Yes $P = 0.0003$
None	6 combo recombinant proteins	6/19	32	Yes $P = 0.018$
6- combo DNAs	6 combo recombinant proteins	5/18	28	Yes $P = 0.03$
6 combo DNAs	rBAP	0/18	0	No
Control VR1012	rBAP	0/20	0	
None	None	0/10	0	

10/30 (33%), 4HWORF1/VR1012, and 1HWORF3/VR1012 each with 8/30 (27%). All the mice in each of the five vaccine groups seroconverted after boosting with respective homologous recombinant protein (data not shown) Antibody isotyping was not done.

Lymphocyte Proliferation Responses and Cytokines Produced

When stimulated with the specific recombinant antigens, splenocytes from mice vaccinated with the partially protective recombinant DNA plasmids (MAP2/VR1012, 18HWORF1/VR1012, 3GDORF3/VR1012, 1HWORF3/VR1012, and 4HWORF1/VR1012) proliferated two- to five-fold more than those obtained from the nonrecombinant control plasmid. Splenocytes cultured from DBA/2 mice vaccinated with the non-protective recombinant plasmids 27HWORF1/VR1012, HSP58/VR1012, 14HWORF1/VR1012, and 14HWORF2/VR1012 showed little or no significant proliferation in response to recombinant antigens (data not shown). After DNA vaccination, IFN-γ was produced in significantly higher amounts in splenocyte cultures of the mice vaccinated with 18HWORF1/VR1012 and 1HWORF3/VR1012, while very little or insignificant amounts were produced in the cultures of mice vaccinated with MAP2/VR1012, 3GDORF3/VR1012, and 4HWORF1/VR1012 (FIG. 1A). After the recombinant protein boost, significantly higher amounts ($P < 0.05$) of IFN-γ were produced in splenocyte cultures of all vaccinated groups in response to stimulation by the respective recombinant proteins compared to levels found in cultures of control mice or no antigen controls (FIG. 1B). Splenocytes of DBA/2 mice that were prime-boosted using the DNA vaccine and respective MAP2, 3GD, or 4HW protein boost produced at least two times higher amounts of IFN-γ than without protein boosting (FIG. 1A and B). There was no increase in the amount of IFN-γ produced by splenocytes of mice boosted with the recombinant 18 HW and 1HW protein, respectively. IL-2 was produced in culture by splenocytes of all five vaccine groups after DNA vaccination and after the recombinant protein boost. IL-6 was also produced in large amounts in cultures of mice in all groups. IL-4 was only detected in the splenocyte cultures of 3GDORF3/VR1012 vaccinated mice, whereas little or no IL-5 was produced. IL-10 was detected in splenocyte cultures of 1HWORF3/VR1012 and 4HWORF1/VR1012 vaccinated mice after DNA vaccination but not after the recombinant protein boost.

DISCUSSION

The data presented here show that the following five *E. ruminantium* genes from clones MAP2, 18HW, 3GD, 1HW, and 4HW administered by DNA vaccination and boosting with a homologous recombinant protein are capable of inducing cellular immune responses associated with partial protection of mice against lethal homologous challenge. Previously, using the *map-1* DNA vaccine, a similar observation was made.[10,11] However, the level of protection against homologous challenge was much higher (67%) compared with heterologous challenge.[11] This observation was not surprising given that the *map-1* gene exhibits significant sequence heterogeneity between *E. ruminantium* strains.[19] Availability of the complete genome sequences (GenBank CR767821, CR925677, CR9225678) of *E. ruminantium* now shows that amino acid sequence homology of the five partially protective vaccine antigens described here was 98% or higher between the Highway, Gardel, and the Welgevonden strains (except for the 3GD, which had 100% homology between Gardel and Welgevonden only) and gives credence for the selection of these antigens to achieve improved cross-protection between different isolates.

The protection with these five DNA vaccines administered as a prime-boost regimen was demonstrated by an increased number of survivors and prolonged average days to death. The immune responses induced by these genes were characterized by the secretion of IFN-γ and IL-2 *in vitro* in lymphocyte proliferation assays on stimulation with specific proteins and *E. ruminantium* organisms and were similar to those induced by DNA vaccination of DBA/2 mice with the *map-1* gene.[10,11] IFN-γ and IL-2 are also produced in *in vitro* proliferation assays by mice infected with *E. ruminantium*, or in mice immunized by infection and treatment.[5] This Th1-type response probably activates macrophages causing them to destroy the intracellular *E. ruminantium* organisms through a cascade of cytokine-mediated effects.[20] The increased secretion of IFN-γ and IL-2 from splenocytes of DNA-vaccinated mice boosted with recombinant antigens correlated with increased survival rates and prolongation of life in these mice. This polarized Th1-type immune response was

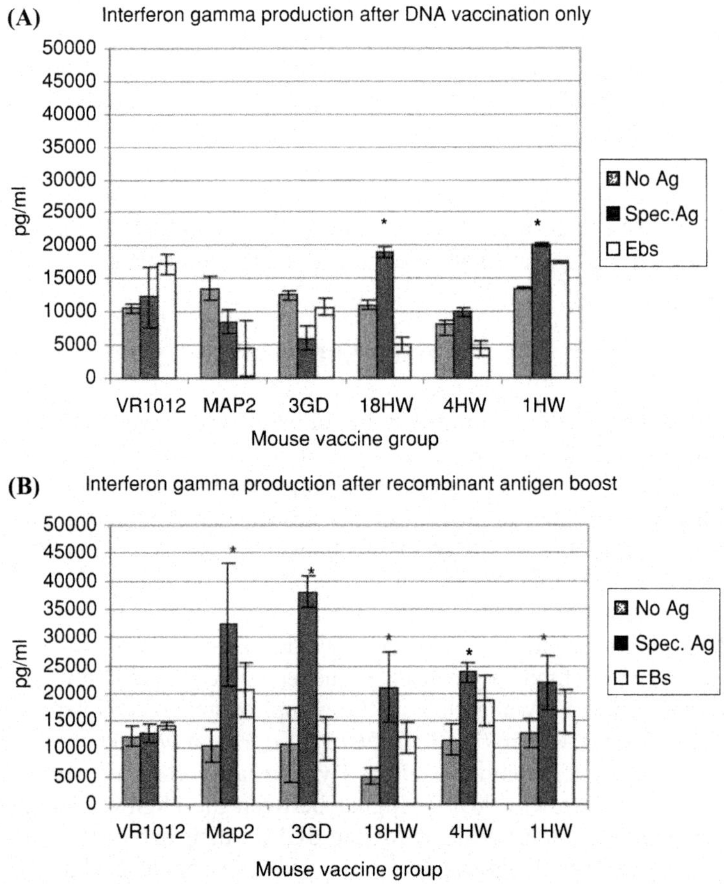

FIGURE 1. (**A**) Production of interferon-gamma in DNA immunized DBA/2 mice after DNA vaccination only. (**B**) Production of IFN-gamma in DBA/2 mice after DNA vaccination was followed by recombinant antigen boost. Mouse spleen cells from DBA/2 mice immunized with various DNA constructs were stimulated *in vitro* with specific homologous antigens, nonspecific antigens or whole EB antigens. IFN-gamma secreted into the supernatants was measured by ELISA (Endogen Inc., USA). The graphs show the mean and one standard deviation for each vaccine group. No Ag, No antigen was added; Spec. Ag, 10 μg of specific recombinant antigen; Ebs, inactivated 10 μg of β propiolactone-inactivated *E. ruminantium* organisms were incubated with splenocytes. *Statistically significant difference from VR1012 control ($P < 0.05$)

further confirmed by the general absence of IL-4, IL-5, and IL-10, which are Th2-type cytokines predominantly associated with antibody responses.[20] However, IL-10 was produced in cultures of splenocytes from mice vaccinated with the 1HW and 4HW DNA vaccines, while IL-4 was produced from vaccination with the 3 GD gene. This would indicate that in addition to the Th1 immune response to these antigens, there was a Th2 component induced. The cytokine IL-10 is anti-inflammatory with an anti-Th1 response effect and is known to exert complex regulatory effects on $CD8^+$ T cells, natural killer cells, and B lymphocytes as well as to suppress cytokine production from macrophages. IL-10 may have been produced to contain the spread of infection.[20] The cultured splenocytes from mice vaccinated with the nonprotective recombinants 14 HW, 27HW, and HSP58 encoding a lipoprotein, an outer membrane protein, and a heat-shock protein, respectively[13] did not proliferate in response to *E. ruminantium* antigens. Very little or no cytokines except IL-6 were produced and this correlated with no protection in the mice. When the *GroEL* and *GroES* genes encoding the 58-kDa heat-shock protein were cloned and the genetic vaccine inoculated into Balb/C and C57BL/6J mice, low levels of protection were induced and none of the inoculated mice survived,[21] indicating that the 58-kDa heat-shock protein is not a good DNA vaccine candidate. A combination DNA vaccine was tested in an attempt to augment protection. However, in this study vaccination with a combination of 6 DNA vaccines did not yield significantly improved protection as compared to the single-gene vaccines. Also, there was little priming effect from the DNA combination. Possible reasons could be suboptimal amounts of each DNA component in the priming vaccination, antigenic interference among the encoded proteins, or poor gene expression. Antigenic interference has been reported when plasmids encoding four different proteins of the human immunodeficiency virus (HIV) were mixed and delivered together, resulting in poor recognition and decreased immune responses.[22] Another reason could be competition for transcription factors leading to decreased or nonexpression, as was observed when combined DNA immunization of three genes that are coexpressed in the HIV resulted in the inhibition of protein expression efficiency and diminished immune response.[23] However, when inactivated whole *E. ruminantium* organisms were used in vaccination alone or with 6 DNA-primed mice, protection was (55% and 50% survival rate, respectively) was superior than when 6 combined recombinant proteins were administered alone or after DNA priming (32% and 28%, respectively). Since these data are preliminary, further evaluation of a combo vaccine approach is required. This evaluation should include monitoring the levels of individual gene expression after DNA vaccination, gene dosage effects, and alternative prime-boosting protocols. The use of molecular adjuvants to enhance the induction of cell-mediated immune responses would be another area requiring attention.

ACKNOWLEDGMENTS

This study was funded by the United States Agency for International Development grant number LAG-1328-G-00-3030-00. We acknowledge the expert technical assistance of Annie Moreland, Jill Palmer, and Camille McCranie.

REFERENCES

1. UILENBERG, G. 1983. Heartwater (*Cowdria ruminantium* infection): current status. *In* Advances in Veterinary Science and Comparative Medicine. C.E. Cornelius, & C.F. Simpson. Eds.: 427–480. Academic Press. New York.
2. UILENBERG, G. *et al.* 1984. Heartwater in the Caribbean. Prev. Vet. Med. **2:** 255–267.
3. DU PLESSIS, J.L. *et al.* 1991. T cell-mediated immunity to *Cowdria ruminantium* in mice: the protective role of Lyt-2^+ T cells. Onderstepoort J. Vet. Res. **58:** 171–179.
4. MWANGI, D.M. *et al.* 1998. Immunization of cattle by infection with *Cowdria ruminantium* elicits T lymphocytes that recognize autologous, infected endothelial cells and monocytes. Infect. Immun. **66:** 1855–1860.
5. BYROM, B. *et al.* 2000. A polarized Thl type immune response to *Cowdria ruminantium* infection is detected in immune DBA/2 mice. J. Parasitol. **86:** 983–992.
6. BYROM, B. *et al.* 2000. CD8 (+) T cell knockout mice are less susceptible to *Cowdria ruminantium* infection than athymic, CD4 (+) T cell knockout, and normal C57BL/6 mice. Vet. Parasitol. **93:** 159–172.
7. MAHAN, S.M. *et al.* 1996. Neutralization of bovine concanavalin-A T cell supernatant-mediated *Cowdria ruminantium* activity with antibodies specific to interferon gamma but not tumor necrosis factor. Parasite Immunol. **18:** 317–324.
8. TOTTE, P. *et al.* 1997. Analysis of T-cell responses in cattle immunized against heartwater by vaccination with killed elementary bodies of *Cowdria ruminantium*. Infect. Immun. **65:** 236–241.
9. MWANGI, D.M. *et al.* 2002. Immunisation of cattle against heartwater by infection with *Cowdria ruminantium* elicits T lymphocytes that recognise major antigenic proteins 1 and 2 of the agent. Vet. Immunol. Immunopathol. **85:** 23–32.
10. NYIKA, A. *et al.* 1998. A DNA vaccine protects mice against the rickettsial agent *Cowdria ruminantium*. Parasite Immunol. **20:** 111–119.
11. NYIKA, A. *et al.* 2002. DNA vaccination with *map1* gene followed by protein boost augments protection against challenge with *Cowdria ruminantium*, the agent of heartwater. Vaccine **20:** 1215–1225.
12. BYROM, B. *et al.* 1993. The development of antibody to *Cowdria ruminantium* in mice and its role in heartwater disease. Rev. Elev. Med. Vet. Pays Trop. **46:** 197–201.
13. BARBET, A.F. *et al.* 2001. A subset of *Cowdria ruminantium* genes important for immune recognition and protection. Gene **275:** 287–298.
14. MAHAN, S.M. *et al.* 1994. Molecular cloning of a gene encoding the immunogenic 21 kDa protein of *Cowdria ruminantium*. Microbiology **140:** 2135–2142.
15. LALLY, N.C. *et al.* 1995. The *Cowdria ruminantium* groE operon. Microbiol. **141:** 2091–2100.
16. SEMU, S. *et al.* 2001. Antibody responses to MAP 1B and other *Cowdria ruminantium* antigens are down regulated in cattle challenged with tick-transmitted heartwater. Clin. Diagn. Lab. Immunol. **8:** 388–396.
17. MAHAN, S.M. *et al.* 1995. Immunisation of sheep against heartwater with inactivated *Cowdria ruminantium*. Res. Vet. Sci. **58:** 146–149.
18. VAN VLIET, A.H.M. *et al.* 1995. Use of a specific immunogenic region on the *Cowdria ruminantium* MAP1 protein in a serological assay. J. Clin. Microbiol. **33:** 2405–2410.

19. REDDY, G.R. *et al.* 1996. Sequence heterogeneity of the major antigenic protein 1 genes from *Cowdria ruminantium* isolates from different geographical areas. Clin. Diagn. Lab. Immunol. **4:** 417–422.
20. JANEWAY, C.A. *et al.* 2001. Immunobiology. The Immune System in Health & Disease, 5th ed. Churchill Livingstone. New York.
21. PRETORIUS, A. *et al.* 2002. Genetic immunization with *Ehrlichia ruminantium* GroEL and GroES homologues. Ann. N.Y. Acad. Sci. **969:** 151–154.
22. HOOPER, J.W. *et al.* 2003. Four-gene-combination DNA vaccine protects mice against a lethal vaccinia virus challenge and elicits appropriate antibody responses in non human primates. Virology **306:** 181–195.
23. KJERRSTROM, A. *et al.* 2001. Interactions of single and combined human immunodeficiency virus type 1(HIV-1) DNA vaccines. Virology **284:** 46–61.

New Findings on Members of the Family *Anaplasmataceae* of Veterinary Importance

YASUKO RIKIHISA

Department of Veterinary Biosciences, College of Veterinary Medicine, The Ohio State University, Ohio 43210, USA

ABSTRACT: Members of the family *Anaplasmataceae* are obligate intracellular Gram-negative bacteria that naturally infect a variety of wild and domestic animal species, the spillover of which may lead to zoonosis. I discuss new findings on members of the family *Anaplasmataceae* of veterinary importance and therefore, I will describe the recent findings on *Neorickettsia risticii* in the trematode and related *Neorickettsia* species. I also will review the recent progress on *Aegyptianella pullorum* and other *Aegyptianella* sp., "*Candidatus* Neoehrlichia mikurensis" and *Anaplasma phagocytophilum* strains in various hosts. The whole genome sequences of two important veterinary pathogens—*Anaplasma marginale,* the bovine anaplasmosis agent, and *Ehrlichia* (formerly *Cowdria*) *ruminantium*, the agent of heartwater of ruminants—have been published. Taken together, these advances in research of the family *Anaplasmataceae* in the veterinary field provide us with insights into the evolution, reservoir, and transmission of these organisms in nature and their pathogenesis in natural and accidental hosts. It is through this work that surveillance, diagnosis, preventive measures, and treatment of ehrlichioses of both animals and humans can be improved.

KEYWORDS: *N. risticii; Ae. pullorum; N. mikurensis; A. phagocytophilum*

INTRODUCTION

Members of the family *Anaplasmataceae* are small Gram-negative pleomorphic cocci that are obligate intracellular bacteria. They replicate in membrane-bound vacuoles (parasitophorous vacuoles) in the cytoplasm of a specific type of host cell of hematopoietic origin and/or in invertebrates.[1,2] These bacteria are not contagious, but are vector-borne, that is, transmitted by ticks or trematodes. Infection of blood cells of domestic and wild animals, and

Address for correspondence: Yasuko Rikihisa, Department of Veterinary Biosciences, College of Veterinary Medicine, The Ohio State University, 1925 Coffey Road, Columbus, OH 43210. Voice: 614-292-9677; fax: 614-292-6473.
e-mail: rikihisa.1@osu.edu

**Ann. N.Y. Acad. Sci. 1078: 438–445 (2006). © 2006 New York Academy of Sciences.
doi: 10.1196/annals.1374.083**

humans with ehrlichial organisms may lead to a clinically apparent illness collectively called ehrlichiosis or anaplasmosis (or rickettsiosis), a febrile systemic illness often accompanied with hematological abnormalities, lymphadenopathy, and elevation of liver enzyme activity. Under the light microscope, ehrlichial organisms are small cocci in the cytoplasm of the host cells that stain dark blue to purple with Romanowsky stain. Microcolonies of ehrlichiae in the host cytoplasm may look like mulberries and thus are called morulae. Small (0.2–0.4 μm), dense forms resembling chlamydial elementary bodies and relatively large, light forms (0.8–1.5 μm) resembling reticular bodies have been noted in cells infected with several species of *Ehrlichia, Neorickettsia*, and *Anaplasma.*[1] On transmission electron microscopy, organisms generally are round but sometimes are highly pleomorphic and found in membrane-lined vacuoles called inclusions. A lack of genes for biosynthesis of lipid A and most genes for biosynthesis of peptidoglycan explains the pleomorphic nature of these bacteria.[3] LPS and peptidoglycan, which are shared by many groups of bacteria,[30] elicit profound innate immune responses in the hosts. Therefore, it is interesting to note that the deletion of genes for biosynthesis of LPS and peptidoglycan in these bacteria likely increases their chance of survival in both vertebrate and invertebrate hosts.

Unlike most other bacteria, *Ehrlichia chaffeensis* and *Anaplasma phagocytophilum* require cholesterol for survival and have evolved to take up cholesterol from the environment.[3] The host cholesterol also is essential for *E. chaffeensis* and *A. phagocytophilum* entry and intracellular infection.[4] Of members of the family *Anaplasmataceae,* the genomes of *Wolbachia pipientis* wMel (1.2 Mb), a *Drosophila* parasitic endosymbiont,[5] the filarial nematode endosymbiont *Wolbachia* sp. *w*Bm,[6] *A. marginale* (1.3 Mb), erythrocyte tropic bovine anaplasmosis agent,[7] *Ehrlichia* (formerly *Cowdria*) *ruminantium* (1.5 Mb), and the agent of heartwater of ruminants,[8] have been published. A draft genome of *E. canis*, the causative agent of canine ehrlichiosis, has been deposited in GenBank (Accession No.: NZ˙AAEJ01000001). Complete genomes of *A. phagocytophilum* (1.5 Mb), *E. chaffeensis* (1.2 Mb), and *Neorickettsia sennetsu* (0.86 Mb) have been published.[30] I will focus my review on the recent progress made in research on members of the family *Anaplasmataceae* of veterinary importance.

NATURAL RESERVOIR AND VECTOR OF *NEORICKETTSIA RISTICII*

Neorickettsia risticii is the causative agent of Potomac horse fever (PHF) or equine monocytic ehrlichiosis.[9] PHF is an acute and potentially fatal equine disease that occurs frequently in the United States, and is occasionally found in Canada, Brazil, Uruguay, and Europe. The symptoms of PHF include depression, anorexia, fever, dehydration, laminitis, abortion, and watery diarrhea.[9] The only effective treatment is the administration of tetracycline, a broad spectrum antibiotic, in the early stages of the disease, and there is no truly effective vaccine available.[9] On the basis of molecular and antigenic similarity between *N. risticii* (formerly called *E. risticii*) and *N. helminthoeca*, trematodes were suspected as the vector of *N. risticii*[10] and the hunt for snails that harbor trematodes infected with *N. risticii* began in the 1990s. *N. risticii* DNA was identified in virgulate cercariae from freshwater snails in the genus *Juga* of the family *Pleuroceridae* in California as well as in virgulate xiphidiocercariae from freshwater snails in the *Elimia* (*Goniobasis*) species, also of the *Pleuroceridae* family, including *Elimia livescens* in Ohio and *E. virginica* in Pennsylvania. Xiphidiocercariae that were identified in Ohio and Pennsylvania, but not those found in California, possessed a stylet that is used to bore through chitinous exoskeletons. These xiphidiocercariae come from digenetic trematodes that utilize insects as their second intermediate hosts. This observation led to the discovery of PCR-positive *N. risticii* DNA in the metacercariae of larval and adult aquatic insects (mayflies, caddisflies, stoneflies, damselflies, and dragonflies). Finally, horizontal transmission of *N. risticii* from trematodes in the metacercaria stage to horses was proven by Koch's postulate, when horses were fed caddisflies from the PHF endemic region. These horses developed clinical signs of PHF and live *N. risticii* were isolated from the horses' blood.[11,12]

Recently, species of this infected trematode and the definitive host of the trematode, where the trematode sexually reproduces, were identified in two Potomac horse fever endemic regions: Northern California and Ohio–Pennsylvania.[13,14] Pools of digenetic trematodes, *Acanthatrium* sp. and *Lecithodendrium* sp. recovered from two *Myotis yumanensis* bats in California, were positive for *N. risticii*. Gravid trematodes infected with *N. risticii* were isolated from the intestines of *Eptesicus fuscus* big brown bats from the PHF endemic region of Pennsylvania and identified as *Acanthatrium oregonense*.[14] *A. oregonense* produces virgulate xiphidiocercariae, and parasitizes pleurocerid snails, caddisfly larvae, stonefly larvae, and bat intestines. Individual eggs isolated from *A. oregonense* were found to be infected with *N. risticii*, indicating that *N. risticii* is transmitted vertically (from adult to egg) in the trematode. This result proved, therefore, for the first time, transovarial transmission of *N. risticii* in the trematode host. Furthermore, *N. risticii* DNA was detected in the blood, liver, or spleen of 23 of 53 *Eptesicus fuscus* and *Myotis lucifugus* bats, suggesting that *N. risticii* also can be transmitted horizontally from trematode to bat.[14] In addition to bats, tree swallows (*Hirundo rustica*) were reported to be infected with *N. risticii*.[13] These results indicate that *A. oregonense* is a natural reservoir and a likely vector of *N. risticii*. Further study of the ecology of *N. risticii* strains and their natural trematode reservoirs will be instrumental in developing effective preventive measures for Potomac horse fever.

AEGYPTIANELLA PULLORUM

In 1928, Carpano[15] described an infectious agent that is seen as intraerythrocytic inclusions in blood smears of domestic fowls in Egypt and named it *Aegyptianella pullorum*. He also suggested that it might be transmitted by the soft tick *Argas (Persicargus) persicus*. The inclusions, which appear only in erythrocytes, are seen as purple intracytoplasmic bodies of 0.3–4 μm in size when stained via the Romanowsky method. Similar inclusions were observed by light microscopy in the red blood cells from various birds in other parts of Africa, Asia, Europe, and South and North America. The ultrastructure of *Ae. pullorum* strongly suggested that this bacterium belongs to the family *Anaplasmataceae*. In fact, *Ae. pullorum* inclusions are separated from the erythrocyte cytoplasm by a single membrane presumably of erythrocyte origin and each inclusion contains between 1 and 26 pleomorphic cocci of 0.25–0.4 μm.[16–18] The organisms are enveloped in wavy outer and inner membranes, and the cytoplasm contains ribosomes and DNA strands.[16–18] We recently determined the 16S rRNA and the *groEL* gene sequences of *Ae. pullorum* derived from Rio Grande wild turkeys in Southern Texas. It was here that the blood smears from an archival specimen that had been previously characterized with high-quality electron micrographs[18] were studied. Our molecular phylogenic analysis revealed that *Ae. pullorum* is related most closely to the genus *Anaplasma* in the family *Anaplasmataceae*.[19]

"*Aegyptianella ranarum*" (order *Rickettsiales*) is known to replicate in the red blood cells of frogs. However, unlike *Ae. pullorum*, this bacterium is a Gram-negative rod, rather than a coccus.[20] We recently obtained sequences of the 16S rRNA and *gyrB* genes of *Ae. ranarum* from a Canadian frog blood specimen from the original author of *Ae. ranarum*, Dr. S. Desser. *In situ* hybridization (with an *Ae. ranarum* 16S rRNA gene PCR product as probe) and electron microscopy confirmed that *Ae. ranarum* forms cytoplasmic inclusions in frog erythrocytes. BLAST comparisons with GenBank 16S rRNA and *gyrB* sequences showed that both *Ae. ranarum* genes were most similar (91% and 67% identity) to those of *Chryseobacterium meningosepticum*, a bacterium in the family *Flavobacteriaceae*. In contrast, *Ae. ranarum* 16S rRNA shared only 61% identity with *Ae. pullorum*.[21] Therefore, *Ae. ranarum* was proposed to be classified as "*Candidatus* Hemobacterium ranarum" in the family *Flavobacteriaceae*.[21] There are additional *Aegyptianella* species that infect the red blood cells of birds, amphibians, and lizards, which remain to be molecularly defined.

"*CANDIDATUS* NEOEHRLICHIA MIKURENSIS"

A novel bacterium that infects laboratory rats was isolated from wild *Rattus norvegicus* rats in Japan.[22] Transmission electron microscopy of the spleen tissue revealed a typical ultrastructure of the members of the family *Anaplasmataceae.* Small cocci surrounded by an inner membrane and a thin rippled outer membrane in the membrane-bound inclusion were found within the cytoplasm of endothelial cells of the rat. Phylogenetic analysis of the 16S rRNA gene sequences of the bacterium found in *R. norvegicus* rats and *Ixodes ovatus* ticks in Japan revealed that the organism represents a new clade in the family *Anaplasmataceae*, which includes the Schotti variant found in *Ixodes ricinus* ticks in the Netherlands, the Ehrlichia-like *Rattus* strain in *R. norvegicus* rats from China, and "*Candidatus* Ehrlichia walkerii" in *Ixodes ricinus* ticks from Italy. The new clade was confirmed by phylogenetic analysis of *groESL* sequences found in *R. norvegicus* rats and *Ixodes ovatus* ticks in Japan. No serological cross-reactivity was detected between this bacterium and members of *Anaplasma* sp., *Ehrlichia* sp., or *Neorickettsia* sp. in the family *Anaplasmataceae.* This new cluster of bacteria was designated as *Candidatus* Neoehrlichia mikurensis.[21] The most well-characterized reference strain is TK4456^{R}.

ANAPLASMA PHAGOCYTOPHILUM STRAINS

In the United States, the majority of human granuloytic anaplasmosis (HGA) cases have originated from the Northeastern and upper Midwestern states—Massachusetts, Connecticut, New York, Minnesota, and Wisconsin—whereas, most cases of equine granulocytic anaplasmosis have been reported in California. White-footed mice, wild deer, and other wild animals were found to be infected with *A. phagocytophilum* strains. Whether humans or animals, *A. phagocytophilum* is transmitted by *Ixodes scapularis* ticks in the Northeast and upper Midwest, and by *Ixodes pacificus* ticks in the Western states. In Europe, *Ixodes ricinus* is considered to be the main vector but other ticks also have been implicated, and *A. phagocytophilum* infection occurs primarily in sheep and cattle and less frequently in humans than in the United States. So far, the 16S rRNA gene sequences of isolates of *A. phagocytophilum* that infect humans are identical. However, divergent 16S rRNA gene sequences were found in ticks and animals. Sequence analysis of *ankA* and *msp4* further revealed the molecular divergence among *A. phagocytophilum* strains.[22–24] The *p44* gene of *A. phagocytophilum* encodes for the immunodominant major outer membrane protein P44s[25] and multiple *p44* homologues have been found in every *A. phagocytophilum* strain examined. *p44* consists of a central hypervariable region and 5′ and 3′ conserved regions.[26] Although the hypervariable regions are quite diverse, phylogenetic analyses of these *p44* sequences are possible.[27] Sequences of 16 distinct *p44*s were obtained from three infected Japanese deer. The Japanese deer *p44*s were quite unique compared to all known *p44* sequences, but retained the conserved *p44* group-specific deduced amino acids observed within the hypervariable region of all sequenced *p44*s.[28] We also compared a specific genomic locus among various

A. phagocytophilum strains, where we found diversity at the *p44-1/18* genomic locus.[26] Intraspecies variation of *p44* genes may contribute to understanding the role of *p44*s in immunoprotection, pathogenesis, and perhaps mammalian host specificity of *A. phagocytophilum*. Our study indicated that *p44-1* and *p44-18* may have evolved at different rates, suggesting that each *p44* may be under different selective pressure.[26] Although the corresponding genomic locus was found in the European sheep isolate, OS, both *p44-1* and *p44-18* were not detected. Thus, the OS and U.S. strains may be quite divergent.

GENOME SEQUENCES OF *E. RUMINANTIUM* AND *A. MARGINALE*

E. ruminantium infects endothelial cells (particularly in the brain) and neutrophils of ruminants and causes a disease called heartwater. Currently, the disease is limited to sub-Saharan Africa and the Caribbean region, where it is transmitted by *Amblyomma* ticks.[2] *A. marginale* infects red blood cells of cattle and causes bovine anaplasmosis throughout the world.[29] Only 62% of the genome of *E. ruminantium* encodes proteins partially due to the large number of tandemly repeated and duplicated sequences.[8] The *msp1* and *msp2* superfamilies including classic *msp1* and *msp2* encoding major surface proteins have been described.[7] Both of these organisms have many pseudogenes that appear to be distinct from pseudogenes in the reductive evolution of *Rickettsia*. The availability of these sequences will facilitate understanding the pathogenesis of heartwater and bovine–ovine anaplasmosis and effective vaccine development for these economically important pathogens.

SUMMARY

The recent ability to apply molecular approaches to *Anaplasmataceae* research has not only advanced our diagnostic capability and knowledge of the evolutionary relationships among members of *Anaplasmataceae* family but also has significantly advanced our understanding of these organisms and the pathogenesis of the diseases they cause. Whole genomes of several members of the family *Anaplasmataceae* have been sequenced and annotated, which has expanded the myriad of research projects that otherwise would not have been possible.

REFERENCES

1. DUMLER, J.S., A.F. BARBET, C.P.J. BEKKER, *et al.* 2001. Reorganization of genera in the families *Rickettsiaceae* and *Anaplasmataceae* in the order *Rickettsiales*: unification of some species of *Ehrlichia* with *Anaplasma*, *Cowdria* with *Ehrlichia*,

and *Ehrlichia* with *Neorickettsia*; Description of six new species combinations; and Designation of *Ehrlichia equi* and "HGE agent" as subjective synonyms of *Ehrlichia phagocytophila*. Int. J. Syst. Evol. Microbiol. **51:** 2145–2165.

2. Rikihisa, Y. 1991. The tribe *Ehrlichieae* and ehrlichial diseases. Clin. Microbiol. Rev. **4:** 286–308.
3. Lin, M. & Y. Rikihisa. 2003. *Ehrlichia chaffeensis* and *Anaplasma phagocytophilum* lack genes for lipid A biosynthesis and incorporate cholesterol for their survival. Infect. Immun. **71:** 5324–5331.
4. Lin, M. & Y. Rikihisa. 2003. Obligate intracellular parasitism by human monocytic and granulocytic ehrlichiosis agents involves caveolae and glycosylphosphatidylinositol-anchored proteins. Cell Microbiol. **5:** 809–820.
5. Wu, M., L.V. Sun, J. Vamathevan, *et al.* 2004. Phylogenomics of the reproductive parasite *Wolbachia pipientis* wMel: a streamlined genome overrun by mobile genetic elements. PLoS Biol. **2:** E69.
6. Foster, J., M. Ganatra, I. Kamal, *et al.* 2005. The *Wolbachia* genome of *Brugia malayi*: endosymbiont evolution within a human pathogenic nematode. PLoS Biol. **3:** e121.
7. Brayton, K.A., L.S. Kappmeyer, D.R. Herndon, *et al.* 2005. Complete genome sequencing of *Anaplasma marginale* reveals that the surface is skewed to two superfamilies of outer membrane proteins. Proc. Natl. Acad. Sci. USA **102:** 844–849.
8. Collins, N.E., J. Liebenberg, E.P. de Villiers, *et al.* 2005. The genome of the heartwater agent *Ehrlichia ruminantium* contains multiple tandem repeats of actively variable copy number. Proc. Natl. Acad. Sci. USA **102:** 838–843.
9. Rikihisa, Y. 2004. Rickettsial diseases. *In* Equine Internal Medicine, 2nd ed. S.M. Reed,W. Bayly & D.C. Sellon Eds.: 96–109 W.B. Saunders. Philadelphia.
10. Wen, B., Y. Rikihisa, S. Yamamoto, *et al.* 1996. Characterization of SF agent, an *Ehrlichia* sp. isolated from *Stellantchasmus falcatus* fluke, by 16S rRNA base sequence, serological, and morphological analysis. Int. J. Syst. Bacteriol. **46:** 149–154.
11. Madigan, J.E., N. Pusterla, E. Johnson, *et al.* 2000. Transmission of *Ehrlichia risticii*, the agent of Potomac horse fever, using naturally infected aquatic insects and helminth vectors: preliminary report. Equine Vet. J. **32:** 275–279.
12. Mott, J., Y. Muramatsu, E. Seaton, *et al.* 2002. Molecular analysis of *Neorickettsia risticii* in adult aquatic insects in Pennsylvania, in horses infected by ingestion of insects, and isolated in cell culture. J. Clin. Microbiol. **40:** 690–693.
13. Pusterla, N., E.M. Johnson, J.S. Chae, *et al.* 2003. Digenetic trematodes, *Acanthatrium* sp. and *Lecithodendrium* sp., as vectors of *Neorickettsia risticii*, the agent of Potomac horse fever. J. Helminthol. **77:** 335–339.
14. Gibson, K.E., Y. Rikihisa, C. Zhang, *et al.* 2005. *Neorickettsia risticii* is vertically transmitted in the trematode *Acanthatrium oregonense* and horizontally transmitted to bats. Environ. Microbiol. **7:** 203–212.
15. Carpano, M. 1928. Piroplasmosis in Egyptian fowls (*Egyptianella pullorum*). Vet. Serv. Bull. Egypt. Min. Agric. Sci. Technol. Serv. **86:** 1–7 (published in 1929).
16. Gothe, R. 1967. Ein Beiträg zür systematischen Stellung von *Aegyptianella pullorum* Carpano 1928. Z. Parasitenkd. **29:** 119–129.
17. Bird, R.G. & P.C.C. Garnham. 1969. *Aegyptianella pullorum* Carpano 1928—fine structure and taxonomy. Parasitology **59:** 745–752.

18. CASTLE, M.D. & B.M. CHRISTENSEN. 1985. Isolation and identification of *Aegyptianella pullorum* (*Rickettsiales, Anaplasmataceae*) in wild turkeys in north America. Avian Dis. **29:** 437–445.
19. RIKIHISA, Y., C. ZHANG & B.M. CHRISTENSEN. 2003. Molecular characterization of *Aegyptianella pullorum* (*Rickettsiales, Anaplasmataceae*). J. Clin. Microbiol. **41:** 5294–5297.
20. DESSER, S.S. 1987. *Aegyptianella ranarum* sp.n. (*Rickettsiales, Anaplasmataceae*): ultrastructure and prevalence in frogs from Ontario. J. Wildl. Dis. **23:** 52–59.
21. ZHANG, C. & Y. RIKIHISA. 2004. Proposal to transfer "*Aegyptianella ranarum*", an intracellular bacterium of frog red blood cells, to the family Flavobacteriaceae as "*Candidatus* Hemobacterium ranarum" comb. nov. Environ. Microbiol. **6:** 568–573.
22. KAWAHARA, M., Y. RIKIHISA, E. ISOGAI, *et al.* 2004. Ultrastructure and phylogenetic analysis of "*Candidatus* Neoehrlichia mikurensis" in the family *Anaplasmataceae* isolated from wild rats and found in *Ixodes ovatus* ticks. Int. J. Syst. Evol. Microbiol. **54:** 1837–1843.
23. DE LA FUENTE, J., R.F. MASSUNG, S.J WONG, *et al.* 2005. Sequence analysis of the msp4 gene of *Anaplasma phagocytophilum* strains. J. Clin. Microbiol. **43:** 1309–1317.
24. MASSUNG, R.F., J.H. OWENS, D. ROSS, *et al.* 2000. Sequence analysis of the ank gene of granulocytic ehrlichiae. J. Clin. Microbiol. **388:** 2917–2922.
25. CHAE, J.S., J.E. FOLEY, J.S. DUMLER, *et al.* 2000. Comparison of the nucleotide sequences of 16S rRNA, 444 Ep-ank, and groESL heat shock operon genes in naturally occurring *Ehrlichia equi* and human granulocytic ehrlichiosis agent isolates from Northern California. J. Clin. Microbiol. **38:** 1364–1369.
26. LIN, Q., Y. RIKIHISA. R.F. MASSUNG, *et al.* 2004. Polymorphism and transcription at the *p44-1/p44-18* genomic locus in *Anaplasma phagocytophilum* strains from diverse geographic regions. Infect. Immunol. **72:** 5574–5581.
27. CASEY, A.N., R.J. BIRTLES, A.D. RADFORD, *et al.* 2004. Groupings of highly similar major surface protein (p44)-encoding paralogues: a potential index of genetic diversity amongst isolates of *Anaplasma phagocytophilum*. Microbiology **150:** 727–734.
28. KAWAHARA, M., Y. RIKIHISA, Q. LIN, *et al.* 2006. Novel genetic variants of *Anaplasma phagocytophilum, Anaplasma bovis, Anaplasma centrale* and a novel *Ehrlichia* sp. in wild deer and ticks on two major islands in Japan. Appl. Env. Microbiol. **72:** 1102–1109.
29. KOCAN, K.M., J. DE LA FUENTE, A.A. GUGLIELMONE, *et al.* 2003. Antigens and alternatives for control of *Anaplasma marginale* infection in cattle. Clin. Microbiol. Rev. **16:** 698–712.
30. HOTOPP, J.C., M. LIN, R. MADUPU, *et al.* 2006. Comparative genomics of emerging human ehrlichiosis agents. PLOS Genetics **2:** e211–216.

Anaplasma phagocytophilum in Ruminants in Europe

ZERAI WOLDEHIWET

Department of Veterinary Pathology, Veterinary Teaching Hospital, University of Liverpool, Leahurst, Neston, South Wirral, CH64 7TE, UK

ABSTRACT: The agent that causes tick-borne fever (TBF) in sheep was first described in 1940, 8 years after the disease was first recognized in Scotland. The same agent was soon shown to cause TBF in sheep and pasture fever in cattle in other parts of the UK, Scandinavia, and other parts of Europe. After the initial use of the name *Rickettsia phagocytophila*, the organism was given the name *Cytoecetes phagocytophila* to reflect its association with granulocytes and its morphological similarity with *Cytoecetes microti*. This name continued to be used by workers in the UK until the recent reclassification of the granulocytic ehrlichiae affecting ruminants, horses, and humans as variants of the same species, *Anaplasma phagocytophilum*. TBF and pasture fever are characterized by high fever, recurrent bacteremia, neutropenia, lymphocytopenia, thrombocytopenia, and general immunosuppression, resulting in more severe secondary infections such as tick pyemia, pneumonic pasteurellosis, listeriosis, and enterotoxemia. During the peak period of bacteremia as many as 90% of granulocytes may be infected. The agent is transmitted transtadially by the hard tick *Ixodes ricinus*, and possibly other ticks. After patent bacteremia, sheep, goats, and cattle become persistently infected "carriers," perhaps playing an important role in the maintenance of infection, in the flock/herd. Little is known about how efficiently ticks acquire and maintain infection in ruminant populations or whether "carrier" domestic ruminants play an important role as reservoirs of infection, but deer, other free-living ruminants, and wild rodents are also potential sources of infection. During the late 1990s serological evidence of infection of humans was demonstrated in several European countries, creating a renewed interest and increased awareness of the zoonotic potential of TBF variants. More recently, a few cases of human granulocytic anaplasmosis (HGA) have been reported in some European countries, but it remains to be established whether the variants causing HGA in Europe are genetically and biologically different from those causing TBF in ruminants. TBF is readily diagnosed by demonstrating intracytoplasmic inclusions in peripheral blood granulocytes or monocytes of febrile animals or by

Address for correspondence: Zerai Woldehiwet, Veterinary Teaching Hospital, Leahurst, Neston, South Wirral, CH64 7TE, UK. Voice: (+44)-151-794-6113; fax: (+44)-151-794-6110.
e-mail: zerai@liverpool.ac.uk

**Ann. N.Y. Acad. Sci. 1078: 446–460 (2006). © 2006 New York Academy of Sciences.
doi: 10.1196/annals.1374.084**

detecting specific DNA by polymerase chain reaction (PCR), and TBF variants of *A. phagocytophilum* can be cultivated in tick cell lines, but the differentiation of TBF variants from HGA variants awaits further investigations.

KEYWORDS: ***anaplasma phagocytophilum*; Europe; ruminants; tick-borne fever**

INTRODUCTION

Variants of the tick-borne bacterium *Anaplasma phagocytophilum* have been recognized as pathogens of domestic and free-living ruminants in Europe for over 70 years.[1] The agent that causes tick-borne fever (TBF) in sheep was first described in 1940,[2] 8 years after the disease was first recognized in Scotland.[3] The same agent was soon shown to cause TBF in sheep and pasture fever in cattle in other parts of the UK,[4] Ireland,[5] Scandinavia[6–8], and other parts of Europe, including the Netherlands,[9] Austria,[10] Switzerland,[11] Spain,[12] and France.[13] After the initial use of the name *Rickettsia phagocytophila,*[14] the organism was given the name *Cytoecetes phagocytophila*[15] to reflect its predilection to granulocytes and its morphological similarity with *Cytoecetes microti.*[16] This name continued to be used by workers in the UK until the recent reclassification of all the granulocytic ehrlichiae that affect ruminants, horses, and humans as variants of the same species, *A. phagocytophilum.*[17] The recent recognition that variants of this bacterium can cause the potentially fatal illnesses in humans, human granulocytic anaplasmosis (HGA), in the USA[18] and in some parts of Europe[19] has created a renewed interest in this organism and resulted in accelerated advances in our knowledge of its biology. It has been known for many years that *A. phagocytophilum* is widespread and common in *Ixodes ricinus* ticks in the UK habitats.[1] However, to date there has been no documented cases of human disease associated with this bacterium in the UK, despite some serological evidence.[20,21] This could suggest that *A. phagocytophilum* maintained in the UK, some parts of Europe, and the USA have different biological characteristics in terms of their pathogenicity and host specificity, and that clinically significant genetic heterogeneity may occur among European isolates.

HOST RANGE AND RESERVOIRS OF INFECTION

Most outbreaks of TBF occur among flocks of sheep and herds of cattle immediately after they have been introduced into tick-infested pastures, but isolated outbreaks have been reported in goats.[22] Among free-living ruminants in the UK, the organism has been detected in feral goats[23] and in red, fallow, and roe deer.[24–25] The organism was also shown to infect a variety of cervids including roe deer, moose, and chamois in Norway,[26–27] Slovenia,[28] Switzerland,[29] and Austria.[30]

Until the recent recognition that free-living rodents do harbor variants of *A. phagocytophilum*, it has been largely thought that TBF variants were maintained in a tick-ruminant cycle, with persistent infections of domestic and free-living ruminants serving as the source of continuous transmission, as no transovarian transmission has been demonstrated.[31] However, the recent demonstration that wood mice (*Apodemus sylvaticus*), yellow-necked shrew (*Apodemus flavicollis*), field voles (*Microtus agrestis*), and bank voles (*Clethrinomyces glareolus*) are competent hosts of *A. phagocytophilum*[32–34] suggests that rodents may also be important reservoirs of infection. However, compared to ruminants, the latter appear to develop low levels of bacteremia and have shorter life cycles. A few reports have also documented some evidence of infection with granulocytic agents in nonruminant domestic animals in Scandinavia and other regions of Europe. The domestic animals in which evidence of infection with *A. phagocytophilum* was reported include cats,[35] dogs,[36–40] and horses.[39,41–44] However, most of the isolates from animals other than ruminants appear to be genetically closer to the HGA variants.

Serologic evidence of human infection was first reported from Switzerland and the UK in 1995[20,45] and this was soon followed by similar reports from other parts of Europe.[46–49] Over the last few years several cases of HGA (previously known as human granulocytic ehrlichiosis [HGE]) have been described in different parts of Europe,[19,50–58] but it remains to be established whether or not TBF variants may be responsible for some or all of these European cases of HGA.

VECTORS

Soon after TBF was described, and before the organism was identified, workers at the Moredun Institute, Edinburgh, carried out a series of experiments to establish its methods of transmission. These studies established that the infectious agent of TBF was transmitted transstadially by *I. ricinus* and that the agent survived in infected ticks for over a year while the tick was awaiting a new host.[31,59] The same workers also speculated that the reported presence of TBF in certain areas of UK, where *I. ricinus* was not present, may indicate ticks other than *I. ricinus*, possibly *Haemophysalis punctata*, may be possible vectors.[60] *I. ricinus* has also been shown to be the most important vector for TBF and pasture fever in other parts of Europe.[8,61] A recent study also reported the detection of variants of *A. phagocytophilum* in *I. persulcatus* in the Baltic regions of Russia.[62] Other recent studies indicate that other ticks such as *I. trianguliceps* may also play an important role in the transmission of *A. phagocytophilum* in rodents,[32,34] but whether or not the strains present in wild rodents are distinct from the TBF variants remains to be elucidated.

The bacteria can survive the moulting process and infect new hosts during the next feeding, but vertical transmission is reported to be nonexistent or inefficient.[31,63] The prevalence of infection in *I. ricinus* ticks may vary from one geographical region to another[64–66] as may the stages of development, with higher infection prevalence in nymphs compared to adults and larvae. Transmission efficiency may be further influenced by other factors including co-feeding, tick density, and anti-tick immunity of the mammalian host.[63]

TARGET CELLS

The first leukocytes to be infected are the eosinophils and neutrophils, with the monocytes being infected at the end of primary bacteremia. During the peak period of bacteremia, as many as 90% of the granulocytes may be infected, with the severity of bacteremia and febrile reaction being influenced by the strain of bacterium and host susceptibility and immune status.[1,8,67] Following acute experimental infection, TBF variants of *A. phagocytophilum* have been shown to cause persistent infection or "carrier" state for up to 2 years following acute experimental infection.[8,68] It remains to be established where the organism multiplies before it is detected in the peripheral blood 2–4 days after experimental infection,[1] and the sites of persistence during in-between periods of recurrent bacteremia remain to be established. However, during acute bacteremia the organism has been demonstrated in the alveolar macrophages, Küpffer cells, and other tissue macrophages.[1,69,70] Moreover, there is some evidence to suggest that the organism may be present in the lungs before its detection in the blood.[71]

IMMUNOSUPPRESSION

Early workers had observed that sheep and cattle infected with *A. phagocytophilum* displayed a range of clinical signs that were attributable to secondary infections. For example, up to 30% of TBF-infected lambs may develop tick pyemia, a crippling lameness and paralysis in tick-infested farms due to infection with *Staphylococcus aureus*.[15,72] When Hudson[4] first described the disease in dairy cattle, the most predominant clinical signs observed were coughing and reduced milk yield. Other workers have observed similar clinical signs in cattle.[8,73] Further studies in the UK and Scandinavia have clearly established that TBF variants of *A. phagocytophilum* are immunosuppressive, resulting in several disease syndromes including tick pyemia,[72,74] abortions,[74–79] pasteurellosis,[80–81] and septicemic listeriosis.[82–83]

Neutrophils form the first line of defence against invading bacteria and fungi and possess a powerful array of cytotoxic enzymes, reactive oxidants, and associated processes, but neutrophils have high rates of spontaneous apoptosis and a very short half-life in the circulation of only 6–12 h.[84–85] In view of this high cytotoxic potential and short half-life, the neutrophil would seem to be an unlikely and unsuitable target for intracellular pathogens such as *A. phagocytophilum*. Earlier studies indicated that neutrophils from infected sheep had reduced capacity for diapedesis *in vivo,*[86] migration *in vitro,*[87] and adherence to glass surfaces.[88] Other studies have shown that the organism adversely affects the phagocytic and bactericidal ability of ovine neutrophils[88–91] and infected neutrophils may be more susceptible to the cytotoxins of *Mannheimia* (formerly *Pasteurella*) *haemolytica.*[92] Recent studies have also shown that the organism inhibits phagosome–lysosome fusion[93] and delays apoptosis of ovine neutrophils *in vivo.*[94] Whilst respiratory burst is upregulated during the early phases of infection, it may be downregulated during the late stages of bacteremia.[91,94] The period of reduced respiratory burst was reported to coincide with a reduction in the number of granulocytes expressing the CD14 epitope.[95] A few studies have shown that some serum factors released during the period of bacteremia may be associated with immunosuppression.[87–88,96–97] Studies in the UK during the 1980s and 1990s showed that during the period of bacteremia, mitogen-driven lymphocyte proliferative responses were reduced[98] and that this may be due to a reduction in the number of CD4+ T cells or their functions and changes in the CD4:CD8 ratio.[99] The nature of the factors which inhibit lymphocyte proliferation and migration have not been investigated fully, but one recent study showed that the expression of the interleukin (IL)-2 receptor CD25 in ovine CD4+ T cells was altered for up to 5 weeks postinfection with *A. phagocytophilum.*[95] In another study an HGA variant was reported to stimulate the production of macrophage inflammatory protein, monocyte chemotactic protein, and IL-8 by HL60 and human bone-marrow cells, at levels likely to suppress hematopoiesis.[100] Detailed information on the mechanisms of immunosuppression and the possible effects of TBF variants on immunoregulatory chemokines, proinflammatory cytokines, and/or anti-inflammatory cytokines await further investigation.

IMMUNITY

Experimental studies have shown that primary infection is followed by a variable degree of resistance to homologous challenge.[8,67–68] Some animals resisted reinfection only for a few months, but others were protected against patent bacteremia after homologous challenge for more than 1 year.[1,68] Resistance was influenced by the strain, the age, and type of host, and by the length of time between primary infection and challenge.[8,67–68,101–103] The role of antibodies and cellular immunity has not been clearly established, but limited experimental studies suggest that both arms of the immune system may be important. One study indicated that the titer of antibodies in peripheral blood may be directly related to the absence of detectable bacteremia following homologous challenge[67] and another study showed that stimulation with fixed infected granulocytes resulted in the proliferation of lymphocytes obtained from primed sheep, indicating CD4 responses.[104]

Early cross-protection studies using European isolates from sheep and cattle suggested a high degree of antigenic diversity amongst different isolates of *A. phagocytophilum.*[68] After showing that primed animals did not resist sequential challenges with several heterologous strains, each causing a reaction, Tuomi[102] concluded that "the extensive immunological heterogeneity observed in this study among randomly isolated strains of tick-borne fever agent appears to be unparalleled by any other species of micro-organisms."

Despite these apparent strain differences based on protection studies, there have been very few attempts at antigenic or genetic differentiation. One study used antibody titers of experimentally infected sheep using homologous and heterologous antigens derived from infected ovine granulocytes to differentiate the Old Sourhope strain from other strains.[103]

In the light of recent evidence of sequential changes in some of the major surface proteins of *Anaplasma* spp. during recurrent bacteremia,[105–106] it remains to be clearly established whether or not these early observations were a true reflection of strain diversity.

PERSISTENCE AND "CARRIER" STATE

It was established early that sheep that had recovered from experimental or natural TBF may continue to harbor the organism for several weeks or even years. In one study, the blood of one sheep was shown to be infective to a naïve sheep 25 months after primary infection.[68] During the same study, samples of blood taken at random from sheep in tick-infested farms at all times of the year were reported to be invariably infective to susceptible sheep, and this was attributed to persistence of infection in "carrier" sheep during the period of no-tick activity. Persistently infected sheep did not develop serious clinical signs, but bacteremia was induced by splenectomy or after treatment with immunosuppressive drugs.[1,68] However, in animals infected under field conditions, the detection of the organism in the peripheral blood appears to be affected by the numbers and frequency of infestation by ticks. This is thought to be either due to an increase in the number of feeding ticks altering the level of bacteremia, or due to frequent reinfections.[63] The increase in the number of feeding ticks could be due to blood loss and/or immunomodulatory effects of tick saliva[107] or due to a reduced mobilization of infected granulocytes at tick-feeding lesions.[108]

DIAGNOSIS

The presence of high fever in animals that have been recently moved into tick-infested pastures is one of the first indications of TBF in sheep.[1,68]

However, the presence of other clinical signs such as tick pyemia in lambs[71] and respiratory signs in cattle[4] affecting several animals and other secondary infections a few days after being introduced to tick-infested pastures are good indicators of TBF. In dairy cattle the other prominent clinical sign is likely to be a significant drop in milk yield.[8,109] The severe leukopenia, particularly the prolonged neutropenia, which accompanies the disease, is also a good indicator of infection.[87,110] Other changes that accompany the disease include thrombocytopenia,[110,111] reduction in the activity of serum alkaline phosphatase and plasma iron and zinc,[112–114] and an increased plasma bilirubin and creatinine.[114] In some cases, abortion storms may occur, particularly when pregnant ewes or cows are moved to tick-infested pastures during the last stages of pregnancy.[75,77,79,115] In young lambs, the main clinical signs are likely to be those of tick pyemia, the most common and serious complication of TBF.[1,74,116] Confirmation of infection is based on the demonstration of typical inclusions in peripheral blood granulocytes and, occasionally, monocytes.[88] The demonstration of specific DNA using primers for various genes is increasingly being used to confirm diagnosis too. TBF variants have been successfully cultivated in tick cell lines[117] but this has not been evaluated as a diagnostic tool.

Following abortion storms, infection in a herd or flock may be retrospectively established by the demonstration of rising antibody titers by complement fixation,[67] counter-current immuneoelectrophoresis,[118] indirect immunofluorescence[119–120] using infected granulocytes as antigens, or by the recently described enzyme-linked immunosorbent assay with antigens derived from infected ovine granulocytes or tick cells.[120] The differentiation of TBF variants from HGA or other variants of *A. phagocytophilum* currently requires the sequencing of amplified fragments of the 16S rRNA gene. Another sequence difference was reported to be present in the *groESL* heat-shock operon.[30] Most interestingly another recent study indicated that variants that infect ruminants in Europe could be differentiated from variants that infect dogs, horses, and humans by sequencing *msp4*, one of the genes encoding major surface proteins.[121] The successful sequencing of the whole genome will no doubt lead to the identification of other genes or gene fragments that could be easily used to identify differences among variants.

CONTROL MEASURES

Nearly half of the >30 million sheep population in the UK is thought to live in hilly, often tick-infested areas, and in one study, an estimated 300,000 lambs were reported to develop TBF followed by tick pyemia annually.[74] Most of the lambs that develop tick pyemia die or are of no economic value. A significant portion of the sheep that develop TBF may also die from other secondary infections, and losses due to TBF-related abortions can be significant. Economic losses due to reduced milk yield and other complications of pasture fever in diary cattle can also be very high. Therefore there is good economic and welfare justification for the control of TBF and pasture fever in ruminants.

Current control strategies are based on the reduction of tick infestations when sheep and cattle are turned out into pastures and the use of long-acting antibiotics as a prophylactic measure given before animals are moved from tick-free environment into tick-infested pastures.[74] The reduction of ticks by regular dipping or pour-on application of synthetic pyrethroids may help to mitigate losses in sheep and particularly in lambs, which are likely to suffer from tick pyemia.[74,122] Another option would be to hold ewes and their lambs in tick-free, fenced pastures until the lambs are 6–7 weeks old. As abortion storms of up to 90% are common in naïve pregnant ewes, pregnant animals should never be moved from tick-free to tick-infested areas.[75–76,78]

The use of long-acting antibiotics is not widely used, but can be effective when strategically used.[74,123] Watt *et al.* reported that the administration of long-acting tetracyclines reduced morbidity and mortality due to tick pyemia by 7.5% and 2.0%, respectively.[123] Treated lambs were also reported to have better weight gains and improved general conditions. This is thought to be due to the prevention of TBF and associated immunosuppression leading to pasteurellosis, colibacillosis, and other conditions.[74] Dipping and treatment with long-acting antibiotics were reported to provide even better results.

The development of prophylactic vaccines against *A. phagocytophilum* is considered to be the most effective strategy for disease control, but no vaccines are currently available. The development of effective vaccines against this intracellular bacterium requires identification of the bacterial components necessary for protective immunity, the development of suitable vaccines containing these immunogenic elements, and effective delivery systems.

REFERENCES

1. WOLDEHIWET, Z. & G.R. SCOTT. 1993. Tick-borne (pasture) fever. *In* Rickettsial and Chlamydial Diseases of Domestic Animals. Z. Woldehiwet & M. Ristic, Eds.: 233–254. Pergamon Press, Oxford.
2. GORDON, W.S., A. BROWNLEE & D.R. WILSON. 1940. Studies on louping ill, tick-borne fever and scrapie. Proceedings of the Third International Congress on Microbiology, New York, pp. 362.
3. MACLEOD, J. 1932. Preliminary studies in tick transmission of louping ill. II. A study of the reaction of sheep to tick infestation. Vet. J. **88:** 276–284.
4. HUDSON, J. R. 1950. The recognition of tick-borne fever as a disease of cattle. Br. Vet. J. **106:** 317.
5. COLLINS, J.D., J. HANNAN, A.R. FERGUSON & J.O. WILSON. 1970. Tick-borne fever in Ireland. Ir. Vet. J. **24:** 162–164.
6. THORSHAUG, K. 1940. Cited by S. Stuen 2003. *Anaplasma phagocytophilum* (formerly *Ehrlichia phagocytophila*) infection in sheep and wild ruminants in Norway: a study on clinical manifestation, distribution and persistence. Doctor Philosophiae thesis, Norwegian College of Veterinary Medicine, Oslo.

8. TUOMI, J. 1967. Experimental studies on bovine tick-borne fever. 1. Clinical and haematological data, some properties of the causative agent, and homologous immunity. Acta Pathol. Microbiol. Scand. **70:** 429–445.
9. BOOL, P.H. & J.S. REINDERS. 1964. Tick-borne fever in cattle in the Netherlands. Tijdschr. Diergeneeskd. **89:** 1519–1527.
10. HINAIDY, H. K. 1973. Zwei neue infektiose Blutkrankheiten des Rindes in Osterreich. Wien. Tierarztl. Monatsschrift. **60:** 364–366.
11. PFISTER, K., A. ROESTI, P.H. BOSS & B. BALSIGER. 1987. *Ehrlichia phagocytophila* als Erreger des "Weidefieber"s im Berner Oberland. Schweizer Archiv fur Tierheilkunde **129:** 343–347.
12. JUSTE, R.A., G.R. SCOTT, E.A. PAXTON & J.L. GELABERT. 1989. Presence of *Cytoecetes phagocytophila* in an atypical disease of cattle in Spain. Vet. Rec. **124:** 636.
13. ARGENTE, G., E. COLLIN & H MORVAN.1992. Ehrlichiosis bovine (fievre des patures): une observation en France. Point Veterinaire **24:** 181–182.
14. FOGGIE, A. 1949. Studies on tick-borne fever. J. Gen. Microbiol. iii: Proceedings, v–vi.
15. FOGGIE, A. 1962. Studies on tick pyemia and tick-borne fever. Symp. Zool. Soc. Lond. **6:** 51–58.
16. TYZZER, E.E. 1938. *Cytoecetes microti*, N.G.H. Sp., a parasite developing in granulocytes and infective for small rodents. Parasitology **30:** 242–257.
17. DUMLER, J.S., A.F. BARBET, C.P.J. BEKKER, *et al.* 2001. Reorganisation of the genera of the families *Rickettsiaceae* and *Anaplasmataceae* in the order *Rickettsiales*: unification of some species of *Ehrlichia* with *Anaplasma, Cowdria* with *Ehrlichia* and *Ehrlichia* with *Neorickettsia*, descriptions of six new combinations and designations of *Ehrlichia equi* and 'HGE agent' as subjective synonyms of *Ehrlichia phagocytophila*. Int. J. Syst. Evol. Microbiol. **51:** 2145–2165.
18. CHEN, S.M., J.S. DUMLER, J.S. BAKKEN & D.H. WALKER. 1994. Identification of a granulocytotropic *Ehrlichia* species as the etiologic agent of human disease. J. Clin. Microbiol. **32:** 589–595.
19. LOTRIC–FURLAN, S., M. PETROVEC, T.A. ZUPANC, *et al.* 1998. Human granulocytic ehrlichiosis in Europe: clinical and laboratory findings for four patients from Slovenia. Clin. Infect. Dis. **27:** 424–428.
20. SUMPTION, K.J., D.J.M. WRIGHT, S.J. CUTLER & B.A.S. DALE. 1995. Human Ehrlichiosis in the UK. Lancet **346:** 1487–1488.
21. THOMAS, D., M. SILLIS, T.J. COLEMAN, *et al.* 1998. Low rates of ehrlichiosis and Lyme borreliosis in English farmworkers. Epidemiol. Infect. **121:** 609–614.
22. GRAY, D., K WEBSTER & J.E. BERRY. 1988. Evidence of louping ill and tick-borne fever in goats. Vet. Rec. **122:** 66.
23. FOSTER, W.N.M. & J.C. GREIG. 1969. Isolation of tick-borne fever from feral goats in New Galloway. Vet. Rec. **85:** 585–586.
24. MCDIARMID, A. 1965. Modern trends in animal health and husbandry: some infectious diseases of free-living wild-life. Br. Vet. J. **121:** 245–257.
25. ALBERDI, M.P., A.R WALKER & K.A. URQUHART. 2000. Field evidence that roe deer (*Capreolus capreolus*) are a natural host for *Ehrlichia phagocytophila*. Epidemiol. Infect. **124:** 315–323.
26. STUEN, S., E.O. ENGVALL, I. VAN DE POL & L.M. SCHOOULS. 2001. Granulocytic ehrlichiosis in a roe deer calf in Norway. J. Wildl. Dis. **37:** 614–616.

27. JENKINS, A., K. HANDELAND, S. STUEN, *et al.* 2001. Ehrlichiosis in a moose calf in Norway. J. Wildl. Dis. **37:** 201–203.
28. PETROVEC, M., A. BIDOVEC, T. AVSIC-ZUPANC, *et al.* 2002. Infection with *Anaplasma phagocytophila* in cervids from Slovenia: evidence of two genotypic lineages. Wien. Klin. Wochenschrift. **114:** 641–647.
29. LIZ, J.S., J.W. SUMNER, K. PFISTER & M. BROSSARD. 2002. PCR detection of and serological evidence of granulocytic ehrlichia; infection in roe deer (*Capreolus capreolus*) and chamois (*Rupicapra rupicapra*). J. Clin. Microbiol. **40:** 892–897.
30. POLIN, H., P. HUFNAGL, R. HAUNSCHMID, *et al.* 2004. Molecular evidence of *Anaplasma phagocytophilum* in *Ixodes ricinus* ticks and wild animals in Austria. J. Clin. Microbiol. **42:** 2285–2286.
31. MACLEOD, J. 1936. Studies on tick-borne fever of sheep. II. Experiment on transmission and distribution of the disease. Parasitology **28:** 320–329.
32. OGDEN, N., K.N.H. BOWN, B.K. HORROCKS, *et al.* 1998. Granulocytic ehrlichia infection in Ixodes ticks and mammals in woodlands and uplands of United Kingdom. Med. Vet. Entomol. **12:** 423–429.
33. LIZ, J.S., L. ANDERES & J.W. SUMNER. 2000. PCR detection of granulocytic ehrlichiae in *Ixodes ricinus* ticks and wild small mammals in western Switzerland. J. Clin. Microbiol. **38:** 1002–1007.
34. BOWN, K.J., M. BENNETT, M. BEGON, *et al.* 2003. Seasonal dynamics of *Anaplasma* (formerly *Ehrlichia) phagocytophila* in a rodent–tick (*Ixodes trianguliceps*) system in the UK. Emerg. Infect. Dis. **9:** 63–70.
35. BJOERSDORFF, A.I., L. SVENDERNIUS, J.H. OWENS & R.F. MASSUNG. 1999. Feline granulocytic ehrlichiosis—a report of new clinical entity and characterization of the infectious agent. J. Small. Anim. Pract. **40:** 20–24.
36. EGENVALL, A.E., A.A. HEDHAMMAR & A.I. BJOERSDORFF. 1997. Clinical features and serology of 14 dogs affected by granulocytic ehrlichiosis in Sweden. Vet. Rec. **140:** 222–226.
37. PUSTERLA, N., J.B. PUSTERLA, P. DEPLAZES, *et al.* 1998. Seroprevalence of *Ehrlichia canis* and of canine granulocytic *Ehrlichia* infection in dogs in Switzerland. J. Clin. Microbiol. **36:** 3460–3462.
38. SHAW, S., M. KENNY, M. DAY, *et al.* 2001. Canine granulocytic ehrlichiosis in the UK. Vet. Rec. **148:** 727–728.
39. ENGVALL, E.O. & A. EGENVALL. 2002. Granulocytic ehrlichiosis in Swedish dogs and horses. Int. J. Med. Microbiol. **291**(Suppl. 33): 100–103.
40. MANNA, L., A. ALBERTI, L.M. PAVONE, *et al.* 2004. First molecular characterization of a granulocytic. Ehrlichia strain isolated from a dog in South Italy. Vet J. **167:** 224–227.
41. PUSTERLA, N., J.B. HUDER, K. FEIGE & H. LUTZ. 1998. Identification of a granulocytic Ehrlichia strain isolated from a horse in Switzerland and comparison with other rickettsiae of the *Ehrlichia phagocytophila* genogroup. J. Clin. Microbiol. **36:** 2035–2037.
42. ARTURSSON, K., A. GUNNARSSON, U.B. WIKSTROM & E.O. ENGVALL. 1999. A serological and clinical follow-up in horses with confirmed equine granulocytic ehrlichiosis. Equine Vet. J. **31:** 473–477.
43. EGENVALL, A., P. FRANZEN, A. GUNNARSSON, *et al.* 2001. Cross-sectional study of the seroprevalence to *Borrelia burgdorferi sensu lato* and granulocytic *Ehrlichia* spp. and demographic, clinical and tick-exposure factors in Swedish horses. Prev. Vet. Med. **49:** 191–208.

44. BERMANN, F., B. DAVOUST, P.E. FOURNIER, *et al.* 2002. *Ehrlichia equi* (*Anaplasma phagocytophila*) infection in an adult horse in France. Vet. Rec. **150:** 787–788.
45. BROUQUI, P., J.S. DULMER, R. LIENHARD, *et al.* 1995. Human granulocytic ehrlichiosis in Europe. Lancet **346:** 782–783.
46. FINGERLE,V., J.L. GOODMAN, R.C. JOHNSON, *et al.* 1997. Human granulocytic ehrlichiosis in southern Germany: increased seroprevalence in high-risk groups. J. Clin. Microbiol. **35:** 3244–3247.
47. LEBECH, A.M., K. HANSEN, P. PANCHOLI, *et al.* 1998. Immunoseroligc evidence of human granulocytic ehrlichiosis in Danish patients with Lyme neuroborreliosis. Scand. J. Infect. Dis. **30:** 173–176.
48. PUSTERLA, N., R. WEBER, C. WOLLFENSBERGER, *et al.* 1998. Serologic evidence of human granulocytic ehrlichiosis in Switzerland. Eur. J. Clin. Microbiol. Infect. Dis. **17:** 207–209.
49. BJOERSDORFF, A., P. BROUQUI, I. ELIASSON, *et al.* 1999. Serological evidence of Ehrlichia infection in Swedish Lyme borreliosis patients. Scand. J. Infect. Dis. **31:** 51–55.
50. PETROVEC, M., S. LOTRIC-FURLAN, T.A. ZUPANC, *et al.* 1997. Human disease in Europe caused by a granulocytic *Ehrlichia* species. J. Clin. Microbiol. **35:** 1556–1559
51. VAN DOBBENBURGH, A., A.P. VAN DAM & E. FIJRIG. 1999. Human granulocytic ehrlichiosis in Western Europe. N. Engl. J. Med. **340:** 1214–1216.
52. OTEO, J.A., J.R. BLANCO, V. DE ARTOLA & V. IBARRA. 2000. First report of human granulocytic ehrlichiosis in southern Europe (Spain). Emerg. Infect. Dis. **6:** 430–431.
53. MISIC-MAJERUS, L.J., N. BUJIC, V MADJARIC & V. JAMES-POJE. 2000. First description of the human granulocytic ehrlichiosis in Croatia. Clin. Microbiol. Infect. **6** (Suppl. 1): 194–195.
54. KARLSSON, U., A. BJOERSDORFF, R.F. MASSUNG & B. CHRISTTENSSON. 2001. Human granulocytic ehrlichiosis—a case in Scandinavia. Scand. J. Infect. Dis. **33:** 805–806.
55. KRISTENSEN, B.E., A. JENKINS, Y. TVETEN, *et al.* 2001. Human granulocytic ehrlichiosis in Norway. Tisskr. Nor. Laegeforen. **121:** 805–806.
56. TYLEWSKA-WIERZBANOWSKA, S., T. CHMIELEWSKI, M. KONDRUSIK, *et al.* 2001. First case of acute human granulocytic ehrlichiosis in Poland. Eur. J. Clin. Microbiol. Infect. Dis. **20:** 196–198.
57. BLANCO, J.R. & J.A. OTEO. 2002. Human granulocytic ehrlichiosis in Europe. Clin. Microbiol. Infect. **8:** 763–772.
58. STRLE, F. 2004. Human granulocytic ehrlichiosis in Europe. Int. J. Med. Microbiol. **293** (Suppl. 37): 27–35.
59. MACLEOD, J. & W.S. GORDON. 1933. Studies on tick-borne fever of sheep. I. Transmission by the tick *Ixodes ricinus* with a description of the disease produced. Parasitology **25:** 273–283.
60. MACLEOD, J. 1962. Ticks and disease in domestic stock in Great Britain. Symp. Zool. Soc. Lond. **6:** 29–50.
61. STUEN, S. 2003. *Anaplasma phagocytophilum* (formerly *Ehrlichia phagocytophila*) infection in sheep and wild ruminants in Norway: a study on clinical manifestation, distribution and persistence. Doctor Philosophiae thesis, Norwegian College of Veterinary Medicine.

62. ALEKSEEV, A.N., H.V. DUBININA, I. VAN DE POL & L.M. SCHOULS. 2001. Identification of *Ehrlichia* and *Borrelia burgdorferi* species in Ixodes ticks in the Baltic regions of Russia. J. Clin. Microbiol. **39:** 2237–2242.
63. OGDEN, N.H., A.N.J. CASEY, N.P. FRENCH, *et al.* 2002. Natural *Ehrlichia phagocytophila* transmission coefficients from sheep "carriers" to *Ixodes ricinus* ticks vary with the numbers of feeding ticks. Parasitology **124:** 127–136.
64. GUY, E.S., S. TASKER & D.H. JOYNSON. 1998. Detection of the agent of human granulocytic ehrlichiosis (HGE) in UK ticks using polymerase chain reaction. Epidemiol. Infect. **121:** 681–683.
65. VON STEDINGK, L.V., M. GURTELSCHMIDT, H.S. HANSON, *et al.* 1997. The human granulocytic ehrlichiosis agent in Swedish ticks. Clin. Microbiol. Infect. **3:** 573–574.
66. ALBERDI, M.P., A.R. WALKER, E.A. PAXTON & K.J. SUMPTION. 1998. Natural prevalence of infection with *Ehrlichia* (*Cytoecetes*) *phagocytophila* of *Ixodes ricinus* ticks in Scotland. Vet. Parasitol. **78:** 203–213.
67. WOLDEHIWET, Z. & G.R. SCOTT. 1982. Immunological studies on tick-borne fever in sheep. J. Comp. Pathol. **92:** 457–467.
68. FOGGIE, A. 1951. Studies on the infectious agent of tick-borne fever. J. Path. Bacteriol. **63:** 1–15.
69. MUNRO, R., A.R. HUNTER, G. MACKENZIE & D.A. MCMARTIN. 1982. Pulmonary lesions in sheep following experimental infection by *Ehrlichia phagocytophila* and *Chlamydia psittaci*. J. Comp. Pathol. **92:** 117–129.
70. LEPIDI, H., J.E. BUNNELL, M.E. MARTIN, *et al.* 2000. Comparative pathology and immunohistology associated with clinical illness after *Ehrlichia phagocytophila*-group infections. Am. J. Trop. Med. Hyg. **62:** 29–37.
71. SNODGRASS, D.R. 1974. Studies on bovine petechial fever and ovine tick-borne fever. Ph.D. thesis, University of Edinburgh, Edinburgh, Scotland.
72. MCEWEN, A.D. 1947. Tick-borne fever in young lambs. Vet. Rec. **59:** 198–201.
73. FOGGIE, A. & C.J. ALLISON. 1960. A note on the occurrence of tick-borne fever in cattle in Scotland with comparative studies of bovine and ovine strains of the organism. Vet. Rec. **72:** 767–770.
74. BRODIE, T.A., P.H. HOLMES & G.M. URQUHART. 1986. Some aspects of tick-borne fever in British sheep. Vet. Rec. **118:** 415–418.
75. JAMIESON, S. 1950. Tick-borne fever as a cause of abortion in sheep. Vet. Rec. **62:** 468–470.
76. STAMP, J.T. & J.A. WATT. 1950. Tick-borne fever as a cause of abortion in sheep—Part I. Vet. Rec. **62:** 465–468.
77. LITTLEJOHN, A.I. 1950. Tick-borne fever as a cause of abortion in sheep. Vet. Rec. **62**: 577–579.
78. JONES, G.L. & I.H. DAVIES. 1995. An abortion storm caused by infection with *Cytoecetes phagocytophila*. Vet. Rec. **136:** 127.
79. GARCIA-PEREZ A.L., J. BARANDIKA, B. OPORTO, *et al.* 2003. *Anaplasma phagocytophila* as an abortifacient agent in sheep farms from northern Spain. Ann. N.Y. Acad. Sci. **990:** 429–432.
80. GILMOUR, N.J.L., T.A. BRODIE & P.H. HOLMES. 1982. Tick-borne fever and pasteurellosis in sheep. Vet. Rec. **111:** 512.
81. OVERAS, J., A. LUND, M.J. ULVUND & H. WALDELAND. 1993. Tick-borne fever as a possible predisposing factor in septicaemic pasteurellosis in lambs. Vet. Rec. **133:** 398.

82. GRONSTOL, H. & M.J. ULVUND. 1977. Listeric septicaemia in sheep associated with tick-borne fever (*Ehrlichiosis ovis*). Acta Vet. Scand. **18:** 575–577.
83. GRONSTOL, H. & J. OVERAS. 1980. Listeriosis in sheep: tick-borne fever used as a model to study predisposing factors. Acta Vet. Scand. **21:** 533–545.
84. SAVILL, J.S., A.H. WYLLIE, J.E. HENSON, *et al.* 1989. Macrophage phagocytosis of aging neutrophils in inflammation. J. Clin. Invest. **83:** 865–875.
85. AKGUL, C., D.A. MOULDING & S.W. EDWARDS. 2001. Molecular control of neutrophil apoptosis. FEBS Lett. **487:** 318–322.
86. FOSTER, W.N.M. & A.E. CAMERON. 1970. Observations on the functional integrity of neutrophil leukocytes infected with tick-borne fever. J. Comp. Pathol. **80:** 487–491.
87. WOLDEHIWET, Z. & G.R. SCOTT. 1982. Tick-borne fever: leukocyte migration inhibition. Vet. Microbiol. **7:** 437–455.
88. WOLDEHIWET, Z. 1987. The effects of tick-borne fever on some functions of polymorphonuclear cells of sheep. J. Comp. Pathol. **97:** 481–485.
89. FOGGIE, A. 1956. The effects of tick-borne fever on the resistance of lambs to staphylococci. J. Comp. Pathol. **66:** 278–285.
90. BATUNGBACAL, M.R. 1995. Effect of *Cytoecetes phagocytophila* on phagocytosis by neutrophils in sheep. Phil. J. Vet. Med. **32:** 70–76.
91. WHIST, S.K., A.K. STORSET & H.J.S. LARSEN. 2002. Functions of neutrophils in sheep experimentally infected with *Ehrlichia phagocytophila*. Vet. Immunol. Immunopathol. **86:** 183–193.
92. WOLDEHIWET, Z., C. DARE C. & S.D. CARTER. 1993. The effects of tick-borne fever on the susceptibility of polymorphonuclear cells of sheep to *Pasteurella haemolytica* cytotoxin. J. Comp. Pathol. **109:** 303–307.
93. GOKCE, H.I., G. ROSS & Z. WOLDEHIWET. 1999. Inhibition of phagosome-lysosome fusion in ovine polymorphonuclear leucocytes by *Ehrlichia (Cytoecetes) phagocytophila*. J. Comp. Pathol. **120:** 369–381.
94. SCAIFE, H., Z. WOLDEHIWET, C.A. HART & S.W. EDWARDS. 2003. *Anaplasma phagocytophilum* reduces neutrophil apoptosis *in vivo.*. Infect. Immun. **71:** 1995–2003.
95. WHIST, S.K., A.K. STORSET, G.M. JOHANSEN & H.J. LARSEN. 2003. Modulation of leukocyte populations and immune responses in sheep experimentally infected with *Anaplasma* (formerly *Ehrlichia*) *phagocytophilum*. Vet. Immunol. Immunopathol. **94:** 163–175.
96. LARSEN, H.J.S., G. OVERAS, H. WALDELAND & G.M. JOHANSEN 1994. Immunosuppression in sheep experimentally infected with *Ehrlichia phagocytophila*. Res. Vet. Sci. **56:** 216–224.
97. GOKCE, I.H. 1998. Studies on the effects of *Ehrlichia* (*Cytoecetes*) *phagocytophila* on some cellular immune responses in sheep. Ph.D. thesis, University of Liverpool, Liverpool, England.
98. WOLDEHIWET, Z. 1987. Depression of lymphocyte responses to mitogens in sheep infected with tick-borne fever. J. Comp. Pathol. **97:** 637–643.
99. WOLDEHIWET, Z. 1991. Lymphocyte subpopulations in peripheral blood of sheep infected with tick-borne fever. Res. Vet. Sci. **51:** 40–43.
100. KLEIN, M.B., S. HU, C.C. CHAO & J.L. GOODMAN J.L. 2000. The agent of human granulocytic ehrlichiosis induces the production of myelosuppressing chemokines without induction of proinflammatory cytokines. J. Infect. Dis. **182:** 200–205.

101. TUOMI, J. 1967. Experimental studies on bovine tick-borne fever. 2. Difference in virulence of strains of cattle and sheep. Acta Pathol. Microbiol. Scand. **70:** 577–589.
102. TUOMI, J. 1967. Experimental studies on bovine tick-borne fever. 3. Immunological strain differences. Acta. Pathol. Microbiol. Scand. **71:** 89–100.
103. WOLDEHIWET, Z. & G.R. SCOTT. 1982. Differentiation of strains of *Cytoecetes phagocytophila*, the causative agent of tick-borne fever, by complement fixation. J. Comp. Pathol. **92:** 475–478.
104. GOKCE, H.I. & Z. WOLDEHIWET. 1999. Lymphocyte responses to mitogens and rickettsial antigens in sheep experimentally infected with *Ehrlichia (Cytoecetes) phagocytophila*. Vet. Parasitol. **83:** 55–64.
105. FRENCH, D.M., W.C. BROWN & G.H. PALMER. 1999. Emergence of *Anaplasma marginale* antigenic variants during persistent rickettsemia. Infect. Immunol. **67:** 5834–5840.
106. ZHI, N., N. OHASHI & Y. RIKIHISA. 1999. Multiple p44 genes encoding major outer membrane proteins are expressed in the human granulocytic ehrlichiosis agent. J. Biol. Chem. **274:** 17828–17836.
107. OGDEN, N.H., A.N.J. CASEY, C. LAWRIE, *et al.* 2002. IgG responses to salivary gland extract of *Ixodes ricinus* ticks vary inversely with resistance in naturally exposed sheep. Med. Vet. Entomol. **16:** 186–192.
108. OGDEN, N.H., A.N.J. CASEY, Z. WOLDEHIWET & N.P. FRENCH. 2003. Transmission of *Anaplasma phagocytophilum* to *Ixodes ricinus* ticks from sheep in the acute and post-acute phase of infection. Infect. Immun. **71:** 2071–2078.
109. VENN, J.A.J. & M.H. WOODFORD. 1956. An outbreak of tick-borne fever in bovines. Vet. Rec. **68:** 132–133.
110. GOKCE, H.I. & Z. WOLDEHIWET. 1999. Differential haematological effects of tick-borne fever in sheep and goats. J. Vet. Med. (B) **46:** 105–115.
111. FOSTER, W.N.M. & A.E. CAMERON. 1969. Thrombocytopaenia in sheep associated with experimental tick-borne fever infection. J. Comp. Pathol. **78:** 251–254.
112. VAN MIERT, A.S.J.P.A.M., C.T.M. VAN DUIN, A.J.H. SCHOTMAN & F.F. FRANSSEN. 1984. Clinical, haematological, and blood biochemical changes in goats after experimental infection with TBF. Vet. Parasitol. **16:** 225–233.
113. BRUN-HANSEN, H., H. GRONSTOL & F. HARDING. 1998. Experimental infection with *Ehrlichia phagocytophila* in cattle. J. Vet. Med. (B) **45:** 193–203.
114. GOKCE, H.I. & Z. WOLDEHIWET. 1999. The effects of *Ehrlichia* (*Cytoecetes*) *phagocytophila* on the clinical chemistry of sheep and goats. J. Vet. Med. (B) **46:** 93–103.
115. WILSON, J.E., A. FOGGIE & M.A. CARMICHAEL. 1964. Tick-borne fever as a cause of abortion and still-births in cattle. Vet. Rec. **76:** 1081–1084.
116. TAYLOR, A.W., H.H. HOLMAN & W.S. GORDON. 1941. Attempts to reproduce the pyemia associated with tick bite. Vet. Rec. **53:** 339–344.
117. WOLDEHIWET, Z., B.K. HORROCKS, H. SCAIFE, *et al.* 2002. Cultivation of an ovine strain of *Ehrlichia phagocytophila* in tick cell culture. J. Comp. Pathol. **127:** 142–147.
118. WEBSTER, K.A. & G.B.B. MITCHELL. 1988. Use of counter immunoelectrophoresis in the detection of antibodies to tick-borne fever. Res. Vet. Sci. **45:** 28–30.
119. PAXTON, E.A. & G.R. SCOTT. 1989. Detection of antibodies to the agent of tick-borne fever by indirect immunofluorescence. Vet. Microbiol. **21:** 133–138.

120. WOLDEHIWET, Z. & B.K. HORROCKS. 2005. Antigenicity of ovine strains of *Anaplasma phagocytophilum* grown in tick cells and ovine granulocytes. J. Comp. Pathol. **132:** 322–328.
121. DE LA FUENTE, J., R.F. MASSUNG, S.J. WONG, *et al.* 2005 Sequence analysis of the *msp4* gene of *Anaplasma phagocytophilum* strains. J. Clin. Microbiol. **43:** 1309–1317.
122. WATSON, W.A., P.R.M. BROWN & J.C. WOOD. 1966. The control of staphylococcal infection (tick pyemia) in lambs by dipping. Vet. Rec. **79:** 101–103.
123. WATT, J.A., W.N.M. FOSTER & A.E. CAMERON. 1968. Benzathine penicillin as a prophylactic in tick pyemia. Vet. Rec. **83:** 507–508.

Epidemiological Survey of *Ehrlichia canis* and Related Species Infection in Dogs in Eastern Sudan

HISASHI INOKUMA,[a] MAREMICHI OYAMADA,[b] BERNARD DAVOUST,[c] MICKAËL BONI,[d] JACQUES DEREURE,[e] BRUNO BUCHETON,[f] AWAD HAMMAD,[g] MALAIKA WATANABE,[b] KAZUHITO ITAMOTO,[b] MASARU OKUDA,[b] AND PHILIPPE BROUQUI[h]

[a]*Obihiro University of Agriculture and Veterinary Medicine, 080-8555 Obihiro, Japan*

[b]*Faculty of Agriculture, Yamaguchi University, 753-8515 Yamaguchi, Japan*

[c]*Direction du Service de Santé en Region Sud-Est, BP16, 69998 Lyon Armeés, France*

[d]*Groupe de Secteurs Vétérinaires Interarmees de Saint Germain en Laye, BP220, 00492 Saint Germain en Laye Armeés, France*

[e]*Laboratoire de Parasitologie, Faculté de Médecine, 34090 Montpellier, France*

[f]*Laboratoire d'Immunologie Parasitaire, Faculté de Médecine, Université de la Méditerranée, Marseille Cedex 5, France*

[g]*Departement of Microbiology and Parasitology, Faculty of Medecine, University of Khartoum, P.O. Box 102, Khartoum, Republic of the Sudan*

[h]*Unité des Rickettsies, Faculté de Médecine, Université de la Méditerranée, Marseille Cédex 5, France*

ABSTRACT: The infection rates of *Ehrlichia canis* and related species in dogs in eastern Sudan were examined using molecular methods. Among 78 dogs examined, 63 (80.8%), 19 (24.4%), and 26 (33.3%) were positive for *E. canis*, *Anaplasma platys*, *Mycoplasma haemocanis*, and "*Candidatus* Mycoplasma haemoparvum," respectively. Among these, 30 dogs were single-positive: 25 for *E. canis*, 2 for *A. platys*, 1 for M. *hemocanis*, and 2 for "*C.* M. haemoparvum." The rest of the dogs (48.7%) were positive for two or more pathogens.

KEYWORDS: *Ehrlichia canis*; *Anaplasma platys*; *Mycoplasma haemocanis*; "*Candidatus* Mycoplasma haemoparvum"; epidemiology; molecular survey

Address for correspondence: Hisashi Inokuma, Obihiro University of Agriculture and Veterinary Medicine, 080-8555 Obihiro, Japan. Voice/fax: +81-155-49-5370.
e-mail: inokuma@obihiro.ac.jp

**Ann. N.Y. Acad. Sci. 1078: 461–463 (2006). © 2006 New York Academy of Sciences.
doi: 10.1196/annals.1374.085**

TABLE 1. Infection rates of *Ehrlichia canis* and related species in dogs in Sudan examined by PCR

	Numbers of dogs among 78
E.canis	63
A.platys	19
M.haemocanis	7
"*C.*M.haemoparuvum"	26
1. Single positive	
E. canis only	25
A. platys only	2
M. haemocanis only	1
"*C.* M.haemoparvum" only	2
2. Dual positive	
E. canis + *A. platys*	10
E. canis + *M. haemocanis*	4
E. canis + "*C.*M. haemoparvum"	16
Other combination	0
3. Triple positive	
E. canis + *A. platys* + "*C.*M. haemoparuvum"	6
E. canis + *M. haemocanis* + "*C.* M.haemoparuvum"	1
Other combination	0
4. All positive	1

Ehrlichioses are emerging tick-borne diseases in both humans and animals. *Ehrlichia canis (E. canis)* is the best-known pathogen that causes canine ehrlichioses. The agent has a worldwide distribution and is transmitted by *Rhipicephalus sanguineus (R. sanguineus)*. The tick species is also suspected as a vector of *Anaplasma platys (A. platys)* and hemoplasma. Reports of such tick-borne infection are mainly from the USA and Europe, but little information is available on canine ehrlichiosis in African countries. And most of epidemiological studies on *Ehrlichia* infection in African countries are performed by using serological tests. [1,6,7] Recently, molecular techniques, including the polymerase chain reaction (PCR) and sequence analysis, have been used for the epidemiological study of canine ehrlichial infections. [2,3] The advantages of the molecular method over other techniques are its higher sensitivity and specificity in the detection of the target pathogens in peripheral blood. Thus the objective of this study was to clarify the infection rates of *E. canis* and related species in dogs in Sudan using molecular methods.

DNA was extracted from the peripheral blood of 78 randomly selected dogs in a village in eastern Sudan from 1997 to 2000. These dogs were allfree-roaming. Detection of *E. canis, A. platys,* and hemoplasma (*Mycoplasma haemocanis* [*M. haemocanis*] and "*Candidatus* Mycoplasma haemoparvum") was attempted using species-specific nested PCR based on the 16S rRNA gene with the method described previously. [3,4] Ticks were removed from the dogs for identification.

The results are shown in TABLE 1. Among 78 dogs examined, 68 (87.2%) dogs were positive for one or more of the pathogens. A total of 63 (80.8%),

19 (24.4%), 7 (9.0%), and 26 (33.3%) dogs were positive for *E. canis, A. platys, M. haemocanis,* and "*C.* M. haemoparvum," respectively. Among these, 30 dogs were single positive; 25 for *E. canis,* 2 for *A. platys*, 1 for *M. haemocanis,* and 2 for "*C.* M. haemoparvum." The rest of the 38 dogs (48.7%) were positive for two or more pathogens. Ten dogs were dually positive for *E. canis* and *A. platys*, 4 for *E. canis* and *M. haemocanis*, and 16 for *E. canis* and "*C.* M. haemoparvum." Another 6 dogs were triplly positive for *E. canis, A. platys* and "*C.* M. haemoparvum," and 1 for *E. canis, M. haemocanis* and "*C.* M. haemoparuvum." The remaining dog was positive for all the pathogens examined. *R. sanguineus* was the most dominant tick species and was detected in 44 dogs, followed by *Amblyomma lepidum (A. lepidum)* (4 dogs) and *Rhipicephalus evertsi evrtsi (R. evertsi evrtsi)* (3 dogs). The infection rates of dogs with *E. canis* in Sudan were extremely higher than those reported in other countries previously. Nearly half the dogs were found to be infected with more than two pathogens, including *E. canis, A. platys,* and hemoplasma. Coinfection of multiple tick-borne pathogens is usually found in dogs. [5,8,9] However, it is the first evidence of coinfection with *E. canis* and hemoplasma, such as *M. haemocanis* and "*C.* M. haemoparvum" in dogs. This coinfection may cause more severe pathogenesis in dogs than single infection. *R. sanguineus* may be responsible for the coinfection. The fact that the agents are potentially zoonotic pathogens should be noted.

REFERENCES

1. BROUQUI, P. *et al.* 1991. Serological evaluation of *Ehrlichia canis* infections in military dogs in Africa and Reunion Island. Vet. Microbiol. **26:** 103–105.
2. INOKUMA, H. *et al.* 2001. Detection of ehrlichial infection by PCR in dogs from Yamaguchi and Okinawa Prefectures, Japan. J. Vet. Med. Sci. **63:** 815–817.
3. INOKUMA, H. *et al.* 2003. Epidemiological survey of *Anaplasma platys* and *Ehrlichia canis* using ticks collected from dogs in Japan. Vet. Parasitol. **115:** 343–348.
4. INOKUMA, H. *et al.* 2004. Molecular survey of *Mycoplasma haemofelis* and '*Candidatus* Mycoplasma haemofelis' in cats in Yamaguchi and surrounding area. J. Vet. Med. Sci. **66:** 1017–1020.
5. KORDICK, S.K. *et al.* 1999. Coinfection with multiple tick-borne pathogens in a Walker hound kennel in North Carolina. J. Clin. Microbiol. **37:** 2631–2638.
6. MATTHEWMAN, L.A. *et al.* 1993. Western blot and indirect fluorescent antibody testing for antibodies reactive with *Ehrlichia canis* in sera from apparently healthy dogs in Zimbabwe. J. S. Afr. Vet. Assoc. **64:** 111–115.
7. PARZY, D. *et al.* 1991. Canine ehrlichiosis in Senegal: human and canine seroepidemiological survey in Dakar. Med. Trop. **51:** 59–63.
8. SUKSAWAT, J. *et al.* 2001. Coinfection with three *Ehrlichia* species in dogs from Thailand and Venezuela with emphasis on consideration of 16S ribosomal DNA secondary structure. J. Clin. Microbiol. **39:** 90–93.
9. SUKSAWAT, J. *et al.* 2001. Serological and molecular evidence of coinfection with multiple vector-borne pathogens in dogs from Thailand. J. Vet. Intern. Med. **15:** 453–462.

Surveys on Seroprevalence of Canine Monocytic Ehrlichiosis among Dogs Living in the Ivory Coast and Gabon and Evaluation of a Quick Commercial Test Kit Dot-ELISA

BERNARD DAVOUST,[a] OLIVIER BOURRY,[b] JOSÉ GOMEZ,[c] LAURENT LAFAY,[d] FANNY CASALI,[d] ERIC LEROY,[b,e] AND DANIEL PARZY[d]

[a]*Direction Régionale du Service de Santé des Armées, BP 16, 69998 Lyon Armées, France*

[b]*Centre International de Recherches Médicales, BP 769, Franceville, Gabon*

[c]*Vétérinaire Indépendant, BP 178, Abidjan 06, Côte d'Ivoire*

[d]*Institut de Médecine Tropicale du Service de Santé des Armées, BP 46, 13998 Marseille Armées, France*

[e]*Institut de Recherche pour le Développement, UR 034, Paris, France*

ABSTRACT: Canine monocytic ehlichiosis (CME), an enzootic disease in Africa, has been studied in canine blood samples (serum). These dogs, without any clinical sign of disease, were living in Abidjan (Ivory Coast) and in several small villages located in northeasst Gabon (Ogooué Ivindo). The results obtained by indirect fluorescent antibody (IFA) test, used as a point of reference, and by a quick test dot-Elisa were compared. Blood samples taken from 390 asymptomatic dogs in 2003 (137 in Ivory Coast and 253 in Gabon) were screened by IFA (antigen from Symbiotics Europe, Lyon) with a positive threshold set at 1/80. Afterwards, CME was detected by the commercial test kit Dot-Elisa in solid phase Snap 3Dx (Idexx, Westbrook, Maine, USA), using recombinant proteins which belong to *Ehrlichia canis*, p30 and p30-1. Using the IFA test, CME seroprevalence in the Ivory Coast is found to be 67.8%. Among 93 Ivorian seropositive blood samples, 76 samples show an antibody titer >1/2560. In Gabon, IFA showed that seroprevalence is only 3.1%. Among 8 seropositive Gabonese dogs, only one sample shows an antibody test titer> 1/2560. Results from the Snap 3Dx test used on 390 blood samples are 100 positive samples and 290 negative ones. Comparison between IFA and Snap test 3DX revealed that the Snap test shows 97.9% specificity,

Address for correspondence: Bernard Davoust, Direction Régionale du Service de Santé des Armées, BP 80, 83800. Toulon Armées, France. Voice: 33-6-82-67-91-83; fax: 33-4-72-00-54-88.
e-mail: bernard.davoust@mageos.com [or] casali.fanny@voila.fr

Ann. N.Y. Acad. Sci. 1078: 464–469 (2006). © 2006 New York Academy of Sciences.
doi: 10.1196/annals.1374.086

93.1% sensitivity, a 94% positive predictive value, a 97.6%, negative predictive value, and 96.6% reliability. In conclusion, CME seroprevalence in Abidjan is very high. Dogs studied for CME were watchdogs, living in kennels, where infection transmitted by *Rhipicephalus sanguineus* seems to be higher in the Gabonese area called Ogooué Ivindo, where semi-stray dogs were subjected to the test. These dogs where carried ticks identified as *Haemaphysalis leachi*, but this kind of tick is not considered as bearing *Ehrlichia canis*. Finally, results of the Snap 3Dx show that it is a simple and reliablemeans for quickly detecting dogs suspected of asymptomatic canine ehrlichiosis.

KEYWORDS: *Ehrlichia canis*; ehrlichiosis; dogs; seroprevalence; dot-ELISA; IFA; Gabon; Ivory Coast

INTRODUCTION

Canine monocytic ehrlichiosis (CME) is a serious infectious disease due to *Ehrlichia canis*. The infection naturally occurs through bites of the brown dog tick *Rhipicephalus sanguineus*. This cosmopolite vector is endophilic and frequently seen in kennels and wall crevices in areas located below 45 degrees latitude.

CME, which is an enzootic disease in Africa, has been researched in dogs' blood samples (serum). The results obtained by indirect fluorescent antibody (IFA) test, used as a point of reference, and by a quick test dot-enzyme-linked immunosorbent assays (ELISA) were compared. [2]

MATERIALS AND METHODS

Dogs

In 2003, 390 dogs were included in this epidemiological survey carried out in two African countries:

1. In the Ivory Coast, 137 dogs belonging to surveillance companies in 16 kennels in Abidjan were studied.
2. In northeast Gabon, 253 dogs belonging to the inhabitants of 15 different localities in the Ogooué-Ivindo region were studied.

All dogs were apparently in good health.

Sample Collection and Pretreatment

A blood sample from the radial vein provided serum that was subsequently deep frozen (–20°C) and transported to a laboratory.

IFA

The use of IFA enabled us to test in accordance with the reference method. The antigen, supplied by the Synbiotics Europe (Lyon), was made up of bacteria grown on MDH cells and formolized. At least 50% of the cells were infected. They were placed on immunofluorescent slides and fixed in cold acetone for 20 min. The technique is classic, with a 20-fold predilution of the serum. Examination was done with an ultraviolet microscope at ×400. The presence of anti-*Ehrlichia* antibodies is detected by a clear fluorescence of intracytoplasmic morulae. A negative serum is either not fluorescent or the entire cytoplasm is only slightly fluorescent. Titers of ≥1/80 were considered to be positive.

Kit Dot-ELISA

Afterward, CME was detected by the commercial test kit dot-ELISA in solid phase Snap 3Dx® (Idexx Laboratories Inc., Westbrook, ME, USA), using recombinant synthetic peptides that belong to *E. canis* p30 and p30-1. To determine the test result the reaction spots are read in the result window. Color (blue) development in the sample spots indicates the presence of *E. canis* antibody in the sample.

RESULTS

Using the IFA test, CME seroprevalence in the Ivory Coast is 67.8% (TABLE 1). Among 93 Ivorian seropositive blood samples, 76 show an antibody titer >1/2 560.

In Gabon, using the IFA test, seroprevalence found is only 3.1%. Among eight seropositive Gabonese dogs, only one sample shows an antibody titer > 1/2 560.

Results from the test Snap 3Dx ® used on 390 blood samples are 100 positive and 290 negative. Comparison (TABLE 2) between IFA and SNAP test 3Dx ® revealed that the snap test shows:

(1) specificity of 97.9% (confidence limits 95.5%–99.2%);
(2) sensitivity of 93.1% (confidence limits 86.2%–97.2%);
(3) positive predictive value of 94%;
(4) negative predictive value of 97.6 %; and
(5) reliability of 96.6%.

The correlation rate is 96.7%.

The test Snap 3Dx ® detected 99% of IFA-positive samples with IFA titers > 1/160.

TABLE 1. Serological results (IFA) of detection of canine monocytic ehrlichiosis in Ivory Coast (Abidjan) and Gabon (Ogooué-Ivindo region)

		Serological Results (IFA) and Antibody Titer									
Location	Number of Dogs	Negative < 1/80	Positive ≥ 1/80	Seroprevalence (%)	1/80	1/ 160	1/ 320	1/ 640	1/ 1 280	1/ 2 560	≥ 1/ 5 120
Ivory Coast (Abidjan)	137	44	93	67,8		8		4	5	6	70
Gabon (Ogooué-Ivindo)	253	245	8	3,1	1	4	1	1		1	
Total	390	289	101	34,9	1	12	1	5	5	7	70

TABLE 2. Comparison of serological assays (IFA and Snap 3Dx ®) to detect *E. canis* antibodies in 390 dogs

		IFA	
		Positive (≥ 1/80)	Negative
Snap 3D x®	Positive	94	6
	Negative	7[a]	283

[a] 1/80 (1), 1/160 (5), 1/640 (1).

CONCLUSION

CME seroprevalence in Abidjan is very high. Dogs studied for CME were watchdogs, living in kennels, where infection transmitted by *R. sanguineus* seems to be higher in the Gabonese area called Ogooué-Ivindo-where semi-stray dogs were subjected to the test. The last mentioned dogs carried ticks identified as *Haemaphysalis leachi*, but this kind of tick is not considered as bearing *E. canis*.

Our evaluation of the dot-ELISA kit confirms its performance. It is faster and easier to use than the IFA "gold standard." Its sensitivity of 93.1% was more effective in our study than in previous trials carried out on 97 serum samples: (89.8%) and on 67 serum samples (71%). [1,3]

This test is recommended for epidemiological surveys. It is possible to compare the results of different surveys. Thus, with this test, seroprevalence is considerably higher in Abidjan than that observed in Mexico (Yucatan)—44.1% (53/120)—and in Brazil: 19.8% (505/2 553). [4,5]

Finally, results of the test Snap 3Dx ® show that it is a simple and reliable means for quickly detecting dogs suspected of asymptomatic canine ehrl ichiosis.

ACKNOWLEDGMENTS

The authors want to thank especially Elisabeth Carme from the Institut de Médecine Tropicale du Service de Santé des Armées and Philippe Parola from the Unité des Rickettsies of Faculté de Médecine de Marseille.

The authors gratefully acknowledge the excellent technical assistance of Loïs Allela and André Délicat at CIRMF. CIRMF is supported by the Government of Gabon, Total-Fina-Elf Gabon, and the Ministère de la Coopération Française. This work was also supported by a Fonds de Solidarité Prioritaire grant from the Ministère des Affaires Etrangères de la France (FSP No. 2002005700).

REFERENCES

1. BELANGER, M., H. SORENSON, M. FRANCE, *et al.* 2002. Comparison of serological detection methods for diagnosis of *Ehrlichia canis* infections in dogs. J. Clin. Microbiol. **40:** 3506–3508.
2. CADMAN, H.S., P.J. KELLY, L.A. MATTHEMAN, *et al.* 1994. Comparison of the dot-blot enzyme-linked immunoassay with immunofluorescence for detecting antibodies to *Ehrlichia canis*. Vet. Rec. **135:** 362.
3. HARRUS, S., A.R. ALLEMAN, H. BARK, *et al.* 2002. Comparison of three enzyme-linked immunosorbent assays with the indirect immunofluorescent antibody test for the diagnosis of canine infection with *Ehrlichia canis*. Vet. Microbiol. **39:** 361–368.
4. LABARTHE, N., M. DE CAMPOS PEREIRA, *et al.* 2003. Serologic prevalence of *Dirofilaria immitis, Ehrlichia canis,* and *Borrelia burgdorferi* infections in Brazil. Vet. Ther. **4:** 67–75.
5. RODRIGUEZ-VIVAS, R.I., R.E. ALBORNOZ & G.M. BOLIO. 2005. *Ehrlichia canis* in dogs in Yucatan, Mexico : seroprevalence, prevalence of infection and associated factors. Vet. Parasitol. **127:** 75–79.

Experimental Infections in Dogs with *Ehrlichia canis* Strain Borgo 89

STÉPHANIE JOURET-GOURJAULT,[a] DANIEL PARZY,[a] AND BERNARD DAVOUST[b]

[a]*Institut de Médecine Tropicale du Service de Santé des Armées, 13998 Marseille Armées, France*

[b]*Direction Régionale du Service de Santé des Armées, BP 16, 69998 Lyon Armées, France*

ABSTRACT: This study attemps to clarify the virulence and the pathogenicity of the Borgo 89 strain of *Ehrlichia canis* isolated from a sick dog in Corsica (France). Four unscathed beagles were intravenously injected with an inoculum of leukocytes infected with the Borgo 89 strain and the animals were examined daily for clinical signs of disease, and blood samples were drawn at frequent intervals for biochemical and hematologic assessment. Serologic (IFI) and PCR assays were also carried out. The results at autopsy are presented in this paper, leading to the conclusion that the Borgo 89 strain has a pathogenicity comparable to that of the known strains. However, the discovery of a case of completely unapparent infection raises the question of a possible individual immunization whose origin remains unexplained.

KEYWORDS: ehrlichiosis; *Ehrlichia canis*; dogs; experimental infections; France

INTRODUCTION

Canine monocytic ehrlichiosis (CME) is a serious infectious disease caused by a gram-negative obligate intracellular bacterium with a tropism for monocytes and macrophages: *Ehrlichia canis (E. canis)*. Discovered more than 70 years ago, this pathogenic agent is widely located geographically as is its vector, the brown tick *Rhipicephalus sanguineus*. Classically, three clinical stages have been differentiated in dogs experimentally infected with *E. canis*: acute, subclinical, and chronic stage.[1] The disease can pass undetected during one of its three successive clinical stages, but it can also be fatal in its acute stage. Clinical signs usually described in CME are fever, depression, anorexia,

Address for correspondence: Bernard Davoust, Direction Régionale du Service de Santé des Armées, BP 16, 69998 Lyon Armées, France. Voice: 33-6-82-67-91-83; fax: 33-4-72 00-54-88.
e-mail: bernard.davoust@mageos.com

Ann. N.Y. Acad. Sci. 1078: 470–475 (2006).
doi: 10.1196/annals.1374.132

weight loss, hemorrhage, gastrointestinal signs such as vomiting or diarrhea, and respiratory disorders and ocular signs.

Canine ehrlichiosis is a common disease in France described essentially in the Mediterranean southeast with a lot of prospective surveys, also in the Rhone Valley, in the southwest and in few temperate regions, such as Lyons or Central Massif. In Borgo, a town on the French Corsica island, an outbreak of CME occurred at a canine military kennel in 1989.[2] This strain of *E. canis* has been contained since that time. It is the first and only insulation of *E. canis* to date carried out in France for study.

The objective of this study was to clarify the virulence and the pathogenicity of this strain with experimental infections in dogs.

MATERIALS AND METHODS

Four unscathed beagle dogs were intravenously injected on day (D) 0 with an inoculum of leukocytes infected with the strain Borgo 89 of *E. canis* isolated from a sick dog in Corsica (France) in 1989. A routine daily physical and clinical examination with recording of appetite, weight, rectal temperature, and any modifications of the general state was carried out, and blood samples drawn for hematological and biochemical analysis (daily between D 10 and D21 and weekly between D0 and D68). Infection with *E. canis* was confirmed by PCR based on the 16S rRNA gene and serology.[3] The presence of specific antibodies against *E. canis* was determined by IFAT.[4] After autopsy of the sick dogs, systematic biopsies of liver, spleen, lungs, kidneys, and mesenteric lymph nodes were obtained for PCR detection and anatomopathologic analysis.

RESULTS

Clinical Signs

The four dogs reacted to inoculation: seroconversion between D7 and D14 and positive PCR starting from D7, but only three dogs presented clinical signs, on average, on D11. Hyperthermia was found in the three cases of acute ehrlichiosis (dogs 1,3, and 4) (FIG. 1). Hyperthermia is an early sign that can be the only apparent symptom. Anorexia and asthenia are not constant, nor are hemorrhagic signs, even when thereis a noticiable drop in blood platelets, only internal petechiae, in particular on the digestive walls, could be observed at autopsy. An adenopathy of the lymph nodes, a hepatomegaly, and a splenomegaly were palpable in the three symptomatic dogs. An oculonasal discharge was noted in two dogs.

The degradation of the general state of dog 1 required the oral intake at D21 of doxycycline 10 mg/kg/day for 28 days.[5] The clinical cure was obtained as

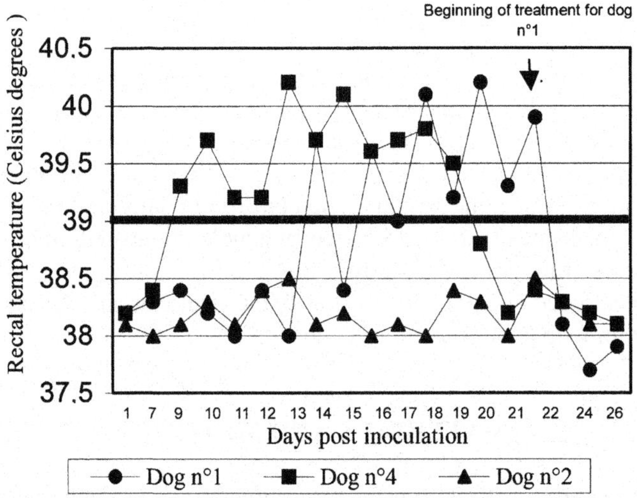

FIGURE 1. Evolution of rectal temperature.

of D30 with return of rectal temperature to that of the first day of treatment and stabilization of the thrombocytes to that of the second day. Dogs 3 and 4 presented chronic infection without treatment.

Dog 2 did not present any symptom in spite of seroconversion and positivity of detection by PCR on an example of a case of bacterial dissemination with no clinical signs, hematologic or inflammatory.

Biological Abnormalities

From a hematologic point of view, pancytopenia is not constant, but thrombocytopenia was present in the three symptomatic dogs (FIG. 2). Thrombocytopenia constitutes the referential biological criterion most constant in the development of the initial infection with *E. canis.*[6] Anemia is also found in our study. Leukopenia is found in dog 1 at the time of the chronic phase of the disease.

Biochemical Modifications

The results of biochemical examinations (hyperbetaglobulinemia and rise in transaminases ALAT and ASAT) and anatomopathologic examination (plasmocytosis, hyperplasy of the cells of Küpfer) indicate that the acute infection in the origin of an inflammatory hepatic attack.[6] The transitory rise in transaminases translates a hepatic necrosis The hypoalbuminemia is described in the three symptomatic dogs and is of renal origin by escape because of lesions of

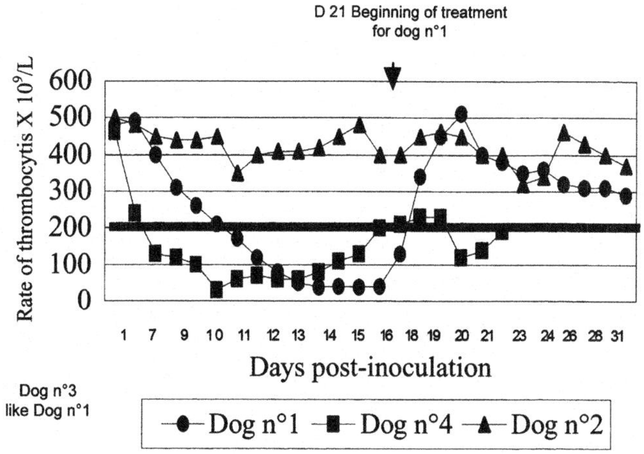

FIGURE 2. Evolution thrombocythemric.

interstitial nephritis as in dog n°3 or of membrane glomerulonephritis as in dog n°4.

Detection by Serologic Study and PCR

Specific antibodies against *E. canis* appeared in serum between D7 and D14 for the four dogs (FIG. 3). They remain high for dogs n°3 and n°4 in chronic ehrlichiosis. Seventeen days after the end of treatment of dog n°1 serologic titer fell by two dilutions. Dog n°2 presents a seroconversion and its titer falls 24 days after inoculation.

Detection by PCR in blood is positive in D7 for dog n°1 and n°2, D9 for dog n°3, and D14 for dog n°4.[3] Twenty days after the end of treatment of dog n°1, a bacteremia is still detected. Therefore, it can be deduced that the treatment does not allow sterilization. Detection by PCR on postmortem tissues is positive on spleen for the four dogs, and also the lymphatic nodes of dog n°2.

DISCUSSION

E. canis Borgo 89 presents a pathogenicity comparable to the known strains of *E. canis*. Our four cases show well the clinical diversity of expression of the disease, which can be severe (dog n°1), subchronic (dogs n°3 and n°4), or asymptomatic (dog n°2). The severity of natural infection probably depends on the duration of infections, opportunistic and secondary infections. But in our study, the four beagle dogs receive the same infectious amount and do not contract other intercurrent diseases. The discovery of a case of completely

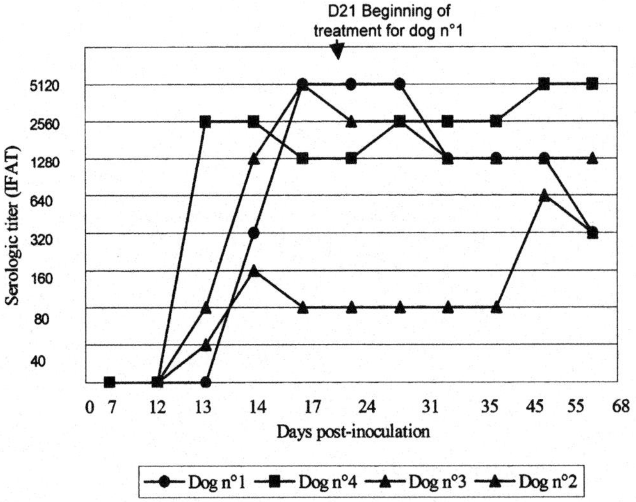

FIGURE 3. Evolution of serologic titer on specific antibodies.

unapparent infection (dog n°2) thus raises the question of a possible individual immunization whose origin remains unexplained. It was the first time that a completely asymptomatic case of canine ehrlichiosis had been described. Thus the prevalence of the natural disease could also be underestimated.

A molecular characterization of this strain is at hand to supplement its description. Its 16S rRNA gene sequence will be aligned with the sequences of other *Ehrlichia* strains. Is it possible to bring it closer to the strain recently isolated in Spain?[7]

ACKNOWLEDGMENTS

The authors want to thank all collaborators on this study, especially Elizabeth Carme and Patrick Barthares from the Institut de Médecine Tropicale du Service de Santé des Armées, Marc Morillon and Michel Chevrier from the Hôpital d'Instruction des Armées Laveran, Thierry Cruel from the Hôpital d'Instruction des Armées Desgenettes, and Jacques Chinchilla from the Secteur Vétérinaire Interarmées de Marseille.

REFERENCES

1. HARRUS, S., T. WANER & T. BARK. 1997. Canine monocytic ehrlichiosis: an update. Comp. Cont. Educ. Pract. Vet. **19:** 431–444.
2. DAVOUST, B., P. BROUQUI, A. RAFFI & D. RAOULT. 1989. L'ehrlichiose canine dans les chenils militaires du Sud-Est: à propos de 14 cas. Point Vet. **21:** 63–67.

3. Mc Bride, J.W., R.E. Corstvet, S.D. Gaunt, *et al.* 1996. PCR detection of acute *Ehrlichia canis* infection in dogs. J. Vet. Diagn. Invest. **8:** 441–447.
4. Neer, T.M., E.B. Breitschwerdt, R.T. Greene, *et al.* 2002. Consensus statement on ehrlichial disease of small animals from the infectious study group of the American College of Veterinary Internal Medicine. J. Vet. Intern. Med. **16:** 309–315.
5. Davoust, B., D. Parzy, E. Vidor, *et al.* 1991. Ehrlichiose canine expérimentale: étude clinique et thérapeutique. Rec. Méd. Vét. **27:** 256–266.
6. Kuehn, N.F. & S.D. Gaunt. 1985. Clinical and hematologic findings in canine ehrlichiosis. J. Am. Vet. Med. Assoc. **186:** 355–358.
7. Aguirre, E., A. Sainz, S. Dunner, *et al.* 2004. First isolation and molecular characterization of *Ehrlichia canis* in Spain. Vet. Parasitol. **125:** 365–372.

Reservoir Competency of Goats for *Anaplasma phagocytophilum*

ROBERT F. MASSUNG,[a] MICHAEL L. LEVIN,[a] NATHAN J. MILLER,[b] AND THOMAS N. MATHER[b]

[a] *Viral and Rickettsial Zoonoses Branch, Centers for Disease Control and Prevention, Atlanta, Georgia, USA*

[b] *Center for Vector-Borne Disease, University of Rhode Island, Kingston, Rhode Island, USA*

ABSTRACT: The susceptibility of goats to infection by *Anaplasma phagocytophilum* (*A. phagocytophilum*) strains Ap-Variant 1 and Ap-ha was assessed by infestation of goats with field-collected *Ixodes scapularis* (*I. scapularis*) ticks. Both strains were infectious in the goat model. These results demonstrate that goats can be used in a laboratory setting to propagate *A. phagocytophilum*. Transmission from an infected goat to a naïve goat by *I. scapularis* tick feeding was used to demonstrate that goats are reservoir-competent for the Ap-Variant 1 strain.

KEYWORDS: *Anaplasma phagocytophilum*; *Ixodes scapularis*; Ap-Variant 1; human granulocytic anaplasmosis

The Ap-Variant 1 strain of *Anaplasma phagocytophilum* (*A. phagocytophilum*) is distinct from strains associated with human disease (Ap-ha strains) and has been detected in *Ixodes scapularis* (*I. scapularis*) ticks in Rhode Island, Connecticut, Maryland, Pennsylvania, and Wisconsin. [1–4] Ap-Variant 1 does not infect mice, and white-tailed deer are likely the primary natural reservoir for Ap-Variant 1. [5] The susceptibility of goats to infection by *A. phagocytophilum* strains Ap-Variant 1 and Ap-ha was assessed by infestation of goats with field-collected *I. scapularis* ticks. Questing adult *I. scapularis* ticks were collected in the spring and fall of 2003 in Rhode Island, at a site where the presence of both Ap-ha and Ap-Variant 1 was detected in prior studies.[3] A representative set of ticks from each collection was individually tested for *A. phagocytophilum* by PCR and DNA sequencing, as previously described, to assess the prevalence of both strains.[4] Ticks were placed in cotton stockinette bags attached to shaved areas on the goat's back or to ears and fed to repletion.

Address for correspondence: Robert F. Massung, Centers for Disease Control and Prevention, 1600 Clifton Rd., MS G-13, Atlanta, GA 30333. Voice: 404-639-1082; fax: 404-639-4436.
e-mail: rfm2@cdc.gov

Ann. N.Y. Acad. Sci. 1078: 476–478 (2006).
doi: 10.1196/annals.1374.088

Goats A and B were each infested with 12 female and 6 male *I. scapularis* from the spring collection in June 2003. Goat C was infested with 6 female ticks from the fall collection in December 2003 and followed by 2 infestations with uninfected laboratory-reared nymphal and larval *I. scapularis* for xenodiagnosis and pathogen propagation at 16 and 24 days, respectively. Four months after molting, the ticks that fed upon goat C were placed on goat E and allowed to feed to repletion. Anticoagulated blood (3.5 mL) collected from goat C at 3 weeks post infection was inoculated by the intravenous route into goat D.

Blood and serum samples were drawn from goats using aseptic techniques. EDTA-treated whole blood samples were collected from each goat and DNA was extracted and tested by PCR for *A. phagocytophilum*. The results of the goat PCR and serological testing are summarized in TABLE 1. Blood samples from goats A and B became PCR-positive on day 11. DNA sequencing of the PCR products showed that both of these goats were infected with the human agent, Ap-ha, which confirmed that goats are susceptible to this agent. In December 2003 goat C was fed upon by 6 female ticks collected from the same site in Rhode Island and the blood of this goat became PCR-positive for *A. phagocytophilum* on day 10 post infestation. DNA sequencing of the PCR products from goat C blood samples identified the infecting agent as Ap-Variant 1. The blood of goat D became PCR-positive on day 7 post injection and DNA sequencing confirmed the infection as Ap-Variant 1. Blood of goat E became PCR-positive on day 5 post infestation, thereby demonstrating the reservoir competency of goats for Ap-Variant 1 and effective tick acquisition and transmission with this host.

Serum samples were sequentially collected from goats C, D, and E, and each of the goats became seropositive when tested using Ap-ha antigen. Goat C became seropositive on day 17, peaked on day 22 with a titer of 1,024, and maintained a 64 titer through day 82. Goat D seroconverted on day 9, showed a peak titer of 256 on day 23, and maintained a titer of 128 through day 60. Goat E became seropositive on day 7, had a peak titer of 512 on day 21, and a titer of 64 on day 91.

TABLE 1. Susceptibility of goats to infections with *A. phagocytophilum* strains Ap-ha and Ap-Variant 1: PCR and serological results

Goat	Strain infected with:[a]	Initial day PCR-positive	Initial day of seroconversion	Maximum titer
A	Ap-ha	11	nt[b]	nt
B	Ap-ha	11	nt	nt
C	Ap-Variant 1	10	17	1/1024
D	Ap-Variant 1	7	9	1/256
E	Ap-Variant 1	5	7	1/512

[a]Determined by PCR and DNA sequencing of 16S rRNA gene.
[b]nt = not tested.

This study is the first report describing an animal species that is reservoir-competent for the Ap-Variant 1 strain. These results also demonstrate that goats can be used in a laboratory setting to propagate the variant strain. This animal model will facilitate studies of the biological and genetic variations that make the Ap-Variant 1 and Ap-ha strains distinct.

REFERENCES

1. BELONGIA, E.A., K.D. REED, P.D. MITCHELL, *et al.* 1997. Prevalence of granulocytic *Ehrlichia* infection among white-tailed deer in Wisconsin. J. Clin. Microbiol. **35:** 1465–1468.
2. COURTNEY, J.W., R.L. DRYDEN, J. MONTGOMERY, *et al.* 2003. Molecular characterization of *Anaplasma phagocytophilum* and *Borrelia burgdorferi* in *Ixodes scapularis* ticks from Pennsylvania. J. Clin. Microbiol. **41:** 1569–1573.
3. MASSUNG, R.F., M.J. MAUEL, J.H. OWENS, *et al.* 2002. Genetic variants of *Ehrlichia phagocytophila*, Rhode Island and Connecticut. Emerg. Infect. Dis. **8:** 467–472.
4. MASSUNG, R.F., K. SLATER, J.H. OWENS, *et al.* 1998. Nested PCR assay for detection of granulocytic ehrlichiae. J. Clin. Microbiol. **36:** 1090–1095.
5. MASSUNG, R.F., T.N. MATHER, R.A. PRIESTLEY & M.L. LEVIN. 2003. Transmission efficiency of the AP-variant 1 strain of *Anaplasma phagocytophila*. Ann. N. Y. Acad. Sci. **990:** 75–79.

An Epidemiological Study on *Anaplasma* Infection in Cattle, Sheep, and Goats in Mashhad Suburb, Khorasan Province, Iran

G.R. RAZMI,[a] K. DASTJERDI,[a] H. HOSSIENI,[a] A. NAGHIBI,[a] F. BARATI,[a] AND M.R. ASLANI[b]

[a]*Department of Pathobiology, School of Veterinary Medicine, Ferdowsi University of Mashhad, Iran*

[b]*Department of Clinical Science, School of Veterinary Medicine, Ferdowsi University of Mashhad, Iran*

ABSTRACT: The prevalence of *Anaplasma* infection was studied in cattle, sheep, and goats in the Mashhad area from 1999 to 2002. A total of 160 cattle from 32 farms and 391 sheep and 385 goats from 77 flocks were clinically examined for the presence of *Anaplasma* spp. in blood smears. The study revealed that 19.37% of cattle were infected with *Anaplasma marginale* and 80.3% of sheep and 38.92% of goats were infected with *Anaplasma ovis*. Prevalence of *Anaplasma* infection between male and female and between different age groups of cattle, sheep, and goats were statistically nonsignificant. Seasonally, the prevalence of *Anaplasma* infection in sheep and goats reached its highest level in summer, while a decrease was observed in autumn, and reached the lowest level in winter. The seasonal prevalence of *Anaplasma* infection in cattle was not significantly different. Symptomatic cases were not observed in any of the cattle, sheep, and goats. The ranges of anaplasmatemia in infected cattle, sheep, and goats were 0.005–0.5%, 0.01–3%, and 0.01–3%, respectively.

KEYWORDS: epidemiology; cattle; sheep; goat; *Anaplasma*

INTRODUCTION

Anaplasmosis of cattle, sheep, and goats is an infectious hemoparasitic disease caused by *Anaplasma* spp. In cattle, severe debility, emaciation, anemia, and jaundice are the major clinical signs. The disease is usually subclinical in sheep and goats. *Anaplasma marginale* is the causative agent in cattle and

Address for correspondence: G.R. Razmi, Department of Pathobiology, School of Veterinary Medicine, Ferdowsi University of Mashhad, P.O. Box: 91775-1793, Iran. Voice: +98-511-6620101; fax: +98-511-6620166.
e-mail: grrazmi@eudoramail.com

**Ann. N.Y. Acad. Sci. 1078: 479–481 (2006). © 2006 New York Academy of Sciences.
doi: 10.1196/annals.1374.089**

wild ruminants, and *Anaplasma ovis* in sheep and goats. Cross-infection and cross-immunity do not occur with *A. ovis*, but subclinical infections with *A. marginale* can occur in sheep and goats.[5] Anaplasmosis in cattle is common in South Africa, Australia, Russia, South America, and the United States, and anaplasmosis of sheep and goats occurs in Africa, Mediterranean countries, Russia, and the United States.[5] Although, *Anaplasma* infection in cattle, sheep, and goats was previously reported from Iran,[1,3,4] no epidemiological studies are available on these animals. The aim of this study was to determine the prevalence of the infection in Mashhad area and to relate the prevalence data to sex, age, and season.

MATERIALS AND METHODS

The study was done in the Mashhad area, capital city of Khorasan province, which in situated in northeast of Iran. About 160 cattle, 391 sheep, and 385 goats were tested to estimate prevalence. Each flock was visited once during 1999 and 2002. First, 5 to 7 animals were randomly sampled from every farm.[8] The selected animals were clinically examined and blood smears were prepared from marginal ear vein. The smears were air-dried, fixed in methanol, and stained in 10% Giemsa solution in phosphate-buffered saline (PBS), pH 7.2.The slides were examined with an oil immersion lens at a total magnification of × 1000. Parasitemia was assessed by counting the number of infected red blood cells on examination of 100 microscopic fields (approximately 100,000 cells). The number infected was then expressed as a percentage. The results of the present study were analyzed by Chi-square test. Significant association was identified when a P value of less than 0.05 was observed.

RESULTS AND DISCUSSION

Among the 160 cattle, 391 sheep, and 385 goats examined, 19.37%, 80.3%, and 47.53% were infected with *Anaplasma*, respectively. The prevalence of *Anaplasma* infection in all age groups and between male and female cattle, sheep, and goats was not significantly different. The seasonally related prevalence of *Anaplasma* infection in sheep and goats was highest in summer ($P < 0.05$), but no difference was found in cattle. No symptom was reported in any of the infected cattle, sheep, and goats. The anaplasmatemia of cattle, sheep, and goats ranged from 0.005% to 0.5% in cattle and from 0.01% to 3% in sheep and goats. Although *A. marginale* and *A. ovis* were previously reported in Iran,[1,3,4] no epidemiological study is available in large and small ruminants. Our data are similar to studies reported from India[4,7] and Israel.[9] In this study, the infected animals had low anaplasmatemia without clinical signs. The results were predictable in sheep and goats,[2,6,10] but observation of clinical anaplasmosis was, however, expected in cattle.[10] It seems that the sample size was not large enough to detect bovine anaplasmosis. In conclusion, Anaplasma infection is common in small and large ruminants in Mashhad area and more studies are needed to understand the epidemiology of anaplasmosis.

ACKNOWLEDGMENTS

We are very grateful to the Khorasan Veterinary Office for providing information about herd sheep and logistic support. The authors would like to acknowledge Mr. G.A.Azari for his technical assistance.

REFERENCES

1. BAZERGANI, T.T., S. RAHBARI, M. NADALIN, *et al.* 1985. A report of some cases of anaplasmosis in cattle and goat. J. Vet. Fac. Univ. Tehran **40:** 94–105. [In Persian, with English abstract].
2. FRIEDHOFF, K.T. 1997. Tick-borne disease of sheep and goats caused by *Babesia, Theileria* or *Anaplasma* spp. Parasitologia **39:** 99–109.
3. HASHEMI-FESHARAKI, R.1997. Tick-borne disease of sheep and goats and their related vectors in Iran. Parasitologia **39:** 115–117.
4. MANICKAM, R. 1987. Epidemiological and clinical observation of acute anaplasmosis in sheep. Indian J. Vet. Med. **7:** 159–160.
5. RAFYI, A. & G. MAGHAMI 1961. Contribution a l'etude de quelques parasites du sang du mouton et de la chevre en Iran et dans les pays voisins. Bull. Off. Int. Epid. **65:** 1769–1783.
6. RADOSTITS, O.M., D.C. BLOOD & C.C. GAY. 1995. Veterinary Medicine. A Textbook of the Disease of Cattle, Sheep, Pig, Goats and Horse. 1146–1150. Bailliere Tindall, London.
7. SINGH, A. & B.S. GILL. 1977. A note of the prevalence of subclinical anaplasmosis in three herds of cattle and buffaloes in Punjab state. Indian J. Anim. Sci. **47:** 224–226.
8. THRUSFIELD, M. 1997.Veterinary Epidemiology:189. Iowa State University Press.
9. YERUHAM, I., A. HADANI, F. GALKER, *et al.* 1992. A field study of haemoparasites in two flock of sheep in Israel. Isr. J. Vet. Med. **47:** 107–111.
10. UILENBERG, G. 1997. General review of tick-borne diseases of sheep and goats world-wide. Parasitologia **39:** 161–165.

Cytokine Gene Expression by Peripheral Blood Leukocytes in Dogs Experimentally Infected with a New Virulent Strain of *Ehrlichia canis*

AHMET UNVER,[a,b] HAIBIN HUANG,[a] AND YASUKO RIKIHISA[a]

[a]*Department of Veterinary Biosciences, College of Veterinary Medicine, The Ohio State University, Columbus, Ohio 43210-1093, USA*

[b] *Department of Microbiology, Faculty of Veterinary Medicine, Kafkas University, 36100 Kars, Turkey*

ABSTRACT: ***Ehrlichia canis (E. canis)*** **is a lipopolysaccharide (LPS)-deficient obligatory intracellular bacterium that causes canine monocytic ehrlichiosis, a chronic febrile disease accompanied with hematological abnormality. This study analyzed temporal expression levels of IL-1β, IL-2, IL-6, IL-8, IFN-γ, and TNF-α mRNA by peripheral blood leukocytes from dogs experimentally infected with a new virulent strain of *E. canis* by using real-time RT-PCR. Relative levels of IL-1β and IL-8 transcripts normalized by the β-actin transcript levels, were significantly upregulated, whereas those of TNF-α and IFN-γ transcripts were only weakly upregulated in all three infected dogs, starting from 2 days up to 52 days post inoculation. The expressions of IL-2 and IL-6 genes were extremely low compared with the positive control (ConA-stimulated canine peripheral blood leukocytes). This study showed that *E. canis* can induce chronic expression of a subset of proinflammatory cytokine genes: balance, timing, and duration of these cytokine generations may contribute to the progression of canine ehrlichiosis.**

KEYWORDS: ***E. canis*; IL-1; IL-8; dog**

INTRODUCTION

Ehrlichia canis (E. canis) is the causative agent of canine monocytic ehrlichiosis (CME) with tropism for monocytes and macrophages. CME has a worldwide distribution with a higher frequency in tropical and subtropical regions.[1] The disease may be manifested by fever, depression, dyspnea, anorexia,

Address for correspondence: Yasuku Rikihisa, Department of Veterinary Biosciences, College of Veterinary Medicine, The Ohio State University, 1925 Coffey Road, Columbus, Ohio 43210. Voice: 614-292-9677; fax: 614-292-6473.
e-mail: rikihisa.1@osu.edu

Ann. N.Y. Acad. Sci. 1078: 482–486 (2006). © 2006 New York Academy of Sciences.
doi: 10.1196/annals.1374.090

hemorrhage, edema, and slight weight loss accompanied with laboratory findings of thrombocytopenia, leukopenia, mild anemia, and hypergammaglobulinemia.[1] The pathological and immunological pathways induced by *E. canis* in dogs are not well defined. Low levels of ehrlichemia and nature of clinical signs in canine ehrlichiosis suggest that the host immune response, including proinflammatory cytokine production, may play a pivotal role in the pathogenesis of ehrlichiosis.[2] Overall, generation of cytokines and chemokines by the members of family *Anaplasmataceae* appears to be different from those by lipopolysaccharide (LPS), corroborating the recent report of absence of genes for biosynthesis of LPS among members of the family *Anaplasmataceae*.[2–4] The involvement of cytokines in pathogenesis of canine ehrlichiosis has not been elucidated. Analyses of cytokines that mediate immune response and inflammation may provide better understanding of pathogenesis of canine ehrlichiosis. The purpose of this study was to analyze several cytokine gene mRNA expressions by PBLs from dogs experimentally infected with a virulent strain of *E. canis* by using quantitative RT-PCR.

MATERIALS AND METHODS

Experimental Infection of Dogs

Seven specific pathogen-free (SPF) female beagle dogs weighing from 6/7 kg to 12/13 kg were used. All dogs were tested free of infection with *E. canis* and other members of the family *Anaplasmataceae*. Three dogs (dog 1, 2, and 3) were inoculated with total of 0.5 mL *E. canis* New Mexico[5]–infected canine blood (0.2 mL intradermally and 0.3 mL subcutaneously). Blood samples were collected periodically from the cephalic vein for indirect immunofluorescent assay (IFA) tests, RT-PCR assays, complete blood counts, and PCR assays. Rectal temperature, appetite, attitude, and any clinical signs were recorded daily and were used to monitor *E. canis* infection status with combination of PCR and IFA as previously described.[6] Dogs 4 and 5 were kept as uninfected controls without any inoculation. Dogs 6 and 7 were inoculated with total of 0.5 mL uninfected canine blood (0.2 mL intradermally and 0.3 mL subcutaneously) and used as inoculation controls.

Real-Time PCR and RT-PCR

Real-time PCR was carried out to determine levels of *E.canis* DNA in canine PBLs for each sample by using Brilliant® SYBR® Green QPCR Core Reagent kit (Stratagene, La Jolla, CA, USA). The forward and reverse primers used in the assay were located at 95–115 and 181–202 nucleotide position of 16S rRNA gene of *E. canis* Oklahoma (GeneBank No: M73221), respectively. To quantify the 16S rRNA gene copy numbers, 389-bp fragment of the 16S rRNA gene was cloned into a PCRII cloning vector (Invitrogen, Carlsbad, CA, USA) and 10-fold serially diluted to generate standard curve.

Total cellular RNA extraction, DNase treatment of RNA, and cDNA synthesis were performed as previously described.[6] Expression of cytokine genes (IL-1 β, IL-2, IL-6, IL-8, IFN-γ, and TNF-α) and β-actin gene was determined by real-time PCR as described by Fujiwara *et al.*[7] To quantify β-actin mRNA levels, 81-bp fragment of canine β-actin gene (GenBank No. Z70044) was cloned into a PCRII cloning vector. As a positive control, PBLs from each dog collected prior to infection were stimulated with concanavalinA (Sigma, St. Louis, MO, USA) (5μg/mL) in RPMI 1640 medium at 37°C for 3 h.

RESULTS

All three dogs inoculated with *E. canis* New Mexico developed signs of an acute severe ehrlichiosis, such as high fever (>40°C), loss of appetite, edema, dehydration, depression and weight loss, and thrombocytopenia (<106 × 10^9 platelets/L, FIG. 1A). The control dogs (dogs 4–7) did not show fever, thrombocytopenia, or any other clinical signs of canine ehrlichiosis during the same period. Temporal patterns of IFA IgG titers using *E.canis* Oklahoma as antigen were similar among three dogs inoculated with *E. canis* New Mexico. The titer started to rise on day 14 PI and reached 1:10,240 on day 56 PI. The IFA titer was negative in all control dogs (dogs 4–7).

Real-time PCR was used to detect 16S rRNA genes of *E. canis* in experimentally infected dogs. β-actin DNA was used to normalize input DNA from canine PBL across specimens. The ratios of *E. canis* 16S rRNA genes to that

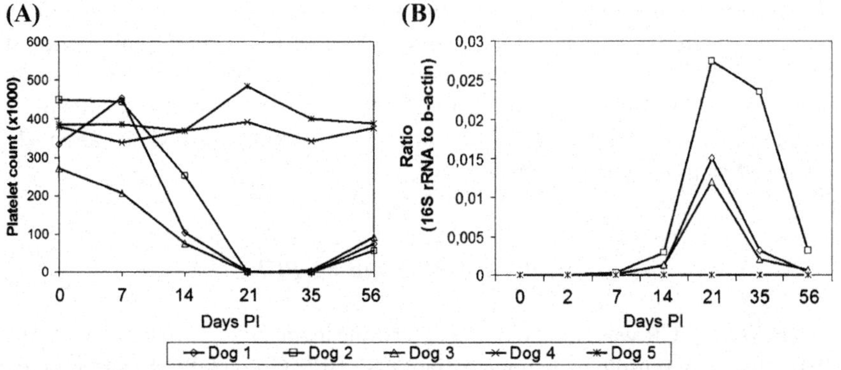

FIGURE 1. Platelet counts (**A**) and ehrlichemia levels (**B**) in the peripheral blood cells from dogs infected with *E. canis* New Mexico (dogs 1–3) and uninfected control dogs (dogs 4 and 5) as determined by real-time PCR.

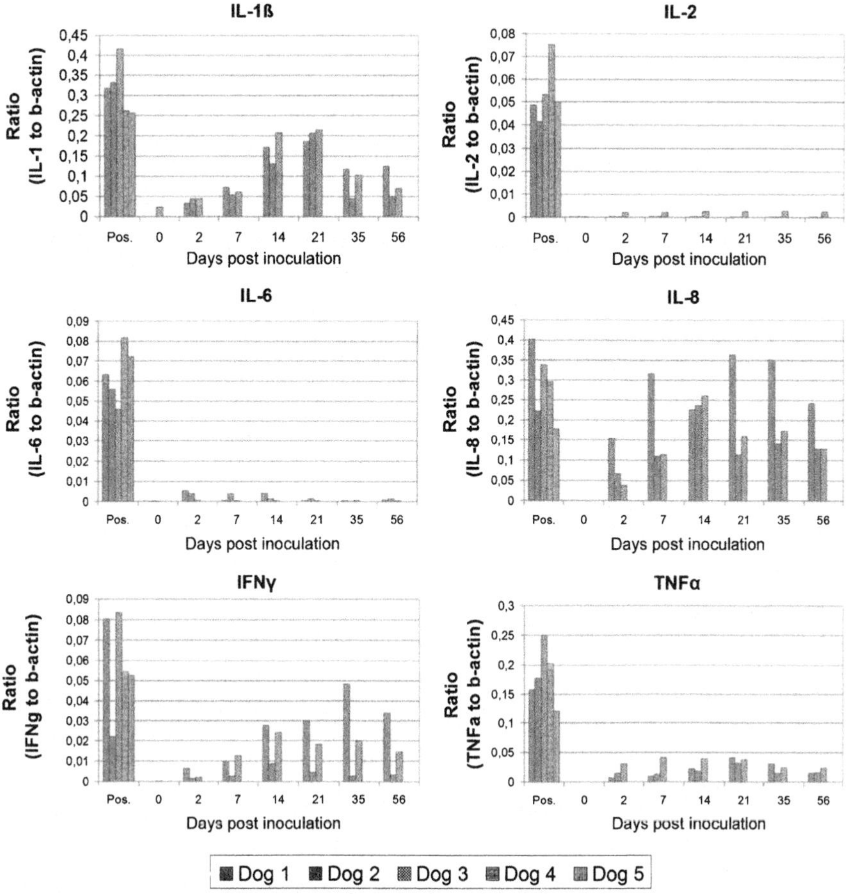

FIGURE 2. Cytokine mRNA expression by PBLs from infected (dogs 1–3) and uninfected (dogs 4 and 5) dogs.

of β-actin DNA were plotted following the time course of infection (FIG. 1B). *E. canis* DNA was never detected in all control dogs (dogs 4–7).

Temporal mRNA levels of six cytokines were examined by real-time RT-PCR in all specimens from day 2 PI (FIG. 2). Rapid and persistent upregulation of IL-1β and IL-8 and relatively low levels of IFN-γ and TNF-α expressions were observed in all three infected dogs (dogs 1–3). None of the cytokine mRNAs was detected in PBLs from uninfected dogs (dogs 4 and 5) and placebo inoculated control dogs (dogs 6 and 7, data not shown). DNA contamination in RNA preparation was negligible, since no PCR products were detected from the RNA samples processed without addition of reverse transcriptase. PBLs from all these dogs were similarly responsive to ConA *in vitro* prior to *E. canis* infection.

DISCUSSION

IL-1β and IL-8 cause inflammatory responses by attracting neutrophils and activating NF-kB-mediated inflammatory cytokine gene expression in a variety of cells.[8,9] High levels of IL-1β and IL-8 expression may be responsible for clinical signs (fever, depression, anorexia, dehydration, and thrombocytopenia) observed in these dogs in this study. IL-1 β and IL-8 are also major cytokines generated by human monocytes upon incubation *in vitro* with *Ehrlichia chaffeensis*.[8] Clinical signs and cytokine responses in dogs infected with the New Mexico strain of *E. canis* were distinct from those of dogs infected with *E. canis* Oklahoma strain,[10] suggesting the association of strain virulence with the pattern of cytokine induction in the host.

REFERENCES

1. BUHLES, W.C., D.L. RUXSOLL & M. RISTIC. 1974. Tropical canine pancytopenia: clinical, haematologic, and serologic, and serologic response of dogs to *Ehrlichia canis* infection, tetracycline therapy, and challenge inoculation. J. Infect. Dis. **130:** 358–367.
2. RIKIHISA, Y. 2003. Mechanisms to create a safe haven by members of the family *Anaplasmataceae*. Ann. N.Y. Acad. Sci. **990**: 548–555.
3. ISMAIL, N., L. SOONG, J.W. MCBRIDE, *et al.* 2004. Overproduction of TNF-alpha by CD8+ type 1 cells and down-regulation of IFN-gamma production by CD4+ Th1 cells contribute to toxic shock-like syndrome in an animal model of fatal monocytotropic ehrlichiosis. J. Immunol. **172:** 1786–1800.
4. LIN, M. & Y. RIKIHISA. 2003. *Ehrlichia chaffeensis* and *Anaplasma phagocytophilum* lack genes for lipid A biosynthesis and incorporate cholesterol for their survival. Infect. Immun. **71**: 5324–5331.
5. FELEK, S., H. HUANG & Y. RIKIHISA. 2003. Sequence and expression analysis of virB9 of the type IV secretion system of *Ehrlichia canis* strains in ticks, dogs, and cultured cells. Infect. Immun. **71:** 6063–6067.
6. UNVER, A., N. OHASHI, T. TAJIMA, *et al.* 2001. Transcriptional analysis of *p30* major outer membrane multigene family of *Ehrlichia canis* in infected dogs and ticks. Infect. Immun. **69:** 6172–6178.
7. FUJIWARA, S., S. YASUNAGA, S. IWABUCHI, *et al.* 2003. Cytokine profiles of peripheral blood mononuclear cells from dogs experimentally sensitized to Japanese cedar pollen. Vet. Immunol. Immunopathol. **93:** 9–20.
8. LEE, E.H. & Y. RIKIHISA. 1997. Anti-*Ehrlichia chaffeensis* antibody complexed with *E. chaffeensis* induces potent proinflammatory cytokine mRNA expression in human monocytes through sustained reduction of IkappaB-alpha and activation of NF-kappaB. Infect. Immun. **65:** 2890–2897.
9. MUKAIDA, N., S.A. KETLINSKY & K. MATSUSHIMA. 2003. İnterleukin-8 and other related chemokines. *In* The Cytokine Handbook, Volume I A.W. Thompson & M.T. Lotze. Eds, Academic Press. San Diego, CA.
10. TAJIMA, T. & Y. RIKIHISA. 2005. Cytokine responses in dogs infected with *Ehrlichia canis* Oklahoma strain. Ann. N.Y. Acad. Sci. **1063:** 429–432.

Serological Evaluation of *Anaplasma phagocytophilum* Infection in Livestock in Northwestern Spain

INMACULADA AMUSATEGUI,[a] ÁNGEL SAINZ,[a]
AND MIGUEL ÁNGEL TESOURO[b]

[a]*College of Veterinary Medicine, Complutense University of Madrid, Madrid, Spain*

[b]*College of Veterinary Medicine, University of León, León, Spain*

ABSTRACT: A total of 1,098 serum samples were analyzed against *Anaplasma phagocytophilum* by immunofluorescent antibody (IFA) test. These serum samples belonged to four different populations distributed throughout two provinces of Galicia (Ourense and Pontevedra) located in northwestern Spain: bovine population (456 samples); ovine population (389 samples); caprine population (207 samples); and equine population (46 serum samples, all from Pontevedra). The seroprevalence against *A. phagocytophilum* within the bovine population was 3.07%. On the other hand, two of 389 (0.51%) sheep and one of 207 (0.48%) goats tested were seropositive, all of them showing low antibody titer. Seroprevalence within the equine population was 6.52% (3/46). Our results reveal the presence of antibodies against *A. phagocytophilum* in livestock from northwestern Spain, mainly in Pontevedra.

KEYWORDS: *Anaplasma phagocytophilum*; Galicia; serology; ruminant; equine

INTRODUCTION

The presence of *Ehrlichia phagocytophila* (proved to be the same microorganism as *Ehrlichia equi* and the human granulocytic ehrlichiosis agent), recently classified as *Anaplasma phagocytophilum (A. phagocytophilum)*,[1] has been demonstrated in livestock from different areas having environmental conditions similar to those of the zone where our work has been performed.[2] Furthermore, a previous study on dogs of this same zone showed the presence of seropositive cases against this agent in Galicia.[3] Seroprevalence studies against an infectious agent in areas where its presence is unknown are the first step

Address for correspondence: Angel Sainz, Dpto. Medicina y Cirugía Animal, Facultad de Veterinaria, UCM, Ciudad Universitaria s/n 28040, Madrid, Spain. Voice: 34-91-394- 3874; fax: 34-91-394-3808.
e-mail: angelehr@vet.ucm.es

**Ann. N.Y. Acad. Sci. 1078: 487–490 (2006). © 2006 New York Academy of Sciences.
doi: 10.1196/annals.1374.091**

not only to establish the possible presence of the agent, but also to estimate its prevalence in that particular area. The aim of the present work is to establish the seroprevalence against *A. phagocytophilum* in livestock from Galicia (northwestern Spain) in order to determine the possible existence of this infection within an area that presents a favorable environment for the development of the vector involved in its transmission (*Ixodes* spp.).

MATERIALS AND METHODS

A total of 1,098 serum samples were analyzed against *A. phagocytophilum* by immunofluorescent antibody (IFA) test (using the HL60 cell line infected by *A. phagocytophilum*). These serum samples belonged to four different populations distributed throughout two provinces of Galicia (Ourense and Pontevedra) located in northwestern Spain: bovine population (456 serum samples, 260 from Ourense and 196 from Pontevedra); ovine population (389 serum samples, 214 from Ourense and 175 from Pontevedra); caprine population (207 serum samples, 123 from Ourense and 84 from Pontevedra); and equine population (46 serum samples, all from Pontevedra).

The sera of all these animals were serially diluted starting in 1/20, the cut-off being 1/40. Thus, cases showing antibody titers ≥1/40 were considered positive.

RESULTS

The seroprevalence against *A. phagocytophilum* within the bovine population was 3.07% (14/456). Highly significant differences ($P < 0.001$) were observed between the percentage of positive cases in Pontevedra, 6.63% (13/196), and in Ourense, 0.38% (1/260). On the other hand, 2 of 389 (0.51%) sheep and 1 of 207 (0.48%) goats tested were seropositive, all of them showing an antibody titer equal to 1/40. Since the three animals were detected in Pontevedra, the rates of positive sheep and goats in this province were 1.14% (2/175) and 1.19% (1/84), respectively. Finally, seroprevalence within the equine population, tested exclusively in Pontevedra, was 6.52% (3/46) (TABLE 1).

TABLE 1. Distribution of the cases depending on the province and the antibody titer

	Pontevedra			Ourense		
	<1/40	1/40	>1/40	<1/40	1/40	>1/40
Bovine ($n = 456$)	183	4	9	259	–	1
Ovine ($n = 389$)	173	2	–	214	–	–
Caprine ($n = 207$)	83	1	–	123	–	–
Equine ($n = 46$)	43	1	2	–	–	–

DISCUSSION

Our results reveal the presence of antibodies against *A. phagocytophilum* in livestock from northwestern Spain, mainly in Pontevedra. In fact, only one cow, located in Ourense, showed an antibody titer over the cutoff. The significant preponderance of positive cases in Pontevedra could be linked to its climate (higher in humidity and warmer in temperature than Ourense) that seems to be particularly favorable for the development of the vectors.[4]

The seroprevalence against *A. phagocytophilum* in horses from Pontevedra was 6.5%, statistically similar to the seroprevalence observed in cattle (6.6%) located in this same province ($P = 0.7675$). On the contrary, the seroprevalence in ovine (1.14%) and caprine (1.19%) populations was very low, and the antibody titer in these three cases was barely 1/40. Although this would lead to the suspicion that *A. phagocytophilum* is not or is scarcely present in sheep and goats from Galicia, previous studies performed in areas with similar environmental conditions have shown the existence of *A. phagocytophilum* in these and other species,[2,5] in good agreement with our findings in cattle and horses from northwestern Spain. In any case, possible cross-reactions with other microorganisms antigenically related to *A. phagocytophilum* must always be considered.[6]

Since these infections are frequently subclinical, they may remain unobserved in spite of their possible endemic presence in certain regions. Nevertheless, even in mild cases economic repercussions may be significant.[7] In such cases the actual health and economic impact caused by these agents can only be assessed through focussed research.

To our knowledge this is the first study related to *A. phagocytophilum* performed in cattle in Galicia. The detection of seropositive cases within the four species tested suggests that this infection may be present in northwestern Spain. Further studies would be necessary in order to establish the real identity of such an infectious agent, its vectors, and its possible pathogenic role.

REFERENCES

1. DUMLER, J.S., A.F. BARBET, *et al.* 2001. Reorganization of genera in the families *Rickettsiaceae* and *Anaplasmataceae* in the order *Rickettsiales*: unification of some species of *Ehrlichia* with *Anaplasma*, *Cowdria* with *Ehrlichia* and *Ehrlichia* with *Neorickettsia*, descriptions of six new species combinations and designation of *Ehrlichia equi* and "HE" agent as subjective synonyms of *Ehrlichia phagocytophila*. Intl. J. Syst. Bacteriol. Evol. Microbiol. **51:** 2145–2165.
2. GARCÍA-PÉREZ, A.L., J. BARANDIKA, *et al.* 2003. *Anaplasma phagocytophila* as an abortifacient agent in sheep farms from northern Spain. Ann. N. Y. Acad. Sci. **990:** 429–432.
3. AMUSATEGUI, I., A. SAINZ & M.A. TESOURO.2001. Seroprevalencia de *Ehrlichia canis*, *Ehrlichia equi* y *Ehrlichia risticii* en perros de Galicia. Presented at the Thirty Sixth National Congress of AVEPA. Barcelona, Spain, November, 2–4.

4. ESTRADA-PEÑA, A. 2003. Las garrapatas del perro en España: especies, distribución, ecología y control. *In* Las Garrapatas del Perro y las Enfermedades Transmitidas en España: Una Panorámica con Aspectos Zoonósicos. A. Estrada-Peña, Ed.: 7–17. Virbac. Barcelona, Spain.
5. OPORTO, B., H. GIL, *et al.* 2003. A survey on *Anaplasma phagocytophila* in wild small mammals and roe deer (*Capreolus capreolus*) in Northern Spain. Ann. N. Y. Acad. Sci. **990:** 98–102.
6. WANER, T., S. HARRUS, *et al.* 2001. Significance of serological testing for ehrlichial diseases in dogs with special emphasis on the diagnosis of canine monocytic ehrlichiosis caused by *Ehrlichia canis*. Vet. Parasitol. **95:** 1–15.
7. STUEN, S., K. BERGSTROM & E. PALMER. 2002. Reduced weight gain due to subclinical *Anaplasma phagocytophilum* (formerly *Ehrlichia phagocytophila*) infection. Exp. Appl. Acarol. **28:** 209–215.

Anaplasma phagocytophilum Infection in Cattle in France

KOTARO MATSUMOTO,[a] GUY JONCOUR,[b] BERNARD DAVOUST,[c] PIERRE-HUGUES PITEL,[d] ALAIN CHAUZY,[e] ERIC COLLIN,[f] HERVÉ MORVAN,[g] NATHALIE VASSALLO,[g] AND PHILIPPE BROUQUI[a]

[a]*Unité des Rickettsies, CNRS UMR 6020 IFR 48, Faculté de Médecine, Marseille, France*

[b]*Groupe Vétérinaire, 26, rue du Cleumeur, 22160 Callac, France*

[c]*Direction Régionale du Service de Santé des Armées, BP 16, 69998 Lyon Armées, France*

[d]*Laboratoire Départemental Frank Duncombe, 1, route de Rosel, 14280 Caen cedex 4, France*

[e]*21140 Semur en Auxois, France*

[f]*22150 Ploeuc sur Lie, France*

[g]*Laboratoire de Développement et d'Analyses, BP 54, 22440 Ploufragan, France*

ABSTRACT: *Anaplasma phagocytophilum* is the agent of pasture fever or tick-borne fever, a disease of ruminants and humans in the United States and in Europe. Although several hundred cases have been suspected to occur in cattle in France, none has yet been microbiologically confirmed. We report the first identification of *A. phagocytophilum* 16S RNA gene sequence in a case of TBF in France. This indicates that the diagnosis of tick-borne fever should be also evoked in cattle exposed to Ixodes ticks in France.

KEYWORDS: *Anaplasma phagocytophilum*; tick-borne fever; cattle diseases; France; 16S RNA

Anaplasma phagocytophilum is an obligate intracellular bacterium and the causative agent of tick-borne fever (TBF) in cattle and sheep, HGE (human anaplasmosis) in the human, and equine ehrlichiosis. This disease has been reported in several countries in Europe.[1–3] In France, although several hundred suspected cases of TBF in cattle have been reported,[4,5] *A. phagocytophilum* has never been identified as the causative agent in those cases using reference methods such as culture, or polymerase chain reaction (PCR), and sequencing.

Address for correspondence: P. Brouqui, Unité des Rickettsies, Faculté de Médecine, 27 Bd. Jean moulin, 13385 Marseille, Cedex 5, France. Voice: 33-49-1-32-43-75; fax: 33-49-1-96-89-38.
e-mail: philippe.brouqui@medecine.univ-mrs.fr

**Ann. N.Y. Acad. Sci. 1078: 491–494 (2006). © 2006 New York Academy of Sciences.
doi: 10.1196/annals.1374.092**

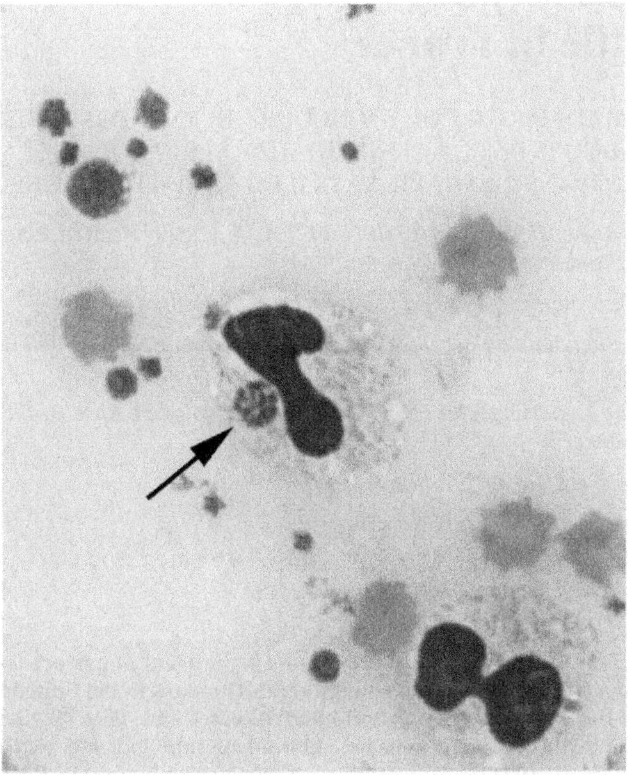

FIGURE 1. Morulae of *Anaplasma phagocytophilum* in white blood cells of cattle (*arrows*).

We report herein the first documented case of *A. phagocytophilum* infection in cattle in France.

In March 2004, a six-year-old charolaise cow from Côte d'Or which was just before its third calving, presented with weariness, hard breathing, and frequent lying-down. After calving, this cow was examined and treated for a peripartum pathology with long-acting benzylpenicillin-dihydrostreptomycin (10mL/100kg). However, 5 days later, the cow presented additional symptoms including a bloody chestnut-colored nasal discharge. Upper respiratory infection was suspected and ceftiofur (2.2mg/kg i. m.) was administered for 4 days. In the following days, the condition of the cow worsened, and the cow presented with coughing and cold edema on distal limb parts, and she did not produce milk anymore. A combination of benzylpenicillin-neomycin and methylprednisolone, and didrophyllin-terpin was prescribed. At this point, clinicians suspected TBF on the basis of the clinical signs.

On day 17 from onset of clinical signs, blood samples were taken for cytohematology, TBF serology, and molecular diagnosis. Serological examination was performed by immunofluorescence assay (IFA) (Focus Diagnostics, Cypress, CA). Molecular detection of *A. phagocytophilum* in blood was performed as described previously.[6] In brief: DNA was extracted from 200 μL of bovine blood using (QIAmp kit Qiagen) and amplified using the primers EHR16SD–EHR16SR.[7] Moreover, the amplification of the 16S rRNA was performed by PCR with a set of primers, EHR16SD–Rp2, and the amplicons were sequenced as described previously.[8]

Blood smear stained with May-Grunwald-Giemsa showed morulae in peripheral mononuclear cells (FIG. 1). The IFA test for *A. phagocytophilum* showed IgG >1:80 dilution. The PCR product was sequenced and identified as *A. phagocytophilum* with a 100% homology with the GenBank reference sequences (915/915, accession number, AY055469).

After *A. phagocytophilum* infection was confirmed by clinical examination, the cow was treated with 10mL/100kg of long-acting oxytetracycline (20%) every 2 days for 8 days. The cow recovered except for milk production. The cow's calf was reared by hand and had no clinical abnormality.

In humans, a confirmed diagnosis of *A. phagocytophilum* infection is based on the criteria described (Atlanta CDCP diagnosis criteria).[9] On the basis of those same criteria, we confirmed *A. phagocytophilum* infection in cattle. *A. phagocytophilum* infection in cattle has been confirmed by molecular biology in several countries in Europe.[1,2,3] Recently, *A. phagocytophilum* infection in a horse in France was also reported.[6] These reports and our study suggest that *A. phagocytophilum* infection in animals is increasingly recognized throughout Europe. Our report indicates that veterinarians should take TBF into consideration in case of febrile disease in cattle occurring in *Ixodes* tick–endemic areas, including France.

REFERENCES

1. PUSTERLA, N., J.B. PUSTERLA, U. BRAUN, *et al.* 1998. Serological, hematologic, and PCR studies of cattle in an area of Switzerland in which tick-borne fever (caused by *Ehrlichia phagocytophila*) is endemic. Clin. Diagn. Lab. Immunol. **5:** 325–327.
2. OGDEN, N.H., A.N. CASEY, N.P. FRENCH, *et al.* 2002. Natural *Ehrlichia phagocytophila* transmission coefficients from sheep "carriers" to *Ixodes ricinus* ticks vary with the numbers of feeding ticks. Parasitology **124:** 127–136.
3. ENGVALL, E.O., B. PETTERSSON, M. PERSSON, *et al.* 1996. A 16S rRNA-based PCR assay for detection and identification of granulocytic *Ehrlichia* species in dogs, horses, and cattle. J. Clin. Microbiol. **34:** 2170–2174.
4. ARGENTÉ, G., E. COLLIN & H. MORVAN. 1992. Ehrlichiose bovine (fièvre des pâtures): une observation en France. Point Vétérinaire. **144:** 89–90.
5. JONCOUR, G., G. ARGENTÉ & L. GUILLOU. 2000. Un épisode d'ehrlichiose dans un troupeau laitier. Bulletin des Groupements Techniques Vétérinaires **5:** 19–24.
6. BERMANN, F., B. DAVOUST, P.E. FOURNIER, *et al.* 2002. *Ehrlichia equi* (*Anaplasma phagocytophila*) infection in an adult horse in France. Vet. Rec. **150:** 787–788.

7. PAROLA, P., V. ROUX, J.L. CAMICAS, *et al.* 2000. Detection of ehrlichiae in African ticks by polymerase chain reaction. Trans. R. Soc. Trop. Med. Hyg. **94:** 707–708.
8. INOKUMA, H., Y. TERADA, T. KAMIO, *et al.* 2001. Analysis of the 16S rRNA gene sequence of *Anaplasma centrale* and its phylogenetic relatedness to other ehrlichiae. Clin. Diagn. Lab. Immunol. **8:** 241–244.
9. BROUQUI, P., F. BACELLAR, G. BARANTON, *et al.* 2004. Guidelines for the diagnosis of tick-borne bacterial diseases in Europe. Clin. Microbiol. Infect. **10:** 1108–1132.

Understanding the Mechanisms of Transmission of *Ehrlichia ruminantium* and Its Influence on the Structure of Pathogen Populations in the Field

NATHALIE VACHIÉRY,[a] MODESTINE RALINIAINA,[b]
FRÉDÉRIC STACHURSKI,[c] HASSANE ADAKAL,[c] SOPHIE MOLIA,[a]
THIERRY LEFRANÇOIS,[a] AND DOMINIQUE MARTINEZ[a]

[a]*CIRAD-EMVT, Domaine Duclos, Prise d'eau 97170, Petit Bourg, Guadeloupe*

[b]*FOFIFA, Antananarivo, Madagascar*

[c]*CIRDES, Bobo Dioulasso, Burkina Faso*

ABSTRACT: The understanding of the structure of *Ehrlichia ruminantium* stock population in the field was highlighted by experiments done in controlled conditions on the goat model. The mixture of strains observed in ticks seemed to be due to simultaneous infections rather than successive infections of the carrier. During a dual infection, the timing of *Ehrlichia ruminantium* circulation of the two stocks in hosts influenced their selection by ticks.

KEYWORDS: heartwater; *Ehrlichia ruminantium*; Amblyomma ticks; structure of population

Heartwater is a ruminant disease present in sub-Saharan Africa, Madagascar, and in some Caribbean islands.[1] The causal agent, *Ehrlichia ruminantium* (ER), is transmitted by *Amblyomma* ticks. Several vaccines such as attenuated vaccine, inactivated vaccine, immunization by infection, and treatment were efficient against homologous virulent strains while conferring poor protection against heterologous strains.[2] The main obstacle to get an efficient vaccine against cowdriosis is the genetic and antigenic diversity of ER. Molecular epidemiology studies conducted in a cowdriosis-endemic region of Africa, the Burkina Faso, during a field evaluation of an inactivated vaccine, have given a description of the distribution of ER genotypes in ticks and hosts. The mechanisms underlying the structuring of populations were studied by conducting investigations on the role of ticks and hosts in ER transmission in controlled conditions.

Address for correspondence: Voice: 0-590-590-255-442; fax: 0-590-590-940-396.
e-mail: nathalie.vachiery@cirad.fr

**Ann. N.Y. Acad. Sci. 1078: 495–497 (2006). © 2006 New York Academy of Sciences.
doi: 10.1196/annals.1374.093**

The inactivated vaccine was evaluated in three villages of Burkina Faso, Lamba, Sara, and Bekuy, and in an experimental station named Banankélédaga. Villages were 20–25 km distant from each other and Banankélédaga was 80 km from the villages. During this field vaccination trial, ticks and brains from reacting and dead sheep were collected. DNA from ticks and brains was submitted to nested polymerase chain reaction (PCR) pCS20 (for ER detection).[3] Positive samples for ER were amplified using the map-1 nested PCR and the amplicons were preferentially digested with Rsa I, Xba I, and Taq I or Msp I to give a restriction fragment length polymorphism (RFLP) pattern.[3] Some map-1 amplicons were sequenced. Interestingly, several *map1* genotypes were shown to circulate in restricted geographical areas without apparently generating a single isolate composed of a mixture of heterologous clones. Seven to 11 map-1 genotypes were present in the four areas—Sara, Bekuy, Lamba, and Banankélédaga. Despite the diversity of genotypes present in the field, only 31 and 18% of brains and ticks were found coinfected, respectively. To understand this situation, experiments were conducted in controlled conditions on goats successively or simultaneously inoculated with two ER stocks. Eleven goats immunized against Gardel survived to homologous virulent challenge. Then, they were submitted to a heterologous challenge with a second strain 2 months to 1 year after Gardel challenge. Seven goats died during the heterologous challenge whereas four animals recovered. Brains and lungs were collected on succumbing animals in order to detect the presence of ER strains. Map-1 nested PCR and RFLP were used to genotype ER strains in host organs as described above. In four of seven brain samples, both strains were detected but the first strain inoculated was generally in lower quantity than the last inoculated strain. In lungs, the last strain inoculated was mainly detected. Ticks were fed on the four recovered goats 2 months after last ER infection. During asymptomatic carrier stage, 37% of the ticks ($N = 32$) fed on these goats became infected, but only with the last stock inoculated. These results suggest that typing in brains reflects the history of successive infection of hosts, whereas typing in ticks gave the last strain infecting the host. More data are needed to confirm these hypotheses.

A second experiment in controlled conditions was done to better characterize the influence of ticks and hosts in ER transmission during coinfection. Two naïve goats were challenged simultaneously with Gardel and Lamba strains. The typing of ER strains circulating in hosts and infecting ticks was done using nested map-1 and RFLP. The percentage of infection of ticks by a genotype was linked to its kinetics of circulation in the blood stream, ticks being preferentially or even exclusively infected by the first genotype circulating.

In conclusion, results in controlled conditions supported field observations: infection with several strains was detected more frequently in brains than in ticks. The mixture of strains observed in ticks in the field seemed to be due to simultaneous infections rather than successive infections of the carrier. Concerning the influence of timing of ER circulation in hosts and the selection of ER genotype by ticks, further experiments will be necessary to validate this hypothesis.

The genetic diversity of ER is associated with an antigenic diversity that is critical for vaccine efficacy. Although *map1* is not a predictor of cross-protection, these studies have shed a new light on the structure of ER stock populations in the field. In addition, various degrees of cross-protection were demonstrated between these heterogeneous populations of *map1* genotypes living in the same area, confirming the need for a better characterization of the diversity for an appropriate formulation of cowdriosis vaccines.

REFERENCES

1. PROVOST, A. & J.D. BEZUIDENHOUT. 1987. The historical background and global importance of heartwater. Onderstepoort J. Vet. Res. **54:** 165–169.
2. JONGEJAN, F. 1991. Protective immunity to heartwater (*Cowdria ruminantium* infection) is acquired after vaccination with *in vitro*-attenuated rickettsiae. Infect. Immunol. **59:** 3243–3246.
3. MARTINEZ, D., N. VACHIÉRY, F. STACHURSKI, *et al.* 2005. Nested PCR for the detection and genotyping of *Ehrlichia ruminantium*: use in genetic diversity analysis. Ann. N.Y. Acad. Sci. **1026:** 106–113.

Incidence of Ovine Abortion by *Coxiella burnetii* in Northern Spain

B. OPORTO, J.F. BARANDIKA, A. HURTADO, G. ADURIZ, B. MORENO, AND A.L. GARCIA-PEREZ

Department of Animal Health, Instituto Vasco de Investigación y Desarrollo Agrario (NEIKER), 48160 Derio, Bizkaia, Spain

ABSTRACT: The infectious causes of ovine abortion occurring in 148 farms in northern Spain between 1999 and 2003 were investigated. Laboratory analysis included microbiological, serological, pathological and molecular techniques. Border disease was diagnosed in 16% of the flocks, toxoplasmosis in 15%, chlamydiosis in 12%, salmonellosis in 10%, Q fever in 3%, miscellaneous infections in 7% (*Yersinia* spp., *Listeria* spp., *Brucella* spp.), and inflammatory lesions compatible with an infectious cause were seen in 7% of the flock. In an additional 1% of the flocks non-infectious causes were identified, and a diagnosis was not reached in 38% of the flocks. When a PCR retrospective study was carried out to investigate the possible implication of *Coxiella burnetii* in the cases without diagnosis, including those with inflammatory lesions, the prevalence of this pathogen increased from 3% up to 9% of the flocks, revealing the importance of this zoonotic pathogen as a small-ruminant abortifacient agent. Placenta was the most commonly positive sample, but other fetal tissues were also of value for *C. burnetii* DNA detection. The present results update information about the situation of abortion in sheep farms in northern Spain, and highlight the relevance of molecular diagnostic tools in routine laboratory analysis of abortions by *C. burnetii.*

KEYWORDS: **sheep abortion; Q fever; *Coxiella burnetii***

Human Q fever is endemic in northern Spain, where the largest series of cases of Q fever pneumonia in the world have been reported.[1] Cattle, sheep, and goats are considered the primary source of transmission to humans, but very few studies have been conducted in Spain and little is known about the quantitative role of *Coxiella burnetii* in ovine abortion in Spain. The aim of this study was to evaluate the importance of *C. burnetii* as an abortifacient in sheep flocks from northern Spain using routine and molecular diagnostic techniques.

Address for correspondence: Ana L. García-Pérez, NEIKER–Instituto Vasco de Investigación y Desarrolo Agrario, Berreaga 1, 48160 Derio (BIZKAIA), Spain. Voice: 34944034312; fax: 34944034310.
e-mail: agarcia@neiker.net

Ann. N.Y. Acad. Sci. 1078: 498–501 (2006).
doi: 10.1196/annals.1374.095

TABLE 1. Diagnostic results in the studied flocks ($n = 148$)

Etiology	Pathogens	No. of farms	%
Bacterial agent	*Chlamydophila abortus*	18	12
	Salmonella abortus ovis	15	10
	Coxiella burnetii	4	3
	Brucella melitensis	4	3
	Listeria ivanovii	4	3
	Yersinia pseudotuberculosis	2	1
	Listeria monocytogenes	1	1
	Bacterial lesions in fetuses	11	7
Protozoal agent	*Toxoplasma gondii*	22	15
Viral agent	Border disease	23	16
Noninfectious	Fetal anomalies	1	1
Not determined	Not determined	58	38

Samples consisting of 188 placentas and 373 fetuses and stillbirths from 148 flocks located in northern Spain were submitted for diagnosis of infectious abortion. In several cases, submissions included vaginal swabs from aborted ewes and maternal blood for bacteriological and serological analyses. Laboratory procedures have been already described elsewhere.[2] In brief: the specific diagnosis of abortion by *C. burnetii* was carried out by means of Stamp staining of placental smears and complement fixation test (CFT) on maternal sera using a commercial antigen (Dade Behring Marburg GmbH, Marburg, Germany) titers $\geq$1:32; were regarded as showing an active infection.

The diagnostic results achieved in the studied flocks are summarized in TABLE 1: viral infection (Border disease) was diagnosed in 16% of the flocks, protozoal infection (*Toxoplasma gondii*) in 15%, and bacterial infection in 33%. Q fever was diagnosed in four flocks (3%) on the basis of the positive Stamp staining of impression smears of the aborted placenta and/or the ewes' serological response ($\geq$1/32). A total of 42% of the maternal sera (16/38) studied in these flocks showed antibodies against *C. burnetii* by CFT. In addition, histological examination of placental and fetal tissues from the flocks with a diagnosis of Q fever revealed suppurative necrotic placentitis in two of them, and significant lesions in the fetuses consisting of multifocal granulomatous or necrotic hepatitis in liver, and, in lung, perivascular mononuclear cuffs. In most of the cases of *C. burnetii*-associated abortions only the placenta is involved and no fetal lesions are generally noted. However some fetuses may develop lesions consistent with those observed in this study.[3] Thus, the granulomatous hepatitis and nonsuppurative interstitial pneumonia, also seen in kids,[3] may be useful in considering differential causes of bacterial abortion.

Interestingly, inflammatory lesions compatible with a bacterial disease, but without bacterial isolation, were seen 11 flocks (7%) (TABLE 1). These

TABLE 2. Comparison of results of different laboratory techniques used for the detection of *C. burnetii* in PCR-positive flocks

Flock	Initial diagnosis	Stamp smear	Bacterial isolation	Maternal serology	Pathology	PCR
1	Q fever	Positive	Negative	≥1/256	Autholysis	Positive
2	Q fever	Positive	Negative	≥1/64	Compatible	Positive
3	Q fever	ND	Negative	≥1/256	NSL	Positive
4	Q fever	Positive	*A. pyogenes*	≥1/256	Compatible	Positive
5	Bacterial abortion	ND	Negative	ND	Compatible	Positive
6	Bacterial abortion	ND	Negative	ND	Compatible	Positive
7	Bacterial abortion	Negative	Negative	Negative	Compatible	Positive
8	Bacterial abortion	Positive	Negative	Negative	Compatible	Positive
9	No diagnosis	Negative	Negative	Negative	NSL	Positive
10	No diagnosis	ND	Negative	Negative	Autholysis	Positive
11	No diagnosis	ND	Negative	Negative	Autholysis	Positive
12	No diagnosis	Negative	Negative	Negative	Autholysis	Positive
13	No diagnosis	Negative	Negative	Negative	NSL	Positive

NOTE: ND = not done, placenta and/or maternal sera not available; NSL = no significant lesions

consisted of suppurative placentitis, periportal necrotic hepatitis, catarrhal bronchopneumonia, and interstitial pneumonia. In one of these cases the placental smears showed bacteria compatible with *C. burnetii*, but maternal serology did not confirm the diagnosis of Q fever (TABLE 2).

As several polymerase chain reaction (PCR)–based methods have been developed recently for the detection of *C. burnetii* in tissues of aborted fetuses, genital swabs, feces, milk, etc.,[4,5] we carried out a PCR assay[4] using primers targeting a trasposon-like repetitive region of the *C. burnetii* genome in order to check retrospectively the presence of *C. burnetii* DNA in: (i) placentae from cases with positive Stamp smears (4 flocks); (ii) placentae and fetal tissues with inflammatory lesions suggestive of infectious abortion but without a confirmative isolation (11 flocks); and (iii) placentae and fetuses without a confirmed diagnosis (58 flocks). A total of 70 placental and 194 fetal tissues were analyzed by PCR. Placenta was the most commonly positive sample, but other fetal tissues were also of value for *C. burnetii* DNA detection, especially in those cases where placenta was not available. A total of 14% of placentae and 17% of fetal samples (lung, spleen, kidney, heart muscle, liver) were positive. Taking into account the PCR results, the prevalence of this pathogen increased from 3% (4/148) to 9% (13/148) of the flocks. The compared performance of the different techniques used in the 13 positive flocks is summarized in TABLE 2. PCR detected DNA of *C. burnetii* in the 4 flocks with a previously confirmed diagnosis of Q fever (100%), in 4 of 11 flocks in which the fetuses showed bacterial lesions (36%), and in 5 of 58 flocks without a confirmed diagnosis (9%). Thus, in 5 flocks *Coxiella* DNA was detected in the absence of a positive result by any of the other techniques (TABLE 2). These results do not confirm *C. burnetii* as the causative agent of the abortion, but indicate that its distribution and prevalence is higher than previously thought. As shown in TABLE 2, CFT did not give a positive result in 7 PCR-positive cases. This and the low seroprevalence found in the four flocks with an initial diagnosis of Q fever (42%) could be due to a lack of sensitivity of the technique[6] or due to the failure of the ewes to develop a humoral response against the bacteria. Moreover, the detection of antibodies against *C. burnetii* can be affected not only by the technique used but also by the nature of the antigen preparations.[6]

In conclusion, these results provide updated information on the situation of abortion in sheep farms in northern Spain, and highlight the relevance of improving diagnostic techniques, including molecular diagnostic tools, in routine laboratory analysis of abortions by *C. burnetii.*

ACKNOWLEDGMENTS

This work had financial support from FIS (G03/057), INIA (RTA02–001), and the Department of Agriculture from the Basque Government.

REFERENCES

1. MAURIN, M. & D. RAOULT. 1999. Q Fever. Clin. Microbiol. Rev. **12:** 518–553.
2. HURTADO, A., G. ADURIZ, B. MORENO, *et al.* 2001. Single tube nested PCR for the detection of *Toxoplasma gondii* in fetal tissues from naturally aborted ewes. Vet. Parasitol. **102:** 17–27.
3. MOORE, J.D., C. BARR, B.M. DAFT & M.T. O'CONNOR. 1991. Pathology and diagnosis of *Coxiella burnetii* infection in a goat herd. Vet. Pathol. **28:** 81–84.
4. BERRI, M., K. LAROUCAU & A. RODOLAKIS. 2000. The detection of *Coxiella burnetii* from ovine genital swabs, milk and fecal samples by the use of a single touchdown polymerase chain reaction. Vet. Microbiol. **72:** 285–293.
5. MASALA, G., R. PORCU, G. SANNA, *et al.* 2004. Occurrence, distribution, and role in abortion of *Coxiella burnetii* in sheep and goats in Sardinia, Italy. Vet. Microbiol. **99:** 301–305.
6. KOVAKOVA, E., J. KAZAR & D. SPANELKOVA. 1998. Suitability of various *Coxiella burnetii* antigen preparations for detection of serum antibodies by various test. Acta Virol. **42:** 365–368.

Detection of *Coxiella burnetii* in Market Chicken Eggs and Mayonnaise

NORIYUKI TATSUMI,[a] ANDREAS BAUMGARTNER,[b] YING QIAO,[a] IKKYU YAMAMOTO,[a] AND KAZUO YAMAGUCHI[c]

[a]*Institute for Preventive Medicine Against Zoonosis (IPMZ), Toyama Prefecture, Japan*

[b]*Food Science Division, Swiss Federal Office of Public Health, Switzerland*

[c]*Division of Functional Genomics Advanced Science Research Center, Kanazawa National University, Kanazawa City, Ishukawa Prefecture, Japan*

ABSTRACT: We tried to detect *C. burnetii* in market chicken eggs and mayonnaise by nested PCR assay. The PCR target was the *com 1* gene of *C. burnetii*. The positive rate for egg and mayonnaise samples was 4.2% and 17.6%, respectively. Direct sequence of some of the positive egg samples shows mutations whereas no mutation was found in the positive mayonnaise samples. The number of molecules of the Q fever agent is estimated at 10^4 to 10^6 per egg, according to our quantitative PCR test.

KEYWORDS: *Coxiella burnetii;* nested PCR; real-time PCR; egg; mayonnaise

OBJECTIVES AND METHODS

Our aim was surveillance of *Coxiella burnetii* (*C. burnetii*) contamination in commercial chicken eggs and egg products.

Methods used in this inquiry included nested PCR, real-time PCR, direct sequencing method, and immunostaining using specific antibody on tissue sections.

RESULTS

We have conducted PCR tests for *C. burnetii*, the agent of Q fever, in a total of 4,252 chicken eggs and 200 bottles of mayonnaise sold in various parts of Japan for the past two years, detecting the *com1* gene of *C. burnetii* in a number of samples. The positive rate of egg samples was 4.2% on average,

Address for correspondence: Nariyuki Tatsumi, the Institute for Preventive Medicine against Zoonosis (IPMZ), 4569-1, Komoridani, Oyabe City, Toyama Prefecture, 932-0133 Japan. Voice: 81-766-69-1116; fax: 81-766-69-7001.
e-mail: info@jinjyuken.gr.jp

**Ann. N.Y. Acad. Sci. 1078: 502–505 (2006). © 2006 New York Academy of Sciences.
doi: 10.1196/annals.1374.096**

TABLE 1. The results of PCR tests for *C. burnetii* in chicken eggs and mayonnaise

Test date	Total number of samples tested	Positive	Positive rate	Mutation	Country of origin
		Eggs			
July 10, 03	300	67	22.3%	8	Japan
Aug. 06, 03	200	9	4.5%	-	Japan
Aug. 14, 03	300	4	2.0%	-	Japan
Sep. 02, 03	400	23	5.7%	4	Japan
Sep. 24, 03	200	9	4.5%	3	Japan
Oct. 10, 03	120	17	14.1%	3	Korea
Oct. 16, 03	12	1	8.3%	0	The Philippines
Jan. 16, 04	200	2	1.0%	1	Japan
Feb. 03, 04	202	7	3.4%	1	Japan
Mar. 06, 04	2,100	21	1.0%	5	Japan
June 25,04[a]	318	19	6.0%	0	Japan
Total	4,252	179	4.2%	25	
		Mayonnaise			
Mar.18, 03	40	14	35.0%	0	Japan
Apr.14, 03	28	7	25.0%	0	Japan
Aug.27, 03	80	11	13.8%	0	Japan
Oct.05, 03	12	2	16.7%	0	Japan
Nov.05, 03	10	1	10.0%	0	Canada
Jun.25, 04*	30	0	0.0%	0	Japan
Total	200	35	17.5%	0	

[a]The tests were conducted in the presence of Dr. Andreas Baumgartner of the Swiss Federal Office of Public Health.

while that of mayonnaise samples was 17.6% (TABLE 1). Sequencing of the PCR products from the positive egg and mayonnaise samples revealed gene sequence identical to that of *C. burnetii.* Mutations were found in 14% of the sequence-confirmed positive egg samples (TABLE 2), whereas no mutation was found in any of the PCR products obtained from the positive mayonnaise samples. The number of bacteria in positive egg samples detected by real-time PCR ranged from 10^4 to 10^6 (TABLE 3). We gave PCR-positive egg samples orally to scid mice. In the immunostaining analysis, tissue sections of their spleens, livers, and intestines were strongly stained with a *C. burnetii*–specific monoclonal antibody. *C. burnetii* DNA was also detected in egg products sold in South Korea, Canada, and the Philippines by PCR tests.

CONCLUSION

C. burnetii was detected in market chicken eggs and mayonnaise in several countries, including Japan. There is a strong possibility that the causative agent of Q fever in the contaminated food is alive.

TABLE 2. Mutations of *C. burnetii com1* gene segments in eggs by direct sequence

	Amino Acid No.																
	97	103	109	116	127	131	136	137	142	151	155	163	164	166	170	175	177
C. burnetii Nine mile	ggc Gly	aca Thr	tgt Cys	aat Asn	aaa Lys	gtt Val	ctg Leu	ccc Pro	caa Gln	tta Leu	aaa Lys	cac His	gac Asp	ctg Leu	gac Asp	gaa Glu	atc Ile
Mutation 1	ggt Gly		cgt Arg														
Mutation 2		gca Ala			aga Arg												
Mutation 3				aac Asn													
Mutation 4					aga Arg												
Mutation 5						gtc Val											
Mutation 6							cag Gln										
Mutation 7								cct Pro	cga Leu								
Mutation 8										cta Leu				ccg Pro			
Mutation 9											aag Lys						
Mutation 10												cat His	aac Asn		aac Asn	gag Glu	
Mutation 11															ggc Gly		
Mutation 12																	aac Asn

NOTE: We have set the following two conditions for determining mutations:
(1) As to a particular base, changes should be found at the same site in mutation screening using a pair of primers.
(2) There is no peak in the curve that is thought to belong to the other base.

TABLE 3. Quantitative PCR for *C. burnetii* in eggs

Sample No.	Number of molecules of *C. burnetii com1* gene	Recognition of bands	Sequence
29	$386^{a} \times 30^{b} \times 9^{c} = 104{,}220^{d}$	+	+
66	$38 \times 30 \times 9 = 10{,}260$	+	+
106	$6{,}181 \times 30 \times 9 = 2{,}208{,}870$	+	+
121	$235 \times 30 \times 9 = 63{,}450$	+	NT
125	$153 \times 30 \times 9 = 41{,}310$	+	+
150	$268 \times 30 \times 9 = 72{,}360$	+	+
164	$109 \times 30 \times 9 = 29{,}430$	+	NT
169	$122 \times 30 \times 9 = 32{,}940$	+	+

NOTE: NT: not tested; quantity of DNA collected was too little to sequence.
[a]Numeric value calculated by standard curve.
[b]1/30 of extracted DNA was used as templates. Therefore, multiply 30.
[c]As 1/9 of one egg yolk was used for DNA extraction, multiply 9.
[d]Number of molecules of *C.burnetii com1* gene contained in one egg yolk.

REFERENCES

1. ZHANG, G.Q. *et al.* 1998. Clinical evaluation of a new PCR assay for detection of *Coxiella bunetii* in human serum samples. *J. Clin. Microbiol.* Jan.: **36:** 77–80.
2. ZHANG, G.Q. *et al.* 1997. Differentiation of *Coxiella burnetii* by sequence analysis of the gene (*com1*) encoding a 27-kDa outer membrane protein. *Microbiol. Immunol.* **41(11):** 871–877.
3. TO, H. *et al.* 1995. Isolation of *Coxiella burnetii* from dairy cattle and ticks, and some characteristics of the isolates in Japan. *Microbiol. Immunol.* **39(9):** 663–671.
4. SOBESLAVSKY, O. *et al.* 1959. Transovular transmission of C. burnetii in the domestic fowl (Gallus gallus domesticus). *J. Hyg. Epidemiol. Microbiol. Immunol.* **III:** 458–464.

Efficacy of Several Anti-Tick Treatments to Prevent the Transmission of *Rickettsia conorii* under Natural Conditions

A. ESTRADA-PEÑA AND J.M. VENZAL BIANCHI

Department of Parasitology, Veterinary Faculty, Miguel Servet 177, 50013-Zaragoza, Spain

ABSTRACT: The efficacy of several anti-tick treatments to prevent the transmission of *Rickettsia* was evaluated under natural conditions of tick pressure in a kennel. Only Amitraz (Preventic) provided total control on transmission (no dogs were infected), whereas with Frontline, Advantix, or Scalibor, the rate of infection varied among the dogs.

KEYWORDS: *Rickettsia conorii*; anti-tick treatment; protection

INTRODUCTION

Rickettsial organisms of the group of Mediterranean group fever are commonly transmitted to dogs and humans through the bite of the brown dog tick, *Rhipicephalus sanguineus (R. sanguineus)*. This tick is abundant in kennels and human constructions around the Mediterranean region. Sites infested by ticks are risky areas for pathogen transmission to dogs. A high parasite reservoir rate is maintained in ticks from feeding on infected animals. The use of antitick treatments seems to be the only effective preventive measure against infection. However, given variable efficacy among acaricide molecules and modes of application, substantial difference is expected between treatments in their ability to break rickettsial transmission to dogs. The purpose of this study was to compare the efficacy of four anti-tick treatments to protect dogs against *Rickettsia conorii* (*R. conorii)* infection.

MATERIALS AND METHODS

A kennel was selected, where previous sampling had recorded the presence of *R. sanguineus* infected with *R. conorii*. The protocol was performed between mid May and early August, 2004. Thirty cross-bred tick-naïve dogs

Address for correspondence: A. Estrada-Peña, Department of Parasitology, Veterinary Faculty, Miguel Servet 177, 50013-Zaragoza, Spain. Voice: +34-976-761-558; fax: +34-976-761-612.
e-mail: aestrada@unizar.es

Ann. N.Y. Acad. Sci. 1078: 506–508 (2006). © 2006 New York Academy of Sciences.
doi: 10.1196/annals.1374.097

(18 male, 12 female), weighing from 15 kg to 21 kg, and determined to be free of rickettsial infection, were included in the study. On day 0 (D0) the animals were randomly allocated to 5 groups of 6 dogs. In the four treatment groups, anti-tick measures included Fipronil Spot-on (Frontline©), Deltamethrin collar (Scalibor©), Permethrin-Imidacloprid Spot-on (Advantix©), and Amitraz collar (Preventic©), respectively. According to manufacturer recommendations, collars were left on the dogs while spot-on treatment was repeated monthly. Another group of dogs remained untreated (controls). From D+14 to D+77, nymphal and adult *R. sanguineus* ticks were counted once a week on each dog. On days D+14, D+28, D+56, and D+84, blood was sampled from each dog and PCR carried out to detect the presence of rickettsial organisms of the Mediterranean group fever. DNA extraction from ticks was performed using the QIAamp tissue kit (Qiagen, Hilden, Germany). The primer set Rp.CS.877-Rp.CS.1258n was used to amplify a 381-bp sequence of the citrate synthase gene.[1] Both positive and negative controls were used. DNA from positive samples was also amplified, using the primer set Rr190.70p-Rr190.602n, which codes for a 532-bp sequence of the 190-kDa surface protein gene.[2,3]

RESULTS AND DISCUSSION

Between D+42 and D+49, the tick burden of individuals in the control group was higher than 100 nymphs—a maximum count of 100 was nevertheless recorded, while an average number of 40 ticks/animal was recorded in the Frontline ©or Scalibor© groups. The lowest tick counts were recorded with the Amitraz collar. Overall nymphal activity sharply decreased as from D+63. Adult tick natural activity decreased between D+35 and D+49, when newly molted nymphs accumulated in the adult population. No dog was found positive for *R. conorii* on D+14 (FIG. 1). Some individuals in the control group

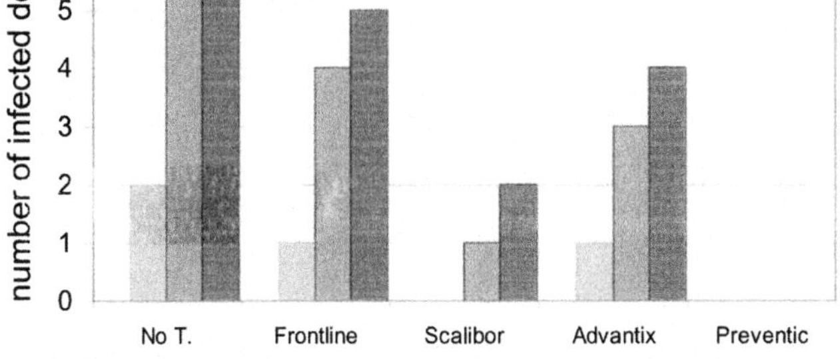

FIGURE 1. Number of dogs with *Rickettsia* at several dogs after treatment (14, 28, 56, and 84) in the different groups of anti-tick treatment. No T. : no treatment.

(2, 33%), Frontline© and Advantix© (1 each, 16%) had become infected by D+28. All dogs in the control group (No T. in the figure) were found positive on D+56. Up to 83% and 66% of dogs in the Frontline© and Advantix© groups, respectively, had become infected on D+84. No dog in the Preventic© group was infected over the study period. This study shows that Amitraz collar is the only anti-tick treatment that provides complete protection against *R. conorii* infection in dogs. This finding is correlated with low tick burden in dogs wearing Amitraz collar and exposed to high brown tick infestation pressure. Considering the figures of tick loads and the percentage of infected animals in this study, it is likely that at midsummer high nymphal tick burden plays a key role in most transmissions, a time when most chemicals exert poor residual activity, below the critical threshold of acaricidal efficacy necessary to break tick-borne disease transmission.

REFERENCES

1. Psaroulaki, A., F.C. Loukaidis & Y. Tselentis. 1999. Detection and identification of the etiological agent of Mediterranean spotted fever (MSF) in two genera of ticks in Cyprus. Trans. R. Soc. Trop. Med. Hyg. **93:** 597–598.
2. Babalis, T., Y. Tselentis, A. Roux, *et al.* 1994. Isolation and identification of a Rickettsial strain related to *Rickettsia massiliae* in Greek ticks. Am. J. Trop. Med. Hyg. **50:** 365–372.
3. Zhang, X.F., M.Y. Fan, J. Chen & D.Z. Bi. 1994. Genotypic identification of seven *Rickettsia conorii* strains. Acta Virol. **38:** 35–37.

Rickettsia spp. in *Ixodes ricinus* Ticks in Bavaria, Germany

R. WÖLFEL,[a] R. TERZIOGLU,[a] J. KIESSLING,[b] S. WILHELM,[b,c] S. ESSBAUER,[a] M. PFEFFER,[a] AND G. DOBLER[a]

[a]*Department of Virology and Rickettsiology, Bundeswehr Institute of Microbiology, Neuherbergstr. 11, 80937 Munich, Germany*

[b]*Institute of Medical Microbiology, Epidemic and Infectious Diseases, LMU Munich, Germany*

[c]*Institute of Animal Hygiene and Veterinary Public Health, University of Leipzig, Leipzig, Germany*

ABSTRACT: This study aims to provide information on the occurrence of spotted fever rickettsiae in *Ixodes ricinus* ticks in southern Germany. A total of 2,141 *I. ricinus* ticks was collected in Bavaria. Pools of 5-10 ticks were studied by a PCR targeting the rickettsial citrate synthase gene *gltA*. The average prevalence rate was 12% (257 of 2,141). Sequencing data exclusively identified *Rickettsia helvetica* DNA. Results and other data demonstrate the possible role of *R. helvetica* in *I. ricinus* as a source of human infections in southern Germany.

KEYWORDS: *Ixodes ricinus*; *Rickettsia helvetica*; Bavaria; Germany

INTRODUCTION

The most prevalent and widely distributed tick species in Germany is the sheep tick, *Ixodes (I.) ricinus*. In many European countries these ticks were found to be infected with spotted fever group (SFG) rickettsiae.[1] In Germany, *Rickettsia (R.) helvetica* has thus far only been identified in ticks from Baden-Wuerttemberg.[2] This study was set up to provide information on the occurrence of SFG rickettsiae in *I. ricinus* ticks in Bavaria, southern Germany.

MATERIALS AND METHODS

A total of 2,141 *I. ricinus* ticks (151 female, 174 male, 1204 nymph, 612 larvae) were collected in six different areas of Bavaria (administrative districts of

Address for correspondence: Roman Wölfel, Department of Virology and Rickettsiology, Bundeswehr Institute of Microbiology, Neuherbergstrasse 11, 80937 Munich, Germany. Fax: +49-89-3168-3292.
e-mail: RomanWoelfel@Bundeswehr.org

Ann. N.Y. Acad. Sci. 1078: 509–511 (2006). © 2006 New York Academy of Sciences.
doi: 10.1196/annals.1374.133

Erlangen, Freyung-Grafenau, Hengersberg, Muehldorf, Munich, Traunstein) by both flagging and from trapped rodents (*Clethrionomys glareolus, Apodemus* spp.). For nucleic acid (NA) extraction all larvae and nymphs were pooled together to a pool size of 10 and a few of the adults were pooled to a pool size of 4 (see TABLE 1). All samples were stored at –80°C until PCR was performed. Ticks were homogenized in 1 mL Eagle's minimal essential medium using a Fast Prep device as described by the supplier (Qbiogene, Heidelberg, Germany). NAs were extracted using QIAamp Viral RNA Mini Kit (Qiagen GmbH, Hilden, Germany) according to the manufacturer's protocol. Samples were screened for the presence of rickettsia DNA by PCR amplification of a fragment of the citrate synthase-encoding gene (*gltA*).[3] A plasmid containing a *gltA*-fragment of *R. rickettsii* was used as positive control in each test. Purified amplicons were submitted to a commercial subcontractor for automated dye-terminator cycle sequencing. Sequences were analyzed by BLAST sequencing analysis in the GenBank database.

RESULTS

All 2,141 ticks (254 pooled, 57 single samples) were examined for *Rickettsia* spp. Using the *gltA* primers, PCR products of rickettsial DNA of 340-bp size were detected in 147 of the 311 studied samples. The positive samples were not clustered to any distinct geographical area of the six regions investigated. Sequencing of the positive *gltA* PCR products identified in all of these *R. helvetica* showed 100% identity. The distribution of positive samples is summarized in TABLE 1.

DISCUSSION

I. ricinus ticks play an important role as the vector of pathogens of medical and veterinary importance. Our results support the occurrence of *R. helvetica*

TABLE 1. Distribution of ticks, tick samples, and pooled samples and distribution of various developing stages in different samples

Parameters	Adult males		Adult females		Nymphs	Larvae	Total
No. of ticks	174		151		1,204	612	2,141
	Pools*	single	Pools*	single	Pools*	Pools*	
No. of samples	35	34	32	23	125	62	311
No. of positive samples	18	4	25	3	63	34	147
No. of ticks in positive samples	72	4	100	3	630	340	1,149
% Positive ticks	12.6–43.7		18.5–68.2		5.2–52.3	5.5–55.5	6.9–53.6

*Pool sizes: for adult ticks = 4; for nymphs and larvae = 10.

in *I. ricinus* in Germany[2] by expanding its geographical range to Bavaria. With the assumption that only one tick in each pooled sample was positive for rickettsial DNA, the minimum infection rate would be 6.9% (147 of 2,141). If, on the other hand, all ticks in the pooled sample were positive, the maximum prevalence would be 53.6% (1,149 of 2,141). We assume that the frequency of positive ticks in the pooled samples can be expected to reflect that of the entire selection of the tick material. Therefore, it is probable to assume that the average prevalence rate corresponds to 12% (257 of 2,141). Data from other parts of Germany and Switzerland show similar infection rates.[2,4] These results imply that transmission of *R. helvetica* to humans may occur quite frequently.

The presence of *R. helvetica* in ticks in Bavaria has a diagnostic importance because this rickettsia species has been recognized as a human pathogen.[1] However, to date, no clinical cases of human *R. helvetica* infections have been reported in Germany. Nevertheless, further studies are needed since contact with *I. ricinus* ticks during outdoor activities is common. Thus more attention should be paid to the incidence of tick-borne rickettsioses.

ACKNOWLEDGMENTS

We thank Kathrin Hartelt (Baden-Wuerttemberg State Health Office) for providing the DNA of *Rickettsia* spp. We acknowledge the skillful assistance of Heike Prabel and Mirko Köhler in field collection of ticks. The views expressed in this article are those of the authors and do not reflect the official policy or position of the German Department of Defense, or the German government.

REFERENCES

1. PAROLA, P. 2004. Tick-borne rickettsial diseases: emerging risks in Europe. Comp. Immunol. Microbiol. Infect. Dis. **27:** 297–304.
2. HARTELT, K., R. OEHME, H. FRANK, *et al.* 2004. Pathogens and symbionts in ticks: prevalence of *Anaplasma phagocytophilum* (*Ehrlichia* sp.), *Wolbachia* sp., *Rickettsia* sp., and *Babesia* sp. in Southern Germany. Int. J. Med. Microbiol. **293** (Suppl 37): 86–92.
3. ROUX, V., E. RYDKINA, M. EREMEEVA & D. RAOULT. 1997. Citrate synthase gene comparison, a new tool for phylogenetic analysis, and its application for the rickettsiae. Int. J. Syst. Bacteriol. **47:** 252–261.
4. BURGDORFER, W., A. AESCHLIMANN, O. PETER, *et al.* 1979. *Ixodes ricinus*: vector of a hitherto undescribed spotted fever group agent in Switzerland. Acta Trop. **36:** 357–367.

The Occurrence of Spotted Fever Group (SFG) Rickettsiae in *Ixodes ricinus* Ticks (Acari: Ixodidae) in Northern Poland

JOANNA STAŃCZAK

Institute of Maritime and Tropical Medicine, Medical University of Gdańsk, Gdańsk, Poland

ABSTRACT: *Ixodes ricinus*, the most commonly observed tick species in Poland, is known vector of microorganisms pathogenic for humans as TBE virus, *Borrelia burgdorferi* s.1., *Anaplasma phagocytophilum* and *Babesia* sp. in this country. Our study aimed to find out whether this tick can also transmit also rickettsiae of the spotted fever group (SFG). DNA extracts from 560 ticks (28 females, 34 males, and 488 nymphs) collected in different wooded areas in northern Poland were examined by PCR for the detection of *Rickettsia* sp., using a primer set RpCS.877p and RpCS.1258n designated to amplify a 381-bp fragment of *gltA* gene. A total of 2.9% ticks was found to be positive. The percentage of infected females and males was comparable (10.5% and 11.8%, respectively) and 6.6–7.6 times higher than in nymphs (1.6%). Sequences of four PCR-derived DNA fragments (acc. no. DQ672603) demonstrated 99% similarity with the sequence of *Rickettsia helvetica* deposited in GenBank. The results obtained suggest the possible role of *I. ricinus* as a source of a microorganism, which recently has been identified as an agent of human rickettsioses in Europe.

KEYWORDS: *Rickettsia* sp.; *Ixodes ricinus;* spotted fever group; rickettsiae; Poland

The spotted fever group (SFG) rickettsiae are small, gram-negative, obligate intracellular bacteria. They are mainly associated with hard ticks (Ixodidae), which can transmit them transstadially and transovarially and serve both as vectors and reservoirs of these pathogens. Vertebrates, including humans, are accidental hosts and acquire infection by a tick bite. From about 30 species of SFG rickettsiae described so far, at least 13 are known to be pathogenic for humans, while the rest of them have been isolated only from arthropods and are often considered as nonpathogenic.[1] The purpose of present investigation was to detect *Rickettsia* spp. in Polish *Ixodes ricinus* as a potential

Address for correspondence: Joanna Stańczak, Institute of Maritime and Tropical Medicine, Medical University of Gdańsk, 9B Powstania Styczniowego Street, 81-519 Gdynia, Poland. Voice: +48-58-69-98-551; fax: +48-58-622-33-54.
e-mail: astan@amg.gda.pl

Ann. N.Y. Acad. Sci. 1078: 512–514 (2006). © 2006 New York Academy of Sciences.
doi: 10.1196/annals.1374.100

vector of *R. helvetica*, which recently has been identified as an agent of human rickettioses.[2,3]

Ixodes ricinus were collected in 2004 by flagging lower vegetation in different wooded areas in northern Poland. DNA was extracted by lysis of crushed ticks in NH_4OH.[4] The polymerase chain reaction (PCR) assay was performed using a primer pair RpCS.877p and RpCS.1258n, derived from the citrate synthetase gene *gltA*, which has conserved regions shared by all known *Rickettsia* species.[5] DNA of *R. slovaca*, kindly provided by Professor Didier Raoult from Unité des Rickettsies, Faculté de Médecine, Université de la Méditerranée, Marseille, France, was used as positive control and double-distilled water as negative controls. The conditions of PCR were as already described.[5] All PCR reactions were carried out in a GeneAmp® PCR System 9700 (Applied Biosystems 850, Foster City, CA). Amplification products were analyzed after electrophoresis in a 2% agarose gel stained with ethidium bromide. DNA bands of 381 bp were considered to be positive results. PCR products of chosen positive samples were purified and sequenced with an ABI Prism 3100 Genetic Analyser (Applied Biosystem 850, Foster City, CA) according to the protocol. Sequences were edited and compared with gene sequences obtained from the GenBank database using NCBI BLAST program.

A total of 560 *I. ricinus,* including 38 females, 34 males, and 488 nymphs, were examined individually and 16 samples (2.9%) were positive for rickettsial DNA. Female ticks were infected in 10.5%, males in 11.8%, and nymphs in 1.6%. Results of our study represent the first demonstration of SFG rickettsiae in Polish *I. ricinus* and confirmed the suggestion that these emerging pathogens are also established in the tick population in Poland.

Sequences of four PCR-derived fragments of *gltA* generated from infected ticks (acc. no. DQ672603) compared with the corresponding sequences of the SFG rickettsiae showed the closest similarities (99%) to *R. helvetica* (accession no. AJ427878 and U59723), with only 1–3 nucleotides difference, respectively. This species has been already isolated from *I. ricinus* in several European countries and probably also infects ticks in Poland. However, for precise species differentiation, additional investigations will be conducted.

Although no clinical cases due to infection with rickettsiae of SFG have been reported from Poland, results of this investigation add them to the list of potentially dangerous pathogens transmitted by ticks in that country. However, further systematic sampling is needed because knowledge of their prevalence in ticks and distribution is insufficient. Moreover, the pathogenicity of different rickettsial species has yet to be determined.

REFERENCES

1. La Scola, B. & D. Raoult. 1997. Laboratory diagnosis of rickettsioses: current approaches to diagnosis of old and new rickettsial diseases. J. Clin. Microbiol. **35:** 2715–2727.

2. FOURNIER, P.E., F. GRUNNENBERGER, B. JAULHAC, *et al.* 2000. Evidence of *Rickettsia helvetica* infection in humans, eastern France. Emerg. Infect. Dis. **6:** 389–392.
3. NILSSON, K., O. LINDQUIST & C. PÅHLSON. 1999. Association of *Rickettsia helvetica* with chronic perimyocarditis is sudden cardiac death. Lancet **354:** 1169–1173.
4. RIJPKEMA, S., D. GOLUBIĊ, M. MOLKENBOER, *et al.* 1996. Identification of four genomic groups of *Borrelia burgdorferi* sensu lato in *Ixodes ricinus* ticks in a Lyme borreliosis endemic region of northern Croatia. Exp. Appl. Acarol. **20:** 23–30.
5. REGNERY, R.L., C.L. SPRUILL & B.D. PLIKYATIS. 1991. Genotypic identification of rickettsiae and estimation of interspecies sequence divergence for portions of two rickettsial genes. J. Bacteriol. **173:** 1576–1589.

Molecular Survey of *Ehrlichia canis* and *Anaplasma phagocytophilum* from Blood of Dogs in Italy

L. SOLANO-GALLEGO, M. TROTTA, L. RAZIA, T. FURLANELLO, AND M. CALDIN

Laboratorio d'Analisi Veterinarie "San Marco," Padova, Italy

ABSTRACT: We investigated the prevalence of *Ehrlichia canis* (*E. canis*) ($n = 601$) and *Anaplasma phagocytophilum* (*A. phagocytophilum*) ($n = 460$) infection by means of real-time PCR from blood of Italian dogs. The prevalence of *E. canis* in northern, central, and southern Italy was 2.9%, 8%, and 9.7%, respectively. The prevalence of *A. phagocytophilum* was 0%.

KEYWORDS: *Ehrlichia canis*; *Anaplasma phagocytophilum*; dog; real-time PCR

INTRODUCTION

Ehrlichia and Anaplasma spp. have been identified within monocytes (*E. canis*), granulocytes (*A. phagocytophilum*) and platelets (*A. platys*) in dogs. The major clinical syndromes of *E. canis* include thrombocytopenia and bleeding diathesis, irreversible bone marrow destruction, and polysystemic immune complex disease. *A. phagocytophilum* causes severe lethargy and weakness, polyarthritis, and central nervous system symptoms. There is limited information regarding canine ehrlichiosis and anaplasmosis in Italy. A few studies have reported *E. canis* seroprevalences ranging from 7% to 47% in dogs from central-south Italy.[1,2] Canine granulocytic anaplasmosis has been previously diagnosed serologically in Italy.[3,4] Recently, the first molecular characterization of an *A. phagocytophilum* strain isolated from a sick dog living in southern Italy has been described.[5] The aim of the study was to investigate the prevalence of *E. canis* and *A. phagocytophilum* by molecular testing from dogs living in Italy.

Address for correspondence: Laia Solano-Gallego, Clinica Veterinaria Privata "San Marco," Via Sorio 114/c, 35141 Padova, Italy. Voice: 00-39-049-8561098; fax: 00-39-02-700-51-8888.
e-mail: laia@sanmarcovet.it

**Ann. N.Y. Acad. Sci. 1078: 515–518 (2006). © 2006 New York Academy of Sciences.
doi: 10.1196/annals.1374.101**

MATERIALS AND METHODS

Blood samples from Italian dogs with clinicopathological findings compatible with tick-borne diseases were submitted to Laboratorio d'Analisi Veterinarie "San Marco" (Padua, Italy) for detection of *E. canis* ($n = 601$) and *A. phagocytophilum* ($n = 460$) DNA by independent quantitative real-time PCR (qPCR) using LightCycler® instrument (Roche, Mannheim, Germany) from January 2003 to May 2005. DNA extraction was performed by the High Pure PCR Template Preparation Kit (Roche). PCR amplification was carried out in a final volume of 10 μL, including 2.5 μL of DNA template, 2 μL of LC fast start DNA master^{PLUS} (Roche) and 0.4 μL of *E. canis* or *A. phagocytophilum* LCSet primers and probes following manufacturer's instructions (TIB MOLBIOL). Thermal cycling comprised an initial denaturation step at 95°C for 8 min, followed by 60 cycles for *E. canis* and 45 cycles for *A. phagocytophilum* of a denaturation at 95°C for 5 sec, annealing at 61°C for 10 sec, and extension at 72°C for 11 sec, with quantification of fluorescence signal. A final melting curve analysis was performed at an initial temperature of 72°C for 30 sec, followed by 95°C for 20 sec and 40°C for 30 sec and continuous heating at 0.5°C/sec to 85°C and a cooling step of 40°C for 1 sec. Negative and positive controls (*E. canis* and *A. phagocytophilum* LCSet positive controls, TIB MOLBIOL) were added in each qPCR run. To determine the absolute and practical sensitivities of each qPCR, *E. canis* and *A. phagocytophilum* DNA synthetic targets (*E. canis* and *A. phagocytophilum* LCSet positive controls) were serially diluted in water and in blood from one healthy dog with 10^{8}copies–10^{-2} copies/μL, respectively. Three replicates of five different concentrations (10^5–10^1/μL) of *E. canis* and *A. phagocytophilum* DNA synthetic targets were tested simultaneously in the same run (intra-assay precision). The interassay precision among five assays was assessed by using *E. canis* and *A. phagocytophilum* DNA synthetic target concentrations run on different days. Variability was reported as the coefficient of variation. The specificity was examined by attempted amplification of DNA from others rickettsial species and related genera (*Brucella canis*, *Borrelia burgdorferi*, *Chlamydia psittaci*, *Rickettsia rickettsii*, *R. conorii*, *E. canis,* and *A. phagocytophilum*).

RESULTS

The primers and probes used were specific for *E. canis* and *A. phagocytophilum*, which was demonstrated by the absence of DNA amplification from other organisms. The absolute and practical detection limits of *E. canis* and *A. phagocytophilum* were 10 DNA synthetic targets/μL. The practical intra-assay variations of C_T values among the replicates were 0.45, 1.28, 1.18, 1.25, and 1.02% for the five different concentrations (10^5–10^1) of *E. canis* and for *A. phagocytophilum* were 0.56, 0.30, 3.47, 0.37, and 2.40%, respectively. The practical interassay variations of C_T values for the *E. canis* DNA concentrations ranging from 10^7 to 10^1 were 5.41, 0.99, 2.02, 4.49, 2.51, 0.36, and 2.38% and for *A. phagocytophilum* were 0.20, 0.39, 3.26, 0.23, 0.42, 0.68, and 2.29%, respectively.

The prevalence of *A. phagocytophilum* in northern (n = 393), central (n = 38), and southern (n = 29) Italy was 0%. The total prevalence of *E. canis* was 6.4% (39/601). The prevalence of *E. canis* in northern, central, and southern Italy was 2.9% (7/238), 8% (16/198), and 9.7% (16/165), respectively. Statistical differences were found between the prevalence of *E. canis* in northern and central Italy and between northern and southern Italy (chi-square = 4.7, $P = 0.028$; chi-square = 7.2, $P = 0.007$, respectively). The prevalence of *E. canis* from November to March over the 2 years studied was 4.6% (10/215), while the prevalence of *E. canis* from April to October was 7.5% (29/386). No statistical differences were found between the two seasonal periods (chi-square = 1.099, $P = 0.29$).

DISCUSSION

The *E. canis* seroprevalence in Sardinia and in central and southern Italy has been shown to be 47%[1] and 13%[2], respectively. Another study detected *E. canis* DNA by PCR–RFLP in 2.6% of dogs with clinical signs compatible with tick-borne disease.[6] The present study indicates that dogs in Italy are frequently infected to *E. canis* (6.4%), and it seems that *E. canis* infection is more prevalent in central and southern regions than in northern Italy.

Despite *A. phagocytophilum* seroprevalence in Italian dogs of 67%,[4] we were unable to detect *A. phagocytophilum* DNA in any of the samples. It is possible that dogs are exposed to *A. phagocytophilum*, and infection is rapidly cleared. Canine anaplasmosis is endemic in northern Europe while the disease was thought to be nonendemic in southern Europe. Recently, *A. phagocytophilum* infection in dogs has been described in southern Europe by means of molecular and serological testing.[5,7] Because of the low numbers of dogs examined in this study from central and southern Italy, further molecular testing in a larger number of dogs is needed to assess the true prevalence of *A. phagocytophilum* in those regions.

Based on our results and clinical observations, many times it is difficult to find a proven etiology in dogs with clinicopathological findings compatible with tick-borne disease. Similar findings are reported in Europe[8], where only 39% of human patients with acute febrile illnesses after tick bite had a proven etiology. For this reason, broad range PCR primers with subsequent sequencing must be applied to detect new possible rickettsial agents in humans and animals.

REFERENCES

1. Cocco, R. *et al.* 2003. Ehrlichiosis and rickettsiosis in a canine population of Northern Sardinia. Ann. N.Y. Acad. Sci. **990:** 126–130.
2. Cuteri, V. *et al.* 2002. *Ehrlichia canis*: indagine sierologica nel cane. Veterinaria **16:** 69–74.
3. Gravino, A.E. *et al.* 1997. Preliminary report of infection in dogs related to *Ehrlichia equi*: description of three cases. N. Microbiol. **20:** 361–363.
4. Furlanello, T. *et al.* 2001. Concurrent coinfections in dogs detected by serology during a survey for *Rickettsia rickettsii*: results from 1093 serum samples collected in Italy. J. Vet. Intern. Med. **15:** 3.
5. Manna, L. *et al.* 2004. First molecular characterization of a granulocytic *Ehrlichia* strain isolated from a dog in South Italy. Vet. J. **167:** 224–227.
6. Di Martino, B. *et al.* 2004. Identificazione diretta di *Ehrlichia canis* mediante una metodica PCR-RFLP. Veterinaria **18:** 39–45.
7. Mylonakis, M.E. *et al.* 2004. Chronic canine ehrlichiosis (*Ehrlichia canis*): a retrospective study of 19 natural cases. J. Am. Anim. Hosp. Assoc. **40:** 174–184.
8. Lotric-Furlan, S. *et al.* 2001. Prospective assessment of the etiology of acute febrile illness after a tick bite in Slovenia. Clin. Infect. Dis. **33:** 503–510.

Spotted Fever Group Rickettsial Infection in Dogs from Eastern Arizona

How Long Has It Been There?

WILLIAM L. NICHOLSON, RONDEEN GORDON, AND LINDA J. DEMMA

Disease Assessment and Epidemiology Team, Viral and Rickettsial Zoonoses Branch, Centers for Disease Control and Prevention, Atlanta, Georgia 30333, USA

ABSTRACT: A serosurvey of free-roaming dogs for antibodies to spotted fever group rickettsiae was conducted using archival samples that had been collected in the White Mountain region of eastern Arizona during a plague study in 1996. Immunoglobulin G antibodies to *Rickettsia rickettsii* (5.1%) and to *R. rhipicephali* (3.6%) were demonstrated, and no cross-reactive samples were identified. This study indicates that *R. rickettsii* was likely present in the dog populations in this area prior to the recognition of human cases of Rocky Mountain spotted fever (RMSF). The role of dogs as short-term reservoirs and primary hosts for the vector tick, *Rhipicephalus sanguineus*, should receive closer attention.

KEYWORDS: *Rickettsia rickettsii*; *Rickettsia rhipicephalus*; spotted fever group rickettsiae; canine serosurvey

INTRODUCTION

Rocky Mountain spotted fever (RMSF) occurs over a wide geographical area in the United States, but most cases are reported from the eastern and central states. Arizona has reported cases of RMSF since at least 1912,[1] but has consistently identified only one or two cases in any year. Cases had been recorded from the White Mountain region in eastern Arizona by the 1920s[2] From 1981–2000, only three cases have been reported from the entire state. The expected tick vectors, *Dermacentor andersoni* and *Dermacentor variabilis*, are not widely distributed in the state, which may account for the low prevalence of human infection.

In 2003, the Centers for Disease Control and Prevention (CDC) identified a case of RMSF in a 14-month-old child residing in the rural eastern White

Address for correspondence: Dr. William L. Nicholson, Viral and Rickettsial Zoonoses Branch, Centers for Disease Control and Prevention, Mail Stop G-13, 1600 Clifton Road, Atlanta, GA 30333, USA. Voice: 404-639-1095; fax: 404-639-4436.
e-mail: wan6@cdc.gov

Ann. N.Y. Acad. Sci. 1078: 519–522 (2006). © 2006 New York Academy of Sciences.
doi: 10.1196/annals.1374.102

Mountain area of Arizona.[3] The case was confirmed by laboratory testing of the serum sediment by polymerase chain reaction (PCR) and nucleotide sequencing to determine the species. This led to an intensive epidemiologic study that is still under way. We were able to identify 13 additional cases and another fatality in this relatively small population (ca. 12,000 people). Thus, the average annual incidence (2002–2004) in this population was over 300 times greater than the national average incidence for the United States. These cases of RMSF in the southeastern United States have stimulated ecological studies of *Rickettsia rickettsii* and tick hosts.

As a part of the epidemiologic study, canine serosurveys were conducted. The prevalence in the area was extremely high in widely separated subdivisions (25–100%) and averaged 62% over the entire region. Seroreactivity to *R. rickettsii* and *R. rhipicephali* were observed and high levels of cross-reactivity may have been present. Because this appeared to have been an emerging infection in Arizona, we conducted a serosurvey on archival canine serum samples that had been collected from this region.

METHODS

Archival specimens were available from a serosurvey of animals for plague antibodies conducted by the CDC and the Arizona Department of Health in 1996. Whole blood samples collected from dogs had been placed on Nobuto blood collection strips. The dried blood was then eluted into phosphate-buffered saline (PBS) and stored as 1/32 dilutions after testing at −70°C. In the Rickettsial Diagnostic Laboratory, samples were thawed and tested at the stored concentration. Any reactive samples were further serially diluted in PBS at pH 7.4. The samples were tested in indirect immunofluorescence assays (IFA) employing both *R. rickettsii* and *R. rhipicephali* antigens. A standard assay format was followed and the conjugate used was fluorescein isothiocyanate (FITC)-labeled goat anti-dog IgG (γ), which reacts specifically with canine IgG.

RESULTS AND DISCUSSION

Of the 353 samples, only 329 contained adequate volume for IFA testing. Seventeen (5.1%) of the 329 dogs tested screened positive for IgG antibodies to *R. rickettsii*, whereas 12 (3.6%) additional dogs were positive for antibodies to *R. rhipicephali*. No samples were positive to both antigens, thus removing cross-reaction as a significant issue in this study. The endpoint titers were low, and only two of the dogs had titers that exceeded 1/32. Both of these animals were positive for antibodies to *R. rickettsii* at titers of 1/128.

The objective of this study was to evaluate canine samples collected in 1995 for exposure to spotted fever group rickettsiae to show historical presence in this area. Although dogs had been known to be susceptible to experimental infection since 1933,[4] the first evidence of naturally occurring RMSF in dogs was reported in 1980.[5] Signs in dogs include sloughing of skin, fever, shifting leg lameness, malaise, bloody vomiting, and diarrhea. The organism can be recovered from the blood for 6–10 days. Thus, dogs may function as short-term reservoirs for the pathogen. Humans could become infected by tick bite or by mucosal contamination following mechanical removal of ticks from dogs. In addition, dogs serve as primary hosts for the brown dog tick, *Rhipicephalus sanguineus*, and they contribute to maintaining large populations of ticks that can feed on animals and humans. Dogs can serve as sentinals for RMSF in human populations, and infections in canines have often been associated with an increased risk for the disease in owners or residents in close proximity. Dogs may serve as transport hosts, carrying infected ticks, which may establish a focus of infection in or near the residence.

Very limited volumes of diluted serum elutions were available. This study would have benefited from additional serum volume so that additional studies could have been conducted. Dilute protein solutions are not stable, and we do not know whether these samples had experienced any prior freeze–thaw cycles. Certainly that could have led to a decreased prevalence and decreased endpoint titers. However, it is interesting to compare these values to our more recently collected seroprevalence and seroreactivity data. We have found seroprevalences that exceed 60% in many subdivisions of the community, and endpoint titers were significantly higher (> 16,384) than those in samples from a decade ago. We have also noted that many animals have significant titers to both *R. rickettsii* and *R. rhipicephali* antigens. Because we did not see cross-reactions between the two rickettsiae in our archived samples, this may suggest a current high level of transmission by both agents, with boosting by either the homologous or heterologous antigen.

Residents noted that the communities have experienced an increase in stray animal activity, particularly stray dogs, in the last 5–6 years. This large increase in the stray animal population could be a contributing factor in the development and maintenance of RMSF within the dogs and ticks in this area. Although *R. sanguineus* is an infrequent human biter, high tick infestations could provide the opportunities for transmission of the pathogen to humans.

CONCLUSION

RMSF was likely present in the dog populations in this area for several years prior to the recognition of human cases. The role of dogs as short-term reservoirs and primary hosts for the vector tick, *R. sanguineus*, should receive closer attention. Tick control efforts focused on the dog will play a valuable role in the control and prevention of RMSF arising from this ecological cycle.

ACKNOWLEDGMENTS

Rondeen Gordon was sponsored by the James A. Ferguson Emerging Infectious Diseases Fellowship. We thank Mira Leslie and the Plague Laboratory at CDC, Ft. Collins and Craig Levy, AZDHS for providing the archival samples.

REFERENCES

1. McClintic, T.B. 1912. Investigations of and tick eradications in Rocky Mountain spotted fever. Public Health Rep. **27:** 732–760.
2. McDade, J.E. & V.F. Newhouse. 1986. Natural history of *Rickettsia rickettsii*. Annu. Rev. Microbiol. **40:** 287–309.
3. Demma, L.J., M.S. Traeger, W.L. Nicholson, *et al.* 2005. Rocky Mountain spotted fever from an unexpected tick vector in Arizona. N. Engl. J. Med. **353:** 587–594.
4. Badger, L.F. 1933. Rocky Mountain spotted fever: susceptibility of the dog and sheep to the virus. Public Health Rep. **48:** 791–795.
5. Lissman, B.A. & J. L. Benach. 1980. Rocky Mountain spotted fever in dogs. J. Am. Vet. Med. Assoc. **176:** 994–995.

Isolation of *Rickettsia rickettsii* and *Rickettsia bellii* in Cell Culture from the Tick *Amblyomma aureolatum* in Brazil

ADRIANO PINTER AND MARCELO B. LABRUNA

Department of Preventive Veterinary Medicine and Animal Health, Faculty of Veterinary Medicine, University of São Paulo, São Paulo 05508-270, Brazil

ABSTRACT: Brazilian spotted fever (BSF) is a highly lethal disease caused by *Rickettsia rickettsii.* In the present study, rickettsial infection was evaluated in 669 *Amblyomma aureolatum* adult ticks collected from naturally infested dogs in Taiaçupeba, a BSF-endemic area in the state of São Paulo. Ten (1.49%) ticks were infected with *Rickettsia bellii,* and 6 (0.89%) ticks were infected with *R. rickettsii.* Both *Rickettsia* species were isolated and established in Vero cell cultures. The *Rickettsia* isolates were characterized by molecular analyses, sequencing fragments of different rickettsial genes. Our results suggest that *A. aureolatum* is an important vector of *R. rickettsii* in Brazil.

KEYWORDS: *Rickettsia rickettsii*; *Rickettsia bellii*; Brazilian spotted fever; *Amblyomma aureolatum*; Brazil

INTRODUCTION

Rickettsia rickettsii and *Rickettsia bellii* have been isolated or detected on several tick species from the Americas.[1,2,3] Whereas *R. rickettsii* is one of the most pathogenic rickettsiae to humans, there is no evidence of human or animal infection by *R. bellii.*

Brazilian spotted fever (BSF), caused by *R. rickettsii*, is an endemic disease in many areas of southeastern Brazil. This disease is also endemic in many parts of the United States, Mexico, Costa Rica, Panama, and Colombia. Tick species that have been recognized as main vectors of *R. rickettsii* are *Dermacentor andersoni* and *Dermacentor variabilis* in the United States, and *Amblyomma cajennense* in Latin America.[1] A reported isolation of *R. rickettsii* from *A. aureolatum* in Brazil during the 1930s[4] has suggested that this tick could also participate as vector of *R. rickettsii.* This suggestion is reinforced by the several reports of human infestation by the adult *A. aureolatum.*[5,6]

Address for correspondence: Marcelo B. Labruna, Faculdade de Medicina Veterinária e Zootecnia, Universidade de São Paulo, Avenue Orlando Marques de Paiva, 87 São Paulo, Brazil 05508-270. Voice: 55-11-3091-1394; fax: 55-011-3091-7928.
e-mail: labruna@usp.br

Ann. N.Y. Acad. Sci. 1078: 523–530 (2006). © 2006 New York Academy of Sciences.
doi: 10.1196/annals.1374.103

The adult *A. aureolatum* feeds most frequently on dogs in rural areas of southeastern Brazil while those at the immature stages seem to feed mostly on passerine birds and some small rodent species.[5,6] The present study evaluated the rickettsial infection in *A. aureolatum* adult ticks collected on dogs from a BSF-endemic area in the state of São Paulo, Brazil.

MATERIAL AND METHODS

Collection of Ticks

This study was conducted in the rural area of Taiaçupeba, Mogi das Cruzes County, state of São Paulo, Brazil (46°11′W, 23°38′S). In this same area, several human lethal cases of BSF have been reported recently.[6] A total of 706 ticks were collected from 21 domestic dogs that lived in nine small farms from December 2000 to November 2002. Details on the study site and tick collection have been reported elsewhere.[6] For the present study, 669 of these ticks were brought alive to the laboratory and further tested.

Hemolymph Test, DNA Extraction, and PCR on Ticks

Each tick specimen was individually tested by the hemolymph test using Gimenez staining for detection of *Rickettsia*-like organisms, as previously described.[2] Right after collection of hemolymph, ticks were frozen at −80°C.

Each frozen tick was placed on dry ice previously covered with an aluminum sheet. Tick legs were extirpated without defrosting the tick, which was immediately returned to the −80°C freezer. DNA was extracted from the legs of each individual tick by the guanidine isothiocyanate–phenol technique, as previously described.[7]

Polymerase chain reaction (PCR) amplification of a rickettsial gene fragment of 381 nucleotides of the citrate synthase gene (*gltA*) was tried on DNA pools of every five ticks, using the primers RpCS.877p and RpCS.1258n (TABLE 1). If a pool demonstrated an expected PCR product, DNA of each tick specimen that composed that pool was individually tested by two new PCRs, one using the primers RpCS.877p and RpCS.1258n, and another using primers Rr190.70p and Rr190.602n (TABLE 1). PCR products of the expected sizes were purified and sequenced as previously described[2], and compared with NCBI Nucleotide BLAST searches.

Isolation of Rickettsiae

Attempts to isolate rickettsiae were performed with two *A. aureolatum* tick specimens: one that was shown to contain a DNA sequence 100% similar to *R. rickettsii* and another that was shown to contain a DNA sequence 100%

TABLE 1. Primer pairs used for amplification of rickettsial genes

Primer pairs	Genes and primers	Primer sequences (5′–3′)	Reference	Position on gene relative to ORF
	gltA			
1	RpCS.877p	GGGGGCCTGCTCACGGCGG	14	638–656
	RpCS.1258n	ATTGCAAAAAGTACAGTGAACA	14	1015–995
2	CS-78	GCAAGTATCGGTGAGGATGTAAT	2	−78–56
	CS-323	GCTTCCTTAAAATTCAATAAATCAGGAT	2	323–296
3	CS-239	GCTCTTCTCATCCTATGGCTATTAT	15	239–263
	CS-1069	CAGGGTCTTCGTGCATTTCTT	15	1069–1049
	17-kDa			
4	17k-5	GCTTTACAAAATTCTAAAAACCATATA	2	−62–34
	17k-3	TGTCTATCAATTCACAACTTGCC	2	+6–464
	ompA			
5	Rr190.70p	ATGGCGAATATTTCTCCAAAA	14	478–499
	Rr190.602n	AGTGCAGCATTCGCTCCCCCT	14	990–969
	ompB			
6	BG1-21	GGCAATTAATATCGCTGACGG	16	696–716
	BG2-20	GCATCTGCACTAGCACTTTC	16	1349–330

similar to *R. bellii* (using the PCR protocol described above for tick legs). Isolation of *R. rickettsii* was performed by inoculating a brain-heart infusion of the infected tick homogenate intraperitoneally in albino guinea pigs (*Cavia aperea porcellus*). Fresh blood from a febrile guinea pig was inoculated onto shell vials containing confluent monolayers of Vero cells and processed as previously described.[2,8] Isolation of *R. bellii* was performed by inoculating the infected tick homogenate directly onto shell vials, as previously described.[2,8]

Infection of the inoculated shell vials was monitored every 3 days by Gimenez staining.[2] If *Rickettsia*-like organisms were visualized, the monolayer of the shell vial was harvested and inoculated into 25 or 150 cm^2 flasks containing a monolayer of confluent uninfected Vero cells. Cell passages of rickettsiae were performed when 90–100% cells were infected, as demonstrated by Gimenez staining. Rickettsial growth in infected Vero cells was observed in three temperatures: 28°, 32°, and 37°C.

Molecular Characterization of the Rickettsial Isolates

For each of the rickettsial isolates, a sample of 100% infected cells from the fourth Vero cell passage was subjected to DNA extraction as described above and thereafter tested by a battery of PCRs targeting fragments of the rickettsial genes *gltA, ompA, ompB,* and the 17-kDa protein gene, using the primer pairs as described in TABLE 1. PCR conditions were as previously described.[2,3] Amplified products were purified and sequenced as previously described[2] and compared with NCBI Nucleotide BLAST searches.

RESULTS

Infection of Ticks by Rickettsiae

By the hemolymph test, a total of 11 (1.64%) *A. aureolatum* ticks contained typical *Rickettsia*-like organisms inside hemocytes, 394 ticks were hemolymph-negative, and 264 ticks yielded inconclusive results because hemolymph cells were completely lost during the washing procedures of the hemolymph test. A total of 16 (2.39%) ticks contained rickettsial DNA by the PCR targeting a fragment of the *gltA* gene, including the 11 hemolymph-positive ticks and 5 hemolymph-inconclusive ticks.

From the 16 *gltA*-positive ticks by PCR, only 6 (0.89% of 669 ticks) were also PCR-positive for a fragment of the *ompA* rickettsial gene. Nucleotide sequences of PCR products of the 10 ticks (1.49%) positive only for the *gltA* gene were identical to each other and 100% (305/305) similar to the corresponding sequence of *R. bellii* in GenBank (U59716). Nucleotide sequences of PCR products of the 6 ticks (0.89%) positive for the *ompA* gene were identical to each other and 100% (392/392) similar to the corresponding sequence of *R. rickettsii* in GenBank (U43804). *R. bellii*-infected ticks were collected during December 2000 (3 specimens), January 2001 (4), June 2001 (1), June 2002 (1), and August 2002 (1). *R. rickettsii*-infected ticks were collected during January 2001 (1 specimen), March 2001 (1), June 2001 (1), March 2002 (1), June 2002 (1), and October 2002 (1). *R. bellii* was detected in 6 female and 4 male ticks, whereas *R. rickettii* was detected in 1 female and 5 male ticks.

Isolation of Rickettsia

R. rickettsii was successfully isolated in Vero cell culture inoculated with guinea-pig-infected blood, whereas *R. bellii* was successfully isolated in Vero cell culture inoculated with tick homogenate. Shell vials inoculated with guinea-pig-infected blood showed typical *Rickettsia*-like organisms by the fifteenth day and were inoculated into 25 cm^2 flasks, which showed 100% infected cells after 6 days, at 32°C. Then infected cells from these flasks were inoculated into 150 cm^2 flasks, which also showed 100% infected cells after 6 days. The *R. rickettsii* isolate displayed stronger cytopathic effect when incubated at 32°C than at 28°C, even though the monolayers underwent 100% infected cells after 6 days at both temperatures. Attempts to propagate the *R. rickettsii* isolate in cells incubated at 37°C never resulted in more than 50% of infected cells.

Shell vials inoculated with *R. bellii*-infected tick homogenate showed typical *Rickettsia*-like organisms by the third day and were inoculated into 25 cm^2 flasks, which showed 100% infected cells after 5 days, at 32°C. Then infected cells from these flasks were inoculated into 150 cm^2 flasks, which also showed 100% infected cells after 5 days. The *R. bellii* isolate displayed a strong cytopathic effect when incubated at 28°C, destroying the monolayer completely in 4 days. However, when cells were incubated at 32°C, there was low cytopathic effect, even when cells were 100% infected in 5 days. Further attempts to propagate the *R. bellii* isolate in cells at 37°C never resulted in more than 10% of infected cells.

Molecular Characterization of the Rickettsial Isolate

DNA of *R. rickettsii*- or *R. bellii*-infected cells at the fourth passage was subjected to PCR targeting the *gltA, ompA, ompB,* and 17-kDa protein genes. For the *R. rickettsii*-infected cells, PCR products of the expected size were obtained with all primer pairs listed in TABLE 1. The *gltA, ompA, ompB,* and 17-kDa partial sequences were 99.7% (1072/1075), 100% (491/491), 100% (529/529), and 100% (435/435), respectively, similar to the corresponding sequences of *R. rickettsii* in GenBank (U59729, U43804, X16353, AY281069). The difference between the *gltA* sequences was due to an insertion or deletion (indel) of the codon CCG at the position 865–867 of the open reading frame (ORF) of the *R. rickettsii gltA* gene. This *Rickettsia* isolated from an *A. aureolatum* tick, named isolate Taiaçu, was genetically identified as *R. rickettsii*.

For *R. bellii*-infected cells, PCR products of the expected sizes were obtained with the *gltA* and 17-kDa primers listed in TABLE 1, but no product was obtained with the *ompA* or *ompB* primers. Fragments of 1053 and 493 nucleotides of the *gltA* and 17-kDa genes, respectively, were sequenced. The *gltA* partial sequence was 100% (1053/1053) similar to the *R. bellii gltA* sequence (U59716), and the 17-kDa partial sequence was 99.8% (477/478) similar to the *R. bellii* 17-kDa gene sequence (AY363702). This *Rickettsia* isolated from an *A. aureolatum* tick, named as isolate Mogi, was genetically identified as *R. bellii*.

The two rickettsial isolates of the present study were successfully established in cell culture and have been deposited as reference strains in our laboratory. GenBank nucleotide sequence accession numbers for the partial sequences generated in this study are DQ115890 for the *gltA* partial sequence of *R. rickettsii* and DQ115891 for the 17-kDa partial sequence of *R. bellii*.

DISCUSSION

The present study showed that *A. aureolatum* ticks collected on dogs from a BSF-endemic area were infected by *R. rickettsii* (0.89%) and *R. bellii* (1.49%). This is the first report of *R. bellii* and the second of *R. rickettsii* in the tick *A. aureolatum*. Decades ago, Gomes[4] isolated *R. rickettsii* from an *A. aureolatum* adult specimen (erroneously referred to as *Amblyomma ovale*[9]) collected from a dog living in a house in São Paulo city, where humans had died of BSF. *R. bellii* has been found in several tick genera, including *Dermacentor, Haemaphysalis, Argas*, and *Ornithodoros* from North America and *Amblyomma* from South America.[2,3]

The prevalence of 0.89% for *R. rickettsii* in *A. aureolatum* ticks is within the range (0.05–1.3%) reported for this *Rickettsia* species in *D. variabilis* ticks in the United States.[1] Our finding of two *Rickettsia* species infecting a single tick population is similar to previous studies that demonstrated some populations of *D. andersoni* and *D. variabilis* to be infected by two or more *Rickettsia* species, including *R. rickettsii* and *R. bellii.*[1] Although it has been observed that a previous infection by one *Rickettsia* species can preclude a tick specimen to become infected by a second *Rickettsia* species,[1,10] the role of *R. bellii* in the natural history of *R. rickettsii* is yet to be determined.

Our findings of *R. rickettsii* infecting ticks at different months throughout the 24-month study period indicate that this bacterium is established in the studied area. The ticks of the present study were collected during another study on seasonal dynamics of *A. aureolatum* ticks,[6] which concluded that adults of *A. aureolatum* parasitize dogs during all months of the years, without any tendency for higher or lower tick activity in any period of the year. As the *R. rickettsii* infection also was shown to be present in ticks during the four seasons of the year, the human and dog population of Taiaçupeba are likely to be exposed to BSF at similar risk levels throughout the year. In fact, the occurrence of human cases of BSF in Taiaçupeba has shown no distinct seasonality.[6] This scenario contrasts to that observed in other BSF-endemic areas, where *A. cajennense* is incriminated as the vector of *R. rickettsii.*[11,12] The ecology of *A. cajennense* in southeastern Brazil is characterized by a one-year generation pattern, with distinct peaks of activity for each of the three parasitic stages.[13] Some studies have correlated the majority of BSF cases with the period of maximum activity of the *A. cajennense* nymphal stage, which is from June to November.[11,12]

The present study brings additional information on the ecology of *R. rickettsii* in Brazil. In addition, it provides the first cell culture isolation of *R. rickettsii* and *R. bellii* from *A. aureolatum* ticks. Both isolated strains are currently being employed for production of antigens for diagnostic tests in our laboratory.

ACKNOWLEDGMENTS

We are very grateful to David H. Walker (University of Texas Medical Branch, USA) for his technical assistance during the present study. This work was supported by Fundação de Amparo a Pesquisa do Estado de São Paulo (FAPESP Grants 01/01218–2 to Adriano Pinter and 03/13872-4 to Marcelo B. Labruna) and Conselho Nacional de Desenvolvimento Científico e Tecnologógico (CNPq scholarship to Marcelo B. Labruna).

REFERENCES

1. BURGDORFER, W. 1988. Ecological and epidemiological considerations of Rocky Mountain spotted fever and scrub typhus. *In* Biology of Rickettsial Diseases. D.H. Walker, Ed.: 33–50. CRC Inc. Boca Raton, FL.
2. LABRUNA, M.B., T. WHITWORTH, M.C. HORTA, *et al.* 2004. *Rickettsia* species infecting *Amblyomma cooperi* ticks from an area in the state of São Paulo, Brazil, where Brazilian spotted fever is endemic. J. Clin. Microbiol. **42:** 90–98.
3. LABRUNA, M.B., T. WHITWORTH, D.H. BOYER, *et al.* 2004. *Rickettsia bellii* and *Rickettsia amblyommii* in *Amblyomma* ticks from the state of Rondonia, Western Amazon. Brazil. J. Med. Entomol. **41:** 1073–1081.
4. GOMES, L.S. 1933. Typho exanthematico de São Paulo. Brasil-Medico **17:** 919–921.
5. GUGLIELMONE, A.A., A. ESTRADA-PENA, A.J. MANGOLD, *et al.* 2003. *Amblyomma aureolatum* (Pallas, 1772) and *Amblyomma ovale* Koch, 1844 (Acari: Ixodidae): hosts, distribution and 16S rDNA sequences. Vet. Parasitol. **113:** 273–288.
6. PINTER, A., R.A. DIAS, S.M. GENNARI, *et al.* 2004. Study of the seasonal dynamics, life cycle, and host specificity of *Amblyomma aureolatum* (Acari: Ixodidae). J. Med. Entomol. **41:** 324–332.
7. SANGIONI, L.A., M.C. HORTA, M.C. VIANNA, *et al.* 2005. Rickettsial infection in animals and Brazilian spotted fever endemicity. Emerg. Infect. Dis. **11:** 265–270.
8. KELLY, P.J., D. RAOULT & P.R. MASON. 1991. Isolation of spotted fever group rickettsias from triturated ticks using a modification of the centrifugation-shell vial technique. Trans. R. Soc. Trop. Med. Hyg. **85:** 397–398.
9. FONSECA, F. 1935. Validade da especie e cyclo evolutivo de *Amblyomma striatum* KOCH, 1844 (Acarina, Ixodidae). Mem. Inst. Butantan **9:** 43–58.
10. MACALUSO, K.R., D.E. SONCENSHINE, S.M. CERAUL, *et al.* 2002. Rickettsial infection in *Dermacentor variabilis* (*Acari:Ixodidae*) inhibits transovarial transmission of a second rickettsia. J. Med. Entomol. **39:** 808–813.
11. DIAS, E. & A.V. MARTINS. 1939. Spotted fever in Brazil. Am. J. Trop. Med. **19:** 103–108.
12. LEMOS, E.R., F.B. ALVARENGA, M.L. CINTRA, *et al.* 2001. Spotted fever in Brazil: a seroepidemiological study and description of clinical cases in an endemic area in the state of São Paulo. Am. J. Trop. Med. Hyg. **65:** 329–334.
13. LABRUNA, M.B., N. KASAI, F. FERREIRA, *et al.* 2002. Seasonal dynamics of ticks (Acari: Ixodidae) on horses in the state of São Paulo Brazil. Vet. Parasitol. **105:** 65–77.
14. REGNERY, R.L., C.L. SPRUILL & B.D. PLIKAYTIS. 1991. Genotypic identification of rickettsiae and estimation of intraspecies sequence divergence for portions of two rickettsial genes. J. Bacteriol. **173:** 1576–1589.
15. LABRUNA, M.B., J.W. MCBRIDE, D.H. BOUYER, *et al.* 2004. Molecular evidence for a spotted fever group Rickettsia species in the tick *Amblyomma longirostre* in Brazil. J. Med. Entomol. **41:** 533–537.
16. EREMEEVA, M., X. YU & D. RAOULT. 1994. Differentiation among spotted fever group rickettsiae species by analysis of restriction fragment length polymorphism of PCR-amplified DNA. J. Clin. Microbiol. **32:** 803–810.

Multiplexed Serology in Atypical Bacterial Pneumonia

FRÉDÉRIQUE GOURIET,[a,b] MICHEL DRANCOURT,[a]
AND DIDIER RAOULT[a]

[a] *Unité des Rickettsies, CNRS UMR 6020, IFR 48, Faculté de Médecine, Université de la Méditerranée, 27 Boulevard Jean Moulin, 13385 Marseille cedex 05, France*

[b] *INODIAG, 52 Avenue de la Tramontane, ZI Athelia IV 13600 la Ciotat, France*

ABSTRACT: Atypical pneumonia is a term applied to lower respiratory tract infections that are not characterized by signs and symptoms of lobar consolidation. This article will discuss the epidemiology, clinical manifestations, and laboratory diagnoses of *Mycoplasma pneumoniae*, *Chlamydia* sp., *Legionella* sp., *Francisella tularensis*, and *Coxiella burnetii*, which are the agents most commonly associated with atypical pneumonia. Because many of these pathogens are intracellular, diagnosis depends upon serological confirmation. The current serological tests used to identify these agents in the etiologic diagnosis of atypical pneumonia are described. Recently, however, it has become possible to make a diagnosis directly in these cases using DNA or protein microarrays. Here, we describe the development of a new, automated technique for simultaneous testing and detection of several pathogens using a multiplexed serology test. This should prove to be a valuable tool for the rapid determination of patient status, allowing effective and efficient postexposure prophylaxis and treatment.

KEYWORDS: *Coxiella burnetii; Francisella tularensis; Legionella* sp.; *Chlamydia pneumoniae*; *Chlamydia psittaci*; *Mycoplasma pneumoniae*; multiplexed serology; immunofluescence

INTRODUCTION

Community-acquired pneumonia (CAP) remains a serious illness that can have a significant impact on patients as well as society. Pneumonia is an infection resulting from microbial invasion of the normally sterile lower respiratory tract and lung tissue. It may be caused by bacteria, viruses, or parasites. Clinically it is characterized by various pulmonary or extrapulmonary signs and symptoms. The risk factors for CAP are: age over 60 years, alcoholism,

Address for correspondence: D. Raoult. Unité des Rickettsies, Faculté de Médecine, 27 Boulevard Jean Moulin, 13385 Marseille, France. Voice: 33-4-91-32-43-75; fax: 33-4-91-38-77-72.
e-mail: Didier.Raoult@medecine.univ-mrs.fr

Ann. N.Y. Acad. Sci. 1078: 530–540 (2006).
doi: 10.1196/annals.1374.104

immunosuppression, asthma, and institutionalization.[1] Pneumonia is a leading cause of morbidity; about 600,000 persons are hospitalized each year due to the effects of pneumonia.[1] In addition, pneumonia is associated with a high mortality rate: approximately 30% in cases of severe CAP.[2] The microbial patterns in CAP differ considerably across studies, depending on the geographic area studied, the patient population included, and the nature and extent of the diagnostic techniques used. The identification of the etiologic agent of pneumonia is difficult and remains unknown in 20–50% of cases.[3] Pathogens like viruses or anaerobes seem to be underestimated. Emerging pathogens such as amoeba-associated bacteria, including *Legionella*-like amoebal pathogens and members of the *Parachlamydiae* genus,[4] are potential causes of community-acquired or ventilation-associated pneumonia. Identification of lung pathogens is a major public health goal. Over 100 different microbial pathogens have been isolated from lung tissue in cases of pneumonia. It is not possible to routinely obtain pulmonary tissue samples, and therefore blood, sputum, and pleural fluid cultures or the results of serological tests are typically used to make an etiologic diagnosis. A microorganism in samples collected from the respiratory tract can be detected and identified by conventional assays including: culture of microorganisms using axenic media, cell culture or a living host, direct microscopic examination or antigen detection, measurement of the host's specific immune response (serology testing), or detection of the micoorganism's specific nucleic acid (polymerase chain reaction [PCR]). The technological revolution in molecular biology has significantly expanded and improved the capabilities of diagnostic microbiology. Originally developed for analysis of whole genome gene expression, DNA microarrays have been used to detect bacteria and viruses, and have considerable potential for application in clinical microbiology.[5,6] Depending on the availability of an appropriate probe set, they enable the detection of several thousand microbial strains, species, genera, or higher clades in a single assay.[7] Recently, there has been an increasing interest and effort devoted to developing biosensor technologies for pathogen identification. In particular, the recent interest in bioterrorism has led to the development of DNA microarrays specifically for this purpose.[8] In the case of CAP, the microarray technique could be used to screen a specimen for a panel of probable pathogens. This could prove especially useful for identifying those pathogens that cause atypical pneumonia and require special media, equipment, biosafety facilities, and technical expertise, for example, *Francisela tularensis, Bacillus anthracis,* and *Coxiella burnetii,* which have been classified as select agents by the Centers for Disease Control and Prevention in Atlanta, and *Chlamydia psittaci*, which has been classified as a Category B bioterrorism agent. In this article we discuss the gold standards and the latest techniques available for the microbiologist to make the diagnosis of atypical pneumonia and we review the most important causative agents: *Mycoplasma pneumoniae, Chlamydia pneumoniae* and *psittaci, C. burnetii, Legionella pneumophila*, and *Francisella tularensis*.

Mycoplasma pneumoniae

Mycoplasma pneumoniae is a slow-growing, motile, aerobe that ferments glucose. *M. pneumoniae* infection occurs both epidemically and endemically in the world in adults and children. In a series of reported CAP cases, *M. pneumoniae* was the most commonly identified etiologic agent in patients treated in an ambulatory setting, accounting for 24% of the cases.[9] *M. pneumoniae* can be transmitted through aerosol droplets or nasal secretions from person to person. The incubation period ranges from 2 to 4 weeks. Infection caused by *M. pneumoniae* usually mimics viral respiratory syndrome; the most common manifestations include sore throat, hoarseness, fever, a cough which is initially nonproductive, headache, chills, coryza, myalgias, and general malaise. Progression to pneumonia occurs in 3% of the cases,[10,11] however, other respiratory syndromes such as bronchitis, bronchiolitis, pharyngitis, and croup also occur. Extrapulmonary manifestations are quite common. These include dermatologic, cardiac, and central nervous system manifestations as well as hemolytic anemia. Elevated cold agglutinins may be an early indication of *M. pneumoniae* infection. For the diagnosis of *M. pneumoniae* infection serological testing is the method of choice; the complement fixation (CF) assay is the reference method. Antibodies are detected by using chloroform-methanol-extracted glycolipid antigens, but this glycolipid antigen is not *M. pneumoniae* -specific and cross-reactions with other microorganims do occur. This test measures primarily the early immunoglobulin (Ig) IgM response, and to a lesser extent that of the subsequent IgG response. The reported sensitivity and specificity is about 90% and 94%, respectively. Due to the poor antibody class detection and the tendency for cross-reactivity, the CF has largely been replaced by enzyme immunoassay (EIA). An indirect immunofluorescence assay (IFA) is also available, although EIA is more sensitive for detecting acute infection. An overall sensitivity of 90% and specificity of 94% were reported for EIA in a large study among adults and children comparing EIA, CF, and IFA.[12] Because of the fastidious growth requirements and long incubation time, culture is not routinely done. Most published studies report superiority of the PCR to culture and serology, but this technique is not widely used in primary care clinics and commercial tests are not yet available.

Chlamydia Sp.

Chlamydia sp. are ubiquitous obligate intracellular bacteria and are common pathogens found in the respiratory tract. *Chlamydia pneumoniae* is a human pathogen. It is transmitted person to person via respiratory droplets and has an incubation period of 7–21 days. The seroprevalence in the adult population is about 50%.[13] Reinfection is common and antibodies do not appear to be protective.[14] *C. pneumoniae* causes acute respiratory disease including pneumonia, bronchitis, sinusitis, and pharyngitis. It is considered responsible for approximately 10% of CAP cases.[15] An association between persistent *C. pneumoniae* and the development and progression of chronic degenerative diseases such as atherosclerosis, neurological disorders, and asthma has been suggested.[13] For the diagnosis of *C. pneumoniae* infection, culture is highly specific, but the bacteria are exceedingly fastidious, slow-growing and require a reference laboratory. Antigen detection, as by EIA or IFA, and molecular detection provide rapid diagnosis without stringent transport requirements. Various tests are available for the serodiagnosis of *Chlamydia* sp. infection: CF, lipopolysachharide (LPS)-based EIA and IFA with a sensitivity of 10% for CF, 72% for EIA, and 88% for IFA.[16] CF detects antibodies against chlamydial LPS and although it is sensitive in primary infection, it is negative in 90% of cases of reinfection. IFA is the most useful and specific tool for the serodiagnosis of *Chlamydia* sp. infection. Acute infection is characterized by a fourfold rise in antibody titers in paired sera taken 3–4 weeks apart. A single IgM antibody titer of $\geq$ 1:16 and a single IgG titer of $\geq$ 1:512 are suggestive of acute infection. IFA allows distinction between IgG and IgM antibodies, which is useful in assessing recent versus remote infection, or distinguishing primary from reinfection. IgM antibodies are rarely produced in reinfections with *C. pneumoniae.*[13]

Chlamydia psittaci is the etiologic agent of ornithosis, a zoonotic infection acquired through exposure to infected birds or through handling of nasal discharge or fecal material from pet birds. Ungulates can also transmit *C. psittaci*. Person to person transmission is rare but has been reported in outbreaks.[14] Similar to that of *C. pneumoniae*, the incubation period is about 7–21 days. Symptomatic illness is manifest as flu-like illness characterized by fever, chills, headache, and less frequently, cough, myalgias, rash, arthralgias, and joint swelling and in the most severe cases, atypical pneumonia. Extrapulmonary signs and symptoms, such as endocarditis and glomerulonephritis, have been described.[14] Diagnosis is usually made by serology; culture is rarely performed as it requires experienced personnel in a specialized laboratory and it represents an important biohazard. CF assays are not species-specific and have a reported sensitivity of 46%.[17] IFA is the method of choice for diagnosis of *C. psittaci* as it can detect species-specific IgG and IgM antibodies. An IgM response of $\geq$1:16 or a fourfold rise in either IgM or IgG is considered diagnostic.[17] PCR assays are not yet commercially available, but they are more sensitive and rapid than serology and allow fingerprinting of the various strains. This is useful for the identification of the source in outbreak investigations and for the confirmation of zoonotic transmission from infected birds and animals.

Q Fever or Coxiella burnetii Infection

Coxiella burnetii is an obligate intracellular microorganism with a coccobacillus form and a Gram-negative cell wall. *C. burnetii* undergoes phase variation (phase I and II) linked to a transition of its LPS. Q fever is a zoonotic disease resulting from a *C. burnetii* infection most often acquired from goat, cattle, or sheep, which are common animal reservoirs of this pathogen. The microorganism reaches a high concentration in placentae of infected animals and aerosolization occurs at the time of parturition. Infection results from the inhalation of aerosolized organisms. Infection due to *C. burnetii* can be divided into acute and chronic varieties. Acute infection follows an incubation period of 2–3 weeks. There are three distinct clinical syndromes of acute Q fever including nonspecific febrile illness, pneumonia, and hepatitis. Chronic Q fever is usually manifest as endocarditis, but occasionally as hepatitis, osteomyelitis, or endovascular infection. The pneumonic form of the illness can range from a mild illness to severe pneumonia requiring assisted ventilation.[18] In most instances the laboratory diagnosis of *C. burnetii* infection is serological. In infected humans the predominant antibody response in acute Q fever is to phase II antigen, while in chronic Q fever the response is primarily to phase I antigen. CF, indirect IFA tests, and in some centers, ELISAs (enzyme-linked immunosorbent assay) are available; however IFA is the gold standard. A fourfold or greater rise in antibody titers between acute and convalescent samples is diagnostic. A phase II IgM titer of $\geq$1:50 or a phase II IgG titer of $\geq$1:200 by IFA is a strong evidence of recent *C. burnetii* infection.[19] Culture using the shell-vial technique is useful for isolation and antibiotic susceptibility assessment.[20] PCR can also be used to amplify *C. burnetii* DNA from samples.[21]

Legionella Pneumophila

Legionella pneumophila is the agent of Legionnaires' disease. *Legionella* sp. are thin, nonspore-forming, uncapsulated, aerobic, Gram-negative bacteria. They are fastidious organisms and do not grow on standard bacteriologic media. They are widespread in nature, usually associated with man-made aquatic environments (hot water tank, shower, public fountains). *Legionella* sp. have been isolated from free-living amoebae. Most human infections are caused by *L. pneumophila* serotypes 1, 4, and 6. *Legionella* culture is the gold standard method for the diagnosis of *Legionella* infection. Several methods have been developed for the direct detection of *Legionella* sp. from patients' specimens, such as immunofluorescence staining and PCR detection. A useful diagnostic tool for detecting *L. pneumophila* serotype 1 is the urinary test. This test is rapid and inexpensive and is both highly sensitive and specific. One of the main disadvantages of a serologic approach is the retrospective nature of the test and, in the case of *Legionella*, the cross-reactivity among *Legionella* species and other bacteria. IFA and ELISA are the most widely used methodologies to measure serological responses. The sensitivity of *Legionella* sp. serology is about 80% and the specificity ranges between 96 and 99%.[22] A fourfold increase in paired sera of antibody titers $\geq$ 1:128 is suggestive of legionellosis. Both IgG and IgM should be evaluated to provide serological diagnosis.

Francisella tularensis

Francisella tularensis is a small pleomorphic Gram-negative bacterium. The bacteria can be transmitted by an arthropod bite, ingestion or inhalation, or direct contact with infected tissue. Tularemia can be a serious disease and is occasionally fatal for humans and animals. The clinical signs and symptoms depend in part on the route of the inoculation. The less common but more severe primary pneumonic form develops after inhalation of the bacteria. This mode of transmission contributes to the classification of this organism as a select agent. The diagnosis of this disease can be difficult because respiratory signs and symptoms may be minimal or absent, and even when they are present they are usually nonspecific. *F. tularensis* is difficult to culture, and because handling this bacterium poses a significant risk of infection to laboratory personnel it requires a high level of biosafety containment equipment. ELISA- and PCR-based methods have been used to detect bacteria in clinical samples, but these methods have not been adequately evaluated for the diagnosis of pneumonic tularemia. Serological tests are frequently used for the diagnosis of tularaemia. There are several methods available, including tube agglutination, microagglutination, IFA, hemagglutination, and ELISA.[23,24] Tube agglutination and microagglutination are the reference method tests. They detect combined IgG and IgM. A fourfold change in titer between acute and convalescent serum specimens with titers of at least 1/160 for tube agglutination and 1/128 for microagglutination are diagnostic for *F. tularensis* infection.

Serological Diagnosis in Atypical Pneumonia

Clearly, then, for atypical bacterial pneumonia, serological testing is the primary diagnostic technique used. By demonstrating seroconversion or rising titers, serology provides indirect evidence for causal relationships between bacteria and disease.

IFA is the most commonly used method to detect antibodies to bacterial agents. It has been used to detect a large panel of infectious agents including *L. pneumophila, C. burnetii, C. pneumoniae* and *psitacci*, and *F. tularensis*. This technique is very effective, easy to perform, and is widely used due to its high sensitivity and specificity. Detection and quantification of antibody reactivity can be accomplished. In addition, various fluorochromes can be used which allow detection of multiple Ig classes such as IgG and IgM. However, reading fluorescence is a subjective process and interpretation of the results is sometimes difficult on account of nonspecific reactions.

The ELISA is a useful technique for epidemiology and the diagnosis of infectious disease and combines simplicity and sensitivity. This technique can be used reliably for screening large numbers of samples for a variety of infectious agents. ELISA is easy to perform and is adaptable. Because the tests are designed to be very sensitive, however, false positive results can occur with varying frequencies, depending on the specificity of the antibody used. Another potential problem with this technique is the high-dose hook effect. This refers to an unexpected fall in signal when the antibody to be assessed is present in very large quantities. The volume of sera required for each test may be a limiting factor when performing multiple tests with a single sample of serum. ELISA is used for *C. pneumoniae*[25] and *M. pneumoniae.*[26]

Complement fixation is a classical technique used in serology; developed in the early 1900s, it is still used in many laboratories. Although CF is a rigorous and standardized procedure it is time consuming, lacks sensitivity, and only detects IgG antibodies. The test requires careful standardization of the amounts of antigens, complement, and sensitive red blood cells. Nevertheless it is a powerful technique used for a large panel of viruses and it was the sole means for detecting antibodies against *M. pneumoniae*[26] for many years.

The agglutination assay is used for the detection of antibodies of various infectious agents including *Treponema pallidum* (TPPA and VRDL), *F. tularensis,* and various viral agents. The microagglutination test is a useful tool for the early and specific serodiagnosis of tularemia and it is used as the reference technique. Both IgG and IgM antibodies can cause agglutination and are detected by this method, but IgM is believed to be several hundred times more efficient. One important limitation of the agglutination assay is the potential for false negative results with high antibody concentrations.

Direct Diagnosis by DNA Microarrays

Direct diagnosis of infectious agents by DNA microarray consists of DNA probes (PCR product or oligonucleotides) bound to a solid support in a prefixed and regular disposition. The target nucleic acid (DNA or RNA) is labelled with a fluorochrome or a radioactive compound.[7] The main advantage of this technique is that thousands of genes can be detected in a single procedure using molecular biological tools. The application of DNA arrays in the field of clinical microbiology is uncommon. Among the specific applications where microarray technology would be beneficial are that the investigation of bacterial pathogenesis, analysis of bacterial evolution and molecular epidemiology, the study of mechanisms of action and resistance to antimicrobial agents, and microbiological diagnosis of infectious diseases. This methodology has been used to detect and identify bacteria in clinical samples.[27] It presents a series of advantages that make it very attractive and in the future it will likely become a valuable tool for clinical diagnosis of infectious diseases in medical laboratories.

Multiplexed Serology

Recently progress in robotic printing technology has allowed the development of high-density nucleic acid and protein arrays that have increased the throughput of a variety of assays. Protein arrays have been used to develop comparative fluorescence assays to measure the concentration of specific proteins and antibodies in complex solutions.[28,29,30] Antigen microarrays have been used for simultaneous determination of antibodies in human sera directed against *Toxoplasma gondii*, rubella virus, cytomegalovirus, and herpes simplex virus types 1 and 2. This technique proved capable of determining the presence or absence of specific IgG and IgM in the sera.[31,32] The microarray assay and commercial ELISA tests were compared and the results obtained showed an overall concordance of >80% between the microarrays and ELISAs. The microarray gave more favorable results when used to detect reactive IgG rather than IgM. Although microarrays represent an emerging and promising technology, they are not yet widely available.

A multiplexed fluorescent covalent microsphere immunoassay has been used for the measurement of specific IgG antibodies against five antigens from organisms defined as biological agents of bioterrorism: *B. anthracis* protective antigen and lethal factor, *Yersinia pestis* F1 and V antigen, *F tularensis*, ricin toxin, and *Staphylococcal* enterotoxin B.[33] Antibody microarrays continue to be developed as a useful tool for multiplexed protein analysis. The benefits of the technology include multiple simultaneous protein measurements, rapid experiments and analysis, quantitative and sensitive detection, and low volume requirements.

Future for Multiplexed Serology

The development of a protein microarray could provide many advantages in the determination of specific etiological pathogens involved in CAP. First, it would provide the ability to select the optimal drug for treatment, which would potentially reduce the occurrence of antibiotic resistance and adverse drug reactions, as well as the cost of antibiotics. This, in conjunction with the technological benefits suggests that multiplexed protein analysis will become an important diagnostic tool in the clinical management of CAP. IFA is currently the reference method for the serological diagnosis of *C. burnettii, Chlamydia* spp., and *Legionella* spp. IFA, which incorporates the whole microbe as the antigen, results in fluorescence that is proportional to antibody concentration; however, the assay is poorly reproducible on account of manual processing and subjective reading of the slides. Multiplexed slides for the serodiagnosis of infectious diseases are available (Euroimmun, Vancouver, B.C., Canada), although they still require the standard manual IFA slide staining, reading, and interpretation techniques. Therefore, even with multiplexing, IFA continues to be a laborious, time- and serum-consuming technique which is poorly reproducible between bench workers and laboratories. With inoDiaG, s.a. (La Ciotat, France) we are currently developing a fully automated IFA, incorporating a calibrated, antigenic microarray, featuring whole microorganism antigens and automated incubation, reading, and interpretation. We used this technique to detect antibodies against several pathogens responsible for atypical pneumonia (*C. burnetii, M. pneumoniae, C. pneumoniae* and *psittaci, L. pneumophila* serotypes 1–6, and *F. tularensis*). Our preliminary results are promising for this multiplex serological test. Standardization, automation, and antigen multiplexing of IFA in cases of atypical pneumonia would allow simultaneous detection of several pathogens including *C. burnetii, M. pneumoniae, C. pneumoniae* and *psittaci, L. pneumophila* serotypes 1–6, and *F. tularensis*. This multiplexed serology test would facilitate the rapid detection of the cause of disease, evaluation of patient status, and development of an effective treatment regimen.

CONCLUSION

Traditional agents of atypical pneumonia constitute a significant percentage of symptomatic CAP. CAP will continue to represent an important threat to patients in the future as the number of patients at risk (i.e., the elderly and those with comorbidity such as diabetes, asthma, and immunosuppression) increase. Accurate rapid diagnostic methods to determine the causative pathogens are needed to allow more specific and directed therapy. Additional tools in diagnostic testing particularly with antigen microarrays would be a suitable format for the serodiagnosis of infectious disease, especially in the case of atypical pneumonia, where the most reliable and commonly used test is serological.

ACKNOWLEDGMENTS

We thank Melanie Ihrig for reviewing the manuscript.

REFERENCES

1. MANDELL, L.A., T.J. MARRIE, R.F. GROSSMAN, *et al.* 2000. Canadian guidelines for the initial management of community-acquired pneumonia: an evidence-based update by the Canadian Infectious Diseases Society and the Canadian Thoracic Society. The Canadian Community-Acquired Pneumonia Working Group. Clin. Infect. Dis. **31:** 383–421.

2. RUIZ, M., S. EWIG, A. TORRES, *et al.* 1999. Severe community-acquired pneumonia: risk factors and follow-up epidemiology. Am. J. Resp. Crit. Care Med. **160:** 923–929.
3. MARRIE, T.J., H. DURANT & L. YATES. 1989. Community-acquired pneumonia requiring hospitalization: 5-year prospective study. Rev. Infect. Dis. **11:** 586–599.
4. MARRIE, T.J., D. RAOULT, B. LA SCOLA, *et al.* 2001. *Legionella*-like and other amoebal pathogens as agents of community-acquired pneumonia. Emerg. Infect. Dis. **7:** 1026–1029.
5. LIU, W.T., A.D. MIRZABEKOV & D.A. STAHL. 2001. Optimization of an oligonucleotide microchip for microbial identification studies: a non-equilibrium dissociation approach. Environ. Microbiol. **3:** 619–629.
6. BORISKIN, Y.S., P.S. RICE, R.A. STABLER, *et al.* 2004. DNA microarrays for virus detection in cases of central nervous system infection. J. Clin. Microbiol. **42:** 5811–5818.
7. BODROSSY, L. & A. SESSITSCH. 2004. Oligonucleotide microarrays in microbial diagnostics. Curr. Opin. Microbiol. **7:** 245–254.
8. BELOSLUDTSEV, Y.Y., D. BOWERMAN, R. WEIL, *et al.* 2004. Organism identification using a genome sequence-independent universal microarray probe set. Biotechniques **37:** 654–658, 660.
9. MARRIE, T.J., R.W. PEELING, M.J. FINE, *et al.* 1996. Ambulatory patients with community-acquired pneumonia: the frequency of atypical agents and clinical course. Am. J. Med. **101:** 508–515.
10. FOY, H.M. 1993. Infections caused by *Mycoplasma pneumoniae* and possible carrier state in different populations of patients. Clin. Infect. Dis. **17** (Suppl. 1): S37–S46.
11. CLYDE, W.A., JR. 1993. Clinical overview of typical *Mycoplasma pneumoniae* infections. Clin. Infect. Dis. **17** (Suppl. 1): S32–S36.
12. ALEXANDER, T.S., L.D. GRAY, J.A. KRAFT, *et al.* 1996. Performance of Meridian ImmunoCard *Mycoplasma* test in a multicenter clinical trial. J. Clin. Microbiol. **34:** 1180–1183.
13. KAUPPINEN, M. & P. SAIKKU. 1995. Pneumonia due to *Chlamydia pneumoniae*: prevalence, clinical features, diagnosis, and treatment. Clin. Infect. Dis. **21** (Suppl. 3): S244–S252.
14. PEELING, R.W. & R.C. BRUNHAM. 1996. *Chlamydiae* as pathogens: new species and new issues. Emerg. Infect. Dis. **2:** 307–319.
15. KUO, C.C., L.A. JACKSON, L.A. CAMPBELL & J.T. GRAYSTON. 1995. *Chlamydia pneumoniae* (TWAR). Clin. Microbiol. Rev. **8:** 451–461.
16. EKMAN, M.R., M. LEINONEN, H. SYRJALA, *et al.* 1993. Evaluation of serological methods in the diagnosis of *Chlamydia pneumoniae* pneumonia during an epidemic in Finland. Eur. J. Clin. Microbiol. Infect. Dis. **12:** 756–760.
17. WONG, K.H., S.K. SKELTON & H. DAUGHARTY. 1994. Utility of complement fixation and microimmunofluorescence assays for detecting serologic responses in patients with clinically diagnosed psittacosis. J. Clin. Microbiol. **32:** 2417–2421.
18. MARRIE, T.J. 2003. *Coxiella burnetii* pneumonia. Eur. Respir. J. **21:** 713–719.
19. DUPONT, H.T., X. THIRION & D. RAOULT. 1994. Q fever serology: cutoff determination for microimmunofluorescence. Clin. Diagn. Lab. Immunol. **1:** 189–196.
20. FOURNIER, P.E., T.J. MARRIE & D. RAOULT. 1998. Diagnosis of Q fever. J. Clin. Microbiol. **36:** 1823–1834.

21. FENOLLAR, F., P.E. FOURNIER & D. RAOULT. 2004. Molecular detection of *Coxiella burnetii* in the sera of patients with Q fever endocarditis or vascular infection. J. Clin. Microbiol. 42: 4919–4924.
22. WILKINSON, H.W., A.L. REINGOLD, B.J. BRAKE, *et al.* 1983. Reactivity of serum from patients with suspected legionellosis against 29 antigens of legionellaceae and Legionella-like organisms by indirect immunofluorescence assay. J. Infect. Dis. **147:** 23–31.
23. SATO, T., H. FUJITA, Y. OHARA & M. HOMMA. 1990. Microagglutination test for early and specific serodiagnosis of tularemia. J. Clin. Microbiol. **28:** 2372–2374.
24. VILJANEN, M.K., T. NURMI & A. SALMINEN. 1983. Enzyme-linked immunosorbent assay (ELISA) with bacterial sonicate antigen for IgM, IgA, and IgG antibodies to *Francisella tularensis*: comparison with bacterial agglutination test and ELISA with lipopolysaccharide antigen. J. Infect. Dis. **148:** 715–720.
25. HERMANN, C., K. GUEINZIUS, A. OEHME, *et al.* 2004. Comparison of quantitative and semiquantitative enzyme-linked immunosorbent assays for immunoglobulin G against *Chlamydophila pneumoniae* to a microimmunofluorescence test for use with patients with respiratory tract infections. J. Clin. Microbiol. **42:** 2476–2479.
26. WAITES, K.B. & D.F. TALKINGTON. 2004. Mycoplasma pneumoniae and its role as a human pathogen [table]. Clin. Microbiol. Rev. **17:** 697–728.
27. MITTERER, G., M. HUBER, E. LEIDINGER, *et al.* 2004. Microarray-based identification of bacteria in clinical samples by solid-phase PCR amplification of 23S ribosomal DNA sequences. J. Clin. Microbiol. **42:** 1048–1057.
28. WILTSHIRE, S., S. O'MALLEY, J. LAMBERT, *et al.* 2000. Detection of multiple allergen-specific IgEs on microarrays by immunoassay with rolling circle amplification. Clin. Chem. **46:** 1990–1993.
29. HAAB, B.B. 2001. Advances in protein microarray technology for protein expression and interaction profiling. Curr. Opin. Drug Discov. Dev. **4:** 116–123.
30. HUANG, R.P., R. HUANG, Y. FAN & Y. LIN. 2001. Simultaneous detection of multiple cytokines from conditioned media and patient's sera by an antibody-based protein array system. Anal. Biochem. **294:** 55–62.
31. MEZZASOMA, L., T. BACARESE-HAMILTON, M. DI CRISTINA, *et al.* 2002. Antigen microarrays for serodiagnosis of infectious diseases. Clin. Chem. **48:** 121–130.
32. BACARESE-HAMILTON, T., L. MEZZASOMA, A. ARDIZZONI, *et al.* 2004. Serodiagnosis of infectious diseases with antigen microarrays. J. Appl. Microbiol. **96:** 10–17.
33. BIAGINI, R.E., D.L. SAMMONS, J.P. SMITH, *et al.* 2005. Simultaneous measurement of specific serum IgG responses to five select agents. Anal. Bioanal. Chem. **382:** 1027–1034.

Isolation of *Anaplasma phagocytophilum* Strain Ap-Variant 1 in a Tick-Derived Cell Line

ROBERT F. MASSUNG,[a] MICHAEL L. LEVIN,[a]
ULRIKE G. MUNDERLOH,[b] DAVID J. SILVERMAN,[c]
MEGHAN J. LYNCH,[b] AND TIMOTHY J. KURTTI[b]

[a] *Viral and Rickettsial Zoonoses Branch, Centers for Disease Control and Prevention, Atlanta, Georgia, USA*

[b] *Department of Entomology, University of Minnesota, Saint Paul, Minnesota, USA*

[c] *School of Medicine, University of Maryland-Baltimore, Baltimore, Maryland, USA*

ABSTRACT: Ten isolates of the Ap-Variant 1 strain of *Anaplasma phagocytophilum* were made in the *Ixodes scapularis (I. scapularis)*–derived cell line, ISE6. Two isolates were obtained from laboratory-infected goats and eight isolates were obtained from field-collected *I. scapularis* ticks. Each isolate showed 16S rRNA sequences identical to those as previously described for the Ap-Variant 1 strain. These are the first tissue culture isolates of the Ap-Variant 1 strain and will allow for further characterization of the biological and antigenic properties of this strain.

KEYWORDS: *Anaplasma phagocytophilum*; *Ixodes scapularis*; Ap-Variant 1; human granulocytic anaplasmosis

Anaplasma phagocytophilum (A. phagocytophilum) causes human granulocytic anaplasmosis, a disease transmitted by *Ixodes scapularis (I. scapularis)* ticks that primarily occurs in the upper midwest and northeast United States.[1] Recently, a strain of *A. phagocytophilum* referred to as Ap-Variant 1 has been described in *I. scapularis* ticks, but this strain has not been associated with human disease.[2–4] Whereas numerous tissue culture isolates from human infections (Ap-ha strains) have been obtained,[5,6] the Ap-Variant 1 strain has never been isolated. The lack of a tissue culture isolate has hindered characterization of Ap-Variant 1 and the goal of this study was to obtain an isolate of this strain *in vitro*.

Address for correspondence: Robert F. Massung, Centers for Disease Control and Prevention, 1600 Clifton Rd., MS G-13, Atlanta, GA 30333. Voice: 404-639-1082; fax: 404-639-4436.
e-mail: rfm2@cdc.gov

Ann. N.Y. Acad. Sci. 1078: 541–544 (2006).
doi: 10.1196/annals.1374.105

Isolation from the blood of two goats infected with Ap-Variant 1 was attempted in a human promyelocytic cell line (HL-60) and an *I. scapularis* tick-derived line (ISE6). One goat had been infected with questing *I. scapularis* ticks collected from Trustom, Rhode Island and the second goat had been infected with blood from the first goat. Isolations directly from partially or fully engorged *I. scapularis* ticks collected from white-tailed deer in Camp Ripley, Minnesota were also attempted in ISE6 cells. Isolation attempts in HL-60 cells were performed as previously described.[7] Goat blood or tick organs were inoculated on ISE6 cells and incubated at 34°C in a candle jar also as previously described.[8] The cultures were examined periodically for the presence of *Anaplasma* by light microscopy using cytocentrifuge-prepared cells fixed in methanol and stained with either Giemsa or Diff-Quik.

Isolates of *A. phagocytophilum* were obtained from each of the two goat bloods and from eight of the tick homogenates in IES6 cells. All isolation attempts in HL-60 cells were unsuccessful. DNA was extracted from each isolate and the 16S rRNA gene was amplified by PCR as previously described.[3,5] Sequencing of these PCR products confirmed that each isolate was the Ap-Variant 1 strain of *A. phagocytophilum* as they had the expected characteristic

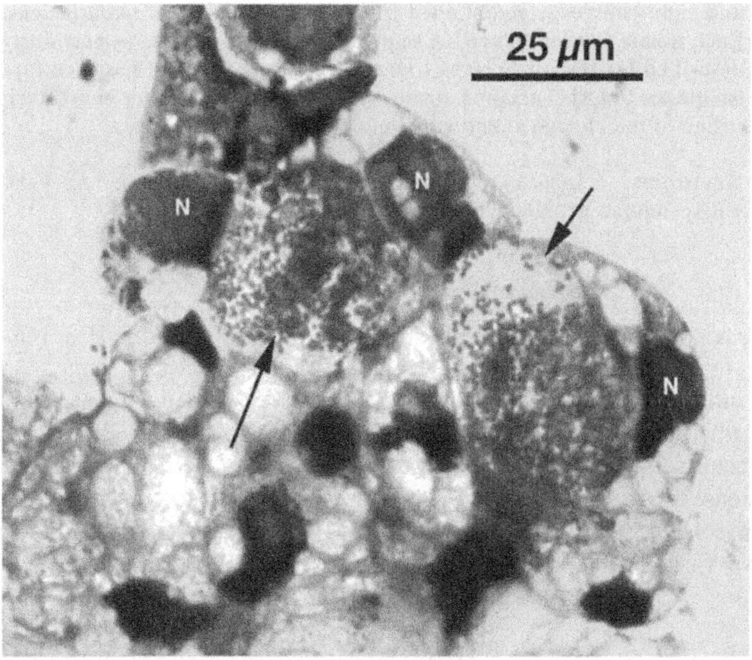

FIGURE 1. Photomicrograph of Giemsa- stained ISE6 cells infected with *A. phagocytophilum* strain Ap-Variant 1. *Arrows* point to large intracellular vacuoles harboring numerous bacteria. N = nucleus.

2-base difference from the Ap-ha strains most commonly associated with human infections. Ap-Variant 1 isolates were maintained by serially transferring infected cell suspensions into fresh ISE6 cell cultures.

Giemsa-stained ISE6 cells infected with Ap-Variant 1 exhibited membrane-bound inclusions in the cytoplasm of infected cells that contain large numbers of bacteria (FIG. 1). Electron microscopy of Ap-Variant 1 in ISE6 cells also showed a cytoplasmic infection characteristic of *A. phagocytophilum* with pleomorphic bacteria in membrane-bound vacuoles, with both electron-dense and electron-lucent organisms.

Ap-Variant 1 strains from Rhode Island have previously been shown to be unable to infect mice, including the white-footed mouse, *Peromyscus leucopus*, which serves as a major reservoir in nature for the Ap-ha strains that cause human disease.[9] To determine whether a similar host tropism may exist among Ap-Variant 1 strains from the upper midwest, a Minnesota tick-derived isolate was used to inoculate a single naïve goat and 5 naïve mice. Attempts to infect mice were unsuccessful, in agreement with previous studies on Ap-Variant 1 from Rhode Island. The Minnesota isolate did infect the goat as its blood became positive by PCR for *A. phagocytophilum* on day 11 and remained PCR positive for 30 days.

In summary, 10 isolates of the novel *A. phagocytophilum* strain Ap-Variant 1 have been obtained. Each isolate has identical 16S rRNA gene sequences, but sequence differences found within the *ank* gene suggests that some natural variation exists among Ap-Variant 1 strains. An Ap-Variant 1 isolate obtained from an *I. scapularis* tick from Minnesota was infectious in a goat but did not infect mice, thus confirming the host reservoir specificity previously described for Ap-Variant 1-infected ticks from Rhode Island.[9]

REFERENCES

1. McQuiston, J.H., C.D. Paddock, R.C. Holman & J.E. Childs. 1999. The human ehrlichioses in the United States. Emerg. Infect. Dis. **5:** 635–642.
2. Belongia, E.A., K.D. Reed, P.D. Mitchell, *et al.* 1997. Prevalence of granulocytic Ehrlichia infection among white-tailed deer in Wisconsin. J. Clin. Microbiol. **35:** 1465–1468.
3. Massung, R.F., K. Slater, J.H. Owens, *et al.* 1998. Nested PCR assay for detection of granulocytic ehrlichiae. J. Clin. Microbiol. **36:** 1090–1095.
4. Massung, R.F., M.J. Mauel, J.H. Owens, *et al.* 2002. Genetic variants of *Ehrlichia phagocytophila*, Rhode Island and Connecticut. Emerg. Infect. Dis. **8:** 467–472.
5. Goodman, J.L., C. Nelson, B. Vitale, *et al.* 1996. Direct cultivation of the causative agent of human granulocytic ehrlichiosis. N. Engl. J. Med. **334:** 209–215.
6. Horowitz, H.W., M.E. Aguero-Rosenfeld, D.F. McKenna, *et al.* 1998. Clinical and laboratory spectrum of culture-proven human granulocytic ehrlichiosis: comparison with culture-negative cases. Clin. Infect. Dis. **27:** 1314–1317.

7. NICHOLSON, W.L., J.A. COMER, J.W. SUMNER, *et al.* 1997. An indirect immunofluorescence assay using a cell culture-derived antigen for detection of antibodies to the agent of human granulocytic ehrlichiosis. J. Clin. Microbiol. **35:** 1510–1516.
8. MUNDERLOH, U.G., S.D. JAURON, V. FINGERLE, *et al.* 1999. Invasion and intracellular development of the human granulocytic ehrlichiosis agent in tick cell culture. J. Clin. Microbiol. **37:** 2518–2524.
9. MASSUNG, R.F., R.A. PRIESTLEY, N.J. MILLER, *et al.* 2003. Inability of a variant strain of *Anaplasma phagocytophilum* to infect mice. J. Infect. Dis. **188:** 1757–1763.

Human Anaplasmosis

The First Spanish Case Confirmed by PCR

J. C. GARCÍA,[a] M. J. NÚÑEZ,[a] B. CASTRO,[a] F. J. FRAILE,[a] A. LÓPEZ,[a] M. C. MELLA,[a] A. BLANCO,[a] C. SIEIRA,[b] E. LOUREIRO,[c] A. PORTILLO,[d] AND J. A. OTEO[d]

[a]*Servicio de Medicina Interna, Hospital Do Salnés, Vilagarcía de Arousa, Spain*

[b]*Laboratorio Clínico, Hospital Do Salnés, Vilagarcía de Arousa, Spain*

[c]*Unidad de Hematología, Hospital Do Salnés, Vilagarcía de Arousa, Spain*

[d]*Área de Enfermedades Infecciosas, Hospitales San Millán-San Pedro-de La Rioja, Logroño, Spain*

ABSTRACT: We report a case of human anaplasmosis (HA) fulfilling the confirmation criteria : epidemiologic data and clinical picture compatible with HA; presence of a morulae within polymorphonuclear leukocyte; and positive PCR assay for *Anaplasma phagocytophilum:* This case report shows the presence of HA in Spain.

KEYWORDS: human anaplasmosis; *Anaplasma phagocytophilum*; Spain

INTRODUCTION

Human anaplasmosis (HA) is considered an emerging zoonoses in the USA and Europe.[1,2] To our knowledge, only one case has been reported in Spain.[3] The first case of HA confirmed by polymerase chain reaction (PCR) in Galicia, Spain, as well as different aspects of the diagnosis, is presented in this work.

CLINICAL OBSERVATION

In May 2002, a 23-year-old man presented at the emergency room with fever, headache, sore throat, malaise, and myalgias. He was discharged with diagnosis of pharyngitis and treated with azithromycin. Two days later he returned to the emergency room with persistent fever, and increased intensity of headache

Address for correspondence: Juan Carlos García, Servicio de Medicina Interna, F.P. Hospital Do Salnés, C/Estromil Ande Rubians, 36600–Vilagarcía de Arousa (Pontevedra), Spain. Voice: + 34-986-568140; fax: +34-986-568001.
e-mail: juan.carlos.garcia.garcia2@sergas.es

Ann. N.Y. Acad. Sci. 1078: 545–547 (2006). © 2006 New York Academy of Sciences.
doi: 10.1196/annals.1374.106

and myalgias located in the back and legs. Previous medical history was unremarkable. He did athletics in the country. Physical examination showed armpit temperature of 39°C, labial herpes, and submandibular painful adenopathy. Spleen was felt. Chest radiographs were normal and abdominal ultrasonography showed slight splenomegaly without focal lesions. Laboratory studies showed leukocyte count of 1,600/mm^3 with 1% band forms, 67% neutrophils, 7% monocytes and 25% lymphocytes, and platelet count of 67,000/mm.[3] The hemoglobin level was normal. AST, ALT, and LDH levels were 47, 52, and 538 U/L, respectively. A urine analysis was normal and Paul–Bunnell test was negative. The patient was interned at the Internal Medicine Service. Twenty-four h later, he developed a nonspecific erythematous rash. Within 72 h, fever, rash, and laboratory findings were persistent. An examination of peripheral blood smear revealed a cytoplasmic inclusion that seemed a morulae in a neutrophil. Blood cultures obtained at admission were negative. Treatment with doxycycline was started empirically and fever disappeared within 48 h. Serologic test results were negative for *Cytomegalovirus*, *Borrellia burgdorferi (B. burgdorferi)*, *Rickettsia conorii (R. conorii)*, *Ehrlichia chaffeensis (E. chaffeensis),* and *Anaplasma phagocytophilum (A. phagocytophilum)*, and indicated immunity for Epstein–Barr virus. An EDTA whole blood sample obtained before starting doxycycline treatment was PCR positive for the *A. phagocytophilum* genogroup. On the 8th day, the patient was asymptomatic and leukocytes and platelets counts were normal. ALT and AST were slightly elevated. He was discharged with treatment of doxycycline for 7 days. Three weeks later the patient was asymptomatic and ALT and AST levels were normal. A year later, a serologic IgG test showed a titer of 1/80 to *A. phagocytophilum.*

DISCUSSION

We present a case of HA fulfilling the confirmation criteria: epidemiological data and clinical picture compatible with HA; presence of morulae within a polymorphonuclear leukocyte; and a positive PCR assay for *A. phagocytophilum.*[4,5] Furthermore, the rapid clinical response to doxycycline is typical of this disease.[1] Rash is uncommon and when it is present, a coinfection with *B. burgdorferi* and other infectious diseases should be ruled out.[1] The visualization of morulae is very uncommon in Europe.[2] This finding has been previously reported only in two cases.[6,7] Initial negative serologic test for *A. phagocytophilum* is frequent in patients with HA and positive PCR assay. On the contrary, a positive PCR test is uncommon in patients with initial positive serologic test for HA agent.[8] This finding shows that PCR assay is sensitive for diagnosis of initial stages of HA. The low IgG titer for HA detected a year later may be due to starting early doxycycline treatment.

This case report shows the presence of HA in Spain. HA should be included in differential diagnosis of clinical pictures with compatible epidemiological context.

REFERENCES

1. BAKKEN, J.S. & J.S. DUMLER. 2000. Human granulocytic ehrlichiosis. Clin. Infect. Dis. **31:** 554–560.
2. BLANCO, J.R. & J.A. OTEO. 2002. Human granulocytic erhlichiosis in Europe. Clin. Microbiol. Infect. **8:** 763–772.
3. OTEO, J.A. & J.R. BLANCO, *et al.* 2000. First report of human granulocytic erhlichiosis from southern Europe (Spain). Emerg. Infect. Dis. **6:** 430–432.
4. WALKER, D.H. 1999. Consensus workshop on diagnosis of human ehrlichiosis. Am. Soc. Rickettsiol. Newslett. **2:** 1–8.
5. BROUQUI, P., F. BACELLAR, *et al.* 2004. Guidelines for the diagnosis of tick-borne diseases in Europe. Clin. Microbiol. Infect. Dis. **10:** 1108–1132.
6. VAN DOBBENBURGH, A., A.P. VAN DAM, *et al.* 1999. Human granulocytic ehrlichiosis in Western Europe. N. Engl. J. Med. **18:** 385–386.
7. REMY, V., Y. HANSMANN, *et al.* 2003. Human anaplasmosis presenting as atypical pneumonitis in France. Clin. Infect. Dis. **37:** 846–848.
8. LOTRIC-FURLAN, S., T. AVSIC-ZUPANC, *et al.* 2001. Clinical and serological follow-up of patients with human granulocytic ehrlichiosis in Slovenia. Clin. Diagn. Lab. Immunol. **8:** 899–903.

Two Cases of Human Granulocytic Ehrlichiosis in Sardinia, Italy Confirmed by PCR

S. MASTRANDREA,[a] M.S MURA,[a] S. TOLA,[b] C. PATTA,[b]
A. TANDA,[b] R. PORCU,[b] AND G. MASALA[b]

[a]*Istituto di Malattie Infettive, University of Sassari, Sassari, Sardinia, Italy*

[b]*Istituto Zooprofilattico Sperimentale della Sardegna, Sassari, Sardinia, Italy*

ABSTRACT: In this work we report the first two cases of human granulocytic ehrlichiosis (HGE) in Sardinia. In early September 2004, a 69-year-old woman (patient 1) was admitted to the Infectious Diseases Institute of Sassari for rickettsiosis like-syndrome: high fever (39.5–40°C), dyspnea, reduced consciousness, vomiting, and cutaneous rash. In late September 2004, a 30-year-old man (patient 2) with high fever was admitted for an evident palmar and oral erythema, edema of the labium, very intense arthralgia, myalgia, and dyspnea. In these two hospitalized patients, the diagnosis was made through indirect IgM and IgG immunofluorescent technique and confirmed by the presence of the specific DNA in the leukocytes. The two patients were *A. phagocytophilum*–PCR positive.

KEYWORDS: *A. phagocytophilum*; granulocytic ehrlichiosis; Sardinia; Italy

INTRODUCTION

Granulocytic ehrlichiae are obligate intracellular bacteria that are transmitted by ticks. Before 2001, they included *Ehrichia phagocytophila (E. phagocytophila)*, *Ehrichia equi* (*E. equi*), and the agent of human granulocytic ehrlichiosis (HGE). Now these three species are unified into a single species reclassified as *Anaplasma phagocytophilum (A. phagocytophilum).*[1] The first case of human disease caused by a granulocytropic ehrlichia species was described in the 1994 in the United States and later in Europe (Slovenia).[2,3]

Two cases of HGE were confirmed in the Friuli Venezia Giulia, an Italian region close to Slovenia.[4] In Sardinia, no human confirmed case of HGE has been reported, except the four cases of clinical and serological diagnosis of

Address for correspondence: Giovanna Masala, Istituto Zooprofilattico Sperimentale della Sardegna, Via Duca degli Abruzzi 8, 07100 Sassari, Sardinia, Italy. Voice: 0792892325; fax: 0792892324.
e-mail: giovanna.masala@izs-sardegna.it

**Ann. N.Y. Acad. Sci. 1078: 548–551 (2006). © 2006 New York Academy of Sciences.
doi: 10.1196/annals.1374.107**

Ehrlichia chaffensis (*E. chaffensis)* infection; the patients were U.S. citizens temporarily living on the U.S. Navy base of La Maddalena.[5]

Here we described two human cases of HGE in Sardinia, in whom diagnosis of HGE was confirmed by PCR.

MATERIALS AND METHODS

Patients

In early September 2004, a 69-year-old woman (patient 1) was admitted to the Infectious Diseases Institute of Sassari for rickettsiosis like-syndrome: high fever (39.5–40°C), dyspnea, reduced consciousness, vomiting, and cutaneous rash. The patient had clinical signs characteristic of atypical pneumonia with radiographically established lung infiltrate. The patient asserted that the symptoms appeared for the first time 10 days earlier. In that period she had been treated at home with paracetamol without any benefit. Laboratory abnormalities were found: leukopenia and thrombocytopenia, elevated liver enzyme levels, erythrocyte edimentation rate (VES), and serum C-reactive protein (PCR) concentration.

In late September 2004, a 30-year-old man (patient 2) with high fever was admitted for an evident palmar and oral erythema, edema of the labium, very intense arthralgia, myalgia, and dyspnea. At the first visit he presented with a high fever (39.5–40°C), which appeared 4 days prior to his admission at the hospital, and clinical characteristics of atypical pneumonia. After a radiographic examination a lung infiltrate was established. This patient also presented laboratory abnormalities: leukopenia and thrombocytopenia, elevated serum transaminases, alterated erythrocyte sedimentation rate (VES), and high concentration of serum C-reactive protein (PCR).

Indirect Fluorescent Antibody Test

Sera from the two patients were diluted 1:40 with further twofold increments up to the dilution 1:320 in PBS (phosphate- buffered saline). Aliquots of 10 μL from each serum were placed in the well containing antigens of HGE (INDX Integrated Diagnostics, Baltimore, USA) or *Rickettsia conorii (R. conorii)* (bioMerieux, Marcy l'Etoile-France). Sera were incubated in a humidified chamber at 37°C for 30 min. After washing in PBS, slides were incubated for 30 min at 37°C with FITC-labeled goat anti-human immunoglobulin (KPL) diluted 1:100. The slides were then washed three times in PBS, counterstained with 1% Evans blue, and examined for fluorescein in a Zeiss microscope. A title of 1:40 was considered positive for IgG and IgM.

DNA Extraction

DNA was extracted from leukocytes, separated from the blood in the buffy coat with phenol-chloroform-isoamyl alcohol, purified following the procedure described by Ausubel *et al.*[6], and used as a template for PCR assays to detect DNA from *A. phagocytophilum*, *E. canis,* and *Rickettsia* spp.

PCR and Sequencing

Detection of the presence of *A. phagocytophilum*, *E canis,* and *Rickettsia* spp. was carried out using a PCR assay targeting specific genes: *16S* rRNA,[7] *p30,*[8] and *ompB*[9], respectively. PCR reactions were performed in a GeneAmp PCR System 9700 (Applied Biosystems, Foster City, CA). PCR products were analyzed on 1.5% agarose gel in 1 × TBE buffer (89 mM Tris, 89 mM boric acid, pH 8.0).

Amplified PCR products were sequenced by a dideoxy-chain-termination reaction using the ALFexpress sequencer (Amersham Buckinghamshire, UK). Nucleotide sequence homology searches were made through the National Center for Biotechnology Information BLAST network service.

RESULTS AND DISCUSSION

Indirect immunofluorescence antibody tests for *A. phagocitophylum* and *Rickettsia* spp. on serum samples taken on patient 1 at the first visit, revealed

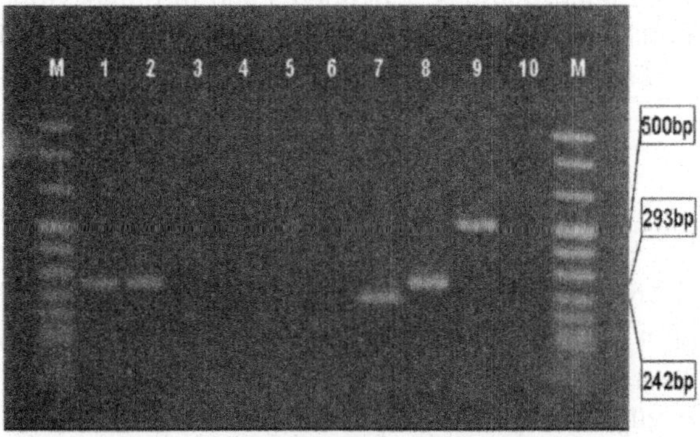

FIGURE 1. Agarose gel electrophoresis of amplified products. *Lane M*: marker VIII (Roche); *lanes* 1–2: amplicons (293 bp) obtained from patient 1and 2 using the primers set of *A. phagocytophilum; lanes* 3–4: no amplicons were obtained with the primer set of *E. canis*; *lanes* 5–6: no amplicons were obtained with the primer set of *Rickettsia* sp.; *lanes* 7–8–9: positive controls of *E. canis, A. phagocytophilum,* and *Rickettsia* spp., respectively; *lane* 10: negative control.

a titer of 1:80 for *Rickettsia* spp.–IgG while – IgM values were in doubt. In patient 2, IFA test revealed a titer of 1:40 for *Rickettsia* spp.–IgG and–IgM; no evidence of antibody response for *A. phagocytophylum* was revealed in both patients. Because of improvement in their clinical conditions, after specific antibiotic therapy (doxycycline 100 mg PO), the patients were discharged after a 10-day hospital stay. DNA amplification of the two human samples, using primers targeting and the *16S* rRNA gene of *A. phagocytophilum*, the *p30* gene of monocytic *Ehrlichia,* and the *omp*B gene of *Rickettsia* spp., revealed a PCR positive for *A. phagocytophilum* only (FIG. 1). Comparison of the sequences of the amplified PCR products to those deposited in the GenBank by using BLAST service, revealed that our samples were 100% identical to a portion of *16S* rRNA gene of *A. phagocytophilum.*

In this study two cases of human ehrlichiosis imputed to *A. phagocytophilum* have been described. This report is the second on the presence of confirmed HGE in humans in Italy,[4] and the first in Sardinia.

REFERENCES

1. DUMLER, J.S., A.F. BARBET, C.P. BEKKER, *et al.* 2001. Reorganization of genera in the families Rickettsiaceae and Anaplasmataceae in the order Rickettsiales: unification of some species of *Ehrlichia* with *Anaplasma*, *Cowdria* with *Ehrlichia* and *Ehrlichia* with *Neorickettsia*, descriptions of six new species combinations and designation of *Ehrlichia equi* and "HGE agent" as subjective synonyms of *Ehrlichia phagocytophilum*. Int. J. Syst. Evol. Microbiol. **51:** 2145–2165.
2. CHEN, S.M., J.S. DUMLER, J.S. BAKKER & D.H. WALKER. 1994. Identification of granulocytotropic *Ehrlichia* species as the ethiologic agent of human disease. J. Clin. Microbiol. **32:** 589–595.
3. PETROVEC, M., F.S. LOTRIC, T.A. ZUPANC, *et al.* 1997. Human disease in Europe caused by granulocytic *Ehrlichia* species. J. Clin. Microbiol. **35:** 1556-1559.
4. RUSCIO, M. & M. CINCO 2003. Human granulocytic ehrlichiosis in Italy. First report on two confirmed cases. Ann. N.Y. Acad. Sci. **990:** 350–352.
5. NUTI, M., D.A. SERAFINI, D. BASSETTI, *et al.* 1998. Ehrlichia infection in Italy. Emerg. Infect. Dis. **4:** 663–665.
6. AUSUBEL, F.M., R. BRENT, R.E. KINGSTON, *et al.* 1993. Current protocols in molecular biology. Wiley/Interscience, New York.
7. KOLBERT, C. 1996. Detection of the agent of human granulocytic ehrlichiosis by PCR. *In* PCR Protocols for Emerging Infectious Diseases. D. Persing, Ed.:106–111 ASM Press, Washington, DC
8. STICH, R.W., Y. RIKIHISA, S.A. EWING, *et al.* 2002. Detection of *Ehrlichia canis* in canine carrier blood and in individual experimentally infected ticks with a p30-based PCR assay. J. Clin. Microbiol. **40:** 540–546.
9. NODA, H., U. MUNDERLOH & T.J. KURTTI. 1997. Endosymbionts of ticks and their relationship to *Wolbachia* spp. and tick-borne pathogens of humans and animals. Appl. Environ. Microbiol. **63:** 3926–3932.

Multiplex Detection of *Ehrlichia* and *Anaplasma* Pathogens in Vertebrate and Tick Hosts by Real-Time RT-PCR

KAMESH R. SIRIGIREDDY, DONALD C. MOCK, AND ROMAN R. GANTA

Department of Diagnostic Medicine/Pathobiology, College of Veterinary Medicine, Kansas State University Manhattan, Kansas 66506, USA

ABSTRACT: Tick-borne rickettsial infections are responsible for many emerging diseases in humans and several vertebrates. These include human infections with *Ehrlichia chaffeensis*, *Ehrlichia ewingii* and *Anaplasma phagocytophilum*. As single or co-infections can result from a tick bite, the availability of a rapid, multiplex molecular test will be valuable for timely diagnosis and treatment. We recently described a muliplex-molecular test that can detect single or co-infections with up to five *Ehrlichia* and *Anaplasma* species. We reported that the test has the sensitivity to identify single infections in the canine host with *E. chaffeensis*, *E. canis*, *E. ewingii*, *A. phagocytophilum*, and *A. platys* and co-infection with *E. canis* and *A. platys*. In this study, ticks were collected from different parts of the state of Kansas during summer months of the year 2003 and tested for the presence of infection using the molecular test. The analysis revealed a minimum of 3.66% of the ticks to be positive for either *E. chaffeensis* or *E. ewingii* in *A. americanum* and *Dermacenter* species. This assay will be valuable in monitoring infections in dogs and ticks, and with minor modifications it can be used for diagnosing infections in people and other vertebrates.

KEYWORDS: *Rickettsia; Ehrlichia; Anaplasma;* multiplex; ticks

INTRODUCTION

Several rickettsial agents of the family *Anaplasmataceae* cause severe, tick-borne pathogen infections in a wide range of vertebrate hosts including humans.[1] They include emerging infections in humans with *Ehrlichia chaffeensis, Ehrlichia ewingii,* and *Anaplasma phagocytophilum* and canine infections with *Ehrlichia canis, E. ewingii, E. chaffeensis, Anaplasma platys,* and *A. phagocytophilum.*[1–4] Coinfections with two or more rickettsiales and other tick-transmitted pathogens are common in vertebrate and tick hosts.[4–13] Because

Address for correspondence: Roman R. Ganta, Department of Diagnostic Medicine/Pathobiology, College of Veterinary Medicine, Kansas State University, 1800 Denison Avenue, Manhattan, KS 66506, USA. Voice: (785)-532-4612; fax: (785)-532-4851.
e-mail: rganta@vet.k-state.edu

Ann. N.Y. Acad. Sci. 1078: 552–556 (2006). © 2006 New York Academy of Sciences.
doi: 10.1196/annals.1374.108

Rickettsiales are able to infect a broad range of hosts and multiple pathogens can coexist in both vertebrate and invertebrate hosts, the availability of a rapid, highly sensitive and specific test that can diagnose one or more pathogens, including coinfections, in a test sample will be valuable for timely diagnosis and treatment. Such a test will be useful for monitoring and controlling the spread of infections from ticks. Recently, we describe the development of a rapid, two-step, species-specific multiplex molecular test to detect one or more infections with three *Ehrlichia* and two *Anaplasma* species.[14] The test protocol includes the magnetic capture-based purification of 16S ribosomal RNA, its enrichment, and specific-pathogen(s) detection by real-time reverse transcription polymerase chain reaction (RT-PCR). The molecular test is used to detect natural infections, including coinfections in dogs and ticks. As the majority of results pertaining to the multiplex molecular test development and its utility for infection monitoring in the canine host is already described recently in our previous paper, we only describe here the tick collection and their analysis.

MATERIALS AND METHODS

Nucleic Acid Isolation

RNA was isolated by using Tri-reagent method as per the manufacturer's protocol (Sigma-Aldrich, St. Louis, MO) and DNA was isolated by the sodium dodecyl sulfate, proteinase K method.[15]

Multiplex Real-Time RT-PCR Assay

Multiplex real-time RT-PCRs were performed in a Smart Cycler system (Cepheid, Sunnyvale, CA). The primers and probes were designed targeting the 16S rRNA region.[14] As the Smart Cycler can detect only up to four different reporter probes, we split the assay into two parts. In the first part, three *Ehrlichia* species—*E. chaffeensis, E. canis*, and *E. ewingii*— were tested and in the second part the two *Anaplasma* species—*A. phagocytophilum* and *A. platys*—were tested. The reaction conditions were optimized for multiplexing by altering the concentrations of enzyme, magnesium probes, and primers.[14] The primers and Tag Man probes were custom-synthesized from Integrated DNA Technologies (Coralville, IA).

Collection of Ticks

Ticks were collected from the northeastern parts of Kansas by using drag-drop method during the summer months of the year 2003.[16] The collected ticks

TABLE 1. Analysis of ticks in Kansas

Total number of ticks processed	246
Number of pools processed (*A. americanum*-52, *Dermacenter* species-21, *Ixodes* species-1)	74
Number of pools positive	9
Minimum % positive	3.66

were sorted and pooled into 74 pools according to the site of collection and species. Each pool consisted of 1–4 ticks. Total RNA was isolated using the Tri-reagent method after the ticks were crushed with plastic homogenizer.

RESULTS

Analysis of Ticks and Their Infection Rates in the State of Kansas

A total of 246 ticks were collected from the northeastern part of the state of Kansas. The ticks were sorted as 74 pools according to species and their collection location. These included 52 pools of *Amblyomma americanum*, 21 pools of *Dermacentor* species, and one pool of *Ixodes* species (TABLE 1). Total RNA isolated from the ticks was tested for the presence of *E. chaffeensis, E. canis, E. ewingii, A. phagocytophilum,* and *A. platys* by real-time multiplex RT-PCR assay. Out of 74 pools analyzed, the test identified 6 pools positive for *E. ewingii* and 3 pools positive for *E. chaffeensis* (TABLE 2). Assuming that there is at least one tick positive in each pool, the minimum infectivity is 3.66%.

DISCUSSION

We described a multiplex molecular test having the ability to detect single or coinfections with up to five tick-transmitted rickettsiales: *E. chaffeensis, E. canis, E. ewingii, A. phacytophilum*, and *A. platys*.[14] The molecular test procedure includes a simplified and rapid magnetic-capture technique to purify

TABLE 2. Infection rates in ticks of Kansas

Tick	County	Positive for	No. of pools
A. americanum	Geary	*E. ewingii*	1
A. americanum	Riley	*E. ewingii*	3
A. americanum	Riley	*E. chaffeensis*	1
Dermacenter sp.	Riley	*E. chaffeensis*	1
A. americanum	Pottawatomie	*E. chaffeensis*	1
A. americanum	Douglas	*E. ewingii*	2

16S rRNA, its enrichment by genera-specific RT-PCR assay, and species-specific pathogen detection by real-time monitoring with species-specific Taq-Man probes. Coinfections with two or more pathogens are also reported in vertebrates and ticks. The ability of this test to detect all known *Ehrlichia* and *Anaplasma* species causing infections in a host, including coinfections, will be valuable for infection monitoring in several vertebrate hosts, including humans. The multiplex molecular test was used to evaluate infections in 95 canine blood samples suspected of ehrlichiosis. The test identified 23 positives, which included 22 for single-pathogen infections of the 5 rickettsials and 1 positive sample for coinfection with *E. canis* and *A. platys*. These results support previous reports that dogs can acquire infections with all the five pathogens included in this assay. Infection rates in canine samples seem similar for *E. canis, E. ewingii,* and *A. platys*. *E. chaffeensis* and *A. phagocytophilum* were detected in fewer samples.[14]

The utility of the assay as an epidemiological tool have been evaluated by analyzing ticks collected from northeastern parts of the state of Kansas. The results showed a minimum infectivity rate of 3.66% in the ticks. In conclusion, we established a multiplex, molecular test that is useful to rapidly diagnose single or coinfections with up to five tick-borne rickettsial pathogens. The test serves as a new tool to monitor infections in dogs and ticks and can be adapted for screening emerging infections in people, cattle, and horses with minor modifications. It is also a useful tool for experimental infection studies of single or coinfections to assess the impact on the disease outcome and also to evaluate vaccines and therapeutics.

ACKNOWLEDGMENTS

This study was supported by the Morris Animal Foundation Grant D01CA-91.

REFERENCES

1. DUMLER, J.S., A.F. BARBET, C.P. BEKKER, *et al.* 2001. Reorganization of genera in the families *Rickettsiaceae* and *Anaplasmataceae* in the order Rickettsiales: unification of some species of *Ehrlichia* with *Anaplasma*, *Cowdria* with *Ehrlichia* and *Ehrlichia* with *Neorickettsia*, descriptions of six new species combinations and designation of *Ehrlichia equi* and "HGE agent" as subjective synonyms of *Ehrlichia phagocytophila*. Int. J. Syst. Evol. Microbiol. **51:** 2145–2165.
2. FRENCH, T.W. & J.W. HARVEY. 1993. Canine infectious cyclic thrombocytopenia (*Ehrlichia platys* infection in dogs). *In* Rickettsial and Chlamydial Diseases of Domestic Animals. Z. Woldehiwet & M. Ristic, Eds.: 195–208. Pergamon Press. Oxford.

3. MURPHY, G.L., S.A. EWING, L.C. WHITWORTH, *et al.* 1998. A molecular and serologic survey of *Ehrlichia canis, E. chaffeensis*, and *E. ewingii* in dogs and ticks from Oklahoma. Vet. Parasitol. **79:** 325–339.
4. PADDOCK, C.D. & J.E. CHILDS. 2003. *Ehrlichia chaffeensis:* a prototypical emerging pathogen. Clin. Microbiol. Rev. **16:** 37–64.
5. KORDICK, S.K., E.B. BREITSCHWERDT, B.C. HEGARTY, *et al.* 1999. Coinfection with multiple tick-borne pathogens in a Walker Hound kennel in North Carolina. J. Clin. Microbiol. **37:** 2631–2638.
6. HOSKINS, J.D., E.B. BREITSCHWERDT, S.D. GAUNT, *et al.* 1988. Antibodies to *Ehrlichia canis, Ehrlichia platys*, and spotted fever group rickettsiae in Louisiana dogs. J. Vet. Intern. Med. **2:** 55–59.
7. HUA, P., M. YUHAI, T. SHIDE, *et al.* 2000. Canine ehrlichiosis caused simultaneously by *Ehrlichia canis* and *Ehrlichia platys*. Microbiol. Immunol. **44:** 737–739.
8. BREITSCHWERDT, E.B., B.C. HEGARTY & S.I. HANCOCK. 1998. Sequential evaluation of dogs naturally infected with *Ehrlichia canis, Ehrlichia chaffeensis, Ehrlichia equi, Ehrlichia ewingii,* or *Bartonella vinsonii.* J Clin. Microbiol. **36:** 2645–2651.
9. LANE, R.S., J.E. FOLEY, L. EISEN, *et al.* 2001. Acarologic risk of exposure to emerging tick-borne bacterial pathogens in a semirural community in northern California. Vector Borne Zoonotic Dis. **1:** 197–210.
10. CHANG, Y.F., V. NOVOSEL, C.F. CHANG, *et al.* 1998. Detection of human granulocytic ehrlichiosis agent and *Borrelia burgdorferi* in ticks by polymerase chain reaction. J. Vet. Diagn. Invest. **10:** 56–59.
11. SEXTON, D.J., G.R. COREY, C. CARPENTER, *et al.* 1998. Dual infection with *Ehrlichia chaffeensis* and a spotted fever group rickettsia: a case report. Emerg. Infect. Dis. **4:** 311–316.
12. MEINKOTH, J.H., S.A. EWING, R.L. COWELL, *et al.* 1998. Morphologic and molecular evidence of a dual species ehrlichial infection in a dog presenting with inflammatory central nervous system disease. J. Vet. Intern. Med. **12:** 389–393.
13. PADDOCK, C.D., S.M. FOLK, G.M. SHORE, *et al.* 2001. Infections with *Ehrlichia chaffeensis* and *Ehrlichia ewingii* in persons coinfected with human immunodeficiency virus. Clin. Infect. Dis. **33:** 1586–1594.
14. SIRIGIREDDY, K.R. & R.R. GANTA. 2005. Multiplex detection of *Ehrlichia* and *Anaplasma* species pathogens in peripheral blood by real-time reverse transcriptase-polymerase chain reaction. J. Mol. Diagn. **7:** 308–316.
15. MANIATIS, T., E.F. FRITSCH & J. SAMBROOK. 1982. Molecular cloning: a laboratory manual. Cold Spring Harbor Laboratory. Cold Spring Harbor, NY.
16. GODDARD, J., J.W. SUMNER, W.L. NICHOLSON, *et al.* 2003. Survey of ticks collected in Mississippi for *Rickettsia, Ehrlichia,* and *Borrelia* species. J. Vector Ecol. **2:** 184–189.

Identification and Characterization of *Coxiella burnetii* Strains and Isolates Using Monoclonal Antibodies

Z. SEKEYOVÁ AND E. KOVÁČOVÁ

Institute of Virology, Slovak Academy of Sciences, Dúbravská Cesta 9, 845 05 Bratislava, Slovak Republic

ABSTRACT: We evaluated 6 monoclonal antibodies (MAbs) for their usefulness in identifying and characterizing recognized laboratory strains as well as field isolates of *Coxiella burnetii*. Five had been generated in response to strain Nine Mile (3 IgM class, 1 IgG class, 1 light chain producers only) and were polypeptide-specific, and 1 was anti-Priscilla (IgG class) and was lipopolysaccharide (LPS)-specific. Initially, the MAbs were used in conjunction with a dot blot assay with which we could differentiate *C. burnetii* from rickettsiae or chlamydiae. Confirmation of the specificity of these MAbs was provided by demonstrating that only *C. burnetii* antigens were recognized by certain combinations of antibodies used for immunoblotting proteins of various *C. burnetii* strains. Subsequently, we characterized antigens of 11 *C. burnetii* field isolates and 3 reference strains by Western blotting with individual MAbs. MAb 921 and 922 (IgG class), MAb 241, 242, 384, 386, 614 (IgM class), and 7A5, 7A1 (light chain) consistently recognized a protein. Staining intensity differed, depending on the strain tested, and there was variability in the size of the antigen immunoreactive with MAb 14H (IgG class, LPS-specific). The most reactive region was at about 249 kD. Variability of reactivities with field isolates was seen in both the distribution of individual bands and their intensities. We conclude that an extensive immunoblotting technique may be useful for *C. burnetii* strain differentiation and routine identification of *C. burnetii* can be accomplished using this MAb-based dot blot assay.

KEYWORDS: *Coxiella burnetii*; monoclonal antibody; differentiation of isolates

INTRODUCTION

Coxiella burnetii is one of the most important aerosol-transmitted bacterial pathogens of livestock and humans and a cause of zoonoses, that is, coxiellosis

Address for correspondence: Zuzana Sekeyová, Institute of Virology, Slovak Academy of Sciences, Dúbravská Cesta 9, 845 05 Bratislava 45, Slovak Republic. Voice: 004212-59-302-433; fax: 004212-5477-4284.
e-mail: viruseke@savba.sk

Ann. N.Y. Acad. Sci. 1078: 557–560 (2006). © 2006 New York Academy of Sciences.
doi: 10.1196/annals.1374.109

in livestock,[1] and Q fever in humans.[2,6,7] Environmental characteristics play a key role in dissemination of this pathogen. Lipopolysaccharide (LPS) and membrane proteins[13] compose the antigenic structures of *Coxiella* spp. and play roles in the development of protective immunity to them. Changes in antigenic structure in phase I of the developmental cycle reflect successful adaptation of this intracellular parasite, namely its capacity to evade host immune response. Complete elimination from an immunocompetent host leads to the phase II developmental stage.[5] As others have done previously,[3,4,8,11,12,14,15] we employed the specificity of monoclonal antibodies (Mabs) as probes to detect antigenic differences. We prepared several Mabs and used them to differentiate numerous isolates. This work revealed unique characteristics of isolates in regard to their Mab binding properties. The results reflected varied structural components and facilitated detection techniques.

MATERIALS AND METHODS

A total of 3 *C. burnetii* strains (Nine Mile, Priscilla, S Q217), and 11 isolates (IXO, Louga, Vologda, Dermacentor, M35, M18, Henzerling, 48, II/1A, No. 9, No. 10) were used in this study. Distinct reservoirs, including human, tick, cattle, sheep, and goats were the sources.

Murine Mabs 921, 922, 241, 242, 384, 386, 614, and 7A5, 7A1 were raised to proteins (anti–Nine Mile Mabs, generated against live whole cells) or to LPS, 14H (anti–Priscilla Mab, generated against formalin-inactivated *C. burnetii*) by techniques described previously.[9,10] Hybridoma cultures that produced *C. burnetii*–reactive antibodies were detected with enzyme-linked immunosorbent assay or by immunofluorescence against whole organisms. For dot blot assays, antigens were applied to PVDF membranes. The immunochemical properties of the outer membrane proteins or to the LPS of a particular isolate were assayed by Western blotting. The antigens were resolved by SDS-PAGE and transferred to the PVDF membranes (Immobilon P membrane, polyvinylidene fluoride, Millipore) as described previously.

RESULTS

Initially, Mabs were tested by a dot blot assay that enabled *C. burnetii* to be differentiated from other bacteria. Confirmation of the specificity of these Mabs was provided by the demonstration that only *C. burnetii* antigens were recognized by a combination of antibodies, using immunoblotting. Antigenic characterization of 3 reference strains and 11 field isolates was performed with individual Mabs by Western blotting. Whereas Mab 921 and 922 (IgG class), Mab 241, 242, 384, 386, 614 (IgM class), and 7A5, 7A1 (light chain) consistently recognized proteins with a molecular weight of approximately

60 kDa and/or 17–22 kDa, staining intensity differed, depending on the strain. There was variability in the size of the antigen immunoreactive with Mab 14H (IgG class, LPS-specific); however, the most reactive region was at about 24–29 kD. Variability of reactivities with field isolates was seen in both the distribution of individual bands and the intensities of those bands.

CONCLUSION

Immunoblotting is useful for *C. burnetii* strain differentiation and for routine identification of *C. burnetii* and can be accomplished using Mab-based dot blot assays.

ACKNOWLEDGMENT

The authors thank Professor Charles H. Calisher, Colorado State University, Fort Collins, Colorado, USA, for reviewing the manuscript. This work was partly supported by Grants 3050 and 5053 of VEGA, Scientific Grant Agency of the Ministry of Education of the Slovak Republic, and the Slovak Academy of Sciences.

REFERENCES

1. BABUDIERI, C. 1959. Q fever Zoonosis. *In* Advances in Veterinary Science. C. A. Brandly & E. L. Tunghen, Eds.: 81–182. Vol. V. Academic Press, Inc. New York NY.
2. FOURNIER, P.E., T.J. MARRIE & D. RAOULT. 1998. Diagnosis of Q fever. J. Clin. Microbiol. **36:** 1823–1834.
3. HOTTA, A., G.Q. ZHANG, M. ANDOH, *et al.* 2004. Use of monoclonal antibodies for analyses of *Coxiella burnetii* major antigens. J. Vet. Med. Sci. **66:** 1289–1291.
4. HOTTA, A., M. KAWAMURA, H. TO, *et al.* 2003. Use of monoclonal antibodies to liposaccharide for antigenic analysis of *Coxiella burnetii*. J. Clin. Microbiol. **41:** 1747–1749.
5. KAZÁR, J., R. BREZINA, S. SCHRAMEK, *et al.* 1974. Virulence, antigenic properties, and physicochemical characteristics of *Coxiella burnetii* strains with different chick embryo yolk sac passage history. Acta Virol. **18:** 434–442.
6. MARRIE, T.J. & D. RAOULT. 2002. Update on Q fever, including Q fever endocarditis. Curr. Clin. Top. Infect. Dis. **22:** 97–124.
7. RAOULT, D., T. MARRIE & J. MEGE. 2005. Natural history and pathophysiology of Q fever. Lancet Infect. Dis. **5:** 219–226.
8. RAOULT, D., J.C. LAURENT & M. MUTILLOD. 1994. Monoclonal antibodies to *Coxiella burnetii* for antigenic detection in cell cultures and paraffin-embedded tissues.Am. J. Clin. Pathol. **101:** 318–320.
9. SEKEYOVÁ, Z., D. THIELE, H. KRAUSS, *et al.* 1996. Monoclonal antibody based differentiation of *Coxiella burnetii* isolates. Acta Virol. **40:** 127–132.

10. SEKEYOVÁ, Z., E. KOVÁOVÁ, J. KAZÁR, *et al.* 1995. Monoclonal antibodies to *Coxiella Burnetii* that cross-react with strain Nine Mile. Clin. Diagn. Lab. Immunol. **2:** 531–534.
11. SESHADRI, R., L.R. HENDRIX & J.E. SAMUEL. 1999. Differential expression of translational elements by life cycle variants of *Coxiella burnetii.* Infect. Immun. **67:** 6026–6033.
12. THIELE, D., M. KARO & H. KRAUSS. 1992. Monoclonal antibody based capture ELISA/ELIFA for detection of *Coxiella burnetii* in clinical specimens. Eur. J. Epidemiol. **8:** 568–574.
13. WILLIAMS, J.C., T.A. HOOVER, D.M. WAAG, *et al.* 1990. Antigenic structure of *Coxiella burnetii*: a comparison of lipopolysaccharide and protein antigens as vaccines against Q fever. Ann. N. Y. Acad. Sci. **590:** 370–380.
14. YU, X. & D. RAOULT. 1994. Serotyping *Coxiella burnetii* isolates from acute and chronic Q fever patients by using monoclonal antibodies. FEMS Microbiol. Lett. **117:** 15–19.
15. ZHANG, G., K. KISS, R. SESHADRI, *et al.* 2004. Identification and cloning of immunodominant antigens of *Coxiella burnetii.* Infect. Immun. **72:** 844–852.

Comparison of Four Commercially Available Assays for the Detection of IgM Phase II Antibodies to *Coxiella burnetii* in the Diagnosis of Acute Q Fever

DIMITRIOS FRANGOULIDIS,[a] ELMAR SCHRÖPFER,[a]
SASCHA AL DAHOUK,[a] HERBERT TOMASO,[a,b]
AND HERMANN MEYER[a]

[a]*Bundeswehr Institute of Microbiology, Munich, Germany*

[b]*Military Hospital Innsbruck, Ministry of Defence, Innsbruck, Austria*

ABSTRACT: Four commercially available serological assays for the detection of IgM phase II antibodies in patients with acute Q fever infection were compared using a panel of 23 serum samples from patients with acute Q fever and 88 control sera from blood donors.

KEYWORDS: Q fever; *Coxiella burnetii*; enzyme-linked immunosorbent assay; immuno-fluorescence assay; serological diagnosis

Coxiella burnetii is the causative agent of Q fever, a worldwide zoonosis. Infection in humans results from inhalation of contaminated aerosols. Although the clinical signs of an acute infection vary, typical symptoms include fever, headache, myalgia, muscle cramps, hepatitis, and respiratory complications. Chronic manifestations may include endocarditis and granulomatous hepatitis.[1] Diagnosis of acute Q fever is based on serological methods because culture of the organism is hazardous and time consuming.[2] Primary binding assays commonly used for measuring antibodies to *C. burnetii* antigens are the complement fixation test (CFT), the indirect immunofluorescent antibody test (IFAT), and the enzyme-linked immunosorbent assay (ELISA). Commercially available assays offer the ability to nonspecialized laboratories to perform a serological diagnosis of acute Q fever and, it is hoped, to improve the identification of this underdiagnosed disease. In our study we compared three indirect IFATs (Focus/Genzyme-Virotech, USA/ Rüsselsheim, Germany; bioMerieux Nürtingen, Germany; and Fuller/Viva-Diagnostics, USA/Cologne Germany)

Address for correspondence: Dimitrios Frangoulidis, Bundeswehr Institute of Microbiology, Neuherbergstr. 11, 80937 Munich, Germany. Voice: +49-89-3168-3869; fax: +49-89-3168-3292.
e-mail: DimitriosFrangoulidis@Bundeswehr.org

Ann. N.Y. Acad. Sci. 1078: 561–562 (2006). © 2006 New York Academy of Sciences.
doi: 10.1196/annals.1374.110

and one ELISA (Virion/Serion, Würzburg, Germany) with respect to their ability to detect IgM phase II antibodies in 23 patients with clinical symptoms of acute Q fever (pneumonia, fever). A total of 88 control sera from blood donors were investigated as well. IgM Phase II test results were considered positive when titers of 1:64 or higher were seen.

The specificity varied from 97.7% (86 out of 88 samples; BioMeriex IFAT) and 98.9% (87/88, Virion/Serion ELISA) to 100% (Focus/Genzyme-Virotech IFAT and Fuller/Viva-Diagnostics IFAT). The sensitivity was 95.65% for the ELISA (22 out of 23 samles) and 100% for all IFATs.

This is the first comparison of commercially available serological assays for the diagnosis of acute Q fever. As has been demonstrated earlier, the ELISA is a useful tool in the diagnosis of acute Q fever and has the advantage of being easy and quick, requiring a single dilution of serum.[3] It also has the potential to be automated. No significant difference was seen between the three IFATs. The immunofluorescence technique remains the reference method to confirm negative and uncertain results.

ACKNOWLEDGMENTS

We thank Alessandra Knoche for technical assistance.

REFERENCES

1. FOURNIER, P., T.J. MARRIE & D. RAOULT. 1999. Diagnosis of Q fever. J. Clin. Microbiol. **36:** 1823–1834.
2. SCOLA, B.L. 2002. Current laboratory diagnosis of Q fever. Semin. Pediatr. Infect. Dis. **13:** 257–262.
3. FIELD, P.R., J.L. MITCHELL, A. SANTIAGO, *et al.* 2000. Comparison of a commercial enzyme-linked immunosorbent assay with immunofluorescence and complement fixation tests for detection of *Coxiella burnetii* (Q fever) immunglobulin M. J. Clin. Microbiol. **38:** 1645–1647.

Evaluation of a Real-Time PCR Assay to Detect *Coxiella burnetii*

SILKE R. KLEE,[a] HEINZ ELLERBROK,[a] JUDITH TYCZKA,[b]
TATJANA FRANZ,[a] AND BERND APPEL[a]

[a]*Robert Koch-Institut, Centre for Biological Safety, Nordufer 20, 13353 Berlin, Germany*

[b]*Institute for Hygiene and Infectious Diseases of Animals, Justus Liebig University of Giessen, Frankfurter Street 85-89, 35392 Giessen, Germany*

ABSTRACT: We evaluated real-time PCR assays for the detection of *C. burnetii* which targets sequences that are present either in one (*icd*) or in several copies (transposase of *IS1111a*) on the chromosome. The assays are highly sensitive, with reproducible detection limits of approximately 10 copies per reaction, at least 100 times more sensitive than capture ELISA, when performed on infected placenta material and specific for *C. burnetii*. The numbers of *IS1111* elements in the genomes of 75 *C. burnetii* isolates were quantified by real-time PCR and proved to be highly variable.

KEYWORDS: *Coxiella burnetii*; real-time PCR; *IS1111* element

INTRODUCTION

For rapid diagnosis of *Coxiella burnetii*, the causative agent of Q fever, we evaluated a real-time polymerase chain reaction (PCR) TaqMan-based assay for two targets, the transposase of insertion sequence (IS) element *IS1111a*, which is present in 20 copies in the genome of the *C. burnetii* Nine Mile RSA493 strain,[1,2] and *icd*, the gene for isocitrate dehydrogenase.[3]

RESULTS

We performed quantitative real-time PCR assays (7700 Sequence Detection System, Applied Biosystems) with 75 *C. burnetii* isolates from all over the world. All isolates contained both the *icd* and the *IS1111* markers, and all other bacterial species tested, including the closely related *Legionella pneumophila*, were negative for both targets, confirming the specificity of the assays.

Address for correspondence: Silke R. Klee, Robert Koch-Institut, Centre for Biological Safety, Nordufer 20, 13353 Berlin, Germany. Voice: +49-0-30-4547-2934; fax: +49-0-30-4547-2110.
e-mail: KleeS@rki.de

Ann. N.Y. Acad. Sci. 1078: 563–565 (2006).
doi: 10.1196/annals.1374.111

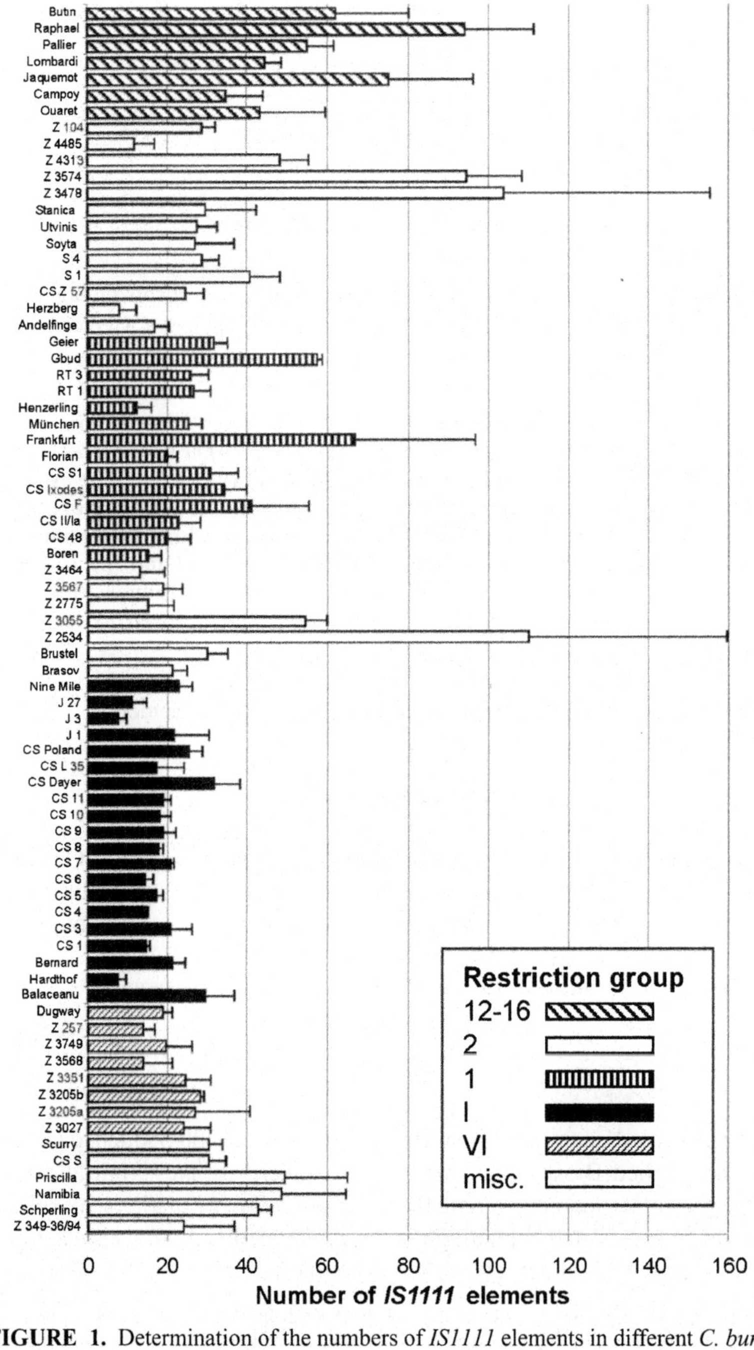

FIGURE 1. Determination of the numbers of *IS1111* elements in different *C. burnetii* isolates. The columns in the figure represent *IS1111* numbers determined from three independent PCR runs; the error bars indicate the standard deviation.

Sensitivity of the assays was examined with dilutions of plasmid standards with cloned inserts of either the *icd* or *IS1111* amplicon and with DNA dilutions of the *C. burnetii* Nine Mile RSA493 strain. Standard curves were established by plotting the starting plasmid copy number against the threshold cycle (C_t), showing a linear quantification over a range from 10^7 to 10 copies and a detection limit of 10 plasmid copies per reaction for both targets. Using dilutions of the *C. burnetii* Nine Mile strain, we were able to detect 10 genome equivalents (GE) per reaction for the *icd* target and 6.5 GE per reaction for the *IS1111* target with 95% probability (probit analysis).

By comparison with plasmid standard curves, the numbers of *IS1111* elements per genome, or per *icd* copy, respectively, were quantified for 75 *C. burnetii* isolates and shown to be very variable (FIG. 1). A correlation of the numbers of IS elements with the restriction group,[4] and hence with the geographical origin, might be possible at least for isolates from groups I and VI (low numbers between 7 and 32) and from groups 12–16 (large numbers between 35 and 94).

Several samples of bovine and ovine placenta material were tested for the presence of *C. burnetii*. Capture enzyme-linked immunosorbent assay (ELISA) and real-time PCR using both targets were able to detect 10^9 coxiellae per gram placenta, whereas 10^7 coxiellae per gram placenta were only detected by real-time PCR, demonstrating that our real-time PCR assays are at least 100 times more sensitive than capture ELISA.

REFERENCES

1. HOOVER, T.A., M.H. VODKIN & J.C. WILLIAMS. 1992. A *Coxiella burnetii* repeated DNA element resembling a bacterial insertion sequence. J. Bacteriol. **174:** 5540–5548.
2. SESHADRI, R., L.T. PAULSEN, J.A. EISEN, *et al.* 2003. Complete genome sequence of the Q-fever pathogen *Coxiella burnetii*. Proc. Natl. Acad. Sci. USA **100:** 5455–5460.
3. NGUYEN, S.V. & K. HIRAI. 1999. Differentiation of *Coxiella burnetii* isolates by sequence determination and PCR-restriction fragment length polymorphism analysis of isocitrate dehydrogenase gene. FEMS Microbiol. Lett. **180:** 249–254.
4. JAGER, C., H. WILLEMS, D. THIELE & G. BALJER. 1998. Molecular characterization of *Coxiella burnetii* isolates. Epidemiol. Infect. **120:** 157–164.

Diagnosis of Acute Q Fever by PCR on Sera during a Recent Outbreak in Rural South Australia

M. TURRA, G. CHANG, D. WHYBROW, G. HIGGINS, AND M. QIAO

Institute of Medical and Veterinary Science, Frome Road, Adelaide, South Australia.

ABSTRACT: Diagnosis of Q fever has largely been dependent upon serology, which may lead to delayed diagnosis as seroconversion can take weeks to develop. During a recent Q fever outbreak (27 patients) in rural South Australia, we compared the diagnostic rate of serology with two separate real-time PCRs, the 27kDa outer membrane protein and the insertion sequence. PCR was positive (on either or both PCR assays) in sera of 67% of the patients. Median time required for making serological diagnosis was 17 days, compared with 4 days by PCR. Q fever PCR is an effective tool in the diagnosis of acute Q fever infection.

KEYWORDS: Q fever PCR; Q fever diagnosis; Q fever serology; rapid detection

OBJECTIVES

The diagnosis of Q fever has largely been dependant upon serological findings. This often leads to delayed diagnosis as seroconversion may take several weeks to develop.[1] During a recent Q fever outbreak in rural South Australia in 2004, we performed real-time polymerase chain reaction (PCR) on serum samples to complement the serological tests used routinely in our laboratory. In this study, the diagnostic rate of serology was compared with combined PCR and serology.

METHODS

During the outbreak, our laboratory received sera from 27 patients with Q fever-like symptoms including fever, myalgia, and headache. A total of

Address for correspondence: Ming Qiao, Consultant Microbiologist/Virologist, Institute of Medical and Veterinary Science, Rundle Mall, Adelaide, South Australia 5000, Australia. Voice: 61-08-8222-3144; fax: 61-08-8222-3543.
e-mail: ming.qiao@imvs.sa.gov.au

Ann. N.Y. Acad. Sci. 1078: 566–569 (2006). © 2006 New York Academy of Sciences.
doi: 10.1196/annals.1374.112

TABLE 1. Primer and probe oligonucleotide sequences

Primer/probe	Sequence
COM1 (forward)	5′-CAAACCGCAGAAAAAGTAGG-3′
COM1 (reverse)	5′-TGGAAGTTATCACGCAGTTG-3′
COM1 Taqman MGB Probe	5′-FAM-CAGGATTATCCATGTCTTT-BHQ-3′
QF-IS-FWD (forward)	5′-ATCTACACGAGACGGG-3′
QF-IS-FWD (reverse)	5′-CTTCAGCTATCGCCTGC-3′
QF-IS-Fluorescein	5′-GCGTGGTGATGGAAGCG-FL-3′
QF-IS-640	5′-LCR640-TGGAGGAGCGAACCATTGG-PHOS-3′

21 of the 27 patients submitted 2 or more samples ($n = 54$); the mean time between paired samples was 12 days (range 4–47). The median age was 46 years (range 16–80). All sera were tested for Q fever antibodies to phase I and II IgM, Ig,A and IgG by indirect immunofluorescence antibody technique, in doubling dilutions starting from 1:10. Titers of $\geq$1:40 were considered positive. A fourfold rise in phase II IgM titers on paired sera was considered a seroconversion.

All sera were tested by two separate PCR assays, one targeting the 27-kDa outer membrane protein (COM1)[2] and the other detecting the insertion sequence (IS1111)[3] (primer/probes sequences are listed in TABLE 1). PCR was performed on 5 μL of the DNA extracted from 200 μL of serum using the Qiagen (Hilden, Germany) DNA blood kit and amplified for 45 cycles using the Roche LightCycler (Basel, Switzerland).

RESULTS

Serology

Seroconversion was detected in all 21 patients with paired sera. Only 2 of 21 (10%) patients had detectable phase II IgM antibody in the first sample. Of the 6 patients for which only a single serum was received, 4 had phase II IgM titers of >1:1280, one had a phase II IgG and IgM of 1:80, and only one patient had no detectable antibody (FIG. 1).

TABLE 2. Comparsion of Q fever serology with PCR result for all samples received

Serology results phase II IgM	COM1 PCR Positive	COM1 PCR Negative	IS1111 PCR Positive	IS1111 PCR Negative	Combined PCR Positive	Combined PCR Negative
Positive ($n = 35$)	2 (6%)	33 (94%)	2 (6%)	33 (94%)	3 (9%)	32 (91%)
Negative ($n = 19$)	11 (58%)	8 (42%)	16 (84%)	3 (16%)	17 (89%)	2 (11%)

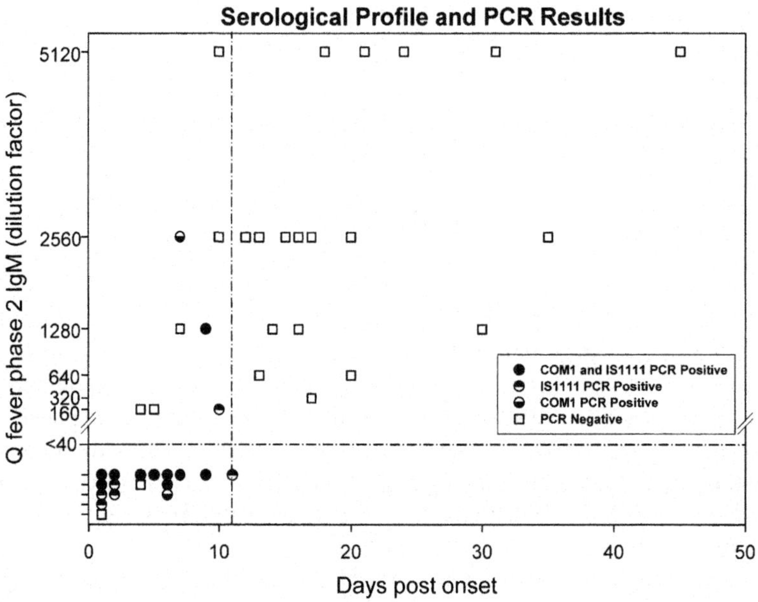

FIGURE 1. Serological profile and PCR result.

PCR

Overall, 17 of 27 patients (63%) were positive by one or both PCR assays. The patient who submitted a single sample only with no detectable antibody was positive by both PCR assays. A total of 12 of 27 patients (44%) were PCR-positive by COM1 PCR, whereas 16 of 27 patients (59%) were IS1111 PCR-positive (TABLE 2). All PCR-positive samples were collected within 11 days post onset (range 1–11 days, FIG. 1). Combined PCR assays were positive in 89% (17/19) of sera with negative phase II IgM antibody. In contrast, only 9% (3/35) of sera with detectable phase II IgM antibody were PCR-positive (TABLE 2, FIG. 1).

DISCUSSION

Early diagnosis of acute Q fever can be made with PCR assays. All PCR-positive results were found within the first 11 days (median 4 days) after the onset of symptoms. In contrast, serological diagnosis was made between 7–35 days (median 17 days) after the onset of symptoms.

PCR positivity was directly related to the absence of phase II IgM antibody ($P < 0.0001$). Of the phase II IgM antibody–negative sera, 17 of 19 (89%) were positive by one or both PCR assays, while only 3 of 35 (9%) antibody-positive sera were PCR-positive. Thus, Q fever DNA detection by PCR provided early diagnosis prior to the appearance of phase II IgM antibody in the majority of patients.

The IS1111 PCR appears to be more sensitive than COM1 PCR (positivity rate 59% vs. 44%). This difference is likely due to the multiple copies of the IS1111 repeats within each genome compared with the single copy of the COM1 gene.

REFERENCES

1. WORSWICK, D. & B.P. MARMION. 1985. Antibody responses in acute and chronic Q fever and in subjects vaccinated against Q fever. J. Med. Microbiol. **19:** 281–296.
2. SCHMEER, N. 1988. Early recognition of a 27 kDa membrane protein (MP27) in *Coxiella burnetii* infected and vaccinated guinea pigs. Zentralbl. Veterinarmed. B. **35:** 338–345.
3. SESHADRI, R. *et al.* 2003. Complete genome sequence of the Q-fever pathogen *Coxiella burnetii*. Proc. Natl. Acad. Sci. USA **100:** 5455–5460.

Evaluation of IgG Antibody Response against *Rickettsia conorii* and *Rickettsia slovaca* in Patients with DEBONEL/TIBOLA

S. SANTIBÁÑEZ,[a] V. IBARRA,[a] A. PORTILLO,[a] J.R. BLANCO,[a] V. MARTÍNEZ DE ARTOLA,[b] A. GUERRERO,[c] AND J.A. OTEO[a]

[a]*Area de Enfermedades Infecciosas, Hospitales San Millán-San Pedro-de La Rioja, Logroño, Spain*

[b]*Servicio de Análisis Clínicos, Hospital Virgen del Camino, Pamplona, Spain*

[c]*Área de Diagnóstico Biológico, Hospital de La Ribera, Valencia, Spain*

ABSTRACT: The aim of the study was to determine the IgG antibody response to spotted fever group *Rickettsia* (SFGR) *R. conorii* and *R. slovaca*, and its specificity and sensitivity in patients with DEBONEL/TIBOLA. A prospective study of 31 patients with DEBONEL was carried out from January 2001 to May 2004. The SFGR serology testing (IgG IFA) for the diagnosis of DEBONEL/TIBOLA showed 61% sensitivity and 100% specificity. The *R. slovaca* antigen allowed the diagnoses in 18 of the 31 patients (58%), and 17 patients (55%) were diagnosed with this disease using *R. conorii* antigen. Therefore, using *R. slovaca* as antigen did not improve the sensitivity of the assay.

KEYWORDS: DEBONEL/TIBOLA; *Rickettsia conorii; Rickettsia slovaca;* indirect immunofluorescence assay (IFA); IgG; Spain

INTRODUCTION

In the last few years, a new tick-borne disease (TBD) with epidemiological, clinical, and microbiological characteristics different from other TBDs has been reported.[1,2] It is known as DEBONEL/TIBOLA and *Rickettsia slovaca* is the main implicated agent.

The aim of this study was to determine the IgG antibody response against spotted fever group *Rickettsia* (SFGR), *R. conorii* and *R. slovaca,* and its specificity and sensitivity for the diagnosis of patients with DEBONEL/TIBOLA.

Address for correspondence: José A. Oteo, Área de Enfermedades Infecciosas, Hospitales San Millán-San Pedro-de La Rioja, Avda de Viana number 1, 26001 Logroño (La Rioja), Spain. Voice: +34-941-297275; fax: +34-941-297267.
e-mail: jaoteo@riojasalud.es

Ann. N.Y. Acad. Sci. 1078: 570–572 (2006).
doi: 10.1196/annals.1374.113

MATERIAL AND METHODS

From January 2001 to May 2004, a prospective study of patients diagnosed with DEBONEL ($n = 31$) was carried out. Diagnosis was made according to criteria of the European Network for Surveillance of Tick-Borne Diseases.[3] For all patients, IgG antibodies against *R. conorii* (BioMérieux, Lyon, France) and *R. slovaca* (antigen slides kindly donated by Dr. Fátima Bacellar, Centro de Estudos de Vectores e Doenças Infecciosas do Instituto Nacional de Saúde, Portugal) were tested using an indirect immunofluorescence assay (IFA) during the acute and convalescence phases (first, third, and sixth month).

Patients who were diagnosed with Lyme borreliosis from January 2003 to October 2004 ($n = 27$) were considered as the control group.

RESULTS AND DISCUSSION

Evidence of recent infection by SFGR was detected in 19 patients with DEBONEL/TIBOLA (61%). A total of 18 of them (95%) seroconverted, against *R. conorii* ($n = 1$), against *R. slovaca* ($n = 1$), or against both species ($n = 16$). Fourfold increase in IgG antibody titers against *R. slovaca* was observed in 1 of the 19 patients (5%). IgG antibodies titers against *R. slovaca* were higher than those against *R. conorii* in 9 cases, being two serial dilutions higher in 5 of them.

Seroconversion or increase of the previous titers was detected during the first month in 17 cases (87%). Only 4 patients (13%) showed seroconversion in the third month. IgG antibodies were observed for at least 6 months.

IgG antibodies against *R. conorii* were detected in two sera (first and third month) of one patient diagnosed with Lyme borreliosis (3.7%). However, this patient had no serological criteria of acute rickettsiosis.

In conclusion, SFGR serology testing (IgG IFA) for the diagnosis of DEBONEL/TIBOLA had 61% sensitivity and 100% specificity. Using *R. slovaca* antigen allowed the serological diagnoses of DEBONEL/TIBOLA in 18 of 31 patients (58%), whereas 17 patients (55%) were diagnosed with this disease using *R. conorii* antigen. Therefore, using *R. slovaca* antigen did not improve the sensitivity of the assay. On account of cross-reactions among SFGR when using IFA assays, and according to our data, commercial IgG substrate slides for *R. conorii* (Bio Mérieux, Lyon, France) are useful markers of rickettsial infection in patients with DEBONEL/TIBOLA.

ACKNOWLEDGMENTS

This study was supported in part by grants from the Fondo de Investigación Sanitaria (FIS PI021810 and G03/057), Ministerio de Sanidad y Consumo, Spain.

REFERENCES

1. RAOULT, D., A. LAKOS, F. FENOLLAR, *et al.* 2002. Spotless rickettsiosis caused by *Rickettsia slovaca* and associated with *Dermacentor* tick. Clin. Infect. Dis. **34:** 1331–1336.
2. OTEO, J.A., V. IBARRA, J.R. BLANCO, *et al.* 2004. *Dermacentor*-borne necrosis erithema and lymphadenopathy: clinical and epidemiological features of a new tick-borne disease. Clin. Microbiol. Infect. **10:** 327–331.
3. BROUQUI, P., F. BACELLAR, G. BARANTON, *et al.* 2004. Guidelines for the diagnosis of tick-borne diseases in Europe. Clin. Microbiol. Infect. Dis. **10:** 1108–1132.

Molecular Typing of Novel *Rickettsia rickettsii* Isolates from Arizona

MARINA E. EREMEEVA, ELIZABETH BOSSERMAN,
MARIA ZAMBRANO, LINDA DEMMA, AND GREGORY A. DASCH

Viral and Rickettsial Zoonoses Branch, CDC, Atlanta, Georgia, USA

ABSTRACT: Seven isolates of *Rickettsia rickettsii* were obtained from a skin biopsy, two whole-blood specimens, and from *Rhipicephalus sanguineus* ticks from eastern Arizona. Molecular typing of seven isolates of *R. rickettsii* and DNA samples from two other *Rh. sanguineus* ticks infected with *R. rickettsii* was conducted by PCR and DNA sequencing of *rompA* and 12 variable-number tandem repeat regions (VNTRs). All DNA specimens from Arizona were identical to each other and to reference human and *Dermacentor andersoni* isolates of *R. rickettsii* from Montana in their rOmpA gene sequences and 10 VNTRs. Two of the twelve VNTRs had differences in the number of repeat sequences in isolates from Arizona compared to those from Montana, thus conferring the novelty of the *Rh. sanguineus*-associated *R. rickettsii*

KEYWORDS: *Rickettsia rickettsii*; VNTR; molecular marker

The state of Arizona has a very low prevalence of Rocky Mountain spotted fever (RMSF) on account of its hot and dry climate and the resultant absence of *Dermacentor variabilis* and *D. andersoni*, classical tick vectors for *Rickettsia rickettsii* in the United States.[1] During an investigation of a cluster of fatal cases of RMSF in eastern Arizona,[2] rickettsial isolates were obtained from an engorged *Rhipicephalus sanguineus* tick from a dog, three questing *Rh. sanguineus* ticks, a patient skin biopsy sample, and two patients whole blood samples (TABLE 1). Isolation was done by centrifugation of samples onto Vero E6 cell monolayers (CRL1587) in 25-cm^2 flasks. Rickettsial growth was monitored by detection of cytotoxic effects on the infected cell monolayers and sustained growth of the isolates by acridine orange staining. Each of the new isolates exhibited a strong cytotoxic phenotype and produced lytic plaques in Vero cell monolayers. All isolates were confirmed to be *R. rickettsii* on the basis of an rOmpA gene fragment (70–701 nt) characterization as previously described.[3] Amplicons from each of the new isolates had 100% sequence homology to the reference sequence of *R. rickettsii* strain Sheila Smith (ATCC VR-149).

Address for correspondence: Marina E. Eremeeva, Mail Stop G-13, 1600 Clifton Road NE, Atlanta, GA 30333, USA. Voice: 404-639-4612; fax: 404-639-4436.
e-mail: MEremeeva@cdc.gov

**Ann. N.Y. Acad. Sci. 1078: 573–577 (2006). © 2006 New York Academy of Sciences.
doi: 10.1196/annals.1374.114**

TABLE 1. Isolates and DNA used in this study

Sample designation	Source	Association with RMSF (host)	Origin of DNA
AZ-1	Engorged *Rh. sanguineus* tick	Sick dog, RMSF not confirmed	Cell culture isolate
AZ-2	Engorged *Rh. sanguineus* tick	Sick dog, RMSF not confirmed	Tick
AZ-3	Human blood	Fatal RMSF	Cell culture isolate
AZ-4	Human blood	Fatal RMSF	Cell culture isolate
AZ-5	Questing *Rh. sanguineus* tick	Environmental sample	Cell culture isolate
AZ-6	Skin biopsy	RMSF	Cell culture isolate
AZ-7	Questing *Rh. sanguineus* tick	Found on human	Tick
AZ-8	Laboratory rabbit engorged *Rh. sanguineus* female tick after oviposition	Flat tick found at the house of a patient deceased from RMSF	Cell culture isolate
AZ-9	Laboratory rabbit engorged *Rh. sanguineus* female tick after oviposition	Flat tick found at the house of a patient deceased from RMSF	Cell culture isolate
SS, Sheila Smith	Human blood, Montana	RMSF	NCBI accession number AADJ01000001
R, Bitterroot	*D. andersoni* tick, Montana	Environmental sample, likely association with fulminant RMSF	Cell culture isolate

(A)

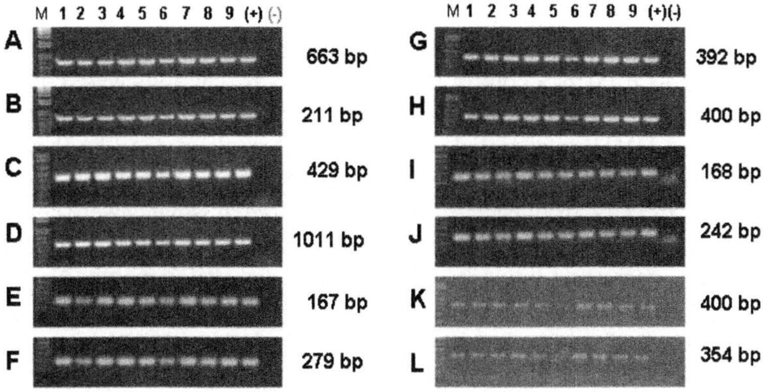

(B)

```
SS      CTACTATAAGGCCTTATGCCTTATGCCTTATTCCTATGCCT
R       CTACTATAAGGCCTTATGCCTTATGCCTTATTCCTATGCCT
AZ-1    CTACTATAAGGCCTTAT--------------TCCTATGCCT
AZ-2    CTACTATAAGGCCTTAT--------------TCCTATGCCT
AZ-3    CTACTATAAGGCCTTAT--------------TCCTATGCCT
AZ-4    CTACTATAAGGCCTTAT--------------TCCTATGCCT
AZ-5    CTACTATAAGGCCTTAT--------------TCCTATGCCT
AZ-6    CTACTATAAGGCCTTAT--------------TCCTATGCCT
AZ-7    CTACTATAAGGCCTTAT--------------TCCTATGCCT
AZ-8    CTACTATAAGGCCTTAT--------------TCCTATGCCT
AZ-9    CTACTATAAGGCCTTAT--------------TCCTATGCCT
                           1      2      3
```

(C)

```
                                                                                          Start ORF104
SS   CCGGAGCCATAGGCGGTGGTGGAGGTGGAGGTGGAGGTGGAAT-[22]-GTGGTGGAGGTGGA------AT-[23]-GTGGTGGAGGTGGAGGTGGAATGTTATTTTCCAAA
R    CCGGAGCCATAGGCGGTGGTGGAGGTGGAGGTGGAGGTGGAAT-[22]-GTGGTGGAGGTGGA------AT-[23]-GTGGTGGAGGTGGAGGTGGAATGTTATTTTCCAAA
AZ-1 CCGGAGCCATAGGCGGTGGTGGAGGTGGAGGTGGAGGTGGAAT-[22]-GTGGTGGAGGTGGAGGTGGAAT-[23]-GTGGTGGAGGTGGAGGTGGAATGTTATTTTCCAAA
AZ-2 CCGGAGCCATAGGCGGTGGTGGAGGTGGAGGTGGAGGTGGAAT-[22]-GTGGTGGAGGTGGAGGTGGAAT-[23]-GTGGTGGAGGTGGAGGTGGAATGTTATTTTCCAAA
AZ-3 CCGGAGCCATAGGCGGTGGTGGAGGTGGAGGTGGAGGTGGAAT-[22]-GTGGTGGAGGTGGAGGTGGAAT-[23]-GTGGTGGAGGTGGAGGTGGAATGTTATTTTCCAAA
AZ-4 CCGGAGCCATAGGCGGTGGTGGAGGTGGAGGTGGAGGTGGAAT-[22]-GTGGTGGAGGTGGAGGTGGAAT-[23]-GTGGTGGAGGTGGAGGTGGAATGTTATTTTCCAAA
AZ-5 CCGGAGCCATAGGCGGTGGTGGAGGTGGAGGTGGAGGTGGAAT-[22]-GTGGTGGAGGTGGAGGTGGAAT-[23]-GTGGTGGAGGTGGAGGTGGAATGTTATTTTCCAAA
AZ-6 CCGGAGCCATAGGCGGTGGTGGAGGTGGAGGTGGAGGTGGAAT-[22]-GTGGTGGAGGTGGAGGTGGAAT-[23]-GTGGTGGAGGTGGAGGTGGAATGTTATTTTCCAAA
AZ-7 CCGGAGCCATAGGCGGTGGTGGAGGTGGAGGTGGAGGTGGAAT-[22]-GTGGTGGAGGTGGAGGTGGAAT-[23]-GTGGTGGAGGTGGAGGTGGAATGTTATTTTCCAAA
Az-8 CCGGAGCCATAGGCGGTGGTGGAGGTGGAGGTGGAGGTGGAAT-[22]-GTGGTGGAGGTGGAGGTGGAAT-[23]-GTGGTGGAGGTGGAGGTGGAATGTTATTTTCCAAA
Az-9 CCGGAGCCATAGGCGGTGGTGGAGGTGGAGGTGGAGGTGGAAT-[22]-GTGGTGGAGGTGGAGGTGGAAT-[23]-GTGGTGGAGGTGGAGGTGGAATGTTATTTTCCAAA
                      1     2     3     4                  5     6     7                 8     9     10
```

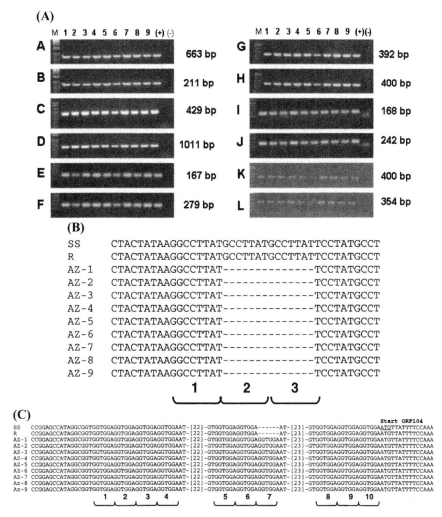

FIGURE 1. PCR detection of 12 VNTR loci in AZ isolates (lanes 1 through 9 correspond to DNA from AZ-1 to AZ-9), (+) DNA from strain Bitterroot, (−) negative control, M, DNA size markers **(A)**. Sequence differences detected in VNTR locus A **(B)** and locus B **(C)**. Repeat sequences (GCCTTAT and GGTGGA, respectively) and their number are identified with the brackets. SS, Sheila Smith; R, Bitterroot.

We performed molecular typing of these *R. rickettsii* isolates to determine whether they differed from patient and tick isolates from western U.S. regions where RMSF is transmitted by *Dermacentor* ticks. The isolates were analyzed for the presence of variable nucleotide tandem repeat (VNTR) sequences that were previously detected by analysis of the complete genome sequence of the Sheila Smith strain of *R. rickettsii* from Montana.[4] The VNTR regions we tested by polymerase chain reaction (PCR) and DNA sequencing were selected on the basis of their size and genome location, and they had 100% repeat sequence identity. Genomic DNA of *R. rickettsii* strain Bitterroot (tick isolate from Montana) was used as a control in all assays. PCR reactions were conducted under standard conditions using Qiagen Taq Master Mix (Qiagen, Valencia, CA). The 12 VNTR sites tested contained 2–7 repeats of 3–87 bp (base pair) length and occurred within 4 open reading frames and 8 intergenic regions. VNTR loci in DNA from each of the new *R. rickettsii* isolates were compared to those found in the reference strains. In total, 998 bp were sampled for each isolate, corresponding to ~0.9% of the *R. rickettsii* genome. Amplicons from DNA from the seven new *R. rickettsii* isolates as well as DNA from two infected ticks (AZ-2, AZ-7) from Arizona were identical in size to amplicons from Sheila Smith and Bitterroot DNA by electrophoresis in 1% agarose gels (FIG. 1A).

DNA sequences of all 12 VNTR were identical in the 9 DNA samples from Arizona and 10 of the VNTR amplicons were identical to sequences in Sheila Smith and Bitterroot isolates. However, 2 of 12 loci had different numbers of repeat units in these samples compared to the tick and patient strains from Montana. Locus A, located within *R. rickettsii* ORF 652, a split gene encoding a probable RND (resistance-nodulation-cell division) efflux transporter, has 3 repeats of 7 bp whereas the same locus in Arizona samples had a single 7 bp sequence (FIG. 1B). Locus B from an intergenic region flanked by *def*3 (ORF 103) and *tlc*1 (ORF 104) contains 10 repeats of 6 bp in all Arizona samples while 9 repeats are found in the reference DNAs from Montana (FIG. 1C).

We describe here the first isolation of *R. rickettsii* from *Rh. sanguineus* ticks in the United States. All isolates of *R. rickettsii* from Arizona were identical to each other in the 13 DNA regions analyzed in this study. However, we demonstrated that 2 of 12 VNTR loci contained different numbers of repeat sequences in the Arizona isolates when compared to 2 isolates from Montana. A total of 4 of the 10 invariant VNTR loci examined in this study exhibited variability in other isolates of *R. rickettsii.*[4] Although the temporal stability of variable VNTR loci in *R. rickettsii* and their degree of association with a particular geographic distribution in the Americas or specific tick host is presently not fully characterized, VNTR and other sequence variants[3] provide novel molecular epidemiologic tools for further investigating the origin of RMSF in Arizona. The identity of the patient and *Rh. sanguineus* tick isolates is consistent with the results of the epidemiologic investigation which implicated *Rh. sanguineus* as the vector that transmitted *R. rickettsii* to these patients.[2]

REFERENCES

1. CHAPMAN, A.S., S.M. MURPHY, L.J. DEMMA, *et al.* 2006. Rocky Mountain spotted fever in the United States, 1997–2002. Vector-Borne Zoonotic Dis. **6:** 170–178.

2. DEMMA, L., M.S. TRAEGER, W.L. NICHOLSON, *et al.* 2005. An outbreak of Rocky Mountain spotted fever in Arizona associated with an unexpected tick vector. N. Engl. J. Med. **353:** 587–594.
3. EREMEEVA, M.E., R.M. KLEMT, L.A. SANTUCCI-DOMOTOR, *et al.* 2003. Genetic analysis of isolates of *Rickettsia rickettsii* that differ in virulence. Ann. N. Y. Acad. Sci. **990:** 717–722.
4. EREMEEVA, M., M. ERDMAN, N. TIOLECO, *et al.* Identification of new genetic markers in *Rickettsia rickettsii* for use in molecular epidemiology and forensics. ASM 2004 Biodefense Meeting. Baltimore, MD, 7–10 March, 2004. (Poster K200).

Ten Years' Experience of Isolation of *Rickettsia* spp. from Blood Samples Using the Shell-Vial Cell Culture Assay

MARIELA QUESADA,[a] ISABEL SANFELIU,[b] NEUS CARDEÑOSA,[a] AND FERRAN SEGURA[a]

[a]*Infectious Disease Program, Department of Internal Medecine, UDIAT Centre Diagnòstic, Corporació Parc Taulí, Sabadell, Barcelona, Spain*

[b]*Microbiology Laboratory, UDIAT Centre Diagnòstic, Corporació Parc Taulí, Sabadell, Barcelona, Spain*

ABSTRACT: Two strategies to improve the efficacy of the shell-vial method for *Rickettsia* were analyzed. Blood samples from 59 patients with Mediterranean spotted fever (MSF) were examined using the shell-vial technique. (i) DNA from positive lenses was obtained when they were contaminated. (ii) Blood sample from one patient was cultured in 17 shell-vials. *R. conorii* was identified in four cases by polymerase chain reaction (PCR)-RFLP. Three of these were obtained from cells adherent to lenses and the fourth one by using total patient blood sample. *Rickettsia* isolation using all blood samples as well as DNA from shell-vial lenses could be useful in the study of rickettsial infections.

KEYWORDS: shell-vial assay; *Rickettsia*; MSF (Mediterranean spotted fever)

INTRODUCTION

The main problem in diagnosing rickettsial disease is because its characteristics of intracellular microorganism makes it difficult to use conventional culture methods. Thus the etiological diagnosis, epidemiological studies, and effectiveness of the treatment are difficult to assess. The specific diagnosis by culture needs a cellular culture medium and the traditional methods are slow and cumbersome and can only be done in special research laboratories. For many years different researchers have adapted the shell-vial culture assay (e.g., viruses) used for the diagnosis of other organisms virus, to *Rickettsia* with various results.[1–6] The shell-vial method combines the centrifugation of

Address for correspondence: Isabel Sanfeliu, Microbiology Laboratory, UDIAT Centre Diagnòstic, Corporació Parc Taulí, Parc Taulí s/n, E-08208 Sabadell, Barcelona, Spain. Voice: +34-93-7458448; fax: +34-93-7162004.
e-mail: isanfeliu@cspt.es

Ann. N.Y. Acad. Sci. 1078: 578–581 (2006). © 2006 New York Academy of Sciences.
doi: 10.1196/annals.1374.115

the sample on the cellular line and the development by immunofluorescence (IFI). In this way we can decrease the time needed for the culture making this method more adaptable to conventional laboratories and allowing in theory a quick diagnosis between 48–72 h. Apart from its possible use for the diagnosis of the disease, this method allows the identification of different *Rickettsia* species for epidemiological studies.[7]

In our laboratory we have been using this method together with specific serological techniques in the study of *Rickettsia* in patient's samples since 1994.

AIM

We presented the results obtained using the shell-vial technique for the detection and isolation of rickettsial strains from blood samples of patients with presumptive diagnosis of Mediterranean spotted fever (MSF). Furthermore, we evaluated two strategies to improve the efficacy of the shell-vial culture method in our laboratory.

PATIENTS AND METHODS

During the period 1994–2003, blood samples from 59 adult patients clinically diagnosed with MSF and with seroconversion to *R. conorii* were cultured by the shell-vial technique using MRC5 cells incubated at 32°C in Eagle's minimal essential medium containing 10% fetal bovine serum and 2 mM L-glutamine (Innogenetics, Barcelona, Spain). From each patient three shell-vial assays were carried out. The time of incubation varied between 4 and 15 days. The supernatants of the shell-vial positive by IFI indirect were inoculated in the same line cells under the same conditions. We used two strategies to improve the results: (i) DNA was obtained from cells adherent to shell-vial lenses positive by IFI whose cultures were contaminated. DNA was extracted by using a QIAamp DNA Mini Kit (Qiagen GmbH, Germany), as recommended by the manufacturer. (ii) In another case, total blood sample (8.5 mL) from one patient in 2004 was cultured in 17 shell-vials in order to increase the sensitivity. Then all the supernatants were cultured. To control the infection Gimenez staining was used.

To identify the SFG rickettsia species seen by IFI and by Gimenez staining, polymerase chain reaction (PCR)/RFLP was used.[8] PCR amplification was performed by using oligonucleotide primer pairs Rp CS877p and Rp CS.1258n generated from the citrate synthase gene of *R. prowazekii*, Rr 190.70p and Rr 190.602n generated from the 190-kDa antigen gene of *R. rickettsii*, and BG1-21 and BG2-20 generated from the 120-kDa antigen gene of *R. rickettsii.*[9,10] PCRs were carried out using a Perkin-Elmer 9600 thermocycler (Roche Diagnostics, Norwalk, CT), according to the protocol described by Regnery *et al.*[10]

The amplified products were analyzed on a 1.8% agarose gel (Bio-Rad Laboratories, Hercules CA) in 0.5 × Tris-borate-EDTA (TBE) buffer (AppliChem GmbH, Darmstat, Germany). As positive control, purified DNA from *R. conorii* (a kind gift from D. Raoult, Unité des Rickettsies CNRS UMR 6020, IFR 48, Faculté de Médecine de Marseille, Université de la Méditerranée, Marseille, France) was used. As a negative control destilled water was included in the same way as the DNA samples.

Amplified products were digested with *Alu*I, *Rsa*I, and *Pst*I restriction endonucleases, according to the protocol described by Eremeeva *et al.*[9] Electrophoretic separation was performed in a gel consisting of 1.8% agarose in TBE buffer.

The DNA fragments were visualized by ethidium bromide staining, and fragment sizes were compared with the sizes from a DNA Molecular Weight Marker VI (Boehringer Mannheim, GmbH, Germany).

RESULTS

From 59 patients (1994–2003) suspected of having MSF, only 5 had positive shell-vial cultures easily recognized by direct IFI staining. Detection of *Rickettsia* spp. was achieved on the first stained shell-vial (day 4) in five cases. One out of five positive shell-vials were not possible to be cultured. Another one was identified by partial *omp*A sequence as *R. conorii* (Malish 7 strain).[1] The remaining three positive shell-vials were contaminated.

Strategy A: Although we could not obtain the strains in three cases because they were contaminated, we could amplify *gtl*A, *omp*A, and *omp*B DNA fragments from DNA extracted from the cells adherent to shell-vial lenses.

Strategy B: A total of 4 of the 17 shell-vials from one patient in 2004 were strongly positive, 6 were weakly positive, and the rest negative. However, when the supernatants were cultured, we isolated *Rickettsia*-like organisms in all cases. We also amplified specific rickettsial DNA fragments. In both strategies, all amplified fragments had the same enzyme restriction profiles as the control *R. conorii*.

CONCLUSIONS

The shell-vial culture is the most suitable method to obtain and identify *Rickettsia* strains from patient blood samples, because it is an easy and fast methodol. To improve its lack of sensibility we can perform PCR of contaminated shell-vials lenses or using up all the extracted blood samples in shell-vial cultures, which in our cases have been shown to be useful. Any way to improve the effectiveness of *Rickettsia* culture will be useful to increase our knowledge of the epidemiologic situation of *Rickettsia* in any country.

REFERENCES

1. CARDEÑOSA, N., V. ROUX, B. FONT, *et al.* 2000. Isolation and identification of two spotted fever group rickettsial strains from patients in Catalonia, Spain. Am. J. Trop. Med. Hyg. **62:** 142–144.
2. MARRERO, M. & D. RAOULT. 1989. Centrifugation-shell-vial technique for rapid detection of Mediterranean spotted fever rickettsia in blood culture. Am. J. Trop. Med. Hyg. **40:** 197–199.
3. ESPEJO-ARENAS, E. & D. RAOULT. 1989. First isolates of *Rickettsia conorii* in Spain using a centrifugation-shell-vial assay. J. Infect. Dis. **159:** 1158–1159.
4. SARDELIC, S., P.E. FOURNIER, V. PUNDA POLIC, *et al.* 2003. First isolation of *Rickettsia conorii* from human blood in Croatia. Croat. Med. J. **44:** 630–634.
5. LA SCOLA, B. & D. RAOULT. 1996. Diagnosis of Mediterranean spotted fever by cultivation of *Rickettsia conorii* from blood and skin samples using the centrifugation-shell-vial technique and by detection of *R. conorii* in circulating endothelial cells: a 6-year follow-up. J. Clin. Microbiol. **34:** 2722–2727.
6. BIRG, M.L., B. LA SCOLA, V. ROUX, *et al.* 1999. Isolation of *Rickettsia prowazekii* from blood by shell-vial cell culture. J. Clin. Microbiol. **37:** 3722–3724.
7. CARDEÑOSA, N., F. SEGURA & D. RAOULT. 2003. Serosurvey among Mediterranean spotted fever patients of a new spotted fever group rickettsial strain (Bar29). Eur. J. Epidemiol. **18:** 351–356.
8. BEATI, L., V. ROUX, A. ORTUNO, *et al.* 1996. Phenotypic and genotypic characterization of spotted fever group Rickettsiae isolated from Catalan *Rhipicephalus sanguineus* ticks. J. Clin. Microbiol. **34:** 2688–2694.
9. EREMEEVA, M., X. YU & D. RAOULT. 1994. Differentiation among spotted fever group rickettsiae species by analysis of restriction fragment length polymorphism of PCR-amplified DNA. J. Clin. Microbiol. **32:** 803–810.
10. REGNERY, R.L., C.L. SPRUILL & B.D. PLIKAYTIS. 1991. Genotypic identification of rickettsiae and estimation of intraspecies sequence divergence for portions of two rickettsial genes. J. Bacteriol. **173:** 1576–1589.

Automated Method Based in VNTR Analysis for Rickettsiae Genotyping

LILIANA VITORINO,[a,b] RITA DE SOUSA,[b] FATIMA BACELLAR,[b] AND LÍBIA ZÉ-ZÉ[a,b]

[a]*Universidade de Lisboa, Faculdade de Ciências, Centro de Genética e Biologia Molecular and Instituto de Ciência Aplicada e Tecnologia, Edifício ICAT, Campus da FCUL, Campo Grande, 1749–016 Lisboa, Portugal*

[b]*CEVDI, Instituto Nacional de Saúde Dr. Ricardo Jorge, 1649-016 Lisboa, Portugal*

ABSTRACT: A genetic locus named Rc-65, which is 5′ adjacent to gene *dks*A and 3′ adjacent to *xer*C gene, has previously been demonstrated to contain a VNTR with high discriminatory power in several rickettsial strains and thus, potentially useful for genetically similar strains identification. In this work, we present an automated molecular identification method based on capillary electrophoresis separation of VNTRs amplicons. The resulting electropherograms were in agreement with the sequence data obtained in a previous work. The presented genotyping method is fast and suitable for full automation, being a powerful tool for epidemiological surveillance in a large number of samples and enables the detection co-infected samples. The combination of other VNTR loci should improve the discriminatory capacity of this typing system, providing greater resolution and contributing to a more accurate VNTR-based assay. To our knowledge, this is the first automated assay for rickettsial strains identification.

KEYWORDS: Mediterranean spotted fever; Rickettsiae; variable number tandem repeats

INTRODUCTION

Tandem repeats (TR) (over 10 nt [nucleotides]) that represent a single locus and show interindividual length polymorphisms are referred to as variable number tandem repeats (VNTRs). The VNTR analysis is a polymerase chain reaction (PCR)-based approach used to genotype several bacteria species, because it is highly discriminatory and reproducible.[1] A VNTR assay for rickettsiae typing was recently described by Fournier *et al.*[2,3] and Vitorino *et al.*[4]

Address for correspondence: Líbia Zé-Zé, CEVDI, Instituto Nacional de Saudé Dr. Ricardo Jorge, INSA, Edifíco LEMES, Av. Padre Cruz, 1649–016 Lisboa, Portugal. Voice: 00351-217-508-126; fax: 00351-217-508-121.

e-mail: libia.zeze@insa.min-saude.pt

Ann. N.Y. Acad. Sci. 1078: 582–586 (2006). © 2006 New York Academy of Sciences. doi: 10.1196/annals.1374.116

We have amplified and sequenced the VNTR locus Rc-65, which is 5′ adjacent to *dks*A gene and 3′ adjacent to *xer*C gene, in order to characterize the repeat diversity within several rickettsiae. This VNTR locus enables us to type different rickettsial species and also distinguishes among "*R. conorii* complex" strains. In this work, we present an automated molecular identification method based on capillary electrophoresis separation of VNTR amplicons. Hence, a fluorescent-labeled primer was used to enable the analysis in an automated DNA-sequencer device. The resulting electropherograms were in agreement with the results previously obtained.[4]

MATERIALS AND METHODS

Strains

Clinical isolates, *Rickettsia conorii* Malish-like Portuguese human rickettsia (PoHuR)1258, *R. conorii* Malish-like PoHuR10710, Israeli tick typhus (ITT) PoHuR1450, ITT PoHuR6647, ITT PoHuR10461, as well as tick isolates, *R. conorii* Malish-like Portuguese tick rickettsia (PoTiR)12, ITT PoTiRb28, *R. slovaca* PoTiR30, and *R. massiliae* PoTir66 were used in this study. These Portuguese isolates were previously identified by serology and *ompA* gene sequencing as described elsewhere.[5] The reference strains, *R. rickettsii, Rickettsia* sp. Bar29, *R. aeschlimannii, R. helvetica, R. australis,* and *R. akari,* provided by D. Raoult (Unité des Ricketsies, Faculté de Médecine, Université de la Méditerranée, Marseille, France) were also included.

VNTR Typing

The PCR reaction was carried out as previously descried,[4] using specific primers Rc-65-Fw, 5′-TTG AGA AGG TTT ATA TCC CAT AG-3′ and Rc-65-Rv, 5′-TAC TAC CGC ATA TCC AAT TAA AAA-3′. The reverse primer was 5′ labeled with dye D3 (Proligo). Fluorescent-labeled amplicons were sized by capillary electrophoresis on an automated DNA sequencer CEQ 2000-XL (Beckman Coulter, Fullerton, CA) using GenomeLabTM DNA Size Standard Kit-600 (Beckman Coulter) as internal standard.

RESULTS AND DISCUSSION

The resulting electropherograms from the VNTR typing assay are in agreement with amplicon sizes assessed by agarose gel electrophoresis and with the sequencing data previously obtained for each rickettsiae.[4] Nevertheless, minor differences (1 bp [base pairs]) in the highest pick can be observed because 3′ adenine addition by the DNA polymerase.

The alleles of Rc-65 locus are easily distinguished, mainly within rickettsiae belonging to the *R. conorii* complex. This approach enables the identification of *R. conorii* Malish-like strains and ITT *Rickettsia* (FIG.1A, 1B, and 1C), which are the causative agents of Mediterranean spotted fever (MSF) in Portugal.[5] ITT Portuguese isolates have two distinct alleles, with three and four repeat units (FIG.1B and 1C, respectively) also distinguished by this approach.

We also investigated whether this assay could be used for the detection of coinfection. Therefore, we amplified a mix sample of *Rickettsia* sp. Bar29 and ITT *Rickettsia* DNA, in the same PCR reaction (FIG.1D). Presently, this methodology is being attempted in several vector species to detect ticks coinfected with different rickettsial species.

Unfortunately, *R. slovaca* has the same allele as some ITT strains, and hence in this case rickettsiae identification is accomplished taking into account the vector species and/or the *omp*A gene sequencing (FIG. 1E and 1F). Likewise, *R. aeschlimannii*, *R. helvetica*, and *Rickettsia* sp. Bar29 cannot be distinguished solely by Rc-65 allele profile, because all have the same allele size (176 bp; data not shown).

This VNTR Rc-65 locus was used to identify *Rickettsia* sp. directly from vector and clinical samples. All the samples were previously screened by *ompA* gene amplification and sequencing. The amplification of rickettsiae Rc-65 locus in tick lysates yielded an amplicon size that corroborates the *ompA* gene identification.

Recently, a new genotype closely related to RpA4 and DnS14, described in Russia[7] and also in Spain,[8] was detected in Portuguese *Dermacentor marginatus* ticks (Vitorino *et al.*, unpublished work). This genotype differs in allele size from *R. slovaca* (FIG. 1E and 1G); hence this methodology enables a straightforward identification of rickettsial strains present in several vectors, even in those infected with other pathogens, such as *Borrelia bugdorferi s.l.* strains (data not shown).

Regarding the clinical samples, previously known to be positive for rickettsiae, the majority of electropherograms were inconclusive, because several picks were observed (data not shown), probably due to cross-contamination. Indeed, the amplification of other genes in these samples is also difficult, and is only achieved by a nested-PCR reaction. In the future, a nested-PCR approach using an additional primer pair will be attempted in these clinical samples. Nevertheless, we could identify *Rickettsia* sp. Bar29-like (FIG. 1I) in a serum sample of a patient with MSF. The tick that bit the patient was also analyzed in this study and was infected with the same strain (FIG. 1H).

In conclusion, this VNTR-based assay provides an automated, reproducible, and discriminatory method for the identification of several rickettsial species. It is also a valuable approach for epidemiological surveillance in a large number of vectors. Further analysis will involve the combination of other VNTR

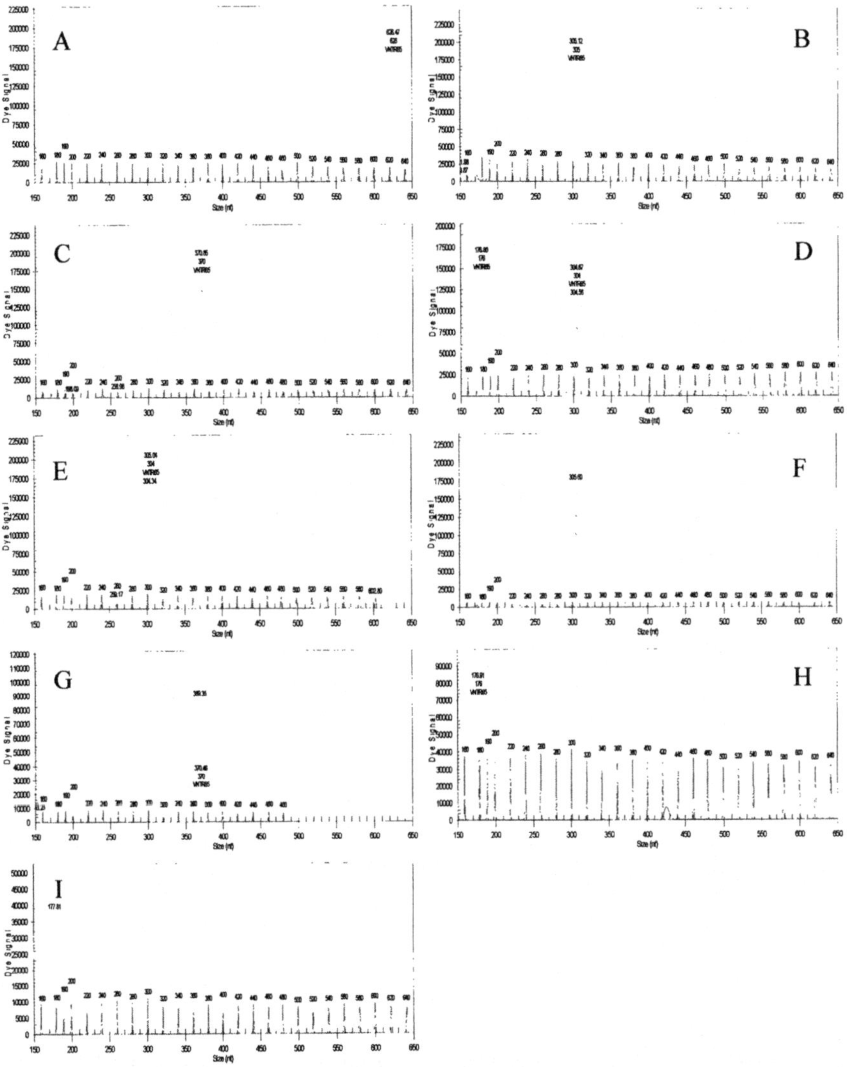

FIGURE 1. Electropherograms obtained from capillary electrophoresis of Rc-65 locus using fluorescence-labeled PCR primers (*gray pick*) obtained with CEQ Genetic Analysis system software (Beckman Coulter). *Black picks* represent the 600-bp size standard (Beckman Coulter). The *y*-axis represents the fluorescence signal and the *x*-axis represents the fragment size in base pairs. (**A**) *R. conorii* PoHuR1258; (**B**) ITT *Rickettsia* PoHuR1450; (**C**) ITT *Rickettsia* PoHuR6647; (**D**) *Rickettsia* sp. *Bar29* and ITT *Rickettsia* PoHuR1450; (**E**) *R. slovaca* detected in *D. marginatus*; (**F**) ITT *Rickettsia* detected in *R. sanguineus*; (**G**) *Rickettsia* sp. RpA4 genotype detected in *Dermacentor marginatus*; (**H**) *Rickettsia* sp. Bar29-like detected in *Rhipicephalus sanguineus*; (**I**) *Rickettsia* sp. Bar29-like, detected in serum from a patient bitten by *R. sanguineus* tick analyzed in electropherogram H.

loci in order to accomplish a multiplex-automated method with higher discriminatory capacity and greater resolution, improving the accuracy and fastness in rickettsial strain identification.

ACKNOWLEDGMENTS

The ICAT/FCUL research team involved in this work strongly acknowledges its scientific leader, Professor Rogério Tenreiro, for the general scientific supervision and engagement in providing the necessary resources and facilities. During the realization of this work L. Vitorino and L. Zé-Zé were the recipients of FCT research grants SFRH/BD/10676/2002 and SFRH/BPD/3653/2000, respectively.

REFERENCES

1. VAN BELKUM, A. *et al.* 1998. Short-sequence DNA repeats in prokaryotic genomes. Microbiol. Mol. Biol. Rev. **62:** 275–293.
2. FOURNIER, P.E., Y. ZHU, H. OGATA & D. RAOULT. 2004. Use of highly variable intergenic spacer sequences for multispacer typing of *Rickettsia conorii* strains. J. Clin. Microbiol. **42:** 5757–5766.
3. ZHU, Y., P.E. FOURNIER, M. EREMEEVA & D. RAOULT. 2005. Proposal to create subspecies of *Rickettsia conorii* based on multi-locus sequence typing and an emended description of *Rickettsia conorii*. BMC Microbiol. **5:** 11.
4. VITORINO, L. *et al.* 2005. Characterization of a tandem repeat polymorphism in *Rickettsia* sp. strains. J. Med. Microbiol. **54:** 833–841.
5. FOURNIER, P.E., V. ROUX & D. RAOULT. 1998. Phylogenetic analysis of the spotted fever *Rickettsiae* by study of the outer surface protein rOmpA. Int. J. Syst. Bacteriol. **48:** 839–849.
6. BACELLAR, F. *et al.* 2003. Boutonneuse fever in Portugal: 1995–2000. Data of state laboratory. Eur. J. Epidemiol. **18:** 275–277.
7. RYDKINA, E. *et al.* 1999. New rickettsiae in ticks collected in territories of the former Soviet Union. Emerg. Infect. Dis. **6:** 811–814.
8. MÁRQUEZ, F.J. *et al.* 2003. Which spotted fever group rickettsia are present in *Dermacentor marginatus* ticks in Spain? Ann. N. Y. Acad. Sci. **990:** 141–142.

Monitoring of Humans and Animals for the Presence of Various Rickettsiae and *Coxiella burnetii* by Serological Methods

E. KOVÁČOVÁ,[a] Z. SEKEYOVÁ,[a] M. TRÁVNIČEK,[b] M.R. BHIDE,[b] S. MARDZINOVÁ,[b] J. ČURLIK,[b] AND D. ŠPANELOVÁ[c]

[a]*Institute of Virology, Slovak Academy of Sciences, Bratislava, Slovak Republic*

[b]*Institute of Veterinary Medicine, Košice, Slovak Republic*

[c]*State Veterinary Institute, Bratislava, Slovak Republic*

ABSTRACT: Serological examination of humans in Slovakia suspected of having rickettsial infections revealed the presence of antibodies to spotted fever group rickettsiae (*R. conorii, R. slovaca*, and *R. typhi*). Of interest is the finding of serological positivity to the newly recognized "IRS" agent. Antibodies to these rickettsiae and to *C. burnetii* were demonstrated also in domestic and hunting dogs and pet animals. These results confirm the occurrence and possible circulation of these rickettsiae and *C. burnetii* in the Slovak Republic.

KEYWORDS: rickettsiae; "IRS 4"; *C. burnetii;* antibodies; ELISA; immunofluorescence test

INTRODUCTION

Infectious diseases caused by *Rickettsia* and *Coxiella burnetii* continue to present public health challenges. Newly described forms of rickettsioses represent distinct entities with unique epidemiological and clinical features which require constant attention. On the basis of serologic examination in Slovakia, only *Rickettsia slovaca, R. typhi,* and *C. burnetii* infections should be considered. However, increasing traveling due to tourism and migration includes the possibility of introducing exotic rickettsioses to this area.[1]

The aim of this study was monitoring of infections with SFG rickettsiae and *C. burnetii* in humans, dogs, and pet animals by enzyme-linked immunosorbent assay (ELISA), immunofluorescence (MIF) test, and cross-absorption test. In the serological tests, Corpuscular antigens of *R. slovaca*, *R. conorii*,

Address for correspondence: E. Kováčová, Institute of Virology, Slovak Academy of Sciences, Bratislava, Slovak Republic. Voice: +4212-59302433; fax: +4212-5477428.
e-mail: virukova@savba.sk

**Ann. N.Y. Acad. Sci. 1078: 587–589 (2006). © 2006 New York Academy of Sciences.
doi: 10.1196/annals.1374.117**

R. typhi, "IRS 4" agent (*Ixodes ricinus* Slovakia)— a newly recognized member of spotted fever group rickettsia — and *C. burnetii* were employed.[2] Dogs were included in our study because they play an important role in the ecology of SFG rickettsiae and *C. burnetii,*[3–6] but little is known about the prevalence of the rickettsial antibodies in our country.[7]

MATERIALS AND METHODS

Human serum samples were obtained from 26 persons professionally exposed to domestic and wild animals and 213 persons suspected of having rickettsial infection or infection of unknown origin. Sera were collected in various districts of Slovakia in 2001–2004. Serum samples from 366 hunting and domestic dogs and pet animals (10 dogs and 15 cats) were collected in eastern and western Slovakia.

All the serum samples were tested by ELISA for reactivity to rickettsiae with *R.conorii, R. slovaca, R. typhi*, and *C. burnetii* as antigens. The ELISA-positive sera were then tested by MIF and eventually confirmed by the cross-absorption test. The testing for the "IRS 4" was done by MIF.

RESULTS

A total of 239 human sera were initially screened by ELISA for IgG and IgM antibodies. Of 56 ELISA-positive sera (23.4%), 7 (12.5.9%) had IgG antibodies to *R. conorii* and 5 (8.9%) to *R. slovaca* and *R. typhi* each in MIF test (the titers varied from 50 to 1,600). Interestingly, antibodies to "IRS 4" were found in 11 (4.6%) sera. Out of those, two sera exhibited titers of 100 and 800 for *R. slovaca* and *R. conorii*, respectively, and one serum had titers of 100 with *R. typhi* and 800 with *R. slovaca*. The positive results for *R. slovaca* with all three sera were confirmed by cross-absorption test. One serum exhibited titers of 100 and 200 for *R. typhi* and "IRS 4," whereas the cross-absorption test confirmed the positive results for "IRS 4" only.

A total of 45 (11.5%) of 391 dog sera had antibodies to rickettsiae and *C. burnetii* in ELISA. In MIF test, out of the 45 ELISA-positive sera, 12 had antibodies to *R. conorii,* and *R. slovaca* and 5 had antibodies to "IRS 4." The cross-absorption test confirmed only 3 of 7 sera positive for *R. slovaca* in MIF test as well as positive for *R. slovaca.* However, no specific antibodies were detected in the sera from pet animals.

CONCLUSIONS

Our results showed that (i) dogs can be naturally infected with various rickettsiae and *C. burnetii*, (ii) a close contact with humans predetermines them as a possible source of SFG rickettsiae and *C. burnetii* infection, and (iii) the presence of antibodies to "IRS 4" in human and animal sera indicates a prevalence of this agent in nature. However, a possible pathogenic threat to public health has not been demonstrated.

ACKNOWLEDGMENTS

The authors are very grateful to Dr. J. Žemla for reviewing the manuscript. The work was supported by the Grants No. 3050 and 5053 of the Scientific Grant Agency of the Ministry of Education of the Slovak Republic and the Slovak Academy of Sciences.

REFERENCES

1. KOVÁČOVÁ, E., D. MOZOLOVÁ, J. BIRČÁK & J. KAZÁR. 2001. An unusual case of rickettsial infection in a Slovak child after return from Dominican Republic. Efermedades Infecciosas y Mikrobiologia. **21**(Suppl.): S138.
2. SEKEYOVÁ, Z., E. KOVÁČOVÁ, P.E. FOURNIER & D. RAOULT. 2003. Isolation and characterization of a new rickettsia from *Ixodes ricinus* ticks collected in Slovakia. Ann. N. Y. Acad. Sci. **990:** 54–56.
3. HORTA, M.C., M.B. LABRUNA, L.A. SANGIONI, *et al.* 2004. Prevalence of antibodies to spotted fever group rickettsiae in humans and domestic animals in a Brazilian spotted fever-endemic area in the state of Sao Paulo, Brazil: serologic evidence for infection by *Rickettsia rickettsii* and another spotted fever group *Rickettsia*. Am. J. Trop. Med. Hyg. **71:** 93–97.
4. SEGURA-PORTA, F., G. DIESTRE-ORTIN, A. ORTUNO-ROMERO, *et al.* 1998. Prevalence of antibodies to spotted fever group rickettsiae in human beings and dogs from an endemic area of Mediterranean spotted fever in Catalonia. Eur. J. Epidemiol. **14:** 395–398.
5. BONI, M., B. DAVOUST, H. TISSOT-DUPONT & D. RAOULT. 1998. Survey of seroprevalence of Q fever in dogs in the southeast of France, French Guyana, Martinique, Senegal and the Ivory Coast. Vet. Microbiol. **64:** 1–5.
6. COMER, J.A., C.D. PADDOCK & J.E. CHILDS. 2001. Urban zoonoses caused by *Bartonella*, *Coxiella*, *Ehrlichia*, and *Rickettsia* species. Vector Borne Zoonotic Dis. **1:** 91–118.
7. KOCIANOVÁ, E., V. LISÁK & M. KOPČOK. 1992. *Coxiella burnetii* and *Chlamydia psittaci* infection in dogs. Vet. Med. (Prague) **37:** 177–183.

Early Diagnosis of Rickettsioses by Electrochemiluminscence

GARY WEN, JERE W. McBRIDE, XIAOFENG ZHANG, AND JUAN P. OLANO

Center for Biodefense and Emerging Infectious Diseases, University of Texas Medical Branch, Galveston, Texas, USA

ABSTRACT: The diagnosis of acute rickettsioses during the acute phase of the disease is challenging. We present preliminary evidence that antigen-capture using streptavidin-coated magnetic beads with biotinylated anti-*rickettsia typhi* rabbit polyclonal antibodies followed by electrochemiluminescent detection with ruthenylated antibodies of the same specificity could be used for the diagnosis of rickettsial diseases in the acute phase.

KEYWORDS: *Rickettsia typhi*; electrochemiluminescence; rickettsioses; ruthenium (II) tris-bipyridine; tripropylamine

BACKGROUND

The gold standard for the diagnosis of rickettsioses is evidence of seroconversion on serum samples taken during the acute and convalescent phases of the disease process.[1,2] The diagnosis of rickettsioses during the acute phase of the disease is difficult. Detection of rickettsial organisms on skin biopsies obtained from patients with rash has been used in some clinical settings but it is relatively invasive and insensitive.[3,4] Detection of rickettsial DNA by polymerase chain reaction (PCR) is both sensitive and specific. However DNA amplification techniques are available only in highly specialized laboratories, remain expensive, and require skilled personnel. Furthermore, culture of rickettsiae is cumbersome and requires both skilled personnel and biosafety level 3 facilities. The development of a fast, accurate, reproducible, highly sensitive and specific, and low-cost alternative technique is therefore needed. Electrochemiluminescence (ECL) is a detection system based on light emission at 622 nm from a ruthenium (II) tris-bipyridine-tagged antibody in the presence of electrical current and tripropylamine (TPA). The system uses magnetic beads

Address for correspondence: Juan P. Olano, Department of Pathology, University of Texas Medical Branch, 301 University Boulevard, R. 0609, Galveston, TX 77555-0609, USA. Voice: 409-772-2870; fax: 409-747-2400.
e-mail: jolano@utmb.edu

Ann. N.Y. Acad. Sci. 1078: 590–594 (2006). © 2006 New York Academy of Sciences.
doi: 10.1196/annals.1374.118

that are coated with streptavidin for the attachment of capturing biotinylated antibodies (B-Ab). Ruthenylated antibodies (R-Ab) are then used to "sandwich" the antigen or pathogen of interest and the magnetic beads are concentrated at the anode and stimulated with current in the presence of TPA. The technique was developed in the early 1990s and has been applied to the detection of certain serum analytes and selected pathogens.[5,6]

MATERIALS AND METHODS

Polyspecific and monospecific (OmpB) polyclonal antibodies against *Rickettsia typhi* were developed in rabbits and mice, respectively. The antibodies were then column-purified, biotinylated, and ruthenylated according to the manufacturer's instructions (Bioveris, Gaithersburg, MD). *R. typhi* was propagated in Vero cells and purified through renografin gradients. Rickettsial stocks were quantified by real-time PCR using primers for the citrate synthase gene. OmpB was extracted hypotonically from purified rickettsiae, resuspended in buffer, and quantified.

ECL Procedure

Different antibody pairs containing the biotin and ruthenium tags were optimized (rabbit–rabbit, mouse–rabbit, and mouse–mouse) using the checkerboard technique at 1, 2, and 4 μg/mL. Antibody pairs were incubated for 30, 60,

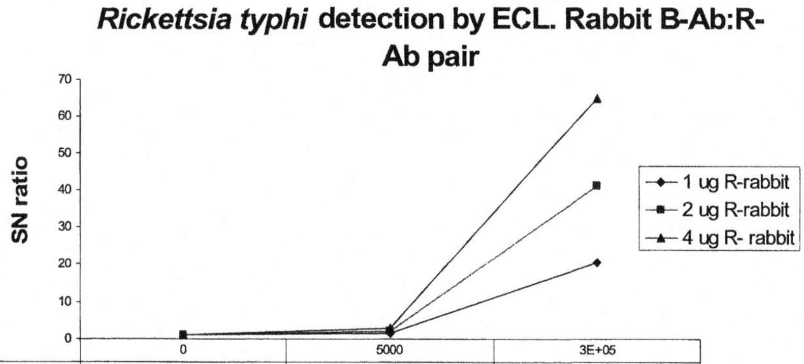

	0	5000	3E+05
1 ug R-rabbit	1	1.5	20.6
2 ug R-rabbit	1	2.2	41.3
4 ug R- rabbit	1	2.9	65.1

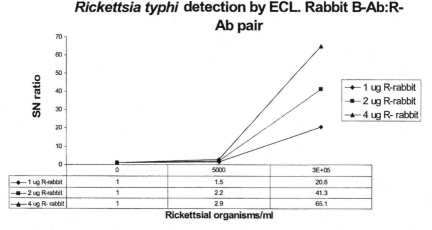

FIGURE 1. Optimization of capturing and detection antibodies using anti–*R. typhi* rabbit polyclonal antibodies. SN ratios improved significantly as the concentration of ruthenylated antibody was increased.

90, and 120 min at room temperature followed by the addition of streptavidin-coated magnetic beads (M270 and M280 Dynabeads, Dynal Biotech, Oslo, Norway) for 30 min. Samples were prepared in 96-well plates (final volume 250 μL) and run through the M1 Analyzer (Bioveris). The analytical sensitivity was determined by using serial dilutions of rickettsial organisms ranging from 0–500,000 organisms/mL. Results were expressed as signal to noise (SN) ratios between negative controls and experimental samples. SN ratios ≥2.0 were considered significant.

RESULTS AND DISCUSSION

Optimization of the capturing antibodies revealed no improvement in SN ratios by increasing the amount of B-Ab in the system. At rickettsial concentrations of 250,000 organisms/mL, increasing the amount of R-Ab from 1 to 2 μg/mL doubled the SN ratio. When the concentration of R-Ab was increased from 1 to 4 μg/mL, the SN ratio tripled. At rickettsial concentrations of 5,000 organisms/mL the SN ratio doubled when R-Ab was increased from 1 to 4 μg/mL (FIG. 1). All subsequent experiments were therefore performed using 1 μg of B-Ab and 4 μg of R-Ab. Optimization of the incubation time also

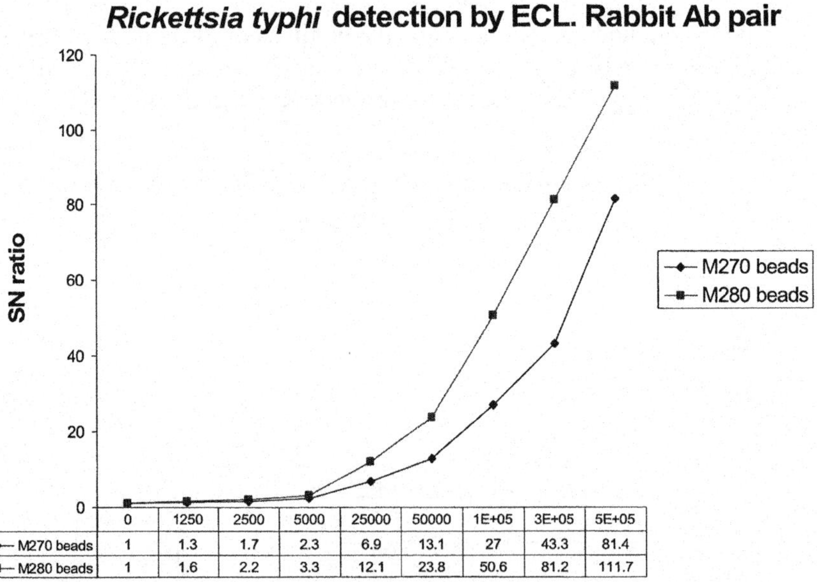

	0	1250	2500	5000	25000	50000	1E+05	3E+05	5E+05
M270 beads	1	1.3	1.7	2.3	6.9	13.1	27	43.3	81.4
M280 beads	1	1.6	2.2	3.3	12.1	23.8	50.6	81.2	111.7

Rickettsial organisms/ml

FIGURE 2. Evaluation of *R. typhi* detection by ECL using different magnetic beads. M280 beads, which have a larger surface area, yielded better SN ratios at low and high concentrations of rickettsiae.

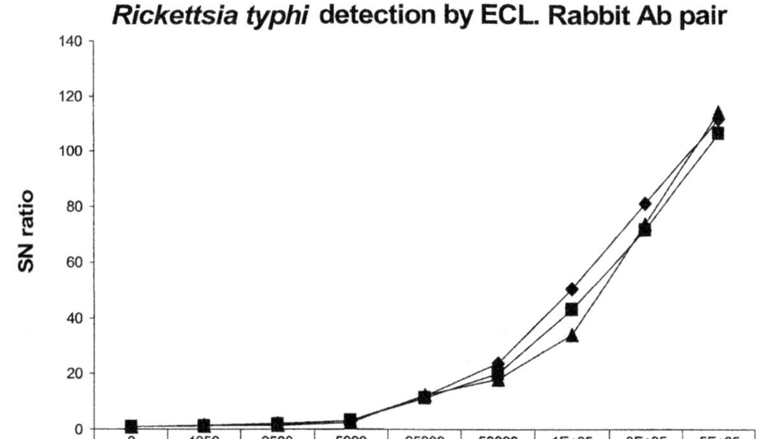

	0	1250	2500	5000	25000	50000	1E+05	3E+05	5E+05
Series1	1	1.6	2.2	3.3	12.1	23.8	50.6	81.2	111.7
Series2	1	1.5	2	3.2	11.4	20.3	43.4	71.8	106.8
Series3	1	1.2	1.8	2.8	12.4	18.1	34.2	73.8	114.4

Rickettsial organisms/ml

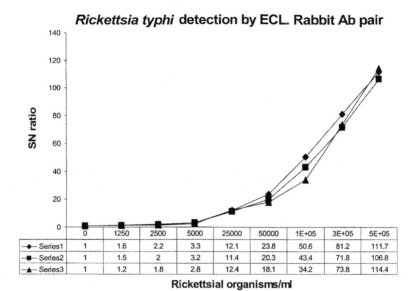

FIGURE 3. Evaluation of the analytical sensitivity of ECL detection of *R. typhi* with rabbit polyclonal antibodies. The lowest detection limit ranges from 2,000–3,000 rickettsial organisms per milliliter. This experiment was performed in triplicate.

revealed improvement in SN ratios up to 90 min of incubation of the antibodies with rickettsial suspensions. No improvement was evident by increasing the incubation time with the magnetic beads beyond 30 min. Optimization of the beads showed that larger beads (M280) increased the SN ratio from 20–50%, depending on the rickettsial concentration (FIG. 2). As to evaluation of the antibody pairs, the best SN ratios were obtained by using the polyclonal rabbit antibody pair or the mouse–rabbit pair in which the rabbit antibody contained the ruthenium tag. The mouse–mouse pair or the rabbit–mouse pair showed SN ratios 10–40 times lower than SN ratios obtained with the other antibody pairs. The analytical sensitivity of the assay for detection of whole rickettsial organisms ranged between 2–3 $\times$ 10^3 organisms/mL (FIG. 3). The analytical sensitivity for the detection of purified OmpB was disappointing because significant SN ratios were obtained at the microgram level. Further experiments are clearly needed to evaluate the clinical usefulness of this novel detection technique. Preliminary data suggest that detection of whole rickettsiae by ECL in the acute phase of the disease could be developed into a useful diagnostic test in clinical practice.

REFERENCES

1. WALKER, D.H., D. RAOULT, J.S. DUMLER & T. MARRIE. 2004. Rickettsial diseases. *In* Harrison's Principles of Internal Medicine. D.L. Kasper, E. Braunward, A.S.

Fauci, S.L. Hauser, D.L. Longo & J.L. Jameson, Eds.: 999–1008. McGraw-Hill. New York, NY.
2. La Scola, B. & D. Raoult. 1997. Laboratory diagnosis of rickettsioses: current approaches to diagnosis of old and new rickettsial diseases. J. Clin. Microbiol. **35:** 2715–2727.
3. Walker, D.H. & B.G. Cain. 1978. A method for specific diagnosis of Rocky Mountain spotted fever on fixed, paraffin-embedded tissue by immunofluorescence. J. Infect. Dis. **137:** 206–209.
4. Walker, D.H., B.G. Cain & P.M. Olmstead. 1978. Laboratory diagnosis of Rocky Mountain spotted fever by immunofluorescent demonstration of *Rickettsia* in cutaneous lesions. Am. J. Clin. Pathol. **69:** 619–623.
5. Yu, H., J.W. Raymonda, T.M. McMahon & A.A. Campagnari. 2000. Detection of biological threat agents by immunomagnetic microsphere-based solid phase fluorogenic- and electro-chemiluminescence. Biosens. Bioelectron. **14:** 829–840.
6. Yu, H. 1996. Enhancing immunoelectrochemiluminescence (IECL) for sensitive bacterial detection. J. Immunol. Methods **192:** 63–71.

Corpuscular Antigenic Microarray for the Serodiagnosis of Blood Culture–Negative Endocarditis

LAURENT SAMSON, MICHEL DRANCOURT, JEAN-PAUL CASALTA, AND DIDIER RAOULT

Unité des Rickettsies CNRS UMR 6020, Faculté de Médecine, IFR 48, Université de la Méditerranée, Marseille, France

ABSTRACT: Blood culture–negative endocarditis is due to fastidious bacteria, including *Coxiella burnetii* and *Bartonella* spp. Diagnosis of such infection relies on serology and microimmunofluorescence is therefore the reference method. We developed a multiplex serology test featuring automatic incubation and reading and incorporating internal controls. Preliminary results indicate that this new serologic test is valuable for the rapid, automated serological diagnosis of blood culture–negative endocarditis.

KEYWORDS: endocarditis; multiplexed serology; immunofluorescence diagnosis

Serological testing using microimmunofluorescence (MIF) proved useful for the indirect diagnosis of blood culture–negative endocarditis due to *Coxiella burnetii* and *Bartonella* spp.[1] MIF is, however, a laborious, time- and serum-consuming technique which is poorly reproducible between bench-workers and laboratories. In order to circumvent MIF limitations, we developed a MIF-based, standardized, automated, and antigen-multiplexed system for the serodiagnosis of endocarditis.

The InoDiaG system for serology consists of standard-format glass slides, spotted with small amounts of biological material and entire bacteria antigens using an automated spotter (Affymetrix, Santa Clara, CA). Each spot contains a single antigen comprising IgG, IgM, Hep-2 cells, and *Staphylococcus aureus* as controls; and *Bartonella henselae, Bartonella quintana*, and *C. burnetii* phase I and II as pathogens potentially responsible for blood culture–negative endocarditis.

The method was: incubation of 1/128 diluted serum samples with spotted antigens at room temperature as well as rinse and incubation with goat anti-

Address for correspondence: Didier Raoult, Unité des Rickettsies, CNRS UMR 6020, IFR 48, Faculté de médecine, Université de la Méditerranée, 27 Bd Jean Moulin, 13385 Marseille Cedex 05, France. Voice: (33)-4-91-38-55-17; fax: (33)-4-91-83-03-90.
e-mail: Didier.Raoult@medecine.univ-mrs.fr

Ann. N.Y. Acad. Sci. 1078: 595–596 (2006). © 2006 New York Academy of Sciences.
doi: 10.1196/annals.1374.119

human IgG : FITC (STAR106F, Serotec, Oxford, U.K.) and goat-anti-human IgM : Texas Red (109-075-129, Jackson ImmunoResearch Laboratories, West Grove, PA). Washing and drying for 5 min were totally automated and performed into a chamber of the automated incubator. The overall process took 1 h. Fluorescence measurements were done by a second apparatus featuring three different filters. Pictures of immunofluorescence were taken with a numeric camera at high resolution in TIF, then transferred to a personal computer for analysis.

A total of 10 human serum sampled in the endocarditis kit as previously described[2] were used to establish cut-off dilutions for the diagnosis of endocarditis using our endocarditis slide. All sera were tested in parallel by MIF for the presence of specific IgG and IgM against *C. burnetii*, and *Bartonella* spp. Five sera were collected from patients with *Bartonella* endocarditis (Ig G > 1 : 800), five from patients with Q fever endocarditis (IgG phase I > 1 : 1600), and five sera from negative control blood donors.

At 470 nm, for a serum dilution of 1:128, a cut-off was established for *C. burnetii* phase I and for *C. burnetii* phase II with a sensitivity and a specificity of 100%. We also determined a cut-off dilution of serum for *B. quintana* and for *B. henselae* with a sensitivity and a specificity of 100%. These results established a proof of concept that automated MIF assay incorporating corpuscular antigenic microarray is a valuable new tool for modern serology.[3]

REFERENCES

1. Rolain, J.M., C. Lecam & D. Raoult. 2003. Simplified serological diagnosis of endocarditis due to *Coxiella burnetii* and *Bartonella*. Clin. Diagn. Lab. Immunol. **10:** 1147–1148.
2. Raoult, D., P.E. Fournier & M. Drancourt. 2004. What does the future hold for clinical microbiology? Nat. Rev. Microbiol. **2:** 151–159.
3. Barrau, K., A. Boulamery, G. Imbert, *et al.* 2004. Causative organisms of infective endocarditis according to host status. Clin. Microbiol. Infect. **10:** 302–308.

Proposal to Create Subspecies of *Rickettsia sibirica* and an Emended Description of *Rickettsia sibirica*

PIERRE-EDOUARD FOURNIER,[a] YONG ZHU,[a] XUEJIE YU,[b] AND DIDIER RAOULT[a]

[a]*Unité des Rickettsies, IFR 48, CNRS UMR 6020, Faculté de Médecine, Université de la Méditerranée, 27 Boulevard Jean Moulin, 13385 Marseille Cedex 05, France*

[b]*Department of Pathology, University of Texas Medical Branch, Galveston, Texas, USA*

ABSTRACT: The *Rickettsia sibirica* species is composed of isolates that are genotypically close but can be classified within two distinct serotypes, that is, *R. sibirica* sensu stricto and *R. sibirica mongolitimonae* (incorrectly named *R. mongolotimonae*). We investigated the possibility of classifying rickettsiae closely related to *R. sibirica* as *R. sibirica* subspecies, as proposed by the *ad hoc* Committee on Reconciliation of Approaches to Bacterial Systematics. For this, we first estimated the genotypic variability by using multilocus sequence typing (MLST), including the sequencing of five genes, and multispacer typing (MST) using three intergenic spacers, of five isolates and three tick amplicons of *R. sibirica* sensu stricto and six isolates of *R. sibirica mongolotimonae*. Then, we selected a representative of each MLST genotype and used mouse serotyping to estimate their degree of taxonomic relatedness. Among the 14 isolates or tick amplicons studied, 2 MLST genotypes were identified: (i) the *R. sibirica* sensu stricto type; and (ii) the *R. sibirica mongolitimonae* type. Representatives of the two MLST types were classified within three MST types and into two serotypes. Therefore, as isolates within the *R. sibirica* species are genotypically homogeneous but show MST genotypic, serotypic, and epidemio-clinical dissimilarities, we propose to modify the nomenclature of the *R. sibirica* species through the creation of subspecies. We propose the names *R. sibirica* subsp. *sibirica* subsp. nov. (type strain = 2-4-6, ATCC VR-541^{T}), and *R. sibirica* subsp. *mongolitimonae* subsp. nov. (type strain = HA-91, ATCC VR-1526^{T}). The description of *R. sibirica* is emended to accommodate the two subspecies.

KEYWORDS: *Rickettsia siberica*; taxonomy; subspecies

Address for correspondence: Pierre-Edouard Fournier, Unité des Rickettsies, IFR 48, CNRS UMR 6020, Faculté de Médecine, Université de la Méditerranée, 27 Boulevard Jean Moulin, 13385 Marseille Cedex 05, France. Voice: 33-04-91-38-55-17; fax: 33-04-91-83-03-90.

e-mail: Pierre-Edouard.Fournier@medecine.univ-mrs.fr

Ann. N.Y. Acad. Sci. 1078: 597–606 (2006).
doi: 10.1196/annals.1374.120

INTRODUCTION

Within the genus *Rickettsia*, since the pioneering work of Philip in 1978,[1] two rickettsial strains were considered as having different serotypes if they exhibited a specificity difference (SPD) of ≥ 3, and some authors have considered this to be a useful guide to speciation.[2,3] However, using genotypic criteria, we have demonstrated that distinct serotypes may not systematically represent distinct species.[2] Regarding rickettsial strains related to *Rickettsia sibirica* opinions divide as to their taxonomic status. These include *R. sibirica* sensu stricto[4] and *R. sibirica mongolitimonae*, incorrectly named "*R. mongolotimonae*" previously.[5] Phylogenetically, these rickettsiae constitute a homogeneous cluster and are distinct from other *Rickettsia* species.[6–8] Initially, we have putatively considered *R. sibirica mongolitimonae* as a new species on the basis of a distinct serotype and specific epidemiological features.[5,9] However, using genotypic criteria, we have recently demonstrated that this rickettsia belongs to the *R. sibirica* species.[2] *R. sibirica* sensu stricto is the causative agent of North Asian tick typhus (also named Siberian tick typhus)[10] in Siberia and western China. It is transmitted to humans mainly by *Dermacentor* sp. ticks.[11] *R. sibirica mongolitimonae* causes LAR (lymphangitis-associated rickettsiosis) in France, Algeria, and South Africa.[11] Although its vector has not been identified in these countries, the latter rickettsia is associated with *Hyalomma* ticks in China and Niger.[5,12] We have recently demonstrated that both diseases are distinguished by specific epidemiological and clinical features.[11]

Recently, we have proposed the creation of subspecies within the *R. conorii* species[13] following the proposal of the *ad hoc* committee on reconciliation of approaches to bacterial systematics[14] that, even if related genetically, bacterial isolates within a given species could be considered as distinct subspecies if they differed phenotypically. In this study, we undertook to examine whether sufficient differences existed among *R. sibirica* isolates that would enable us to classify them as subspecies.

MATERIALS AND METHODS

All rickettsial isolates tested and their original sources are listed in TABLE 1. Rickettsial isolates for which sequences were not available were grown in Vero cell monolayers at 32°C in 150- cm^2 cell culture flasks as previously described.[15] When available, we used gene sequences present in GenBank. The accession numbers for these sequences are reported at the end of this article. Missing sequences were obtained in this study. Genomic DNA was extracted from rickettsial suspensions and three *Dermacentor nuttali* ticks collected in Russia as previously described.[13] Multilocus sequence typing (MLST) using the 16S rDNA, *glt*A, *omp*A, *omp*B, and *sca*4 genes, and multispacer typing (MST) using the *dks*A/*xer*C, *mpp*A/*pur*C, and *rpm*E/tRNA-fMet intergenic

TABLE 1. Rickettsial isolates included in the study

Strain (ATCC number)	Original source	Geographical origin	Strain provided by	Reference
R. sibirica sensu stricto				
246 (VR-151^T)	*D. nuttali*	Russia	ATCC	Ref. 22
Gornyi 37/89	*D. nuttali*	Russia	S. Shpynov*	
M	*D. marginatus*	Armenia	M. Eremeeva¶	Ref. 23
BJ-90	*D. sinicus*	China	X. Yu†	Ref. 5
R. sibirica mongolotimonae				
HA-91 (VR-1526^T)	*Hy. asiaticum*	China	X. Yu†	Ref. 5
URRMTMFEe13	Human	France	‡	Ref. 11
URRMTMe19	Human	France	‡	Ref. 11
URRMTMFe30	Human	France	‡	Ref. 11
URRMTMFEe37	Human	France	‡	Ref. 11
URRMTMFEe65	Human	France	‡	Ref. 11

*Omsk Research Institute of Natural Foci Infections, Omsk, Russia.
¶Centers for Disease Control, Atlanta, GA, USA.
†University of Texas Medical Branch, Galveston, TX, USA.
‡Strains cultivated in our laboratory from human eschar biopsies.

spacers were performed using the previously described primers and polymerase chain reaction (PCR) conditions.[11,16] For each gene and spacer, the pairwise sequence similarity among studied isolates and tick amplicons was estimated using the multisequence alignment program CLUSTAL W.[17] Then, we selected for mouse serotyping two representative isolates of the *sibirica* MLST genotype, that is, isolates 246, and BJ-90, as well as two representative isolates of the *mongolitimonae* genotype, *that is,* HA-91 and URRMTMFEe65. Mouse serotyping was performed and SPDs were calculated as previously described.[13]

RESULTS

MLST Genotyping

All five genes studied were amplified from all 14 isolates and ticks tested. All negative controls remained negative. Two MLST types were identified. Within the first type, which we named *sibirica* type, all *R. sibirica* isolates and tick amplicons exhibited identical sequences for each of the five genes tested except isolate BJ-90. The latter isolate showed subtle nucleotide substitutions in four of them. Degrees of pairwise similarity in nucleotide sequence between isolates BJ-90 and 246 were of 99.8, 100, 99.8, 99.7, and 99.9% for the 16S rDNA, *glt*A, *omp*A, *omp*B, and *sca*4 genes, respectively. Within the *mongolitimonae* type, all isolates showed identical sequences of the five genes studied except the URRMTMFEe65 isolate, which showed subtle nucleotide substitutions in four of them. In this isolate, the pairwise nucleotide similarity with isolate HA-91 for 16S rDNA, *glt*A, *omp*A, *omp*B, and *sca*4 genes was 100, 99.9, 99.9, 99.9, and 99.5%, respectively. As representative isolates for each MLST genotype, we have chosen isolates 246 and BJ-90 for the *sibirica* genotype, and isolates HA-91 and URRMTMFEe65 for the *mongolitimonae* genotype. Between the two MLST genotypes, the degree of pairwise nucleotide similarity was of 100, 99.8, 99.0, 99.6, and 99.4% for the 16S rRNA, *glt*A, *omp*A, *omp*B, and *sca*4 genes, respectively. The dendrograms obtained using the three different tree-building analysis methods showed similar organization for the five rickettsiae studied. Representatives of the two MLST genotypes were grouped into two monophyletic groups (FIG. 1).

MST Genotyping

All *R. sibirica* sensu stricto isolates and tick amplicons exhibited identical spacer sequences. They showed unique 232-bp (base pair) *dks*A/*xer*C (type Q, GenBank accession number DQ008262), 153-bp *mpp*A/*pur*C (type J, DQ008283), and 200-bp *rpm*E/tRNA-fMet (type E, DQ008247) spacer sequences. All *R. sibirica mongolitimonae* isolates exhibited identical spacer sequences except isolate URRMTMFEe65. Isolate HA-91 showed unique 232-bp *dks*A/*xer*C (type R, DQ008263), 153-bp *mpp*A/*pur*C (type K, DQ008284), and 200-bp *rpm*E/tRNA-fMet (type F, DQ008248) spacer sequences. Isolate URRMTMFEe65 showed unique 232-bp *dks*A/*xer*C (type S, GenBank accession number DQ008264) and 153-bp *mpp*A/*pur*C (type L, DQ008285), but an *rpm*E/tRNA-fMet spacer sequence identical to that of isolate HA-91. When combining the genotypes obtained from each of the three spacers into a MST genotype, three distinct MST genotypes were identified. The MST genotypes so obtained were different from those previously published for *R. conorii.*[16] The phylogenetic tree inferred from the concatenated spacer sequences showed an organization similar to that obtained from MLST when *R. conorii* subsp. *conorii* strain Malish was used as an outgroup (FIG. 2).

Mouse Serotyping

SPD values are presented in TABLE 2. The antibody titers obtained in homologous sera were at least onefold higher than those observed in heterologous sera. Two serotypes were identified, that is, isolates of *R. sibirica* sensu stricto and isolates of *R. sibirica mongolitimonae*.

DISCUSSION

We propose to create subspecies within the *R. sibirica* species to accommodate both the genotypic homogeneity and the serotypic, MST genotypic,

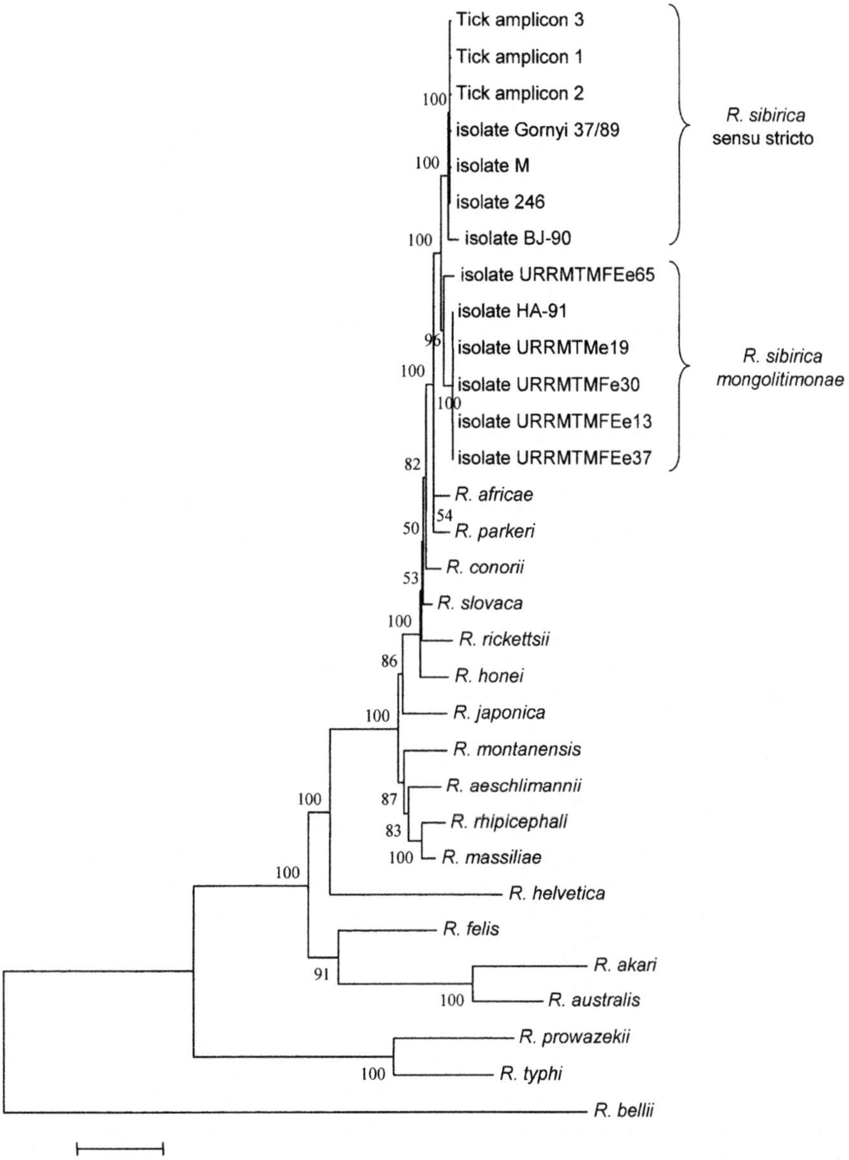

FIGURE 1. Unrooted phylogenetic tree derived from the comparison of concatenated sequences of the16S rDNA, *glt*A, *omp*B, and *sca*4 genes from 19 validated species of the *Rickettsia* genus. *R. canadensis*, in which neither *omp*B nor *sca*4 could be amplified, was not included. *omp*B and *sca*4 sequences from *R. bellii* were retrieved from the sequence of its genome (GenBank accession No. NC_007940). The analysis was conducted using the neighbor-joining method and the Kimura 2 parameter. Bootstrap values are indicated at the nodes of the phylogenetic tree. The scale bar indicates a 2% nucleotide sequence divergence.

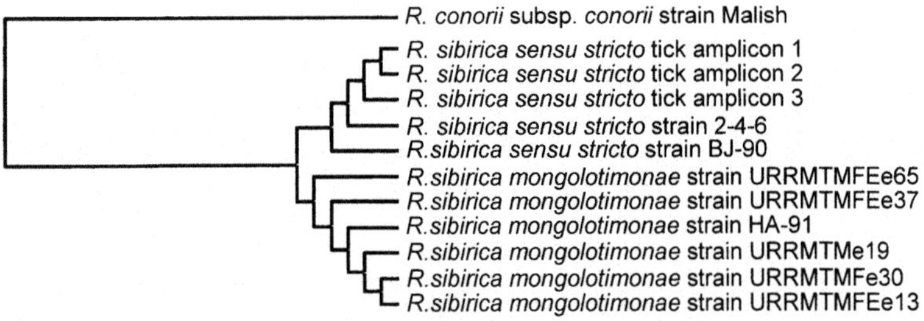

FIGURE 2. Unrooted phylogenetic trees derived from the comparison of concatenated sequences of the *dks*A/*xer*C, *mpp*A/*pur*C, and *rpm*E/tRNA-fMet intergenic spacers using the parsimony method.

geographic, and pathogenic diversities of rickettsia strains within this species.

Species definition within the *Rickettsia* genus is subject to controversy. Rickettsiologists have agreed to use mouse serotyping as one criterion,[1] and we have used this classification scheme as a basis for the development of genotypic criteria at the species level.[2] However, as confusion still existed about the taxonomic status of members of the *R. conorii* complex, we have proposed the creation of subspecies of *R. conorii.*[13] A similar confusion existed among rickettsial isolates related to *R. sibirica*.

In the present study, we have demonstrated, using MLST, that rickettsiae closely related to *R. sibirica* sensu stricto isolate 246 are very homogeneous and can be classified within two genotypes. Except for isolate BJ-90, which showed subtle mutations in four genes, all studied *R. sibirica* sensu stricto isolates and tick amplicons had identical sequences for the five genes studied and thus constituted a reliable taxon. Among isolates and tick amplicons of *R. sibirica mongolitimonae*, the degree of MLST sequence similarity was of 100% except for the URRMTMFEe65 isolate, which is genotypically slightly different from *R. sibirica mongolitimonae* isolate HA-91 in four genes.[11] Among representatives of these MLST genotypes, the minimal pairwise similarity in the nucleotide sequences of 16S rDNA, *glt*A, *omp*A, *omp*B, and *sca*4 genes

TABLE 2. Microimmunofluorescence antibody titers and SPDs from reciprocal cross-reactions of mouse antisera to members of the *R. sibirica* species

	MIF antibody titer (SPD) of mouse antisera to:[a]			
Antigens	Strain 246	Strain BJ-90	Strain HA-91	Strain URMTMFEe65
Strain 246	1,024	512 (1)	128 (5)	128 (6)
Strain BJ-90	1,024 (1)	1,024	128 (5)	128 (5)
Strain HA-91	256 (5)	256 (5)	1,024	1,024 (0)
Strain URRMTMFEe65	128 (6)	128 (5)	512 (0)	512

[a]Titers are the reciprocals of the highest dilution of antisera that gave a positive reaction.

between the rickettsial isolates tested and *R. sibirica* was 99.9, 99.8, 99.0, 99.6, and 99.4%, respectively. Thus, all isolates and tick amplicons tested fulfilled the criteria to belong to the *R. sibirica* species[2] and formed a phylogenetically strongly supported group distinct from other *Rickettsia* species (FIG. 1). In addition, MST confirmed that the two MLST genotypes consistently form two phylogenetic clusters (FIG. 2), and mouse serotyping identified representatives of these two clusters as distinct serotypes.

Thus, we demonstrated a discrepancy between the high level of conservation in nucleotide sequence of these rickettsiae, and phenotypic specificities including serotyping, epidemiological, and clinical characteristics. Therefore, according to the accepted criteria of the *ad hoc* Committee on Reconciliation of Approaches to Bacterial Systematics, which allows the creation of subspecies for genetically close isolates that express phenotypic differences,[14] we are justified in creating new subspecies for these organisms.

Emendation of the Description of **R. sibirica** *(Bell 1960)*

The characteristics of this taxon are similar to those described by Weiss and Moulder for the species.[18] The species contains two subspecies. The minimal degree of pairwise nucleotide similarity of the *R. sibirica* species is that of 99.8% for 16S rDNA, 99.4% for *glt*A, 98.3% for the 5′ end of *omp*A, 98.6% for *omp*B, and 99.3% for *sca*4, by comparison with isolate 246.

Description of **R. sibirica** *subsp.* **sibirica** *subsp. nov.*

R. sibirica subsp. *sibirica* (si.bi'ri.ca N. L. fem. adj. *sibirica*, from Siberia, where the tick providing the first isolate was collected): The characteristics are the same as those of the species. It is transmitted to humans through the bite of *Dermacentor marginatus, D. nuttali, D. silvarum, D. pictus, D. sinicus, D. auratus, Haemaphysalis concinna, Hyalomma wellingtoni,* and *Hy. yeni* ticks and grows in Vero cells at 32°C in antibiotic-free Minimal Essential Medium supplemented with 2% fetal calf serum and 2 mg/m L-glutamine.

The type strain is 246, ATCC VR-151^T, which was isolated in Russia by Golinevitch in 1948. This type is the most common and representative of the isolates of *R. sibirica* subsp. *sibirica* subsp. nov. The nucleotide sequence of the 16S rRNA, *glt*A, *omp*A, *omp*B, and *sca*4 genes have been deposited in the GenBank database under the accession numbers, L36218,[19] U59734,[15] U43807,[20] AF123722,[7] and AF155057,[8] respectively. The nucleotide sequence of the 16S rRNA, *glt*A, *omp*A, *omp*B, and *sca*4 genes of isolate BJ-90 have been deposited in the GenBank database under the accession numbers AF178036, AF178035, AF179365,[21] AF123715,[7] and AF151725,[8] respectively. To be classified as *R. sibirica* subsp. *sibirica* subsp. nov., a rickettsial isolate should exhibit a minimal pairwise nucleotide sequence similarity of 99.8, 100, 99.8, 99.7, and 99.9% for the 16S rDNA, *glt*A, *omp*A, *omp*B, and *sca*4 genes, respectively, with those of *R. sibirica* isolate 246, ATCC VR-151^T. *R. sibirica* subsp. *sibirica* subsp nov. strain 246 has also been deposited in the Collection de Souches de l'Unité des Rickettsies (CSUR), Marseille, France.

Description of R. sibirica *subsp.* mongolitimonae *subsp. nov.*

R. sibirica subsp. *mongolitimonae* (mon.go.li.ti.mo'nae. N. L. fem. n., from latinized Timone [Timone Hospital, Marseille], N.L. mongolitimonae of Mongolia and the Timone Hospital): The characteristics are the same as those of the species. It is associated with *Hy. asiaticum* ticks in China, and with *H. truncatum* ticks in Niger. Vector for transmission to humans is unknown. It grows in Vero cells at 32°C in antibiotic-free Minimal Essential Medium supplemented with 2% fetal calf serum and 2 mg/mL L-glutamine.

The type strain is HA-91, ATCC VR-1526^T, which was isolated from a *Hy. asiaticum* tick collected in China in 1991. The nucleotide sequence of the 16S rRNA, *glt*A, *omp*A, *omp*B, and *sca*4 genes have been deposited in the GenBank database under the accession numbers L36219,[19] U59731,[15] U43796,[20] AF123715,[7] and AY331397,[2] respectively. The nucleotide sequence of the 16S rRNA, *glt*A, *omp*A, *omp*B, and *sca*4 genes of *R. sibirica* subsp. *mongolitimonae* subsp. nov. strain URMTMFEe65 have been deposited in the GenBank database under the accession numbers DQ097085, DQ097081, DQ097082, DQ097083, and DQ097084, respectively. To be classified as *R. sibirica* subsp. *mongolitimonae* subsp. nov., a rickettsial strain should exhibit a minimal pairwise nucleotide sequence similarity of 100, 99.9, 99.9, 99.9, and 99.5% for the 16S rDNA, *glt*A, *omp*A, *omp*B, and *sca*4 genes, respectively, with those of *R. sibirica* subsp. *mongolitimonae* subsp. nov., isolate HA-91. *R. sibirica* subsp. *mongolitimonae* subsp. nov. isolate HA-91 has also been deposited in the CSUR.

ACKNOWLEDGMENTS

We are grateful to Dr. S. Shpynov for *D. nuttali* ticks.

REFERENCES

1. PHILIP, R.N., E.A. CASPER, W. BURGDORFER, *et al.* 1978. Serologic typing of rickettsiae of the spotted fever group by micro-immunofluorescence. J. Immunol. **121:** 1961–1968.
2. FOURNIER, P.E., J.S. DUMLER, G. GREUB, *et al.* 2003. Gene sequence-based criteria for the identification of new *Rickettsia* isolates and description of *Rickettsia heilongjiangensis* sp. nov. J. Clin. Microbiol. **41:** 5456–5465.

3. KELLY, P.J. & P.R. MASON. 1990. Serological typing of spotted fever group rickettsia isolates from Zimbabwe. J. Clin. Microbiol. **28:** 2302–2304.
4. BELL, E.J. & H.G. STOENNER. 1960. Immunologic relationships among the spotted fever group of rickettsias determined by toxin neutralisation tests in mice with convalescent animal serums. J. Immunol. **84:** 171–182.
5. YU, X., Y. JIN, M. FAN, *et al.* 1993. Genotypic and antigenic identification of two new strains of spotted fever group rickettsiae isolated from China. J. Clin. Microbiol. **31:** 83–88.
6. FOURNIER, P.E., V. ROUX & D. RAOULT. 1998. Phylogenetic analysis of spotted fever group rickettsiae by study of the outer surface protein rOmpA. Int. J. Syst. Bacteriol. **48:** 839–849.
7. ROUX, V. & D. RAOULT. 2000. Phylogenetic analysis of members of the genus *Rickettsia* using the gene encoding the outer-membrane protein rOmpB (ompB). Int. J. Syst. Evol. Microbiol. **50:** 1449–1455.
8. SEKEYOVA, Z., V. ROUX & D. RAOULT. 2001. Phylogeny of *Rickettsia* spp. inferred by comparing sequences of 'gene D', which encodes an intracytoplasmic protein. Int. J. Syst. Evol. Microbiol. **51:** 1353–1360.
9. RAOULT, D., P. BROUQUI & V. ROUX. 1996. A new spotted-fever-group rickettsiosis. Lancet **348:** 412.
10. LYSKOVTSEV, M.M. 1968. Tickborne rickettsiosis. Miscellaneous Publ. Entomol. Soc. Am. 42–140.
11. FOURNIER, P.E., F. GOURIET, P. BROUQUI, *et al.* 2005. Lymphangitis-associated rickettsiosis, a new rickettsiosis caused by *Rickettsia sibirica* mongolotimonae: seven new cases and review of the literature. Clin. Infect. Dis **40:** 1435–1444.
12. PAROLA, P., H. INOKUMA, J.L. CAMICAS, *et al.* 2001. Detection and identification of spotted fever group Rickettsiae and Ehrlichiae in African ticks. Emerg. Infect. Dis. **7:** 1014–1017.
13. ZHU, Y., P.E. FOURNIER, M. EREMEEVA & D. RAOULT. 2005. Proposal to create subspecies of *Rickettsia conorii* based on multi-locus sequence typing and an emended description of *Rickettsia conorii*. BMC. Microbiol. **5:** 11.
14. WAYNE, L.G., D.J. BRENNER, R.R. COLWELL, *et al.* 1987. Report of the ad hoc Committee on Reconciliation of Approaches to Bacterial Systematics. Int. J. Syst. Bacteriol. **37:** 463–464.
15. ROUX, V., E. RYDKINA, M. EREMEEVA & D. RAOULT. 1997. Citrate synthase gene comparison, a new tool for phylogenetic analysis, and its application for the rickettsiae. Int. J. Syst. Bacteriol. **47:** 252–261.
16. FOURNIER, P.E., Y. ZHU, H. OGATA & D. RAOULT. 2004. Use of highly variable intergenic spacer sequences for multispacer typing of *Rickettsia conorii* strains. J. Clin. Microbiol. **42:** 5757–5766.
17. THOMPSON, J.D., D.G. HIGGINS & T.J. GIBSON. 1994. CLUSTAL W: improving the sensitivity of progressive multiple sequence alignment through sequence weighting, position-specific gap penalties and weight matrix choice. Nucleic Acids Res. **22:** 4673–4680.
18. WEISS, E. & J.W. MOULDER. 1984. Order I *Rickettsiales*, Gieszczkiewicz 1939. *In* Bergey's Manual of Systematic Bacteriology, Vol. 1. N.R. Krieg & J.G. Holt, Eds.: 687–703. Williams & Wilkins. Baltimore.
19. ROUX, V. & D. RAOULT. 1995. Phylogenetic analysis of the genus *Rickettsia* by 16S rDNA sequencing. Res. Microbiol. **146:** 385–396.
20. ROUX, V., P.E. FOURNIER & D. RAOULT. 1996. Differentiation of spotted fever group rickettsiae by sequencing and analysis of restriction fragment length

polymorphism of PCR amplified DNA of the gene encoding the protein rOmpA. J. Clin. Microbiol. **34:** 2058–2065.
21. ZHANG, J.Z., M.Y. FAN, X.J. YU & D. RAOULT. 2000. Phylogenetic analysis of the Chinese isolate BJ-90. Emerg. Infect. Dis. **6:** 432–433.
22. ZDRODOVSKII, P.F. 1949. Systematics and comparative characterization of endemic rickettsioses. Zhur. Mikrobiol. Epidemiol. **10:** 19–28.
23. EREMEEVA, M.E., N.M. BALAYEVA, V.F. IGNATOVICH & D. RAOULT. 1993. Proteinic and genomic identification of spotted fever group rickettsiae isolated in the former USSR. J. Clin. Microbiol. **31:** 2625–2633.

Comparison of Immune Response against *Orientia tsutsugamushi*, a Causative Agent of Scrub Typhus, in 4-Week-Old and 10-Week-Old Scrub Typhus-Infected Laboratory Mice Using Enzyme-Linked Immunosorbent Assay Technique

KRIANGKRAI LERDTHUSNEE,[a] SUPAK JENKITKASEMWONG,[b] SUCHEERA INSUAN,[a] WARISA LEEPITAKRAT,[a] TAWEESAK MONKANNA,[a] NITTAYA KHLAIMANEE,[a] WEERAYUT CHAREONSONGSERMKIJ,[a] SURACHAI LEEPITAKRAT,[a] KWANTA CHAYAPHUM,[a] AND JAMES. W. JONES[a]

[a]*Department of Entomology, USAMC-AFRIMS, Bangkok, Thailand*

[b]*Department of Biology, Faculty of Science, Chulalongkorn University, Bangkok, Thailand*

ABSTRACT: Using the indirect enzyme-linked immunosorbent assay (ELISA) technique, we compared the immune response of specific antibodies (IgM and IgG) in serum samples of scrub typhus-infected and non-infected mice against *Orientia sutsugamushi* (a causative agent of scrub typhus). Two different age groups (4-week-old and 10-week-old) of ICR laboratory mice were infected with *O. tsutsugamushi* by the animal passage procedure. Serum samples were detected for scrub typhus-specific antibodies using ELISA technique. Results from determining the presence of IgM and IgG in the serum samples obtained from scrub typhus-infected mice showed that the IgG was first detected on day-13 after the initial infection in both the 4-week-old and 10-week-old mice. The IgG titer levels of both groups were not significantly different. Although the presence of IgM in the in serum samples obtained from scrub typhus-infected mice was first detected on day 13 in the 4-week-old mice and on day-12 in the 10-week-old mice, the IgM titer in the 4-week-old mice was slightly lower than in the 10-week-old mice. Clinical observations of the scrub typhus-infected mice showed that the older mice become ill on day 9 whereas the younger mice exhibited the symptoms on day 12.

Address for correspondence: Kriangkrai Lerdthusnee, Department of Entomology, USAMC-AFRIMS, 315/6 Rajvithi Road, Phya Thai, Bangkok 10400, Thailand. Voice: 66-2-644-4888; fax: 66-2-354-7885.

e-mail: kRiangkRAil@afrims.org

Ann. N.Y. Acad. Sci. 1078: 607–612 (2006). © 2006 New York Academy of Sciences.
doi: 10.1196/annals.1374.121

Considering the earlier detection and slightly higher level of specific IgM antibody, it could be interpreted that the older mice may appear to have responded against *O. tsutsugamushi* faster than the younger mice.

KEYWORDS: scrub typhus-infected mice; immune response; ELISA technique

INTRODUCTION

Scrub typhus is an acute febrile vector-borne/infectious disease, caused by a Gram-negative, rod-shaped bacterium known as *Orientia tsutsugamushi* (an obligate intracellular parasite that lives in mite [the primary reservoir] and later in rodents and/or humans). Scrub typhus, has occurred and threatened people throughout Asia and Australia, extending from the Far East to the Middle East and to Asian Pacific regions. *O. tsutsugamushi* is transmitted to vertebrate hosts (wild rodent and human) by the bites of larval mites (known as chiggers) of genus *Leptotrombidium* (i.e., *Leptotrombidium imphalum, L. chiangraiensis,* and *L. deliense*) that live on the animal hosts. In the chigger, *O. tsutsugamushi* bacterium is transmitted transovarially (from adult female to egg) and transstadially (from egg to larval stages and to adult stage). Scrub typhus in humans generally occurs after exposure to areas with scrub vegetation, habitats of infected rodents, and chigger mites.[1–3] Clinical symptoms of this disease are high fever, a skin rash, a severe headache, a wound "eschar" at the bite site, and swelling of the lymph glands, liver and spleen. In severe cases, it can lead to organ failure from second infection resulting in respiratory failure and death. Typically, scrub typhus can be easily treated with antibiotics, — doxycycline, chloramphenical and tetracycline—which normally result in rapid and complete recovery. However, the most appropriate and/or effective way to prevent from scrub typhus infection when encroaching the scrub typhus high-risk areas is to avoid chigger bites using personal protection measures such as insect repellents. Currently, there is no vaccine for scrub typhus available.

The objective of this study is to compare immune responses of scrub typhus-infected 4-week-old (young) and 10-week-old (older) ICR (CD-1®, Charles River Laboratories) laboratory mice against *O. tsutsugamushi.*

MATERIALS AND METHODS

(i) Scrub typhus-infected *Leptotrombidium* chigger mite colonies were maintained by the "animal passage procedure" through the ICR laboratory mice. Two groups of 4-week-old (the "young mice" group, $n = 14$ mice/group) and 10-week-old (the "old mice" group, $n = 13$ mice/group) mice underwent scrub typhus-infected chigger-feeding procedure (by putting scrub typhus-infected chiggers on the mouse's ears) for 3 consecutive days. Daily observations and clinical symptoms were recorded.

Serum samples were collected daily from both groups, started from day 1 after completion of 3-day-chigger feeding, and continued to day 30 and/or whenever the animals exhibited signs of illness, body weight loss exceeded 10%, or the animal died. Experiments in both groups were duplicated.

(ii) Scrub typhus-specific antibodies detection for both immunoglobulin M (a primary antibody response against pathogen, IgM) and immunoglobulin G (a secondary antibody response against pathogen, IgG), using r56 Karp strain recombinant antigen, was performed on both the young and the old mice using the indirect enzyme-linked immunosorbent assay (ELISA) technique.[4–8] An ELISA technique is considered as the gold standard for detecting the presence of *O. tsutsugamushi* antigens. A screen at titers of 1:50 was first performed; sera positive at that level were then serially diluted. Any titers greater than 1:800 for IgG or 1:1,600 for IgM were considered positive for scrub typhus infection.

(iii) Experiments in all animals used in this study are strictly conducted under the Animal Use Protocol written specifically for this procedure and are approved by the AFRIMS-IACUC committee. Animals were observed for symptoms and signs of sickness for 30 days. Any mouse that became ill was euthanized by CO_2.

RESULTS

(i) *Clinical observations*: The scrub typhus-infected mice started to show signs of illness on day 9 for the 10-week-old mice and on day-12 for the 4-week-old mice.

(ii) *Scrub typhus-specific IgM detection*: . The positive IgM values in serum samples were first detected on day 13 in the young mice and on day 12 in the old mice (FIG. 1). The optical density (OD) levels of the IgM titers (a concentration of 1:150) of the young mice were slightly lower than those of the old mice group (FIG. 3).

(iii) *Scrub typhus-specific IgG detection*: Positive IgG serum samples of both groups (FIG. 2). The optical density (OD) levels of the IgG titers (at a concentration of 1:50) of both groups were not significantly different (FIG. 4). were first detected on day-13

CONCLUSION

(i) The 10-week-old mice became ill earlier than the 4-week-old mice.

(ii) Older mice can produce scrub typhus-specific IgM earlier and at a higher level than the younger ones.

(iii) Both young and old mice can produce scrub typhus-specific IgG almost at the same time (on day 13) and levels (as shown on the OD values).

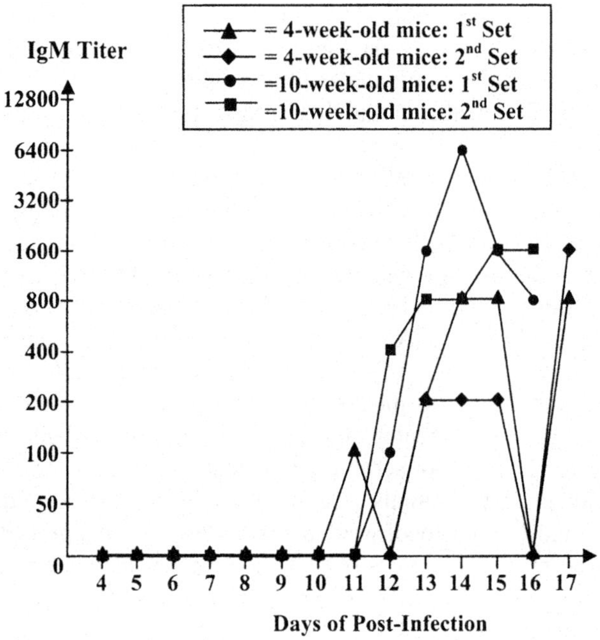

FIGURE 1. Positive IgM in serum samples were first detected on day 13 in the 4-week-old group and on day 12 in the 10-week-old group.

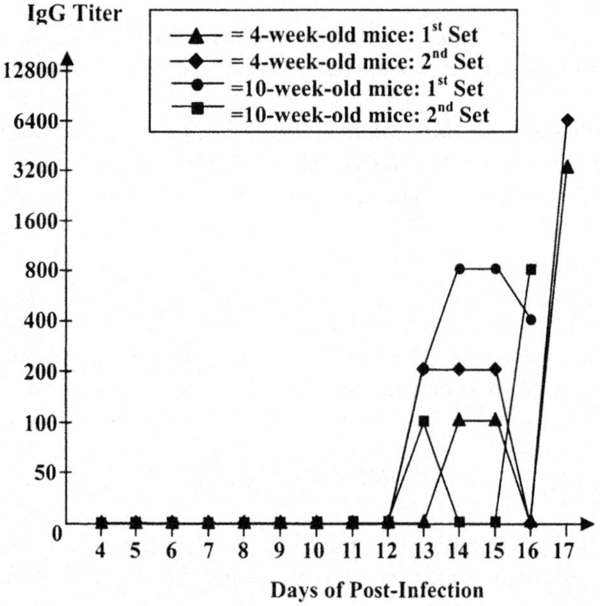

FIGURE 2. Positive IgG serum samples of both the 4-week-old and 10-week-old mice were first detected on day 13.

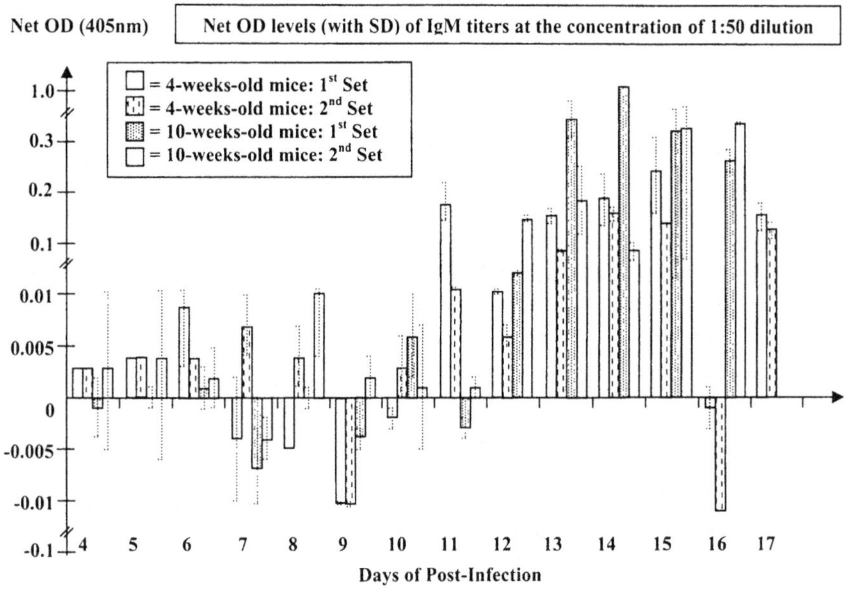

FIGURE 3. Net OD levels (with standard deviation) of IgM titers of the 4-week-old mice were slightly lower than those of the 10-week-old mice.

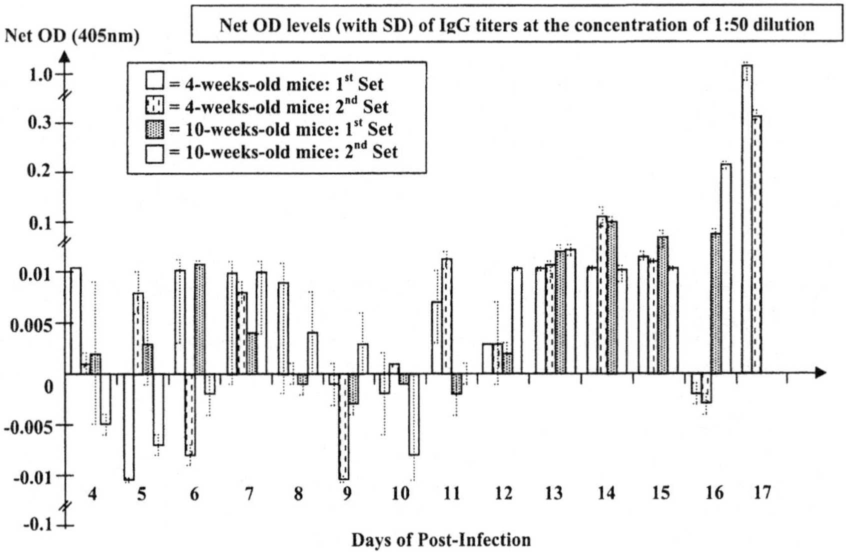

FIGURE 4. Net OD levels (with SD) of IgG titers of the 4-week-old and 10-week-old mice showed no significant difference.

Considering an earlier detection and slightly higher level of specific IgM exhibited in the older mice, it could be assumed as well as interpreted that the Older mice might be able to immunologically respond against *O. tsutsugamushi* faster than the younger mice.

Results obtained from this study will be useful in setting up any criteria for determining whether or not to choose young or older ICR laboratory mice in the evaluation procedures of the scrub typhus-vaccine candidates (currently being developed) by either or both needle challenging (intraperitoneal inoculation of *O. tsutsugamushi*) or natural challenging (by the biting of scrub typhus-infected *Leptotrombidium* chiggers) of vaccinated and unvaccinated ICR laboratory mice.

ACKNOWLEDGMENTS

Funding for this project was provided by the Military Infectious Diseases Research Program of the U.S. Army Medical Research and Materiel Command, Fort Detrick, MD, U.S.A.

REFERENCES

1. LERDTHUSNEE, K., N. KHLAIMANEE, T. MONKANNA, *et al.* 2002. Efficiency of *Leptotrombidium* chiggers at transmitting *Orientia tsutsugamushi* to laboratory mice. J. Med. Entomol. **39:** 521–525.
2. LERDTHUSNEE, K., B. KHUNTIRAT, W. LEEPITAKRAT, *et al.* 2003. Scrub typhus: vector competence of *Leptotrombidium chiangraiensis* chiggers and transmission efficacy and isolation of *Orientia tsutsugamushi*. Ann. N. Y. Acad. Sci. **990:** 25–35.
3. COLEMAN, R.E., T. MONKANNA, K.J. LINTHICUM, *et al.* 2003. Occurrence of *Orientia tsutsugamushi* in small mammals from Thailand. Am. J. Trop. Med. Hyg. **69:** 519–524.
4. DASCH, G.A., S. HALLE, A.L. BOURGEOIS, *et al.* 1979. ELISA for antibodies against scrub typhus. J. Clin. Microbiol. **9:** 38–48.
5. SUWANABUN, N., C. CHOURIYAGUNE, C. EAMSILA, *et al.* 1997. Evaluation of an enzyme-linked immunosorbent assay in Thai scrub typhus patients. Am. J. Trop. Med. Hyg. **56:** 38–43.
6. CHING, W.M., H. WANG, C. EAMSILA, *et al.* 1998. Expression and refolding of truncated recombinant major outer membrane protein antigen (r56) of *Orientia tsutsugamushi* and its use in enzyme-linked immunosorbent assays. Clin. Diagn. Lab. Immunol. **5:** 519–526.
7. COLEMAN, R.E., V. SANGKASUWAN, N. SUWANABUN, *et al.* 2002. Comparative evaluation of selected diagnostic assay for the detection of IgG and IgM antibody to *Orientia tsutsugamushi* in Thailand. Am. J. Trop. Med. Hyg. **67:** 497–503.
8. JANG, W.J., M.S. HUH, K.H. PARK, *et al.* 2003. Evaluation of an Immunoglobulin M capture enzyme-linked immunosorbent assay for diagnosis of *Orientia tsutsugamushi* infection. Clin. Diagn. Lab. Immunol. **10:** 394–398..

Methods of Isolation and Cultivation of New Rickettsiae from the Nosoarea of the North Asian Tick Typhus in Siberia

I.E. SAMOYLENKO,[a] L.V. KUMPAN,[b] S.N. SHPYNOV,[a,c] A.S. OBERT,[d,e] O.V. BUTAKOV,[f] AND N.V. RUDAKOV[a,c]

[a] *Omsk Research Institute of Natural Foci Infections, Omsk, Russia*

[b] *Omsk Regional Centre of the State Sanitary and Epidemiologic Supervision, Omsk, Russia*

[c] *Omsk State Medical Academy, Omsk, Russia*

[d] *Institute for Water and Ecological Problems*

[e] *Altai State Medical University, Prospekt Lenina 40, Barnaul 656099, Russia*

[f] *Ust-Tarsk Hospital, Novosibirsk, Russia*

KEYWORDS: **new rickettsial genotypes; isolation; cultivation**

Identification of spotted fever group (SFG) rickettsiae and development of approaches to the study of rickettsial population structure are real problems for modern rickettsiology. We need to mention that a significant number of "new" rickettsiae are revealed by genotyping in carriers without isolating the strain. Pathogenicity of the majority of newly revealed agents (including genotypes of the *R. massiliae* group—R.sp.RpA4, R.sp.DnS14, and R.sp.DnS28)[1] is not established yet.

Modeling of the natural cycle of metamorphosis of ticks is not new in rickettsiae and has been widely used for 60–70 years;[2,3] however, combining modern methods of research, such as the use of monoclonal antibodies and genetic identification, might shed light in this area. We used an updated method of Tagiltzev *et al.* (1990)[4] for cultivation of SFG rickettsiae, and also for the study of their biological characteristics at different stages of metamorphosis.

Inasmuch as the tick is a natural habitat for SFG rickettsiae, experimental modeling of the metamorphosis of carriers (EMNCMc) allows us to carry out the study of variant rickettsiae in a population, including its nonvirulent part. Eight strains of SFG rickettsiae not cultivated in guinea pigs and poorly cultivated with embryonated eggs were isolated by this method. Using PCR

Address for correspondence: I.E. Samoylenko, Omsk Research Institute of Natural Foci Infections, 644080, Prospekt Mira, 7 Omsk, Russia.
e-mail: irinasam59@mail.ru

**Ann. N.Y. Acad. Sci. 1078: 613–616 (2006). © 2006 New York Academy of Sciences.
doi: 10.1196/annals.1374.035**

amplification and sequencing, these agents were identified by S. Shpynov as genotypes R.sp.RpA4 (four strains), R.sp.DnS14 (one strain), and R.sp.DnS28 (three strains). Six of them were deposited in the Russian museum of rickettsial cultures. The strains of researched genotypes were successfully kept in laboratory conditions in ticks for 4 generations (R.sp.DnS14) and 6–7 generations (R.sp.RpA4 and R.sp.DnS28).

Using MEMNCMc, we investigated the efficiency of the transovarial (TOT) and transstadial (TST) transmission of rickettsiae of three new genotypes: R.sp.RpA4, R.sp.DnS14, and R.sp.DnS28. TOTs made up 43.0–100% before feeding larvae and 86.4–100% after feeding; TSTs made up 90.0–100% and 95.0–100%, respectively.[5]

The phenomenon of increase of concentration of rickettsiae after feeding of hungry carriers has also been established. On the basis of the results of one-factor dispersive analysis, authentic distinctions (P <0001) between accumulated rickettsiae in larvae and nymphs of experimental lines before and after feeding were revealed.

We also investigated the localization of rickettsiae in organs of carriers by smear-prints from hemolymph, salivary glands, malpighian vessels, and genital glands of the ticks containing rickettsiae of genotypes R.sp.RpA4 and R.sp.DnS28. The rickettsiae have been found in all investigated organs, but in various concentrations.

Analysis of the antigenic structure of rickettsial genotypes R.sp.RpA4 and R.sp.DnS28 revealed a change of antigenic properties of rickettsiae in the metamorphic process of carriers. Low-expression OmpA was obvious in rickettsiae contained in larvae of ticks. Modification of the antigenic structure of rickettsiae depends on stages of the vector's metamorphosis and makes it possible to suppose existence of "antigenic mimicry," which permits rickettsiae in larvae and nymphs to be alive in the same carriers while feeding on blood.[6]

Interference of microorganisms spreads greatly in nature. Our experimental study demonstrated that those infected with R.sp.RpA4, mainly, were infected with the virulent *R. sibirica* unsuccessfully.[7,8]

Culturing cells is a prospective method for isolation and study of SFG rickettsiae, especially those with low virulence. We studied the possibility of culturing rickettsiael strains of genotypes R.sp.DnS24, R.sp.DnS28, R.sp.RpA4, *R. slovaca*, and *Anaplasma* sp. Omsk (GenBank AY 649325), and also of primary isolation of rickettsiae of new genotypes from carriers with the use of Vero and HEP-2 cultures.[5]

We cultured rickettsiae R.sp.DnS24, R.sp.DnS28, and R.sp.RpA4 for 16 passages with moderate accumulation and with moderate cytopathogenicity, essential distinctions in a level of accumulation of the agents in the given cultures of cells. It is not marked. Vero cells also were used for accumulation of a biomass of rickettsiae genotypes R.sp.DnS24, R.sp.DnS28, and R.sp.RpA4 and the subsequent preparation of soluble antigens. Antigens were prepared according to P.F. Zdrodovskii's method and E.M. Golinevich's recommendations

(1972)[9] and tested for specificity and sensitivity in cross reactions. With commercial sera against *R. sibirica*, and *R. prowazekii*, negative results were obtained. Differences in antigenic structure of rickettsiae of new genotypes in reaction with sera of patients infected by Siberian tick typhus are shown. It is necessary to continue studying sera of patients by immunoblot and genetic research of material to specify the role of rickettsiae in new genotypes in an infectious pathology.

We also show primary isolation of rickettsiae, different from *R. sibirica*, from vectors. For this reason, infection of cultures of Vero cells with suspensions from individual exemplars of ticks *I. persulcatus* collected in *Candidatus* rickettsia tarasevichiae[10] has been carried out and has provided about 40 isolates of rickettsiae.[11] Some were recognized as *Candidatus* rickettsia tarasevichiae (Shpynov *et al.*, nonpublished data, 2005).

Thus, we proved that at least two models, EMNCMc and cultures of Vero and HEP-2 cells, are adequate for primary isolation, cultivation, and study of biological and genetic characteristics of rickettsiae of new genotypes R.sp.DnS14, R.sp.DnS28, R.sp.RpA4, and *Candidatus* rickettsia tarasevichiae, not cultivated in traditional models (guinea pigs, chicken embryos). We have been keeping a collection of rickettsiae of new genotypes, using those models.

ACKNOWLEDGMENTS

We are grateful to Professor Irina Tarasevich for stimulating this investigation, to Professor David Walker for kindly providing monoclonal antibodies, and to Professor Didier Raoult for giving an opportunity to carry out genetic research in Unité des Rickettsies, Université de la Méditerranée.

REFERENCES

1. RYDKINA, E., V. ROUX, N. FETISOVA *et al.* 1999. New *Rickettsiae* in ticks collected in territories of the former Soviet Union. Emerg. Infect. Dis. **5:** 811–814.
2. KORSHUNOVA, O.S. 1967. Ticks Ixodoidae and *Rickettsia sibirica* (*Dermacentroxenus sibiricus*): field and experimental investigations: 86–103. Moscow: Medicine.
3. GROKHOVSKAYA, I.M. & V.E. SIDOROV. 1967. Ticks Ixodoidea and *Dermacentroxenus sibiricus* (experimental investigations): 107–125. Moscow. Medicine.
4. TAGILTZEV, A.A., L.N. TARASEVICH, I.I. BOGDANOV & V.V. YAKIMENKO. 1990. The study of nesting-barrow complex's arthropods in transmission of arboviruses infections. Tomsk.
5. SAMOILENKO, I.E., N.V. RUDAKOV, S.N. SHPYNOV, *et al.* 2003. Study of biological characteristics of spotted fever group rickettsial genotypes RpA4, DnS14 and DnS28. Ann. N.Y. Acad. Sci. **990:** 612–616.
6. RUDAKOV, N.V., S.N. SHPYNOV, I.E. SAMOILENKO & M.A. TANKIBAEV. 2003. Ecology and epidemiology of spotted fever group rickettsiae and new data from their study in Russia and Kazakhstan. Ann. N.Y. Acad. Sci. **990:** 12–24.

7. SHPYNOV, S., P. PAROLA, N. RUDAKOV, *et al.* 2001. Detection and identification of spotted fever group rickettsiae in *Dermacentor* ticks from Russia and Central Kazahstan. Eur. J. Clin. Microbiol. Inf. Dis. **20:** 903–905.
8. RUDAKOV, N.V., I.E. SAMOILENKO, V.V. YAKIMENKO, *et al.* 1999. The re-emergence of Siberian tick typhus: field and experimental observations. *In* Rickettsiae and Rickettsial Diseases in the Turn of the Third Millenium: 269–273. Elsevier. Marseille.
9. ZDRODOVSKII, P.F. & E.M. GOLINEVITCH. 1972. Rickettsiae and Rickettsial Diseases. 3rd ed. Moscow. Medicine.
10. SHPYNOV, S., P.-E. FOURNIER, N. RUDAKOV & D. RAOULT. 2003. "*Candidatus Rickettsia tarasevichiae* " in *Ixodes persulcatus* ticks collected in Russia. Ann. N.Y. Acad. Sci. **990:** 162–172.
11. SAMOYLENKO, I.E., L.V. KUMPAN, S.N. SHPYNOV, *et al.* 2005. Methods of isolation and cultivation of rickettsiae of "new genotypes" from nosoarea of the North Asian tick typhus in Siberia. 4th int. conf. on Rickettsiae and Rickettsial Diseases: book of abstracts. 185. Logroño (La Rioja). Spain.

Validation of a *Rickettsia prowazekii*–Specific Quantitative Real-Time PCR Cassette and DNA Extraction Protocols Using Experimentally Infected Lice

PATRICK J. ROZMAJZL,[a] LINDA HOUHAMDI,[b] JU JIANG,[a] DIDIER RAOULT,[b] AND ALLEN L. RICHARDS[a]

[a]*Naval Medical Research Center, Silver Spring, Maryland, USA*

[b]*Faculte de Medecine – Timone, Marseille, France*

ABSTRACT: The presence of *R. prowazekii* in lice using *R. prowazekii*-specific qPCR cassettes showed the changes of the rickettsial burden in the lice post-infection.

KEYWORDS: PCR studies in lice; DNA extraction in lice; Rickettsia prowazekii

OBJECTIVE

The efficacy of a *Rickettsia* genus–specific and a *Rickettsia prowazekii*–specific quantitative real-time PCR (qPCR) cassette was assessed using experimentally infected human body lice. Subsequently, various DNA isolation protocols were compared using the *R. prowazekii*–specific cassette to determine which one yielded greatest qPCR sensitivity.

METHODS

Human body lice (*Pediculus humanus humanus*) were infected with *R. prowazekii* by feeding on a New Zealand white rabbit made bacteremic by infusion of cultured *R. prowazekii* as previously described.[1] In brief: *R. prowazekii* isolated from L929 cell culture were used to establish a 20-mL inoculum with a final concentration of 105 organisms per milliliter. Lice (n = 200) were allowed to feed on the shaved abdomen of the rabbit (R1) during infusion of the

Address for correspondence: Patrick J. Rozmajzl, NMRC-IDD/RDD, 503 Robert Grant Ave., Silver Spring, MD 20910-7500.
e-mail: RozmajzlP@nmrc.navy.mil

**Ann. N.Y. Acad. Sci. 1078: 617–619 (2006). © 2006 New York Academy of Sciences.
doi: 10.1196/annals.1374.036**

inoculum. Subsequently, infected lice were fed once daily on a nurse rabbit (R2 or R3, alternated daily). Death of infected lice routinely occurred by day 6 post infection. Lice were sampled on days 0–6 and frozen at –70°C. DNA was extracted by trituration and boiling of individual lice in water or Prepman Ultra (Applied Biosystems, Foster City, CA, USA). Alternatively, DNA was isolated from triturated lice using Chelex (Biorad, Hercules, CA, USA) or Qiagen spin columns (Qiagen, Valencia, CA, USA) per manufacturers' instructions.

Rickettsia genus–specific[2] and *R. prowazekii*–specific[3] qPCR reactions were performed using LightCycler equipment and software (Roche, Alameda, CA, USA). Reaction mixes were formulated by rehydration of lyophilized OmniMix HS beads (Cepheid, Sunnyvale, CA). All reactions were performed in a final volume of 25 μL, with template DNA volumes of 1 μL, 0.2 mM dNTPs, 5 mM MgCl2 1.5 U TaKaRa Taq polymerase, and 25 mM HEPES buffer. For the *Rickettsia* genus cassette, primers were added to a final concentration of 0.3 μM and probe to a final concentration of 0.4 μM. In the case of the *Rickettsia prowazekii*–specific cassette, primers and probe were added to a final concentration of 0.2 μM and 0.3 μM, respectively. Thermal cycling parameters for all assays were as follows: 94°C for 3 min followed by 50 cycles of 94°C for 5 sec and 60°C for 30 sec. Samples demonstrating a logarithmic increase in fluorescence beyond a minimum default threshold value were identified as positive and assigned a cycle threshold (CT) value equal to that cycle number at which the increase in fluorescence exceeded the threshold.

RESULTS

Initial studies assaying the efficacy of qPCR cassettes specific for *Rickettsia* genus and *R. prowazekii* using infected lice sampled at day 6 demonstrated that both assays are efficacious using rickettsial DNA extracted from infected lice in water by simple boiling in the presence or absence of trituration (data not shown). Preliminary studies using DNA extracted by simple boiling from day 0–infected lice, however, demonstrated negative results.

In order to increase assay sensitivity in day 0–infected lice, three DNA isolation protocols were used to isolate DNA from potential inhibitors of PCR. DNA isolated by Qiagen spin columns, Chelex, or by boiling in Prepman Ultra demonstrated dramatically improved sensitivity at day 0 (FIG. 1). Lower cycle threshold (CT) values are indicative of greater concentrations of organismal DNA. Although *Rickettsia*-specific DNA was detected by qPCR from preparations using each isolation protocol, on the basis of CT values, the qPCR results using DNA isolated by Prepman Ultra and Qiagen protocols were generally comparable to one another and superior to those obtained using DNA isolated by Chelex for each day examined.

Given the experimental design, examination of the qPCR results allowed for an assessment of the dynamics of rickettsial infection in lice. All three

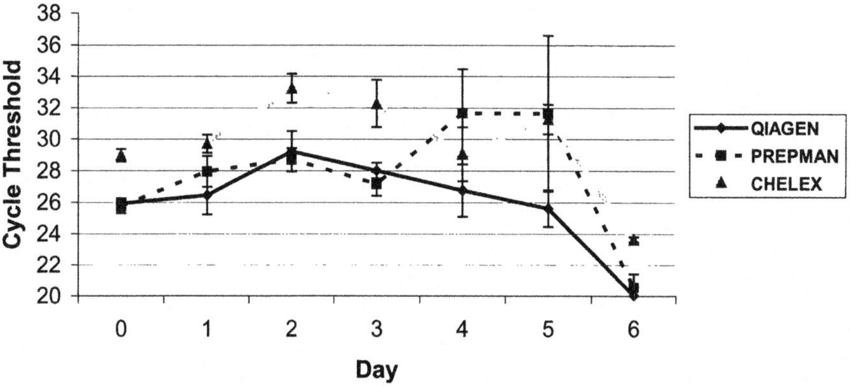

FIGURE 1. qPCR cycle threshold (CT) ± S.D. versus day post infection for three DNA isolation protocols. Lower CT values indicate higher gene copy numbers.

protocols demonstrated an initial decrease in rickettsia load followed by a dramatic increase on day 6, coinciding with universal louse death.

CONCLUSIONS

The *Rickettsia* genus–specific and *R. prowazekii*–specific qPCR cassettes, in general, reliably detect and enumerate *R. prowazekii* present in human body lice. The use of any of the three DNA isolation protocols studied here demonstrated improved sensitivity in day 0–infected lice. qPCR data using DNA isolated by means of Prepman Ultra are generally comparable to those using DNA isolated via Qiagen columns and superior to those using DNA isolated by Chelex. Given that the Prepman Ultra protocol requires the fewest manipulations, this method may be considered preferable for field use. The rickettsial burden in infected lice demonstrates an initial decrease in total number of rickettsia followed by robust increase coincident with louse death.

REFERENCES

1. HOUHAMDI, L., P.E. FOURNIER, R. FAND, *et al.* 2002. An experimental model of human body louse infection with *Rickettsia prowazekii*. J. Inf. Dis. **186**: 1639–1646.
2. JIANG, J., T.C. CHAN, J.J. TEMENAK, *et al.* 2004. Development of a quantitative real-time polymerase chain reaction assay specific for *Orientia tsutsugamushi*. Am. J. Trop. Med. Hyg. **70**: 351–356.
3. JIANG, J., J.J. TEMENAK & A.L. RICHARDS. 2003. Real-time PCR duplex assay for *Rickettsia prowazekii* and *Borrelia recurrentis*. Ann. N. Y. Acad. Sci. **990**: 302–310.

Index of Contributors

Erratum

In [1], the following error was published on page 1.

JAMES N. IHLE,[b] WILLIAM THIERFELDER, STEPHAN TEGLUND, DIMITRIOS STRAVAPODIS, DEMIN WANG, JIAN FENG, AND EVAN PARGANAS

The text was incorrect and should have read:

JAMES N. IHLE,[b] WILLIAM THIERFELDER, STEPHAN TEGLUND, DIMITRIOS STRAVOPODIS, DEMIN WANG, JIAN FENG, AND EVAN PARGANAS

The spelling of the 4th author's last name should be STRAVOPODIS

We apologize for this error.

Reference

1. IHLE, J.N. *et al.* 1998. Signaling by the Cytokine Receptor Superfamily. Ann. N. Y. Acad. Sci. **865:** 1–9. DOI: 10.1111/j.1749-6632.1998.tb11157.x

CPSIA information can be obtained at www.ICGtesting.com
Printed in the USA
LVOW04s0908300815

452091LV00023B/1106/P